INTRODUCTION TO QUANTUM MECHANICS
Schrödinger Equation and Path Integral

INTRODUCTION TO QUANTUM MECHANICS
Schrödinger Equation and Path Integral

Harald J W Müller-Kirsten

University of Kaiserslautern, Germany

World Scientific

NEW JERSEY · LONDON · SINGAPORE · BEIJING · SHANGHAI · HONG KONG · TAIPEI · CHENNAI

Published by

World Scientific Publishing Co. Pte. Ltd.

5 Toh Tuck Link, Singapore 596224

USA office: 27 Warren Street, Suite 401-402, Hackensack, NJ 07601

UK office: 57 Shelton Street, Covent Garden, London WC2H 9HE

British Library Cataloguing-in-Publication Data
A catalogue record for this book is available from the British Library.

INTRODUCTION TO QUANTUM MECHANICS:
Schrödinger Equation and Path Integral

ISBN 981-256-691-0
ISBN 981-256-692-9 (pbk)

Printed in Singapore by World Scientific Printers (S) Pte Ltd

Contents

18 Anharmonic Oscillator Potentials   379

19 Singular Potentials   435

Preface

With the discovery of quantization by Planck in 1900, quantum mechanics is now more than a hundred years old. However, a proper understanding of the phenomenon was gained only later in 1925 with the fundamental Heisenberg commutation relation or phase space algebra and the associated uncertainty principle. The resulting Schrödinger equation has ever since been the theoretical basis of atomic physics. The alternative formulation by Feynman in terms of path integrals appeared two to three decades later. Although the two approaches are basically equivalent, the Schrödinger equation has found much wider usefulness, particularly in applications, presumably, in view of its simpler mathematics. However, the realization that solutions of classical equations, notably in field theory, play an important role in our understanding of a large number of physical phenomena, intensified the interest in Feynman's formulation of quantum mechanics, so that today this method must be considered of equal basic significance. Thus there are two basic approaches to the solution of a quantum mechanical problem, and an understanding of both and their usefulness in respective domains calls for their application to exemplary problems and their comparison. This is our aim here on an introductory level.

Throughout the development of theoretical physics two types of forces played an exceptional role: That of the restoring force of simple harmonic motion proportional to the displacement, and that in the Kepler problem proportional to the inverse square of the distance, i.e. Newton's gravitational force like that of the Coulomb potential. In the early development of quantum mechanics again oscillators appeared (though not really those of harmonic type) in Planck's quantization and the Coulomb potential in the Bohr model of the hydrogen atom. Again after the full and proper formulation of quantum mechanics with Heisenberg's phase space algebra and Born's wave function interpretation the oscillator and the Coulomb potentials provided the dominant and fully solvable models with a large number of at least approximate applications. To this day these two cases of interaction with nonresonant spectra feature as the standard and most important illustrative examples in any treatise on quantum mechanics and — excepting various kinds of square well and rectangular barrier potentials — leave the student sometimes puzzled about other potentials that he encounters soon thereafter, like periodic potentials, screened Coulomb potentials and maybe singular potentials, but also about complex energies that he encounters in a parallel course on nuclear physics. Excluding spin, any problem more complicated is frequently dispensed with by referring to cumbersome perturbation methods.

Diverse and more detailed quantum mechanical investigations in the second half of the last century revealed that perturbation theory frequently does permit systematic procedures (as is evident e.g. in Feynman diagrams in quantum electrodynamics), even though the expansions are mostly asymptotic. With various techniques and deeper studies, numerous problems could, in fact, be treated to a considerable degree of satisfaction perturbatively. With the growing importance of models in statistical mechanics and in field theory, the path integral method of Feynman was soon recognized to offer frequently a more general procedure of enforcing first quantization instead of the Schrödinger equation. To what extent the two methods are actually equivalent, has not always been understood well, one problem being that there are few nontrivial models which permit a deeper insight into their connection. However, the aforementioned exactly solvable cases, that is the Coulomb potential and the harmonic oscillator, again point the way: For scattering problems the path integral seems particularly convenient, whereas for the calculation of discrete eigenvalues the Schrödinger equation. Thus important level splitting formulas for periodic and anharmonic oscillator potentials (i.e. with degenerate vacua) were first and more easily derived from the Schrödinger equation. These basic cases will be dealt with in detail by both methods in this text, and it will be seen in the final chapter that potentials with degenerate vacua are not exclusively of general interest, but arise also in recently studied models of large spins.

The introduction to quantum mechanics we attempt here could be subdivided into essentially four consecutive parts. In the **first part**, Chapters 1 to 14, we recapitulate the origin of quantum mechanics, its mathematical foundations, basic postulates and standard applications. Our approach to quantum mechanics is through a passage from the Poisson algebra of classical Hamiltonian mechanics to the canonical commutator algebra of quantum mechanics which permits the introduction of Heisenberg and Schrödinger pictures already on the classical level with the help of canonical transformations. Then the Schrödinger equation is introduced and the two main exactly solvable cases of harmonic oscillator and Coulomb potentials are treated in detail since these form the basis of much of what follows. Thus this first part deals mainly with standard quantum mechanics although we do not dwell here on a large number of other aspects which are treated in detail in the long-established and wellknown textbooks.

In the **second part**, Chapters 15 to 20, we deal mostly with applications depending on perturbation theory. In the majority of the cases that we treat we do not use the standard Rayleigh–Schrödinger perturbation method but the systematic perturbation procedure of Dingle and Müller which is introduced in Chapter 8. After a treatment of power potentials, the chapter thereafter deals with Yukawa potentials, and their eigenvalues. This is followed by the important case of the cosine or Mathieu potential for which the perturbation method was originally developed, and the behaviour of the eigenvalues is discussed in both weak and strong coupling domains with formation of bands and their asymptotic limits. The solution of this case — however in nonperiodic form — turns out to be a prerequisite for the complete solution of the Schrödinger equation for the singular potential $1/r^4$ in Chapter 19, which is presumably the only such singular case permitting complete solution and was achieved only recently. The earlier Chapter 17 also contains a brief description of a similar treatment of the elliptic or Lamé potential. The following Chapter then deals with Schrödinger potentials which represent essentially anharmonic oscillators. The most prominent examples here are the double well potential and its inverted form. Using perturbation theory, i.e. the method of matched asymptotic expansions with boundary conditions (the latter providing the so-called nonperturbative effects), we derive respectively the level-splitting formula and the imaginary energy part for these cases for arbitrary states. In the final chapter of this part we discuss the large order behaviour of the perturbation expansion with particular reference to the cosine and double well potentials.

In **part three** the path integral method is introduced and its use is illustrated by application to the Coulomb potential and to the derivation of the Rutherford scattering formula. Thereafter the concepts of instantons, periodic instantons, bounces and sphalerons are introduced and their relevance in quantum mechanical problems is discussed (admittedly in also trespassing the sharp dividing line between quantum mechanics and simple scalar field theory). The following chapters deal with the derivation of level splitting formulas (including excited states) for periodic potentials and anharmonic oscillators and — in the one-loop approximation considered — are shown to agree with those obtained by perturbation theory with associated boundary conditions. We also consider inverted double wells and calculate with the path integral the imaginary part of the energy (or decay width). The potentials with degenerate minima will be seen to re-appear throughout the text, and the elliptic or Lamé potential — here introduced earlier as a generalization of the Mathieu potential — re-appears as the potential in the equations of small fluctuations about the classical configurations in each of the basic cases (cosine, quartic, cubic). All results are compared with those obtained by perturbation theory, and whenever available also with the results of WKB calculations, this comparison on a transparent level being one of the main aims of this text.

The introduction of collective coordinates of classical configurations and the fluctuations about these leads to constraints. Our **fourth and final part** therefore deals with elementary aspects of the quantization of systems with constraints as introduced by Dirac. We then illustrate the relevance of this in the method of collective coordinates. In addition this part considers in more detail the region near the top of a potential barrier around the configuration there which is known as a sphaleron. The physical behaviour there (in the transition region between quantum and thermal physics) is no longer controlled by the Schrödinger equation. Employing anharmonic oscillator and periodic potentials and re-obtaining these in the context of a simple spin model, we consider the topic of transitions between the quantum and thermal regimes at the top of the barrier and show that these may be classified in analogy to phase transitions in statistical mechanics. These considerations demonstrate (also with reference to the topic of spin-tunneling and large-spin behaviour) the basic nature also of the classical configurations in a vast area of applications.

Comparing the Schrödinger equation method with that of the path integral as applied to identical or similar problems, we can make the following observations. With a fully systematic perturbation method and with applied boundary conditions, the Schrödinger equation can be solved for practically any potential in complete analogy to wellknown differential equations of mathematical physics, except that these are no longer of hypergeometric type. The particular solutions and eigenvalues of interest in physics are — as a rule — those which are asymptotic expansions. This puts Schrödinger equations with e.g. anharmonic oscillator potentials on a comparable level with, for instance, the Mathieu equation. The application of path integrals to the same problems with the same aims is seen to involve a number of subtle steps, such as limiting procedures. This method is therefore more complicated. In fact, in compiling this text it was not possible to transcribe anything from the highly condensed (and frequently unsystematic) original literature on applications of path integrals (as the reader can see, for instance, from our precise reference to unavoidable elliptic integrals taken from Tables). An expected observation is that — ignoring a minor deficiency — the WKB approximation is and remains the most immediate way to obtain the dominant contribution of an eigenenergy, it is, however, an approximation whose higher order contributions are difficult to obtain. Nonetheless, we also consider at various points of the text comparisons with WKB approximations, also for the verification of results.

In writing this text the author considered it of interest to demonstrate the parallel application of both the Schrödinger equation and the path integral to a selection of basic problems; an additional motivation was that a sufficient understanding of the more complicated of these problems had been achieved only in recent years. Since this comparison was the guide-line in writing the text, other topics have been left out which are usually found in books on quantum mechanics (and can be looked up there), not the least for permitting a more detailed and hopefully comprehensible presentation here. Throughout the text some calculations which require special attention, as well as applications and illustrations, are relegated to separate subsections which — lacking a better name — we refer to as Examples.

The line of thinking underlying this text grew out of the author's association with Professor R. B. Dingle (then University of Western Australia, thereafter University of St. Andrews), whose research into asymptotic expansions laid the ground for detailed explorations into perturbation theory and large order behaviour. The author is deeply indebted to his one-time supervisor Professor R. B. Dingle for paving him the way into this field which — though not always at the forefront of current research (including the author's) — repeatedly triggered recurring interest to return to it. Thus when instantons became a familiar topic it was natural to venture into this with the intent to compare the results with those of perturbation theory. This endeavour developed into an unforeseen task leading to periodic instantons and the exploration of quantum-classical transitions. The author has to thank several of his colleagues for their highly devoted collaboration in this latter part of the work over many years, in particular Professors J.-Q. Liang (Taiyuan), D. K. Park (Masan), D. H. Tchrakian (Dublin) and Jian-zu Zhang (Shanghai). Their deep involvement in the attempt described here is evident from the cited bibliography.*

H. J. W. Müller-Kirsten

---

*In the running text references are cited like e.g. Whittaker and Watson [283]. For ease of reading, the references referred to are never cited by mere numbers which have to be identified e.g. at the end of a chapter (after troublesome turning of pages). Instead a glance at a nearby footnote provides the reader immediately the names of authors, e.g. like E. T. Whittaker and G. N. Watson [283], with the source given in the bibliography at the end. As a rule, formulas taken from Tables or elsewhere are referred to by number and/or page number in the source, which is particularly important in the case of elliptic integrals which require a relative ordering of integration limits and parameter domains, so that the reader is spared difficult and considerably time-consuming searches in a source (and besides, shows him that each such formula here has been properly looked up).

Chapter 1

Introduction

1.1 Origin and Discovery of Quantum Mechanics

The observation made by Planck towards the end of 1900, that the formula he had established for the energy distribution of electromagnetic black body radiation was in agreement with the experimentally confirmed Wien- and Rayleigh–Jeans laws for the limiting cases of small and large values of the wave-length λ (or λT) respectively is generally considered as the discovery of quantum mechanics. Planck had arrived at his formula with the assumption of a distribution of a countable number of infinitely many oscillators. We do not enter here into detailed considerations of Planck, which involved also thermodynamics and statistical mechanics (in the sense of Boltzmann's statistical interpretation of entropy). Instead, we want to single out the vital aspect which can be considered as the discovery of quantum mechanics. Although practically every book on quantum mechanics refers at the beginning to Planck's discovery, very few explain in this context what he really did in view of involvement with statistical mechanics.

A "perfectly black body" is defined to be one that absorbs all (thermal) radiation incident on it. The best approximation to such a body is a cavity with a tiny opening (of solid angle $d\Omega$) and whose inside walls provide a diffuse distribution of the radiation entering through the hole with the intensity of the incoming ray decreasing rapidly after a few reflections from the walls. Thermal radiation (with wave-lengths $\lambda \sim 10^{-5}$ to 10^{-2} cm at moderate temperatures T) is the radiation emitted by a body (consisting of a large number of atoms) as a result of the temperature (as we know today as a result of transitions between a large number of very closely lying energy levels). Kirchhoff's law in thermodynamics says that in the case of equilibrium, the amount of radiation absorbed by a body is equal to the amount the body emits. Black bodies as good absorbers are therefore also good emitters, i.e. radiators. The (equilibrium) radiation of the black body can be determined experimentally by sending radiation into a cavity surrounded by a heat bath at temperature T, and then measuring the increase in temperature of the heat bath.

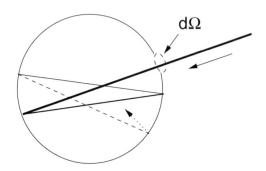

Fig. 1.1 Absorption in a cavity.

Let us look at the final result of Planck, i.e. the formula (to be explained)

$$\bar{u}(\nu, T) = 2\frac{4\pi\nu^2}{c^3}\left(\frac{x}{e^x - 1}\right)kT, \quad \text{where} \quad x = \frac{h\nu}{kT} = \frac{hc}{k\lambda T}. \qquad (1.1)$$

Here $\bar{u}(\nu, T)d\nu$ is the mean energy density (i.e. energy per unit volume) of the radiation (i.e. of the photons or photon gas) in the cavity with both possible directions of polarization (hence the factor "2") in the frequency domain $\nu, \nu + d\nu$ in equilibrium with the black body at temperature T. In Eq. (1.1) c is the velocity of light with $c = \nu\lambda$, λ being the wave-length of the radiation. The parameters k and h are the constants of Boltzmann and Planck:

$$k = 1.38 \times 10^{-23} J\,K^{-1}, \quad h = 6.626 \times 10^{-34} J\,s.$$

How did Planck arrive at the expression (1.1) containing the constant h by treating the radiation in the cavity as something like a gas? By 1900 two theoretically-motivated (but from today's point of view incorrectly derived) expressions for $\bar{u}(\nu, T)$ were known and tested experimentally. It was found that one expression agreed well with observations in the region of small λ (or λT), and the other in the region of large λ (or λT). These expressions are:
(1) *Wien's law:*

$$\bar{u}(\nu, T) = C_1\nu^3 e^{-C_2\nu/T}, \qquad (1.2)$$

and the

(2) *Rayleigh–Jeans law:*

$$\bar{u}(\nu, T) = 2\frac{4\pi\nu^2}{c^3}C_3 T, \tag{1.3}$$

C_i, C_2, C_3 being constants.
Considering Eq. (1.1) in regions of x "small" (i.e. $\exp(x) \simeq 1+x$) and "large"
($\exp(-x) \ll 1$), we obtain:

$$\bar{u}(\nu, T) \simeq 2\frac{4\pi\nu^2}{c^3}kT, \quad (x \text{ small}),$$

$$\bar{u}(\nu, T) \simeq 2\frac{4\pi\nu^2}{c^3}e^{-x}h\nu, \quad (x \text{ large}).$$

We see, that the formulas (1.2) and (1.3) are contained in Eq. (1.1) as approximations. Indeed, in the first place Planck had tried to find an expression linking both, and he had succeeded in finding such an expression of the form

$$\bar{u}(\nu, T) = \frac{a\nu^3}{e^{b\nu/T} - 1},$$

where a and b are constants. When Planck had found this expression, he searched for a derivation. To this end he considered Boltzmann's formula $S = k \ln W$ for the entropy S. Here W is a number which determines the distribution of the energy among a discrete number of objects, and thus over a discrete number of admissible states. This is the point, where the

Fig. 1.2 Distributing quanta (dots) among oscillators (boxes).

discretization begins to enter. Planck now imagined a number N of oscillators or N oscillating degrees of freedom, every oscillator corresponding to an eigenmode or eigenvibration or standing wave in the cavity and with mean energy U. Moreover Planck assumed that these oscillators do not absorb or emit energy continuously, but — here the discreteness appears properly — only in elements (quanta) ϵ, so that W represents the number of possible ways of distributing the number $P := NU/\epsilon$ of energy-quanta ("photons", which are indistinguishable) among the N indistinguishable oscillators at temperature T, $U(T)$ being the average energy emitted by one oscillator. We visualize the N oscillators as boxes separated by $N-1$ walls, with the quanta represented schematically by dots as indicated in Fig. 1.2. Then W is given by

$$W = \frac{(N+P-1)!}{(N-1)!P!}. \tag{1.4}$$

With the help of *Stirling's formula*[*]

$$\ln N! \simeq N \ln N - N + O(0), \quad N \to \infty,$$

and the second law of thermodynamics $((\partial S/\partial U)_V = 1/T)$, one obtains (cf. Example 1.1)

$$U = \frac{\epsilon}{e^{\epsilon/kT} - 1} \tag{1.5}$$

as the mean energy emitted or absorbed by an oscillator (corresponding to the classical expression of $2 \times kT/2$, as for small values of ϵ). Agreement with Eq. (1.2) requires that $\epsilon \propto \nu$, i.e.

$$\epsilon = h\nu, \quad h = \text{const.} \tag{1.6}$$

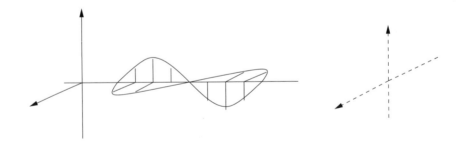

Fig. 1.3 Comparing the polarization modes with those
of a 2-dimensional oscillator.

We now obtain the energy density of the radiation, $\bar{u}(\nu, T)d\nu$, by multiplying U with the number $n_\nu d\nu$ of modes or oscillators per unit volume with frequency ν in the interval $\nu, \nu + d\nu$, i.e. with

$$n_\nu d\nu = 2 \times \frac{4\pi \nu^2}{c^3} d\nu, \tag{1.7}$$

---

[*]See e.g. I. S. Gradshteyn and I. M. Ryzhik [122], formula 8.343(2), p. 940, there not called Stirling's formula, as in most other Tables, e.g. W. Magnus and F. Oberhettinger [181], p.3. The Stirling formula or approximation will appear frequently in later chapters.

where the factor 2 takes the two possible mutually orthogonal linear directions of polarization of the electromagnetic radiation into account, as indicated in Fig.1.3. We obtain the expression (1.7) for instance, as in electrodynamics, where we have for the electric field

$$\mathbf{E} \propto e^{i\omega t} \sum_{\kappa} \mathbf{e}_{\kappa} \sin \kappa_1 x_1 \sin \kappa_2 x_2 \sin \kappa_3 x_3$$

with the boundary condition that at the walls $\mathbf{E} = 0$ at $x_i = 0, L$ for $i = 1, 2, 3$ (as for ideal conductors). Then $L\kappa_i = \pi n_i$, $n_i = 1, 2, 3, \ldots$,

$$L^2 \kappa^2 = \pi^2 \mathbf{n}^2,$$

where[†]

$$\kappa^2 = \left(\frac{2\pi\nu}{c}\right)^2, \quad \text{so that} \quad \left(\frac{2\nu L}{c}\right)^2 = \mathbf{n}^2.$$

The number of possible modes (states) is equal to the volume of the spherical octant (where $n_i > 0$) in the space of $n_i, i = 1, 2, 3$. The number with frequency ν in the interval $\nu, \nu + d\nu$, i.e. $n_\nu d\nu$ per unit volume, is given by

$$
\begin{aligned}
d\mathcal{N} &= \frac{d\mathcal{N}}{d\nu} d\nu \equiv n_\nu d\nu = \frac{d}{d\nu}\left[\frac{1}{8}\frac{4}{3}\pi \left(\frac{2\nu L}{c}\right)^3 / L^3\right] d\nu \\
&= \frac{1}{8}\frac{4}{3}\pi \frac{8}{c^3} 3\nu^2 d\nu = \frac{4\pi\nu^2}{c^3} d\nu,
\end{aligned}
$$

as claimed in Eq. (1.7). We obtain therefore

$$\bar{u}(\nu, T) = U n_\nu = 2\frac{4\pi\nu^2}{c^3}\frac{h\nu}{e^{h\nu/kT} - 1}. \tag{1.8}$$

This is *Planck's formula* (1.1). We observe that $\bar{u}(\nu, T)$ has a maximum which follows from $d\bar{u}/d\lambda = 0$ (with $c = \nu\lambda$). In terms of λ we have

$$\bar{u}(\lambda, T)d\lambda = \frac{8\pi}{\lambda^4}\frac{hc/\lambda}{e^{hc/\lambda kT} - 1} d\lambda,$$

so that the derivative of $\bar{u}$ implies (x as in Eq. (1.1))

$$\left(1 - \frac{x}{5}\right)e^x = 1.$$

The solutions of this equation are

$$x_{\text{max}} = 4.965 \quad \text{and} \quad x_{\text{min}} = 0.$$

[†]From the equation $\left(\frac{1}{c^2}\frac{\partial^2}{\partial t^2} - \nabla^2\right)\mathbf{E} = 0$, so that $-\omega^2/c^2 + \kappa^2 = 0, \omega = 2\pi\nu$.

The first value yields

$$\lambda_{\max} T = \frac{hc}{4.965k} = \text{const.}$$

This is Wien's *displacement law*, which had also been known before Planck's discovery, and from which the constant h *can be determined* from the known value of k.

Later it was realized by H. A. Lorentz and Planck that Eq. (1.8) could be derived much more easily in the context of statistical mechanics. If an oscillator with thermal weight or occupation probability $\exp(-nx)$ can assume only discrete energies $\epsilon_n = nh\nu$, $n = 0, 1, 2, \ldots$, then (with $x = h\nu/kT$) its mean energy is

$$
\begin{aligned}
U &= \frac{\sum_{n=0}^{\infty} nh\nu e^{-nx}}{\sum_{n=0}^{\infty} e^{-nx}} = -h\nu \frac{d}{dx} \ln \sum_{n=0}^{\infty} e^{-nx} \\
&= -h\nu \frac{d}{dx} \ln \frac{1}{1 - e^{-x}} = h\nu \frac{(1 - e^{-x})}{(1 - e^{-x})^2} e^{-x} \\
&= \frac{h\nu}{e^x - 1}.
\end{aligned}
\tag{1.9}
$$

We observe that for $T \to 0$ (i.e. $x \to \infty$) the mean energy vanishes ($0 < U \leq \infty$). Thus we have a rather complicated system here, that of an oscillation system at absolute temperature $T \neq 0$. One expects, of course, that it is easier to consider first the case of $T = 0$, i.e. the behaviour of the system at zero absolute temperature. Since temperature originates through contact with other oscillators, we then have at $T = 0$ independent oscillators, which can assume the discrete energies $\epsilon_n = nh\nu$. We are not dealing with the linear harmonic oscillator familiar from mechanics here, but one can expect an analogy. We shall see later that in the case of this *linear harmonic oscillator* the energies E_n are given by

$$
E_n = \left(n + \frac{1}{2}\right) h\nu \equiv \left(n + \frac{1}{2}\right) \hbar\omega, \quad \hbar = \frac{h}{2\pi}, \quad n = 0, 1, 2. \ldots
\tag{1.10}
$$

Thus here the so-called *zero point energy* appears, which did not arise in Planck's consideration of 1900.

One might suppose now, that we arrive at quantum mechanics simply by discretizing the energy and thus by postulating — following Planck — for the harmonic oscillator the expression (1.10). However, such a procedure leads to contradictions, which can not be eliminated without a different approach. We therefore examine such contradictions next.

Example 1.1: Mean energy of an oscillator

In Boltzmann's statistical mechanics the entropy S is given by the following expression (which we cite here with no further explanation) $S = k \ln W$, where k is Boltzmann's constant and W is the number of times P indistinguishable elements of energy ϵ can be distributed among N indistinguishable oscillators, i.e.

$$W = \frac{(N + P - 1)!}{(N - 1)!P!}, \quad \text{and} \quad P = \frac{UN}{\epsilon}.$$

Show with the help of Stirling's formula that the mean energy U of an oscillator is given by

$$U = \frac{\epsilon}{\exp(\epsilon/kT) - 1}.$$

Solution: Inserting W into Boltzmann's formula and using $\ln N! \simeq N \ln N - N$, we obtain

$$S = k[\ln(N + P - 1)! - \ln(N - 1)! - \ln P!] \simeq kN\left[\left(1 + \frac{U}{\epsilon}\right)\ln\left(1 + \frac{U}{\epsilon}\right) - \frac{U}{\epsilon}\ln\frac{U}{\epsilon}\right].$$

The second law of thermodynamics says

$$\left(\frac{\partial S}{\partial U}\right)_V = \frac{1}{T}.$$

For a single oscillator the entropy is $s = S/N$, so that

$$\frac{1}{T} = \left(\frac{\partial s}{\partial U}\right)_V = k\frac{\partial}{\partial U}\left[\left(1 + \frac{U}{\epsilon}\right)\ln\left(1 + \frac{U}{\epsilon}\right) - \frac{U}{\epsilon}\ln\frac{U}{\epsilon}\right] = \frac{k}{\epsilon}\ln\left(\frac{\epsilon}{U} + 1\right),$$

i.e.

$$U = \frac{\epsilon}{\exp(\epsilon/kT) - 1},$$

which for $\epsilon/kT \to 0$ becomes

$$U \simeq \frac{\epsilon}{1 + \frac{\epsilon}{kT} - 1} \simeq kT.$$

This means U is then the classical expression resulting from the mean kinetic energy per degree of freedom, $kT/2$, for 2 degrees of freedom.

1.2 Contradicting Discretization: Uncertainties

The far-reaching consequences of Planck's quantization hypothesis were recognized only later, around 1926, with Heisenberg's discovery of the uncertainty relation. In the following we attempt to incorporate the above discretizations into classical considerations* and consider for this reason so-called *thought experiments* (from German "Gedankenexperimente"). We

---

*This is what was effectively done before 1925 in Bohr's and Sommerfeld's atomic models and is today referred to as "*old quantum theory*".

shall see that we arrive at contradictions. As an example[†] we consider the
linear harmonic oscillator with energy

$$E = \frac{1}{2}m\dot{x}^2 + \frac{1}{2}m\omega^2 x^2. \tag{1.11}$$

The classical equation of motion

$$\frac{dE}{dt} = \dot{x}(m\ddot{x} + m\omega^2 x) = 0$$

permits solutions $x = A\cos(\omega t + \delta)$, so that

$$E = \frac{1}{2}m\omega^2 A^2,$$

where A is the maximum displacement of the oscillation, i.e. at $\dot{x} = 0$.
We consider first this case of velocity and hence momentum precisely zero,
and investigate the possibility to fix the amplitude. If we replace E by the
discretized expression (1.10), i.e. by $E_n = (n + 1/2)\hbar\omega$, we obtain for the
amplitude A

$$A \longrightarrow A_n = \sqrt{\frac{2\hbar}{m\omega}}\sqrt{n + \frac{1}{2}}. \tag{1.12}$$

Thus the amplitude can assume only these definite values. We now perform
the following thought experiment. We give the oscillator initially an amplitude which is not contained in the set (1.12), i.e. for instance an amplitude
$\tilde{A}$ with

$$A_n < \tilde{A} < A_{n+1}.$$

Energy conservation then requires that the oscillator has to oscillate all the
time with this (according to Eq. (1.12) nonpermissible) amplitude. In order
to be able to perform this experiment, the difference

$$\triangle A = A_{n+1} - A_n$$

must not be too small, i.e. the difference

$$\triangle A = \sqrt{\frac{2\hbar}{m\omega}}\left(\sqrt{n + \frac{3}{2}} - \sqrt{n + \frac{1}{2}}\right) \simeq \sqrt{\frac{2\hbar}{m\omega}}\frac{2}{4\sqrt{n}}[1 + O(1/n)].$$

For $m = 2kg$, $\hbar = 1 \times 10^{-34}\,J\,s$, $\omega = 1s^{-1}$, we obtain

$$\triangle A \simeq \frac{10^{-17}}{2\sqrt{n}}[1 + O(1/n)] \quad \text{meter}.$$

[†]H. Koppe [152].

This distance is even less than what one would consider as a certain "diameter" of the electron ($\sim 10^{-15}$ meter). Thus it is even experimentally impossible to fix the amplitude $\tilde{A}$ of the oscillator with the required precision. Since A is the largest value of x, where $\dot{x} = 0$, we have the problem that for a given definite value of $m\dot{x}$, i.e. zero, the value of $x = A$ can not be determined, i.e. given the energy of Eq. (1.10), it is *not possible* to give the oscillator at the same time at a definite position a definite momentum.

The above expression (1.10) for the energy of the harmonic oscillator, which we have not established so far, has the further characteristic of possessing the "*zero-point energy*" $\hbar\omega/2$, the smallest energy the oscillator can assume, according to the formula. Let us now consider the oscillator as a pendulum with frequency ω in the gravitational field of the Earth.[‡] Then

$$\omega^2 = \frac{g}{l}, \tag{1.13}$$

where l is the length of the pendulum. Thus we can vary the frequency ω by varying the length l. This can be achieved with the help of a pivot, attached to a movable frame as indicated in Fig. 1.4. The resultant of the tension in the string of the pendulum, **R**, always has a nonnegative vertical component. If the pivot is moved downward, work is done against this vertical component of **R**; in other words, the system receives additional energy. However, there is one case, in which for a very short interval of time, δt, the pendulum is at angle $\theta = 0$. Reducing in this short interval of time the length of the pendulum (by an appropriately quick shift of the pivot) by a factor of 4, the frequency of the oscillator is doubled, without supplying it with additional energy. Thus the energy

$$E_n = \left(n + \frac{1}{2} \right) \hbar\omega \quad \text{becomes} \quad \left(n + \frac{1}{2} \right) \hbar\, 2\omega,$$

without giving it additional energy. This is a self-evident *contradiction*. This means — if the quantum mechanical expression (1.10) is valid — we cannot simultaneously fix the energy (with energy conservation), as well as time t to an interval $\delta t \to 0$.[§]

The source of our difficulties in the considerations of these two examples is that in both cases we try to incorporate the discrete energies (1.10) into the framework of classical mechanics without any changes in the latter. Thus the theory with discrete energies must be very different from classical mechanics with its continuously variable energies.

[‡]H. Koppe [152].
[§]See also Example 1.3.

It is illuminating in this context to consider the linear oscillator in phase space (q, p) with

$$E = \frac{p^2}{2m} + \frac{1}{2}m\omega^2 q^2 = \text{const.} \tag{1.14}$$

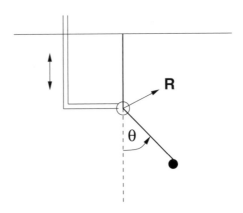

Fig. 1.4 The pendulum with variable length.

This equation is that of an ellipse as a comparison with the Cartesian form

$$\frac{x^2}{a^2} + \frac{y^2}{b^2} = 1$$

reveals immediately. Evidently the ellipses in the (q, p)-plane have semi-axes of lengths

$$a = \sqrt{\frac{2E}{m\omega^2}}, \quad b = \sqrt{2mE}. \tag{1.15}$$

Inserting here (1.10), we obtain

$$a_n = \sqrt{\frac{2(n + 1/2)\hbar\omega}{m\omega^2}}, \quad b_n = \sqrt{2m(n + 1/2)\hbar\omega}. \tag{1.16}$$

We see that for $n = 0, 1, 2, \ldots$ only certain ellipses are allowed. The area enclosed by such an ellipse is (note A earlier amplitude, now means area)

$$A_n = \pi a_n b_n = \frac{2\pi E_n}{\omega} = 2\pi\hbar\left(n + \frac{1}{2}\right), \tag{1.17a}$$

or

$$\oint p \, dq = 2\pi\hbar\left(n + \frac{1}{2}\right). \tag{1.17b}$$

In the first of the examples discussed above the contradiction arose as a consequence of our assumption that we could put the oscillator initially at any point in phase space, i.e. at some point which does not belong to one of the allowed ellipses. In the second example we chose $n = 0$ and thus restricted ourselves to the innermost orbit. However, we also assumed we would know at which point of the orbit the pendulum could be found.

Thus in attempting to incorporate the discrete quantization condition into the context of classical mechanics we see, that a system cannot be localized with arbitrary precision in phase space, in other words the area $\triangle A$, in which a system can be localized, is not nought. We can write this area

$$\triangle A \geq A_{n+1} - A_n \overset{(1.17a)}{=} 2\pi\hbar,$$

since the system cannot be "between" A_{n+1} and A_n. Since $\triangle A$ represents an element of area of the (q, p)-plane, we can write more precisely

$$\triangle p \triangle q \geq 2\pi\hbar. \tag{1.18}$$

This relation, called the *Heisenberg uncertainty relation*, implies that if we wish to make q very precise by arranging $\triangle q$ to be very small, the complementary uncertainty in momentum, $\triangle p$, becomes correspondingly large and extends over a large number of quantum states, as — for instance — in the second example considered above and illustrated in Fig. 1.5.

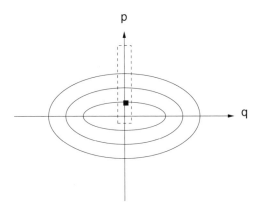

Fig. 1.5 Precise q implying large uncertainty in p.

Thus we face the problem of formulating classical mechanics in such a way that by some kind of extension or generalization we can find a way to quantum mechanics. Instead of the deterministic Newtonian mechanics — which for a given precise initial position and initial momentum of a system yields the precise values of these for any later time — we require a formulation answering the question: If the system is at time $t = 0$ in the area defined by the limits

$$0 \leq q \leq q + \triangle q, \quad 0 \leq p \leq p + \triangle p,$$

what can be said about its position at some later time $t = T$? The appropriate formulation does not yet have anything to do with quantum mechanics; however, it permits the transition to quantum mechanics, as we shall see. Before we continue in this direction, we return once again briefly to the historical development, and there to the ideas leading to *particle-wave duality*.[¶]

1.3   Particle-Wave Dualism

The wave nature of light can be deduced from the phenomenon of interference, as in a double-slit experiment, as illustrated in Fig. 1.6.

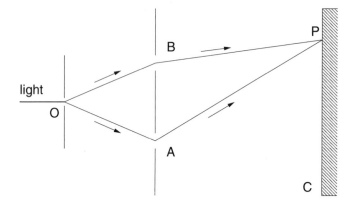

Fig. 1.6 Schematic arrangement of the double-slit experiment.

Light of wave-length λ from a source point O can reach point P on the observation screen C either through slit A or through slit B in the diaphragm placed somewhere in between. If the difference of the path lengths OBP, OAP is $n\lambda, n \in \mathbb{Z}$, the wave at P is re-inforced by superposition and one observes a bright spot; if the difference is $n\lambda/2$, the waves annul each other and one observes a dark spot. Both observations can be understood by a wave propagation of light. The *photoelectric effect*, however, seems to suggest a corpuscular nature of light. In this effect[*] light of frequency ν is sent onto a metal plate in a vacuum, and the electrons ejected by the light from the plate are observed by applying a potential difference between this plate and another one. The energy of the observed electrons depends only on ν and

---

[¶]See also M.–C. Combourieu and H. Rauch [58].

[*]This is explained in experimental physics; we therefore do not enter into a deeper explanation here.

the number of such *photo-electrons* on the intensity of the incoming light. This is true even for very weak light. Einstein concluded from this effect, that the energy in a light ray is transported in the form of localized packets, called *wave packets*, which are also described as *photons* or *quanta*. Indeed the Compton effect, i.e. the elastic scattering of light, demonstrates that photons can be scattered off electrons like particles. Thus whereas Planck postulated that an oscillator emits or absorbs radiation in units of $h\nu = \hbar\omega$, Einstein went further and postulated that radiation consists of discrete quanta.

Thus light can be attributed a wave nature but also a corpuscular, i.e. particle-like, nature. In the interference experiment light behaves like a wave, but in the photoelectric effect like a stream of particles. One could try to play a trick, and use radiation which is so weak that it can transport only very few photons. What does the interference pattern then look like? Instead of bands one observes a few point-like spots. With an increasing number of photons these spots become denser and produce bands. Thus the interference experiment is always indicative of the wave nature of light, whereas the photoelectric effect is indicative of its particle-like nature. Without going into further historical details we add here, that it was Einstein in 1905 who attributed a momentum p to the light quantum with energy $E = h\nu$, and both he and Planck attributed to this the momentum

$$p = \frac{h\nu}{c} = \frac{h}{\lambda}. \tag{1.19}$$

The hypothesis that every freely moving nonrelativistic microscopic particle with energy E and momentum $\mathbf{p}$ can be attributed a plane harmonic matter wave $\psi(\mathbf{r}, t)$ was put forward much later, i.e. in 1924, by de Broglie.[†] This wave can be written as a complex function

$$\psi(\mathbf{r}, t) = A e^{i\mathbf{k}\cdot\mathbf{r} - i\omega t},$$

where $\mathbf{r}$ is the position vector, and ω and $\mathbf{k}$ are given by

$$E = \hbar\omega, \quad \mathbf{p} = \hbar\mathbf{k}.$$

Thus particles also possess a wave-like nature. It is wellknown that this was experimentally verified by Davisson and Germer [64], who demonstrated the existence of electron waves by the observation of diffraction fringes instead of intensity distributions in appropriate experiments.

[†]L. de Broglie [39].

1.4   Particle-Wave Dualism and Uncertainties

We saw above that we can observe the wave nature of light in one type of experiment, and its particle-like nature in another. We cannot observe both types simultaneously, i.e. the wave-like nature together with the particle-like nature. Thus these wave and particle aspects are complementary, and show up only under specific experimental situations. In fact, they exclude each other. Every attempt to single out either of these aspects, requires a modification of the experiment which rules out every possibility to observe the other aspect.[‡] This becomes particularly clear, if in a double-slit experiment the detectors which register outcoming photons are placed immediately behind the diaphragm with the two slits: A photon is registered only in one detector, not in both — hence it cannot split itself. Applying the above uncertainty principle to this situation, we identify the attempt to determine which slit the photon passes through with the observation of its position coordinate q. On the other hand the observation of the interference fringes corresponds to the observation of its momentum p.[§] Since the reader will ask himself what happens in the case of a single slit, we consider this case in Example 1.2.

Example 1.2: The Single-Slit Experiment
Discuss the uncertainties of the canonical variables in relation to the diffraction fringes observed in a single-slit experiment.

Solution: Let light of wave-length λ fall vertically on a diaphragm S_1 with slit AB as shown schematicaly in Fig. 1.7.

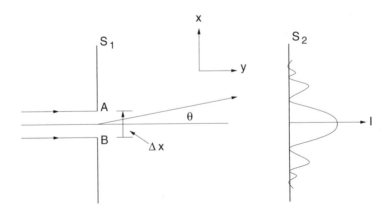

Fig. 1.7 Schematic arrangement of the single-slit experiment.

On the screen S_2 one then observes a diffraction pattern of alternately bright and dark fringes, in the

---

[‡]See, for instance, the discussion in A. Messiah [195], Vol. I, Sec. 4.4.4.

[§]Considerable discussion can be found in A. Rae [234].

figure indicated by maxima and minima of the light intensity I. As remarked earlier, the fringes are formed by interference of rays traversing different paths from the source to the observation screen. Before we enter into a discussion of uncertainties, we derive an expression for the intensity I. Since the derivation is not of primary interest here, we resort to a (still somewhat cumbersome) trick justification, which however can also be obtained in a rigorous way.[¶] We subdivide the distance $AB = \triangle x$ into N equal pieces $AP_1, P_1P_2, \ldots$, as indicated in Fig. 1.8.

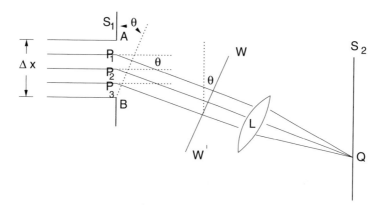

Fig. 1.8 The wave-front WW'.

We consider rays deflected by an angle θ with wave-front WW' and bundled by a lense L and focussed at a point Q on the screen S_2. Since WW' is a wave-front, all points on it have the same phase, so that light sent out from a source at Q reaches every point on WW' at the same time and across equal distances. Hence a phase difference at Q can be attributed to different path lengths from $P_1, P_2, \ldots$ to WW'. Considering two paths from neighbouring points P_i, P_j along AB, the difference in their lengths is $\triangle x \sin\theta / N$. In the case of a wave having the shape of the function

$$\sin kr = \sin \frac{2\pi}{\lambda} r,$$

this implies a phase difference given by

$$\delta_N = \frac{2\pi}{\lambda} \frac{\triangle x}{N} \sin\theta. \tag{1.20}$$

Just as we can represent an amplitude r having phase θ by a vector $\mathbf{r}$, i.e. $\mathbf{r} \to |\mathbf{r}| \exp(i\theta)$, we can similarly imagine the wave at Q, and this means its amplitude and phase, as represented by a vector, and similarly the wave of any component of the ray passing through $AP_1, P_1P_2, \ldots$. If we represent their effects at Q by vectors of equal moduli but different directions, their sum is the resultant OP_N as indicated in Fig. 1.9. In the limit $N \to \infty$ the N vectors produce the arc of a circle. The angle δ between the tangents at the two ends is the phase difference of the rays from the edges of the slit:

$$\delta = 2\alpha = \lim_{N\to\infty} N\delta_N = \frac{2\pi}{\lambda} \triangle x \sin\theta. \tag{1.21}$$

If all rays were in phase, the amplitude, given by the length of the arc OQ, would be given by the chord OQ. Hence we obtain for the amplitude A at Q if A_0 is the amplitude of the beam at the slit:

$$A = A_0 \frac{\text{length of chord } OQ}{\text{length of arc } OQ} = A_0 \frac{2a\sin\alpha}{a\,2\alpha} = A_0 \frac{\sin\alpha}{\alpha}. \tag{1.22}$$

[¶]S. G. Starling and A. J. Woodall [260], p. 664. For other derivations see e.g. A. Brachner and R. Fichtner [32], p. 52.

The intensity at the point Q is therefore

$$I_\theta = I_0 \left(\frac{\sin \alpha}{\alpha} \right)^2,$$

where from Eq. (1.21)

$$\alpha = \frac{\pi}{\lambda} \triangle x \sin \theta = \frac{k}{2} \triangle x \sin \theta.$$

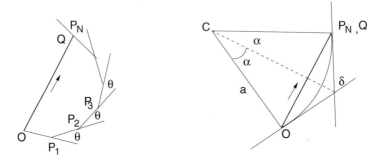

Fig. 1.9 The resultant OP_N of N equal vectors with varying inclination.

Thus the intensity at the point Q is

$$I_\theta = I_0 \frac{\sin^2(k \triangle x \sin \theta / 2)}{(k \triangle x \sin \theta / 2)^2}. \tag{1.23}$$

The maxima of this distribution are obtained for

$$\frac{1}{2} k \triangle x \sin \theta = (2n + 1) \frac{\pi}{2}, \quad \text{i.e. for } \triangle x \sin \theta = (2n + 1) \frac{\pi}{k} = (2n + 1) \frac{\lambda}{2} \tag{1.24a}$$

and minima for

$$\frac{1}{2} k \triangle x \sin \theta = n\pi, \quad \text{i.e. for } \triangle x \sin \theta = n\lambda. \tag{1.24b}$$

The maxima are not exactly where only the numerator assumes extremal values, since the variable also occurs in the denominator, but nearby.

We return to the single-slit experiment. Let the light incident on the diaphragm S_1 have a sharp momentum $p = h/\lambda$. When the ray passes through the slit the position of the photon is fixed by the width of the slit $\triangle x$, and afterwards the photon's position is even less precisely known. We have a situation which — for the observation on the screen S_2 is a past (the uncertainty relation does not refer to this past with $p_x = 0$, rather to the position and momentum later; for the situation of the past $\triangle x \triangle p$ is less than h). The above formula (1.23) gives the probability that after passing through the slit the photon appears at some point on the screen S_2. This probability says, that the photon's momentum component p_x after passing through the slit is no longer zero, but indeterminate. It is not possible to predict at which point on S_2 the photon will appear (if we knew this, we could derive p_x from this). The momentum uncertainty in the direction x can be estimated from the geometry of Fig. 1.10, where θ is the angle in the direction to the first minimum:

$$\triangle p_x = 2p_x = 2p \sin \theta = \frac{2h}{\lambda} \sin \theta. \tag{1.25}$$

From Eq. (1.24b) we obtain for the angle θ in the direction of the first minimum

$$\triangle x \sin \theta = \lambda,$$

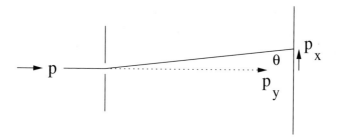

Fig. 1.10 The components of momentum p.

so that

$$\triangle x \triangle p_x = 2h.$$

If we take the higher order minima into account, we obtain $\triangle x \triangle p_x = 2nh$, or

$$\triangle x \triangle p_x \geq h.$$

We see that as a consequence of the indeterminacy of position and momentum, one has to introduce probability considerations. The limiting value of the uncertainty relation does not depend on how we try to measure position and momentum. It does also not depend on the type of particle (what applies to electromagnetic waves, applies also to particle waves).

1.4.1 Further thought experiments

Another experiment very similar to that described above is the attempt to localize a particle by means of an idealized microscope consisting of a single lense. This is depicted schematically in Fig. 1.11.

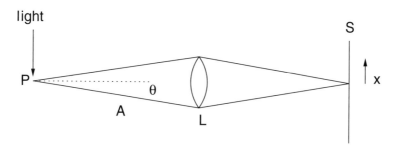

Fig. 1.11 Light incident as shown.

The resolving power of a lense L is determined by the separation $\triangle x$ of the first two neighbouring interference fringes, i.e. the position of a particle is at best determinable only up to an uncertainty $\triangle x$. Let θ be one half of the angle as shown in Fig. 1.11, where P is the particle. We allow light to fall in the direction of $-\mathbf{x}$ on the particle, from which it is scattered. We assume a quantum of light is scattered from P through the lense L to S where it is focussed and registered on a photographic plate. For the resolving power $\triangle x$ of the lense one can derive a formula like Eqs. (1.24a), (1.24b) . This is derived in books on optics, and hence will not be verified here, i.e.[||]

$$\triangle x \simeq \frac{\lambda}{2\sin\theta}. \tag{1.26a}$$

The precise direction in which the photon with momentum $p = h/\lambda$ is scattered is not known. However, after scattering of the photon, for instance along PA in Fig. 1.11, the uncertainty in its x-component is

$$\triangle p_x = 2p\sin\theta = \frac{2h}{\lambda}\sin\theta \tag{1.26b}$$

(prior to scattering the x-components of the momenta of the particle and the photon may be known precisely). From Eqs. (1.26a), (1.26b) we obtain again

$$\triangle x\, \triangle p_x \sim h.$$

The above considerations lead to the question of what kind of physical quantities obey an uncertainty relation. For instance, how about momentum and kinetic energy T? Apparently there are "*compatible*" and "*incompatible*" quantities, the latter being those subjected to an uncertainty relation. If the momentum p_x is "sharp", meaning $\triangle p_x = 0$, then also $T = p_x^2/2m$ is sharp, i.e. T and p_x are compatible. In the case of angular momentum $\mathbf{L} = \mathbf{r} \times \mathbf{p}$, we have

$$|\mathbf{L}| = |\mathbf{r}||\mathbf{p}'| = rp',$$

where $p' = p\sin\phi$. As one can see, r and p' are perpendicular to each other and thus can be sharp simultaneously. If p' lies in the direction of x, we have

$$\triangle x\, \triangle p' \geq h,$$

where now $\triangle x = r\triangle\varphi$, φ being the azimuthal angle, i.e.

$$r\triangle\varphi\, \triangle p' \geq h, \quad \text{i.e.} \quad \triangle L\, \triangle\varphi \geq h.$$

Thus the angular momentum $\mathbf{L}$ is not simultaneously exactly determinable with the angle φ. This means, when $\mathbf{L}$ is known exactly, the position of the object in the plane perpendicular to $\mathbf{L}$ is totally indeterminate.

Finally we mention an uncertainty relation which has a meaning different from that of the relations considered thus far. In the relation $\triangle x\, \triangle p_x \geq 0$ the

---

[||]See, for instance, N. F. Mott, [199], p. 111. In some books the factor of "2" is missing; see, for instance, S. Simons [251], p. 12.

quantities $\triangle x, \triangle p_x$ are uncertainties at one and the same instant of time, and x and p_x cannot assume simultaneously precisely determined values. If, however, we consider a *wave packet*, such as we consider later, which spreads over a distance $\triangle x$ and has *group velocity* $v_G = p/m$, the situation is different. The energy E of this wave packet (as also its momentum) has an uncertainty given by

$$\triangle E \approx \frac{\partial E}{\partial p} \triangle p = v_G \triangle p.$$

The instant of time t at which the wave packet passes a certain point x is not unique in view of the wave packet's spread $\triangle x$. Thus this time t is uncertain by an amount

$$\triangle t \approx \frac{\triangle x}{v_G}.$$

It follows that

$$\triangle t \triangle E \approx \triangle x \triangle p \geq h. \tag{1.27}$$

Thus if a particle does not remain in some state of a number of states for a period of time longer than $\triangle t$, the energy values in this state have an indeterminacy of $|\triangle E|$.

1.5 Bohr's Complementarity Principle

Vaguely expressed the complementarity principle says that two canonically conjugate variables like position coordinate x and the the associated canonical momentum p of a particle are related in such a way that the measurement of one (with uncertainty $\triangle x$) has consequences for the measurement of the other. But this is essentially what the uncertainty relation expresses. Bohr's complementarity principle goes further. Every measurement we are interested in is performed with a macroscopic apparatus at a microscopic object. In the course of the measurement the apparatus interferes with the state of the microscopic object. Thus really one has to consider the combined system of both, not a selected part alone. The uncertainty relation shows: If we try to determine the position coordinate with utmost precision all information about the object's momentum is lost — precisely as a consequence of the disturbance of the microscopic system by the measuring instrument. The so-called Kopenhagen view, i.e. that of Bohr, is expressed in the thesis that the microscopic object together with the apparatus determine the result of a measurement. This implies that if a beam of light or electrons is passed through a double-slit (this being the apparatus in this case) the photons or electrons behave like waves precisely because under these observation conditions they are waves, and that on the other hand, when observed in a counter, they behave like a stream of particles because under these conditions they are particles. In fact, without performance of some measurement (e.g. at some electron) we cannot say anything about the object's existence. The Kopenhagen view can also be expressed by saying that a quantity is real, i.e. physical, only when it is measured, or — put differently — the properties of a quantum system (e.g. whether wave-like or corpuscular) depend on the method of observation. This is the domain of conceptual difficulties which we do not enter into in more detail here.[*]

1.6   Further Examples

Example 1.3: The oscillator with variable frequency

Consider an harmonic oscillator (i.e. simple pendulum) with time-dependent frequency $\omega(t)$.
(a) Considering the case of a monotonically increasing frequency $\omega(t)$, i.e. $d\omega/dt > 0$, from ω_0 to ω', show that the energy E' satisfies the following inequality

$$E_0 \leq E' \leq \frac{\omega'^2}{\omega_0^2} E_0, \tag{1.28}$$

where E_0 is its energy at deflection angle $\theta = \theta_0$. Compare the inequality with the quantum mechanical zero point energy of an oscillator.
(b) Considering the energy of the oscillator averaged over one period of oscillation (for slow, i.e. adiabatic, variation of the frequency) show that the energy becomes proportional to ω. What is the quantum mechanical interpretation of the result?

Solution: (a) The equation of motion of the oscillator of mass m and with variable frequency $\omega(t)$ is

$$m\ddot{x} + m\omega^2(t)x = 0,$$

where, according to the given conditions,

$$\frac{d\omega}{dt} \geq 0, \quad \omega = \omega_0 \text{ at } t = 0, \quad \omega = \omega' \text{ at } t = T,$$

i.e. $\omega(t)$ grows monotonically. Multiplying the equation of motion by $\dot{x}$ we can rewrite it as

$$\frac{d}{dt}\left[\frac{1}{2}m\dot{x}^2 + \frac{1}{2}m\omega^2(t)x^2\right] - \frac{1}{2}mx^2\frac{d\omega^2}{dt} = 0.$$

The energy of the oscillator is

$$E = \frac{1}{2}m\dot{x}^2 + \frac{1}{2}m\omega^2(t)x^2, \quad \text{so that} \quad \frac{dE}{dt} = \frac{1}{2}mx^2\frac{d\omega^2}{dt} \geq 0, \tag{1.29}$$

where we used the given conditions in the last step. On the other hand, dividing the equation of motion by ω^2 and proceeding as before, we obtain

$$\frac{\dot{x}}{\omega^2}[m\ddot{x} + m\omega^2(t)x] = 0, \quad \text{i.e.} \quad \frac{d}{dt}\left[\frac{1}{\omega^2}\frac{1}{2}m\dot{x}^2 + \frac{1}{2}mx^2\right] - \frac{1}{2}m\dot{x}^2\frac{d}{dt}\left(\frac{1}{\omega^2}\right) = 0,$$

---

[*]See e.g. A. Rae [234]; P. C. W. Davies and J. R. Brown [65].

or

$$\frac{d}{dt}\left(\frac{E}{\omega^2}\right) = \frac{1}{2}m\dot{x}^2\frac{d}{dt}\left(\frac{1}{\omega^2}\right) = -\frac{m\dot{x}^2}{\omega^3}\frac{d\omega}{dt} \leq 0, \tag{1.30}$$

where the inequality again follows as before. We deduce from the last relation that

$$\frac{1}{\omega^2}\frac{dE}{dt} - \frac{E}{\omega^4}\frac{d\omega^2}{dt} \leq 0, \quad \text{i.e.} \quad \frac{1}{E}\frac{dE}{dt} \leq \frac{1}{\omega^2}\frac{d\omega^2}{dt}. \tag{1.31}$$

Integrating we obtain

$$\int_{E_0}^{E'}\frac{dE}{E} \leq \int_{\omega_0^2}^{\omega'^2}\frac{d\omega^2}{\omega^2}, \quad \text{i.e.} \quad [\ln E]_{E_0}^{E'} \leq [\ln \omega^2]_{\omega_0^2}^{\omega'^2}, \quad \text{i.e.} \quad \frac{E'}{E_0} \leq \frac{\omega'^2}{\omega_0^2},$$

or

$$E' \leq \frac{\omega'^2}{\omega_0^2}E_0.$$

Next we consider the case of the harmonic oscillator as a simple pendulum in the gravitational field of the Earth with

$$\ddot{\theta} + \omega_0^2\theta \simeq 0, \quad \omega_0^2 = \frac{g}{l},$$

and we assume that — as explained in the foregoing — the length of the pendulum is reduced by one half so that

$$\omega'^2 = 2\frac{g}{l} = 2\omega_0^2.$$

Then the preceding inequality becomes

$$E' \leq 2E_0.$$

In shortening the length of the pendulum we apply energy (work against the tension in the string), maximally however E_0. Only in the case of the instantaneous reduction of the length at $\theta = 0$ (the pivot does not touch the string!) no energy is added, so that in this case $E' = E_0$, i.e.

$$E_0 \leq E' \leq 2E_0.$$

We can therefore rewrite the earlier inequality as

$$E_0 \leq E' \leq \frac{\omega'^2}{\omega_0^2}E_0.$$

Just as the equality on the left applies in the case of an instantaneous increase of the frequency (shortening of pendulum string), so the equality on the right applies to $\theta = \theta_{\max}$. In *classical physics* we have

$$E = \frac{1}{2}m\dot{x}^2 + \frac{1}{2}m\omega^2x^2.$$

If no energy is added, but ω^2 is replaced by $2\omega^2$, then x changes, and also $\dot{x}$, i.e. x becomes shorter and $\dot{x}$ becomes faster. The quantum mechanical expression for the energy of the oscillator in its ground state is the zero point energy $E = \hbar\omega/2$. Here in *quantum physics* we cannot change ω without changing E. This means if we double ω instantaneously (i.e. in a time interval $\Delta t \to 0$) without addition of energy (to $\hbar\omega/2$), then the result $E' = \hbar\omega$ is incorrect by $\Delta E = \hbar\omega/2$. We cannot have simultaneously $\Delta t \to 0$ and error $\Delta E = 0$.

(b) The classical expression for E contains ω quadratically, the quantum mechanical expression is linear in ω. We argue now that we can obtain an expression for $E_{\text{classical}}$ by assuming that $\omega(t)$ varies very little (i.e. "adiabatically") within a period of oscillation of the oscillator, T. Classical mechanics is deterministic (i.e. the behaviour at time t follows from the equation of motion and the initial conditions); hence for the consideration of a single mass point there is no reason for an averaging over a period, unless we are not interested in an exact value but, e.g. in the average

$$\left\langle \frac{1}{2}mx^2 \right\rangle = \frac{1}{T}\int_0^T \frac{1}{2}mx^2(t)dt. \tag{1.32}$$

If ω is the frequency of $x(t)$, i.e. $x(t) \propto \cos\omega t$ or $\sin\omega t$ depending on the initial condition, then $\dot{x}^2(t) = \omega^2 x^2$ and hence

$$\left\langle \frac{1}{2}m\omega^2 x^2 \right\rangle = \left\langle \frac{1}{2}m\dot{x}^2 \right\rangle = \frac{1}{2}E$$

(as follows also from the virial theorem). If we now insert in the equation for dE/dt, i.e. in Eq. (1.29), for $mx^2/2$ the mean value

$$\left\langle \frac{1}{2}mx^2 \right\rangle = \frac{1}{2}\frac{E}{\omega^2},$$

we obtain

$$\frac{dE}{dt} = \left\langle \frac{1}{2}mx^2 \right\rangle \frac{d\omega^2}{dt} = \frac{E}{2\omega^2}\frac{d\omega^2}{dt}, \quad \text{or} \quad \frac{dE}{E} = \frac{1}{2}\frac{d\omega^2}{\omega^2} = \frac{d\omega}{\omega},$$

and hence

$$\frac{E}{\omega} = \text{const.}$$

In quantum mechanics with $E = \hbar\omega(n+1/2)$ this implies $\hbar(n+1/2) = \text{const.}$, i.e. $n = \text{const.}$ This means, with slow variation of the frequency the system remains in state n. This is an example of the so-called *adiabatic theorem of Ehrenfest*, which formulates this in a general form.[†]

Example 1.4: Angular spread of a beam
A dish-like aerial of radius R is to be designed which can send a microwave beam of wave-length $\lambda = 2\pi\hbar/p$ from the Earth to a satellite. Estimate the angular spread θ of the beam.

Solution: Initially the photons are restricted to a transverse spread of length $\triangle x = 2R$. From the uncertainty relation we obtain the uncertainty $\triangle p_x$ of the transverse momentum p_x as $\triangle p_x \simeq \hbar/2R$. Hence the angle θ is given by

$$\theta \simeq \frac{\triangle p_x}{p} = \frac{\hbar}{2R}\left(\frac{\lambda}{2\pi\hbar}\right) = \frac{\lambda}{4\pi R}.$$

[†]See e.g. L. Schiff [243], pp. 25 - 27.

Chapter 2

Hamiltonian Mechanics

2.1 Introductory Remarks

In this chapter we first recapitulate significant aspects of the Hamiltonian formulation of classical mechanics. In particular we recapitulate the concept of Poisson brackets and re-express Hamilton's equations of motion in terms of these. We shall then make the extremely important observation that these equations can be solved on the basis of very general properties of the Poisson bracket, i.e. without reference to the original definition of the latter. This observation reveals that classical mechanics can be formulated in a framework which permits a generalization by replacing the c-number valued functions appearing in the Poisson brackets by a larger class of quantities, such as matrices and operators. Thus in this chapter we attempt to approach quantum mechanics as far as possible within the framework of classical mechanics. We shall see that we can even define such concepts as Schrödinger and Heisenberg pictures in the purely classical context.

2.2 The Hamilton Formalism

In courses on classical mechanics it is shown that Hamilton's equations can be derived in a number of ways, e.g. from the Lagrangian with a Legendre transform or with a variational principle from the Hamiltonian $H(q_i, p_i)$, i.e.

$$\delta \int_{t_1}^{t_2} \left[\sum_i p_i \dot{q}_i - H(q_i, p_i) \right] dt = 0,$$

where now (different from the derivation of the Euler-Lagrange equations) the momenta p_i and coordinates q_i are treated as independent variables. As is wellknown, one obtains the Hamilton equations*

$$\dot{q}_i = \frac{\partial H}{\partial p_i}, \quad \dot{p}_i = -\frac{\partial H}{\partial q_i}. \tag{2.1}$$

In this Hamilton formalism it is wrong to consider the momentum p_i as $m\dot{q}_i$, i.e. as mass times velocity. Rather, p_i has to be considered as an independent quantity, which can be observed directly at time t, whereas the velocity requires observations of space coordinates at different times, since

$$\dot{q}_i = \lim_{\delta t \to 0} \frac{q_i(t + \delta t) - q_i(t)}{\delta t}.$$

Real quantities which are directly observable are called *observables*. A system consisting of several mass points is therefore described by a number of such variables, which all together describe the *state* of the system. All functions $u(q_i, p_i)$ of q_i, p_i are therefore again observables. Compared with an arbitrary function $f(q_i, p_i, t)$, the entire time-dependence of observables $u(q_i, p_i)$ is contained implicitly in the canonical variables q_i and p_i. The total time derivative of u can therefore be rewritten with the help of Eqs. (2.1) as

$$\frac{d}{dt} u(q_i, p_i) = \sum_i \left(\frac{\partial u}{\partial q_i} \dot{q}_i + \frac{\partial u}{\partial p_i} \dot{p}_i \right) = \sum_i \left(\frac{\partial u}{\partial q_i} \frac{\partial H}{\partial p_i} - \frac{\partial u}{\partial p_i} \frac{\partial H}{\partial q_i} \right). \tag{2.2}$$

If we have only one degree of freedom ($i = 1$), this expression is simply a functional determinant. One now defines as (nonrelativistic) *Poisson bracket* the expression[†]

$$\{A, B\} := \sum_i \left(\frac{\partial A}{\partial q_i} \frac{\partial B}{\partial p_i} - \frac{\partial A}{\partial p_i} \frac{\partial B}{\partial q_i} \right). \tag{2.3}$$

With this definition we can rewrite Eq. (2.2) as

$$\frac{du}{dt} = \{u, H\}. \tag{2.4}$$

This equation is, in analogy with Eqs. (2.1), the equation of motion of the observable u. One can verify readily that Eq. (2.4) contains as special cases the Hamilton Eqs. (2.1). We can therefore consider Eq. (2.4) as the generalization of Eqs. (2.1). It suggests itself therefore to consider more closely the properties of the symbols (2.3). The following properties can be verified:

---

*See e.g. H. Goldstein [114], chapter VII.

[†]As H. Goldstein [114] remarks at the end of his chapter VIII, the standard reference for the application of Poisson brackets is the book of P. A. M. Dirac [75], chapter VIII. It was only with the development of quantum mechanics by Heisenberg and Dirac that Poisson brackets gained widespread interest in modern physics.

(1) Antisymmetry:
$$\{A, B\} = -\{B, A\}, \tag{2.5a}$$

(2) linearity:

$$\{A, \alpha_1 B_1 + \alpha_2 B_2\} = \alpha_1\{A, B_1\} + \alpha_2\{A, B_2\}, \tag{2.5b}$$

(3) complex conjugation (note: observables are real, but could be multiplied by a complex number):

$$\{A, B\}^* = \{A^*, B^*\}, \tag{2.5c}$$

(4) product formation:

$$\{A, BC\} = \{A, B\}C + B\{A, C\}, \tag{2.5d}$$

(5) Jacobi identity:

$$\{A, \{B, C\}\} + \{B, \{C, A\}\} + \{C, \{A, B\}\} = 0. \tag{2.5e}$$

The first three properties are readily seen to hold. Property (2.5d) is useful in calculations. As long as we are concerned with commuting quantities, like here, it is irrelevant whether we write

$$\{A, B\}C \quad \text{or} \quad C\{A, B\}.$$

Later we shall consider noncommuting quantities, then the ordering is taken as in (2.5d) above.

If we evaluate the Poisson brackets for q_i, p_i, we obtain the *fundamental Poisson brackets*. These are

$$\{q_i, q_k\} = 0, \quad \{q_i, p_k\} = \delta_{ik}, \quad \{p_i, p_k\} = 0. \tag{2.6}$$

We can now show, that the very general Eq. (2.4), which combines the Hamilton equations, can be solved solely with the help of the properties of Poisson brackets and the fundamental Poisson brackets (2.6), in other words without any reference to the original definition (2.3) of the Poisson bracket. If, for example, we wish to evaluate $\{A, B\}$, where A and B are arbitrary observables, we expand A and B in powers of q_i and p_i and apply the above rules until only the fundamental brackets remain. Since Eqs. (2.6) give the values of these, the Poisson bracket $\{A, B\}$ is completely evaluated. As an example we consider a case we shall encounter again and again, i.e. that of the linear harmonic oscillator. The original definition of the Poisson bracket will not be used at all. In the evaluation one should also note that the fact that q_i and p_i are ordinary real number variables and that $H(q, p)$ is an ordinary function is also irrelevant. Since constants are also irrelevant in this context, we consider as Hamiltonian the function

$$H(q, p) = \frac{1}{2}(p^2 + q^2). \tag{2.7}$$

According to Eq. (2.4) we have for $u = q, p$,

$$\dot{q} = \{q, H\}, \tag{2.8a}$$

and

$$\dot{p} = \{p, H\}. \tag{2.8b}$$

We insert (2.7) into (2.8a) and use the properties of the Poisson bracket and Eqs. (2.6). Then we obtain:

$$\begin{aligned}
\dot{q} &= \left\{q, \frac{1}{2}(p^2 + q^2)\right\} = \frac{1}{2}(\{q, p^2\} + \{q, q^2\}) \\
&= \frac{1}{2}(\{q, p\}p + p\{q, p\}) \\
&= p.
\end{aligned} \tag{2.9}$$

Similarly we obtain from Eq. (2.8b)

$$\dot{p} = -q. \tag{2.10}$$

From Eqs. (2.9) and (2.10) we deduce

$$\ddot{q} = \dot{p} = -q, \quad \ddot{q} + q = 0,$$

and so

$$\ddot{q} = -q, \quad \dddot{q} = -\dot{q}, \quad \ddddot{q} = q, \dots,$$

from which we infer that

$$q(t) = q_0 \cos t + p_0 \sin t, \tag{2.11a}$$

or

$$q(t) = q_0 + p_0 t - \frac{1}{2} q_0 t^2 - \frac{1}{3!} p_0 t^3 + \dots. \tag{2.11b}$$

In classical mechanics one studies also *canonical transformations*. These are transformations

$$q_i \longrightarrow Q_i = Q_i(q, p, t), \quad p_i \longrightarrow P_i = P_i(q, p, t), \tag{2.12}$$

for which the new coordinates are also *canonical*, which means that a Hamilton function $K(Q, P)$ exists, for which Hamilton's equations hold, i.e.

$$\dot{Q}_i = \frac{\partial K}{\partial P_i}, \quad \dot{P}_i = -\frac{\partial K}{\partial Q_i}. \tag{2.13}$$

We write the reversal of the transformation (2.12)

$$Q_i \longrightarrow q_i = q_i(Q, P, t), \quad P_i \longrightarrow p_i = p_i(Q, P, t). \tag{2.14}$$

With the help of the definition (2.3) we can now express the Poisson bracket $\{A, B\}$ of two observables A and B in terms of either set of canonical variables, i.e. as

$$\{A, B\}_{q,p} \quad \text{or as} \quad \{A, B\}_{Q,P}.$$

One can then show that

$$\{A, B\}_{q,p} = \{A, B\}_{Q,P}, \tag{2.15a}$$

provided the transformation $q, p \leftrightarrow Q, P$ is canonical in the sense defined above. The proof requires the invariance of the fundamental Poisson brackets under canonical transformations, i.e. that (dropping the subscripts therefore)

$$\{p_i, p_k\} = 0, \quad \{q_i, p_k\} = \delta_{ik}, \quad \{q_i, q_k\} = 0,$$
$$\{P_i, P_k\} = 0, \quad \{Q_i, P_k\} = \delta_{ik}, \quad \{Q_i, Q_k\} = 0. \tag{2.15b}$$

The proof of the latter invariance is too long to be reproduced in detail here but can be found in the book of Goldstein.[‡] Hence in Example 2.1 we verify only Eq. (2.15a). Example 2.2 below contains a further illustration of the use of Poisson brackets, and Example 2.3 deals with the relativistic extension.

In classical mechanics we learned yet another important aspect of the Hamilton formalism: We can inquire about that particular canonical transformation, which transforms q_i, p_i back to their constant initial values, i.e. those at a time $t = 0$. Of course, this transformation is described precisely by the equations of motion but we shall not consider this in more detail here.

Example 2.1: Canonical invariance of Poisson bracket

Assuming the invariance of the fundamental Poisson brackets under canonical transformations $q_j = q_j(Q, P), p_j = p_j(Q, P)$, verify that the Poisson bracket of two observables A and B is invariant, i.e. Eq. (2.15a).

[‡]H. Goldstein [114], chapter VIII.

Solution: Using the definition of the Poisson bracket applied to A and B we have

$$
\begin{aligned}
\{A, B\}_{q,p} &= \sum_j \left(\frac{\partial A}{\partial q_j} \frac{\partial B}{\partial p_j} - \frac{\partial A}{\partial p_j} \frac{\partial B}{\partial q_j} \right) \\
&= \sum_{j,k} \left[\frac{\partial A}{\partial q_j} \left(\frac{\partial B}{\partial Q_k} \frac{\partial Q_k}{\partial p_j} + \frac{\partial B}{\partial P_k} \frac{\partial P_k}{\partial p_j} \right) - \frac{\partial A}{\partial p_j} \left(\frac{\partial B}{\partial Q_k} \frac{\partial Q_k}{\partial q_j} + \frac{\partial B}{\partial P_k} \frac{\partial P_k}{\partial q_j} \right) \right] \\
&= \sum_k \left[\{A, Q_k\}_{q,p} \frac{\partial B}{\partial Q_k} + \{A, P_k\}_{q,p} \frac{\partial B}{\partial P_k} \right]
\end{aligned}
$$

Replacing here A by Q and B by A, we obtain

$$
\{Q_k, A\}_{q,p} = \sum_j \left[\{Q_k, Q_j\}_{q,p} \frac{\partial A}{\partial Q_j} + \{Q_k, P_j\}_{q,p} \frac{\partial A}{\partial P_j} \right] \overset{(2.15b)}{=} \frac{\partial A}{\partial P_k}.
$$

Replacing in the above A by P and B by A, we obtain analogously

$$
\{P_k, A\}_{q,p} = -\frac{\partial A}{\partial Q_k}.
$$

Inserting both of these results into the first equation, we obtain as claimed by Eq. (2.15a)

$$
\{A, B\}_{q,p} = \sum_k \left(\frac{\partial A}{\partial Q_k} \frac{\partial B}{\partial P_k} - \frac{\partial A}{\partial P_k} \frac{\partial B}{\partial Q_k} \right) = \{A, B\}_{Q,P}.
$$

Example 2.2: Solution of Galilei problem with Poisson brackets

Consider the Hamiltonian for the free fall of a mass point m_0 in the gravitational field (linear potential),

$$
H = \frac{1}{2m_0} p^2 + m_0 g q
$$

and solve the canonical equations with Poisson brackets for initial conditions $q(0) = q_0, p(0) = p_0$.

Solution: The solution can be looked up in the literature.[§]

Example 2.3: Relativistic Poisson brackets

By extending q_i, p_i to four-vectors (in a $(1 + 3)$-dimensional space) define relativistic Poisson brackets.

Solution: Relativistically we have to treat space and time on an equal footing. Thus we extend q and p to space-time vectors $(t, q/c)$ and (E, pc), their product $Et - qp$ being relativistically invariant. Thus whenever q and p are multiplied, we have $-Et$. The relativistic Poisson bracket (subscript r) therefore becomes

$$
\{u, F\}_r = \left\{ \frac{\partial u}{\partial q} \frac{\partial F}{\partial p} - \frac{\partial u}{\partial p} \frac{\partial F}{\partial q} \right\} - \left\{ \frac{\partial u}{\partial t} \frac{\partial F}{\partial E} - \frac{\partial u}{\partial E} \frac{\partial F}{\partial t} \right\}.
$$

Consider

$$
F = H(q, p) - E(t).
$$

[§]See P. Mittelstaedt [197], p. 236.

(This is, of course, numerically zero, but partial derivatives of F do not vanish, since H is expressed as a function of q and p, and E as a function of t). Then

$$\{u, H(q,p) - E(t)\}_r = \frac{\partial u}{\partial q}\frac{\partial H}{\partial p} - \frac{\partial u}{\partial p}\frac{\partial H}{\partial q} + \frac{\partial u}{\partial t}\frac{\partial E(t)}{\partial E} = \frac{\partial u}{\partial t} + \frac{\partial u}{\partial q}\dot{q} + \frac{\partial u}{\partial p}\dot{p} = \frac{du}{dt}.$$

Hence

$$\frac{du}{dt} = \{u, H(q,p) - E(t)\}_r.$$

Relativistically we really should have $du/d\tau$, where $d\tau$ is the difference of proper time given by

$$(d\tau)^2 = (dt)^2 - \frac{dq^2}{c^2}, \quad \frac{d\tau}{dt} = \sqrt{1 - \frac{1}{c^2}\left(\frac{dq}{dt}\right)^2}, \quad \text{with} \quad \frac{du}{d\tau} = \frac{du}{dt}\frac{1}{\sqrt{1 - \dot{q}^2/c^2}}.$$

2.3 Liouville Equation, Probabilities

2.3.1 Single particle consideration

We continue to consider classical mechanics in which the canonical coordinates q_i, p_i are the coordinates of some mass point m_i, and the space spanned by the entire set of canonical coordinates is described as its *phase space*. But now we consider a system whose phase space coordinates are not known precisely. Instead we assume a case in which we know only that the system is located in some particular domain of phase space. Let us assume that at some initial time t_0 the system may be found in a domain $G_0(q,p)$ around some point q_0, p_0, and at time $t > t_0$ in a domain $G_1(q,p)$. Of course, it is the equations of motion which lead from $G_0(q,p)$ to $G_1(q,p)$. Since Hamilton's equations give a continuous map of one domain onto another, also boundary points of one domain are mapped into boundary domains of the other, so that $G_0(q_0, p_0) = G_1(q,p)$, i.e. if q_0, p_0 is a point on G_0, one obtains G_1 with

$$q = q(q_0, p_0, t_0; t), \quad p = p(q_0, p_0, t_0; t).$$

We distinguish in the following between two kinds of probabilities. We consider first the *a priori weighting* or *a priori probability*, g, which is the probability of a particle having a coordinate q between q and $q + \triangle q$ and a momentum p between p and $p + \triangle p$. This probability is evidently proportional to $\triangle q \triangle p$, i.e.

$$g \propto \triangle q \triangle p. \tag{2.16}$$

For example, in the case of the linear oscillator with energy E given by Eq. (1.14) and area A of the phase space ellipse given by Eq. (1.17a), we have

$$A = \oint dp\, dq = \frac{2\pi E}{\omega} \quad \text{and hence} \quad g \propto \frac{dE}{\omega}.$$

If g depended on time t it would be dynamical and would involve known information about the particle. Thus g must be independent of t, as is demonstrated by *Liouville's theorem* in Example 2.4; in view of this independence it can be expressed in terms of the conserved energy E. Example 2.5 thereafter provides an illustration of the a priori weighting expressed in terms of energy E.

Example 2.4: Liouville's theorem

Show that $\triangle q \triangle p$ is independent of time t, which means, this has the same value at a time t_0, as at a time $t_0' \neq t_0$.

Solution: We consider

$$\frac{d}{dt}\ln(\triangle q \triangle p) = \frac{d(\triangle q)}{dt}\frac{1}{\triangle q} + \frac{d(\triangle p)}{dt}\frac{1}{\triangle p}.$$

Here $d(\triangle q)/dt$ is the rate at which the q-walls of the phase space element move away from the centre of the element,

$$\dot{q} + \frac{\partial \dot{q}}{\partial q}\frac{\triangle q}{2} \quad \text{to the right} \quad \text{and} \quad \dot{q} - \frac{\partial \dot{q}}{\partial q}\frac{\triangle q}{2} \quad \text{to the left.}$$

Hence from the difference:

$$\frac{d(\triangle q)}{dt} = \frac{\partial \dot{q}}{\partial q}\triangle q, \quad \text{and similarly} \quad \frac{d(\triangle p)}{dt} = \frac{\partial \dot{p}}{\partial p}\triangle p,$$

and with with Hamilton's equations (2.1):

$$\frac{d}{dt}\ln(\triangle q \triangle p) = \frac{\partial \dot{q}}{\partial q} + \frac{\partial \dot{p}}{\partial p} = \frac{\partial^2 H}{\partial q \partial p} - \frac{\partial^2 H}{\partial p \partial q} = 0.$$

Example 2.5: A priori weighting of a molecule

If the rotational energy of a diatomic molecule with moment of inertia I is

$$E = \frac{1}{2I}\left(p_\theta^2 + \frac{p_\phi^2}{\sin^2\theta}\right), \quad 1 = \frac{p_\theta^2}{2IE} + \frac{p_\phi^2}{2IE\sin^2\theta},$$

in spherical polar coordinates, the (p_θ, p_ϕ)-curve for constant E and θ is — as may be seen by comparison with Eqs. (1.14) to (1.15) — an ellipse of area $\oint dp_\theta dp_\phi = 2\pi IE\sin\theta$. Show that the total volume of phase space covered for constant E is $8\pi^2 IE$, and hence $g \propto 8\pi^2 IdE$.

Solution: Integrating over the angles we have

$$\int_{\phi=0}^{\phi=2\pi}\int_{\theta=0}^{\theta=\pi} 2I\pi E\sin\theta d\theta d\phi = 8\pi^2 IE.$$

Hence $g \propto 8\pi^2 IdE$.

2.3.2 Ensemble consideration

We now assume a large number of identical systems — the entire collection
is called an *ensemble* — all of whose initial locations are possible locations of
our system in the neighbourhood of the point q_0, p_0. Thus we assume a large
number of identical sytems, whose positions in phase space are characterized
by points. We consider the totality of these systems which is described by a
density of points ρ (number dn of points per infinitesimal volume) in phase
space, i.e. by

$$\frac{dn}{dqdp} = \rho(q, p, t), \qquad dqdp = \prod_i dq_i dp_i. \qquad (2.17)$$

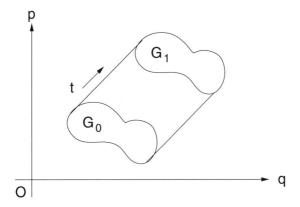

Fig. 2.1 The system moving from domain G_0 to domain G_1.

Thus dn is that number of systems which at time t are contained in the
domain $q, q + dq; p, p + dp$. The total number of systems N is obtained by
integrating over the whole of phase space, i.e.

$$\int dn = \int \rho(q, p, t) dqdp = N. \qquad (2.18)$$

With a suitable normalization we can write this

$$\int W(q, p, t) dqdp = 1, \qquad W = \frac{1}{N} \rho(q, p, t). \qquad (2.19)$$

Thus W is the *probability to find the system at time t at q, p*. Since W has
the dimension of a reciprocal action, it is suggestive to introduce a factor
$2\pi\hbar$ with every pair $dpdq$ without, however, leaving the basis of classical
mechanics! Hence we set

$$\int W(q, p, t) \frac{dqdp}{2\pi\hbar} = 1 \qquad (2.20)$$

We can consider $2\pi\hbar$ as a unit of area in (here the $(1+1)$-dimensional) phase space.

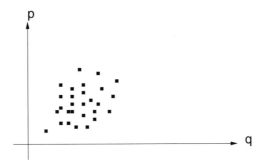

Fig. 2.2 The ensemble in phase space.

We are now interested in how n or W changes in time, i.e. how the system moves about in phase space. The equation of motion for n or W is the so-called *Liouville equation*. In order to derive this equation, we consider the domain G in Fig. 2.3 and establish an equation for the change of the number of points or systems in G in the time interval dt. In doing this, we take into account, that in our consideration no additional points are created or destroyed.

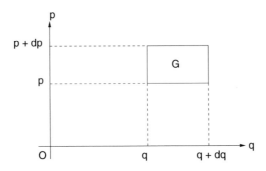

Fig. 2.3 The region G.

The number of points at time $t + dt$ in domain G, i.e. $\rho(q, p, t + dt)dqdp$, is equal to the number in G at time t plus the number that went into G in the time interval dt minus the number that left G in the time interval dt, i.e. — if $v_q(q, p)$ and $v_p(q, p)$ denote the velocities in directions q and p —

$$\rho(q, p, t + dt)dqdp - \rho(q, p, t)dqdp$$
$$= \rho(q, p, t)dp\left(\frac{dp}{dt}\right)_{q,p} dt - \rho(q + dq, p, t)dp\left(\frac{dp}{dt}\right)_{q+dq,p} dt + \cdots,$$

and thus

$$\rho(q, p, t + dt)dqdp - \rho(q, p, t)dqdp$$
$$= \rho(q, p, t)v_q(q, p)dtdp - \rho(q + dq, p, t)v_q(q + dq, p)dtdp$$
$$+ \rho(q, p, t)v_p(q, p)dtdq - \rho(q, p + dp, t)v_p(q, p + dp)dtdq.$$

Dividing both sides by $dqdpdt$ this becomes

$$\frac{\rho(q, p, t + dt) - \rho(q, p, t)}{dt}$$
$$= \frac{\rho(q, p, t)v_q(q, p) - \rho(q + dq, p, t)v_q(q + dq, p)}{dq}$$
$$+ \frac{\rho(q, p, t)v_p(q, p) - \rho(q, p + dp, t)v_p(q, p + dp)}{dp},$$

or

$$\frac{\partial \rho}{\partial t} = -\frac{\partial}{\partial q}[\rho v_q(q, p)] - \frac{\partial}{\partial p}[\rho v_p(q, p)].$$

However,

$$v_q(q, p) = \dot{q} = \frac{\partial H}{\partial p}, \quad v_p(q, p) = \dot{p} = -\frac{\partial H}{\partial q},$$

so that

$$\frac{\partial \rho}{\partial t} = -\frac{\partial}{\partial q}\left(\rho \frac{\partial H}{\partial p}\right) + \frac{\partial}{\partial p}\left(\rho \frac{\partial H}{\partial q}\right) = -\frac{\partial \rho}{\partial q}\frac{\partial H}{\partial p} + \frac{\partial \rho}{\partial p}\frac{\partial H}{\partial q} = \{H, \rho\}.$$

Hence

$$\frac{\partial \rho}{\partial t} = \{H, \rho\}. \tag{2.21}$$

This is the Liouville equation which describes the motion of the ensemble or, put differently, the probable motion of the system under consideration. Comparison of Eq. (2.21) with Eq. (2.4) shows that ρ and u satisfy very similar equations. With Eqs. (2.19), (2.20) and (2.21) we can also write

$$\frac{\partial W(q, p, t)}{\partial t} = \{H(q, p), W(q, p, t)\} \quad \text{with} \quad \int W(q, p, t)\frac{dqdp}{2\pi\hbar} = 1. \tag{2.22}$$

The generalization to n degrees of freedom is evident: The volume element of phase space is

$$\prod_i \left(\frac{dq_i dp_i}{2\pi\hbar}\right) := \left(\frac{dqdp}{2\pi\hbar}\right)^n \quad \text{with} \quad \int W(q, p, t)\left(\frac{dqdp}{2\pi\hbar}\right)^n = 1, \tag{2.23}$$

where

$$W(q,p,t)\left(\frac{dqdp}{2\pi\hbar}\right)^n$$

is the probability for the system to be at time t in the volume $q, q+dq; p, p+dp$.
We deduce from the Liouville equation the important consequence that

$$\frac{d\rho(q,p,t)}{dt} = 0, \tag{2.24}$$

since the total derivative is made up of precisely the partial derivatives contained in Eq. (2.24). Equation (2.24) implies that ρ is a constant in time, and hence that equal phase space volumes contain the same number of systems, and this means — since these systems are contained in a finite part V of phase space — that

$$\frac{d}{dt}\int_V \rho dV = \frac{dN}{dt} = 0.$$

We have in particular, since no systems are created or destroyed, that

$$dV \propto \left(\frac{dq_0 dp_0}{2\pi\hbar}\right)^n = \left(\frac{dqdp}{2\pi\hbar}\right)^n, \tag{2.25}$$

if q_0, p_0 are the initial values of q, p (cf. Example 2.4). Thus in Fig. 2.1 the area G_0 is equal to the area G_1.

2.4 Expectation Values of Observables

Let $u = u(q,p)$ be an observable. We define as *expectation value* of $u(q,p)$ the following expression:

$$\langle u \rangle = \int u(q,p)W(q,p,t)\left(\frac{dqdp}{2\pi\hbar}\right)^n. \tag{2.26}$$

With Eq. (2.4), i.e.

$$\frac{du(q,p)}{dt} = \{u, H\},$$

we described the time variation of the observable $u(q,p)$. We now inquire about the time variation of the expectation value $\langle u \rangle$ of u. We shall see that we have two possibilities for this, i.e. for

$$\frac{d}{dt}\langle u \rangle = \frac{d}{dt}\int u(q,p)W(q,p,t)\left(\frac{dqdp}{2\pi\hbar}\right)^n. \tag{2.27}$$

The first and most immediate possibility is — as indicated – that the density or probability $W(q,p,t)$ depends *explicitly* on time t (if determined at a fixed point in phase space), and the time variation $d\langle u \rangle / dt$ is attributed to the fact that it is this probability (that $u(q, p)$ assumes certain values) that depends explicitly on time. Then Eq. (2.27) becomes

$$
\begin{aligned}
\frac{d}{dt}\langle u \rangle &= \int u(q, p) \frac{\partial}{\partial t} W(q, p, t) \left(\frac{dqdp}{2\pi\hbar} \right)^n \\
&= \int u(q, p) \{ H(q, p), W(q, p, t) \} \left(\frac{dqdp}{2\pi\hbar} \right)^n,
\end{aligned}
\tag{2.28}
$$

where we used Eq. (2.22).

However, we can also employ a more complicated consideration.[¶] Solving the equations of motion for q, p, we can express these in terms of their initial values q_0, p_0, i.e. at $t = 0$, so that

$$
q = g(q_0, p_0, t), \quad p = f(q_0, p_0, t),
\tag{2.29}
$$

and hence

$$
u(q, p) \equiv u(q, p, 0) = u(g(q_0, p_0, t), f(q_0, p_0, t), 0) \equiv u_0(q_0, p_0, t).
\tag{2.30}
$$

The distribution of the canonical variables is given by $W(q, p, t)$. Thus we can write, since $W \propto \rho$ is constant in time according to Eq. (2.24):

$$
\begin{aligned}
W(q, p, t) &= W(g(q_0, p_0, t), f(q_0, p_0, t), t) \\
&= W(q_0, p_0, 0) \equiv W_0(q_0, p_0) \text{ at time } t = 0,
\end{aligned}
\tag{2.31}
$$

i.e. W is the density in the neighbourhood of a given point in phase space and has an *implicit* dependence on time t. With these expressions we obtain for the expectation value $\langle u \rangle_0$:

$$
\langle u \rangle_0 = \int u_0(q_0, p_0, t) W_0(q_0, p_0) \left(\frac{dq_0 dp_0}{2\pi\hbar} \right)^n.
\tag{2.32}
$$

In this expression the time t is contained explicitly in the observable $u(q, p) \equiv u_0(q_0, p_0, t)$. We expect, of course, that

$$
\langle u \rangle = \langle u \rangle_0.
\tag{2.33}
$$

We verify this claim as follows.

Reversing Eq. (2.29), we have

$$
q_0 = \tilde{g}(q, p, t), \quad p_0 = \tilde{f}(q, p, t),
\tag{2.34}
$$

[¶] See also H. Goldstein [114], Sec. 8.8.

so that on the other hand with Eq. (2.29)

$$q \equiv g(\tilde{g}(q,p,t), \tilde{f}(q,p,t), t), \quad p \equiv f(\tilde{g}(q,p,t), \tilde{f}(q,p,t), t). \tag{2.35}$$

Inserting these expressions into u_0 we obtain

$$
\begin{aligned}
u_0(q_0, p_0, t) \quad &= \quad u_0(\tilde{g}(q,p,t), \tilde{f}(q,p,t), t) \\
&\overset{(2.30)}{=} \quad u(g(\tilde{g}(q,p,t), \tilde{f}(q,p,t), t), f(\tilde{g}(q,p,t), \tilde{f}(q,p,t), t), 0) \\
&\overset{(2.35),(2.30)}{=} \quad u(q,p,0) \equiv u(q,p).
\end{aligned} \tag{2.36}
$$

Moreover, (cf. Eq. (2.31))

$$W_0(q_0, p_0) = W(q, p, t). \tag{2.37}$$

Inserting Eqs. (2.36) and (2.37) and (2.25) into Eq. (2.32), we obtain

$$\langle u \rangle_0 = \int u(q,p) W(q,p,t) \left(\frac{dq\,dp}{2\pi\hbar} \right)^n = \langle u \rangle,$$

as had to be shown.

Taking now the total time derivative of $\langle u \rangle_0$, Eq. (2.32), we obtain an expression which is different from that in Eq. (2.28), i.e.

$$
\begin{aligned}
\frac{d}{dt} \langle u \rangle_0 \quad &= \quad \frac{d}{dt} \int u_0(q_0, p_0, t) W_0(q_0, p_0) \left(\frac{dq_0\,dp_0}{2\pi\hbar} \right)^n \\
&= \quad \int \frac{\partial u_0(q_0, p_0, t)}{\partial t} W_0(q_0, p_0) \left(\frac{dq_0\,dp_0}{2\pi\hbar} \right)^n.
\end{aligned} \tag{2.38}
$$

We deal with the partial derivative with the help of Eq. (2.30):

$$
\begin{aligned}
\frac{\partial u_0(q_0, p_0, t)}{\partial t} \quad &= \quad \left(\frac{\partial u}{\partial q} \right)_{\substack{q=g(q_0,p_0,t), \\ p=f(q_0,p_0,t)}} \frac{dq}{dt} + \left(\frac{\partial u}{\partial p} \right)_{\substack{q=g(q_0,p_0,t), \\ p=f(q_0,p_0,t)}} \frac{dp}{dt} \\
&= \quad \left(\frac{\partial u}{\partial q} \right)_{\substack{q=g(q_0,p_0,t), \\ p=f(q_0,p_0,t)}} \frac{\partial H}{\partial p} - \left(\frac{\partial u}{\partial p} \right)_{\substack{q=g(q_0,p_0,t), \\ p=f(q_0,p_0,t)}} \frac{\partial H}{\partial q} \\
&= - \quad (\{H(q,p), u(q,p)\})_{\substack{q=g(q_0,p_0,t), \\ p=f(q_0,p_0,t)}}
\end{aligned} \tag{2.39}
$$

Substituting this into Eq. (2.38) we obtain

$$\frac{d}{dt} \langle u \rangle_0 = - \int (\{H(q,p), u(q,p)\})_{\substack{q=g(q_0,p_0,t), \\ p=f(q_0,p_0,t)}} W_0(q_0, p_0) \left(\frac{dq_0\,dp_0}{2\pi\hbar} \right)^n. \tag{2.40}$$

Here we perform the transformation (2.34) and use (2.25) and (2.31), so that

$$\frac{d}{dt}\langle u\rangle_0 = -\int \{H(q,p), u(q,p)\} W(q,p,t) \left(\frac{dqdp}{2\pi\hbar}\right)^n. \qquad (2.41)$$

This expression contains $\{H, u\}W$ instead of $u\{H, W\}$ in Eq.(2.28). However, from the properties of the Poisson bracket, we obtain

$$\{H, uW\} = \{H, u\}W + u\{H, W\}. \qquad (2.42)$$

The phase-space integral of a Poisson bracket like

$$I := \int \{H, uW\} \left(\frac{dqdp}{2\pi\hbar}\right)^n$$

vanishes under certain conditions. Consider

$$\{H, uW\} = \sum_i \left[\frac{\partial H}{\partial q_i}\frac{\partial(uW)}{\partial p_i} - \frac{\partial H}{\partial p_i}\frac{\partial(uW)}{\partial q_i}\right]$$

$$= \sum_i \left[\frac{\partial}{\partial p_i}\left(\frac{\partial H}{\partial q_i}uW\right) - \frac{\partial}{\partial q_i}\left(\frac{\partial H}{\partial p_i}uW\right)\right]. \qquad (2.43)$$

If for all i:

$$\lim_{p_i\to\pm\infty}\frac{\partial H}{\partial q_i}uW = 0, \qquad \lim_{q_i\to\pm\infty}\frac{\partial H}{\partial p_i}uW = 0, \qquad (2.44)$$

(which is reasonable since the density vanishes at infinity), we obtain zero after partial integration of I and hence from Eqs. (2.41) and (2.42) the relation

$$\frac{d}{dt}\langle u\rangle_0 = \int u(q,p)\{H(q,p), W(q,p,t)\} \left(\frac{dqdp}{2\pi\hbar}\right)^n \qquad (2.45)$$

in agreement with Eq. (2.28). Alternatively we could deduce from Eqs. (2.28) and (2.45) the Liouville equation.

The considerations we just performed demonstrate that we have two ways of treating the time-dependence: The explicit time-dependence can either be contained in the probability W or in the (transformed) observables. In the first case, described as "*Schrödinger picture*", the observable u is treated as a function $u(q,p)$, and the time variation of $\langle u(q,p)\rangle$ is attributed to the probability $W(q,p,t)$ of u assuming certain values q and p. This time-dependence is described by the Liouville equation

$$\frac{\partial W(q,p,t)}{\partial t} = \{H(q,p), W(q,p,t)\}.$$

In the other case, called *"Heisenberg picture"*, the probability of the initial values $W_0(q_0, p_0)$ is assumed,[||] and the explicit time-dependence is transferred into the correspondingly transformed observables $u_0(q_0, p_0, t)$. The equation of motion is then that of an observable, i.e. (cf. Eq. (2.4))

$$\frac{du(q, p)}{dt} = \{u(q, p), H(q, p)\},$$

the reason being that — since q_0, p_0 are constant initial values — we have

$$
\begin{aligned}
\frac{\partial u_0(q_0, p_0, t)}{\partial t} &= \frac{du_0(q_0, p_0, t)}{dt} \\
&\overset{(2.30)}{=} \frac{du(q, p)}{dt} \\
&\overset{(2.4)}{=} \{u(q, p), H(q, p)\}.
\end{aligned}
\tag{2.46}
$$

We thus also recognize the connection between the Liouville equation, as the equation of motion of an ensemble or of a probability distribution on the one hand, and the equation of motion (2.4) of an observable on the other.

2.5 Extension beyond Classical Mechanics

With the above considerations we achieved a general formulation of classical mechanics. This formulation deals with observables u representing physical quantities, and probabilities W, which describe the state of a system. The time dependence of the expectation values can be dealt with in two different ways, as we observed, which are described as "Schrödinger picture" and "Heisenberg picture"— all this on a purely classical level but with the use of canonical transformations.[**]

These considerations point the way to a generalization which results if we permit u and W to belong to a more general class of mathematical quantities. Thus we arrive at a more general theory if we define u and W with the following properties:

(a) An addition is defined between the quantities, for which the axioms of a commutative group apply.

(b) As usual a muliplication by a complex number is defined.

(c) A multiplication of the quantities among themselves is defined, which, does not have to be commutative, but does satisfy the usual associative and distributive laws.

[||]This means: Only in the Heisenberg picture $\partial W/\partial t = 0$.

[**]The author learned this approach from lectures of H. Koppe [152] at the university of Munich around 1964.

Quantities satisfying these properties define a linear algebra, so that we consider this next. We shall then interpret as "*canonical quantization*" the procedure which allocates to each of the fundamental Poisson brackets (2.6) a so-called "*commutator*" $[A, B] := AB - BA$ in the following way:

$$\{q, p\} = 1 \qquad \longrightarrow \qquad -\frac{i}{\hbar}[q, p] = 1,$$

$$\{q, q\} = 0 \qquad \longrightarrow \qquad \frac{i}{\hbar}[q, q] = 0,$$

$$\{p, p\} = 0 \qquad \longrightarrow \qquad \frac{i}{\hbar}[p, p] = 0. \qquad (2.47)$$

One verifies readily that the commutator relations are satisfied by the differential operator representations

$$q_x \to x, \quad p_x \to -i\hbar\frac{\partial}{\partial x}.$$

In view of our later considerations, and to be able to correlate these with the above classical considerations, it is helpful to introduce already at this stage some additional terminology. The quantity corresponding to W in quantum mechanics is the so-called "*statistical operator*", also called "*density matrix*". Moreover, for a phase-space integral we define the word or symbol "*trace*", also written "Tr", by

$$\text{Trace } u := \int u(q, p) \left(\frac{dq\, dp}{2\pi\hbar} \right)^n. \qquad (2.48)$$

In matrix theory the symbol "trace" has a well-defined meaning. Introducing it here assumes, therefore, that its use here implies the essential properties it has in matrix theory. That this is the case, can be deduced from the following characteristic properties of a trace, which all apply to (2.48):
(a) $\text{Tr } u* = (\text{Tr } u)^* \equiv \overline{\text{Tr } u}$,
(b) $\text{Tr}(\alpha u + \beta u) = \alpha \text{Tr } u + \beta \text{Tr } v$,
(c) $\text{Tr } (u^*u) > 0$, if $u \neq 0$,
(d) $\text{Tr}(uv) = \text{Tr } (vu)$.
Thus we can write the expectation value of an observable u (cf. Eq. (2.26))

$$\langle u \rangle = \text{Tr}(uW). \qquad (2.49)$$

With these considerations we have reviewed aspects of classical particle mechanics in as close an approach to quantum mechanics as seems possible. In Chapter 9 we attempt a corresponding approach for classical systems with a wave-like nature, i.e. electrodynamics, and, excepting the Poisson brackets, obtain corresponding results — as one would envisage in view of the expected particle-wave duality in quantum mechanics. However, it will be shown that in electrodynamics, the Poisson brackets require modification to Dirac brackets, since gauge fixing (i.e. a constraint) has to be taken into account. These aspects will be considered in Chapter 27. Thus we can now proceed to prepare the ground for the extension of classical mechanics into an operator formulation.

Chapter 3

Mathematical Foundations of Quantum Mechanics

3.1 Introductory Remarks

In Chapter 2 we investigated the algebraic structure of classical Hamiltonian mechanics. We found that the Poisson algebra permits extensions to non-c-number formulations, which turn out to be those of the theory today known as quantum mechanics. In this chapter we therefore introduce important basic mathematical concepts of this non-c-number mechanics, i.e. quantum mechanics: The Hilbert space as the space of state vectors representing the states of a physical system, and selfadjoint operators in this space as representatives of observables, i.e. of measurable quantities, with the canonical commutation relations or Heisenberg algebra defining the basic product relations. These somewhat abstract considerations — although later in many cases not referred back to — are a necessary prerequisite for the formulation of a mechanics which is not of the c-number type as classical mechanics. We also introduce in this chapter the concepts of linear functionals and distributions so that we can make free use of the delta distribution and similar objects in later chapters.

3.2 Hilbert Spaces

We first recapitulate some fundamental concepts of linear algebra and begin with the axioms defining a linear vector space. Building upon this, we can define the Hilbert space as the space of states of a physical system.

A set $\mathcal{M} = \{\psi_i\}$ is called a *linear vector space* on the set of numbers $\mathbb{K} \in \{\mathbb{C}\}$, if the elements ψ_i of $\mathcal{M}$ satisfy the usual axioms of addition and multiplication by complex numbers, i.e.

$$
\begin{aligned}
\psi_i + \psi_j &= \psi_j + \psi_i, \\
(\psi_i + \psi_j) + \psi_k &= \psi_i + (\psi_j + \psi_k), \\
\psi_i + \emptyset &= \psi_i, \\
\psi_i + (-\psi_i) &= \emptyset, \\
(\emptyset : \text{null element} \ &\in \ \mathcal{M}),
\end{aligned}
\tag{3.1}
$$

and with complex numbers α and β:

$$
\begin{aligned}
\alpha(\psi_i + \psi_j) &= \alpha\psi_i + \alpha\psi_j, \\
(\alpha + \beta)\psi_i &= \alpha\psi_i + \beta\psi_i, \\
\alpha(\beta\psi_i) &= (\alpha\beta)\psi_i, \quad (\alpha, \beta \in \mathbb{K}), \\
\mathbb{1}\psi_i &= \psi_i.
\end{aligned}
\tag{3.2}
$$

Vectors $\psi_1, \psi_2, \ldots, \psi_n$ are said to be *linearly dependent*, if numbers $\alpha_i, i = 1, 2, \ldots, n$ exist (not all zero), so that

$$
\sum_{i=1}^{n} \alpha_i \psi_i = 0.
\tag{3.3}
$$

If all $\alpha_i = 0$, the vectors $\psi_1, \psi_2, \ldots, \psi_n$ are said to be *linearly independent*. If $n + 1$ elements $\psi_i \in \mathcal{M}$ are linearly dependent, and n is the smallest such number, n is called the *dimension* of $\mathcal{M}$. In each case n linearly independent vectors are said to form a *basis*, if every vector $\psi \in \mathcal{M}$ can be associated with numbers $c_i, i = 1, 2, \ldots, n$, such that

$$
\psi = \sum_{i=1}^{n} c_i \psi_i.
\tag{3.4}
$$

The vector space $\mathcal{M}$ is said to be a *metric vector space* or a *pre-Hilbert space*, if any two elements ψ_1, ψ_2 of this space can be associated with a complex number (ψ_1, ψ_2) called *inner product*, $(\psi_1, \psi_2) : \mathcal{M} \times \mathcal{M} \to \mathbb{K}$, with the properties ($\alpha_i \in \mathbb{K}$):

$$
(\psi_2, \psi_1) = (\psi_1, \psi_2)^* \quad \text{(hermiticity)}, \tag{3.5a}
$$

$$
(\psi, \alpha_1\psi_1 + \alpha_2\psi_2) = \alpha_1(\psi, \psi_1) + \alpha_2(\psi, \psi_2), \tag{3.5b}
$$

$$
(\psi, \psi) \begin{cases} > 0 \ \text{ if } \psi \neq 0, \\ = 0 \ \text{ if } \psi = 0, \end{cases} \tag{3.5c}
$$

where the asterix * means complex conjugation. The first two properties imply

$$(\alpha_1\psi_1 + \alpha_2\psi_2, \psi) = \alpha_1^*(\psi_1, \psi) + \alpha_2^*(\psi_2, \psi), \tag{3.6}$$

i.e. linearity in the second component, antilinearity in the first component (also described as *sesquilinearity*, meaning one-and-a-halffold linearity).

The *norm* of the vector ψ (pre-Hilbert space norm) is defined as

$$||\psi|| := (\psi, \psi)^{1/2}. \tag{3.7}$$

The *distance* between two vectors $\psi_1, \psi_2 \in \mathcal{M}$ is defined by

$$d(\psi_1, \psi_2) := ||\psi_1 - \psi_2||. \tag{3.8}$$

In addition the following relations hold in a metric space $\mathcal{M}$ for $\psi_1, \psi_2 \in \mathcal{M}$:

$$|(\psi_1, \psi_2)| \leq ||\psi_1|| \cdot ||\psi_2|| \ \text{(Schwarz inequality)}, \tag{3.9a}$$

$$||\psi_1 + \psi_2|| \leq ||\psi_1|| + ||\psi_2|| \ \text{(triangle inequality)}, \tag{3.9b}$$

$$||\psi_1 + \psi_2||^2 = ||\psi_1||^2 + ||\psi_2||^2, \ \text{if} \ (\psi_1, \psi_2) = 0 \ \text{(Pythagoras theorem)}, \tag{3.9c}$$

$$||\psi_1 + \psi_2||^2 + ||\psi_1 - \psi_2||^2 = 2||\psi_1||^2 + 2||\psi_2||^2 \ \text{(parallelogram equation)}, \tag{3.9d}$$

$$||\psi_1|| = \sup_{||\psi_2||=1} |(\psi_1, \psi_2)|. \tag{3.9e}$$

We restrict ourselves here to some remarks on the verification of these well-known relations. In order to verify Eq. (3.9a) we start from $\psi = \psi_1 + \lambda\psi_2 \in \mathcal{M}$ for arbitrary λ and $\psi_2 \neq 0$:

$$(\psi_1 + \lambda\psi_2, \psi_1 + \lambda\psi_2) \geq 0,$$

which we can write

$$\begin{aligned} 0 \ \leq \ & (\psi_1, \psi_1) + \lambda^*(\psi_2, \psi_1) + \lambda(\psi_1, \psi_2) + |\lambda|^2(\psi_2, \psi_2) \\ = \ & ||\psi_1||^2 + 2\Re(\psi_1, \lambda\psi_2) + |\lambda|^2||\psi_2||^2 \\ = \ & \left|\lambda||\psi_2|| + \frac{(\psi_2, \psi_1)}{||\psi_2||}\right|^2 - \frac{|(\psi_2, \psi_1)|^2}{||\psi_2||^2} + ||\psi_1||^2. \end{aligned} \tag{3.10}$$

For $\psi_2 \neq 0$ we set

$$\lambda = -\frac{(\psi_2, \psi_1)}{||\psi_2||^2},$$

so that

$$0 \leq -\frac{|(\psi_2, \psi_1)|^2}{||\psi_2||^2} + ||\psi_1||^2,$$

or

$$|(\psi_2, \psi_1)| \le ||\psi_1|| \cdot ||\psi_2||, \tag{3.11}$$

thus verifying Eq. (3.9a). In verifying the triangle inequality we use this result (3.11), beginning with $\lambda = 1$ in the second line of Eq. (3.10):

$$\begin{aligned}
||\psi_1 + \psi_2||^2 &= ||\psi_1||^2 + ||\psi_2||^2 + 2\Re(\psi_1, \psi_2) \\
&\le ||\psi_1||^2 + ||\psi_2||^2 + 2|(\psi_2, \psi_1)| \\
&\le ||\psi_1||^2 + ||\psi_2||^2 + 2||\psi_2|| \cdot ||\psi_1|| \\
&= (||\psi_1|| + ||\psi_2||)^2,
\end{aligned}$$

so that

$$||\psi_1 + \psi_2|| \le ||\psi_1|| + ||\psi_2||, \tag{3.12}$$

We omit here the verification of the remaining relations (3.9c), (3.9d), (3.9e). We also omit the verification of the following properties of the norm of a vector $\psi_1 \in \mathcal{M}$ with $\psi_2 \in \mathcal{M}, \alpha \in \mathbb{K}$:

$$\begin{aligned}
||\psi_1|| &\ge 0, \\
||\psi_1|| &= 0 \Leftrightarrow \psi_1 = 0, \\
||\psi_1 + \psi_2|| &\le ||\psi_1|| + ||\psi_2||, \\
||\alpha\psi|| &= |\alpha| \cdot ||\psi||
\end{aligned} \tag{3.13}$$

If for a vector $\psi \in \mathcal{M}$:

$$||\psi|| = 1,$$

the vector is said to be *normalized.*[*] Two vectors $\psi_1, \psi_2 \in \mathcal{M}$ are said to be *orthogonal* if

$$(\psi_1, \psi_2) = 0.$$

Examples of metric vector spaces:
(1) Let $\mathcal{M}$ be the set of all column vectors

$$v = (v_i) = \begin{pmatrix} v_1 \\ v_2 \\ \vdots \end{pmatrix}$$

with complex numbers v_i, for which

$$||v||^2 := \sum_{i=1}^{\infty} |v_i|^2 < \infty.$$

---

[*]Not all the wave functions we consider in the following and in later chapters are automatically normalized to 1; hence verification in each case is necessary.

Then we define

$$v + w := (v_i) + (w_i) := (v_i + w_i),$$

etc. with inner product

$$(v, w) := \sum_{i=1}^{\infty} v_i^* w_i.$$

The Schwarz inequality is then

$$|(v, w)| \le \sqrt{\sum_{i=1}^{\infty} |v_i|^2 \sum_{j=1}^{\infty} |w_j|^2},$$

and so on.

(2) Let $\mathcal{M} = \mathcal{L}^2$ be the set of all complex-valued integrable functions $f(\mathbf{x})$ on $S \subset \mathbb{R}^3$ (in the sense of Lebesgue) for which

$$\int_{S \subset \mathbb{R}^3} |f(\mathbf{x})|^2 d^3x < \infty.$$

With the scalar product

$$(f, g) = \int_S f(\mathbf{x})^* g(\mathbf{x}) d^3x,$$

the space $\mathcal{L}^2$ is not yet a metric vector space although for (cf. Eq. (3.5c))

$$f(\mathbf{x}) = 0 \Rightarrow (f, f) = 0.$$

But this applies also in the case of any function which is nonzero only on a set of measure zero, i.e.

$$f(\mathbf{x}) = \begin{cases} f_0 & \text{for } x = 0, \\ 0 & \text{otherwise.} \end{cases}$$

In order to avoid this difficulty, all square-integrable functions f which are "almost everywhere equal", i.e. which differ solely on a set of measure zero, are combined to an *equivalence class* $[f]$ (with space L^2), and one defines addition and multiplication by complex numbers with respect to these classes. Elements of the classes are then called *representatives* of these classes. Then L^2 is the space of all these equivalence classes, for which the scalar product, which satisfies relations (3.5a), (3.5b), (3.5c), is defined by

$$([f], [g]) := \int_{\mathcal{M}} f(\mathbf{x})^* g(\mathbf{x}) d^3x,$$

and

$$||[f]|| = 0 \Rightarrow [f] = [0],$$

where $[0]$ is defined as the class of all functions which are almost everywhere zero. This means that functions that differ only on a pointset of Lebesgue measure zero are looked at as identical. Unless necessary, we ignore in the following mostly for simplicity the distinction between $\mathcal{L}^2$ and L^2. Convergence of sequences in Hilbert space is then called convergence "almost everywhere".

With the help of the concept of a norm we can introduce the concepts of convergence and point of accumulation.

Definition: A sequence $\{\psi_n\} \in \mathcal{M}$ is said to converge (strongly) towards $\psi \in \mathcal{M}$, if the distance $||\psi - \psi_n||$ tends towards zero, i.e.

$$\lim_{n \to \infty} ||\psi - \psi_n|| = 0. \tag{3.14}$$

The vector ψ is then called *point of accumulation*. The point of accumulation does not have to be an element of $\mathcal{M}$. If $\mathcal{M}$ contains all of its points of accumulation, the set $\mathcal{M}$ is said to be *closed*.

A normalized vector space $\mathcal{M}$ which with every convergent sequence contains a vector towards which the sequence converges, is said to be *complete*, i.e. if $\psi_n \in \mathcal{M}$ with

$$\lim_{m,n \to \infty} ||\psi_n - \psi_m|| = 0$$

(called *Cauchy sequence*), there is a $\psi \in \mathcal{M}$ with

$$\psi = \lim_{n \to \infty} \psi_n, \quad \text{i.e.} \quad \lim_{n \to \infty} ||\psi - \psi_n|| = 0. \tag{3.15}$$

Every *finite-dimensional* vector space (on $\mathbb{K}$) is complete in the sense of the concept of convergence defined above (so that completeness does not have to be demanded separately). In order to see this, we consider the convergent sequence

$$\psi_\alpha = \sum_{i=1}^{n} C_{\alpha i} \psi_i, \quad C_{\alpha i} \in \mathbb{K}, \tag{3.16}$$

where $\psi_1, \ldots, \psi_n \in \mathcal{M}$ constitute a basis in $\mathcal{M}$. Then (according to Pythagoras)

$$||\psi_\alpha - \psi_\beta||^2 = ||\sum_{i=1}^{n} (C_{\alpha i} - C_{\beta i}) \psi_i||^2 = \sum_{i=1}^{n} |C_{\alpha i} - C_{\beta i}|^2, \tag{3.17}$$

a relation also known as *Parseval equation*. The convergence of the sequence ψ_α implies the convergence of the sequence $\{C_{\alpha i}\}$ towards a number C_i. Then for the vector

$$\psi = \sum_{i=1}^{n} C_i \psi_i$$

we have

$$||\psi_\alpha - \psi|| = \sum_{i=1}^{n} |C_{\alpha i} - C_i|^2, \qquad (3.18)$$

i.e. that the sequence of the vectors ψ_α converges towards ψ. We thus arrive at the definition of a Hilbert space.

Definition: An infinitely dimensional, metric vector space, which is also complete with regard to (strong) convergence, is called a *Hilbert space* $\mathcal{H}$.

The given definition of a Hilbert space is that usually given in mathematics.[†] In physics this is generally supplemented by the requirement that the space be *separable*, i.e. of a countably infinite dimensionality. Naturally Hilbert spaces with a countable basis are the simplest.

We supplement the above by referring to the concept of a *dense set or subset* $\mathcal{M}$ of $\mathcal{H}$. A subset $\mathcal{M}$ of $\mathcal{H}$ is said to be dense in $\mathcal{H}$, if to every $f \in \mathcal{H}$ there exists a sequence of vectors $f_n, f_n \in \mathcal{M}$, so that $f_n \to f$, i.e. f_n converges strongly to f, implying that every vector $f \in \mathcal{H}$ can be approximated arbitrarily precisely. We consider next some examples.

Examples of Hilbert spaces:
(1) The *hyperspherical functions* $Y_{l,m}(\theta, \varphi)$ define a complete set of basis functions on the unit sphere. Any function $f(\theta, \varphi)$ with

$$\int |f(\theta, \varphi)|^2 d\Omega < \infty$$

can be written as a convergent series

$$f(\theta, \varphi) = \sum_{l=0}^{\infty} \sum_{m=-l}^{l} C_{l,m} Y_{l,m}(\theta, \varphi). \qquad (3.19)$$

For completeness we recall here the definition

$$Y_{l,m}(\theta, \varphi) = \frac{(-1)^m}{2^l l!} \left[\frac{(2l+1)(l-m)!}{4\pi(l+m)!} \right]^{1/2} e^{im\varphi} \sin^m \theta \left(\frac{d}{d\theta} \right)^{l+m} (\cos^2 \theta - 1)^l,$$

[†] See e.g. N. I. Achieser and L. M. Glasman [3].

so that

$$Y_{0,0} = \frac{1}{\sqrt{4\pi}}, \quad Y_{1,0} = \sqrt{\frac{3}{4\pi}}\cos\theta, \quad Y_{1,\pm 1} = \mp\sqrt{\frac{3}{8\pi}}e^{\pm i\varphi}\sin\theta.$$

A Hilbert space contains a complete set of orthonormal basis vectors or a corresponding sequence precisely then if it is separable.

(2) On the space $L^2(0, 2\pi)$, i.e. on the space of square-integrable functions on the interval $[0, 2\pi]$, an orthonormal system, called the *trigonometric system*, is defined by the functions

$$\frac{1}{\sqrt{2\pi}}e^{inx}, \quad \pm n = 0, 1, 2, 3, \ldots . \tag{3.20}$$

(3) On the space $L^2(a, b)$, where (a, b) is an arbitrary but finite interval, a complete but not orthonormal system is given by

$$1, x, x^2, \ldots .$$

In order to obtain the orthonormalized system, one employs the orthogonalization procedure of E. Schmidt.[‡] The sequence of polynomials thus obtained consists of the *Legendre polynomials* which are defined on the interval $-1 \leq x \leq 1$. These polynomials are defined as follows:

$$P_0(x) = 1, \quad P_n(x) = \frac{1}{2^n n!}\frac{d^n}{dx^n}(x^2 - 1)^n, \quad n = 1, 2, \ldots . \tag{3.21}$$

These polynomials satisfy the following normalization conditions, i.e. are orthogonal but are not normalized to 1:

$$\int_{-1}^{1} P_m(x)P_n(x)dx = 0, \quad (m \neq n), \tag{3.22a}$$

and

$$\int_{-1}^{1} [P_n(x)]^2 dx = \frac{2}{2n + 1}. \tag{3.22b}$$

(4) By orthogonalization of the following functions

$$e^{-x^2/2}, \quad xe^{-x^2/2}, \quad x^2 e^{-x^2/2}, \ldots,$$

one obtains the following functions defined on the space $L^2(-\infty, \infty)$:

$$\phi_n(x) = (-1)^n e^{x^2/2}\frac{d^n}{dx^n}e^{-x^2} = H_n(x)e^{-x^2/2} \tag{3.23a}$$

[‡]This procedure is wellknown in analysis, and hence will not be elaborated on here.

with

$$\int_{-\infty}^{\infty} \phi_n(x)\phi_m(x) = 2^n n!\sqrt{\pi}\delta_{nm}, \qquad (3.23b)$$

where $H_n(x)$ is the *Hermite polynomial* of the n-th degree (which we do not go into in more detail at this stage).[§]

3.3 Operators in Hilbert Space

Having defined the admissible states of a physical system as the vectors which span a Hilbert space, the next step is to introduce quantities representing operations in this space. This is our objective in this section. We begin with a number of definitions.

Definition: Let $\mathcal{D}_A$, the domain of definition, be a (dense) subspace of the Hilbert space $\mathcal{H}$. Then one defines as a *linear operator* A on $\mathcal{H}$ the mapping

$$A : \mathcal{D}_A \to \mathcal{H} \qquad (3.24a)$$

with $(\alpha, \beta \in \mathbb{C}, \ \psi_1, \psi_2 \in \mathcal{D}_A \subset \mathcal{H})$

$$A(\alpha\psi_1 + \beta\psi_2) = \alpha(A\psi_1) + \beta(A\psi_2). \qquad (3.24b)$$

Definition: One defines as *norm* (i.e. operator norm) of the operator A the quantity

$$\|A\| := \sup_{\psi \in \mathcal{D}_A \backslash \{0\}} \frac{\|A\psi\|}{\|\psi\|}. \qquad (3.25)$$

Definition: An operator A is said to be *bounded*, if its norm is finite, i.e.

$$\|A\| < \infty.$$

Definition: Two operators $A : \mathcal{D}_A \to \mathcal{H}$ and $B : \mathcal{D}_B \to \mathcal{H}$ are said to be *equal* if and only if

$$A\psi = B\psi, \text{ for every } \psi \in \mathcal{D}_A \text{ and } \mathcal{D}_A = \mathcal{D}_B.$$

Example: An example of a linear operator is given by the differential operator

$$D := \frac{d}{dx} \text{ with } \mathcal{D}_D = \left\{ \psi \in L^2, \frac{d\psi}{dx} \in L^2 \right\}. \qquad (3.26)$$

Definition: We define the operations of *addition and multiplication* of operators by the relations

$$(A + B)\psi := A\psi + B\psi, \ \forall \psi \in \mathcal{D}_{A+B} = \mathcal{D}_A \cap \mathcal{D}_B, \qquad (3.27a)$$

[§]Hermite polynomials are dealt with in detail in Chapter 6.

$$(AB)\psi := A(B\psi), \quad \forall \psi \in \mathcal{D}_B, \ B\psi \in \mathcal{D}_A. \tag{3.27b}$$

Definition: We define operators called *commutators* as follows: Let $A : \mathcal{D}_A \to \mathcal{H}$, and $B : \mathcal{D}_B \to \mathcal{H}$ be linear operators in the Hilbert space $\mathcal{H}$. Then we define as *commutator* the expression

$$[A, B] := AB - BA \text{ with } \mathcal{D}_{[A,B]} = \mathcal{D}_A \cap \mathcal{D}_B \to \mathcal{H}. \tag{3.28}$$

Definition: If $A : \mathcal{D}_A \to \mathcal{H}$ for $\psi \in \mathcal{D}_A \backslash \{0\}$, then ψ is called *eigenvector* with respect to the *eigenvalue* $\lambda \in \mathbb{C}$ if and only if

$$A\psi = \lambda\psi. \tag{3.29}$$

Very important for our purposes are the concepts of *adjoint* and *self-adjoint operators*.

Definition: Let $A : \mathcal{D}_A \to \mathcal{H}$ and $\psi \in \mathcal{D}_A \subset \mathcal{H}$. Then $A^\dagger$ is called *adjoint operator* of A if for $A^\dagger \phi := \phi^\dagger, \phi \in \mathcal{D}_{A^\dagger}$, the following relation holds:

$$(A^\dagger \phi, \psi) = (\phi, A\psi). \tag{3.30}$$

Definition: The operator $A : \mathcal{D}_A \to \mathcal{H}$, is said to be *symmetric* if and only if

$$(\phi, A\psi) = (A\phi, \psi), \quad \forall \, \phi, \psi \in \mathcal{D}_A \subset \mathcal{D}_{A^\dagger}. \tag{3.31}$$

Definition: The operator A, $A : \mathcal{D}_A \to \mathcal{H}$, is said to be *hermitian* or *selfadjoint* if and only if

$$A = A^\dagger, \quad \text{i.e.} \ \ \mathcal{D}_A = \mathcal{D}_{A^\dagger} \ \text{ and } \ A^\dagger\phi = A\phi. \tag{3.32}$$

One can verify that:

$$(A^\dagger)^\dagger = A, \tag{3.33a}$$

$$(A + B)^\dagger = A^\dagger + B^\dagger, \tag{3.33b}$$

$$(\lambda A)^\dagger = \lambda^* A^\dagger, \tag{3.33c}$$

$$(AB)^\dagger = B^\dagger A^\dagger, \tag{3.33d}$$

$$(A^{-1})^\dagger = (A^\dagger)^{-1}. \tag{3.33e}$$

We are now in a position to construct relations between operators A and B in $\mathcal{H}$, which correspond to the relations (2.5a) to (2.5e) of Poisson brackets, however, we omit their verification here:

$$[A, B] = -[B, A], \tag{3.34a}$$

$$[A, \alpha_1 B_1 + \alpha_2 B_2] = \alpha_1[A, B_1] + \alpha_2[A, B_2], \tag{3.34b}$$

$$[A, B]^\dagger = -[A^\dagger, B^\dagger], \tag{3.34c}$$

$$[A, BC] = [A, B]C + B[A, C], \tag{3.34d}$$

$$[A, [B, C]] + [B, [C, A]] + [C, [A, B]] = 0. \tag{3.34e}$$

The last relation is again called a *Jacobi identity*. Comparison of Eq. (3.34c) with Eq. (2.5c) requires here in the definition of the corresponding quantity the introduction of a factor "i" (see Eq. (2.47)).

As important examples we consider the following operators.

(1) $q_j : \mathcal{D}_{q_j} \to L^2(\mathbb{R}^3)$.

We write sometimes the application of the operator q_j to $\phi \in L^2(\mathbb{R}^3)$: $(q_j\phi)(\mathbf{x})$, and we define

$$(q_j\phi)(\mathbf{x}) := x_j\phi(\mathbf{x}). \tag{3.35}$$

We can read this equation as an eigenvalue equation: Operator q_j applied to the vector ϕ yields the eigenvalue x_j multiplied by ϕ, in the present case on $\mathbb{R}^3$. Since

$$\mathcal{D}_{q_j} = \{\phi \in L^2(\mathbb{R}^3) : q_j\phi \in L^2(\mathbb{R}^3)\},$$

we have

$$\mathcal{D}_{q_j} = \mathcal{D}_{q_j{}^\dagger}.$$

Furthermore, for instance for $\psi, \phi \in L^2(\mathbb{R}^1)$, we have

$$\begin{aligned}
(\psi, q\phi) &= \int \psi^*(x)(q\phi)(x)dx \overset{(3.35)}{=} \int \psi^*(x)x\phi(x)dx \\
&= \int x\psi^*(x)\phi(x)dx = \int \{(q\psi)(x)\}^*\phi(x)dx \\
&= (q\psi, \phi). \tag{3.36}
\end{aligned}$$

Since for the adjoint operator $A^\dagger$ of A:

$$(\psi, A\phi) = (A^\dagger\psi, \phi),$$

it follows (with $\mathcal{D}_{q_j} = \mathcal{D}_{q_j{}^\dagger}$ from above) that $q_j = q_j{}^\dagger$.

(2) $p_j : \mathcal{D}_{p_j} \to L^2(\mathbb{R}^3)$ defined by

$$(p_j\phi)(\mathbf{x}) := -i\hbar\frac{\partial}{\partial x_j}\phi(\mathbf{x}). \tag{3.37}$$

In this case we have

$$\mathcal{D}_{p_j} = \left\{\phi \in L^2(\mathbb{R}^3) : \phi \text{ continuous}, \frac{\partial\phi}{\partial x_j} \in L^2(\mathbb{R}^3)\right\} = \mathcal{D}_{p_j{}^\dagger}.$$

Here for $\psi, \phi \in L^2(\mathbb{R}^1)$:

$$
\begin{aligned}
(\psi, p\phi) &= \int \psi^*(x)(p\phi)dx \overset{(3.37)}{=} -i\hbar \int \psi^*(x)\frac{d}{dx}\phi(x)dx \\
&= -i\hbar \left[\psi^*(x)\phi(x)|_{-\infty}^{\infty} - \int \frac{d}{dx}\psi^*(x)\phi(x)dx \right] \\
&= -\int \left(i\hbar\frac{d}{dx}\psi(x) \right)^* \phi(x)dx = \int (p\psi)^*(x)\phi(x)dx \\
&= (p\psi, \phi) \equiv (p^\dagger\psi, \phi),
\end{aligned}
\tag{3.38}
$$

so that $p^\dagger = p$.

Something similar applies in the case of the following operator which represents classically the kinetic energy T:

(3) $T : \mathcal{D}_T \to L^2(\mathbb{R}^3)$

and

$$
(T\phi)(\mathbf{x}) := -\frac{\hbar^2}{2m}\triangle\phi(\mathbf{x}) = \left(\frac{1}{2m}\sum_{j=1}^{3}p_j{}^2 \right)\phi(\mathbf{x}).
\tag{3.39}
$$

As a further example we consider the commutator.

(4) Let the commutator be the mapping

$$
[p_j, q_k] : \mathcal{D}_{[p_j, q_k]} \to L^2(\mathbb{R}^3).
$$

Then for $\phi \in L^2(\mathbb{R}^3)$:

$$
\begin{aligned}
[p_j, q_k]\phi(\mathbf{x}) &= (p_j q_k - q_k p_j)\phi(\mathbf{x}) = (p_j q_k \phi)(\mathbf{x}) - (q_k p_j \phi)(\mathbf{x}) \\
&= -i\hbar\left(\frac{\partial}{\partial x_j}x_k\phi(\mathbf{x}) - x_k\frac{\partial}{\partial x_j}\phi(\mathbf{x}) \right) = -i\hbar\delta_{jk}\phi(\mathbf{x}),
\end{aligned}
$$

i.e. formally

$$
[p_j, q_k] = -i\hbar\delta_{jk}.
\tag{3.40}
$$

The following commutators which define the *Heisenberg algebra*

$$
[p_j, q_k] = -i\hbar\delta_{jk}, \quad [p_j, p_k] = 0, \quad [q_j, q_k] = 0
$$

are called *canonical quantization conditions* with respect to a theory whose classical version possesses the fundamental Poisson brackets

$$
\{p_j, q_k\} = \delta_{jk}, \quad \{p_j, p_k\} = 0, \quad \{q_j, q_k\} = 0.
$$

The simplest example to consider is the harmonic oscillator. We postpone this till later (Chapter 6). We add, that the quantization must always be carried out on Cartesian coordinates. Moreover, the above relations assume that the three degrees of freedom are independent, i.e. there are no constraints linking them. Systems with constraints can be handled, but require a separate treatment.¶

3.4 Linear Functionals and Distributions

We now introduce the concept of a *continuous linear functional* on a so-called *test function space.*‖ Our aim is here, to provide a frame in which the formal Dirac bra- and ket-formalism to be developed later finds its mathematical justification. We require in particular the *delta distribution* and *Fourier transformations*.

A subset of a Hilbert space $\mathcal{H}$ is called a *linear manifold* $\mathcal{D}$, if along with any two elements $\phi_1, \phi_2 \in \mathcal{D} \subset \mathcal{H}$ this also contains the linear combination of these, i.e.

$$\alpha_1 \phi_1 + \alpha_2 \phi_2 \in \mathcal{D}, \quad \alpha_1, \alpha_2 \in \mathbb{C}. \tag{3.41}$$

A *linear functional* $[f]$ in the Hilbert space $\mathcal{H}$ is a mapping of the manifold $\mathcal{D}$ into the set of complex numbers, i.e.

$$[f] : \mathcal{D} \to \mathbb{C}$$

and is written

$$[f]\langle\phi\rangle \equiv f\langle\phi\rangle := \int_{\mathbb{R}^n} f(\mathbf{x})\phi(\mathbf{r})d\mathbf{x}, \quad (\forall \phi \in \mathcal{D}) \tag{3.42}$$

with the property of linearity, i.e.

$$f\langle\phi_1 + \phi_2\rangle = f\langle\phi_1\rangle + f\langle\phi_2\rangle,$$

(i.e. the expression (3.42) is antilinear in the first component). If $f(\mathbf{x})$ is for instance a locally integrable function (i.e. for a compact set $\in \mathbb{R}^n$) like $\exp(i\mathbf{k} \cdot \mathbf{x})$, then it is clear that the function $\phi(\mathbf{x})$ has to decrease sufficiently fast at infinity, so that the expression in Eq. (3.42) exists. Functions which provide this are called *test functions* (see also later). Instead of the Hilbert space we therefore consider now a *space of test functions* (vector space of test functions), and on this space linear functionals.

Definition: The *compact support* of a continuous function $\phi : \mathbb{R}^n \to \mathbb{C}$ is defined to be the compact (i.e. closed and bounded) set of points outside

---

¶See Chapter 27.

‖We follow here to some extent W. Güttinger [127].

that of $\phi = 0$. Test functions ϕ with compact support are exactly zero outside their support; they define the space $D(\mathbb{R}^n)$.

A different class of test functions ϕ consists of those which together with all of their derivatives $|D^n \phi|$ fall off at infinity faster than any inverse power of $|\mathbf{x}|$. These test functions are called "*rapidly decreasing test functions*" and constitute the space $S(\mathbb{R}^n)$:

$$
\begin{aligned}
D(\mathbb{R}^n) &:= \{\phi \in C^\infty(\mathbb{R}^n \to \mathbb{C}) : \text{ support of } \phi \text{ compact}\}, \\
S(\mathbb{R}^n) &:= \{\phi \in C^\infty(\mathbb{R}^n \to \mathbb{C}) : |\mathbf{x}|^m |D^n \phi| \text{ bounded}, \ m, n, \ldots \in \mathbb{N} \geq 0\}.
\end{aligned}
$$
$$(3.43)$$

Definition: *Distributions* $f\langle\phi\rangle$ are defined to be the linear functionals on $D(\mathbb{R}^n)$ and *tempered distributions* the linear functionals on $S(\mathbb{R}^n)$. A subset of distributions can obviously be identified with ordinary functions, which is the reason why distributions are also called "*generalized functions*".

3.4.1   Interpretation of distributions in physics

It is possible to attribute a physical meaning to a functional of the form

$$
f\langle\phi\rangle := \int dx f(x)\phi(x). \tag{3.44}
$$

In order to perform a measurement at some object, one observes and hence measures the reaction of this object to some tests. If we describe the object by its density distribution $f(x)$, like e.g. mass, and that of its testing object by $\phi(x)$, then the product $f(x)\phi(x)$ describes the result of the testing procedure at a point x, since

$$
f(x)\phi(x) = 0, \quad \text{provided } f(x) \neq 0 \text{ and } \phi(x) \neq 0,
$$

i.e. if object and testing object do not meet. The expression then describes the result of the testing procedure in the entire space. If we perform the testing procedure with different testing objects and hence with different test functions $\phi_i(x), i = 1, 2, \ldots$, we obtain as a result for the entire space a set of different numbers which correspond to the individual $\phi_i(x)$. These ϕ-dependent numbers are written as in Eq. (3.44):

$$
f\langle\phi\rangle = \int f(x)\phi(x)dx.
$$

If $f\langle\phi\rangle = 0$ for every continuously differentiable function $\phi(x)$, then $f(x) = 0$. In general one expects that a knowledge of the numbers $f\langle\phi\rangle$ and the test functions $\phi(x)$ permits one to characterize the function $f(x)$ itself, provided the set of test functions is complete. In this way one arrives at a new concept of functions: Instead of its values $y = f(x)$ the function f is now determined by its action on all the test functions $\phi(x)$. One refers to this as a *functional*: The functional f associates a number $f\langle\phi\rangle$ with every test function ϕ. Thus the functional is the mapping of a space of functions into the space of numbers. $f\langle\phi\rangle$ is the value of the functional at the "point" ϕ. With this concept of a functional we can define quantities which are not functions in the sense of classical analysis. As an example we consider in the following the so-called "delta distribution".

3.4.2 Properties of functionals and the delta distribution

The *delta distribution* is defined as the functional $\delta\langle\phi\rangle$ which associates with every test function $\phi(x)$ a number, in this case the value of the test function at $x = 0$, i.e.

$$\delta\langle\phi\rangle = \phi(0), \tag{3.45a}$$

where according to Eq. (3.44)

$$\delta\langle\phi\rangle = \int \delta(x)\phi(x)dx. \tag{3.45b}$$

The result of the action of the "delta function" $\delta(x)$ on the test function $\phi(x)$ is the number $\phi(0)$. The notation $\int \delta(x)\phi(x)dx$ is to be understood only symbolically. The example of the delta function shows that a function does not have to be given in order to allow the definition of a functional.

In order to insure that in the transition from a function $f(x)$ defined in the classical sense to its corresponding functional $f\langle\phi\rangle$ no information about f is lost, i.e. to insure that $f\langle\phi\rangle$ is equivalent to $f(x)$, the class of test functions must be sufficiently large. Thus, if the integral $\int f(x)\phi(x)dx$ is to exist also for a function $f(x)$ which grows with x beyond all bounds, the test functions must decrease to zero sufficiently fast for large x, exactly how fast depending on the given physical situation. In any case the space of test functions must contain those functions $\phi(x)$ which vanish outside a closed and bounded domain, since these correspond to the possibility to measure mass distributions which are necessarily restricted to a finite domain. Furthermore, for the integral (3.44) to exist, the test functions must also possess a sufficiently regular behaviour. For these reasons one demands continuous differentiability of any arbitrary order as in the case of $S(\mathbb{R}^n)$ above. Certain continuity properties of the function $f(x)$ should also be reflected in the associated functional. The reaction of a mass distribution $f(x)$ on a test object $\phi(x)$ is the weaker, the weaker $\phi(x)$ is. Thus it makes sense to demand that if a sequence $\{\phi_i(x)\}$ of test functions and the sequences resulting from the latter's derivatives of arbitrary order, $\phi_i^{(n)}(x)$, converge uniformly towards zero, that then also the sequence of numbers $\langle\phi_i\rangle$ converges towards zero. We refrain, however, from entering here into a deeper discussion of convergence properties in the space of test functions.

The *derivative of a distribution* $f\langle\phi\rangle$, indicated by a prime, is defined by the following equation

$$f'\langle\phi\rangle := -f\langle\phi'\rangle. \tag{3.46}$$

This definition suggests itself for the following reason. If we associate with the function $f(x)$ the distribution

$$f\langle\phi\rangle = \int_{-\infty}^{\infty} dx f(x)\phi(x),$$

then the derivative $f'(x)$ becomes the functional

$$f'\langle\phi\rangle = \int_{-\infty}^{\infty} dx f'(x)\phi(x) = -f\langle\phi'\rangle, \tag{3.47}$$

as in Eq. (3.44). Partial integration of this integral then yields, in view of the conditions $\phi(\pm\infty) = 0$,

$$f'\langle\phi\rangle = [f(x)\phi(x)]_{-\infty}^{\infty} - \int_{-\infty}^{\infty} dx f(x)\phi'(x) = -f\langle\phi'\rangle,$$

as in Eq. (3.47). Equation (3.46) defines the derivative of the functional $f\langle\phi\rangle$ even if there is no function $f(x)$ which defines the functional. For instance in the case of the delta distribution we have

$$\delta'\langle\phi\rangle = -\delta\langle\phi'\rangle = -\phi'(0) \tag{3.48}$$

according to Eq. (3.45a). Formally one writes, of course,

$$\int_{-\infty}^{\infty} dx \delta'(x)\phi(x) = [\delta(x)\phi(x)]_{-\infty}^{\infty} - \int_{-\infty}^{\infty} dx \delta(x)\phi'(x)$$
$$= -\delta\langle\phi'\rangle = -\phi'(0). \tag{3.49}$$

For an infinitely often differentiable function $g(x)$ one has apparently

$$(g \cdot f)\langle\phi\rangle = f\langle g\phi\rangle, \tag{3.50}$$

so that

$$(x\delta)\langle\phi\rangle = \delta\langle x\phi\rangle = [x\phi]_{x=0} = 0 \times \phi(0) = 0, \tag{3.51}$$

or

$$\int x\delta(x)\phi(x)dx = 0, \qquad x\delta(x) = 0. \tag{3.52}$$

Thus *formally* one can operate with the delta distribution or delta function in much the same way as with a function of classical analysis. As a further example we consider the relation

$$f(x)\delta(x) = f(0)\delta(x). \tag{3.53}$$

According to Eq. (3.50) we have

$$
\begin{aligned}
\int f(x)\delta(x)\phi(x)dx &= \delta\langle f\phi\rangle = [f(x)\phi(x)]_{x=0} = f(0)\phi(0) \\
&= f(0)\delta\langle\phi\rangle = \int f(0)\delta(x)\phi(x)dx,
\end{aligned}
$$

as claimed in Eq. (3.53). Formal differentiation of the relation (3.53) yields

$$f(x)\delta'(x) = f(0)\delta'(x) - f'(x)\delta(x). \tag{3.54}$$

One can convince oneself that this formal relation follows also from the defining equation (3.46). In particular for $f(x) = x$ we obtain the useful relation

$$x\delta'(x) = -\delta(x). \tag{3.55}$$

A very important relation for applications is the *Heaviside* or *step function* $\theta(x)$ which is defined as follows:

$$\theta(x) = \begin{cases} 1 & \text{for } x > 0, \\ 0 & \text{for } x < 0. \end{cases} \tag{3.56}$$

From Eq. (3.56) we can deduce the relation

$$\theta'(x) = \delta(x). \tag{3.57}$$

For the verification we associate with the step function the following functional

$$\theta(x)\langle\phi\rangle = \int_{-\infty}^{\infty} \theta(x)\phi(x)dx = \int_{0}^{\infty} \phi(x)dx.$$

For the derivative we have according to Eq. (3.46)

$$
\begin{aligned}
\theta'(x)\langle\phi\rangle &= -\theta(x)\langle\phi'\rangle = -\int_{0}^{\infty} dx\phi'(x) = \phi(0) - \phi(\infty) = \phi(0) \\
&= \delta(x)\langle\phi\rangle,
\end{aligned}
$$

or symbolically $\theta'(x) = \delta(x)$, i.e. Eq. (3.57).

After the introduction of the delta function it is customary to consider briefly *Fourier transforms*, i.e. in the case of one dimension the integral relations

$$f(x) = \frac{1}{\sqrt{2\pi}} \int_{-\infty}^{\infty} dk\, g(k) e^{-ikx} \equiv F^{-1} g(k),$$

$$g(k) = \frac{1}{\sqrt{2\pi}} \int_{-\infty}^{\infty} dx\, f(x) e^{ikx} \equiv F f(x). \tag{3.58}$$

We assume here some familiarity with these integral relations. It is clear that the existence of these integrals assumes significant restrictions on the functions $f(x), g(k)$. As a formal relation in the sense of the theory of distributions we deduce from Eq. (3.58) the important formal integral representation of the delta function, i.e. the relation

$$\delta(x) = \frac{1}{2\pi} \int_{-\infty}^{\infty} dk\, e^{ikx} = \delta(-x). \tag{3.59}$$

One can see this as follows. According to Eq. (3.58)

$$f(0) = \frac{1}{\sqrt{2\pi}} \int dk\, g(k) \quad \text{and} \quad g(k) = \frac{1}{\sqrt{2\pi}} \int dx\, f(x) e^{ikx}.$$

Inserting the second relation into the first, we obtain

$$f(0) = \int dx\, f(x) \left(\frac{1}{2\pi} \int dk\, e^{ikx} \right),$$

and comparison with Eqs. (3.45a) and (3.45b) yields (3.59).

We close this topic with a comment. The singular delta distribution was introduced by Dirac in 1930. The rigorous mathematical theory which justifies the formal use of Dirac's delta function was only later developed by mathematicians, in particular L. Schwartz. Thus today the singular delta distribution is written as a regular distribution, i.e. as an integral operator, by writing, for instance,

$$\delta_{\mathbf{a}} = \int_{\mathbb{R}^n} \delta(\mathbf{x} - \mathbf{a}) \phi(\mathbf{x}) d\mathbf{x} = \phi(\mathbf{a}) \ \forall\, \phi \in S(\mathbb{R}^n), \tag{3.60}$$

and one derives from this that the delta function $\delta(\mathbf{x})$ has the (impossible) properties of being (1) zero everywhere except at $\mathbf{x} = 0$ where it increases so enormously that (2) the integral over $\delta(\mathbf{x})$ is unity:

$$\int_{\mathbb{R}^n} \delta(\mathbf{x}) d\mathbf{x} = 1.$$

Chapter 4

Dirac's Ket- and Bra-Formalism

4.1 Introductory Remarks

In this chapter we introduce the (initially position or momentum space representation-independent) notation of Dirac* which is of considerable practicability. We also introduce those properties of the notation which make the calculations with states and operators amenable to simple manipulations. The notation will be used extensively in later chapters.

The integral representation of the delta function discussed in Chapter 3 can be considered as a formal orthogonality relation for harmonic waves $\exp(ikx)$ which finds its rigorous justification in the context of distribution theory. Thus we have — again for simplicity here for the one-dimensional case —

$$\delta(x - x') = \frac{1}{2\pi} \int_{-\infty}^{\infty} dk e^{ikx} e^{-ikx'} \tag{4.1a}$$

or

$$\delta(k - k') = \frac{1}{2\pi} \int_{-\infty}^{\infty} dx e^{ikx} e^{-ik'x}. \tag{4.1b}$$

Since k and similarly x here assume a continuum of values, Eqs.(4.1a) and (4.1b) are described as normalization conditions of the continuum functions $\exp(ikx)$ as distinct from orthogonality conditions for functions depending on integers m, n as, for instance, the trigonometric functions

$$u_m(x) = \cos(mx), \quad v_m(x) = \sin(mx),$$

---

*See P. A. M. Dirac [75].

for which

$$\frac{1}{\pi} \int_{-\pi}^{\pi} dx u_m(x) u_n(x) = \delta_{mn}, \tag{4.2a}$$

$$\frac{1}{\pi} \int_{-\pi}^{\pi} dx v_m(x) v_n(x) = \delta_{mn}, \tag{4.2b}$$

$$\frac{1}{\pi} \int_{-\pi}^{\pi} dx u_m(x) v_n(x) = 0. \tag{4.2c}$$

In the following we shall therefore "orthogonalize" continuum states represented by vectors of a Hilbert space which is no longer separable (i.e. which has at least a subset whose vectors are characterized by a continuous parameter) in the sense of the relation (4.1a) to a delta function instead to a Kronecker delta as in Eq. (4.2a). With this *formal use* of the delta function we can manipulate continuum states easily in much the same way as discrete states which implies an enormous simplification of numerous calculations.[†]

The Fourier transforms introduced in Chapter 3 permit an additional important observation: We have several possibilities to *represent* vectors in Hilbert space, since the Fourier transform describes the transformation from one representation to another (" configuration" or "position space representation" ↔ "momentum space representation"). A position space Schrödinger wave function $\psi(\mathbf{x})$, which we shall consider in detail in numerous examples later, corresponds to the representation of the vector ψ of a vector space (as representative of a state of the system under consideration) in the position space representation, i.e. as a function of coordinates $\mathbf{x}$. In the momentum space represenation the vector ψ is the Fourier-transformed $\tilde{\psi}(\mathbf{k})$ of $\psi(\mathbf{x})$; this representation will be used in particular in Chapter 13.

4.2   Ket and Bra States

In the following we introduce the notation of Dirac.[‡] We define ket-vectors as elements of a linear vector space and bra-vectors as those of an associated *dual vector space*. The syllables *"bra"* and *"ket"* are those of the word *"bracket"*. The spaces are linear vector spaces, but not necessarily separable Hilbert spaces, unless we are dealing with an entirely discrete system. More as an excercise than as a matter of necessity we recall in the following some considerations of Sec. 3.1, expressed, however, in the notation of Dirac. Hopefully this partial overlap is instructive.

[†]Formally this means that we have to go to an *extended Hilbert space*, which contains also states with infinite norm, see e.g. A. Messiah [195], Vol. I, Sec. 7.1.3.

[‡]P. A. M. Dirac [75].

In order to achieve a representation-independent formulation we assign to every state of the system under consideration a vector, called ket-vector and designated $|\rangle$, in a vector space $\mathcal{V}$. In the symbol $|u\rangle$, for example, u is an index of the spectrum, i.e. a discrete index in the case of the discrete spectrum, or a continuous index in the case of a continuous spectrum. The linearity of the vector space implies, that if ξ is a continuous index, and $\lambda(\xi)$ an arbitrary complex function, then with $|\xi\rangle$,

$$|w\rangle = \int \lambda(\xi)|\xi\rangle d\xi \qquad (4.3)$$

is also an element of $\mathcal{V}$. As mentioned, the space can be finitely or infinitely dimensional. In the vector space $\mathcal{V}$ of ket-vectors a set of basis vectors can be defined, so that every vector can be expressed as a linear combination of these basis vectors.

With the vector space $\mathcal{V}$ of ket-vectors we can associate a *dual space* $\bar{\mathcal{V}}$ of bra-vectors, which are written $\langle\chi|$. The bra-vector $\langle\chi|$ is defined by the linear function

$$\langle\chi|\{|u\rangle\}$$

of ket-vectors $|u\rangle$. For a certain ket-vector $|u\rangle$ the quantity χ has the value $\langle\chi|u\rangle$, which is, in general, a complex number. For a better understanding of the concept of the dual space, we recall first the difference between a linear operator and a linear functional. According to definition, the linear operator which acts on a ket-vector $|u\rangle \in \mathcal{V}$ yields again a ket-vector $|u'\rangle \in \mathcal{V}$. On the other hand, a linear functional χ is an operation, which assigns every ket-vector $|u\rangle \in \mathcal{V}$ linearly a complex number $\langle\chi|u\rangle$, i.e.

$$\chi : \quad |u\rangle \in \mathcal{V} \rightarrow \chi(|u\rangle)$$

with

$$\chi(\lambda_1|u_1\rangle + \lambda_2|u_2\rangle) = \lambda_1\chi(|u_1\rangle) + \lambda_2\chi(|u_2\rangle).$$

The functionals defined on the ket-vectors $|u\rangle \in \mathcal{V}$ define the vector space $\bar{\mathcal{V}}$ called *dual space of* $\mathcal{V}$. An element of this vector space, i.e. a functional χ is symbolized by the bra-vector $\langle\chi|$.[§] Then $\langle\chi|u\rangle$ is the number that results by allowing the linear functional $\langle\chi| \in \bar{\mathcal{V}}$ to act on the ket-vector $|u\rangle$, i.e.

$$\chi(|u\rangle) = \langle\chi|u\rangle.$$

One should compare this with our earlier definition of the functional f on the space of test functions ϕ. We wrote this functional

$$f\langle\phi\rangle \equiv f(|\phi\rangle).$$

[§]By construction every ket-vector is assigned an appropriate bra-vector. The reverse is not true in general. See e.g. C. Cohen–Tannoudji, B. Diu and F. Laloa [53], p. 112.

Furthermore we have:

$$\langle \chi | = 0, \quad \text{if} \quad \chi | u \rangle = 0 \quad \forall \text{ kets } |u\rangle \in \mathcal{V}.$$

Also: $\langle \chi_1 | = \langle \chi_2 |$, if $\langle \chi_1 | u \rangle = \langle \chi_2 | u \rangle$ for all $|u\rangle \in \mathcal{V}$. The dimension of $\bar{\mathcal{V}}$ is the same as the dimension of $\mathcal{V}$.

We consider next the important case that a unique antilinear relation called *conjugation* exists between the kets $|u\rangle \in \mathcal{V}$ and the bras $\langle \chi | \in \bar{\mathcal{V}}$, i.e. that the vector $|u\rangle \in \mathcal{V}$ is associated with a vector in $\bar{\mathcal{V}}$, which we write $\langle u |$. Because of the anti-linearity the conjugate bra-vector associated with the ket-vector

$$|v\rangle = \lambda_1 |1\rangle + \lambda_2 |2\rangle \tag{4.4a}$$

is

$$\langle v | = \lambda_1^* \langle 1 | + \lambda_2^* \langle 2 |, \tag{4.4b}$$

or

$$|v\rangle = \int \lambda(\xi) |\xi\rangle d\xi \tag{4.4c}$$

and

$$\langle v | = \int \lambda^*(\xi) \langle \xi | d\xi. \tag{4.4d}$$

One defines as the *scalar product* or *inner product* of $|u\rangle \in \mathcal{V}$ and $|v\rangle \in \mathcal{V}$ the c-number $\langle v | u \rangle$. This expression is linear with respect to $|u\rangle$ and anti-linear with respect to $|v\rangle$. As in our earlier considerations we demand for $\langle v | u \rangle$ the following properties:

(1) $\langle u | v \rangle = \langle v | u \rangle^*$, $\langle \xi | u \rangle = \langle u | \xi \rangle^*$,

(2) $\langle u | u \rangle \geq 0$, the norm squared, $\forall |u\rangle$ (in the case of $|\xi\rangle$ with infinite norm!),

(3) $\langle u | u \rangle = 0$, if $|u\rangle = 0$ (see (3.5c)).

It follows that we can now write the *Schwarzian inequality* (3.9a):

$$|\langle u | v \rangle|^2 \leq \langle u | u \rangle \langle v | v \rangle. \tag{4.5}$$

Two ket-vectors $|u\rangle, |v\rangle \in \mathcal{V}$ are said to be *orthogonal*, if $\langle u | v \rangle = 0$.

4.3 Linear Operators, Hermitian Operators

A *linear operator* A in the space of ket-vectors $\mathcal{V}$ acts such that

$$A|u\rangle = |v\rangle \in \mathcal{V}$$

with $A = 0$ if $A|u\rangle = 0$ for every vector $|u\rangle \in \mathcal{V}$. Also $A = B$ if for every vector $|u\rangle \in \mathcal{V}$: $\langle u|A|u\rangle = \langle u|B|u\rangle$. Then the following rules apply:

$$A + B = B + A, \quad AB \neq BA.$$

Multiplication by unity, i.e. $\mathbb{1}|u\rangle = |u\rangle$, defines the *unit* or *identity operator.*

Two operators A and B are said to be *inverse* to each other, if

$$|v\rangle = A|u\rangle \quad \text{and} \quad |u\rangle = B|v\rangle,$$

i.e. if

$$AB = \mathbb{1}, \quad BA = \mathbb{1}.$$

The inverse of A is written A^{-1}. This does not always exist. The *inverse of a product* is

$$(AB)^{-1} = B^{-1}A^{-1}.$$

The operator A, defined as a linear operator on $\mathcal{V}$, also acts on the dual space $\bar{\mathcal{V}}$, as we can see as follows. Let the following bra-vector be given: $\langle \chi| \in \bar{\mathcal{V}}$. Then $\langle \chi|(A|u\rangle)$ is a linear functional of $|u\rangle \in \mathcal{V}$ (since A is linear). Furthermore, let $\langle \eta| \in \bar{\mathcal{V}}$ be the bra-vector defined by this functional. Then one says: $\langle \eta|$ results from the action of A on $\langle \chi|$, and one writes:

$$\langle \eta| = \langle \chi|A.$$

It follows that

$$\langle \eta|u\rangle = (\langle \chi|A)|u\rangle \stackrel{!}{=} \langle \chi|(A|u\rangle) \equiv \langle \chi|A|u\rangle.$$

In this way the operations of operators $A, B, \dots$ on bra-vectors are similar to those on ket-vectors.

Finally we define the *tensor-* or *Kronecker-* or *direct product of vector spaces.* Let $|u\rangle^{(1)}$ be vectors of the space $\mathcal{V}_1$ and similarly $|u\rangle^{(2)}$ the vectors of the space $\mathcal{V}_2$. The product of such vectors is written:

$$|u^{(1)}u^{(2)}\rangle \equiv |u\rangle^{(1)}|u\rangle^{(2)}.$$

This product is commutative. Furthermore linearity applies, i.e. with

$$|u\rangle^{(1)} = \lambda_1|v\rangle^{(1)} + \lambda_2|w\rangle^{(1)},$$

we have

$$|u^{(1)}u^{(2)}\rangle = \lambda_1|v^{(1)}u^{(2)}\rangle + \lambda_2|w^{(1)}u^{(2)}\rangle$$

and so on. The product vectors $|u^{(1)}u^{(2)}\rangle$ span a new vector space, the space $\mathcal{V}_1 \otimes \mathcal{V}_2$. If $\mathcal{V}_i$ has the dimension $\mathcal{N}_i$, the space $\mathcal{V}_1 \otimes \mathcal{V}_2$ has the dimension

$\mathcal{N}_1\mathcal{N}_2$. Let $A^{(1)}$ be an operator in the space $\mathcal{V}_1$ and $A^{(2)}$ an operator in the space $\mathcal{V}_2$, i.e.

$$A^{(1)}|u\rangle^{(1)} = |v\rangle^{(1)}.$$

Every such operator is then also associated with an operator, here designated with the same symbol, in the product space, i.e.

$$A^{(1)}|u^{(1)}\rangle|u^{(2)}\rangle = |v^{(1)}u^{(2)}\rangle.$$

Every operator $A^{(1)}$ commutes with every operator $A^{(2)}$, i.e.

$$[A^{(1)}, A^{(2)}] = 0,$$

since

$$A^{(1)}A^{(2)}|u^{(1)}\rangle|u^{(2)}\rangle = |v^{(1)}u^{(2)}\rangle = A^{(2)}A^{(1)}|u^{(1)}\rangle|u^{(2)}\rangle.$$

Assuming now that $|v\rangle \in \mathcal{V}$ and $\langle u|A \in \bar{\mathcal{V}}$ are conjugate vectors as explained above, then $|v\rangle$ is a linear function of $|u\rangle$, i.e. we write

$$|v\rangle = A^\dagger|u\rangle, \quad \text{and} \quad \langle v| = \langle u|A.$$

Next we construct the following scalar products:

$$\langle t|v\rangle = \langle t|A^\dagger|u\rangle, \quad \langle v|t\rangle = \langle u|A|t\rangle.$$

Since we demanded the scalar products to have the property

$$\langle t|v\rangle = \langle v|t\rangle^*,$$

we obtain the conjugate relation

$$\langle t|A^\dagger|u\rangle = \langle u|A|t\rangle^* \tag{4.6}$$

for all $|u\rangle, |t\rangle \in \mathcal{V}$. If $\langle v| = \langle u|A$, the operator $A^\dagger$ is called *hermitian conjugate* or *adjoint operator of A* (cf. also Sec. 3.3). As a consequence we arrive at the following properties (e.g. by replacing in the above the bra-vector $\langle t|$ by $\langle t|B^\dagger$)

$$(AB)^\dagger = B^\dagger A^\dagger, \tag{4.7a}$$

$$(cA)^\dagger = c^* A^\dagger, \tag{4.7b}$$

$$(A + B)^\dagger = A^\dagger + B^\dagger, \tag{4.7c}$$

$$(|u\rangle\langle v|)^\dagger = |v\rangle\langle u|. \tag{4.7d}$$

Examples of operator relations:
(1) $(AB|u\rangle\langle v|C)^\dagger = C^\dagger|v\rangle\langle u|B^\dagger A^\dagger.$

(2) The conjugate bra-vector of

$$AB|u\rangle\langle v|C|w\rangle \quad \text{is} \quad \langle w|C^\dagger|v\rangle\langle u|B^\dagger A^\dagger.$$

The linear operator H is called *hermitian*, if $H = H^\dagger$ and *anti-hermitian*, if $H = -H^\dagger$.

Every operator A can be written as the sum of two such parts:

$$A = \underbrace{\frac{1}{2}(A + A^\dagger)}_{\text{hermitian}} + \underbrace{\frac{1}{2}(A - A^\dagger)}_{\text{anti-hermitian}} \quad .$$

The *product of two hermitian operators* is not necessarily hermitian, but only if these commutators commute, i.e. if $H = H^\dagger$ and $K = K^\dagger$, then $(HK)^\dagger = HK$ only if

$$K^\dagger H^\dagger = KH, \quad \text{i.e.} \quad [H, K] = 0.$$

The *commutator of two hermitian operators* is anti-hermitian:

$$\begin{aligned}
[H, K]^\dagger &= [HK - KH]^\dagger = K^\dagger H^\dagger - H^\dagger K^\dagger \\
&= KH - HK = -[H, K].
\end{aligned}$$

The separation of HK into hermitian and anti-hermitian parts is therefore

$$HK = \frac{1}{2}(HK + KH) + \frac{1}{2}[H, K].$$

Obviously the quantity $|a\rangle\langle a|$ is an hermitian operator, for we had

$$(|u\rangle\langle v|)^\dagger = |v\rangle\langle u|,$$

so that

$$(|a\rangle\langle a|)^\dagger = |a\rangle\langle a|.$$

The operator H is said to be *positive definite*, if H is hermitian and $\langle u|H|u\rangle \geq 0$ for all ket-vectors $|u\rangle$.

An *important theorem* is the following. An hermitian operator A has the properties (a) that all its eigenvalues are real, and (b) that the eigenvalues with respect to $|u\rangle$ are equal to those with respect to $\langle u|$ and vice versa. Verification: Since $A = A^\dagger$ and $A|u\rangle = a|u\rangle$, it follows that

$$\langle u|A|u\rangle = a\langle u|u\rangle, \quad \text{or} \quad \langle u|A|u\rangle^* = \langle u|A^\dagger|u\rangle = \langle u|A|u\rangle,$$

so that $\langle u|A|u\rangle$ is real, as is also $\langle u|u\rangle$. It follows that a is real. Analogous considerations apply in the case of bra-vectors $\langle u|$.

Moreover, since a is real, it follows from $A|u\rangle = a|u\rangle$ that $\langle u|A = a\langle u|$ or the other way round.

Orthogonality: Two eigenvectors of some hermitian operator A belonging to different eigenvalues are orthogonal, as we can see as follows. Assuming

$$A|u\rangle = a|u\rangle, \quad \langle v|A = b\langle v|, \quad a \neq b,$$

we have

$$0 = \langle v|A|u\rangle - \langle v|A|u\rangle = a\langle v|u\rangle - b\langle v|u\rangle = (a - b)\langle v|u\rangle,$$

so that

$$\langle v|u\rangle = 0, \quad \text{since } a \neq b.$$

A continuous spectrum can immediately be included by passing to the extended Hilbert space, which includes vectors of infinite norm. The above theorem remains unchanged, but the continuum vectors are normalized to a delta function, i.e.

$$\langle n|n'\rangle = \delta_{nn'}, \quad \langle n|\nu'\rangle = 0, \quad \langle \nu|\nu'\rangle = \delta(\nu - \nu').$$

The entire set of ket-vectors spans the *extended Hilbert* space. If these form a complete system, then the hermitian operator defined on this space is an *observable.*

Next we introduce the very useful concept of *projection operators*. Let S be a subspace of the Hilbert space $\mathcal{H}$, and let S^* be its complement. Then every vector $|u\rangle \in \mathcal{H}$ can be written

$$|u\rangle = |u_s\rangle + |u_{s*}\rangle \quad \text{with} \quad |u_s\rangle \in S, |u_{s*}\rangle \in S^*, \quad \langle u_s|u_{s*}\rangle = 0.$$

The operator P_S which projects onto S is defined by the properties:

$$P_S|u\rangle = |u_s\rangle = P_S|u_s\rangle, \quad \text{i.e.} \quad P_S|u_{s*}\rangle = 0. \tag{4.8}$$

The projection operator has the following important properties:

$$P_S \text{ is hermitian, i.e. } \langle u|P_S = \langle u_s|. \tag{4.9a}$$

This follows from observing that

$$\langle u|P_S|v\rangle = \langle u|v_s\rangle = \langle u_s|v_s\rangle = \langle u_s|v\rangle,$$

so that $\langle u|P_s = \langle u_s|$.

$$P_S \text{ is } idempotent, \text{ i.e. } P_S{}^2 = P_S. \tag{4.9b}$$

This property follows from the observation that

$$P_S^2|u\rangle = P_S|u_s\rangle = |u_s\rangle = P_S|u\rangle, \quad |u\rangle \text{ arbitrary,}$$

so that

$$(P_S^2 - P_S)|u\rangle = 0, \quad \text{i.e. } P_S^2 = P_S.$$

$$\text{The only eigenvalues of } P_S \text{ are : } 0 \text{ and } 1. \tag{4.9c}$$

This property can be seen as follows. Let p be an eigenvalue of the projection operator P, i.e. $P|p\rangle = p|p\rangle$. Then

$$0 = (P^2 - P)|p\rangle = (p^2 - p)|p\rangle, \quad \text{i.e. } p^2 - p = 0,$$

and hence $p = 0, 1$.

$$P_S \text{ is projector onto } S, \ (1 - P_S) \text{ is projector onto } S^*. \tag{4.9d}$$

Thus the vectors $P|u\rangle, (1 - P)|u\rangle$ are orthogonal.

Example: The vector $|a\rangle$ normalized to 1 with $\langle a|a\rangle = 1$ spans a 1-dimensional subspace. The projection $|u_a\rangle$ of an arbitrary vector $|u\rangle$ onto this subspace is $|u_a\rangle = |a\rangle\langle a|u\rangle$, as can be seen as follows. Set $|u\rangle = |u_a\rangle + |u_{a*}\rangle$ with $\langle a|u_{a*}\rangle = 0$, $|u_a\rangle = c|a\rangle$. Then $\langle a|u\rangle = \langle a|u_a\rangle + \langle a|u_{a*}\rangle = \langle a|u_a\rangle$ and so $\langle a|u\rangle = \langle a|u_a\rangle$ and hence $\langle a|u\rangle = c\langle a|a\rangle = c$, i.e. $c = \langle a|u\rangle$, so that

$$|u_a\rangle = \langle a|u\rangle|a\rangle = |a\rangle\langle a|u\rangle.$$

The quantity $|a\rangle\langle a|$ is called *elementary projector*.

Obviously

$$P_N = \sum_{n=1}^{N} |n\rangle\langle n|$$

is the projection operator onto the N-dimensional subspace S_N, if the set of vectors $|1\rangle, |2\rangle, \ldots, |N\rangle$ is a set of orthonormalized states.

Very similarly we can construct *operators which project onto the subspace spanned by the continuum states*. Let a continuum state be written $|\xi\rangle$, where ξ is a continuous index. If for these (by construction)

$$\langle \xi'|\xi\rangle = \delta(\xi - \xi'),$$

the projector P_2 onto a subspace S_2 is

$$P_2 = \int_{\xi_1}^{\xi_2} |\xi\rangle d\xi \langle \xi|,$$

and hence

$$P_2|u\rangle = \int_{\xi_1}^{\xi_2} |\xi\rangle d\xi \langle\xi|u\rangle.$$

Numerous properties of the projection operators P_i are self-evident. In the case of the infinite-dimensional Hilbert space with a countable number of orthonormalized basis vectors, the operator P is correspondingly

$$P = \sum_{n=1}^{\infty} |n\rangle\langle n|.$$

In the case of continuum states (which are no longer countable) one has correspondingly as projector of all states in a domain $\xi \in [\xi_2, \xi_1]$

$$P = \int_{\xi_1}^{\xi_2} d\xi \, |\xi\rangle\langle\xi|,$$

or in the case of the differential domain $d\xi$ of ξ:

$$dP = \int_{\xi}^{\xi+d\xi} d\xi' \, |\xi'\rangle\langle\xi'|.$$

This latter operator is called *differential projection operator*.

4.4 Observables

Operators which play a particular role in quantum mechanics are those called *observables*, which we introduced earlier. Observables are representatives of measurable quantities.

Let us assume that A is an hermitian operator with a completely discrete spectrum (as for instance in the case of the harmonic oscillator). The eigenvalues λ_i then form a discrete sequence with associated eigenvectors $|u_i\rangle \in \mathcal{H}$. Let $|u_1\rangle, |u_2\rangle, \ldots$ be a system of basis vectors in this space. In general it is possible, that one and the same eigenvalue λ_i is associated with several eigenfunctions, which are orthogonal to each other (i.e. are linearly independent). If there are r such vectors, one says the *degree of degeneracy is* r.[¶] The projector onto the subspace with eigenvalue λ_i can then be written

$$P_i = \sum_r |u_i, r\rangle\langle u_i, r|.$$

[¶]An example is provided by the case of the hydrogen atom, i.e. Coulomb potential; see Eq. (11.114c). However, in this case the spectrum also has a continuous part. Degeneracy will be discussed at various points in this text. See e.g. Sec. 8.6, Examples 8.1 and 11.6, and Eq. (11.114b).

The dimension of this subspace is that of the degree of degeneracy, and

$$AP_i = \lambda_i P_i, \quad (A - \lambda_i)P_i = 0.$$

If $\lambda_i \neq \lambda_{i'}$, then $P_i P_{i'} = 0$. Let us set

$$P_A := \sum_i P_i.$$

If A is an observable and if the spectrum is purely discrete, the projection of this operator is the entire space, i.e.

$$P_A \equiv \sum_i P_i \equiv \sum_{i,r} |u_i, r\rangle\langle u_i, r| = \mathbb{1}. \tag{4.10}$$

This expression is known as *completeness relation* or closure relation, or also as *subdivision of unity* or *of the unit operator*. The operators P_i are linearly independent. The uniqueness of the expression (4.10) follows from the fact that

$$|u\rangle = \sum_i P_i |u\rangle.$$

Applying the operator A to the projector (4.10), we obtain

$$A \sum_i P_i = A, \quad A \sum_i P_i = A \sum_i |u_i\rangle\langle u_i| = \sum_i \lambda_i |u_i\rangle\langle u_i|,$$

i.e.

$$\sum_i \lambda_i P_i = A = \sum_i \lambda_i |u_i, r\rangle\langle u_i, r|, \tag{4.11}$$

i.e. the operator A is completely determined by specification of the eigenvalues λ_i and the eigenvectors $|u_i, r\rangle$ (the convergence, which we do not enter into here, is always implied). Together with the orthogonality condition, i.e. in the case of degeneracy with

$$\langle u_i, r | u_{i'}, r'\rangle = \delta_{ii'}\delta_{rr'},$$

the relation (4.10) expresses the completeness of the orthonormal system. Applied to an arbitrary vector $|u\rangle \in \mathcal{H}$ Eq. (4.10) gives the linear combination

$$|u\rangle = \sum_{i,r} |u_i, r\rangle\langle u_i, r | u\rangle.$$

It follows that the norm squared is given by

$$\langle u | u \rangle = \langle u | P_A | u \rangle = \sum_{i,r} \langle u | u_i, r\rangle\langle u_i, r | u\rangle = \sum_{i,r} |\langle u_i, r | u\rangle|^2. \tag{4.12}$$

This is the expression called *Parseval equation* which we encountered with Eq. (3.17) previously, now however, written in Dirac's notation. If the spectrum of an observable A consists of discrete as well as continuous parts, we have the corresponding generalizations. Let ν be the continuous parameter which characterizes the continuum. Then we denote by $|u_\nu\rangle$ the eigenvector of A whose eigenvalue is $\lambda(\nu)$, i.e. (note: $\langle u_\nu|A|u_{\nu'}\rangle$ is the *matrix representation* of A)

$$A|u_\nu\rangle = \lambda(\nu)|u_\nu\rangle.$$

The ket-vectors $|u_\nu\rangle$ are orthonormalized to a delta function, as explained earlier, i.e.

$$\langle u_\nu|u_{\nu'}\rangle = \delta(\nu - \nu').$$

Then the operator P_A which expresses the completeness of the entire system of eigenvectors is (all continuum states assumed to be in the interval (ν_1, ν_2))

$$P_A \equiv \sum_i |u_i\rangle\langle u_i| + \int_{\nu_1}^{\nu_2} d\nu |u_\nu\rangle\langle u_\nu| = \mathbb{1}. \tag{4.13}$$

For an arbitrary vector $|u\rangle$ of the appropriately *extended Hilbert space* we then have

$$|u\rangle = P_A|u\rangle = \sum_i |u_i\rangle\langle u_i|u\rangle + \int_{\nu_1}^{\nu_2} d\nu |u_\nu\rangle\langle u_\nu|u\rangle,$$

and hence

$$\langle u|u\rangle = \langle u|P_A|u\rangle = \sum_i |\langle u_i|u\rangle|^2 + \int_{\nu_1}^{\nu_2} d\nu |\langle u_\nu|u\rangle|^2,$$

and

$$A = AP_A = \sum_i |u_i\rangle\langle u_i|\lambda_i + \int_{\nu_1}^{\nu_2} \lambda(\nu)|u_\nu\rangle\langle u_\nu| d\nu.$$

In a similar way we can handle functions $f(A)$ of an observable A, for instance exponential functions. One defines as action of $f(A)$ on, for instance, the eigenvector $|u_i\rangle$:

$$f(A)|u_i\rangle = f(\lambda_i)|u_i\rangle.$$

Then

$$f(A) = f(A)P_A = \sum_i f(\lambda_i)|u_i\rangle\langle u_i| + \int_{\nu_1}^{\nu_2} f(\lambda(\nu))|u_\nu\rangle\langle u_\nu| d\nu. \tag{4.14}$$

This relation expresses an arbitrary function $f(A)$ of an observable A in terms of the eigenfunctions of this operator. In the differential operator representation the problem to establish such a relation is known as the *Sturm–Liouville problem*.

Finaly we recapitulate the following theorem from linear algebra: Two observables A and B commute, i.e. their commutator vanishes, $[A, B]$, if and only if they possess at least one common system of basis vectors. Extending this we can say: A sequence of observables $A, B, C, \ldots$, which all commute with one another, form a *complete set of commuting observables*, if (a) they all commute pairwise, and (b) if they possess a uniquely determined system of basis vectors. We shall return to this theorem, here presented without proof, again later.

4.5 Representation Spaces and Basis Vectors

We began by considering ket-vectors $|u\rangle$ in which u is a parameter of the (energy) spectrum, which is discrete in the case of discrete energies and continuous in the case of scattering states (with continuous energies). In the following we consider more generally ket-vectors $|\psi\rangle$ as representatives of the physical states, where the symbol ψ is already indicative of the Schrödinger wave function ψ with energy E.

For many practical purposes the use of the Fourier transform is unavoidable. Therefore we want to re-express the Fourier transform (3.58) in terms of ket- and bra-vectors. First of all we can rewrite the integral representation of the delta function, Eq. (3.59), as a formal orthonormality condition. In the one-dimensional case we write

$$\langle x|x'\rangle = \delta(x - x'), \tag{4.15}$$

where the vectors $\{|x\rangle\}$ are to be a complete set of basis vectors of a linear vector space in its position space representation F_x, i.e.

$$\int dx|x\rangle\langle x| = \mathbb{1}, \quad |x\rangle \in F_x. \tag{4.16}$$

Correspondingly we also have a complete set of basis vectors $\{|k\rangle\}$ of an associated vector space F_k, for which the completeness relation is

$$\int dk|k\rangle\langle k| = \mathbb{1}, \quad |k\rangle \in F_k. \tag{4.17}$$

The Fourier transform provides the transition from one representation and basis to the other. Since both expressions (4.15) and (4.16) represent subdivisions of the unit operator, we can rewrite Eq. (4.15):

$$\delta(x - x') = \langle x|x' \rangle = \langle x|\mathbb{1}|x' \rangle = \langle x| \int dk |k\rangle\langle k|x' \rangle. \qquad (4.18)$$

According to Eq. (3.59) or Eq. (4.1a), this expression has to be identified with

$$\frac{1}{2\pi} \int dk e^{ikx} e^{-ikx'},$$

i.e.

$$\langle x|k \rangle = \frac{1}{\sqrt{2\pi}} e^{ikx}, \quad \langle k|x' \rangle = \frac{1}{\sqrt{2\pi}} e^{-ikx'} = \langle x'|k \rangle^*. \qquad (4.19)$$

Comparison with the orthonormalized system of trigonometric functions (4.2a) etc. shows that these expressions are the corresponding continuum functions (the continuous parameter k replaces the discrete index n). The vectors $|x\rangle$ and $|k\rangle \in F$ are not to be confused with the vectors $|u\rangle$ or $|\psi\rangle \in \mathcal{H}$, which are representatives of the states of our physical system. Rather $|x\rangle$ and $|k\rangle$ serve as basis vectors in the representation spaces F_x, F_k. The representation of a state vector $|\psi\rangle \in \mathcal{H}$ in position space F_x is the mapping of the vector $|\psi\rangle$ into the complex numbers $\langle x|\psi \rangle$, called *wave function*, i.e.

$$\psi(x) := \langle x|\psi \rangle \; : \; \mathcal{H} \to \mathbb{C}. \qquad (4.20)$$

The representation of the corresponding bra-vector $\langle \psi| \in \bar{\mathcal{H}}$ in the position space F_x is correspondingly written

$$\langle \psi|x \rangle = \langle x|\psi \rangle^*.$$

The Fourier representation

$$\psi(x) = \frac{1}{\sqrt{2\pi}} \int e^{ikx} \tilde{\psi}(k) dk \qquad (4.21)$$

provides the ket-vector $|\psi\rangle$ in the k-space representation,

$$\tilde{\psi}(k) := \langle k|\psi \rangle.$$

Obviously we obtain this by inserting a complete system of basis vectors of F_k, i.e.

$$\langle x|\psi \rangle = \int dk \langle x|k \rangle\langle k|\psi \rangle. \qquad (4.22)$$

All of these expressions can readily be transcribed into cases with a higher dimensional position space.

Chapter 5

Schrödinger Equation and Liouville Equation

5.1 Introductory Remarks

In this chapter we remind ourselves of the measuring process briefly alluded to in Chapter 1 and of the necessity to recognize the abstraction of a physical situation that we perform, if we do not take into account the complementary part of the universe. Thus we distinguish between so-called *pure states* and *mixed states* of a system and thereby introduce the concept of *density matrix* and that of the *statistical operator*. We postulate the (time-dependent) *Schrödinger equation*[*] (with the Hamiltonian as the time-development operator) and obtain the quantum mechanical analogue of the Liouville equation. Finally we introduce briefly the canonical distribution of statistical mechanics and show that with this the density matrix can be calculated like the Green's function of a Schrödinger equation (inverse temperature replacing time), the calculation of which for specific cases is the subject of Chapter 7.

5.2 The Density Matrix

We shall establish a matrix ρ, which is the quantum mechanical analogue of the classical probability density $\rho(q, p, t)$. This will then also clarify the role of ρ as an operator in the quantum mechanical analogy with the Liouville equation. We mentioned earlier (in Sec. 1.5) that in general and in reality we ought to consider the system under consideration together or in interaction with the rest of the universe, i.e. with the complementary system or

---

[*]E. Schrödinger [244].

the measuring instruments. The actual and real state of a system is then a superposition of two types of states, $|i\rangle, |a\rangle$. The states $|i\rangle \in \mathcal{H}$, where i represents collectively all quantum numbers of the state of the system under consideration, are called *"pure states"*. These are the quantum mechanical states which are also referred to as *micro-states*. The states $|a\rangle \in \mathcal{H}'$ on the other hand are the corresponding states of the complementary part of the universe. The difficulty we have to face is the impossibility to specify precisely the stationary state of our limited system. For an exact treatment we would have to express the actual state $|\psi\rangle$ of the system as a supersposition not only of the states $|i\rangle, \dots$ but also of the states of the complementary set of the universe, $|a\rangle, \dots$, that is, we have to deal with a so-called *"mixed state"*, which in Dirac's notation we can write as

$$|\psi\rangle = \sum_{a,i} C_{a,i} |a\rangle |i\rangle \ \in \mathcal{H} \otimes \mathcal{H}'. \tag{5.1}$$

We assume that the states are orthonormal and in their respective subspaces also complete, i.e. we have

$$\langle a|b\rangle = \delta_{ab}, \quad \langle i|j\rangle = \delta_{ij}, \quad \langle a|i\rangle = 0 = \langle j|b\rangle, \tag{5.2}$$

as well as

$$\sum_i |i\rangle\langle i| = \mathbb{1} \ \text{ in the subspace of pure states} \tag{5.3a}$$

and

$$\sum_a |a\rangle\langle a| = \mathbb{1} \ \text{ in the complementary subspace.} \tag{5.3b}$$

The quantum mechanical expectation value of an observable A, i.e. of an operator that acts only on the states of the system we are interested in, is then given by

$$\langle A \rangle \ := \ \langle \psi|A|\psi\rangle = \sum_{a,b;i,j} \langle j|\langle b|A|a\rangle|i\rangle C_{ai} C_{bj}{}^*$$

$$= \ \sum_{a,b;i.j} \langle j|A|i\rangle \delta_{ba} C_{ai} C_{bj}{}^* = \sum_{ij} \langle j|A|i\rangle \rho_{ij} \tag{5.4}$$

where

$$\rho_{ij} \equiv \langle i|\rho|j\rangle = \sum_a C_{ai} C_{aj}{}^*. \tag{5.5}$$

Here ρ is an hermitian matrix, called the *density matrix*.[†] Since ρ is hermitian, the matrix can be diagonalized by a transition to a new set of states

---

[†]The hermiticity can be demonstrated as follows, where T means transpose: $\rho = C^T C^*$, $\rho^\dagger = (C^T C^*)^\dagger = C^T C^* = \rho$.

$|i'\rangle \in \mathcal{H}$. This transition to a new set of basis vectors in which the matrix becomes diagonal, can be achieved with the help of the completeness relation of the vectors $|i\rangle$:

$$|i'\rangle = \sum_i |i\rangle\langle i|i'\rangle \quad \in \mathcal{H}, \tag{5.6}$$

i.e. (see below)

$$\langle i'|\rho|i''\rangle = \omega_{i'}\delta_{i'i''} = \text{real}, \tag{5.7}$$

or

$$\rho = \sum_{i',i''} |i'\rangle\langle i'|\rho|i''\rangle\langle i''| = \sum_{i'} \omega_{i'}|i'\rangle\langle i'|, \tag{5.8}$$

since then

$$\langle j'|\rho|j''\rangle = \sum_{i'} \omega_{i'}\langle j'|i'\rangle\langle i'|j''\rangle = \sum_{i'} \omega_{i'}\delta_{i'j'}\delta_{i'j''} = \omega_{j'}\delta_{j'j''}, \tag{5.9}$$

is diagonal. In the following we write simply $|i\rangle$ instead of $|i'\rangle$. Thus the pure states $|i\rangle$ form in the subspace of the space of states which is of interest to us, a complete orthonormal system with properties (5.2) and (5.3a). The operator $P_i \equiv |i\rangle\langle i|$ projects a state $|\phi\rangle \in \mathcal{H}$ onto $|i\rangle$, i.e.

$$P_i|\phi\rangle = |i\rangle\langle i|\phi\rangle = \langle i|\phi\rangle|i\rangle \tag{5.10}$$

and is therefore, as we saw earlier, a *projection operator*. This projection operator represents a quantum mechanical probability as compared with the numbers ω_i which represent classical probabilities.[‡]

We can specify the state of our system, as we discussed previously, in a specific representation, e.g. in the position state representation or x-representation, where x represents collectively the entire set of position coordinates of the (particle) system. In this sense we have

$$\rho(x', x) := \langle x'|\rho|x\rangle = \sum_i \omega_i \langle x'|i\rangle\langle i|x\rangle \equiv \sum_i \omega_i \, i(x')i^*(x). \tag{5.11}$$

In Eq. (5.4) we had the expression

$$\langle A\rangle = \sum_{i,j} \rho_{ij}\langle j|A|i\rangle \quad \text{with} \quad \rho_{ij} = \sum_a C_{ai}C_{aj}^*. \tag{5.12}$$

[‡]In his reformulation of the statistical mechanics developed by Boltzmann, Gibbs introduced the concept of ensemble, which today is a fundamental term in statistical mechanics. This ensemble is interpreted as a large set of identical systems, which are all subjected to the same macroscopic boundary and subsidiary conditions, but occupy in general different microscopic, i.e. quantum states. The ensemble can be represented as a set of points in phase space. If the ensemble corresponding to the system under consideration consists of Z elements (i.e. seen macroscopically of Z copies of the system), then $(\omega_i/\sum_i \omega_i)Z$ is the number of ensemble elements in the pure state $|i\rangle$. Thus the real system is in the pure state $|i\rangle$ with probability $\omega_i/\sum_i \omega_i$.

We now define the *operator* ρ by the relation (cf. Eq. (5.5) above)

$$\langle i|\rho|j\rangle = \rho_{ij}. \tag{5.13}$$

Then the expectation value of the observable A is (cf. Eq. (5.4))

$$\langle A\rangle = \sum_{ij}\langle j|A|i\rangle\langle i|\rho|j\rangle = \sum_{j}\langle j|A\rho|j\rangle, \tag{5.14}$$

which we can write

$$\langle A\rangle = \mathrm{Tr}(\rho A). \tag{5.15}$$

In particular for $A = \mathbb{1}$, we have

$$\mathrm{Tr}(\rho) = 1, \quad \mathrm{Tr}\left(\sum_{i}\omega_{i}|i\rangle\langle i|\right) = 1. \tag{5.16}$$

Thus

$$\sum_{j}\langle j|\sum_{i}\omega_{i}|i\rangle\langle i|j\rangle = 1, \quad \text{or} \quad \sum_{i}\omega_{i}\langle i|i\rangle = \sum_{i}\omega_{i} = 1. \tag{5.17}$$

Hence the sum of the classical probabilities is 1. We can also show that

$$\omega_{i} \geq 0. \tag{5.18}$$

To this end we set

$$A \rightarrow A_{i'} := |i'\rangle\langle i'| \quad (\text{no summation over } i').$$

Substituting this into Eq. (5.14) we obtain

$$\langle A_{i'}\rangle = \sum_{ij}\langle j|i'\rangle\langle i'|i\rangle\langle i|\rho|j\rangle = \sum_{i,j}\delta_{ji'}\delta_{ii'}\rho_{ij} = \rho_{i'i'} \stackrel{(5.7)}{=} \omega_{i'}.$$

On the other hand, according to Eq. (5.4),

$$\langle A_{i'}\rangle := \langle\psi|A_{i'}|\psi\rangle = \langle\psi|i'\rangle\langle i'|\psi\rangle = |\langle\psi|i'\rangle|^2 \geq 0.$$

Hence $\omega_{i'} \geq 0$, as claimed by Eq. (5.18).

We recall that the elements of ρ originate from the coefficients C_{ai} in

$$|\psi\rangle = \sum_{a,i}C_{ai}|a\rangle|i\rangle = \sum_{i}\left(\sum_{a}C_{ai}|a\rangle\right)|i\rangle. \tag{5.19}$$

We see therefore that ρ defines the *mixed states*. The effect of the interaction of the system under consideration with the surroundings, which is contained in the coefficients C_{ai}, thus enters these states. If all ω_i but one are nought, i.e. the only remaining one is 1, one says, the system is in a pure state; in all other cases the state of the system is a mixed state. In the following we are mostly concerned with considerations of systems in pure states, but we see here, that these are actually not sufficiently general, since they do not take into account the (interaction with the) rest of the universe. (Question: Is the state of the universe a pure state?)

5.3 The Probability Density $\rho(\mathbf{x}, t)$

The expression (5.8), i.e.

$$\rho = \sum_i \omega_i |i\rangle \langle i|,$$

defines the *statistical operator*, in which the numbers ω_i represent *classical probabilities* and the operator $|i\rangle\langle i|$ *quantum mechanical probabilities* or *weights*. We now consider the latter expression. For the case

$$\omega_j = \begin{cases} 1 & \text{for } j = i, \\ 0 & \text{for } j \neq i, \end{cases} \tag{5.20}$$

we have

$$\rho = |i\rangle\langle i| \equiv P_i. \tag{5.21}$$

Equation (5.20) expresses that the system can be found with probability 1 in the state $|i\rangle$, and in the other states with probability zero. Thus we have a system in a pure state, not in a mixture of states. We observed above, that — different from ω_i — P_i represents a quantum mechanical probability. We see this more clearly by going to the position space representation, i.e. by representing the vector $|i\rangle$ by the *wave function* $i(x,t) \equiv \langle x|i\rangle$,[§] $\{|x\rangle\}$ is a complete system of basis vectors in the position representation space F_x with

$$\int dx |x\rangle\langle x| = \mathbb{1}, \quad \langle x|x'\rangle = \delta(x - x'). \tag{5.22}$$

Then

$$\langle x|P_i|x\rangle = \langle x|i\rangle\langle i|x\rangle = |\langle x|i\rangle|^2. \tag{5.23}$$

Instead of $i(x,t)$ we now write $\psi(x,t)$ and instead of $\langle x|P_i|x\rangle$ we now write $\rho(x,t)$, so that

$$\rho(x,t) = |\psi(x,t)|^2. \tag{5.24}$$

[§]In the following $\langle x|i\rangle$ is always to be understood as $\langle x|i(t)\rangle$.

The relation (5.22) can also be written $|\langle x|i\rangle|^2 = \langle i|x\rangle\langle x|i\rangle$, and hence the integral over all space

$$\int_{\mathbb{R}^1} dx \rho(x,t) = \int dx \langle i|x\rangle\langle x|i\rangle = \langle i|i\rangle = 1. \tag{5.25}$$

Thus the particle whose state is described by $|i\rangle$, or the wave function $\psi(x,t)$, can be found with probability 1 in the space $\mathbb{R}^1$. The generalization to a three-dimensional position space is self-evident:

$$\rho(\mathbf{x},t) = |\psi(\mathbf{x},t)|^2. \tag{5.26}$$

5.4   Schrödinger Equation and Liouville Equation

We encountered the classical form of the Liouville equation in Sec. 2.3. We now want to obtain the quantum mechanical analogue. To this end we have to differentiate the density matrix ρ, i.e. the expression (5.8), with respect to time. This differentiation requires the time derivative of a state vector $|i\rangle \in \mathcal{H}$. *We therefore postulate* first an equation for the time dependence of a state vector, which is chosen in such a way that the desired analogy is obtained. We shall convince ourselves later that the postulated equation is sensible (in fact, the Schrödinger equation is always postulated in some way, it is not "derived"). Hence we *postulate*: The time development of a state $|j\rangle \in \mathcal{H}$ is given by the equation

$$i\hbar\frac{\partial}{\partial t}|j\rangle = H|j\rangle, \tag{5.27}$$

where H is the Hamilton operator of the system. Equation (5.27) is known as the *Schrödinger equation.*[¶] The equation which is the conjugate of Eq. (5.27) is

$$-i\hbar\langle j|\overleftarrow{\frac{\partial}{\partial t}} = \langle j|H, \tag{5.28}$$

Differentiating Eq. (5.8) and multiplying by $i\hbar$, we obtain

$$i\hbar\frac{\partial\rho}{\partial t} = i\hbar\frac{\partial}{\partial t}\sum_j \omega_j|j\rangle\langle j| = i\hbar\sum_j \frac{\partial\omega_j}{\partial t}|j\rangle\langle j| + i\hbar\sum_j \omega_j\left(\frac{\partial}{\partial t}|j\rangle\right)\langle j|$$
$$+i\hbar\sum_j \omega_j|j\rangle\left(\frac{\partial}{\partial t}\langle j|\right). \tag{5.29}$$

[¶]More precisely this equation should be called Schrödinger equation of motion (since H is fixed, but the linear operator states are moving — even if the system is observably stationary), and the time-independent equation should correspondingly be called Schrödinger wave equation (i.e. the equation of motion with states represented in a system for which the q's or here x's are diagonal).

Inserting here Eqs. (5.27) and (5.28) we obtain (see also Eq. (5.31) below)

$$
\begin{aligned}
i\hbar\frac{\partial\rho}{\partial t} &= \sum_j \omega_j H|j\rangle\langle j| - \sum_j \omega_j|j\rangle\langle j|H + i\hbar\sum_j \frac{\partial\omega_j}{\partial t}|j\rangle\langle j| \\
&= \sum_j \omega_j(HP_j - P_jH) + i\hbar\sum_j \frac{\partial\omega_j}{\partial t}|j\rangle\langle j| \\
&= [H,\rho] + i\hbar\sum_j \frac{\partial\omega_j}{\partial t}|j\rangle\langle j|, \tag{5.30}
\end{aligned}
$$

where in statistical equilibrium

$$
\sum_j \frac{\partial\omega_j}{\partial t}|j\rangle\langle j| = 0. \tag{5.31}
$$

The eigenvalues ω_i of ρ determine the fraction of the total number of systems of the ensemble describing the actual system which occupy the state $|i\rangle$. With Eqs. (5.27), (5.28) the states are considered as time-dependent; such a formulation is described as *Schrödinger picture*. The Schrödinger picture and the alternative Heisenberg picture will later be considered separately. We thus obtain the operator form of the *Liouville equation*, i.e. the relation

$$
i\hbar\frac{\partial\rho}{\partial t} = [H,\rho]. \tag{5.32}
$$

Comparing Eq. (5.32) with its classical counterpart (2.21), we observe that here on the left hand side we have the total time derivative, not the partial derivative. The reason for this is that in addition to the "classical probabilities", ρ also contains "quantum mechanical probabilities" (i.e. both ω_i and $|i\rangle\langle i|$). The correspondence to the classical case is obtained with the substitution

$$
\frac{1}{i\hbar}[\dots,\dots] \to \{\dots,\dots\} \tag{5.33}
$$

and $d\rho/dt = 0$ (cf. Eq. (2.24)), $\partial\rho/\partial t \neq 0$.

We can now consider the Schrödinger equation (5.27) in the position space representation:

$$
i\hbar\langle x|\frac{\partial}{\partial t}|j\rangle = \langle x|H|j\rangle = \langle x|H(p_i,x_i)|j\rangle, \tag{5.34}
$$

i.e. (cf. Eq. (3.37))

$$
i\hbar\frac{\partial}{\partial t}\langle x|j\rangle = H\left(-i\hbar\frac{\partial}{\partial x_i},x_i\right)\langle x|j\rangle,
$$

or with $\langle x|j\rangle \rightarrow \psi(\mathbf{x}, t)$:

$$i\hbar\frac{\partial}{\partial t}\psi(\mathbf{x}, t) = H\left(-i\hbar\frac{\partial}{\partial \mathbf{x}}, \mathbf{x}\right)\psi(\mathbf{x}, t). \tag{5.35}$$

Here we can look at the Hamilton operator as a *"time generator"*, since the solution of Eq. (5.35) can be written[||]

$$\psi(\mathbf{x}, t) = \exp\left[\frac{1}{i\hbar}Ht\right]\psi(\mathbf{x}, 0) \quad \text{or} \quad |\psi\rangle_t = |\psi\rangle_{t=0}\exp\left[\frac{1}{i\hbar}Ht\right]. \tag{5.36}$$

The initial wave function or time-independent wave function $\psi(\mathbf{x}, 0)$ can be expanded in terms of a complete set of eigenvectors $|E_n\rangle$ of the Hamilton operator H, i.e. we write

$$\psi(\mathbf{x}, 0): \quad \langle x|j\rangle_{t=0} = \sum_n \langle x|E_n\rangle\langle E_n|j\rangle_{t=0}, \tag{5.37}$$

where

$$H\left(-i\hbar\frac{\partial}{\partial x_i}, x_i\right)\langle x|E_n\rangle = E_n\langle x|E_n\rangle, \quad \text{or} \quad H|E_n\rangle = E_n|E_n\rangle. \tag{5.38}$$

This equation is described as the *time-independent Schrödinger equation.*

5.4.1 Evaluation of the density matrix

As a side-remark with regard to statistical mechanics we can make an interesting observation at this point, namely that the density matrix satisfies an equation analogous to the time-dependent Schrödinger equation — not, however, with respect to time t, but with respect to the parameter $\beta = 1/kT$, T meaning temperature, appearing in the Boltzmann distribution.[**] In view of the close connection between time- and temperature-dependent Green's functions, which we shall need later, it is plausible to refer to this equation at this stage. For a mixture of states $|i\rangle$ of the system under consideration caused by the rest of the universe we have $\omega_i \neq 1$. In expression (5.8) for ρ, i.e. in $\rho = \sum_i \omega_i|i\rangle\langle i|$, the state $|i\rangle$ is still time-dependent. With the help of Eq. (5.36) however, we can replace these states by corresponding time-independent states, since the time-dependence cancels out (the exponential functions involve the same operator H but with signs reversed; this is an application of the *Baker–Campbell–Hausdorff formula* which we deal with in

---

[||]Later, in Sec. 10.4, we shall describe as *"time development operator"* the exponentiated operator given by $U(t, t_0) = \exp[-iH(t - t_0)/\hbar]$.

[**]R. P. Feynman [94].

Example 5.1). Hence we have (with $\langle E_i|E_j\rangle = \delta_{ij}$) in what may be called the energy representation

$$\rho = \sum_i \omega_i |E_i\rangle\langle E_i|. \tag{5.39}$$

Without proof we recall here that in the so-called *canonical distribution* the weight factors ω_i (similar to those of the Boltzmann distribution) are such that (cf. Eq. (1.9) with $E_n \to nh\nu$)

$$\omega_i \propto e^{-\beta E_i}, \quad \beta = 1/kT, \quad \text{or} \quad \omega_i = \frac{e^{-\beta E_i}}{\sum_i e^{-\beta E_i}}, \tag{5.40}$$

so that $\sum_i \omega_i = 1$. In the position space representation Eq. (5.39) therefore becomes

$$\langle x|\rho|x'\rangle = \sum_i \omega_i \langle x|E_i\rangle\langle E_i|x'\rangle, \tag{5.41}$$

or with $\phi_i(x) := \langle x|E_i\rangle$ this is

$$\rho(x, x') = \sum_i \omega_i \phi_i(x)\phi_i^*(x'). \tag{5.42}$$

Inserting here Eq. (5.40), we obtain

$$\rho(x, x') \to \rho(x, x'; \beta) = \frac{\sum_i e^{-\beta E_i}\phi_i(x)\phi_i^*(x')}{\sum_i e^{-\beta E_i}}. \tag{5.43}$$

Since $H\phi_i(x) = E_i\phi_i(x)$, we can rewrite ρ, i.e. the expression

$$\rho = \frac{\sum_i e^{-\beta E_i}|E_i\rangle\langle E_i|}{\sum_i e^{-\beta E_i}}, \tag{5.44}$$

also as (see below)

$$\rho = \frac{e^{-\beta H}}{(\text{Tr}e^{-\beta H})}, \tag{5.45}$$

since on the one hand

$$\text{Tr}(e^{-\beta H}) = \sum_i \langle E_i|e^{-\beta H}|E_i\rangle = \sum_i e^{-\beta E_i}\langle E_i|E_i\rangle = \sum_i e^{-\beta E_i},$$

and on the other hand

$$e^{-\beta H} = e^{-\beta H}\sum_i |E_i\rangle\langle E_i| = \sum_i e^{-\beta E_i}|E_i\rangle\langle E_i|,$$

so that ρ is given by Eq. (5.45). We now rewrite the factor $\exp(-\beta H)$ in Eq. (5.45) without normalization as

$$\rho_N(\beta) := e^{-\beta H}. \tag{5.46}$$

In the *energy representation* this expression is

$$\rho_{Nij}(\beta) := \langle E_i | \rho_N(\beta) | E_j \rangle = \langle E_i | e^{-\beta H} | E_j \rangle = e^{-\beta E_i} \delta_{ij} \tag{5.47}$$

with

$$\rho_{Nij}(0) = \delta_{ij}, \quad \rho_N(0) = \mathbb{1}. \tag{5.48}$$

Differentiating Eq. (5.47) with respect to β, we obtain

$$\frac{\partial \rho_{Nij}(\beta)}{\partial \beta} = -\delta_{ij} E_i e^{-\beta E_i} = -E_i \rho_{Nij}(\beta). \tag{5.49}$$

In the position or configuration space representation Eq. (5.46) is

$$\rho_N(x, x'; \beta) := \langle x | \rho_N(\beta) | x' \rangle = \langle x | e^{-\beta H} | x' \rangle, \quad \text{Tr}(\rho_N) = \int dx \rho_N(x, x; \beta). \tag{5.50}$$

Differentiating this equation with respect to β, we obtain

$$\begin{aligned}
\frac{\partial \rho_N(x, x'; \beta)}{\partial \beta} &= -\langle x | H e^{-\beta H} | x' \rangle = -H\left(-i\hbar \frac{\partial}{\partial x}, x\right) \langle x | e^{-\beta H} | x' \rangle \\
&\equiv -H_x \rho_N(x, x'; \beta),
\end{aligned} \tag{5.51}$$

where the subscript x indicates that H acts on x of $\rho_N(x, x'; \beta)$. Equation (5.51) is seen to be very similar to the Schrödinger equation (5.35). With Eqs. (5.45) and (5.46) we can write the expectation value of an observable A:

$$\langle A \rangle \equiv \text{Tr}\langle \rho A \rangle = \frac{\text{Tr}(\rho_N A)}{(\text{Tr}\rho_N)}. \tag{5.52}$$

Hence this expectation value can now be evaluated with a knowledge of ρ_N. The quantity ρ_N is obtained from the solution of the Schrödinger-like equation (5.51). This solution is the *Green's function* or *kernel* $K(x, x'; \beta)$, whose calculation for specific cases is a topic of Chapter 7 (see Eq. (7.55)).

Example 5.1: Baker–Campbell–Hausdorff formula
Verify that if A and B are operators, the following relation holds:

$$\exp(A)\exp(B) = \exp\left(A + B + \frac{1}{2}[A, B] + \frac{1}{12}[A, [A, B]] - \frac{1}{12}[B, [B, A]] + \cdots\right).$$

Solution: The relation can be verified by expansion of the left hand side.[††]

[††]For a more extensive discussion see R. M. Wilcox [284].

Chapter 6

Quantum Mechanics of the Harmonic Oscillator

6.1 Introductory Remarks

In the following we consider in detail the quantization of the linear harmonic oscillator. This is a fundamental topic since the harmonic oscillator, like the Coulomb potential, can be treated exactly and therefore plays an important role in numerous quantum mechanical problems which can be solved only with the help of perturbation theory. The importance of the harmonic oscillator can also be seen from comments in the literature such as:* *"...the present form of quantum mechanics is largely a physics of harmonic oscillators."* Quite apart from this basic significance of the harmonic oscillator, its quantization here in terms of quasi-particle creation and annihilation operators also points the direction to the quantization of field theories with creation and annihilation of particles, which is often described as "second quantization" as distinct from the quantization of quantum mechanics which is correspondingly described as "first quantization" (thus one could visualize the free electromagnetic field as consisting of two mutually orthogonal oscillators at every point in space). An alternative method of quantization of the harmonic oscillator — by consideration of the Schrödinger equation as the equation of Weber or parabolic cylinder functions — is discussed in Chapter 8 (see Eqs. (8.51) to (8.53)). However, we shall see on the way, that the oscillator, can — in fact — be quantized in quite a few different ways. Since the spectrum of the harmonic oscillator is entirely discrete, its associated space of state vectors also provides the best illustration of a properly separable Hilbert space.

---

*Comment of Y. S. Kim and M. E. Noz [149].

6.2   The One-Dimensional Linear Oscillator

In the case of the one-dimensional, linear harmonic oscillator the Hamilton operator is given by the expression

$$H = \frac{1}{2m_0}p^2 + \frac{1}{2}m_0\omega^2 q^2.$$

The quantization of this oscillator is achieved with the three postulated canonical quantization conditions, of which the only non-trivial one here is

$$[q, p] = i\hbar \tag{6.1}$$

for the hermitian operators q, p.[†] The first problem is to determine the eigenvalues and eigenfunctions of H. This can be done in a number of ways. For instance we can go to the special configuration or position space representation given by

$$q \rightarrow x, \quad p \rightarrow -i\hbar\frac{\partial}{\partial x},$$

and then to the time-independent Schrödinger equation $H\psi(x) = E\psi(x)$, in the present case

$$\frac{\hbar^2}{2m_0}\frac{d^2\psi}{dx^2} + \left(E - \frac{1}{2}m_0\omega^2 x^2\right)\psi = 0, \tag{6.2}$$

which for $m_0 = 1/2, \hbar = 1, \omega = 1$ reduces to the *parabolic cylinder equation*. This equation would then have to be solved as a second order differential equation; the boundary conditions on the solutions $\psi(x)$ would be square integrability over the entire domain of x from $-\infty$ to ∞. However, we can also proceed in a different way, which is — in the first place — representation-independent. To this end we introduce first of all the dimensionless quantities

$$\sqrt{\frac{m_0\omega}{\hbar}}q, \quad \frac{p}{\sqrt{m_0\omega\hbar}},$$

by setting

$$A := \frac{1}{\sqrt{2}}\left(\sqrt{\frac{m_0\omega}{\hbar}}q + \frac{i}{\sqrt{m_0\omega\hbar}}p\right), \tag{6.3}$$

and

$$A^\dagger = \frac{1}{\sqrt{2}}\left(\sqrt{\frac{m_0\omega}{\hbar}}q - \frac{i}{\sqrt{m_0\omega\hbar}}p\right), \tag{6.4}$$

[†]Note that we assume it is obvious from the context whether q and p are operators or c-numbers; hence we do not use a distinguishing notation. Also observe that H does not contain terms like pq or qp, so that there is no ambiguity due to commutation. When such a term arises one has to resort to some definition, like taking half and half, which is called Weyl ordering.

(p, q are hermitian). With the help of Eq. (6.1) we obtain immediately the commutation relation

$$[A, A^\dagger] = \mathbb{1}. \tag{6.5}$$

Re-expressing q and p in terms of $A, A^\dagger$, we obtain

$$q = \sqrt{\frac{\hbar}{m_0\omega}} \frac{A + A^\dagger}{\sqrt{2}} \tag{6.6}$$

and

$$p = \sqrt{m_0\omega\hbar} \frac{A - A^\dagger}{i\sqrt{2}}. \tag{6.7}$$

Inserting these expressions for p and q into H we obtain

$$H = \frac{1}{2}\hbar\omega(A^\dagger A + AA^\dagger) = \hbar\omega\left(A^\dagger A + \frac{1}{2}\right). \tag{6.8}$$

The eigenstates of H are therefore essentially those of

$$N := A^\dagger A. \tag{6.9}$$

We observe first that if $|\alpha\rangle$ is a normalized eigenvector of N with eigenvalue α, i.e. if

$$A^\dagger A|\alpha\rangle = \alpha|\alpha\rangle, \tag{6.10}$$

then

$$\alpha = \langle\alpha|A^\dagger A|\alpha\rangle = ||A|\alpha\rangle||^2 \geq 0. \tag{6.11}$$

Thus the eigenvalues are real and non-negative. We now use the relation (3.34d):

$$[A, BC] = [A, B]C + B[A, C], \quad \text{and} \quad [AB, C] = A[B, C] + [A, C]B, \tag{6.12}$$

in order to obtain the following expressions:

$$[A^\dagger A, A] = [A^\dagger, A]A = -A, \quad [A^\dagger A, A^\dagger] = A^\dagger[A, A^\dagger] = A^\dagger. \tag{6.13}$$

From these relations we obtain

$$(A^\dagger A)A = A(A^\dagger A - 1), \tag{6.14a}$$

$$(A^\dagger A)A^\dagger = A^\dagger(A^\dagger A + 1). \tag{6.14b}$$

From Eq. (6.14a) we deduce for an eigenvector $|\alpha\rangle$ of $A^\dagger A$, i.e. for

$$A^\dagger A|\alpha\rangle = \alpha|\alpha\rangle,$$

the relation

$$(A^\dagger A)A|\alpha\rangle = A(A^\dagger A - 1)|\alpha\rangle = A(\alpha - 1)|\alpha\rangle = (\alpha - 1)A|\alpha\rangle. \qquad (6.15)$$

Thus $A|\alpha\rangle$ is eigenvector of $A^\dagger A$ with eigenvalue $(\alpha - 1)$, unless $A|\alpha\rangle = 0$. Similarly we obtain from Eq. (6.14b) :

$$(A^\dagger A)A^\dagger|\alpha\rangle = A^\dagger(A^\dagger A + 1)|\alpha\rangle = A^\dagger(\alpha + 1)|\alpha\rangle = (\alpha + 1)A^\dagger|\alpha\rangle. \qquad (6.16)$$

This means, $A^\dagger|\alpha\rangle$ is eigenvector of $A^\dagger A$ with eigenvalue $(\alpha + 1)$, unless $A^\dagger|\alpha\rangle = 0$. The norm of $A|\alpha\rangle$ is

$$||A|\alpha\rangle||^2 = \langle\alpha|A^\dagger A|\alpha\rangle = \alpha\langle\alpha|\alpha\rangle = \alpha, \qquad (6.17)$$

or

$$||A|\alpha\rangle|| = \sqrt{\alpha}.$$

Similarly

$$||A^\dagger|\alpha\rangle||^2 = \langle\alpha|AA^\dagger|\alpha\rangle = \langle\alpha|1 + A^\dagger A|\alpha\rangle = (\alpha + 1)\langle\alpha|\alpha\rangle = \alpha + 1, \qquad (6.18)$$

or

$$||A^\dagger|\alpha\rangle|| = \sqrt{\alpha + 1}.$$

Next we consider the vector $A^2|\alpha\rangle$. In this case we have

$$
\begin{aligned}
(A^\dagger A)A^2|\alpha\rangle &= A(A^\dagger A - 1)A|\alpha\rangle = A(A^\dagger AA - A)|\alpha\rangle \\
&\overset{(6.14a)}{=} A(AA^\dagger A - 2A)|\alpha\rangle = A(A\alpha - 2A)|\alpha\rangle \\
&= (\alpha - 2)A^2|\alpha\rangle, \qquad\qquad (6.19)
\end{aligned}
$$

i.e. $A^2|\alpha\rangle$ is eigenvector of $A^\dagger A$ with eigenvalue $(\alpha - 2)$, unless $A^2|\alpha\rangle = 0$. If we continue like this and consider the vectors

$$A^n|\alpha\rangle \neq 0 \text{ for all } n,$$

we find that $A^n|\alpha\rangle$ is eigenvector of $A^\dagger A$ with eigenvalue $(\alpha - n)$. This would mean that for sufficiently large values of n the eigenvalue would be negative. However, in view of Eq. (6.11) this is not possible, since this equation implies that the eigenvalues cannot be negative. Thus for a certain value of $n \geq 0$, we must have

$$(a) \qquad A^n|\alpha\rangle \neq 0, \quad \text{but} \quad (b) \qquad A^{n+1}|\alpha\rangle = 0. \qquad (6.20)$$

Let

$$|\alpha - n\rangle := \frac{A^n|\alpha\rangle}{||A^n|\alpha\rangle||} \qquad (6.21)$$

be a normalized eigenvector of $A^\dagger A$ with eigenvalue $(\alpha - n)$, so that

$$\langle \alpha - n | \alpha - n \rangle = \frac{\langle \alpha | (A^n)^\dagger A^n | \alpha \rangle}{||A^n |\alpha\rangle||^2} = \frac{||A^n |\alpha\rangle||^2}{||A^n |\alpha\rangle||^2} = 1.$$

Replacing in Eq. (6.17) α by $(\alpha - n)$, we obtain

$$\alpha - n = ||A|\alpha - n\rangle||^2 \overset{(6.21)}{=} \left|\left| \frac{A^{n+1}|\alpha\rangle}{||A^n |\alpha\rangle||} \right|\right|^2. \tag{6.22}$$

With relation (b) of Eq (6.20) we obtain from this that the right hand side vanishes, i.e. that

$$\alpha - n = 0, \quad \text{or} \quad \alpha = n \geq 0. \tag{6.23}$$

Hence the eigenvalues of the operator $N := A^\dagger A$ are nonnegative integers. For $\alpha = n = 0$ we deduce from (b) of (6.20) that

$$A|0\rangle = 0. \tag{6.24}$$

This is a very important relation which we can use as definition of the state vector $|0\rangle$. The state $|0\rangle$ is called *ground state* or *vacuum state*. In the following manipulations we always use the commutator (6.5) to shift operators A to the right which then acting on $|0\rangle$ give zero. From relation (6.18) we obtain for $\alpha = n$:

$$||A^\dagger |n\rangle||^2 = n + 1, \tag{6.25}$$

so that

$$A^\dagger |n\rangle \neq 0 \quad \text{for all } n.$$

In particular we have

$$A^\dagger |0\rangle \neq 0$$

and

$$||A^\dagger |0\rangle||^2 = \langle 0|AA^\dagger|0\rangle = \langle 0|1 + A^\dagger A|0\rangle = \langle 0|0\rangle = 1. \tag{6.26}$$

Moreover,

$$\begin{aligned} ||A^\dagger A^\dagger |0\rangle||^2 &= \langle 0|AAA^\dagger A^\dagger|0\rangle = \langle 0|A(A^\dagger A + 1)A^\dagger|0\rangle \\ &= \langle 0|AA^\dagger AA^\dagger + AA^\dagger|0\rangle = \langle 0|2(A^\dagger A + 1)|0\rangle = 2. \end{aligned} \tag{6.27}$$

According to (b) of (6.20) we have for $\alpha = n = 1$:

$$A^2|1\rangle = 0.$$

But

$$A^2 A^\dagger |0\rangle = A(AA^\dagger)|0\rangle = A(A^\dagger A + 1)|0\rangle = 0, \tag{6.28}$$

and

$$
\begin{aligned}
A^2 A^\dagger A^\dagger |0\rangle &= A(A^\dagger A + 1)A^\dagger |0\rangle = (AA^\dagger AA^\dagger + AA^\dagger)|0\rangle \\
&= \{AA^\dagger(A^\dagger A + 1) + A^\dagger A + 1\}|0\rangle \\
&= (AA^\dagger + 1)|0\rangle = (A^\dagger A + 2)|0\rangle \\
&= 2|0\rangle \neq 0.
\end{aligned} \tag{6.29}
$$

Hence we obtain

$$
|1\rangle \propto A^\dagger |0\rangle,
$$

and in view of Eq. (6.26) the equality

$$
|1\rangle = A^\dagger |0\rangle. \tag{6.30}
$$

Similarly we find

$$
|2\rangle \propto A^\dagger A^\dagger |0\rangle,
$$

and in view of Eq. (6.29), i.e.

$$
\langle 0|A^2 (A^\dagger)^2|0\rangle = 2,
$$

we have

$$
|2\rangle = \frac{1}{\sqrt{2}} A^\dagger A^\dagger |0\rangle. \tag{6.31}
$$

In general we have

$$
|n\rangle = \frac{1}{\sqrt{n!}} (A^\dagger)^n |0\rangle \tag{6.32}
$$

(arbitrary phase factors which are not excluded by the normalization have been put equal to 1).

The states $|n\rangle$ thus defined are orthonormal, as we can see as follows. According to Eq. (6.32) we have

$$
\langle n|m\rangle = \frac{1}{\sqrt{n!m!}} \langle 0|A^n (A^\dagger)^m |0\rangle. \tag{6.33}
$$

But using Eqs.(6.11), (6.5) and again (6.11) we have

$$
[A, (A^\dagger)^2] = [A, A^\dagger]A^\dagger + A^\dagger[A, A^\dagger] = 2A^\dagger,
$$

and

$$
[A, (A^\dagger)^3] = [A, (A^\dagger)^2]A^\dagger + (A^\dagger)^2[A, A^\dagger] = 2A^\dagger A^\dagger + (A^\dagger)^2 \times 1 = 3(A^\dagger)^2,
$$

and in general

$$
[A, (A^\dagger)^n] = n(A^\dagger)^{n-1}, \tag{6.34}
$$

so that

$$\langle 0|A^n(A^\dagger)^m|0\rangle = \langle 0|A^{n-1}A^{\dagger m}A|0\rangle + \langle 0|A^{n-1}m(A^\dagger)^{m-1}|0\rangle$$
$$= 0 + m\langle 0|A^{n-1}(A^\dagger)^{m-1}|0\rangle.$$

By repeated application of this relation on itself it follows that (since this is nonzero only for $n = m$)

$$\langle 0|A^n(A^\dagger)^m|0\rangle = n(n-1)\ldots 1\delta_{nm} = n!\delta_{nm}. \tag{6.35}$$

Inserting this result into Eq. (6.33), we obtain

$$\langle n|m\rangle = \delta_{nm} \tag{6.36}$$

We also deduce from Eq. (6.32):

$$A^\dagger|n\rangle = \frac{1}{\sqrt{n!}}(A^\dagger)^{n+1}|0\rangle = \sqrt{n+1}|n+1\rangle \tag{6.37}$$

and with Eq. (6.34)

$$A|n\rangle = \frac{1}{\sqrt{n!}}A(A^\dagger)^n|0\rangle = \frac{1}{\sqrt{n!}}[(A^\dagger)^nA + n(A^\dagger)^{n-1}]|0\rangle = \sqrt{n}|n-1\rangle \tag{6.38}$$

and

$$A^\dagger A|n\rangle = \sqrt{n}A^\dagger|n-1\rangle = n|n\rangle. \tag{6.39}$$

With Eq. (6.8) we obtain therefore

$$H|n\rangle = \hbar\omega\left(A^\dagger A + \frac{1}{2}\right)|n\rangle = \hbar\omega\left(n + \frac{1}{2}\right)|n\rangle, \tag{6.40}$$

i.e. the eigenvalues of the Hamiltonian H are

$$E_n = \hbar\omega\left(n + \frac{1}{2}\right), \quad n = 0, 1, 2, \ldots. \tag{6.41}$$

The contribution $\hbar\omega/2$ is called *zero point energy*. In view of the properties (6.34) and (6.35) the operators $A^\dagger$ and A are called respectively *raising* and *lowering operators* or also *shift operators*. In view of the same properties A is also called *annihilation operator* and $A^\dagger$ *creation operator* of so-called "*quasi-particles*", whose number is given by the integer eigenvalue of the *number operator* $N = A^\dagger A$. Here, in quantum mechanics, we do not have creation or annihilation of any real particles as in field theory (hence the word "quasi-particle"). The terminology is, however, chosen in close analogy to the postulated "second quantization relations" of field theory.

6.3   The Energy Representation of the Oscillator

The *energy representation*, also called *Heisenberg representation*, is defined by projection of operators on eigenfunctions of energy, and — in fact — we have used this already previously (see e.g. Eq. (5.39)). The representation of the Hamiltonian in this energy representation is

$$\langle n|H|n'\rangle = E_n\delta_{nn'}, \quad E_n = \hbar\omega\left(n+\frac{1}{2}\right), \quad n = 0,1,2,\ldots. \tag{6.42}$$

We can deduce the energy representation of the operators $A, A^\dagger$ from the relations (6.38) and (6.39):

$$
\begin{aligned}
\langle n'|A^\dagger|n\rangle &= \sqrt{n+1}\langle n'|n+1\rangle = \sqrt{n+1}\delta_{n',n+1}, \\
\therefore \langle n+1|A^\dagger|n\rangle &= \sqrt{n+1}, \quad \text{other elements zero,}
\end{aligned} \tag{6.43}
$$

and similarly

$$
\begin{aligned}
\langle n'|A|n\rangle &= \sqrt{n}\langle n'|n-1\rangle = \sqrt{n}\delta_{n',n-1}, \\
\therefore \langle n-1|A|n\rangle &= \sqrt{n}, \quad \text{other elements zero.}
\end{aligned} \tag{6.44}
$$

In matrix form this means

$$
A^\dagger = \begin{pmatrix}
0 & 0 & 0 & 0 & \cdot \\
\sqrt{1} & 0 & 0 & 0 & \cdot \\
0 & \sqrt{2} & 0 & 0 & \cdot \\
0 & 0 & \sqrt{3} & 0 & \cdot \\
0 & 0 & 0 & \sqrt{4} & \cdot \\
\cdot & \cdot & \cdot & \cdot & \cdot
\end{pmatrix}, \quad
A = \begin{pmatrix}
0 & \sqrt{1} & 0 & 0 & 0 & \cdot \\
0 & 0 & \sqrt{2} & 0 & 0 & \cdot \\
0 & 0 & 0 & \sqrt{3} & 0 & \cdot \\
0 & 0 & 0 & 0 & \sqrt{4} & \cdot \\
0 & 0 & 0 & 0 & 0 & \cdot \\
\cdot & \cdot & \cdot & \cdot & \cdot & \cdot
\end{pmatrix}.
\tag{6.45}
$$

Correspondingly we obtain with Eqs. (6.6) and (6.7) the energy representation of the operators q and p, i.e.

$$
q = \sqrt{\frac{\hbar}{2m_0\omega}}\begin{pmatrix}
0 & \sqrt{1} & 0 & 0 & \cdot \\
\sqrt{1} & 0 & \sqrt{2} & 0 & \cdot \\
0 & \sqrt{2} & 0 & \sqrt{3} & \cdot \\
0 & 0 & \sqrt{3} & 0 & \cdot \\
0 & 0 & 0 & \sqrt{4} & \cdot \\
\cdot & \cdot & \cdot & \cdot & \cdot
\end{pmatrix},
$$

$$
p = i\sqrt{\frac{m_0\hbar\omega}{2}}\begin{pmatrix}
0 & -\sqrt{1} & 0 & 0 & \cdot \\
\sqrt{1} & 0 & -\sqrt{2} & 0 & \cdot \\
0 & \sqrt{2} & 0 & -\sqrt{3} & \cdot \\
0 & 0 & \sqrt{3} & 0 & \cdot \\
0 & 0 & 0 & \sqrt{4} & \cdot \\
\cdot & \cdot & \cdot & \cdot & \cdot
\end{pmatrix}. \tag{6.46}
$$

It is an instructive excercise to check by direct calculation that Eqs. (6.1) and (6.5) are also satisfied as matrix equations.

6.4 The Configuration Space Representation

We saw that the eigenstates are given by Eq. (6.32). Correspondingly the position space representation is given by the wave function

$$\phi_n(x) := \langle x|n \rangle. \tag{6.47}$$

The ground state wave function $\phi_0(x)$ is defined by Eq. (6.24),

$$A|0\rangle = 0,$$

i.e. by (cf. Eq. (6.3))

$$\sqrt{\frac{m_0\omega}{2\hbar}}\left(q + \frac{i}{m_0\omega}p\right)|0\rangle = 0. \tag{6.48}$$

Applying from the left the bra-vector $\langle x|$ and remembering that

$$\langle x|p|\phi \rangle = -i\hbar\frac{d}{dx}\langle x|\phi \rangle,$$

we obtain

$$\sqrt{\frac{m_0\omega}{2\hbar}}\left(x + \frac{\hbar}{m_0\omega}\frac{d}{dx}\right)\langle x|0\rangle = 0. \tag{6.49}$$

This is a simple differential equation of the first order with solution

$$\langle x|0\rangle = Ce^{-m_0\omega x^2/2\hbar}.$$

The normalization constant C is determined by the condition[‡]

$$1 = \langle 0|0\rangle = \int_{-\infty}^{\infty} dx\langle 0|x\rangle\langle x|0\rangle = |C|^2 \int_{-\infty}^{\infty} e^{-m_0\omega x^2/\hbar}dx = |C|^2\sqrt{\frac{\pi\hbar}{m_0\omega}},$$

so that

$$C = e^{i\theta}\left(\frac{m_0\omega}{\pi\hbar}\right)^{1/4}.$$

We choose the arbitrary phase θ to be zero. Hence

$$\phi_0(x) \equiv \langle x|0\rangle = \left(\frac{m_0\omega}{\pi\hbar}\right)^{1/4}e^{-m_0\omega x^2/2\hbar}. \tag{6.50}$$

[‡]Recall $\int_{-\infty}^{\infty} dx\, e^{-w^2x^2/2} = \sqrt{2\pi/w^2}$.

This is therefore the ground state wave function of the one-dimensional harmonic oscillator. In order to obtain the wave functions of higher states, it suffices according to Eq. (6.32) to apply the appropriate number of creation operators $A^\dagger$ to the vacuum state $|0\rangle$, i.e.

$$\phi_n(x) := \langle x|n\rangle = \frac{1}{\sqrt{n!}}\langle x|(A^\dagger)^n|0\rangle. \tag{6.51}$$

Now

$$\langle x|A^\dagger = \sqrt{\frac{m_0\omega}{2\hbar}}\langle x|\left(q - \frac{i}{m_0\omega}p\right) = \sqrt{\frac{m_0\omega}{2\hbar}}\left(x - \frac{\hbar}{m_0\omega}\frac{d}{dx}\right)\langle x|,$$

so that

$$\begin{aligned}
\langle x|n\rangle &= \frac{1}{\sqrt{n!}}\left(\frac{m_0\omega}{2\hbar}\right)^{n/2}\left(x - \frac{\hbar}{m_0\omega}\frac{d}{dx}\right)^n\langle x|0\rangle \\
&= \frac{1}{\sqrt{n!}}\left(\frac{m_0\omega}{2\hbar}\right)^{n/2}\left(x - \frac{\hbar}{m_0\omega}\frac{d}{dx}\right)^n\left(\frac{m_0\omega}{\pi\hbar}\right)^{1/4}e^{-m_0\omega x^2/2\hbar}. \tag{6.52}
\end{aligned}$$

Setting

$$\alpha = \sqrt{\frac{\hbar}{m_0\omega}}$$

we have

$$\langle x|n\rangle = \frac{1}{\pi^{1/4}\sqrt{\alpha\, n!}2^{n/2}}\left(\frac{x}{\alpha} - \alpha\frac{d}{dx}\right)^n e^{-x^2/2\alpha^2}. \tag{6.53}$$

Setting

$$\xi = \frac{x}{\alpha},$$

and

$$H_n(\xi) = e^{\xi^2/2}\left(\xi - \frac{d}{d\xi}\right)^n e^{-\xi^2/2}, \tag{6.54}$$

we have

$$\phi_n(x) = \frac{e^{-\frac{1}{2}\left(\frac{x}{\alpha}\right)^2}H_n\left(\frac{x}{\alpha}\right)}{\pi^{1/4}\sqrt{\alpha\, 2^n\, n!}}. \tag{6.55}$$

The functions $H_n(\xi)$ are obviously polynomials of the n-th degree; they are called *Hermite polynomials*. These polynomials are real for real arguments and have the symmetry property

$$H_n(-\xi) = (-1)^n H_n(\xi), \quad\text{so that}\quad \phi_n(-x) = (-1)^n\phi_n(x). \tag{6.56}$$

We have as an operator identity the relation

$$e^{\xi^2/2}\frac{d}{d\xi}e^{-\xi^2/2} = \frac{d}{d\xi} - \xi, \tag{6.57a}$$

i.e.

$$\left(e^{\xi^2/2}\frac{d}{d\xi}e^{-\xi^2/2}\right)\psi(\xi) = \left(\frac{d}{d\xi} - \xi\right)\psi(\xi)$$

for some function $\psi(\xi)$. It follows that

$$
\begin{aligned}
H_1(\xi) &= -e^{\xi^2/2}\left(\frac{d}{d\xi} - \xi\right)e^{-\xi^2/2}\\
&= -e^{\xi^2/2}\left(e^{\xi^2/2}\frac{d}{d\xi}e^{-\xi^2/2}\right)e^{-\xi^2/2}\\
&= -e^{\xi^2}\frac{d}{d\xi}e^{-\xi^2} = 2\xi.
\end{aligned}
$$

Again as an operator relation we have

$$
\begin{aligned}
e^{\xi^2/2}\frac{d^2}{d\xi^2}e^{-\xi^2/2} &= e^{\xi^2/2}\frac{d}{d\xi}\left[e^{-\xi^2/2}\frac{d}{d\xi} - \xi e^{-\xi^2/2}\right]\\
&= e^{\xi^2/2}\left[e^{-\xi^2/2}(-\xi)\frac{d}{d\xi} + e^{-\xi^2/2}\frac{d^2}{d\xi^2}\right.\\
&\qquad\left. -\xi e^{-\xi^2/2}\frac{d}{d\xi} - e^{-\xi^2/2} + \xi^2 e^{-\xi^2/2}\right]\\
&= -2\xi\frac{d}{d\xi} + \frac{d^2}{d\xi^2} - 1 + \xi^2\\
&= \left(\frac{d}{d\xi} - \xi\right)^2. \tag{6.57b}
\end{aligned}
$$

Generalizing this result and inserting it into Eq. (6.54) we find that (observe that there is no factor 2 in the arguments of the exponentials!)[§]

$$H_n(\xi) = (-1)^n e^{\xi^2}\frac{d^n}{d\xi^n}e^{-\xi^2}. \tag{6.58a}$$

In particular we have

$$
\begin{aligned}
H_0(\xi) &= 1,\\
H_1(\xi) &= 2\xi,\\
H_2(\xi) &= (2\xi)^2 - 2,\\
H_3(\xi) &= (2\xi)^3 - 6(2\xi),\\
H_4(\xi) &= (2\xi)^4 - 12(2\xi) - 12. \tag{6.58b}
\end{aligned}
$$

[§]This definition — apart from the usual one — is also cited e.g. by W. Magnus and F. Oberhettinger [181], p. 81.

The differential equation of the *Hermite polynomials* $H_n(\xi)$ defined in this way is

$$\left(\frac{d^2}{d\xi^2} - 2\xi \frac{d}{d\xi} + 2n \right) H_n(\xi) = 0. \tag{6.58c}$$

Here special care is advisable since the Hermite polynomials defined above were motivated by the harmonic oscillator and hence are those frequently used by physicists. Mathematicians often define Hermite polynomials $He_n(\xi)$ as[¶]

$$He_n(\xi) = (-1)^n e^{\xi^2/2} \frac{d^n}{d\xi^n} (e^{-\xi^2/2}) \quad \text{or} \quad H_n(\xi) = 2^{n/2} He_n(\sqrt{2}\xi). \tag{6.59a}$$

Then

$$
\begin{aligned}
He_0(\xi) &= 1, \\
He_1(\xi) &= \xi, \\
He_2(\xi) &= \xi^2 - 1, \\
He_3(\xi) &= \xi^3 - 3\xi, \\
He_4(\xi) &= \xi^4 - 6\xi^2 + 3.
\end{aligned} \tag{6.59b}
$$

The differential equation obeyed by the *Hermite polynomials* $He_n(\xi)$ defined in this way is

$$\left(\frac{d^2}{d\xi^2} - \xi \frac{d}{d\xi} + n \right) He_n(\xi) = 0. \tag{6.59c}$$

The *recurrence relations* satisfied by these Hermite polynomials (the second will be derived later in Example 6.3) are (as given in the cited literature and with a prime meaning derivative)

$$
\begin{aligned}
He_{n+1}(\xi) &= \xi He_n(\xi) - He_n'(\xi), \\
He_n(\xi) &= He_{n+1}(\xi) + n He_{n-1}(\xi), \\
He_n'(\xi) &= n He_{n-1}(\xi).
\end{aligned} \tag{6.59d}
$$

The normalization of Hermite polynomials is given by the relations:

$$\int_{-\infty}^{\infty} d\xi\, H_n(\xi) H_m(\xi) e^{-\xi^2} = 2^n\, n! \sqrt{\pi}\, \delta_{nm}, \tag{6.60a}$$

and

$$\int_{-\infty}^{\infty} d\xi\, He_n(\xi) He_m(\xi) e^{-\xi^2/2} = n! \sqrt{2\pi}\, \delta_{nm}. \tag{6.60b}$$

[¶]See for instance W. Magnus and F. Oberhettinger [181], pp. 80, 81.

The relation between Hermite polynomials and parabolic cylinder functions $D_n(\xi)$ — which we use frequently here — is given by[||]

$$
\begin{aligned}
D_n(\xi) &= (-1)^n e^{\xi^2/4} \frac{d^n}{d\xi^n} e^{-\xi^2/4} \\
&= e^{-\xi^2/4} He_n(\xi) = e^{-\xi^2/4} 2^{-n/2} H_n(\xi/\sqrt{2}), \\
D_n(\sqrt{2}\xi) &= e^{-\xi^2/2} 2^{-n/2} H_n(\xi).
\end{aligned}
\tag{6.61}
$$

Their normalization is therefore given by

$$
\begin{aligned}
\int_{-\infty}^{\infty} d\xi\, D_n^2(\xi) &= \int_{-\infty}^{\infty} d\xi\, e^{-\xi^2/2} He_n^2(\xi) = n!\sqrt{2\pi}, \\
&= \frac{\sqrt{2}}{2^n} \int_{-\infty}^{\infty} d\xi\, e^{-\xi^2} H_n^2(\xi).
\end{aligned}
\tag{6.62}
$$

The following function is generally described as *generating function* of Hermite polynomials $H_n(\xi)$:

$$
F(\xi, s) = \exp[\xi^2 - (\xi - s)^2].
\tag{6.63}
$$

The meaning is that if we expand $F(\xi, s)$ as a Taylor series,

$$
F(\xi, s) = F(\xi, 0) + \frac{s}{1!}\left(\frac{\partial F}{\partial s}\right)_{s=0} + \dots,
$$

we obtain

$$
\begin{aligned}
\left.\frac{\partial F}{\partial s}\right|_{s=0} &= 2(\xi - s)e^{\xi^2 - (\xi - s)^2}\Big|_{s=0} = 2\xi = H_1(\xi), \\
\left.\frac{\partial^2 F}{\partial s^2}\right|_{s=0} &= \frac{\partial}{\partial s}\left[2(\xi - s)e^{\xi^2 - (\xi - s)^2}\right]\Big|_{s=0} \\
&= [4(\xi - s)^2 - 2]e^{\xi^2 - (\xi - s)^2}\Big|_{s=0} \\
&= 2(2\xi^2 - 1) = (2\xi)^2 - 2 = H_2(\xi),
\end{aligned}
$$

and so on. Thus the generating function can be written

$$
F(\xi, s) = \sum_{n=0}^{\infty} \frac{H_n(\xi)}{n!} s^n.
\tag{6.64}
$$

[||]See e.g. W. Magnus and F. Oberhettinger [181], p. 93.

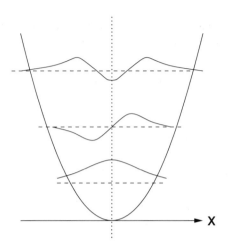

Fig. 6.1 The comparative behaviour of the lowest three wave functions
of the harmonic oscillator.

The comparative behaviour of the wave functions of the three lowest states
of the harmonic oscillator is shown in Fig. 6.1. We see that the lowest state is
symmetric, i.e. has parity +1, and we see that even and odd states alternate.
Furthermore we observe that the wave function of the n-th state has n zeros
in finite domains of the variable x. We also see from the eigenvalue that the
energy of the lowest (i.e. ground) state is $\hbar\omega/2$. In classical mechanics the
energy of the oscillator can be zero, i.e. for $x(t) = 0$ and $p(t) = m_0\dot{x}(t) = 0$.
But this would mean that both $x(t)$ and $p(t)$ would assume simultaneously
the "sharp" values zero. In fact we can use the uncertainty relation $\triangle x \triangle p_x \simeq$
$\hbar/2$, in order to estimate the lowest energy of the harmonic oscillator. With

$$\triangle p_x = m_0 \triangle v_x = m_0 \omega \triangle x$$

we have

$$\triangle x \simeq \frac{\hbar/2}{\triangle p_x} = \frac{\hbar/2}{m_0 \omega \triangle x}, \quad \triangle x = \sqrt{\frac{\hbar/2}{m_0 \omega}}, \quad \triangle p_x = \sqrt{m_0 \omega \hbar/2},$$

and thus

$$
\begin{aligned}
E_0 &\simeq \frac{1}{2m_0}(\triangle p_x)^2 + \frac{1}{2}m_0\omega^2(\triangle x)^2 \\
&= \frac{m_0\omega\hbar/2}{2m_0} + \frac{1}{2}\frac{m_0\omega^2\hbar/2}{m_0\omega} \\
&= \frac{1}{2}\hbar\omega.
\end{aligned}
$$

The *zero point energy* plays an important role in atomic oscillators, like two-atomic molecules, and is directly observable. This observability can be taken as direct proof of the uncertainty relation. It is clear that the uncertainties are not due to deficiencies of measuring instruments (we are concerned here with a system in the pure state $|0\rangle$). Finally we point out that the Hamiltonian of the harmonic oscillator can be diagonalized in many different ways. In Example 6.1 we demonstrate how the eigenvalues may be obtained by the method of *"poles at infinity"*. The analogous Example 6.2 contains only the answers since the evaluation of the integrals proceeds as in Example 6.1.

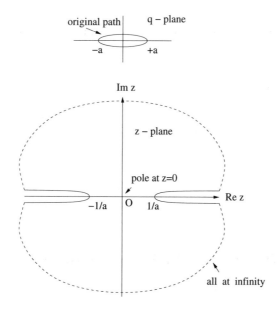

Fig. 6.2 The original path and the transformed path at infinity.

Example 6.1: Eigenvalues obtained by contour integration
Use the corrected Bohr–Sommerfeld–Wilson quantization condition (cf. Sec. 14.4) to obtain by contour integration the eigenvalues of the harmonic oscillator.

Solution: Consider the simple harmonic oscillator of natural frequency of vibration ν, mass m_0, and with potential $V(q) = 2\pi^2 m_0 \nu^2 q^2$ and total energy $E = p^2/2m_0 + V(q)$. According to "old quantum theory" with integer n, in corrected version instead with $n^* = n + 1/2$,

$$\oint_{\text{classical orbit}} pdq = n^*h, \quad \text{i.e.} \quad 2\pi m_0 \nu \oint_{\text{classical orbit}} \sqrt{a^2 - q^2}dq = n^*h, \quad E = 2\pi^2 m_0 \nu^2 a^2.$$

Here q goes from 0 to a, back to 0, to $-a$, to 0, which is the complete orbit. The kinetic energy is always positive or zero. In the plane of complex q the integral has no poles for finite values of q. But it has a pole at infinity. Thus, setting $q = 1/z$, we obtain the integral

$$\oint \sqrt{a^2 - \frac{1}{z^2}} \frac{dz}{z^2} = \oint \frac{dz}{z^3} \sqrt{a^2 z^2 - 1}.$$

This integral has a triple pole at $z = 0$ (i.e. in q at infinity). We can evaluate the integral with the help of the *Cauchy formula* (the superscript meaning derivative)

$$\int_C dz \frac{f(z)}{(z-a)^m} = 2\pi i \frac{f^{(m-1)}(a)}{(m-1)!}. \tag{6.65}$$

Hence we obtain

$$\oint \frac{dz}{z^3} \sqrt{a^2 z^2 - 1} = \frac{2\pi i}{2!} \left[\frac{\partial^2}{\partial z^2} \sqrt{a^2 z^2 - 1} \right]_{z=0} = \pi i a^2 \left[\frac{\partial}{\partial z} \frac{z}{\sqrt{a^2 z^2 - 1}} \right]_{z=0} = \frac{\pi i a^2}{\sqrt{-1}} = \pi a^2.$$

It follows that

$$2\pi m_0 \nu (\pi a^2) = n^* h = E/\nu, \quad \text{i.e.} \quad E = n^* h\nu = \left(n + \frac{1}{2} \right) h\nu.$$

Example 6.2: Contour integration along a classical orbit
Setting $q = 1/z$, verify that

$$I_1 = \oint dq \frac{\sqrt{1-q^2}}{aq+b} = -2\pi i \left[\frac{\partial}{\partial z} \left(\frac{\sqrt{z^2-1}}{a+bz} \right) \right]_{z=0} - 2\pi i \frac{\sqrt{a^2-b^2}}{a^2} = \frac{2\pi}{a^2} (b - \sqrt{b^2 - a^2}),$$

$$I_2 = \oint dq \left[-a^2 + \frac{2b}{q} - \frac{c^2}{q^2} \right]^{1/2} = \frac{2\pi}{a}(ac-b) \quad \text{and} \quad I_3 = \oint \frac{qdq}{[-a^2 + \frac{2b}{q} - \frac{c^2}{q^2}]^{1/2}} = \frac{\pi}{a^5}(3b - a^2 c^2).$$

Solution: The solution proceeds as in Example 6.1. With $a^2 = -E, 2b = Ze^2, c^2 = (l + 1/2)^2$ (natural units, $m_0 = 1/2$) the integral I_2 can be checked to give the eigenvalues for the Coulomb potential $-Ze^2/q$ (2 turning points), i.e. $-E = Z^2 e^4/4(l + n + 1)^2$ (cf. Eq. (11.114a)).

6.5 The Harmonic Oscillator Equation

In the above we considered mainly the energy representation of the harmonic oscillator and encountered the Hermite polynomials. In the following we investigate in more detail the important differential equation of the harmonic oscillator. After removal of a Gaussian or WKB exponential, we start with a *Laplace transform ansatz* for the remaining part of the solution, and obtain with this all properties of the solution. In Chapter 11 we perform similar calculations for the radial equation of the Coulomb potential — as in Example 11.3, for instance — without deriving all properties of the solutions, since these two cases, harmonic oscillator and Coulomb interaction, are the most basic and exactly solvable cases in quantum mechanics.

6.5.1 Derivation of the generating function

From the above we know that the energy is quantized; hence we consider the following differential equation in which $n = 0, 1, 2, \ldots$ but here we leave this latter property open for determination later:

$$\frac{d^2\psi}{dt^2} + \left[n + \frac{1}{2} - \frac{1}{4}t^2\right]\psi = 0. \tag{6.66}$$

First we have to remove the t^2-term, since this can give rise to exponentially increasing behaviour (at most a factor t can be tolerated). Hence we set (with the chosen sign in the exponential)

$$\psi(t) = y(t)e^{-t^2/4} \quad \text{with} \quad \frac{d^2y(t)}{dt^2} - t\frac{dy(t)}{dt} + ny(t) = 0, \tag{6.67}$$

the latter being the differential equation of Hermite polynomials $He_n(t)$ for integral values of n.** We make the Laplace transform ansatz†† (with limits to be decided later)

$$y(t) = \int s(p)e^{-pt}dp. \tag{6.68}$$

Then (with partial integration)

$$t\frac{dy}{dt} = -t\int ps(p)e^{-pt}dp = [ps(p)e^{-pt}] - \int \frac{d}{dp}(ps(p))e^{-pt}dp,$$

$$\frac{d^2y}{dt^2} = \int p^2 s(p)e^{-pt}dp.$$

Substituting these expressions into Eq. (6.67) we obtain

$$-[ps(p)e^{-pt}] + \int \left[p^2 s(p) + \frac{d}{dp}(ps(p)) + ns(p)\right]e^{-pt}dp = 0.$$

This has to be true for all values of t. We demand that

$$C(p) := -[ps(p)e^{-pt}] = 0, \quad \text{(this determines the limits)},$$

so that the integrand of the remaining integral vanishes, i.e.

$$\frac{d}{dp}(ps(p)) = -(p^2 + n)s(p), \quad \frac{1}{ps(p)}\frac{d}{dp}(ps(p)) = -p - \frac{n}{p}.$$

Integrating this expression, we obtain

$$\ln(ps(p)) = -\frac{1}{2}p^2 - n\ln p \quad \text{or} \quad s(p) \propto \frac{e^{-p^2/2}}{p^{n+1}}, \tag{6.69}$$

** See e.g. W. Magnus and F. Oberhettinger [181], p. 80.

†† Thus — ignoring details — $\int dp f(p)e^{ipx}$ is the Fourier transform of $f(p)$, $\int dp f(p)e^{-px}$ the Laplace transform of $f(p)$, and $\int dp f(p)p^n$ the Mellin transform of $f(p)$. E.g. (cf. Eq. (11.111)) the Mellin transform of e^{-p} is $n!$.

and hence

$$y(t) = \int \frac{e^{-pt-p^2/2}}{p^{n+1}} dp \quad \text{with} \quad C(p) = \left[\frac{e^{-pt-p^2/2}}{p^n}\right] = 0 \qquad (6.70)$$

between the two limits of integration.

We now investigate possibilities to satisfy the boundary condition $C(p) = 0$. (1) This will be satisfied if $|p| \to \pm\infty$. Hence a possible path of integration is from $-\infty$ to ∞. (2) If $n < 0$, we can choose a path from zero to infinity (the condition will vanish at $p = 0$). (3) We can choose a contour once around $p = 0$ in the complex p-plane. For single-valuedness, n must then be an integer, since otherwise (with $p = |p| \exp(i\theta)$) we will get part of the solution from 0 to infinity, as indicated in Fig. 6.3.

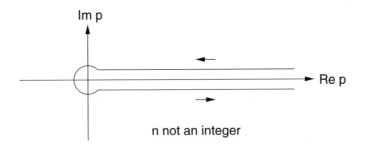

Fig. 6.3 The path when n is not an integer.

We can show that if $|p|$ is allowed to be large, then for large t we have $y(t) \sim \exp(t^2/2)$. To see this, consider the exponential in the integrand. This exponential assumes a maximum value when

$$pt + \frac{1}{2}p^2 = \text{minimal, i.e. when } p = -t$$

(which is possible, of course). This can be attained if $|p|$ is allowed to be large (t can be of either sign). The value of the exponential at this point is $\sim \exp(t^2/2)$, so that $y(t) \sim \exp(t^2/2)$. The quantum mechanical wave function is then (cf. Eq. (6.67))

$$\psi_n(t) \propto e^{-t^2/4} y(t) \sim e^{t^2/4}.$$

But this is not permissible, so that p cannot go to infinity. Thus the only solution left which does not involve large values of $|p|$ is that with $\oint_{p=0}$ for n an integer. Therefore we take

$$y(t) \quad \propto \quad \frac{1}{2\pi i} \oint \frac{e^{-pt-p^2/2}}{p^{n+1}} dp \quad \text{for } n \text{ integral and}$$

$$\therefore y(t) \quad = \quad \text{coefficient of } p^n \text{ in } e^{-pt-p^2/2} \qquad (6.71)$$

with Cauchy's residue theorem. The exponential function here is described as the *generating function* of the *Hermite polynomials* $He_n(t)$. The Hermite polynomial $He_n(t)$ is now defined such that the coefficient of t^n is unity. In $y(t)$ above, the term of highest degree in t in the exponential corresponds to taking all p's from $\exp(-pt)$. Therefore the coefficient of $(pt)^n$ in $\exp(-pt - p^2/2)$ is $(-1)^n/n!$. The Hermite polynomial is therefore given by

$$He_n(t) = \frac{(-1)^n n!}{2\pi i} \oint \frac{e^{-pt-p^2/2}}{p^{n+1}} dp. \tag{6.72}$$

We now obtain the polynomial form with Cauchy's residue theorem as

$$
\begin{aligned}
He_n(t) &= (-1)^n n! \times \text{ coefficient of } p^n \text{ in } e^{-pt-p^2/2} \\
&= (-1)^n n! \times \text{ coefficient of } p^n \text{ in } \sum_{\nu=0}^{\infty} \frac{(-pt)^\nu}{\nu!} e^{-p^2/2} \\
&= (-1)^n n! \times \text{ coefficient of } p^{n-\nu} \text{ in } \sum_{\nu=0}^{\infty} \frac{(-t)^\nu}{\nu!} e^{-p^2/2} \\
&= (-1)^n n! \sum_{\nu=0,\, n-\nu \text{ even}}^{\infty} \frac{(-t)^\nu}{\nu!} \frac{(-1)^{(n-\nu)/2}}{2^{(n-\nu)/2} \left(\frac{n-\nu}{2}\right)!}. \tag{6.73}
\end{aligned}
$$

Hence we obtain the polynomial of degree n

$$He_n(t) = n! \sum_{\text{see below}} \frac{t^\nu}{(-2)^{(n-\nu)/2} \, \nu! \left(\frac{n-\nu}{2}\right)!}, \tag{6.74}$$

where $\nu = 0, 2, 4, \ldots$ if n is even and $\nu = 1, 3, 5, \ldots$ if n is odd. We can also obtain the derivative form of this Hermite polynomial. We have

$$
\begin{aligned}
He_n(t) &= (-1)^n \frac{n!}{2\pi i} \oint_{(p=0)} dp \frac{e^{-pt-p^2/2}}{p^{n+1}} \\
&= (-1)^n \frac{n!}{2\pi i} e^{t^2/2} \oint_{(p=0)} dp \frac{e^{-(p+t)^2/2}}{p^{n+1}}. \tag{6.75}
\end{aligned}
$$

Setting $q = p + t$, this becomes

$$
\begin{aligned}
He_n(t) &= (-1)^n \frac{n!}{2\pi i} e^{t^2/2} \oint_{(q=t)} dq \frac{e^{-q^2/2}}{(q-t)^{n+1}} \\
&= (-1)^n \frac{n!}{2\pi i} e^{t^2/2} \frac{1}{n!} \left(\frac{\partial}{\partial t}\right)^n \oint_{(q=t)} dq \frac{e^{-q^2/2}}{(q-t)} \\
&= e^{t^2/2} \left(-\frac{\partial}{\partial t}\right)^n e^{-t^2/2}, \tag{6.76}
\end{aligned}
$$

which we recognize as agreeing with Eq. (6.59a). We leave the consideration of the recurrence relation of Hermite polynomials (of paramount importance in the perturbation method of Sec. 8.7 that we use throughout this text) and their orthogonality and normalization to Examples 6.3 and 6.4. Example 6.5 is a further application of the method for the evaluation of integrals that occur frequently in practice (e.g. in radiation problems, for the evaluation of expectation values, and elsewhere, as in the normalization of asymptotic expansions of Mathieu functions).

Example 6.3: Recurrence relation of Hermite polynomials
Show that

$$tHe_n(t) = (n, n+1)He_{n+1}(t) + (n, n-1)He_{n-1}(t), \quad (n, n+1) = 1, \quad (n, n-1) = n.$$

Solution: We multiply the generating function by a factor t and perform a partial integration on the factor $\exp(-pt)$ contained in the following integral (the differentiation of the other factor leading to two contributions), and we obtain then:

$$\oint t\frac{e^{-pt-p^2/2}}{p^{n+1}}dp = \left[-\frac{e^{-pt-p^2/2}}{p^{n+1}}\right] - (n+1)\oint \frac{e^{-pt}e^{-p^2/2}}{p^{n+2}}dp - \oint \frac{e^{-pt}e^{-p^2/2}}{p^n}dp.$$

Here the bracketed expression vanishes in view of our condition (6.70). Re-interpreting the remaining integrals with Eq. (6.75) as Hermite polynomials, we obtain

$$\frac{tHe_n(t)}{(-1)^n n!} = -\frac{He_{n-1}(t)}{(-1)^{n-1}(n-1)!} - \frac{(n+1)He_{n+1}(t)}{(-1)^{n+1}(n+1)!},$$

i.e. $tHe_n(t) = nHe_{n-1}(t) + He_{n+1}(t).$ (6.77)

Starting from $He_0(t) = 1$, $He_1(t) = t$, one can construct the tower of polynomials $He_n(t)$.

Example 6.4: Orthonormality of Hermite polynomials
Establish the orthogonality and normalization of Hermite polynomials.

Solution: We have to evaluate

$$I = \int_{-\infty}^{\infty} He_n(t)He_m(t)e^{-t^2/2}dt \cdot \text{ using } He_n(t) = (-1)^n\frac{n!}{2\pi i}\oint \frac{e^{-pt-p^2/2}}{p^{n+1}}dp.$$

It follows that

$$I = (-1)^{n+m}\frac{n!m!}{(2\pi i)^2}\oint_{p=0}\oint_{q=0}dpdq\frac{e^{-(p^2+q^2)/2}}{p^{n+1}q^{m+1}}\int_{t=-\infty}^{\infty}e^{-(p+q)t-t^2/2}dt.$$

The integral with respect to t is a Gauss integral[‡‡] and yields

$$e^{pq}e^{(p^2+q^2)/2}\int_{-\infty}^{\infty}e^{-\{t+(p+q)\}^2/2}dt = e^{pq}e^{(p^2+q^2)/2}\sqrt{2\pi},$$

[‡‡] $\int_{-\infty}^{\infty}dte^{-t^2/2} = \sqrt{2\pi}.$

and hence

$$I = (-1)^{n+m} \frac{\sqrt{2\pi}n!m!}{(2\pi i)^2} \oint_{p=0} \oint_{q=0} \frac{e^{pq}}{p^{n+1}q^{m+1}} \, dp dq.$$

But

$$\frac{1}{2\pi i} \oint \frac{e^{pq}}{q^{m+1}} dq = \text{coefficient of } q^m \text{ in } e^{pq} = \frac{p^m}{m!}.$$

Therefore

$$
\begin{aligned}
I &= (-1)^{n+m} \frac{\sqrt{2\pi}n!}{2\pi i} \underbrace{\oint_{p=0} \frac{dp}{p^{n-m+1}}}_{=2\pi i \text{ if } n=m, \text{ and } 0 \text{ otherwise}} \\
&= \sqrt{2\pi}(-1)^{n+m}n! \text{ if } n = m, \text{ zero otherwise} \\
&= \sqrt{2\pi}n!\delta_{nm}.
\end{aligned}
\tag{6.78}
$$

Hence the normalized wave function of Eq. (6.66) is

$$\psi_n(t) = He_n(t)e^{-t^2/4} \times \frac{1}{\sqrt{n!\sqrt{2\pi}}}.$$

Example 6.5: Generalized power integrals with Hermite polynomials
Establish the following general formula for Hermite polynomials:

$$
\begin{aligned}
I : &= \int_0^\infty t^{2s} He_n(t)He_{n+2r}(t)e^{-t^2/2} dt \\
&= \sqrt{\frac{\pi}{2}} \frac{n!(2s)!(n+2r)!}{2^{s-r}} \sum_{\nu=0}^{s-r} \frac{2^\nu}{\nu!(2r+\nu)!(n-\nu)!(s-r-\nu)!}.
\end{aligned}
\tag{6.79}
$$

Solution: From the above we know that $He_n(t)$ is the coefficient of p^n in $(-1)^n n! \exp(-pt-p^2/2)$. It follows therefore that the following integral (with an exponential $\exp(-\alpha t)$ from whose expansion we pick later the power of t) is given by

$$\int_{-\infty}^\infty e^{-\alpha t} He_n(t)He_{n+2r}(t)e^{-t^2/2} dt = n!(n+2r)! \times \text{ coefficient of } p^n q^{n+2r} \text{ in } f(p,q), \tag{6.80}$$

where

$$
\begin{aligned}
f(p,q) &= e^{-(p^2+q^2)/2} \int_{-\infty}^\infty e^{-(p+q+\alpha)t-t^2/2} dt \\
&= e^{\alpha^2/2}e^{pq}e^{\alpha(p+q)} \int_{-\infty}^\infty e^{-(t+p+q+\alpha)^2/2} dt = \sqrt{2\pi}e^{\alpha^2/2}e^{q\alpha}e^{p(q+\alpha)}.
\end{aligned}
\tag{6.81}
$$

The coefficient of p^n in $f(p,q)$ is

$$\frac{\sqrt{2\pi}}{n!} e^{\alpha^2/2}e^{q\alpha}(q+\alpha)^n = \frac{\sqrt{2\pi}}{n!}e^{\alpha^2/2}e^{q\alpha} \sum_{\mu=0}^n \alpha^{n-\mu} \frac{n!}{\mu!(n-\mu)!}q^\mu.$$

Next, the coefficient of $q^{n+2r-\mu}$ in $\exp(q\alpha)$ is $\alpha^{n+2r-\mu}/(n+2r-\mu)!$, so that

the coefficient of $p^n q^{n+2r}$ in $f(p,q)$ is the product $\sqrt{2\pi}e^{\alpha^2/2} \sum_{\mu=0}^n \frac{\alpha^{2(n+r-\mu)}}{\mu!(n-\mu)!(n+2r-\mu)!}.$

$$\tag{6.82}$$

It now follows that — in later equations with $\nu := (n - \mu), \mu = (n - \nu)$ —

$$J := \frac{\int_{-\infty}^{\infty} e^{-\alpha t} He_n(t) He_{n+2r}(t) e^{-t^2/2} dt}{\int_{-\infty}^{\infty} He_n^2(t) e^{-t^2/2} dt} = (n+2r)! e^{\alpha^2/2} \sum_{\mu=0}^{n} \frac{\alpha^{2(n+r-\mu)}}{\mu!(n-\mu)!(n+2r-\mu)!}$$

$$= (n+2r)! e^{\alpha^2/2} \sum_{\nu=0}^{n} \frac{\alpha^{2(r+\nu)}}{\nu!(n-\nu)!(2r+\nu)!}. \tag{6.83}$$

It follows that the coefficient of $\alpha^{2s}/(2s)!$ in this expression J is the quantity K given by

$$K := \frac{\int_{-\infty}^{\infty} t^{2s} He_n(t) He_{n+2r}(t) e^{-t^2/2} dt}{\int_{-\infty}^{\infty} He_n^2(t) e^{-t^2/2} dt}.$$

Now, the coefficient of

$$\alpha^{2(s-r-\nu)} \text{ in } e^{\alpha^2/2} \text{ is } \frac{1}{2^{s-r-\nu}(s-r-\nu)!}.$$

As we observed, K is the coefficient of $\alpha^{2s}/(2s)!$ in the expression J, so that now with integration limits from 0 to ∞

$$\begin{aligned}
K &= \text{coefficient of } \frac{\alpha^{2s}}{(2s)!} \text{ in } J \\
&= \text{coefficient of } \frac{\alpha^{2s}}{(2s)!} \text{ in } (n+2r)! e^{\alpha^2/2} \sum_{\nu=0}^{n} \frac{\alpha^{2(r+\nu)}}{\nu!(n-\nu)!(2r+\nu)!} \\
&= \text{coefficient of } \frac{\alpha^{2s}}{(2s)!} \text{ in } (n+2r)! \sum_{i=0}^{\infty} \frac{(\alpha^2/2)^i}{i!} \sum_{\nu=0}^{n} \frac{\alpha^{2(r+\nu)}}{\nu!(n-\nu)!(2r+\nu)!} \\
&= \text{coefficient of } \frac{\alpha^{2s}}{(2s)!} \text{ in } \sum_{\nu=0}^{n} \frac{(n+2r)!}{(s-r-\nu)!} \frac{\alpha^{2(s-r-\nu)}}{2^{s-r-\nu}} \frac{\alpha^{2(r+\nu)}}{\nu!(n-\nu)!(2r+\nu)!} \\
&= \frac{(2s)!(n+2r)!}{2^{s-r}} \sum_{\nu=0}^{s-r} \frac{2^{\nu}}{\nu!(2r+\nu)!(n-\nu)!(s-r-\nu)!}.
\end{aligned}$$

Multiplying both sides by the value of the normalization integral, i.e. by $n!(\pi/2)^{1/2}$, we obtain the result (6.79). As an example we obtain for the integral from 0 to ∞ (also re-expressing the integral in terms of parabolic cylinder functions $D_n(t)$):

$$\int_0^{\infty} t^4 He_n(t) He_{n+2}(t) e^{-t^2/2} dt \equiv \int_0^{\infty} t^4 D_n(t) D_{n+2}(t) dt = \sqrt{2\pi}(2n+3)(n+2)!.$$

This result can be verified by inserting in the integral for $t^4 He_n(t)$ the linearized expression obtained with the help of the recurrence relation (6.77), i.e. the relation

$$\begin{aligned}
t^4 He_n(t) &= (n, n+1)(n+1, n+2)(n+2, n+3)(n+3, n+4) He_{n+4}(t) \\
&+ [(n, n+1)(n+1, n+2)(n+2, n+3)(n+3, n+2) \\
&+ (n, n+1)(n+1, n+2)(n+2, n+1)(n+1, n+2) \\
&+ (n, n+1)(n+1, n)(n, n+1)(n+1, n+2) \\
&+ (n, n-1)(n-1, n)(n, n+1)(n+1, n+2)] He_{n+2}(t) + \cdots.
\end{aligned}$$

From Example 6.3 we know that up-going coefficients are 1 and down-going coefficients $(n, n-1)$ are given by the first index, i.e. n. Hence

$$\begin{aligned}
t^4 He_n(t) &= He_{n+4}(t) + [(n+3) + (n+2) + (n+1) + n] He_{n+2}(t) + \cdots \\
&= He_{n+4}(t) + 2(2n+3) He_{n+2}(t) + \cdots.
\end{aligned}$$

The result then follows with half the normalization and orthogonality integrals (6.78).

Chapter 7

Green's Functions

7.1 Introductory Remarks

In this chapter we consider *Green's functions* of Schrödinger equations in both time-dependent and time-independent forms. In particular we derive the time-dependent Green's functions for the case of a *free particle* and for that of a *particle in an harmonic potential*. The first of these cases will re-appear later, in Chapter 21 (with a different calculation), in Feynman's path integral as the *kernel* or *free particle propagator*, and is thus an important point to be noted here. The other example of the oscillator enables us to evaluate corresponding expectation values of observables in the canonical distribution introduced in Chapter 5. This example also offers a convenient context to introduce the *inverted oscillator potential*, which is normally not discussed. With reference to this case, we consider the *sojourn time* of a quantum mechanical particle at a point, which is a point of unstable equilibrium for a classical particle.

7.2 Time-dependent and Time-independent Cases

In the case of Green's functions we distinguish between those which are time-dependent, which we write here $K(x, x'; t)$, and those which are time-independent, which we write $G(x, x')$.

We consider first *time-independent Green's functions*. These are the Green's functions of the time-independent Schrödinger equation or, more generally, of a differential equation of second order. Let

$$H\Psi = E\Psi \tag{7.1}$$

be the time-independent Schrödinger equation which has to be solved. We assume a Hamiltonian of the form

$$H = H_0 + H_I, \tag{7.2}$$

where H_0 consists, for instance, of the kinetic part $p^2/2m_0$ and some part of the potential, e.g. αq^2, and H_I is some other contribution, e.g. a perturbation part like βq^4. The time-independent Green's function $G(x, x')$ is defined by the equation*

$$(H_0 - E^{(0)})_x G(x, x') = \delta(x - x'), \tag{7.3}$$

where

$$E^{(0)} = \lim_{H_I \to 0} E.$$

The boundary conditions which $G(x, x')$ has to satisfy correspond to those of Ψ (to insure e.g. square integrability, i.e. appropriately decreasing behaviour at infinity). We can obtain an expression for $G(x, x')$ by recalling the completeness relation, i.e. the relation

$$\sum_i |i\rangle\langle i| = \mathbb{1},$$

which has the specific representation

$$\sum_i \langle x|i\rangle\langle i|x'\rangle = \delta(x - x') \quad \text{or} \quad \sum_i \psi_i(x)\psi_i^*(x') = \delta(x - x'), \tag{7.4}$$

where the functions $\psi_i(x)$ are solutions of

$$H_0\psi_i(x) = E_i\psi_i(x). \tag{7.5}$$

We can readily see that $G(x, x')$ can be written[†]

$$G(x, x') = \sum_i \frac{\psi_i(x)\psi_i^*(x')}{E_i - E^{(0)}} \quad \text{for} \quad E^{(0)} \neq E_i. \tag{7.6}$$

This holds since

$$(H_0 - E^{(0)})_x G(x, x') = \sum_i \frac{(H_0 - E^{(0)})_x \psi_i(x)\psi_i^*(x')}{E_i - E^{(0)}} = \delta(x - x').$$

---

*For simplicity we consider the one-dimensional case, since the generalization to higher dimensions is self-evident, although the explicit calculations can be much more involved.
[†]The case of $E^{(0)}$ equal to some E_i is considered later, cf. Eq. (7.13).

We note that the Green's function possesses simple poles in the plane of complex $E^{(0)}$, and that the residue at a pole is determined by the wave function belonging to the eigenvalues E_i.

The *time-dependent Green's function* $K(x, x'; t)$ (K for "kernel") is defined as solution of the equation

$$i\hbar \frac{\partial}{\partial t} K(x, x'; t) = H_0 K(x, x'; t) \tag{7.7}$$

with the initial condition

$$K(x, x'; 0) = \delta(x - x'). \tag{7.8}$$

This Green's function can be seen to be given by the following expansion in terms of a complete set of states (see verification below)

$$K(x, x'; t) = \sum_i e^{E_i t / i\hbar} \psi_i(x) \psi_i^*(x'), \tag{7.9}$$

since

$$
\begin{aligned}
i\hbar \frac{\partial}{\partial t} K(x, x'; t) &= \sum_i E_i e^{E_i t / i\hbar} \psi_i(x) \psi_i^*(x') = H_0 \sum_i e^{E_i t / i\hbar} \psi_i(x) \psi_i^*(x') \\
&= H_0 K(x, x'; t).
\end{aligned}
$$

We see that the initial condition (7.8) is satisfied as a consequence of the completeness relation of the wave functions $\psi_i(x)$.

We can see the *significance of the time-independent Green's function* for the complete problem as follows. We rewrite the complete time-independent Schrödinger equation $H\Psi = E\Psi$ as

$$(H_0 - E^{(0)})\Psi = (E - E^{(0)} - H_I)\Psi. \tag{7.10}$$

The contribution on the right hand side is the so-called perturbation contribution, which consists of the perturbation part H_I of the Hamiltonian and the corresponding correction $E - E^{(0)}$ of the eigenvalue. For the solution Ψ we can immediately write down the homogeneous integral equation (again to be verified below)

$$\Psi(x) = \int dx' G(x, x')(E - E^{(0)} - H_I(x'))\Psi(x'). \tag{7.11}$$

We can check this by considering

$$
\begin{aligned}
(H_0 - E^{(0)})_x \Psi(x) &= \int dx' (H_0 - E^{(0)})_x G(x, x')(E - E^{(0)} - H_I(x'))\Psi(x') \\
&\overset{(7.3)}{=} \int dx' \delta(x - x')(E - E^{(0)} - H_I(x'))\Psi(x') \\
&= (E - E^{(0)} - H_I(x))\Psi(x).
\end{aligned}
$$

Can we add on the right hand side of Eq. (7.11) an inhomogeneous term which is a solution of the non-perturbed part of the Schrödinger equation? If we keep in mind the anharmonic oscillator with square integrable wave functions $\Psi(x)$, then at best such a contribution would be something like $c\psi_i(x)$. But if we assume that $E^{(0)} \neq E_i$ for all i, the function $\psi_i(x)$ would not be a solution of

$$(H_0 - E^{(0)})\psi = 0,$$

and it is not possible to add an inhomogeneous contribution. This will be different, if we assume that $E^{(0)} = E_j$. But then the Green's function (7.6) is not defined. This difficulty[‡] can be circumvented by demanding from the beginning that

$$0 = \int dx' \psi_j^*(x')(E - E_j - H_I(x'))\Psi(x'), \qquad (7.12)$$

i.e. that

$$(E - E_j - H_I(x'))\Psi(x')$$

is a vector in $\mathcal{H}$ which is orthogonal to ψ_j. In this case the perturbation is restricted to that subspace of the Hilbert space which is orthogonal to ψ_j. Instead of Eq. (7.6) we would have

$$G(x, x') = \sum_{i, i \neq j} \frac{\psi_i(x)\psi_i^*(x')}{E_i - E^{(0)}}, \qquad E^{(0)} = E_j. \qquad (7.13)$$

Equations of the form of Eq. (7.11) are called (homogeneous or inhomogeneous) *Fredholm integral equations*. In cases where an inhomogeneous contribution is given, it is possible to solve the integral equation by an iterative perturbation procedure (see later: Born approximation). In general, however, integral equations are more difficult to solve than the corresponding differential equations. All such methods of solution are based on the analogy of Eq. (7.11) with a system of linear equations of the form

$$y_i = M_{ij}y_j. \qquad (7.14)$$

We can now see the significance of the time-dependent Green's function $K(x, x'; t)$ for the complete problem as follows. The time-dependent Schrödinger equation

$$i\hbar \frac{\partial}{\partial t}|\psi\rangle_t = H|\psi\rangle_t \qquad (7.15)$$

[‡]This problem and its circumvention can be formulated as a theorem; see E. Merzbacher [194], 2nd ed., Sec. 17.2 (not contained in the first edition!). The theorem is also known as Fredholm alternative; see B. Booss and D. D. Bleecker [33]. Examples in various contexts have been investigated by L. Maharana, H. J. W. Müller–Kirsten and A. Wiedemann [183].

permits *stationary states* $|E_n\rangle$ defined by

$$|\psi\rangle_t = |E_n\rangle e^{E_n t/i\hbar}, \quad H|E_n\rangle = E_n|E_n\rangle. \tag{7.16}$$

As usual, we assume that the set of states $|E_n\rangle$ is a complete set of eigenvectors of the Hamiltonian H (in the energy representation or energy basis) so that (cf. Eq. (6.42))

$$|\psi\rangle_{t=0} = \sum_n |E_n\rangle\langle E_n|\psi\rangle_{t=0}. \tag{7.17}$$

Thus at time $t = 0$ the state $|\psi\rangle_t$ is a linear superposition of the vectors $|E_n\rangle$ with coefficients $\langle E_n|\psi\rangle_{t=0}$, i.e. (7.17) is the initial condition of $|\psi\rangle_t$. The solution of Eq. (7.15) which contains (7.17) as initial condition is obviously

$$|\psi\rangle_t = \sum_n |E_n\rangle\langle E_n|\psi\rangle_{t=0} e^{E_n t/i\hbar}, \tag{7.18}$$

or, in different formulations,

$$\langle x|\psi\rangle_t = \sum_n \int dx' \langle x|E_n\rangle\langle E_n|x'\rangle\langle x'|\psi\rangle_{t=0} e^{E_n t/i\hbar},$$

$$\psi(x,t) = \sum_n \int dx' \psi_n(x)\psi_n^*(x')\psi(x',0) e^{E_n t/i\hbar}. \tag{7.19}$$

We can write this expression also as[§]

$$\psi(x,t) = \int dx' K(x,x';t)\psi(x',0), \quad t > 0, \tag{7.20}$$

where

$$K(x,x';t) = \sum_n e^{E_n t/i\hbar}\psi_n(x)\psi_n^*(x'), \quad t > 0, \tag{7.21}$$

is the time-dependent Green's function,[¶] which obviously satisfies the initial condition

$$K(x,x';0) = \delta(x - x'). \tag{7.22}$$

According to Eq. (7.20) the Green's function $K(x,x';t)$ describes the evolution of the wave function from its initial value $\psi(x,0)$. Comparison of Eq. (7.7) with Eq. (5.51) shows that we obtain a very analogous expression for the density matrix $\rho_N(\beta)$ as for the Green's function $K(x,x';t)$ with the

---

[§]Note that when $K(x,x';t)$ is known, e.g. for the oscillator potential, this relation provides the probability density $|\psi(x,t)|^2$. We use this in Sec. 7.5.2 for the computation of the sojourn time.

[¶]See e.g. E. Merzbacher [194], 2nd ed., p.158.

difference that $\beta = 1/kT$ plays the role of *it*. In the following we shall derive the Green's function for the case of the harmonic oscillator; we obtain then also the corresponding density matrix.

As a consequence of the above considerations one wants to know the connection between the time-dependent and the time-independent Green's functions, i.e. between (with $E^{(0)} \equiv E$)

$$G(x, x') \equiv G_E(x, x') = \sum_n \frac{\psi_n(x)\psi_n^*(x')}{E_n - E}$$

and

$$K(x, x'; t) = \sum_n e^{E_n t/i\hbar} \psi_n(x)\psi_n^*(x'), \quad t > 0.$$

We see that $G \equiv G_E$ depends on E. We therefore consider the following integral with $\epsilon > 0$:

$$I(t) := -i \int_C \frac{dE}{2\pi} e^{Et/i\hbar} G_{E+i\epsilon}(x, x')\theta(t) \tag{7.23}$$

along the contour C in the plane of complex E as shown in Fig. 7.1. Inserting for $G_{E+i\epsilon}$ the expression above, we obtain

$$I(t) := -i \int_C \frac{dE}{2\pi} e^{Et/i\hbar} \sum_n \frac{\psi_n(x)\psi_n^*(x')}{E_n - E - i\epsilon}\theta(t). \tag{7.24}$$

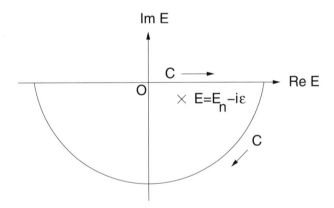

Fig. 7.1 The contour of integration.

With Cauchy's residue theorem we obtain

$$I(t) := \sum_n e^{E_n t/i\hbar} \psi_n(x)\psi_n^*(x')\theta(t) = K(x, x'; t) \tag{7.25}$$

in agreement with the time-dependent Green's function $K(x, x'; t)$.

7.3   The Green's Function of a Free Particle

The *time-dependent Green's function* of a free particle which we now derive
is an important quantity and will reappear later in Feynman's path integral
method. It is clear that it is nontrivial to solve an equation like Eq. (7.7).
We therefore consider first the simplest case with

$$H_0 = \frac{p^2}{2m_0} \rightarrow -\frac{\hbar^2}{2m_0}\frac{\partial^2}{\partial x^2}. \tag{7.26}$$

This is the case of a free particle with mass m_0, which is moving in one space
dimension. In this case the Green's function is the solution of the equation

$$i\hbar\frac{\partial}{\partial t}K(x,x';t) = -\frac{\hbar^2}{2m_0}\frac{\partial^2}{\partial x^2}K(x,x';t). \tag{7.27}$$

An equation of this type — called of the type of a *diffusion equation* — can
be solved with an ansatz. Thus we try the ansatz, A and B being constants,

$$K(x,x';t) = \frac{A}{\sqrt{t}}e^{-B(x-x')^2/t}. \tag{7.28}$$

In this case we have

$$\frac{\partial K}{\partial t} = \left[-\frac{A}{2t^{3/2}} + \frac{A}{t^{1/2}}\frac{B(x-x')^2}{t^2}\right]e^{-B(x-x')^2/t} \tag{7.29}$$

and

$$\frac{\partial K}{\partial x} = \frac{A}{t^{1/2}}\left[-\frac{2B(x-x')}{t}\right]e^{-B(x-x')^2/t},$$

$$\frac{\partial^2 K}{\partial x^2} = \left[-\frac{2AB}{t^{3/2}} + \frac{4AB^2(x-x')^2}{t^{5/2}}\right]e^{-B(x-x')^2/t}. \tag{7.30}$$

Inserting Eqs. (7.29) and (7.30) into Eq. (7.27), and identifying coefficients
of the same powers of t on both sides, we obtain

$$i\hbar\left(-\frac{A}{2}\right) = -\frac{\hbar^2}{2m_0}(-2AB), \quad i\hbar(AB) = -\frac{\hbar^2}{2m_0}(4B^2A), \quad \text{i.e.} \quad B = \frac{m_0}{2i\hbar}. \tag{7.31}$$

The constant A has to be chosen such that

$$K(x,x';0) = \delta(x-x'), \quad \text{i.e.} \quad \int dx K(x,x';0) = 1.$$

For parameter values such that the following integral exists, we have

$$\int dx\, K(x,x';t) = \frac{A}{\sqrt{t}} \int e^{-B(x-x')^2/t}\, dx = \frac{A}{\sqrt{t}}\left(\frac{\sqrt{\pi}}{\sqrt{B/t}}\right) = \frac{A\sqrt{\pi}}{\sqrt{B}}.$$

This is 1 provided

$$A = \sqrt{\frac{B}{\pi}} = \sqrt{\frac{m_0}{2i\hbar\pi}}. \tag{7.32}$$

It follows that

$$K(x,x';t) = \sqrt{\frac{m_0}{2i\hbar\pi t}}\, e^{-m_0(x-x')^2/2i\hbar t}. \tag{7.33}$$

Can we demonstrate that this expression can also be obtained from Eq.(7.21), i.e. from

$$K(x,x';t) = \sum_n e^{E_n t/i\hbar}\psi_n(x)\psi_n^*(x'), \quad t > 0. \tag{7.34}$$

For a free particle moving in the one-dimensional domain $|x| \le L \to \infty$, we have to make the replacements

$$\sum_n \to \int dk, \quad \psi_n(x) \to \psi_k(x) = \frac{e^{ikx}}{\sqrt{2\pi}}, \quad E_n \to \frac{(\hbar k)^2}{2m_0}, \tag{7.35a}$$

so that

$$K(x,x';t) = \int \frac{dk}{2\pi} e^{(\hbar k)^2 t/2m_0 i\hbar} e^{ik(x-x')}. \tag{7.35b}$$

We set

$$\alpha = i\frac{\hbar t}{2m_0}, \quad \beta = i(x - x'). \tag{7.36}$$

Then — provided that the parameters assume values such that the intergral exists —

$$
\begin{aligned}
K(x,x';t) &= \frac{1}{2\pi}\int_{-\infty}^{\infty} dk\, e^{-\alpha k^2 + \beta k} = \frac{1}{2\pi}e^{\beta^2/4\alpha}\int_{-\infty}^{\infty} dk\, e^{-(\sqrt{\alpha}k - \beta/2\sqrt{\alpha})^2} \\
&= \frac{1}{2\pi}e^{\beta^2/4\alpha}\int \frac{dz}{\sqrt{\alpha}} e^{-z^2} = \frac{\sqrt{\pi}}{2\pi\sqrt{\alpha}}e^{\beta^2/4\alpha} \\
&\stackrel{(7.36)}{=} \left(\frac{m_0}{2\pi i t\hbar}\right)^{1/2} e^{-m_0(x-x')^2/2it\hbar}
\end{aligned}
\tag{7.37}
$$

in agreement with Eq. (7.33).$^{\|}$ We can insert the expression for $K(x,x';t)$ into Eq. (7.20) and can then obtain $\psi(x,t)$ — for instance for a wave packet given at time $t = 0$ of the form

$$\psi(x,0) \propto e^{-\alpha x^2 + ik_0 x}. \tag{7.38}$$

$^{\|}$See the excercise in E. Merzbacher [194], 1st ed., p. 158.

The result (7.33), (7.37) will later be obtained by a different method — see Eq. (21.25) — in the context of Feynman's path integral method.

7.4 Green's Function of the Harmonic Oscillator

The next most obvious case to consider is that of a particle subjected to the harmonic oscillator potential. We consider the one-dimensional harmonic oscillator with Hamilton operator H_0,

$$H_0 = \frac{1}{2m_0}p^2 + \frac{1}{2}m_0\omega^2 q^2. \qquad (7.39)$$

In this case the *time-dependent Green's function* K of Eq. (7.7) is the solution of

$$i\hbar\frac{\partial}{\partial t}K(x,x';t) = -\frac{\hbar^2}{2m_0}\frac{\partial^2}{\partial x^2}K(x,x';t) + \frac{1}{2}m_0\omega^2 x^2 K(x,x';t). \qquad (7.40)$$

We now set

$$\xi = \sqrt{\frac{m_0\omega}{\hbar}}x, \quad f = \frac{1}{2}\hbar\omega\frac{it}{\hbar}. \qquad (7.41)$$

Then Eq. (7.40) becomes

$$\frac{2}{\omega\hbar}i\hbar\frac{\partial}{\partial t}K(x,x';t) = -\frac{\hbar^2}{2m_0}\frac{2}{\omega\hbar}\frac{\partial^2}{\partial x^2}K(x,x';t) + \frac{m_0\omega}{\hbar}x^2 K(x,x';t),$$

i.e.

$$-\frac{\partial}{\partial f}K(x,x';f) = -\frac{\partial^2}{\partial \xi^2}K(x,x';f) + \xi^2 K(x,x';f) \qquad (7.42)$$

with the initial condition (7.8), i.e.

$$K(x,x';0) = \delta(x-x') \quad \text{at} \quad f = 0.$$

We rewrite this initial condition in terms of ξ and use for $a = $ const. the relation*

$$\delta(x) = |a|\delta(ax), \qquad (7.43)$$

so that

$$K(\xi,\xi';f)|_{f=0} = \sqrt{\frac{m_0\omega}{\hbar}}\delta(\xi-\xi'). \qquad (7.44)$$

In the domain of small values of x (near the minimum of the potential) the Hamiltonian H_0 is dominated by the kinetic energy, i.e. in this domain

---

*Recall that $\delta(x) = \frac{1}{2\pi}\int e^{ikx}dk$, i.e. $\delta(ax) = \frac{1}{2\pi}\int e^{ikax}dk = \frac{1}{|a|}\delta(x)$.

the particle behaves almost like a free particle, and we expect $K(x, x'; t)$ to become similar to expression (7.33), i.e.

$$K(\xi, \xi'; f) \simeq \sqrt{\frac{m_0 \omega}{4\pi \hbar f}} e^{-(\xi - \xi')^2 / 4f} \quad \text{for} \quad \xi \neq \xi'. \qquad (7.45)$$

The same approximation is also valid for large energies E and for t or f small (near zero) in view of the relation

$$\triangle E \triangle t \sim \hbar. \qquad (7.46)$$

If we interpret Eq. (7.45) in this sense, that is, as the limiting case $f \to 0$, it is suggestive to attempt for K the following ansatz[†]

$$K(\xi, \xi'; f) \propto \exp[-\{a(f)\xi^2 + b(f)\xi + c(f)\}] \qquad (7.47)$$

with

$$a(0) \to \frac{1}{4f}, \quad b(0) \to -\frac{\xi'}{2f}, \quad c(0) \to \frac{\xi'^2}{4f}. \qquad (7.48)$$

We insert this ansatz for K into Eq. (7.42) and obtain the equation (with $a' \equiv da/df$ etc.)

$$a'\xi^2 + b'\xi + c' = (1 - 4a^2)\xi^2 - 4ab\xi + 2a - b^2. \qquad (7.49)$$

Identifying coefficients on both sides, we obttain

$$a' = 1 - 4a^2, \qquad (7.50a)$$

$$b' = -4ab, \qquad (7.50b)$$

$$c' = 2a - b^2. \qquad (7.50c)$$

Integrating Eq. (7.50a) we obtain

$$a = \frac{1}{2}\coth 2(f - f_0) = \frac{1}{2\tanh 2(f - f_0)}.$$

To ensure that the expression (7.47) becomes (7.45) in accordance with (7.48), we must have $f_0 = 0$, so that for $f \to 0$:

$$a \to \frac{1}{4f}, \quad \text{i.e.} \quad a = \frac{1}{2}\coth 2f. \qquad (7.51)$$

[†]See R. P. Feynman [94], p. 50.

Correspondingly we obtain from integration of Eq. (7.50b)

$$b(f) = \frac{A}{\sinh 2f}, \quad A \text{ independent of } f.$$

To ensure that in accordance with Eq. (7.48) $b(0) = -\xi'/2f$, we must have $A = -\xi'$, i.e.

$$b(f) = -\frac{\xi'}{\sinh 2f}. \tag{7.52}$$

Finally Eq. (7.50c) yields for $c(f)$, with A, B independent of f,

$$c(f) = \frac{1}{2}\ln(\sinh 2f) + \frac{1}{2}A^2\coth 2f - \ln B.$$

In order to satisfy Eq. (7.48), i.e. $c(0) = \xi'^2/4f$, and to ensure that we obtain the prefactor of Eq. (7.45), we must have (besides $A = -\xi'$)

$$B = \sqrt{\frac{m_0\omega}{2\pi\hbar}}.$$

Inserting $a(f), b(f), c(f)$ into Eq. (7.47) we obtain

$$K = \frac{B}{\sqrt{\sinh 2f}} \exp\left[-\left\{\frac{1}{2}\xi^2\coth 2f + \frac{A\xi}{\sinh 2f} + \frac{1}{2}A^2\coth 2f\right\}\right], \tag{7.53}$$

or, if we return to x, x', t,

$$K(x, x'; t) = \sqrt{\frac{m_0\omega}{2\pi\hbar i \sin(\omega t)}} \exp\left[-\frac{m_0\omega}{2\hbar i \sin(\omega t)}\{(x^2 + x'^2)\cos\omega t - 2xx'\}\right]. \tag{7.54}$$

For $t \to 0$ this expression goes over into the expression (7.33) for the Green's function of a free particle, as one can verify.[‡] With this result we have another important quantity at our disposal, as we shall see in the following, in particular for the derivation of the sojourn time in Sec. 7.5.2.

Comparing the Eqs. (7.27), (7.40) of the time-dependent Green's function with Eq. (5.51) for the density matrix $\rho_N(x, x'; \beta)$, we can use $K(x, x'; t)$ to obtain this element (x, x') of the density matrix ρ_N (with respect to the canonical distribution with $\beta = 1/kT$):

$$\rho_N(x, x'; \beta) = \sqrt{\frac{m_0\omega}{2\pi\hbar \sinh(\hbar\omega/kT)}}$$

$$\times \exp\left[-\frac{m_0\omega}{2\hbar \sinh(\hbar\omega/kT)}\left\{(x^2 + x'^2)\cosh\frac{\hbar\omega}{kT} - 2xx'\right\}\right]. \tag{7.55}$$

[‡]For an alternative derivation and further discussion see also B. Felsager [91], p. 174.

With this expression we can evaluate (cf. Eq. (5.52)) the expectation value of an observable A in the canonical distribution (i.e. at temperature T):

$$\langle A \rangle = \mathrm{Tr}\langle \rho A \rangle = \frac{\mathrm{Tr}(\rho_N A)}{\mathrm{Tr}\rho_N}.$$

For instance we have with Eq. (5.52):

$$\langle q^2 \rangle = \frac{\mathrm{Tr}\langle \rho_N q^2 \rangle}{\mathrm{Tr}\rho_N} = \frac{\int x^2 \rho_N(x, x; \beta) dx}{\int \rho_N(x, x; \beta) dx}. \tag{7.56a}$$

Thus for $\langle q^2 \rangle = \mathrm{Tr}\langle \rho q^2 \rangle = \sum_i \langle i | \rho q^2 | i \rangle$ we obtain:

$$
\begin{aligned}
\langle q^2 \rangle &= \sum_i \int \int \int dx dx' dx'' \langle i | x \rangle \langle x | \rho | x' \rangle \langle x' | q^2 | x'' \rangle \langle x'' | i \rangle \\
&= \sum_i \int \int \int dx dx' dx'' \langle x'' | i \rangle \langle i | x \rangle \langle x | \rho | x' \rangle \langle x' | q^2 | x'' \rangle \\
&= \int \int \int dx dx' dx'' \langle x'' | x \rangle \langle x | \rho | x' \rangle \langle x' | q^2 | x'' \rangle \\
&= \int \int \int dx dx' dx'' \delta(x'' - x) \langle x | \rho | x' \rangle \langle x' | q^2 | x'' \rangle,
\end{aligned}
$$

i.e. we verify the relation:

$$
\begin{aligned}
\langle q^2 \rangle &= \int \int dx dx' \langle x | \rho | x' \rangle \langle x' | q^2 | x \rangle = \int \int dx dx' \langle x | \rho | x' \rangle \langle x' | x \rangle x^2 \\
&= \int dx \rho(x, x; \beta) x^2.
\end{aligned}
$$

Inserting into Eq. (7.56a) the expression (7.55), we obtain[§]

$$\langle q^2 \rangle = \frac{\hbar}{2m_0\omega} \coth\frac{\hbar\omega}{2kT} \xrightarrow{T \to 0} \frac{\hbar}{2m_0\omega}. \tag{7.56b}$$

What is the meaning of this expression? At temperature T the fraction (cf. Eq. (5.40))[¶]

$$\omega_i = \frac{e^{-\beta E_i}}{\sum_j e^{-\beta E_j}} = \frac{1}{1 + \sum_{j \neq i} e^{-\beta(E_j - E_i)}} \tag{7.57}$$

of the number of systems of the ensemble occupies the quantum mechanical state i. Thus the system is in a mixed state and the expectation value

---

[§]Cf. R. P. Feynman [94], p. 52; we skip the algebra here.
[¶]For $T \to 0$: $\omega_0 \to 1, \omega_{j>0} \to 0$.

(7.56b) is that with respect to this mixed state (whose cause is the finite temperature T). If we consider the system in the pure state $|i\rangle$, which means in the oscillator state $|i\rangle$ with eigenenergy $\hbar\omega(i+1/2)$, the expression for $\langle q^2\rangle$ would be:[||]

$$\langle q^2\rangle_i = \frac{\int\langle i|x\rangle x^2\langle x|i\rangle dx}{\int\langle i|x\rangle\langle x|i\rangle dx}, \qquad \langle q^2\rangle_0 = \frac{\hbar}{2m_0\omega}. \tag{7.58}$$

Next we explore the connection between the explicit form of K and the latter's expansion in terms of a complete set of states. We return to the Green's function (7.54). We assume $t > 0$ and $t \to \infty$, and we replace ω by $\omega - i\epsilon, \epsilon > 0$. In this case the Green's function $K(x, x'; t)$ of Eq. (7.54) is

$$K(x, x'; t)\Big|_{\substack{t\to\infty \\ \omega\to\omega-i\epsilon}}$$

$$= \left(\frac{m_0\omega/\pi\hbar}{e^{i\omega t} - e^{-i\omega t}}\right)^{1/2} \exp\left[-\frac{m_0\omega}{\hbar e^{i\omega t}}\left\{(x^2 + x'^2)\frac{1}{2}e^{i\omega t} - 2xx'\right\}\right]\Bigg|_{\substack{\omega\to\omega-i\epsilon \\ t\to\infty}}$$

$$\simeq \sqrt{\frac{m_0\omega}{\pi\hbar}}e^{-i\omega t/2}\exp\left[-\frac{m_0\omega}{2\hbar}(x^2 + x'^2)\right]$$

$$= e^{\hbar\omega/2i\hbar t}\sqrt{\frac{m_0\omega}{\pi\hbar}}\exp\left[-\frac{1}{2}\frac{m_0\omega}{\hbar}x^2\right]\exp\left[-\frac{1}{2}\frac{m_0\omega}{\hbar}x'^2\right]$$

$$= e^{E_0/i\hbar t}\phi_0(x)\phi_0(x') \quad \text{for} \quad t > 0, \quad E_0 = \frac{1}{2}\hbar\omega. \tag{7.59a}$$

This is the first and hence dominant term of the expression (7.21), i.e. of

$$K(x, x'; t) = \sum_n e^{E_n t/i\hbar}\phi_n(x)\phi_n(x'). \tag{7.59b}$$

For $t > 0$ and $E_n = (n + 1/2)\hbar\omega$ the factor $\exp(E_n t/i\hbar)$ is

$$\exp\left[\left(n + \frac{1}{2}\right)\frac{\hbar\omega t}{i\hbar}\right] \overset{\text{here}}{=} \exp\left[-i\left(n + \frac{1}{2}\right)(\omega - i\epsilon)t\right]$$

$$= e^{-(n+1/2)\epsilon t}e^{-i(n+1/2)\omega t}.$$

[||] With the normalized ground state wave function of the harmonic oscillator given by Eq. (6.50) we obtain, setting $\alpha = m_0\omega/\hbar$,

$$\langle q^2\rangle_0 = \left(\frac{\alpha}{\pi}\right)^{1/2}\int_{-\infty}^\infty dx\, x^2 e^{-\alpha x^2} = -\left(\frac{\alpha}{\pi}\right)^{1/2}\frac{d}{d\alpha}\int_{-\infty}^\infty dx e^{-\alpha x^2}$$

$$= -\left(\frac{\alpha}{\pi}\right)^{1/2}\frac{d}{d\alpha}\left(\frac{\pi}{\alpha}\right)^{1/2} = \frac{1}{2\alpha} = \frac{\hbar}{2m_0\omega}.$$

For t large (i.e. to infinity) the contribution with $n = 0$ dominates, as in Eq. (7.59a).

In a very analogous manner we obtain the solution of the equation for the density matrix, i.e. of

$$-\frac{\partial \rho_N}{\partial \beta} = H_x \rho_N,$$

as

$$\rho_N(x, x'; \beta) = \sum_n e^{-\beta E_n} \phi_n(x) \phi_n^*(x') \xrightarrow{T \to 0} e^{-\beta E_0} \phi_0(x) \phi_0^*(x'). \qquad (7.60)$$

7.5   The Inverted Harmonic Oscillator

We encountered the inverted harmonic oscillator already in some examples. Considered classically, a particle placed at the maximum of the inverted oscillator potential (which is classically a position of unstable equilibrium) will stay there indefinitely. However quantum mechanically in view of the uncertainties in position and momentum, the particle will stay there only for a finite length of time T. In the following we want to calculate (more precisely estimate) with the help of the Green's function the time interval T which a pointlike particle can stay at the maximum of the potential before it rolls down as a result of the quantum mechanical uncertainties. We first introduce the concept of a wave packet and then use the particular form of a wave packet in order to describe the state of the particle at time $t = 0$, and with this we estimate the sojourn time T. In Example 7.1 we estimate T semiclassically.

7.5.1   Wave packets

The simplest type of wave is the so-called plane wave or monochromatic wave of frequency ω represented by the expression

$$\exp[i(\mathbf{k} \cdot \mathbf{r} - \omega t)], \quad |\mathbf{k}| = k. \qquad (7.61)$$

The word "plane" implies that the points of constant phase $\varphi := \mathbf{k} \cdot \mathbf{r} - \omega t$ at $t = $ const. lie on surfaces, which are planes. The wavelength λ is given by

$$\lambda = \frac{2\pi}{k}.$$

The *phase velocity* v_φ, i.e. the velocity of planes of equal phase, is defined as

$$v_\varphi = \frac{\omega}{k}. \qquad (7.62)$$

Every frequency ω belongs to a definite (particle) energy E (cf. the fundamental postulate on matter waves in Chapter 2):

$$E = \hbar\omega, \quad \omega = \omega(\mathbf{k}). \tag{7.63}$$

The relation $\omega = \omega(\mathbf{k})$ is known as *dispersion* or *dispersion law*.

A *wave packet* is defined as a superposition of plane waves with almost equal wave vectors $\mathbf{k}$, i.e.

$$\psi(\mathbf{r}, t) = \int f(\mathbf{k}')e^{i(\mathbf{k}'\cdot\mathbf{r}-\omega't)}d\mathbf{k}', \tag{7.64}$$

where $f(\mathbf{k}')$ differs substantially from zero only near $\mathbf{k}' = \mathbf{k}$. The wave packet describes a wave of limited extent. If we assume for $f(\mathbf{k}')$ a Gauss distribution, i.e.

$$e^{-\alpha(\mathbf{k}-\mathbf{k}')^2}, \quad \alpha > 0,$$

then $\psi(\mathbf{r}, t)$ is the spatial Fourier transform of this Gauss distribution — as we know, essentially again a Gauss curve.

We now ask: How and under what conditions does the time variation of the function $\psi(\mathbf{r}, t)$ describe the motion of a classical particle? For reasons of simplicity we restrict ourselves here to the one-dimensional case, i.e. to the function ψ given by

$$\psi(x, t) = \int_{-\infty}^{\infty} f(k')e^{i(k'x-\omega't)}dk'. \tag{7.65}$$

Let

$$f(k') = |f(k')|e^{i\alpha} \quad \text{and} \quad \varphi := k'x - \omega't + \alpha.$$

One defines as *centre of mass of the wave packet* that value of x for which

$$\frac{d\varphi}{dk'} = 0, \quad \text{i.e.} \quad x - t\frac{d\omega'}{dk'} + \frac{d\alpha}{dk'} = 0,$$

or

$$x = t\frac{d\omega'}{dk'} - \frac{d\alpha}{dk'}. \tag{7.66}$$

The centre of mass determines the particular phase, for which $|\psi|$ assumes its largest value:

$$
\begin{aligned}
\psi^*(x, t)\psi(x, t) &= \int_{-\infty}^{\infty}\int_{-\infty}^{\infty} |f(k')||f(k'')|e^{i(\varphi(k')-\varphi(k''))}dk'dk'' \\
&= \int_{-\infty}^{\infty}\int_{-\infty}^{\infty} |f(k')||f(k'')|[\cos(\varphi(k') - \varphi(k'')) \\
&\qquad\qquad + i\sin(\varphi(k') - \varphi(k''))]dk'dk''.
\end{aligned}
$$

This expression assumes its maximal real value when

$$\varphi(k') = \varphi(k'') = \text{const.},$$

i.e. as claimed for

$$\frac{\partial \varphi}{\partial k'} = 0. \qquad (7.67)$$

The centre of mass moves with uniform velocity v_g called *group velocity* of the waves $\exp[i(kx - \omega t)]$. For

$$E = \hbar \omega, \quad p = \hbar k$$

the group velocity is equal to the particle velocity v:

$$v_g = \frac{d\omega}{dk} = \frac{d\hbar\omega}{d\hbar k} = \frac{dE}{dp} = \frac{\partial E}{\partial p}.$$

We can also argue the other way round and say: By identifying $v_g = v$, we obtain the de-Broglie relation $p = \hbar k$. The three-dimensional generalization is evidently

$$\mathbf{v}_g = \frac{\partial \omega}{\partial \mathbf{k}} \equiv \text{grad}_k\, \omega, \quad \mathbf{v} = \frac{\partial E}{\partial \mathbf{p}} \equiv \text{grad}_p\, E. \qquad (7.68)$$

It is instructive to consider at this point the following examples.

Example 7.1: Fourier transform of a Gauss function
Calculate the Fourier transform of the spatial Gauss function

$$e^{-ax^2}, \quad a > 0.$$

Solution: The function to be calculated is the integral

$$g(k) = \int_{-\infty}^{\infty} dx\, e^{-ax^2} e^{ikx}. \qquad (7.69)$$

In general one uses the theory of functions for the evaluation of this integral. In the present case, however, we can use a simpler method. Differentiation of $g(k)$ yields

$$g'(k) = \frac{i}{-2a} \int_{-\infty}^{\infty} dx\, e^{-ax^2} (-2ax) e^{ikx} = -\frac{i}{2a} \int_{-\infty}^{\infty} dx\, \frac{d}{dx}\left(e^{-ax^2}\right) e^{ikx}.$$

With partial integration we obtain from this

$$g'(k) = \frac{i}{2a} \int_{-\infty}^{\infty} ike^{ikx} e^{-ax^2}\, dx = -\frac{k}{2a} g(k).$$

Thus the expression $g(k)$ is solution of the following first order differential equation

$$g'(k) + \frac{k}{2a} g(k) = 0.$$

Simple integration yields

$$g(k) = ce^{-k^2/4a},$$

where $c = g(0)$ is a constant. Since

$$g(0) = \int_{-\infty}^{\infty} dx e^{-ax^2} = \sqrt{\frac{\pi}{a}},$$

we obtain

$$g(k) = \sqrt{\frac{\pi}{a}} e^{-k^2/4a}. \tag{7.70}$$

With the help of this example we can obtain some useful representations of the delta function or distribution as in the next example.

Example 7.2: Representations of the delta distribution
Use Eqs. (7.69) and (7.70) to verify the following representations of the delta distribution:

$$\delta(x) = \lim_{\epsilon \to 0} \frac{e^{-x^2/\epsilon^2}}{\epsilon\sqrt{\pi}} \tag{7.71}$$

and

$$\delta(x) = \lim_{\epsilon \to 0} \frac{1}{\pi} \frac{\epsilon}{\epsilon^2 + x^2}. \tag{7.72}$$

Solution: From Eqs. (7.69) and (7.70) we obtain

$$\delta(k) = \lim_{a \to 0} \frac{1}{2\pi} \int_{-\infty}^{\infty} dx e^{-ax^2} e^{ikx} = \lim_{a \to 0} \frac{1}{2\sqrt{\pi a}} e^{-k^2/4a} = \lim_{\epsilon \to 0} \frac{e^{-k^2/\epsilon^2}}{\epsilon\sqrt{\pi}}.$$

The second important example can be verified by immediate integration. We have

$$\frac{1}{2\pi} \int_{-\infty}^{\infty} dk e^{-\epsilon|k|} e^{ikx} = \frac{1}{\pi} \int_0^{\infty} dk e^{-\epsilon k} \cos kx = \frac{1}{\pi} \frac{\epsilon}{\epsilon^2 + x^2} \quad (\epsilon > 0). \tag{7.73}$$

From Eq. (7.73) we obtain the requested representation of the delta distribution with

$$\delta(x) = \lim_{\epsilon \to 0} \frac{1}{2\pi} \int_{-\infty}^{\infty} dk e^{-\epsilon|k|} e^{ikx} = \lim_{\epsilon \to 0} \frac{1}{\pi} \frac{\epsilon}{\epsilon^2 + x^2}.$$

With the help of Eq. (7.43) one can verify the following important relation[*]

$$\delta[(x - a)(x - b)] = \frac{1}{|a - b|} [\delta(x - a) + \delta(x - b)] \quad \text{for} \quad a \neq b. \tag{7.74}$$

Example 7.3: The uncertainty relation for Gaussian wave packets
For the specific Gauss wave packet of Example 7.1 verify the uncertainty relation

$$\triangle x \triangle k = 8 \ln 2. \tag{7.75}$$

---

[*]E.g. with partial fraction decomposition, or see H. J. W. Müller–Kirsten [215], Appendix A.

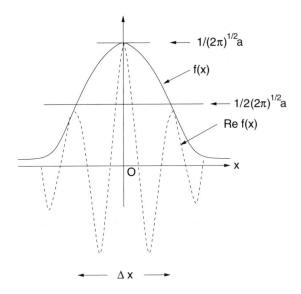

Fig. 7.2 The Gaussian curve.

Solution: In Fig. 7.2 we sketch the behaviour of the Gaussian function

$$f(x) = \frac{e^{-x^2/2a^2} e^{-ik_0 x}}{\sqrt{2\pi}a}. \tag{7.76}$$

The uncertainty $\triangle x$ is defined to be the width of the curve at half the height of the maximum, i.e. where $|f(x)| = \max|f(x)|/2$. A simple calculation yields

$$\triangle x = 2\sqrt{2\ln 2}\,a. \tag{7.77a}$$

According to Eq. (7.70) the Fourier transform of $f(x)$ is

$$g(k) = e^{-a^2(k-k_0)^2/2}. \tag{7.77b}$$

The breadth $\triangle k$ of the curve $g(k)$ around $k = k_0$, where $g(k) = \max|g(k)|/2$, is

$$\triangle k = 2\sqrt{2\ln 2}\,\frac{1}{a}. \tag{7.78}$$

It follows that the product of the uncertainties is $\triangle x \triangle k = 8\ln 2$. Thus a sharp maximum of the function $f(x)$, i.e. when $\triangle x$ is very small, requires according to Eq. (7.77a) small values of a, and hence implies large values of $\triangle k$.

Thus a slim maximum of the curve of $|f(x)|$ leads to a very broad maximum of the Fourier transform $|g(k)|$. In quantum mechanics the square of the modulus of the wave function $\psi(x, t)$ is a measure of the probability to find the particle at time t at the position x. The wave function corresponds to the function $f(x)$ in the above considerations (e.g. at time $t = 0$), the width $\triangle x$ is a measure of the uncertainty of the probability. Physically

$p = \hbar k$ is the canonical momentum associated with x. The function $g(k)$ is therefore described as momentum space representation of the wave function $f(x)$. The result (7.75) implies therefore: The more precise the coordinate of a microscopic particle is determined, the less precise is the determination of its associated momentum. In the limit $a \to 0$ we obtain from Eqs. (7.71) and (7.76)

$$\lim_{a \to 0} |f(x)| = \lim_{a \to 0} \frac{e^{-x^2/2a^2}}{\sqrt{2\pi}a} = \delta(x).$$

This means the particle is localized at $x = 0$. In the same limit the momentum uncertainty $\triangle k$ grows beyond all bounds, which means all values of the momentum are equally probable. In the reverse case a sharp localization of the particle in momentum space implies a correspondingly large uncertainty of its spatial coordinate.

7.5.2 A particle's sojourn time T at the maximum

The very instructive topic of this subsection has been explored in detail in a paper by Barton.* We obtain the Green's function K_{io} for the inverted harmonic oscillator from Eq. (7.54) with the substitution $\omega \to i\omega$. For reasons of simplicity in the following we set in addition† $m_0 = 1, \omega = 1, \hbar = 1$, so that

$$K_{io}(x, x'; t) = \frac{1}{\sqrt{2\pi i \sinh t}} \exp\left[\frac{i}{2 \sinh t}\{(x^2 + x'^2) \cosh t - 2xx'\}\right]. \quad (7.79)$$

According to Eq. (7.20) the wave packet at time $t > 0$ is obtained from that at time $t = 0$ with the relation

$$\psi(x, t) = \int dx' K(x, x'; t)\psi(x', 0). \quad (7.80)$$

We assume that initially, i.e. at time $t = 0$, the wave packet has its centre of mass at the origin and has the following Gaussian shape as the pure initial state:

$$\psi(x, 0) = \frac{1}{\sqrt{\pi^{1/2}b}} e^{-x^2/2b^2}. \quad (7.81)$$

---

*G. Barton [15]. As G. Barton mentions at the beginning of his paper, many supposedly elementary problems request the calculation of the sojourn time of a quantum system near a classically unstable equilibrium configuration. Drastic idealizations are then required in order to reformulate the classical situation into that of a quantum system. Thus G. Barton recalls that W. E. Lamb at Oxford set the following problem in an examination in 1957: *"A pencil is to be balanced so as to stand upright on its point on a horizontal surface. Estimate the maximum length of time compatible with quantum limitations before the pencil falls over..."*.

†In Eq. (7.54) these parameters appear in the combinations $m_0\omega/\hbar$ (dimension: length^{-2}) and ωt.

Inserting (7.81) and (7.79) into (7.80) we obtain

$$\psi(x,t) = \int_{-\infty}^{\infty} dx' \frac{\exp\left[\frac{i}{2\sinh t}\{(x^2 + x'^2)\cosh t - 2xx'\} - \frac{x'^2}{2b^2}\right]}{\sqrt{2\pi i \sinh t}\sqrt{\pi^{1/2} b}}. \quad (7.82)$$

Setting

$$A^2 := \frac{1}{b^2} - i\frac{\cosh t}{\sinh t}, \quad \text{and} \quad AB := \frac{i}{\sinh t}, \quad (7.83)$$

we can rewrite ψ as

$$\psi(x,t) = \int_{-\infty}^{\infty} dx' \frac{\exp\left[-\frac{1}{2}(Ax' + Bx)^2 + \frac{1}{2}(B^2 + i\coth t)x^2\right]}{\sqrt{2\pi i \sinh t}\sqrt{\pi^{1/2} b}}. \quad (7.84)$$

With some algebra one can show that

$$\frac{1}{2}(B^2 + i\coth t) = \frac{1}{2}\left(\frac{i}{A\sinh t}\right)^2 + \frac{i}{2}\frac{\cosh t}{\sinh t} \stackrel{(7.83)}{=} \frac{\sinh^2 t + (i/2b^2)\sinh 2t}{2A^2 \sinh^2 t} \quad (7.85)$$

and

$$A^2 \sinh^2 t \stackrel{(7.83)}{=} \frac{\sinh^2 t}{b^2} - \frac{i}{2}\sinh 2t,$$

so that (this defining $C(t)$ and the imaginary part φ on the right)

$$\frac{1}{2}(B^2 + i\coth t) = \frac{\sinh^2 t + (i/2b^2)\sinh 2t}{2\left[(1/b^2)\sinh^2 t - (i/2)\sinh 2t\right]} \equiv -\frac{1}{2C^2(t)} + i\varphi. \quad (7.86)$$

Evaluating the Gaussian integral of Eq. (7.84) we obtain[‡]

$$\psi(x,t) = \frac{\sqrt{2\pi}}{A\sqrt{2\pi i \sinh t}\sqrt{\pi^{1/2} b}} e^{-x^2/2C^2 + i\varphi x^2}.$$

In the further calculations it is convenient to set

$$\tan 2\theta = b^2 \coth t \quad \text{and} \quad \sqrt{A + iB} := Re^{i\theta}. \quad (7.87)$$

Then

$$R^4 = A^2 + B^2 = \frac{\sinh^2 t}{b^4} + \cosh^2 t. \quad (7.88)$$

[‡] $\int_{-\infty}^{\infty} dx e^{-x^2/2\alpha^2} = \sqrt{2\pi}\alpha.$

The real part of Eq. (7.86) has

$$C^2(t) := b^2 \cosh^2 t + \frac{1}{b^2} \sinh^2 t. \tag{7.89}$$

In taking the modulus of $\psi(x, t)$ the phase φ drops out and one has

$$|\psi(x, t)|^2 = \frac{\exp[-x^2/C^2(t)]}{\pi^{1/2}C(t)} \tag{7.90}$$

with the expected limiting behaviour

$$\lim_{t \to 0} |\psi(x, t)|^2 = \frac{\exp[-x^2/b^2]}{\pi^{1/2}b}.$$

In the following we choose $b = 1$.

The probability for the particle to be at time t still within the distance l away from the origin is plausibly given by

$$Q(l, t) = \int_{-l}^{l} dx |\psi(x, t)|^2 = \frac{2}{\sqrt{\pi}} \int_0^{\xi_0(t)} d\xi \, e^{-\xi^2}, \quad \xi = \frac{x}{b}, \tag{7.91}$$

where

$$\xi_0(t) = \frac{l}{C(t)}. \tag{7.92}$$

The *sojourn time* T is now defined as the mean time T given by

$$T := \int_0^\infty dt \, Q(l, t) \equiv \int_0^\infty t \left(-\frac{\partial Q}{\partial t} dt \right), \tag{7.93}$$

where $-(\partial Q/\partial t)\delta t$ is the probability that the particle leaves the vicinity of the origin in the interval δt around t (partial integration leads from the second expression back to the first). We have therefore

$$T = \int_{t=0}^\infty dt \frac{2}{\sqrt{\pi}} \int_0^{\xi_0(t)=l/C(t)} d\xi \, e^{-\xi^2}, \tag{7.94}$$

or, since

$$\frac{d}{dy} \int_0^{g(y)} dx f(x) = g'(y) f(g(y)),$$

the derivative

$$\frac{dT}{dl} = \int_{t=0}^\infty dt \frac{2}{\sqrt{\pi}} \frac{1}{C(t)} \exp\left[-\frac{l^2}{C^2(t)} \right]. \tag{7.95}$$

For $b = 1$ we have

$$C^2(t) = 1 + 2\sinh^2 t.$$

We set

$$\eta := \frac{l}{C(t)}, \quad \frac{d\eta}{dt} = -\eta\frac{C'(t)}{C(t)}, \tag{7.96}$$

so that

$$\frac{dT}{dl} = \frac{1}{l}\int_{t=0}^{\infty}\frac{2}{\sqrt{\pi}}dt\eta e^{-\eta^2} = \frac{1}{l}\int_{\eta=0}^{l}\frac{2}{\sqrt{\pi}}d\eta\frac{C(t)}{C'(t)}e^{-\eta^2}. \tag{7.97}$$

Now,

$$C(t) = \frac{l}{\eta} = \sqrt{1 + 2\sinh^2 t}, \quad \sinh^2 t = \frac{1}{2}\left(\frac{l^2}{\eta^2} - 1\right)$$

and $(\cosh t = \sqrt{1 + \sinh^2 t})$

$$C'(t) = \frac{2\sinh t\sqrt{1 + \sinh^2 t}}{C(t)} = \frac{\sqrt{2}\eta}{l}\sqrt{\frac{l^2}{\eta^2} - 1}\sqrt{1 + \frac{1}{2}\left(\frac{l^2}{\eta^2} - 1\right)}.$$

From this we obtain

$$\frac{C(t)}{C'(t)} = \frac{l^2}{\sqrt{(\eta^2 - l^2)(l^2 - \eta^2)}}. \tag{7.98}$$

We insert this into Eq. (7.97) and obtain

$$\frac{dT}{dl} = \frac{2l}{\sqrt{\pi}}\int_{\eta=0}^{l}\frac{d\eta}{\sqrt{(l^2 - \eta^2)(l^2 + \eta^2)}}e^{-\eta^2},$$

or with $\xi = \eta/l$:

$$\frac{dT}{dl} = \frac{2}{\sqrt{\pi}}\int_0^1\frac{d\xi e^{-l^2\xi^2}}{\sqrt{(1 - \xi^2)(1 + \xi^2)}} \simeq \frac{2}{\sqrt{\pi}}\int_0^1 d\xi e^{-l^2\xi^2}$$

$$= \frac{2}{\sqrt{\pi}}\frac{1}{l}\int_0^l dx e^{-x^2} = \frac{2}{\sqrt{\pi}}\frac{1}{l}\int_0^{x'=\sqrt{2}l}\frac{dx'}{\sqrt{2}}e^{-\frac{x'^2}{2}}.$$

Using the following expansion of the integral*

$$\frac{1}{\sqrt{2\pi}}\int_{-x}^{x}e^{-t^2/2}dt \approx 1 - \sqrt{\frac{2}{\pi}}\frac{e^{-x^2/2}}{x}\left[1 + O\left(\frac{1}{x^2}\right)\right],$$

we obtain

$$\frac{dT}{dl} \simeq \frac{1}{l} + O(e^{-l^2}) \tag{7.99}$$

---

*See for instance H. B. Dwight [81], p.136.

for l sufficiently large. It follows that

$$T \simeq \ln l. \tag{7.100}$$

Re-introducing the dimensional parameters we had set equal to 1, this is

$$T \simeq \frac{1}{\omega} \ln(l\sqrt{m_0\omega/\hbar}). \tag{7.101}$$

A more detailed evaluation of the constant can be found in the paper of Barton cited above. The following Example 7.4, also motivated by this paper, is instructive in revealing the basic quantum mechanics involved in the finiteness of the sojourn time.

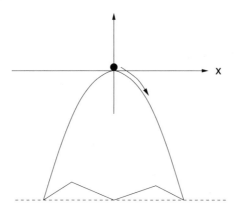

Fig. 7.3 How stable is the particle on top of the eggshell in quantum mechanics?

Example 7.4: The sojourn time calculated semiclassically

A tiny ball of mass m_0 is placed at the apex of an upright egg. Calculate with the help of a semiclassical consideration the (quantum mechanically limited) maximal length of time which passes before the ball is observed to roll down. Explain the parameters entering the calculation.

Solution: In the case of the inverted oscillator (representing the egg) we have in the usual notation the classical Hamiltonian

$$H = \frac{p^2}{2m_0} - \frac{1}{2}m_0\omega^2 x^2$$

with time derivatives

$$\dot{x} = \frac{\partial H}{\partial p} = \frac{p}{m_0}, \quad m_0\ddot{x} = \dot{p} = -\frac{\partial H}{\partial x} = m_0\omega^2 x.$$

The classical equation of motion is therefore[†]

$$\ddot{x} - \omega^2 x = 0 \tag{7.102}$$

[†]As a matter of interest we add that in the theory of a free scalar field x, this equation is the equation of motion of this scalar field with negative mass-squared. Such a field arises in the spectrum of states in string field theories and is there called a *"tachyon"*. The equation there describes the classical *"rolling down"* of this tachyon. See e.g. B. Zwiebach [294], p. 238.

with $\dot{x}^2 = \omega^2 x^2 + \text{const.}$ and solution

$$x = A \cosh \omega t + B \sinh \omega t, \quad \dot{x} = A\omega \sinh \omega t + B\omega \cosh \omega t.$$

We ask: At what time $T \gg 0$ does the ball reach the point with horizontal coordinate $x = l$. Thus we set

$$l = A \cosh \omega T + B \sinh \omega T \simeq \frac{1}{2}(A + B)e^{\omega T}.$$

Let A and $B\omega$ be the values of x and $\dot{x}$ at time $t = 0$, i.e.

$$x(0) = A, \quad \dot{x}(0) = B\omega,$$

so that

$$l = \frac{1}{2}\left[x(0) + \frac{\dot{x}(0)}{\omega}\right]e^{\omega T} = \frac{1}{2}\left[x(0) + \frac{p(0)}{m_0\omega}\right]e^{\omega T}.$$

We assume now that at time $t = 0$ the ball is placed at the point $x(0)$ with momentum $p(0)$. Quantum mechanically $x(0), p(0)$ have to be replaced by the positions of the spatial and momentum maxima of a wave packet. Thus quantum mechanically $x(0), p(0)$ cannot be determined with arbitrary precision, but instead are subject to an uncertainty relation of the form

$$\triangle x \triangle p \geq \frac{1}{2}\hbar,$$

where $\triangle x, \triangle p$ are defined by the *mean square deviations* given by (see Eq. (10.5))

$$\triangle x = \sqrt{\langle x^2 \rangle - \langle x \rangle^2}, \quad \triangle p = \sqrt{\langle p^2 \rangle - \langle p \rangle^2}.$$

For a symmetric state like the usual ground state of the harmonic oscillator we have (since the wave function is an even function, so that e.g. the expectation value of x is the integral with an odd integrand)

$$\langle x \rangle = 0 = \langle p \rangle,$$

so that

$$\triangle x \triangle p = \sqrt{\langle x^2 \rangle \langle p^2 \rangle}.$$

In our semiclassical consideration we set therefore

$$x(0) = \sqrt{\langle x^2 \rangle}, \quad p(0) = \sqrt{\langle p^2 \rangle},$$

and for simplicity we take (cf. Eq. (7.102))

$$\langle p^2 \rangle := m_0^2 \omega^2 \langle x^2 \rangle,$$

so that

$$x(0) + \frac{p(0)}{m_0\omega} = 2\sqrt{\langle x^2 \rangle}.$$

It follows that

$$l = \sqrt{\langle x^2 \rangle}e^{\omega T}.$$

For a minimal uncertainty (hence the factor of 2 in the following) we then have

$$\frac{1}{2}\hbar = \triangle x \triangle p = \sqrt{\langle x^2 \rangle \langle p^2 \rangle} = m_0\omega\langle x^2 \rangle, \quad \text{i.e.} \quad \langle x^2 \rangle = \frac{\hbar}{2m_0\omega}.$$

Hence

$$l = \sqrt{\frac{\hbar}{2m_0\omega}}e^{\omega T},$$

and so

$$T = \frac{1}{\omega}\ln\left(\sqrt{\frac{2m_0\omega}{\hbar}}l\right). \tag{7.103}$$

Here l is a largely arbitrary but macroscopic length like the length of the power of resolution of a microscope — so to speak, the smallest macroscopic length. For $\hbar \to 0$ the time $T \to \infty$ in agreement with our classical expectation.

Chapter 8

Time-Independent Perturbation Theory

8.1 Introductory Remarks

The Schrödinger equation can be solved exactly (i.e. in closed form) only for very few potentials. It follows that in general one depends on some approximation procedure which is usually described as a perturbation method. In the case of the harmonic oscillator potential αx^2 the Schrödinger equation can be solved exactly with ease, but this is no longer the case for an anharmonic oscillator potential like $\alpha x^2 + \beta x^4$. The perturbation method generally described in textbooks — and frequently called *Rayleigh–Schrödinger perturbation theory* — consists in assuming power series expansions for the wave function Ψ and the eigenvalue E in terms of a parameter like β which is assumed to be small, i.e. one would set

$$\Psi = \psi^{(0)} + \beta \psi^{(1)} + \beta^2 \psi^{(2)} + \cdots ,$$
$$E = E^{(0)} + \beta E^{(1)} + \beta^2 E^{(2)} + \cdots$$

(β can also be thought of as a kind of "book-keeping" parameter in retaining corresponding powers of some kind of expansion). Frequently even the calculation of the next to leading contributions $\psi^{(1)}, E^{(1)}$ is already a bigg problem. For mathematical purists the question of convergence of the series is an even bigger challenge. In general perturbation series do not converge. An example permitting convergent perturbation series is provided by the trigonometric potential $\cos 2x$ with one-dimensional Schrödinger equation given by

$$\psi'' + [E - 2h^2 \cos 2x]\psi = 0.$$

In the case of this equation, which is known as the *Mathieu equation*, the expansions in ascending powers of the parameter h^2 can indeed be shown to have a definite radius of convergence,[*] i.e. the expansions

$$\psi = \psi^{(0)} + h^2\psi^{(1)} + h^4\psi^{(2)} + \cdots ,$$
$$E = E^{(0)} + h^2 E^{(1)} + h^4 E^{(2)} + \cdots .$$

However, the expansions in descending powers of h^2, e.g.

$$E = a_{-2}h^2 + a_{-1}h + a_0 + \frac{a_1}{h} + \frac{a_2}{h^2} + \cdots ,$$

are in the strict mathematical sense divergent. Actually expansions of this type which are ubiquitous in physics are so-called *asymptotic series* which were originally also described as *semi-convergent series* in view of the decreasing behaviour of their first few terms. In the following we explain the difference between convergent and asymptotic series, explore the latter in somewhat more detail and finally consider methods for deriving perturbation solutions of the Schrödinger equation. It will be seen later (e.g. at the end of Chapter 26) — and by comparing the results of WKB, perturbation theory and path integral methods — that the light-minded way in which perturbation theory is sometimes discarded is not justified.

8.2   Asymptotic Series versus Convergent Series

Before we actually define asymptotic series we illustrate some of their characteristic properties by considering specific examples which demonstrate also how they differ from convergent series. The series

$$e^{-x} = 1 - \frac{x}{1!} + \frac{x^2}{2!} - \cdots = \sum_{n=0}^{\infty} \frac{(-x)^n}{n!} \tag{8.1}$$

of the exponential function is well known to converge absolutely for all real and complex values of x. This convergence can be shown with D'Alembert's ratio test since[†]

$$\lim_{n\to\infty} \left| \frac{x}{n} \right| = 0 < 1. \tag{8.2}$$

It is interesting to compare the behaviour of the series (8.1) and its terms with the behaviour of the following series and its terms:

$$f(x) = 1 - \frac{1!}{x} + \frac{2!}{x^2} - \cdots + (-1)^n \frac{n!}{x^n} + R_n(x), \tag{8.3}$$

---

[*]See e.g. J. Meixner and F. W. Schäfke [193].

[†]For further details, concerning also the uniform convergence of the exponential series, see e.g. E. T. Whittaker and G. N. Watson [283], p. 581.

where $R_n(x)$ is the remainder sum. We observe in the first place that since

$$\lim_{n \to \infty} \left| \frac{n}{x} \right| = \infty \qquad (8.4)$$

for every value of x, the series (8.3) diverges for every value of x. However, in spite of this the series (8.3) can still be used to obtain almost correct values of $f(x)$ for x sufficiently large. We can see this as follows. Normally a divergent series is characterized by an ever increasing behaviour of its terms as in the case of $x = 1$, i.e.

$$f(1) = 1 - 1! + 2! - 3! + \cdots .$$

However, for larger values of x, e.g. for $x = 3$, we have

$$
\begin{aligned}
f(3) &= 1 - \frac{1}{3} + \frac{2}{9} - \frac{6}{27} + \frac{24}{81} - \frac{120}{243} + \frac{720}{729} - \frac{5040}{2187} + \cdots , \\
&= 1 - 0.33333\ldots + 0.22222\ldots - 0.22222\ldots \\
&\quad + 0.29629\ldots - 0.4938\ldots + 0.9876\ldots - 2.376\ldots + \cdots ,
\end{aligned}
$$

and we see that the moduli of successive terms first decrease and then, after reaching $6/27 = 0.222$, begin to increase again, and in fact increase indefinitely. The theory of asymptotic expansions claims that if the expansion is truncated at the least term, the partial sum of terms up to and including the least term yields a reasonably precise value of the function at that point with an error of the order of the first term of the remainder. It is evident that the larger the value of $|x|$, the better the approximation obtained. Thus e.g.

$$f(1000) = 1 - 0.001 + 0.000\,002\,00 - 0.000\,000\,06 + \cdots .$$

Comparing the asymptotic series (8.3) with the convergent series (8.1), we observe that the individual terms of the asymptotic series have the form of a factorial divided by the power of some parameter which is large. This type of behaviour is characteristic of the terms of an asymptotic expansion as we shall see in more detail in Chapter 20.

As a second example we consider two series expansions of the *gamma function* or factorial $\Gamma(z) = (z-1)!$. From the Weierstrass product which defines this function[‡] one obtains the series

$$\ln(z-1)! = -\gamma(z-1) + \sum_{n=1}^{\infty} \left[\frac{z-1}{n} - \ln\left(1 + \frac{z-1}{n}\right) \right], \qquad (8.5)$$

[‡]Cf. E. T. Whittaker and G. N. Watson [283], p. 235.

where $\gamma = 0.5772157\ldots$. Here the series on the right is an absolutely and uniformly convergent series of the analytic function.[§] However, for $\ln(z-1)!$ one can also obtain the *Stirling series*

$$\ln(z-1)! = \ln(2\pi)^{\frac{1}{2}} - z + \left(z - \frac{1}{2}\right)\ln z$$
$$+ \ln\left(1 + \frac{1}{12z} + \frac{1}{288z^2} - \frac{139}{51840z^3} + \cdots\right). \qquad (8.6)$$

Here the series on the right is an asymptotic series which is particularly useful for large values of z but is readily checked to yield very good approximations for values as small as 2 or so. Applications of Stirling's series can be found in all areas of the physical sciences (particularly in statistical problems), whereas applications of the convergent series (8.5) are practically unknown. This latter observation hints already at the importance of asymptotic series in applications.

It is inherent in the nature of an approximation in a physical problem that in deciding between dominant or primary effects and those of secondary importance, the dominant approximation is, in fact, the leading term of an asymptotic expansion. Thus one can say that the vast majority of expansions of this type in physics is asymptotic and not convergent. This is not appreciated by mathematical purists. It is interesting to observe the comments of various authors on this point. Thus Merzbacher[¶] says: "*Simple perturbation theory applies when these eigenvalues and eigenfunctions can be expanded in powers of ϵ (at least in the sense of an asymptotic expansion) in the hope that for practical calculations only the first few terms of the expansions need be considered.*" This is a very clear statement which does not try to pretend that a perturbation expansion would have to be convergent. On the other hand Schiff[‖] says: "*We assume that these two series (for Ψ and E) are analytic for ϵ between zero and one, although this has not been investigated except for a few simple problems*". It seems that Schiff tries to cling to the idea that a proper series has to be analytic, meaning convergent. In many respects Messiah aims at more rigour in his arguments. It is therefore not surprising that he[**] says: "*If the perturbation ϵV is sufficiently small, it is sensible to make the assumption, that E and $|\psi\rangle$ (i.e. ψ) can be represented by rapidly converging power series in ϵ.*" These "rapidly converging power series" in physical contexts are most likely very rare cases. In fact, at a later point in his treatise Messiah admits that the expansions are mostly

---

[§]Cf. E. T. Whittaker and G. N. Watson [283], p. 236.
[¶]E. Merzbacher [194], p. 371.
[‖]L. Schiff [243], p. 152.
[**]A. Messiah [195], Vol. I, Sec. 16.1.1 between Eqs. (16.41) and (16.5).

asymptotic. He says,[††] somewhat afraid to say so himself and therefore with reference to investigations of T. Kato: "*Indeed the perturbation expansion is in most cases an asymptotic expansion ....*" Of these three authors the first, Merzbacher, seems to be closest to the truth and does not attempt to give the impression that the series has to converge. In fact, there is no need for such bias. We know from books on Special Functions that all of these functions, solutions of second order differential equations, which have been studied in great detail, possess asymptotic expansions which have for a long time been important and accepted standard results of mathematics.

8.2.1 The error function and Stokes discontinuities

Another extremely instructive example which illustrates the nature — and in addition the origin — of asymptotic expansions is the *error function* $\phi(x)$ defined by the integral

$$\phi(x) = \frac{2}{\sqrt{\pi}} \int_0^x e^{-t^2} \, dt. \tag{8.7}$$

This function has been considered in detail in the book of Dingle,[*] and we follow some of the considerations given there. Replacing the exponential in Eq. (8.7) by its power series and integrating term by term one obtains the absolutely convergent series

$$\phi(x) = \frac{2x}{\sqrt{\pi}} \sum_{n=0}^{\infty} \frac{(-x^2)^n}{n!(2n+1)}. \tag{8.8}$$

Considering for the time being only real and positive values of x, we can rewrite Eq. (8.7) as

$$\phi(x) = \frac{2}{\sqrt{\pi}} \left[\int_0^\infty e^{-t^2} \, dt - \int_x^\infty e^{-t^2} \, dt \right] = 1 - \frac{2}{\sqrt{\pi}} \int_x^\infty e^{-t^2} \, dt. \tag{8.9}$$

The integral is expected to be dominated by the behaviour of the exponential at the lower limit. It is therefore suggestive to write

$$\phi(x) = 1 - \frac{2}{\sqrt{\pi}} e^{-x^2} \int_x^\infty e^{-(t^2-x^2)} \, dt. \tag{8.10}$$

Changing the variable of integration to

$$u = t^2 - x^2, \quad du = 2t \, dt, \tag{8.11}$$

[††]A. Messiah [195], Vol. II, p. 198 (German edition).
[*]R. B. Dingle [70].

we obtain

$$\phi(x) = 1 - \frac{2}{\sqrt{\pi}} e^{-x^2} \int_0^\infty e^{-u} \frac{du}{2(u+x^2)^{\frac{1}{2}}},$$

i.e.

$$\phi(x) = 1 - \frac{e^{-x^2}}{x\sqrt{\pi}} \int_0^\infty e^{-u} \left(1 + \frac{u}{x^2}\right)^{-\frac{1}{2}} du. \tag{8.12}$$

The binomial expansion

$$\left(1 + \frac{u}{x^2}\right)^{-\frac{1}{2}} = \sum_{n=0}^\infty \binom{-\frac{1}{2}}{n} \left(\frac{u}{x^2}\right)^n = \sum_{n=0}^\infty \frac{(-\frac{1}{2})!}{n!(-\frac{1}{2}-n)!} \left(\frac{u}{x^2}\right)^n \tag{8.13}$$

presupposes that

$$\left|\frac{u}{x^2}\right| < 1.$$

Using the *reflection formula*

$$(z-1)!(-z)! = \frac{\pi}{\sin \pi z} \tag{8.14}$$

for $z = n + \frac{1}{2}$, we have

$$\left(-\frac{1}{2} - n\right)! = \frac{\pi}{(n-\frac{1}{2})! \sin\{\pi(n+\frac{1}{2})\}} = \frac{\pi}{(n-\frac{1}{2})!(-1)^n}.$$

Then, since $(-\frac{1}{2})! = \sqrt{\pi}$, we obtain

$$\left(1 + \frac{u}{x^2}\right)^{-\frac{1}{2}} = \sum_{n=0}^\infty \frac{\sqrt{\pi}(n-\frac{1}{2})!(-1)^n}{n!\pi} \left(\frac{u}{x^2}\right)^n,$$

i.e.

$$\left(1 + \frac{u}{x^2}\right)^{-\frac{1}{2}} = \frac{1}{\sqrt{\pi}} \sum_{n=0}^\infty \frac{(n-\frac{1}{2})!}{n!} \left(-\frac{u}{x^2}\right)^n. \tag{8.15}$$

For the domain $|u/x^2| < 1$ we can insert this expansion into Eq. (8.12). Then

$$\begin{aligned}
\phi(x) &= 1 - \frac{e^{-x^2}}{x\sqrt{\pi}} \left[\int_0^{x^2} \frac{e^{-u}}{\sqrt{\pi}} \sum_{n=0}^\infty \frac{(n-\frac{1}{2})!}{n!} \left(-\frac{u}{x^2}\right)^n du \right. \\
&\qquad\qquad \left. + \int_{x^2}^\infty e^{-u} \left(1 + \frac{u}{x^2}\right)^{-\frac{1}{2}} du\right] \\
&= 1 - \frac{e^{-x^2}}{x\sqrt{\pi}} \left[\int_0^\infty \frac{e^{-u}}{\sqrt{\pi}} \sum_{n=0}^\infty \frac{(n-\frac{1}{2})!}{n!} \left(-\frac{u}{x^2}\right)^n du + \int_{x^2}^\infty e^{-u} du \right. \\
&\qquad\qquad \left. \left\{\left(1 + \frac{u}{x^2}\right)^{-\frac{1}{2}} - \frac{1}{\sqrt{\pi}} \sum_{n=0}^\infty \frac{(n-\frac{1}{2})!}{n!} \left(-\frac{u}{x^2}\right)^n\right\}\right]. \tag{8.16}
\end{aligned}$$

We can evaluate the first integral with the help of the integral representation of the *gamma function*, i.e.

$$\Gamma(n) = (n-1)! = \int_0^\infty e^{-t}t^{n-1}dt. \tag{8.17}$$

Then

$$\begin{aligned}
\phi(x) &= 1 - \frac{e^{-x^2}}{x\pi}\sum_{n=0}^\infty \frac{(n-\frac{1}{2})!}{(-x^2)^n} - \frac{e^{-x^2}}{x\sqrt{\pi}}\int_{x^2}^\infty due^{-u} \\
&\quad \times \left[\left(1+\frac{u}{x^2}\right)^{-\frac{1}{2}} - \frac{1}{\sqrt{\pi}}\sum_{n=0}^\infty \frac{(n-\frac{1}{2})!}{n!}\left(-\frac{u}{x^2}\right)^n\right]
\end{aligned} \tag{8.18}$$

and we write

$$\phi(x) \simeq 1 - \frac{e^{-x^2}}{x\pi}\sum_{n=0}^\infty \frac{(n-\frac{1}{2})!}{(-x^2)^n}, \quad |\arg x| < \frac{1}{2}\pi, \tag{8.19}$$

where "$\simeq$" means *asymptotically equal to* which in turn means that the right hand side of Eq. (8.19) approximates $\phi(x)$ the better the larger x is. Frequently "$=$" is written instead of "$\simeq$".

We now see how the asymptotic expansion (8.19) originates. The expansion (8.19) implies either that we ignore the remainder or the correction term given by the integral in Eq. (8.18) or that we insert the binomial expansion (8.15) into (8.12) ignoring the fact that the latter is valid (i.e. convergent) only in the restricted domain $u < x^2$. Whichever way we look at the result, the fact that we effectively use a binomial expansion beyond its circle of convergence implies that the resulting series is divergent so that even if the first few terms decrease in magnitude the later terms will increase eventually as a reflection of this procedure.

As in the case of the gamma function, the asymptotic expansion of the error function is, in general, much more useful than the convergent expansion. We can actually understand the deeper reason for this in practice. In an asymptotic expansion the first few terms successively decrease in magnitude, whereas in a convergent expansion the terms first increase in magnitude, and in fact so slowly, that a large number of terms has to be summed in order to obtain a reasonable approximation of the quantity concerned.

In the foregoing discussion of the error function we assumed that x was real and positive, i.e. $\arg x = 0$. We now want to relax this condition and allow for phases $\arg x \neq 0$. It is fairly clear that the above considerations remain unaffected for

$$|\arg x| < \frac{1}{2}\pi,$$

since for this range of phases the decreasing nature of the exponential is
maintained. The situation becomes critical, however, when $|\arg x| = \pi/2$.
We therefore consider this case in detail, i.e. we set $x = iy$ in Eq. (8.7) and
obtain

$$\phi(iy) = \frac{2}{\sqrt{\pi}} \int_0^{iy} e^{-t^2} dt = \frac{2i}{\sqrt{\pi}} \int_0^y e^{s^2} ds, \qquad (8.20)$$

where we set $t = is$. Proceeding along lines similar to those above we write

$$\phi(iy) = \frac{2i}{\sqrt{\pi}} e^{y^2} \int_0^y e^{(s^2 - y^2)} ds. \qquad (8.21)$$

Changing the variable of integration to

$$v = y^2 - s^2, \qquad dv = -2s\,ds = -2\sqrt{y^2 - v}\,ds, \qquad (8.22)$$

we obtain

$$\phi(iy) = \frac{i}{\sqrt{\pi}} e^{y^2} \int_0^{y^2} e^{-v} \frac{dv}{\sqrt{y^2 - v}}. \qquad (8.23)$$

Throughout the range of integration $0 < v < y^2$ the integrand is real. We
can therefore rewrite the integral as

$$\phi(iy) = \frac{i}{y\sqrt{\pi}} e^{y^2} \int_0^{y^2} e^{-v} \frac{1}{\sqrt{1 - (v/y^2)}} dv. \qquad (8.24)$$

We can read off the binomial expansion of the factor in the integrand from
Eq. (8.15); thus

$$\frac{1}{\sqrt{1 - (v/y^2)}} = \frac{1}{\sqrt{\pi}} \sum_{n=0}^{\infty} \frac{(n - \frac{1}{2})!}{n!} \left(\frac{v}{y^2}\right)^n \qquad (8.25)$$

with $|v/y^2| < 1$, and we can write $\phi(iy)$:

$$\phi(iy) = \frac{ie^{y^2}}{y\pi} \int_0^{y^2} e^{-v} \sum_{n=0}^{\infty} \frac{(n - \frac{1}{2})!}{n!} \left(\frac{v}{y^2}\right)^n dv. \qquad (8.26)$$

If we want to proceed with the evaluation of the integral in order to arrive at
an expansion of $\phi(iy)$, we can try to proceed with Eq. (8.26) or else return

to Eq. (8.24). If we proceed with Eq. (8.26) we can write[†]

$$
\phi(iy) = \frac{ie^{y^2}}{y\pi} \int_0^\infty e^{-v} \sum_{n=0}^\infty \frac{(n-\frac{1}{2})!}{n!} \left(\frac{v}{y^2}\right)^n dv
$$

$$
- \frac{ie^{y^2}}{y\pi} \int_{y^2}^\infty e^{-v} \sum_{n=0}^\infty \frac{(n-\frac{1}{2})!}{n!} \left(\frac{v}{y^2}\right)^n dv
$$

$$
= \frac{ie^{y^2}}{y\pi} \sum_{n=0}^\infty \frac{(n-\frac{1}{2})!}{y^{2n}} - \frac{ie^{y^2}}{y\pi} \int_{y^2}^\infty e^{-v} \sum_{n=0}^\infty \frac{(n-\frac{1}{2})!}{n!} \left(\frac{v}{y^2}\right)^n dv.
$$

$$(8.27)$$

Then

$$
\phi(iy) \simeq \frac{ie^{y^2}}{y\pi} \sum_{n=0}^\infty \frac{(n-\frac{1}{2})!}{y^{2n}}, \quad \arg(x=iy) = \frac{1}{2}\pi. \tag{8.28}
$$

We observe that this expansion differs significantly from (8.19); i.e. for $|\arg x| < \pi/2$ and for $\arg x = \pi/2$ the error function $\phi(x)$ possesses different asymptotic expansions, which will be studied in more detail below. Returning to Eqs. (8.27) and (8.28) we observe that the expansion (8.28) ignores the integral contribution of Eq. (8.27). These integral contributions are of order $1/y$ since the integral behaves like e^{-y^2} and is multiplied by a factor e^{+y^2} in front. However, for large values of y, e^{y^2} is much larger than something of order $1/y$, and the larger y, the better the approximation expressed by Eq. (8.28) because then the correction term $\propto \int_{y^2}^\infty$ becomes smaller and smaller.

Suppose now we consider Eq. (8.24). We know that for $v < y^2$ the integrand is real, and so the integral is real. We have therefore

$$
\phi(iy) = \frac{i}{y\sqrt{\pi}} e^{y^2} \mathcal{R} \int_0^{y^2} e^{-v} \left(1 - \frac{v}{y^2}\right)^{-\frac{1}{2}} dv
$$

$$
= \frac{i}{y\sqrt{\pi}} e^{y^2} \mathcal{R} \left[\int_0^\infty - \int_{y^2}^\infty\right] e^{-v} \left(1 - \frac{v}{y^2}\right)^{-\frac{1}{2}} dv, \tag{8.29}
$$

where $\mathcal{R}$ means "*real part*". For $v > y^2$ the second integral is seen to be imaginary and thus drops out in taking the real part; i.e. we obtain

$$
\phi(iy) = \frac{i}{y\sqrt{\pi}} e^{y^2} \mathcal{R} \int_0^\infty e^{-v} \left(1 - \frac{v}{y^2}\right)^{-\frac{1}{2}} dv. \tag{8.30}
$$

[†]Cf. Eq. (8.17): $\int_0^\infty e^{-v} v^n dv = n!$.

If we insert here the expansion (8.25) and integrate from 0 to ∞, we use the expansion outside its circle of convergence and hence obtain a divergent expansion. This result is identical with that obtained previously, since (8.30) can be written

$$
\begin{aligned}
\phi(iy) &= \frac{i}{y\sqrt{\pi}} e^{y^2} \int_0^{y^2} e^{-v} \left(1 - \frac{v}{y^2}\right)^{-\frac{1}{2}} dv \\
&= \frac{ie^{y^2}}{y\pi} \int_0^{y^2} e^{-v} \sum_{n=0}^{\infty} \frac{(n-\frac{1}{2})!}{n!} \left(\frac{v}{y^2}\right)^n dv,
\end{aligned}
\tag{8.31}
$$

which is (8.26). With a similar type of reasoning one can show that the same expansion, i.e. (8.28), is obtained for $\phi(x)$ with $\arg x = -\pi/2$. We have therefore found that

$$
\phi(x) \simeq 1 - \frac{e^{-x^2}}{x\pi} \sum_{n=0}^{\infty} \frac{(n-\frac{1}{2})!}{(-x^2)^n} \quad \text{for } |\arg x| < \frac{\pi}{2},
\tag{8.32}
$$

and

$$
\phi(x) \simeq -\frac{e^{-x^2}}{x\pi} \sum_{n=0}^{\infty} \frac{(n-\frac{1}{2})!}{(-x^2)^n} \quad \text{for } |\arg x| = \frac{\pi}{2}.
\tag{8.33}
$$

The sudden disappearance of "1" at $|\arg x| = \pi/2$ hints at something like a discontinuity of $\phi(x)$ at $\arg x = \pi/2$ which was discovered by Stokes and is therefore known as a *Stokes discontinuity*.[*] We observe that this Stokes discontinuity is a *property of the asymptotic expansion but not of the function itself*. For the sake of completeness of the above example we continue the phase to $|\arg x| > \pi/2$. Since (cf. Eq. (8.8))

$$
\phi(x) = -\phi(-x)
$$

we see that

$$
\phi(x) = -1 - \frac{e^{-x^2}}{x\pi} \sum_{n=0}^{\infty} \frac{(n-\frac{1}{2})!}{(-x^2)^n}
\tag{8.34}
$$

for $\pi/2 < |\arg x| < 3\pi/2$. Looking at Eqs. (8.31) and (8.33) we see that the Stokes discontinuities occur at those phases for which these expansions (i.e. (8.31), (8.33)) *would* have

- a *maximally* increasing exponential e^{+x^2}, and
- higher order terms of the associated series all have the same sign.

The corresponding phase lines are called *Stokes rays*.

---

[*]Cf. R. B. Dingle [70], Chapter 1.

It should be observed that the phase of the Stokes ray is the phase at which an exponential is not just increasing but maximally increasing. Thus if we set $x = x_R + ix_I$, we have

$$e^{-x^2/2} = e^{-(x_R^2 - x_I^2 + 2ix_R x_I)/2}.$$

The exponential is maximally increasing for $x_R = 0$, i.e. for arg $x = \pi/2$. We also observe that the expansion for $|\arg x| = \pi/2$ is half the sum of the expansions on either side of the Stokes ray, i.e. in the sum of the right hand sides of Eqs. (8.32) and (8.34) the contributions $1, -1$ cancel out, and

$$
\phi(x)\Big|_{\arg x = \pi/2} = \frac{1}{2}\left\{ \phi(x)\Big|_{\arg x < \pi/2} + \phi(x)\Big|_{\arg x > \pi/2} \right\}
$$
$$
= -\frac{e^{-x^2}}{x\pi} \sum_{n=0}^{\infty} \frac{(n - \frac{1}{2})!}{(-x^2)^n}. \tag{8.35}
$$

8.2.2   Stokes discontinuities of oscillator functions

Dingle[†] has formulated rules which permit one to continue an asymptotic series across a Stokes discontinuity without the necessity of a separate calculation of the asymptotic series in the new domain. These rules, though, apply predominantly to asymptotic series of the solutions of second order differential equations, and so may not be universally applicable. Later we shall make extensive use of *parabolic cylinder functions* $D_\nu(x)$ which are solutions of the equation

$$
\frac{d^2 y}{dx^2} + \left[\nu + \frac{1}{2} - \frac{1}{4}x^2 \right] y = 0 \tag{8.36}
$$

which is of the type of a Schrödinger equation for an *harmonic oscillator potential*.[‡] In view of the considerable importance of the harmonic oscillator and the associated parabolic cylinder functions in later chapters we consider this case now in more detail. Equation (8.36) has an exponentially decreasing solution and an exponentially increasing solution for zero phase of x, which can be written (for ν not restricted to integral values)

$$
\begin{aligned}
D_\nu^{(1)}(x) &= x^\nu \phi_\nu(x), \\
D_\nu^{(2)}(x) &= x^{-\nu-1}\psi_\nu(x) = x^{-\nu-1}\phi_{-\nu-1}(\pm ix),
\end{aligned} \tag{8.37}
$$

[†]Cf. R. B. Dingle [70], Chapter 1, pp. 9 - 13.
[‡]Cf. Eq. (6.2). See also Tables of Special Functions, e.g. W. Magnus and F. Oberhettinger [181], p. 91.

where[§]

$$\phi_\nu(x) \quad = \quad \frac{e^{-\frac{1}{2}x^2}}{(-\nu - 1)!} \sum_{i=0}^{\infty} \frac{(2i - \nu - 1)!}{i!(-2x^2)^i},$$

$$\psi_\nu(x) \quad = \quad \frac{e^{\frac{1}{2}x^2}}{\nu!} \sum_{i=0}^{\infty} \frac{(2i + \nu)!}{i!(2x^2)^i}. \qquad (8.38)$$

We observe that one solution follows from the other by making the replacements[¶]

$$x \to \pm ix, \quad \nu \to -\nu - 1.$$

The equation (8.36) is invariant under these replacements. We also observe that the late (large i) terms of the series in (8.38) are of the type

$$\sim \frac{(2i)!}{i!(-2x^2)^i} \sim \frac{i!}{(-2x^2)^i}$$

and so have the form of a factorial divided by a power which (as discussed previously) is the behaviour typical of asymptotic series.

We consider $D_\nu^{(1)}(x)$. At $\arg x = \pi/2$ the exponential factor becomes an increasing exponential and late terms of the series have the same sign. Thus $D_\nu^{(1)}(x)$ (i.e. the asymptotic series (8.37), (8.38)) has a Stokes ray at $\arg x = \pi/2$. Hence

$$D_\nu^{(1)}(x) = x^\nu \phi_\nu(x), \quad 0 \leq \arg x < \frac{\pi}{2} \qquad (8.39)$$

(as in the example of the error function). The continuation of (8.39) onto the Stokes ray and from there into the neighbouring domain is determined by the following rules which will not be established here:[||]

- *Dingle's rule* (1): On reaching $\arg x = \pi/2$, the series of $D_\nu^{(1)}(x)$ develops an additive contribution (the discontinuity) which is $\pi/2$ out of phase with $x^\nu \phi_\nu(x)$ and proportional to the associated function, i.e. $x^{-\nu-1}\psi_\nu(x)$, with a real proportionality factor, i.e.

$$D_\nu^{(1)}(x) \quad = \quad x^\nu \phi_\nu(x) + \frac{1}{2}\alpha e^{i\pi/2} e^{i\nu\pi/2} |x|^{-\nu-1}\psi_\nu(x)$$

$$= \quad x^\nu \phi_\nu(x) + \frac{1}{2}\alpha e^{i\pi(\nu+1)} x^{-\nu-1}\psi_\nu(x), \quad \arg x = \frac{\pi}{2}, \qquad (8.40)$$

[§]Cf. R. B. Dingle [70], Chapter 1, Eq. (14).

[¶]Such symmetries are extensively exploited in the perturbation method of Sec. 8.7, as can most easily be seen in the case of the Mathieu potential, Chapter 17.

[||]Cf. R. B. Dingle [70], Chapters I (in particular p. 9) and XXI.

for $\arg x = \pi/2$, where in the first line on the right hand side the second term contains the real proportionality constant $\alpha/2$, the extra phase factor $e^{i\pi/2}$ and the phase $e^{i\nu\pi/2}$ of $x^\nu \phi_\nu(x)$ (so that as the rule requires the added contribution is $\pi/2$ out of phase with the first). The value of $D_\nu^{(1)}(x)$ on the other side of the Stokes ray, i.e. beyond $\arg x = \pi/2$, is given by

● *Dingle's rule* (2): One half of the discontinuity appears on reaching the Stokes ray, another half of the discontinuity appears on leaving the Stokes ray on the other side. Thus for $\arg x > \pi/2$ we have

$$
\begin{aligned}
D_\nu^{(1)}(x) &= x^\nu \phi_\nu(x) + 2\frac{1}{2}\alpha e^{i\pi(\nu+1)}x^{-\nu-1}\psi_\nu(x) \\
&= x^\nu \phi_\nu(x) + \alpha(-x)^{-\nu-1}\psi_\nu(x), \quad \frac{\pi}{2} < \arg x < \pi. \quad (8.41)
\end{aligned}
$$

In (8.40), (8.41) the Stokes multiplier α is, as yet an unknown real constant. It is determined by continuing the asymptotic series to $\arg x = \pi$ and demanding that the result be real since $D_\nu^{(1)}(x)$ is real when x is real. We therefore proceed to continue (8.41) to $\arg x = \pi$. On reaching π the part containing $\psi_\nu(x)$, i.e.

$$
e^{+\frac{1}{2}x^2} = e^{\frac{1}{2}(x_R^2 - x_I^2 + 2ix_R x_I)},
$$

is maximally exponentially increasing (i.e. for $x_I = 0$) and therefore possesses a Stokes discontinuity there. Applying Dingle's rule (1) (to the Stokes discontinuity of $\psi_\nu(x)$) we obtain on the ray

$$
\begin{aligned}
D_\nu^{(1)}(x) &= x^\nu \phi_\nu(x) + \alpha\left\{(-x)^{-\nu-1}\psi_\nu(x) + \frac{1}{2}\beta e^{i\pi/2}e^{i\pi\nu}|x|^\nu \phi_\nu(x)\right\} \\
&= \left(e^{i\pi\nu} + \frac{1}{2}i\alpha\beta\right)(-x)^\nu \phi_\nu(x) + \alpha(-x)^{-\nu-1}\psi_\nu(x), \quad (8.42)
\end{aligned}
$$

with $|x| = (-x)$ for $\arg x = \pi$. Applying rule (2) we obtain

$$
D_\nu^{(1)}(x) = \left(e^{i\pi\nu} + \frac{2}{2}i\alpha\beta\right)(-x)^\nu \phi_\nu(x) + \alpha(-x)^{-\nu-1}\psi_\nu(x) \quad (8.43)
$$

for $\pi < \arg x < 3\pi/2$. Since $D_\nu^{(1)}(x)$ is real when x is real, we must have at $\arg x = \pi$ in the dominant factor on the right of Eq. (8.42)

$$
\Im\left(e^{i\pi\nu} + \frac{1}{2}i\alpha\beta\right) = 0
$$

i.e. $\sin(\pi\nu) = -\alpha\beta/2$ or

$$
\alpha\beta = -2\sin(\pi\nu). \quad (8.44)
$$

Now α and β can still be functions of ν, i.e.

$$\alpha(\nu)\beta(\nu) = -2\sin(\pi\nu) = -2\sin\pi(-\nu - 1) \tag{8.45}$$

Thus the right hand side of this equation remains unchanged if ν is replaced by $-\nu - 1$. Hence the left hand side must have the same property, i.e.

$$\alpha(\nu)\beta(\nu) = \alpha(-\nu - 1)\beta(-\nu - 1) \tag{8.46}$$

For a given function $\alpha(\nu)$ this equation determines $\beta(\nu)$. We can see that the equation is satisfied by

$$\beta(\nu) = \pm\alpha(-\nu - 1), \quad \beta(-\nu - 1) = \pm\alpha(\nu). \tag{8.47}$$

Equation (8.45) is therefore given by

$$\alpha(\nu)\alpha(-\nu - 1) = -2\sin\pi\nu. \tag{8.48}$$

We compare this with the *reflection formula* (8.14), i.e.

$$(z)!(-z - 1)! = \frac{\pi}{\sin\pi(z + 1)} = -\frac{\pi}{\sin\pi z},$$

or

$$\frac{\sqrt{2\pi}}{(z)!}\frac{\sqrt{2\pi}}{(-z - 1)!} = -2\sin\pi z. \tag{8.49}$$

Comparing this with Eq. (8.48) we can set

$$\alpha(\nu) = \frac{\sqrt{2\pi}}{(-\nu - 1)!}, \quad \beta(\nu) = \frac{\sqrt{2\pi}}{\nu!}, \tag{8.50a}$$

or

$$\alpha(\nu) = \frac{\sqrt{2\pi}}{\nu!}, \quad \beta(\nu) = \frac{\sqrt{2\pi}}{(-\nu - 1)!} \tag{8.50b}$$

In order to decide which case is relevant we observe from Eq. (8.42) that α multiplies $\psi_\nu(x)$, i.e. $D_\nu^{(1)}(x)$ is (cf. Eq. (8.42)) for $\arg x = \pi$ (in the second line of Eq. (8.42) we insert from Eq. (8.45) that $\alpha\beta = -2\sin(\pi\nu)$)

$$
\begin{aligned}
D_\nu^{(1)}(x) &= \cos\pi\nu(-x)^\nu\phi_\nu(x) + \alpha(\nu)(-x)^{-\nu-1}\psi_\nu(x) \\
&\overset{(8.38)}{=} \cos\pi\nu\frac{(-x)^\nu e^{-\frac{1}{2}x^2}}{(-\nu - 1)!}\sum_{i=0}^{\infty}\frac{(2i - \nu - 1)!}{i!(-2x^2)^i} \\
&\quad + \frac{\alpha(\nu)(-x)^{-\nu-1}e^{\frac{1}{2}x^2}}{\nu!}\sum_{i=0}^{\infty}\frac{(2i + \nu)!}{i!(2x^2)^i}, \quad \arg x = \pi. \tag{8.51}
\end{aligned}
$$

We choose (8.50a) because this enables us to define normalizable parabolic cylinder functions, i.e. solutions of Eq. (8.36), which vanish at $x = \pm\infty$. Then (in the second step using $(-\nu - 1)!\nu! = -\pi/\sin(\pi\nu)$)

$$
\begin{aligned}
D_\nu^{(1)}(x) &= \cos\pi\nu \frac{(-x)^\nu e^{-\frac{1}{2}x^2}}{(-\nu - 1)!} \sum_{i=0}^\infty \frac{(2i - \nu - 1)!}{i!(-2x^2)^i} \\
&+ \frac{\sqrt{2\pi}(-x)^{-\nu-1}e^{\frac{1}{2}x^2}}{(-\nu - 1)!\nu!} \sum_{i=0}^\infty \frac{(2i + \nu)!}{i!(2x^2)^i} \\
&= \cos\pi\nu(-x)^\nu e^{-\frac{1}{2}x^2} \sum_{i=0}^\infty \frac{(2i - \nu - 1)!}{(-\nu - 1)!i!} \frac{1}{(-2x^2)^i} \\
&- \sqrt{\frac{2}{\pi}} \sin\pi\nu(-x)^{-\nu-1}e^{\frac{1}{2}x^2} \sum_{i=0}^\infty \frac{(2i + \nu)!}{i!(2x^2)^i}
\end{aligned}
\tag{8.52}
$$

($\arg x = \pi$). We observe that for $\nu = n = 0, 1, 2, \ldots$ the second contribution vanishes and we obtain

$$
D_n^{(1)}(x) = x^n e^{-\frac{1}{2}x^2} \sum_{i=0}^\infty \frac{(2i - n - 1)!}{(-n - 1)!i!} \frac{1}{(-2x^2)^i} = (-1)^n D_n^{(1)}(-x) \tag{8.53}
$$

We have therefore obtained in a natural way the *quantization of the harmonic oscillator*. It should be noted that in Eq. (8.53) the factor $(-n - 1)!$ is to be understood in association with $(2i - n - 1)!$ in the numerator, i.e.

$$
\frac{(2i - n - 1)!}{(-n - 1)!} = \frac{n!}{(n - 2i)!}
$$

So far we have been considering the asymptotic series expansion of $D_\nu^{(1)}(x)$ (cf. (8.35), (8.36)) in the domain

$$
0 \le \arg x \le \pi.
$$

Proceeding along similar lines we can continue the expansion into the domain $\pi \le \arg x \le 2\pi$. In a similar way we can examine $D_\nu^{(2)}(x)$ and its Stokes discontinuities (at $\arg x = 0, \pi$). For details we refer again to the book of Dingle, Chapter I.

8.3   Asymptotic Series from Differential Equations

In the case of the *error function* we obtained the asymptotic expansion from the integral representation of the function, i.e. Eq. (8.7). However, the error

function can also be obtained as the solution of a second order differential
equation, i.e.

$$\frac{d^2\phi}{dx^2} + 2x\frac{d\phi}{dx} = 0 \tag{8.54}$$

with the boundary conditions

$$\lim_{x\to\pm\infty} \phi(x) = \pm 1. \tag{8.55}$$

The large x asymptotic expansion can also be obtained from the differential
equation. Instead of dealing with Eq (8.54) we consider again the important
equation of the harmonic oscillator, i.e. Eq. (8.36), and demonstrate, how
series of the type (8.37), (8.38) can be obtained. We write Eq. (8.36) in
the form (coefficient of $-x^2$ here chosen to be $\frac{1}{4}$; previously we had $\frac{1}{2}$, cf.
Eq. (8.36))

$$\frac{d^2y}{dx^2} + x^2 Q(x)y = 0, \tag{8.56}$$

where

$$Q(x) = -\frac{1}{2} + \frac{\nu + \frac{1}{2}}{x^2}. \tag{8.57}$$

We then set

$$y = e^{S(x)} \tag{8.58}$$

so that the equation becomes

$$\frac{d^2S}{dx^2} + \left(\frac{dS}{dx}\right)^2 + x^2 Q(x) = 0. \tag{8.59}$$

Since x^2 is the highest power of x appearing in $x^2Q(x)$ we set

$$S(x) = \frac{1}{2}S_0 x^2 + S_1 x + S_2 \ln x + \frac{S_3}{x} + \frac{S_4}{x^2} + \cdots. \tag{8.60}$$

Inserting (8.60) into (8.59) and equating to zero the coefficients of powers
of $x^2, x, x^0 \ldots$, we obtain equations from which the coefficients $S_0, S_1, S_2, \ldots$
can be determined. Thus

$$-\frac{1}{4} + S_0^2 = 0,$$
$$2S_0 S_1 = 0,$$
$$S_0 + 2S_0 S_2 + \nu + \frac{1}{2} = 0,$$

and so on. Solving these equations we obtain

$$S_0 = \pm\frac{1}{2}, \quad S_1 = 0, \quad S_2 = \begin{cases} -(\nu+1) \\ +\nu. \end{cases}$$

Then

$$y = e^{S(x)} = e^{\pm\frac{1}{4}x^2} x^{S_2} \left[1 + O\left(\frac{1}{x^2}\right)\right] \tag{8.61}$$

(the dominant terms of (8.35) and (8.36) have correspondingly $e^{\pm x^2/2}$), and it is clear that higher order terms follow accordingly.

The expansions (8.19), (8.37) with (8.38), etc. are asymptotic series in a variable x. In the following we are mostly concerned with asymptotic series in some parameter. Nonetheless we shall need asymptotic expansions like (8.37), (8.38) in a variable for the solutions of approximated differential equations. In particular we shall need the asymptotic expansions of *parabolic cylinder functions* in various domains. The parabolic cylinder function normally written $D_\nu(x)$ is frequently defined via the *Whittaker function* $W_{\kappa,\mu}(x)$ which is a solution of *Whittaker's equation*

$$\frac{d^2W}{dx^2} + \left[-\frac{1}{4} + \frac{\kappa}{x} + \frac{1-\mu^2}{x^2}\right]W = 0.$$

Thus[**]

$$D_\nu(x) \equiv 2^{\frac{1}{4}+\frac{\nu}{2}} x^{-\frac{1}{2}} W_{\frac{1}{4}+\frac{\nu}{2},-\frac{1}{4}}\left(\frac{x^2}{2}\right) = 2^{\frac{\nu}{2}} e^{-\frac{x^2}{4}} \left[\frac{(-\frac{1}{2})!}{(\frac{-\nu-1}{2})!}\right.$$

$$\times {}_1F_1\left(-\frac{\nu}{2},\frac{1}{2};\frac{x^2}{2}\right) + \frac{x}{\sqrt{2}} \frac{(-\frac{3}{2})!}{(-\frac{\nu}{2}-1)!} {}_1F_1\left(\frac{1-\nu}{2},\frac{3}{2};\frac{x^2}{2}\right)\right], \tag{8.62}$$

where ${}_1F_1(a,b;z)$ is a *confluent hypergeometric function*, i.e.

$${}_1F_1(a,b;z) = 1 + \frac{a}{b}\frac{z}{1!} + \frac{a(a+1)}{b(b+1)}\frac{z^2}{2!} + \cdots,$$

which has the unit circle as its circle of convergence. The derivation of the asymptotic series from integral representations (cf. Whittaker and Watson [283]) is now somewhat elaborate (although in principle the same as before). We, therefore, do not enter here into their derivation and simply quote the result: With $|x| \gg 1$, $|x| \gg |\nu|$ and

(a) for $|\arg x| < 3/4\pi$:

$$D_\nu(x) \simeq e^{-\frac{x^2}{4}} x^\nu \left[1 - \frac{\nu(\nu-1)}{2x^2} + \frac{\nu(\nu-1)(\nu-2)(\nu-3)}{2.4x^2} + \cdots\right], \tag{8.63}$$

[**]W. Magnus and F. Oberhettinger [181], p.91.

(b) for $5/4\pi > \arg x > \pi/4$:

$$
D_\nu(x) \simeq e^{-\frac{x^2}{4}} x^\nu \left[1 - \frac{\nu(\nu-1)}{2x^2} + \frac{\nu(\nu-1)(\nu-2)(\nu-3)}{2.4x^2} + \cdots \right]
$$
$$
- \frac{\sqrt{2\pi}}{(-\nu-1)!} e^{i\pi\nu} e^{\frac{x^2}{4}} x^{-\nu-1} \left[1 + \frac{(\nu+1)(\nu+2)}{2x^2} \right.
$$
$$
\left. + \frac{(\nu+1)(\nu+2)(\nu+3)(\nu+4)}{2.4x^2} + \cdots \right], \tag{8.64}
$$

(c) for $-\pi/4 > \arg x > -5\pi/4$:

$$
D_\nu(x) \simeq e^{-\frac{x^2}{4}} x^\nu \left[1 - \frac{\nu(\nu-1)}{2x^2} + \cdots \right] - \frac{\sqrt{2\pi}}{(-\nu-1)!} e^{-i\nu\pi}
$$
$$
\times e^{\frac{x^2}{4}} x^{-\nu-1} \left[1 + \frac{(\nu+1)(\nu+2)}{2x^2} + \cdots \right]. \tag{8.65}
$$

We observe that the series of (a) and (b) have the common domain of validity

$$
\frac{1}{4}\pi < \arg x < \frac{3}{4}\pi. \tag{8.66}
$$

In this domain $\Im x > |\Re x|$ and so

$$
e^{\frac{x^2}{4}} = e^{\frac{1}{4}(\Re x)^2 - \frac{1}{4}(\Im x)^2} e^{\frac{i}{2}(\Re x)(\Im x)}
$$

decreases exponentially, whereas $e^{-x^2/4}$ increases exponentially. Thus the series expansion (a) in this case is multiplied by an increasing exponential and the asymptotic equality "$\simeq$" means an exponentially decreasing contribution has been ignored. Thus there is no contradiction[††] between (a) and (b).

8.4   Formal Definition of Asymptotic Expansions

Finally we introduce the formal definition of an asymptotic expansion. If $S_n(x)$ is the sum of the first $n+1$ terms of the divergent series expansion

$$
a_0 + \frac{a_1}{x} + \frac{a_2}{x^2} + \cdots + \frac{a_n}{x^n} + \cdots
$$

of a function $f(x)$ for a given range of $\arg x$, then the series is said to be an asymptotic expansion of $f(x)$ in that range if for a fixed value of n

$$
\lim_{|x| \to \infty} x^n |f(x) - S_n(x)| = 0 \tag{8.67}
$$

[††]Cf. E. T. Whittaker and G. N. Watson [283], top of p. 349.

although

$$\lim_{n \to \infty} |x^n \left[f(x) - S_n(x) \right]| = \infty \tag{8.68}$$

The definition (8.68) is (effectively) simply a statement of the observations made at the beginning. The definition is due to Poincaré (1866). A critical discussion of the definition can be found in the book by Dingle [70], Chapter I, in particular with regard to possible exponentially small contributions to the expansion for which e.g.

$$\lim_{|x| \to \infty} x^n e^{-|x|} = 0 \tag{8.69}$$

and the consequent nonunique definition of the expansion. In fact, the divergent nature of asymptotic expansions together with a vagueness of definition (which is avoidable as explained by Dingle) frequently tempt mathematically prejudiced purists to turn away from asymptotic expansions, thereby (unknowingly) discarding the vast majority of expansions which have been used in applications to physics. It may be noted that Whittaker and Watson's internationally aclaimed text on *"Modern Analysis"*, first published in 1902, possesses a whole chapter on asymptotic expansions.

8.5 Rayleigh–Schrödinger Perturbation Theory

Rayleigh–Schrödinger perturbation theory is the usual perturbation theory which one can describe as textbook perturbation theory as distinguished from other less common methods. The equation to be solved is the equation

$$H\Psi = E\psi \quad \text{with} \quad H = H_0 + \epsilon V. \tag{8.70}$$

It is assumed that the spectrum $\{E_n^{(0)}\}$ and the eigenfunctions $\{\psi_n\}$ of H_0 are known with

$$H_0 \psi_n = E_n^{(0)} \psi_n \tag{8.71}$$

and the orthonormality

$$\delta_{mn} = \langle \psi_m | \psi_n \rangle = \int dx \psi_m^*(x) \psi_n(x). \tag{8.72}$$

For the eigenvalue E near E_n and the eigenfunction Ψ near ψ_n we assume power expansions in terms of the parameter ϵ, which is assumed to be small, i.e. we set

$$
\begin{aligned}
E \to E_n &= E_n^{(0)} + \epsilon E_n^{(1)} + \epsilon^2 E_n^{(2)} + \cdots, \\
\Psi \to \Psi_n &= \psi_n + \epsilon \psi_n^{(1)} + \epsilon^2 \psi_n^{(2)} + \cdots.
\end{aligned} \tag{8.73}
$$

We insert these expansions into Eq. (8.70) and equate on both sides the coefficients of the same powers of ϵ. We obtain then the following set of equations:

$$H\psi_n = E_n^{(0)}\psi_n, \tag{8.74}$$

$$H_0\psi_n^{(1)} + V\psi_n = E_n^{(0)}\psi_n^{(1)} + E_n^{(1)}\psi_n, \tag{8.75}$$

$$H_0\psi_n^{(2)} + V\psi_n^{(1)} = E_n^{(0)}\psi_n^{(2)} + E_n^{(1)}\psi_n^{(1)} + E_n^{(2)}\psi_n. \tag{8.76}$$

The first equation is that of the unperturbed problem. The second equation can be rewritten as

$$(H_0 - E_n^{(0)})\psi_n^{(1)} = (E_n^{(1)} - V)\psi_n. \tag{8.77}$$

The states $|\psi_n\rangle$ of the unperturbed problem span the Hilbert space $\mathcal{H}$. As a consequence the state $|\Psi\rangle$ is a vector in this space, as are also the states $|\psi_n^{(1)}\rangle, |\psi_n^{(2)}\rangle, \ldots$. We therefore write these vectors as superpositions of the basis vectors provided by the unperturbed problem, i.e. we write in the position space representation

$$\psi_n^{(1)} = \sum_{i \neq n} a_i^{(1)}\psi_i. \tag{8.78}$$

A contribution to ψ_n can be combined with the unperturbed part of the wave function Ψ, so that we choose

$$\int dx\psi_n^*\psi_n^{(1)} = 0, \quad \text{i.e.} \quad (\psi_n, \psi_n^{(1)}) = 0.$$

We insert the ansatz (8.78) into Eq. (8.77) and obtain

$$(H_0 - E_n^{(0)})\sum_{i \neq n} a_i^{(1)}\psi_i = (E_n^{(1)} - V)\psi_n. \tag{8.79}$$

We multiply this equation from the left by ψ_n^* and integrate over x. With the help of the orthonormality condition (8.72) we obtain

$$\int_{-\infty}^{\infty} dx\psi_n^*(H_0 - E_n^{(0)})\sum_{i \neq n} a_i^{(1)}\psi_i = \int_{-\infty}^{\infty} dx\psi_n^*(E_n^{(1)} - V)\psi_n,$$

i.e.

$$\int_{-\infty}^{\infty} dx\psi_n^* \sum_i (E_i^{(0)} - E_n^{(0)})\psi_i = \int dx\psi_n^*(E_n^{(1)} - V)\psi_n,$$

or

$$0 = E_n^{(1)} - \int dx(\psi_n^* V\psi_n).$$

and so

$$E_n^{(1)} = \langle n|V|n \rangle \equiv (\psi_n, V\psi_n). \tag{8.80}$$

In order to obtain $\psi_n^{(1)}$, i.e. the coefficients $a_i^{(1)}$, we return to Eq. (8.79). We multiply the equation by $\psi_m^*, m \neq n$ and integrate:

$$\int_{-\infty}^{\infty} dx \psi_m^* \sum_i a_i^{(1)}(E_i^{(0)} - E_n^{(0)})\psi_i = \int_{-\infty}^{\infty} dx \psi_m^*(E_m^{(1)} - V)\psi_n,$$

i.e.

$$a_m^{(1)}(E_m^{(0)} - E_n^{(0)}) = -(\psi_m, V\psi_n),$$

and hence

$$a_m^{(1)} = -\frac{(\psi_m, V\psi_n)}{E_m^{(0)} - E_n^{(0)}}, \quad m \neq n. \tag{8.81}$$

It follows that to the first order in the parameter ϵ:

$$E_n = E_n^{(0)} + \epsilon(\psi_n, V\psi_n) + O(\epsilon^2), \tag{8.82}$$

$$\Psi_n = \psi_n + \epsilon \sum_{i \neq n} \frac{(\psi_i, V\psi_n)}{E_n^{(0)} - E_i^{(0)}}\psi_i + O(\epsilon^2). \tag{8.83}$$

We observe that Eq. (8.83) makes sense only if

$$E_i^{(0)} \neq E_n^{(0)} \quad \text{for all } i \neq n.$$

Thus the procedure forbids the equality of any $E_i^{(0)}, i \neq 0$, with $E_n^{(0)}$; i.e. the procedure forbids degeneracy and we can consider here only *nondegenerate eigenstates*.

It is clear that we can proceed similarly in the calculation of the higher order contributions to the expansions of E_n and Ψ_n. In order to appreciate the structure of these expansions it is instructive to go one step further and to derive the next order contribution. We therefore consider Eq. (8.76) which we rewrite as

$$(H_0 - E_n^{(0)})\psi_n^{(2)} = (E_n^{(1)} - V)\psi_n^{(1)} + E_n^{(2)}\psi_n. \tag{8.84}$$

Setting

$$\psi_n^{(2)} = \sum_{i \neq n} a_i^{(2)}\psi_i \tag{8.85}$$

(excluding $i = n$ for reasons discussed above), multiplying Eq. (8.84) from the left by ψ_m^* and integrating over the entire domain of x we obtain

$$\int_{-\infty}^{\infty} dx \psi_m^* \sum_{i \neq n} a_i^{(2)}(E_i^{(0)} - E_n^{(0)})\psi_i$$

$$= \int_{-\infty}^{\infty} dx \psi_m^*(E_n^{(1)} - V)\psi_n^{(1)} + E_n^{(2)} \int_{-\infty}^{\infty} dx \psi_m^*\psi_n.$$

Setting $m = n$ we obtain with $(\psi_n, \psi_n^{(1)}) = 0$

$$0 = -(\psi_n, V\psi_n^{(1)}) + E_n^{(2)},$$

i.e.

$$E_n^{(2)} = \left(\psi_n, V\sum_{i\neq n}\frac{(\psi_i, V\psi_n)}{E_n^{(0)} - E_i^{(0)}}\psi_i\right) = \sum_{i\neq n}\frac{(\psi_n, V\psi_i)(\psi_i, V\psi_n)}{E_n^{(0)} - E_i^{(0)}}. \qquad (8.86)$$

For $m \neq n$ we obtain from Eq. (8.85) together with (8.79)

$$a_m^{(2)}(E_m^{(0)} - E_n^{(0)}) = \int_{-\infty}^{\infty} dx\psi_m^*(E_n^{(1)} - V)\psi_n^{(1)}$$

$$= \int_{-\infty}^{\infty} dx\psi_m^*(E_n^{(1)} - V)\sum_{j\neq 0}a_j^{(1)}\psi_j = E_n^{(1)}a_m^{(1)} - \sum_{j\neq 0}a_j^{(1)}(\psi_m, V\psi_j).$$

Using (8.81) we obtain

$$a_m^{(2)} = -\frac{(\psi_m, V\psi_n)E_n^{(1)}}{(E_m^{(0)} - E_n^{(0)})^2} + \sum_{j\neq 0}\frac{(\psi_m, V\psi_j)(\psi_j, V\psi_n)}{(E_m^{(0)} - E_n^{(0)})(E_j^{(0)} - E_n^{(0)})}. \qquad (8.87)$$

Hence

$$\psi_n^{(2)} = \sum_{k\neq n}\left[-\frac{(\psi_k, V\psi_n)(\psi_n, V\psi_n)}{(E_k^{(0)} - E_n^{(0)})^2} + \sum_{j\neq n}\frac{(\psi_k, V\psi_j)(\psi_j, V\psi_n)}{(E_k^{(0)} - E_n^{(0)})(E_j^{(0)} - E_n^{(0)})}\right]\psi_k. \qquad (8.88)$$

Inserting Eq. (8.86) into the expression for E_n we obtain

$$\begin{aligned}
E_n &= E_n^{(0)} + \epsilon(\psi_n, V\psi_n) + \epsilon^2\sum_{i\neq n}(\psi_n, V\psi_i)\frac{1}{E_n^{(0)} - E_i^{(0)}}(\psi_i, V\psi_n) \\
&\quad + \epsilon^3\sum_{j\neq n}\sum_{i\neq j}(\psi_n, V\psi_j)\frac{1}{E_n^{(0)} - E_j^{(0)}}(\psi_j, V\psi_i)\frac{1}{E_j^{(0)} - E_i^{(0)}}(\psi_i, V\psi_n) \\
&\quad + \cdots,
\end{aligned} \qquad (8.89)$$

where we guessed the term of $O(\epsilon^3)$. Inserting the expression (8.88) into ψ_n we obtain similarly

$$\begin{aligned}
\Psi_n &= \psi_n + \epsilon\sum_{i\neq n}\frac{(\psi_i, V\psi_n)}{E_n^{(0)} - E_i^{(0)}}\psi_i + \epsilon^2\sum_{k\neq n}\left[-\frac{(\psi_k, V\psi_n)(\psi_n, V\psi_n)}{(E_k^{(0)} - E_n^{(0)})^2}\right. \\
&\quad \left. + \sum_{j\neq n}\frac{(\psi_k, V\psi_j)(\psi_j, V\psi_n)}{(E_n^{(0)} - E_k^{(0)})(E_n^{(0)} - E_j^{(0)})}\right]\psi_k \\
&\quad + \cdots.
\end{aligned} \qquad (8.90)$$

The expressions (8.89), (8.90) can be found in just about any book on quantum mechanics, sometimes expressed in terms of projection operators which one can construct from the states $|\psi_m\rangle$ (see e.g. Merzbacher[194]). An ugly feature of the expansion (8.90) is the term with only one summation $\sum$ in the contribution of $O(\epsilon^2)$. One would prefer to lump it together with the other contribution. The functions $\psi_n(x)$ are eigenfunctions of some second-order differential equation. These functions generally obey one or more recurrence relations which are effectively equivalent to the differential equation in the sense of difference equations. These recurrence relations allow one to re-express expressions like

$$x^m \psi_n$$

as linear contributions of ψ_i with constant coefficients, i.e. in the form

$$\sum_i c_i \psi_{n+i}.$$

Adopting this procedure for the perturbing potential $V(x)$ we can set

$$V(x)\psi_n(x) = \sum_i c_i \psi_{n+i}. \tag{8.91}$$

The coefficients c_i follow from the recurrence relations. We then obtain

$$(\psi_k, V\psi_n) = \sum_i c_i(\psi_k, \psi_{n+i}) = \sum_i c_i \delta_{k,n+i} = c_{k-n}.$$

We can now rewrite $\psi_n^{(2)}$ as

$$\psi_n^{(2)} = \sum_{k\neq n}\left[-\frac{c_{k-n}c_0}{(E_k^{(0)} - E_n^{(0)})^2} + \sum_{j\neq n}\frac{c_{k-j}c_{j-n}}{(E_k^{(0)} - E_n^{(0)})(E_j^{(0)} - E_n^{(0)})}\right]\psi_k.$$

The sum $\sum_{j\neq n}$ includes a term with $j = k$ which is exactly cancelled by the other contribution. We can therefore rewrite $\psi_n^{(2)}$:

$$\psi_n^{(2)} = \sum_{k\neq n}\sum_{j\neq n,k}\frac{c_{k-j}c_{j-n}}{(E_k^{(0)} - E_n^{(0)})(E_j^{(0)} - E_n^{(0)})}\psi_k. \tag{8.92}$$

The perturbation theory described in Sec. 8.7 is based on this procedure.

8.6   Degenerate Perturbation Theory

It is clear from the expressions obtained above for E_n and Ψ_n that no two eigenvalues $E_n^{(0)}$, all n, are allowed to be equal since otherwise the expansion becomes undefined. Equality of eigenvalues, i.e. degeneracy, however, is not uncommon. For example the magnetic states enumerated by the magnetic quantum number $m, m = l, l - 1, \ldots, -l$, in a central force problem all have the same energy unless the system is placed in a magnetic field. In order to deal with the problem of degeneracy we consider the simplest case, i.e. that of *double degeneracy* $E_n^{(0)} = E_m^{(0)}$, so that

$$E_n = E_n^{(0)} + \epsilon E_n^{(1)} + \epsilon^2 E_n^{(2)} + \cdots , \tag{8.93a}$$

$$E_m = \underbrace{E_n^{(0)}}_{=E_m^{(0)}} + \epsilon E_m^{(1)} + \epsilon^2 E_m^{(2)} + \cdots . \tag{8.93b}$$

The lowest order eigenfunctions belonging to E_n, E_m can now be linear combinations of ψ_n, ψ_m. Hence we write

$$\Psi_n = c_{11}\psi_n + c_{12}\psi_m + \epsilon \psi_n^{(1)} + \epsilon^2 \psi_n^{(2)} + \cdots , \tag{8.94a}$$

$$\Psi_m = c_{21}\psi_n + c_{22}\psi_m + \epsilon \psi_m^{(1)} + \epsilon^2 \psi_m^{(2)} + \cdots \tag{8.94b}$$

with $(\psi_n, \psi_m) = 0 = (\psi_m, \psi_n)$. We now insert Eqs. (8.93a), (8.93b), (8.94a), (8.94a) into Eq. (8.70), i.e.

$$(H_0 + \epsilon V)\Psi_n = E_n\Psi_n, \tag{8.95a}$$

and

$$(H_0 + \epsilon V)\Psi_m = E_m\Psi_m. \tag{8.95b}$$

Proceeding as before we obtain

$$\begin{aligned} H_0(c_{11}\psi_n + c_{12}\psi_m) &= E_n^{(0)}(c_{11}\psi_n + c_{12}\psi_m) \\ H_0(c_{21}\psi_n + c_{22}\psi_m) &= E_n^{(0)}(c_{21}\psi_n + c_{22}\psi_m) \end{aligned} \tag{8.96}$$

and

$$(H_0 - E_n^{(0)})\psi_n^{(1)} = (E_n^{(1)} - V)(c_{11}\psi_n + c_{12}\psi_m), \tag{8.97a}$$

$$(H_0 - E_n^{(0)})\psi_m^{(1)} = (E_m^{(1)} - V)(c_{21}\psi_n + c_{22}\psi_m), \tag{8.97b}$$

with $\psi_n \neq \psi_m$. Equations (8.96) are effectively the known equations of ψ_n, ψ_m and so yield no new information. Equations (8.97a) and (8.97b) are inhomogeneous equations. We now multiply these equations from the left by ψ_l^* and integrate. Then

$$(\psi_l, (H_0 - E_n^{(0)})\psi_n^{(1)}) = E_n^{(1)}(\psi_l, c_{11}\psi_n + c_{12}\psi_m)$$
$$-(\psi_l, c_{11}V\psi_n + c_{12}V\psi_m), \quad (8.98a)$$

and

$$(\psi_l, (H_0 - E_n^{(0)})\psi_m^{(1)}) = E_m^{(1)}(\psi_l, c_{21}\psi_n + c_{22}\psi_m)$$
$$-(\psi_l, c_{21}V\psi_n + c_{22}V\psi_m). \quad (8.98b)$$

Setting $l = n, m$, we obtain with $(\psi_n, \psi_n^{(1)}n) = 0 = (\psi_m, \psi_m^{(1)})$ the equations

$$(\psi_n, H_0\psi_n^{(1)}) = c_{11}[E_n^{(1)} - (\psi_n, V\psi_n)] - c_{12}(\psi_n, V\psi_m),$$

$$(\psi_m, H_0\psi_n^{(1)}) - E_n^{(0)}(\psi_m, \psi_n^{(1)}) = c_{12}[E_n^{(1)} - (\psi_m, V\psi_m)] - c_{11}(\psi_m, V\psi_n),$$

$$(\psi_n, H_0\psi_m^{(1)}) - E_m^{(0)}(\psi_n, \psi_m^{(1)}) = c_{21}[E_m^{(1)} - (\psi_n, V\psi_n)] - c_{22}(\psi_n, V\psi_m),$$

$$(\psi_m, H_0\psi_m^{(1)}) = c_{22}[E_m^{(1)} - (\psi_m, V\psi_m)] - c_{21}(\psi_m, V\psi_n). \quad (8.99)$$

But

$$(\psi_n, H_0\psi_n^{(1)}) = \left(\psi_n, H_0 \sum_{i \neq n} a_i^{(1)}\psi_i\right) = \left(\psi_n, \sum_{i \neq n} a_i^{(1)} E_i^{(0)}\psi_i\right) = 0,$$

and

$$(\psi_m, H_0\psi_n^{(1)}) - E_n^{(0)}(\psi_m, \psi_n^{(1)})$$
$$= \left(\psi_m, H_0 \sum_{i \neq n} a_i^{(1)}\psi_i\right) - E_n^{(0)}\left(\psi_m, \sum_{i \neq n} a_i^{(1)}\psi_i\right)$$
$$= \left(\psi_m, \sum_{i \neq n} a_i^{(1)} E_i^{(0)}\psi_i\right) - E_n^{(0)} a_m^{(1)} = a_m^{(1)} E_m^{(0)} - E_n^{(0)} a_m^{(1)} = 0,$$

since $E_m^{(0)} = E_n^{(0)}$ (degeneracy). Similarly

$$(\psi_n, H_0\psi_m^{(1)}) - E_m^{(0)}(\psi_n, \psi_m^{(1)}) = a_n^{(1)} E_n^{(0)} - E_m^{(0)} a_n^{(1)} = 0,$$

and

$$(\psi_m, H_0\psi_m^{(1)}) = 0.$$

Equations (8.99) can now be rewritten as

$$\begin{pmatrix} (\psi_n, V\psi_n) & (\psi_n, V\psi_m) \\ (\psi_m, V\psi_n) & (\psi_m, V\psi_m) \end{pmatrix} \begin{pmatrix} c_{11} \\ c_{12} \end{pmatrix} = E_n^{(1)} \begin{pmatrix} c_{11} \\ c_{12} \end{pmatrix} \quad (8.100a)$$

and

$$\left(\begin{array}{cc} (\psi_n, V\psi_n) & (\psi_n, V\psi_m) \\ (\psi_m, V\psi_n) & (\psi_m, V\psi_m) \end{array} \right) \left(\begin{array}{c} c_{21} \\ c_{22} \end{array} \right) = E_m^{(1)} \left(\begin{array}{c} c_{21} \\ c_{22} \end{array} \right). \qquad (8.100b)$$

These matrix equations have nontrivial solutions if and only if

$$\left| \begin{array}{cc} (\psi_n, V\psi_n) - E_n^{(1)} & (\psi_n, V\psi_m) \\ (\psi_m, V\psi_n) & (\psi_m, V\psi_m) - E_n^{(1)} \end{array} \right| = 0. \qquad (8.101)$$

This secular equation determines the first order correction $E_n^{(1)}$. The equations (8.100a) determine the coefficients c_{11}, c_{12}. Similarly Eqs. (8.100b) determine c_{21}, c_{22}. This completes the derivation of the first order perturbation corrections if an unperturbed eigenvalue is doubly degenerate. An example which emphasizes the significance of the spatial dimensionality in this context is treated in Example 8.1.

Example 8.1: Do discrete spectra permit degeneracy?
Show that in the case of a one-dimensioanl Schrödinger equation with discrete point spectrum there is no degeneracy.

Solution: We consider the Schrödinger equation with potential V in the simplified form

$$-\psi'' + (V - E)\psi = 0.$$

Let ψ and ϕ be two eigenfunctions belonging to the same eigenvalue E. Then

$$\frac{\psi''}{\psi} = V - E = \frac{\phi''}{\phi}.$$

Hence

$$0 = \psi''\phi - \phi''\psi = \frac{d}{dx}(\psi'\phi - \phi'\psi), \quad \text{i.e.} \quad (\psi'\phi - \phi'\psi) = \text{const.}.$$

At $x = \pm\infty$ the constant is zero. It follows that ψ and ϕ are linearly dependent, i.e.

$$\frac{\psi'}{\psi} = \frac{\phi'}{\phi}, \quad \ln\psi = \ln\phi + \text{const.}, \quad \psi = c\phi.$$

Hence there is no degeneracy, just as there is no degeneracy in the case of the one-dimensional harmonic oscillator (Chapter 6). However, there is degeneracy in the case of the hydrogen atom with Coulomb potential (Chapter 11) which is based on three-dimensional spherical polar coordinates r, θ, φ. If one wants to perform a perturbation calculation in such a case, the degenerate perturbation theory must be used, i.e. the secular determinant. Nonetheless we shall see that the Schrödinger equation of the hydrogen-like problem can also be separated in some other orthogonal coordinates, in particular in parabolic coordinates (see Sec. 11.6.3). It is then possible that a special perturbation, like that providing the *Stark effect* (proportional to $r\cos\theta$), splits the problem into two independent one-dimensional systems. A one-dimensional system, as we saw, cannot be degenerate because there is only one quantum number corresponding to one definite energy and hence cannot be degenerate. In this particular case therefore non-degenerate perturbation theory can be used (cf. Example 11.5).

8.7 Dingle–Müller Perturbation Method

We have seen above that although the procedure of calculating higher order contributions of perturbation expansions is in principle straightforward, one quickly arrives at clumsy expressions which do not reveal much about the general structure of a higher order perturbation term. It is therefore natural to inquire about a procedure which permits one to generate successive orders of a perturbation expansion in a systematic way, and such that one can even obtain a recurrence relation for the coefficients of the expansion. In the following we describe such a procedure as first applied to the strong coupling case of the cosine potential or Mathieu equation.* In various versions this method is used throughout this book, in particular in applications to periodic potentials, anharmonic oscillator potentials and screened Coulomb potentials. Applications of the method are, of course, also possible in areas not of immediate relevance here.† Since in general a solution is valid only in a limited region of the variable, the method has to be applied in each domain separately, and the solutions then have to be matched in their regions of overlap. Thus the method is *a systematic method of matched asymptotic expansions.*

We consider an equation of the form

$$D\phi + (E - V)\phi = 0, \tag{8.102}$$

where D is a second order differential operator and E a parameter. It is immaterial here whether D does or does not contain a first order derivative — in fact, with some restrictions on the function V, a potential, the considerations given below would be valid even if D contained only the derivative of the first order. We assume that V can be expressed as the sum of two terms, i.e.

$$V = V_0 + v, \tag{8.103}$$

such that if E is written correspondingly

$$E = E_0 + \epsilon, \tag{8.104}$$

Eq. (8.102) can be written

$$D\phi + (E_0 - V_0)\phi = (v - \epsilon)\phi. \tag{8.105}$$

The subdivisions (8.103), (8.104) are chosen such that the equation

$$D\phi + (E_0 - V_0)\phi = 0 \tag{8.106}$$

---

*This method was developed in R. B. Dingle and H. J. W. Müller [73]. See also R. B. Dingle [72] and H. J. W. Müller [210].

†For instance spheroidal functions can be studied with this method; cf. H. J. W. Müller [216].

is exactly solvable and has the eigenvalues E_m and eigenfunctions ϕ_m, where m is an additional integral or near-integral parameter (depending on boundary conditions). Then if

$$|\epsilon| \ll |E| \quad \text{and} \quad |v - \epsilon| \ll |V_0 - E_0|$$

(the latter in a restricted domain), the function

$$\phi^{(0)} = \phi_m \tag{8.107}$$

represents a first approximation to the solution ϕ of Eq. (8.102) with $E \simeq E_m$. The next step is to use the recursion formulae of the functions ϕ_m in order to re-express $(v - \epsilon)\phi_m$ as a linear combination of functions ϕ_{m+s}, s an integer:

$$(v - \epsilon)\phi_m = \sum_s (m, m+s)\phi_{m+s} \tag{8.108}$$

with coefficients $(m, m+s)$ which are determined by the recursion formulae. We write the coefficients "$(m, m+s)$" in order to relate them to "steps" from m to $m+s$. Considering ϕ_{m+s} we have, since

$$D\phi_m + (E_m - V_0)\phi_m = 0,$$

the equation

$$D\phi_{m+s} + (E_{m+s} - V_0)\phi_{m+s} = 0, \tag{8.109}$$

and so

$$\underbrace{D\phi_{m+s} + (E_m - V_0)\phi_{m+s}}_{\equiv (D+E_m-V_0)\phi_{m+s}} = (E_m - E_{m+s})\phi_{m+s}. \tag{8.110}$$

If we write the next contribution $\phi^{(1)}$ to $\phi^{(0)}$ in the form

$$\phi^{(1)} = \sum_s c_s \phi_{m+s}, \tag{8.111}$$

then (see below)

$$\phi^{(1)} = \sum_{s \neq 0} \frac{(m, m+s)}{(E_m - E_{m+s})}\phi_{m+s}. \tag{8.112}$$

This result is obtained as follows. Consider

$$(D + E_m - V_0)(\phi^{(0)} + \phi^{(1)}) = (D + E_m - V_0)\phi^{(1)}$$
$$= (D + E_m - V_0)\sum_s c_s\phi_{m+s} = \sum_s c_s(E_m - E_{m+s})\phi_{m+s}$$
$$= (v - \epsilon)\phi^{(0)} = \sum_s (m, m+s)\phi_{m+s}, \tag{8.113}$$

and so

$$s \neq 0: \qquad c_s = \frac{(m, m+s)}{E_m - E_{m+s}}. \tag{8.114}$$

To insure that $\phi = \phi^{(0)} + \phi^{(1)} + \cdots$ is a solution of Eq. (8.102) to order (1) of the perturbation $(v - \epsilon)$, the coefficient of the term $(m, m)\phi_m$ in (8.113) must vanish. Thus

$$(m, m) = 0 \quad \text{(first approximation)}. \tag{8.115}$$

Repeating this procedure with $\phi^{(1)}$ instead of $\phi^{(0)}$ we obtain the next contribution

$$\phi^{(2)} = \sum_{s \neq 0} \sum_{r \neq 0} \frac{(m, m+s)(m+s, m+r)}{(E_m - E_{m+s})(E_m - E_{m+r})} \phi_{m+r} \tag{8.116}$$

together with

$$(m, m) + \sum_{s \neq 0} \frac{(m, m+s)(m+s, m)}{(E_m - E_{m+s})} = 0 \quad \text{(second approximation)}. \tag{8.117}$$

This procedure can be repeated indefinitely. We then find altogether

$$\phi = \phi^{(0)} + \phi^{(1)} + \phi^{(2)} + \cdots = \phi_m + \sum_{s \neq 0} \frac{(m, m+s)}{(E_m - E_{m+s})} \phi_{m+s}$$

$$+ \sum_{s \neq 0} \sum_{r \neq 0} \frac{(m, m+s)(m+s, m+r)}{(E_m - E_{m+s})(E_m - E_{m+r})} \phi_{m+r} + \cdots \tag{8.118}$$

together with

$$0 = (m, m) + \sum_{s \neq 0} \frac{(m, m+s)(m+s, m)}{(E_m - E_{m+s})}$$

$$+ \sum_{s \neq 0} \sum_{r \neq 0} \frac{(m, m+s)(m+s, m+r)(m+r, m)}{(E_m - E_{m+s})(E_m - E_{m+r})} + \cdots . \tag{8.119}$$

The first expansion determines the solution ϕ (apart from an overall normalization constant), and the second expansion is an equation from which ϵ and hence the eigenvalue E is determined (i.e. Eq. (8.119) is effectively the secular equation). We observe that the coeffcients of expansions (8.118), (8.119) follow a definite pattern and can therefore be constructed in a systematic way. The factors $(E_m - E_{m+s})$ in the denominators result from the right hand side of Eq. (8.110) and are therefore related to the inverse of a

Green's function, as will be seen later. The systematics of the coefficients suggests that one can construct the coefficients of an arbitrary order of this perturbation theory. This can in fact be done and will be dealt with later. One should note that in this perturbation theory every order in the perturbation parameter also contains contributions of higher orders; this is — so to speak — the price paid for obtaining a clear systematics which is essential, for instance, for the investigation of the behaviour of the late terms of the perturbation series. A further conspicuous difference between this method and the usual Rayleigh–Schrödinger perturbation method is that the first approximation of the expansion, i.e. ϕ_m in the above, does not occur in any of the later contributions. This means the solution is not normalized. Imposing normalization later one obtains the normalization constant as an asymptotic expansion, and when the solution is multiplied by this factor one obtains the contributions involving ϕ_m in the higher order terms (for an explicit demonstration of normalization see Example 17.2). An additional aspect of the method is the full exploitation of the symmetries of the differential equation, as already mentioned after Eq. (8.38).

An explicit application of the method (not including matching) seems suitable at this point, since it will be used extensively in later chapters. The following Example 8.2 therefore illustrates the use of this perturbation method in the derivation of the eigenvalues of the *Gauss potential*. The eigenvalue expansion obtained is an asymptotic expansion, and is typical of a large number of cases, as will be seen in later chapters. Other applications can be found in the literature.[‡]

Example 8.2: Eigenenergies of the Gauss Potential[§]
Use the above perturbation theory for the calculation of the eigenvalues $E = k^2$ of the radial Schrödinger equation

$$\left[\frac{d^2}{dr^2} + k^2 - \frac{l(l+1)}{r^2} - V(r)\right]\psi(r) = 0 \text{ with Gauss potential } V(r) = -g^2 e^{-a^2 r^2}$$

for large values of g^2. The result is the following expansion, in which $q = 4n + 3, n = 0, 1, 2, \ldots$, i.e.

$$
\begin{aligned}
k^2 + g^2 \;=\; & ga(2l+q) - \frac{a^2}{2^4}[3(q^2+1) + 4(3q-1)l + 8l^2] \\
& -\frac{a^3}{3 \cdot 2^8 g}[q(11q^2+1) + 2(33q^2 - 6q + 1)l + 24(5q-1)l^2 + 64l^3] \\
& -\frac{a^4}{3 \cdot 2^{15} g^2}[4(85q^4 + 2q^2 - 423) + l(2720q^3 - 71q^2 + 32q + 2976) \\
& +32l^2(252q^2 - 12q + 64) + 256l^3(41q - 9) + 4096l^4] + O(1/g^3).
\end{aligned} \tag{8.120}
$$

[‡]The potential $\lambda r^2/(1 + gr^2)$, for instance, has been treated by R. S. Kaushal [146].
[§]H. J. W. Müller [211].

Solution: We expand the exponential and rewrite the Schrödinger equation in the form

$$\frac{d^2\psi}{dr^2} + \left(k^2 - \frac{l(l+1)}{r^2} + g^2 - g^2 a^2 r^2\right)\psi = -g^2 \sum_{i=2}^{\infty} \frac{(-a^2)^i}{i!} r^{2i}\psi.$$

Here we change the independent variable to $z = \sqrt{2ga}r$. Then

$$\frac{d^2\psi}{dz^2} + \left(\frac{k^2 + g^2}{2ga} - \frac{l(l+1)}{z^2} - \frac{1}{4}z^2\right)\psi = -\frac{1}{2}\sum_{i=2}^{\infty} \left(\frac{a}{g}\right)^{i-1} \frac{(-z^2/2)^i}{i!}\psi. \tag{8.121}$$

In the limit $|g| \to \infty$ we can neglect the right hand side and write the solution $\psi \to \psi_0$. We then set

$$\psi_0(z) = z^{l+1}e^{-z^2/4}X_0(z) \quad \text{and} \quad S = \frac{1}{2}z^2.$$

Now the function $X_0(z)$ satisfies the *confluent hypergeometric equation* (cf. Secs. 11.6 and 16.2)

$$S\frac{d^2X_0}{dS^2} + (b - S)\frac{dX_0}{dS} - aX_0 = 0,$$

where

$$a = \frac{1}{2}\left(l + \frac{3}{2}\right) - \frac{(k^2 + g^2)}{4ga} \quad \text{and} \quad b = l + \frac{3}{2}.$$

The solution $X_0(S) = \Phi(a, b; S)$ is a confluent hypergeometric function. The solution

$$\psi_0(z) = z^{l+1}e^{-z^2/4}\Phi\left(a, b; \frac{1}{2}z^2\right)$$

is a normalizable function if $a = -n$ for $n = 0, 1, 2, \ldots$. Setting $q = 4n + 3$ implies $k^2 + g^2 = ga(2l + q)$. Therefore in the general case we may set

$$k^2 + g^2 = ga(2l + q) - 2a^2\triangle, \tag{8.122}$$

where $\triangle$ is of order $1/g$. We now insert this equation into Eq. (8.121), multiply the equation by one-half and set $h = -a/g$. Then the equation is

$$D_q\psi = \left[-\frac{1}{2}\triangle h + \frac{1}{4}\sum_{i=2}^{\infty} \frac{h^{i-1}}{i!}\left(\frac{1}{2}z^2\right)^i\right]\psi \quad \text{with} \quad D_q \equiv \frac{1}{2}\left[\frac{d^2}{dz^2} + l + \frac{1}{2}q - \frac{l(l+1)}{z^2} - \frac{1}{4}z^2\right]. \tag{8.123}$$

The first approximation (note the last expression as convenient notation)

$$\psi \simeq \psi^{(0)} = \psi_q(z) = z^{l+1}e^{-z^2/4}\Phi\left(a, b; \frac{1}{2}z^2\right) \equiv \psi(a)$$

leaves unaccounted for on the right hand side of Eq. (8.123) the contribution

$$R^{(0)} = \left[-\frac{1}{2}\triangle h + \frac{1}{4}\sum_{i=2}^{\infty} \frac{h^{i-1}}{i!}\left(\frac{1}{2}z^2\right)^i\right]\psi_q(z).$$

The recurrence relation for the functions $\psi(a)$ follows from that for confluent hypergeometric functions:

$$\frac{1}{2}z^2\psi(a) = (a, a+1)\psi(a+1) + (a, a)\psi(a) + (a, a-1)\psi(a-1),$$

where

$$(a, a+1) = a = -\frac{1}{4}(q-3), \quad (a,a) = b - 2a = l + \frac{1}{2}q, \quad (a, a-1) = a - b = -\frac{1}{4}(q+3) - l.$$

In general

$$\left(\frac{1}{2}z^2\right)^m \psi(a) = \sum_{j=-m}^{m} S_m(a, a+j)\psi(a+m).$$

The coefficients $S_m(a, a+r)$ satisfy the following recurrence relation:

$$\begin{aligned} S_m(a, a+r) &= S_{m-1}(a, a+r-1)(a+r-1, a+r) + S_{m-1}(a, a+r)(a+r, a+r) \\ &\quad + S_{m-1}(a, a+r+1)(a+r+1, a+r) \end{aligned}$$

with the boundary conditions $S_0(a,a) = 1, S_0(a, a+i) = 0$ for $i \neq 0, S_m(a, a+r) = 0$ for $|r| > m$. The above remainder $R^{(0)}$ can now be rewritten as

$$R^{(0)} = \sum_{i=0}^{\infty} h^{i+1} \sum_{j=-(i+2)}^{i+2} [a, a+j]_{i+1} \psi(a+j),$$

where

$$[a, a]_1 = -\frac{1}{2}\triangle + \frac{1}{4\cdot 2!}S_2(a,a), \quad [a, a+j]_{i+1} = \frac{1}{4(i+2)!}S_{i+2}(a, a+j)$$

for i and j not zero simultaneously. Now we make the following important observation that

$$D_q\psi(a+j) = j\psi(a+j).$$

This equation shows that a term $\mu\psi(a+j)$ on the right hand side of Eq. (8.123) and so in $R^{(0)}$ can be taken care of by adding to the previous approximation the contribution $\mu\psi(a+j)/j$ except, of course, when $j = 0$. Hence the next order contribution to $\psi^{(0)}$ becomes

$$\psi^{(1)} = \sum_{i=0}^{\infty} h^{i+1} \sum_{j=-(i+2), j\neq 0}^{i+2} \frac{[a, a+j]_{i+1}}{j}\psi(a+j) \quad \text{with} \quad h[a,a]_1 = 0,$$

the latter determining to that order the eigenvalue. Now proceeding as explained in the text one obtains finally the expansion

$$\psi = \psi^{(0)} + \psi^{(1)} + \psi^{(2)} + \cdots$$

along with the expansion from which $\triangle$ is determined, i.e.

$$\begin{aligned} 0 &= h[a,a]_1 + h^2\Big[[a,a]_2 + \frac{[a, a+2]_1}{2}[a+2, a]_1 + \frac{[a, a+1]_1}{1}[a+1, a]_1 \\ &\quad + \frac{[a, a-1]_1}{-1}[a-1, a]_1 + \frac{[a, a-2]_1}{-2}[a-2, a]_1\Big] + O(h^3). \end{aligned}$$

It is clear that the various contributions can be obtained from a consideration of "allowed steps". In fact, one can write down recursion relations for the coeffcients of powers of h^i as will be shown later in a simpler context. If we now evaluate the various coefficients in the last expansion we obtain the result (8.120). As a check on the usefulness of the perturbation method employed here, one could now use the Rayleigh–Schrödinger method and rederive the result (8.120).

Chapter 9

The Density Matrix and Polarization Phenomena

9.1 Introductory Remarks

We saw in the foregoing chapters that in considering a particle system which is classically described by the solution of Newton's equation* is quantum mechanically described by states which in the position or configuration space representation are given by the solutions of the Schrödinger wave equation. In continuation of our earlier search for an approach to quantum mechanics as close as possible to classical mechanics by looking for the generalizability of classical mechanics — as already attempted in Chapter 2 — it is reasonable to consider also classical systems with a wave-like nature, i.e. light, and this means electrodynamics. We therefore begin with the appropriate classical considerations with a view to generalization, and we shall then see that these lead to analogous results as in Chapter 2, except for the canonical algebra which has to be formulated with *Dirac brackets* replacing *Poisson brackets*, as we shall see in Chapter 27.

9.2 Reconsideration of Electrodynamics

In this chapter we consider some aspects of classical electrodynamics with a view to their generalization in the direction of quantum mechanics.

We know from classical electrodynamics that a planar light wave with propagation or wave vector $\mathbf{k}$ and frequency ω is represented by a vector

---

*So far our considerations were restricted to one-particle systems; the generalization to many particles, although not exactly straight-forward, proceeds along similar lines. Regrettably this has to be left out here in view of lack of space.

potential $\mathbf{A}(\mathbf{r}, t)$ given by

$$\mathbf{A}(\mathbf{r}, t) = \Re \mathbf{A}_0 e^{i(\mathbf{k} \cdot \mathbf{r} - \omega t)} \tag{9.1}$$

which satisfies the transversality or Lorentz condition

$$\mathbf{k} \cdot \mathbf{A} = 0. \tag{9.2}$$

In general the constant $\mathbf{A}_0$ is complex, i.e.

$$\mathbf{A}_0 = \mathbf{A}_{0R} + i\mathbf{A}_{0I}. \tag{9.3}$$

It follows therefore that

$$\mathbf{A}(\mathbf{r}, t) = \mathbf{A}_{0R} \cos(\mathbf{k} \cdot \mathbf{r} - \omega t) - \mathbf{A}_{0I} \sin(\mathbf{k} \cdot \mathbf{r} - \omega t)$$

and at $\mathbf{r} = 0$:

$$\mathbf{A}(0, t) = \mathbf{A}_{0R} \cos(\omega t) + \mathbf{A}_{0I} \sin(\omega t). \tag{9.4}$$

This is the equation of an ellipse as indicated in Fig. 9.1.

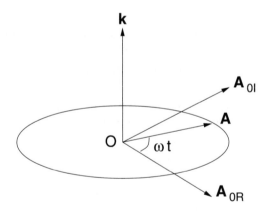

Fig. 9.1 The vector potential ellipse.

When $\mathbf{A}_{0R}$ is parallel to $\mathbf{A}_{0I}$ the ellipse becomes a straight line and the wave is said to be linearly polarized. When $\mathbf{A}_{0R}$ is perpendicular to $\mathbf{A}_{0I}$ and $|\mathbf{A}_{0R}| = |\mathbf{A}_{0I}|$ the ellipse becomes a circle and the field is said to be circularly polarized. In the latter case one has the possibility of

$$(\mathbf{A}_{0R} \times \mathbf{A}_{0I}) \cdot \mathbf{k} > 0, \tag{9.5}$$

which is described as right polarization as compared with left polarization for which

$$(\mathbf{A}_{0R} \times \mathbf{A}_{0I}) \cdot \mathbf{k} < 0. \tag{9.6}$$

As a consequence of the Lorentz condition (9.2) the vector potential $\mathbf{A}$ is completely determined by its two (say (x, y)) components $\mathbf{A}_1, \mathbf{A}_2$. We set therefore

$$\mathbf{A}_0 = \mathbf{A}_1 + \mathbf{A}_2, \quad \text{with} \quad \mathbf{k} \cdot \mathbf{A}_1 = 0 = \mathbf{k} \cdot \mathbf{A}_2. \tag{9.7}$$

We consider now measurements with polarized light,[†] in which $\mathbf{k}$ and ω remain unchanged, but the polarization (state) and the intensity can change (the latter by absorption) from

$$\{A_i^{(0)}\} \longrightarrow \{A_i^{(1)}\}.$$

In view of the linearity of the Maxwell equations, the connection must be linear, i.e.

$$A_i^{(1)} = \sum_{k=1,2} F_{ik} A_k^{(0)} \tag{9.8}$$

with coefficients F_{ik}.

In classical electrodynamics, we recall, the fields $\mathbf{E}$ and $\mathbf{B}$ are observables, not, however, the vector potential $\mathbf{A}$. Here we are interested in observable properties of light waves — like, for instance, the intensity of a wave. The intensity I of a light wave with amplitude coefficients $A_1^{(0)}, A_2^{(0)}$ is the sum of the moduli, i.e.

$$I = A_1^{(0)*} A_1^{(0)} + A_2^{(0)*} A_2^{(0)} \equiv (\mathbf{A}^{(0)}, \mathbf{A}^{(0)}) \equiv |\mathbf{A}^{(0)}|^2. \tag{9.9}$$

The intensity is characterized by the fact that it can be looked at as the "*expectation value*" $\langle 1 \rangle$ of the unit matrix $\mathbb{1}_{2\times2}$, i.e.

$$I \equiv \langle \mathbb{1}_{2\times2} \rangle := \sum_{i,k} A_i^{(0)*} \mathbb{1}_{ik} A_k^{(0)}. \tag{9.10}$$

We give the initial wave the label "0" and subject this to an experiment, represented by a filter "F", which leads to an outgoing wave which we give the label "1". Then their connection is given by Eq. (9.8). Observable quantities can only be those which are invariant under translations and rotations. From the vectors

$$\mathbf{A}^{(0)}, \mathbf{A}^{(1)}$$

we can construct only the following scalar products with this invariance:

$$(\mathbf{A}^{(0)}, \mathbf{A}^{(0)}), \quad (\mathbf{A}^{(1)}, \mathbf{A}^{(0)}), \quad (\mathbf{A}^{(0)}, \mathbf{A}^{(1)}), \quad (\mathbf{A}^{(1)}, \mathbf{A}^{(1)}). \tag{9.11}$$

[†]This means we consider the polarization, i.e. the direction of polarization, determined with the help of a polarizer, as e.g. a calcite crystal or a film of nitrocellulose. See for instance A. Rae [234], p. 17.

Only the second last of these combinations describes the experiment, i.e.

$$(\mathbf{A}^{(0)}, \mathbf{A}^{(1)}) = \sum_{i,k} A_i^{(0)*} F_{ik} A_k^{(0)}, \tag{9.12}$$

i.e. the connection between the initial polarization and the final polarization. We write the observable quantity as

$$\langle F \rangle := \sum_{i,k} A_i^* F_{ik} A_k. \tag{9.13}$$

This consideration of translation and rotation invariant quantities is somewhat outside the framework of our earlier arguments. But we can convince ourselves that we achieve exactly the same as with our earlier consideration of observables. To this end we observe that as a consequence of Eq. (9.13) the quantity F must be an hermitian matrix and thus representative of an observable, so that $\langle F \rangle$ can be looked at as the expectation value of an observable F. Thus, if $\langle F \rangle$ is an observable quantity, it must be real and hence

$$\langle F \rangle = \langle F \rangle^*.$$

Then according to Eq. (9.13):

$$\sum_{i,k} A_i^* F_{ik} A_k = \sum_{i,k} A_i F_{ik}^* A_k^* = \sum_{i,k} A_k^* F_{ki}^{*T} A_i = \sum_{ik} A_i^* F_{ik}^{*T} A_k,$$

and hence

$$F = F^\dagger. \tag{9.14}$$

We now introduce a matrix ρ — called *density matrix* — which is defined by the relation

$$\rho_{ik} := A_i A_k^*, \qquad \rho_{ki} = A_k A_i^*. \tag{9.15}$$

This matrix is hermitian since

$$(\rho_{ik})^\dagger = (A_i A_k^*)^\dagger = (A_i^* A_k)^T = (A_k A_i^*)^T = (\rho_{ki})^T = (\rho_{ik}).$$

We use definition (9.15) in order to rewrite Eq. (9.13). Thus

$$\langle F \rangle = \sum_{i,k} F_{ik} \rho_{ki} = \sum_i (F\rho)_{ii}, \tag{9.16}$$

i.e.

$$\langle F \rangle = \text{Tr}(F\rho). \tag{9.17}$$

In particular we have

$$I = \text{Tr}(\rho). \tag{9.18}$$

The vector $\mathbf{A}$ describes a particular ray of light. But light, like that of a bulb, is in general unpolarized, i.e. it is a statistical admixture of light rays whose polarization vectors are uniformly distributed over all directions, i.e. "*incoherently*". This admixture is described by introducing matrices ρ. In the present case of a polarized beam of light we have two possible, mutually orthogonal polarization directions x and y described by the matrices (see also below)

$$\rho_x = \begin{pmatrix} 1 & 0 \\ 0 & 0 \end{pmatrix} \quad \text{and} \quad \rho_y = \begin{pmatrix} 0 & 0 \\ 0 & 1 \end{pmatrix}. \tag{9.19}$$

In the case of an admixture with equal portions of 50%, we have

$$\rho = \frac{1}{2} \begin{pmatrix} 1 & 0 \\ 0 & 0 \end{pmatrix} + \frac{1}{2} \begin{pmatrix} 0 & 0 \\ 0 & 1 \end{pmatrix} = \begin{pmatrix} \frac{1}{2} & 0 \\ 0 & \frac{1}{2} \end{pmatrix}. \tag{9.20}$$

In this case there is no preferential direction. The so-called pure case, i.e. that of a *pure state*, is given by a single polarized wave.[‡] For a light-wave travelling in the direction of z, we can describe the wave function of the state polarized in the direction of x or y respectively by

$$|\varphi\rangle_1 = \begin{pmatrix} 1 \\ 0 \end{pmatrix} \quad \text{and} \quad |\varphi\rangle_2 = \begin{pmatrix} 0 \\ 1 \end{pmatrix}.$$

A special state is represented by

$$\begin{pmatrix} a \\ b \end{pmatrix} = a \begin{pmatrix} 1 \\ 0 \end{pmatrix} + b \begin{pmatrix} 0 \\ 1 \end{pmatrix} \tag{9.21a}$$

with

$$|a|^2 + |b|^2 = 1. \tag{9.21b}$$

This state is still a pure state. The density matrix for the *pure state* (9.21a) follows from Eq. (9.15) with $A_1 = a, A_2 = b$ as

$$(\rho_{ik}) = (A_i A_k^*) = \begin{pmatrix} aa^* & ab^* \\ ba^* & bb^* \end{pmatrix}. \tag{9.22}$$

For $a = 1, b = 0$ and $a = 0, b = 1$ we obtain the expressions (9.19). For a 45° polarized wave we have

$$a = \frac{1}{\sqrt{2}}, \quad b = \frac{1}{\sqrt{2}},$$

[‡]The classical polarization vector or wave vector $\mathbf{A}$ corresponds in the quantum mechanical case to the wave function ψ.

and

$$\rho_{45°} = \begin{pmatrix} \frac{1}{2} & \frac{1}{2} \\ \frac{1}{2} & \frac{1}{2} \end{pmatrix}. \tag{9.23}$$

For a $135°$ polarized wave we have

$$a = -\frac{1}{\sqrt{2}}, \quad b = \frac{1}{\sqrt{2}},$$

and

$$\rho_{135°} = \begin{pmatrix} \frac{1}{2} & -\frac{1}{2} \\ -\frac{1}{2} & \frac{1}{2} \end{pmatrix}. \tag{9.24}$$

The following two 50% admixtures yield the same effect:

$$\rho = \frac{1}{2}\rho_x + \frac{1}{2}\rho_y = \begin{pmatrix} \frac{1}{2} & 0 \\ 0 & \frac{1}{2} \end{pmatrix},$$

$$\rho = \frac{1}{2}\rho_{45°} + \frac{1}{2}\rho_{135°} = \begin{pmatrix} \frac{1}{2} & 0 \\ 0 & \frac{1}{2} \end{pmatrix}. \tag{9.25}$$

9.3   Schrödinger and Heisenberg Pictures

As in Chapter 2 we now consider on the purely classical basis the (analogues of the) Schrödinger and Heisenberg picture descriptions. Equation (9.8) describes the relation between the initial and final polarization vectors $\mathbf{A}^{(0)}$ and $\mathbf{A}^{(1)}$ respectively in a measurement of the observable F, in which the intensity $I = \mathrm{Tr}(\rho)$ is not conserved (e.g. as a consequence of absorption). We now inquire about the the way ρ changes in the process of the transmission of the light-wave through an apparatus. We shall see that the matrices appearing in the description of polarization properties of waves possess the same generalizability as the particle considerations of classical mechanics. We set in analogy to Eq. (9.8)

$$A_i^{(1)} = \sum_k R_{ik} A_k^{(0)}, \quad R = R(\alpha), \tag{9.26}$$

where α is a parameter whose variation describes the continuous variation of the polarization from its initial state labeled with "0" and hence parameter value α_0 (analogous to t or β in our earlier considerations of the density matrix) to the final state with label "1". Then according to Eq. (9.15)

$$\begin{aligned} \rho_{ik}^{(1)} &= A_i^{(1)} A_k^{(1)*} = \sum_{j,m} R_{ij} A_j^{(0)} R_{km}^* A_m^{(0)*} \\ &= \sum_{j,m} R_{ij} \rho_{jm}^{(0)} R_{mk}^\dagger, \end{aligned}$$

i.e.

$$\rho^{(1)} = R\rho^{(0)}R^\dagger. \tag{9.27}$$

The measurement of an observable F in the new state with superscript "1" then implies according to Eq. (9.17)

$$\begin{aligned}\langle F \rangle^{(1)} &= \text{Tr}(F\rho^{(1)}) = \text{Tr}(FR\rho^{(0)}R^\dagger) \\ &= \text{Tr}(R^\dagger F R\rho^{(0)}) = \text{Tr}(F'\rho^{(0)}),\end{aligned} \tag{9.28}$$

where

$$F' = R^\dagger F R. \tag{9.29}$$

We can interpret the result either as

$$\langle F \rangle^{(1)} = \text{Tr}(F\{R\rho^{(0)}R^\dagger\}) \tag{9.30}$$

or as

$$\langle F \rangle^{(1)} = \text{Tr}(\{R^\dagger F R\}\rho^{(0)}). \tag{9.31}$$

In the case of Eq. (9.30) the dependence on α is contained in

$$\rho^{(1)}(\alpha) = R(\alpha)\rho^{(0)}(\alpha_0)R^\dagger(\alpha) \tag{9.32}$$

(i.e. in the state functional) in analogy to the time dependence of $W(p, q, t)$ in Chapter 2 in what we described as the Schrödinger picture there, the observable F remaining unchanged. In the case of Eq. (9.31), on the other hand, the dependence on α is contained in

$$F' = R^\dagger(\alpha)FR(\alpha), \tag{9.33}$$

i.e. the observable is transformed, here this contains the dependence on α, and $\rho^{(0)}$ remains unchanged. The two cases are completely equivalent and can therefore be described as *Schrödinger picture* and *Heisenberg picture* representations.

9.4 The Liouville Equation

It is natural to go one step further and to derive the equation analogous to the Liouville equation. The *Liouville equation* describes the motion of an equal number of systems in phase space elements of the same size, in other words without creation or annihilation of systems. The corresponding condition here is that no absorption of the wave takes place, i.e. that the initial intensity I_0 is equal to the final state intensity I_1, i.e. that

$$I_0 = \text{Tr}(\rho^{(0)}) = \text{Tr}(\rho^{(1)}) = I_1, \tag{9.34}$$

i.e.

$$\text{Tr}(\rho^{(0)}) = \text{Tr}(\rho^{(1)}) = \text{Tr}(R\rho^{(0)}R^\dagger) = \text{Tr}(R^\dagger R\rho^{(0)}),$$

so that

$$R^\dagger R = \mathbb{1}. \tag{9.35}$$

Thus the matrix R has to be unitary. For an infinitesimal variation of the state of polarization we have

$$R(\alpha) = \mathbb{1} + M d\alpha. \tag{9.36}$$

In view of Eq. (9.35) we have

$$\mathbb{1} = R^\dagger R = (\mathbb{1} + M^\dagger d\alpha)(\mathbb{1} + M d\alpha) \simeq \mathbb{1} + (M^\dagger + M)d\alpha,$$

i.e. the matrix M is anti-Hermitian:

$$M^\dagger = -M. \tag{9.37}$$

We set therefore

$$M^\dagger := iH, \quad H = H^\dagger. \tag{9.38}$$

For

$$\rho^{(1)}(\alpha) = \rho^{(0)}(\alpha_0 + d\alpha)$$

follows from Eq. (9.32) in the Schrödinger picture that

$$\begin{aligned}
\rho^{(0)}(\alpha_0 + d\alpha) &= R(\alpha)\rho^{(0)}(\alpha_0)R^\dagger(\alpha) \\
&= (\mathbb{1} - iH d\alpha)\rho^{(0)}(\alpha_0)(\mathbb{1} + iH d\alpha) \\
&= \rho^{(0)}(\alpha_0) - i[H, \rho^{(0)}(\alpha_0)]d\alpha,
\end{aligned}$$

or (with $\rho^{(0)} \to \rho$)

$$\frac{\partial \rho}{\partial \alpha} = -i[H, \rho]. \tag{9.39}$$

If the dependence on α is not contained in $\rho^{(1)}$ but in F' (cf. Eq. (9.33)) we obtain correspondingly in the Heisenberg picture the equation

$$\frac{\partial F}{\partial \alpha} = +i[H, F]. \tag{9.40}$$

Equations (9.39) and (9.40) can be considered as quasi-equations of motion. Comparison of these equations with Eqs. (2.21) and (2.46) suggests to define *Poisson brackets* of matrices by the correspondence

$$\{A, B\} := i[A, B]. \tag{9.41}$$

We see therefore that the 2×2 matrices entering the description of polarization properties of waves possess the same generalizability as the particle considerations of classical mechanics in Chapter 2. In Chapter 27 we shall see that the Poisson brackets here are actually Dirac brackets because the gauge fixing condition (9.2) has to be taken into account.

Chapter 10

Quantum Theory: The General Formalism

10.1 Introductory Remarks

In this chapter we are concerned with the formulation of quantum mechanics in general. Thus in particular we establish the uncertainty relation for observables in general, we establish their Heisenberg equations of motion, the Schrödinger, Heisenberg and interaction pictures, and consider time-dependent perturbation theory.

10.2 States and Observables

Every state of a physical system is represented by a certain ket-vector $|u\rangle$ which is an element of the corresponding Hilbert space $\mathcal{H}$. In keeping with our earlier notation the expectation values of dynamical variables are defined as follows:

Postulate: The expectation value or mean value of an arbitrary function $F(A)$ of an observable A in the case of a system in the state $|u\rangle \in \mathcal{H}$ is defined as the expression

$$\langle F(A)\rangle = \langle u|F(A)|u\rangle. \tag{10.1}$$

This expectation value remains unchanged if $|u\rangle$ is replaced by $e^{i\alpha}|u\rangle$ with α real.

Thus the expectation value with respect to the vector describing the state in the Hilbert space does not depend on this phase factor. To insure that the expectation value of the unit vector or identity operator is 1, we must have

$$\langle u|u \rangle = 1. \tag{10.2}$$

Naturally one can also write

$$\langle F(A) \rangle = \frac{\langle u|F(A)|u \rangle}{\langle u|u \rangle}.$$

The first step in the investigation of a quantum mechanical system is to specify its dynamical variables A. We can describe a system as one with a classical analogue if the Hamilton operator of the quantum system is obtained from the Hamilton function of classical mechanics by the correspondence

$$x_i \rightarrow \hat{q}_i, \quad p_i \rightarrow \hat{p}_i,$$

where $i = 1, 2, 3$ represent the three Cartesian coordinate directions of x, y, z and for definiteness here — which will not be maintained throughout — the "hat" denotes that the quantity is an operator. Such a correspondence with classical mechanics is not always possible,[*] a conspicuous exception being for instance a spin system. Here, however, we assume this now, and assume that $A = A(p, q)$. The phase space coordinates q_i, p_i are called *fundamental observables*. For these the following rules of *canonical quantization* are postulated:

Postulate: The operators $\hat{q}_i, \hat{p}_i$ are postulated to obey the following *fundamental commutation relations* (again leaving out "hats"):

$$[q_i, q_j] = 0, \quad [p_i, p_j] = 0, \quad [q_i, p_j] = i\hbar\delta_{ij}. \tag{10.3}$$

These fundamental commutators determine the commutators of arbitrary observables $A = A(p, q)$, such as for instance of the angular momentum operators L_i. In order to avoid multivaluedness one always starts from Cartesian coordinates in position or configuration space.

10.2.1 Uncertainty relation for observables A, B

We first establish the following theorem: If A and B are observables and hence hermitian operators obeying the commutation relation

$$[A, B] = i\hbar, \tag{10.4}$$

---

[*]Even in those cases where this is possible the operator correspondence may not be unique. E.g. classically $q_i p_i = p_i q_i$, which clearly does not hold in the case of operators.

then their uncertainties $\triangle A, \triangle B$ defined by the *mean square deviation*

$$\triangle C = (\langle C^2 \rangle - \langle C \rangle^2)^{1/2}, \quad C = A, B, \dots, \tag{10.5}$$

are subject to the following inequality called *uncertainty relation*:

$$\triangle A \triangle B \geq \frac{1}{2}\hbar. \tag{10.6}$$

In proving the relation we first set

$$\hat{A} := A - \langle A \rangle, \quad \hat{B} := B - \langle B \rangle, \tag{10.7a}$$

so that

$$\langle \hat{A} \rangle = \langle A \rangle - \langle A \rangle = 0, \quad \langle \hat{B} \rangle = 0. \tag{10.7b}$$

Then

$$\begin{aligned}
[\hat{A}, \hat{B}] = \hat{A}\hat{B} - \hat{B}\hat{A} &= (A - \langle A \rangle)(B - \langle B \rangle) - (B - \langle B \rangle)(A - \langle A \rangle) \\
&= [A, B] = i\hbar, \tag{10.8}
\end{aligned}$$

and, using Eq. (10.7b),

$$\begin{aligned}
\triangle \hat{A} &= (\langle \hat{A}^2 \rangle - \langle \hat{A} \rangle^2)^{1/2} \overset{(10.7a)}{=} (\langle A^2 + \langle A \rangle^2 - 2A\langle A \rangle \rangle - \langle \hat{A} \rangle^2)^{1/2} \\
&= (\langle A^2 \rangle + \langle A \rangle^2 - 2\langle A \rangle^2 - 0)^{1/2} \\
&= (\langle A^2 \rangle - \langle A \rangle^2)^{1/2} = \triangle A. \tag{10.9}
\end{aligned}$$

Since $\hat{A} = A - \langle A \rangle$, we have

$$\hat{A}^2 = A^2 - 2A\langle A \rangle + \langle A \rangle^2, \quad \langle \hat{A}^2 \rangle = \langle A^2 \rangle - \langle A \rangle^2 = (\triangle A)^2. \tag{10.10}$$

Thus

$$\triangle A = \triangle \hat{A} = \langle \hat{A}^2 \rangle^{1/2}, \quad \triangle B = \triangle \hat{B} = \langle \hat{B}^2 \rangle^{1/2}. \tag{10.11}$$

Hence

$$(\triangle A)^2 (\triangle B)^2 = \langle \hat{A}^2 \rangle \langle \hat{B}^2 \rangle = \langle u|\hat{A}^2|u \rangle \langle u|\hat{B}^2|u \rangle, \tag{10.12}$$

where $|u\rangle \in \mathcal{H}$ is a ket-vector representing the state of the system. Applying to the expression (10.12) the *Schwarz inequality* (cf. Eqs. (3.9a), (4.5)) (for $\hat{A}|u\rangle, \hat{B}|u\rangle \in \mathcal{H}$ as vectors) we obtain (for an application see Example 10.1)

$$\begin{aligned}
(\triangle A)^2 (\triangle B)^2 &\equiv \langle u|\hat{A}^2|u \rangle \langle u|\hat{B}^2|u \rangle \equiv ||\hat{A}|u\rangle||^2 ||\hat{B}|u\rangle||^2 \\
&\geq |\langle u|\hat{A}\hat{B}|u \rangle|^2. \tag{10.13}
\end{aligned}$$

Next we separate the product $\hat{A}\hat{B}$ into its hermitian and antihermitian components[†] and use Eq. (10.4). Then

$$\hat{A}\hat{B} = \frac{1}{2}(\hat{A}\hat{B} + \hat{B}\hat{A}) + \frac{1}{2}(\hat{A}\hat{B} - \hat{B}\hat{A}) \stackrel{(10.8)}{=} \frac{1}{2}(\hat{A}\hat{B} + \hat{B}\hat{A}) + \frac{1}{2}i\hbar. \qquad (10.14)$$

Hence

$$\langle u|\hat{A}\hat{B}|u\rangle = \left\langle \frac{1}{2}(\hat{A}\hat{B} + \hat{B}\hat{A}) \right\rangle + \frac{1}{2}i\hbar, \qquad (10.15)$$

where, since A and B are *hermitian*,

$$\begin{aligned}
\langle \hat{A}\hat{B} + \hat{B}\hat{A} \rangle &\equiv \langle u|\hat{A}\hat{B} + \hat{B}\hat{A}|u\rangle = \langle u|(\hat{A}\hat{B} + \hat{B}\hat{A})^{\dagger}|u\rangle^{*} \\
&= \langle u|(\hat{A}\hat{B} + \hat{B}\hat{A})|u\rangle^{*}. \qquad (10.16)
\end{aligned}$$

This means

$$\langle \hat{A}\hat{B} + \hat{B}\hat{A} \rangle$$

is real. We can therefore rewrite Eq. (10.15) as

$$\langle u|\hat{A}\hat{B}|u\rangle = R + iI, \quad R = \frac{1}{2}\langle \hat{A}\hat{B} + \hat{B}\hat{A} \rangle, \quad I = \frac{1}{2}\hbar. \qquad (10.17)$$

From this we obtain

$$|\langle u|\hat{A}\hat{B}|u\rangle|^{2} = R^{2} + I^{2},$$

and with Eq. (10.13)

$$(\triangle A)^{2}(\triangle B)^{2} \geq R^{2} + I^{2},$$

i.e.

$$\triangle A \, \triangle B \geq I = \frac{1}{2}\hbar, \qquad (10.18)$$

as had to be shown. The equal-to-sign applies in the case when (a) it applies in the Schwarz inequality, and (b) when

$$\langle \hat{A}\hat{B} + \hat{B}\hat{A} \rangle = 0.$$

The condition (a) is satisfied when, for a complex constant c,

$$\hat{A}|u\rangle = c\hat{B}|u\rangle, \qquad (10.19)$$

and hence for condition (b)

$$\begin{aligned}
0 &= \langle u|\hat{A}\hat{B} + \hat{B}\hat{A}|u\rangle = \langle u|\hat{A}\hat{B}|u\rangle + \langle u|\hat{B}\hat{A}|u\rangle \\
&= c^{*}\langle u|\hat{B}\hat{B}|u\rangle + c\langle u|\hat{B}\hat{B}|u\rangle = (c^{*} + c)\langle u|\hat{B}^{2}|u\rangle.
\end{aligned}$$

[†]See Sec. 4.3.

Thus $c = -c^*$, i.e. $\Re c = 0$, implying that the vector $|u\rangle$ has to satisfy the following equation:

$$\hat{A}|u\rangle = i(\Im c)\hat{B}|u\rangle \qquad (10.20)$$

or, in view of Eq. (10.7a),

$$(A - \langle A\rangle)|u\rangle = i(\Im c)(B - \langle B\rangle)|u\rangle.$$

Example 10.1: Angular momentum and uncertainties

Show that in the case of angular momentum operators $L_i, i = x, y, z$, the following uncertainty relation holds:

$$(\triangle L_i)^2(\triangle L_j)^2 \geq \frac{1}{4}\hbar^2\langle L_k\rangle^2.$$

What does the relation imply for states $|u\rangle \equiv |lm\rangle$, for which one component of $\mathbf{L}$ is "sharp"?

Solution: Applying Eq. (10.13) to operators $\hat{L}_i \equiv L_i$, we obtain

$$(\triangle L_i)^2(\triangle L_j)^2 \geq |\langle u|L_iL_j|u\rangle|^2, \qquad \langle u|L_iL_j|u\rangle \equiv \Re + i\Im.$$

Separating the product into symmetric and antisymmetric parts, we have

$$L_iL_j = \frac{1}{2}(L_iL_j + L_jL_i) + \frac{1}{2}[L_i, L_j] = \frac{1}{2}(L_iL_j + L_jL_i) + \frac{1}{2}i\hbar\epsilon_{ijk}L_k,$$

and

$$\Re = \frac{1}{2}\langle u|L_iL_j + L_jL_i|u\rangle, \qquad \Im = \frac{1}{2}\hbar\langle u|\epsilon_{ijk}L_k|u\rangle,$$

so that

$$|\langle u|L_iL_j|u\rangle|^2 = \Re^2 + \Im^2 \geq \Im^2 = \frac{1}{4}\hbar^2\langle L_k\rangle^2.$$

Hence

$$(\triangle L_x)^2(\triangle L_y)^2 \geq \frac{1}{4}\hbar^2\langle L_z\rangle^2, \quad (\triangle L_y)^2(\triangle L_z)^2 \geq \frac{1}{4}\hbar^2\langle L_x\rangle^2, \quad (\triangle L_z)^2(\triangle L_x)^2 \geq \frac{1}{4}\hbar^2\langle L_y\rangle^2.$$

For states $|u\rangle$, such that L_z is "sharp", i.e. $\triangle L_z = 0$ (these are states $|lm\rangle$ with $L_z|lm\rangle = m|lm\rangle$), it follows that

$$\langle L_x\rangle^2 = 0, \quad \langle L_y\rangle^2 = 0, \quad \langle L_z\rangle^2 \neq 0,$$

implying

$$\langle L_x\rangle = 0, \quad \langle L_y\rangle = 0, \quad \text{i.e. } \langle L_x \pm iL_y\rangle \equiv \langle u|L_\pm|u\rangle = 0.$$

Thus in this case L_x, L_y are not "sharp", i.e. $\triangle L_x \neq 0, \triangle L_y \neq 0$.

10.3 One-Dimensional Systems

We now consider one-dimensional systems with a classical analogue as described earlier. The observables like angular momentum are functions of p and q. For these we have in the one-dimensional case the relation

$$[p, q] = -i\hbar.$$

Since this says that p and q do not commute (as operators), they do not have a common system of eigenvectors, as remarked earlier. This is why the operator p requires a representation in the space of eigenvectors of q, i.e. for $q|x\rangle = x|x\rangle$ we have

$$p|x\rangle = |x\rangle \left(-i\hbar \frac{\overleftarrow{\partial}}{\partial x} \right).$$

Now let A be an observable, i.e. $A = A(p,q)$. Then (see below) we can derive for A the following commutator equations

$$[q, A(q,p)] = i\hbar \frac{\partial A(q,p)}{\partial p} \qquad (10.21\text{a})$$

and

$$[p, A(q,p)] = -i\hbar \frac{\partial A(q,p)}{\partial q}. \qquad (10.21\text{b})$$

We obtain these equations as follows. We have (cf. Eq. (3.34d))

$$[A, BC] = [A, B]C + B[A, C].$$

Using $[q,p] = i\hbar$ we obtain therefore

$$[q, p^2] = [q, p]p + p[q, p] = 2i\hbar p = i\hbar \frac{\partial}{\partial p}(p^2).$$

Proceeding in this way we obtain

$$[q, p^n] = i\hbar \frac{\partial}{\partial p}(p^n). \qquad (10.22)$$

Similarly we obtain

$$[p, q^2] = [p, q]q + q[p, q] = -2i\hbar q = -i\hbar \frac{\partial}{\partial q}(q^2)$$

and

$$[p, q^n] = -i\hbar \frac{\partial}{\partial q}(q^n). \qquad (10.23)$$

Equations (10.21a) and (10.21b) now follow with the assumption that $A(q,p)$ can be expanded as a power series in q and p.

Next we have a closer look at the operator q. We demonstrate that
(a) its eigenvectors have infinite norm (in the sense of a delta function),
(b) its spectrum is necessarily continuous, and
(c) the spectrum is not degenerate.

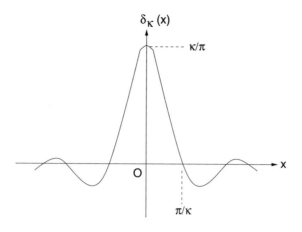

Fig. 10.1 The delta function for $\kappa \to \infty$.

We begin with (a). We have

$$\langle x|x'\rangle = \int_{-\infty}^{\infty} dk\langle x|k\rangle\langle k|x'\rangle \overset{(4.19)}{=} \frac{1}{2\pi}\int_{-\infty}^{\infty} dk e^{ikx}e^{-ikx'}$$
$$= \lim_{\kappa\to\infty}\delta_\kappa(x-x') = \delta(x-x'), \tag{10.24}$$

where

$$\delta_\kappa(x-x') = \frac{1}{2\pi}\int_{-\kappa}^{\kappa} dk e^{ik(x-x')} = \frac{\sin\kappa(x-x')}{\pi(x-x')}. \tag{10.25}$$

It is shown in Fig. 10.1 how the delta function arises in the limit $\kappa \to \infty$. We thus see explicitly that the vectors $|x\rangle$ have infinite norm (in the sense of delta function normalization).

For the cases (b) and (c) we consider the operator

$$U(a) = e^{-ipa/\hbar}. \tag{10.26}$$

Here p is an observable, i.e. an operator, and a a c-number. Since $p = p^\dagger$, it follows that

$$U^\dagger(a) = e^{ipa/\hbar} = U(a)^{-1} = U(-a), \tag{10.27}$$

so that

$$U(a)U^\dagger(a) = \mathbb{1}, \tag{10.28}$$

i.e. U is unitary. Replacing in Eq. (10.21a) A by U, we obtain

$$[q,U] = i\hbar\frac{\partial U}{\partial p} = aU.$$

i.e.

$$qU = Uq + aU = U(q + a),$$

or

$$qU|x\rangle = U(q + a)|x\rangle = (x + a)U|x\rangle. \tag{10.29}$$

This means: $U|x\rangle$ is eigenvector of q with eigenvalue $x + a$. Since this holds for any arbitrary value of a in $(-\infty, \infty)$, this means: With a unitary transformation (which leaves the matrix elements unaffected[‡] and hence the physically observable quantities) one can pass to any arbitrary eigenvalue in the domain $(-\infty, \infty)$. Thus all these values belong to the spectrum of the operator q, which means that the spectrum of this operator is a continuum. The eigenvectors have the same norm as $|x\rangle$:

$$\langle x|U^\dagger U|x'\rangle = \langle x|x'\rangle = \delta(x - x').$$

Evidently every eigenvalue has only one eigenvector, i.e. the spectrum is not degenerate.

In an analogous way by defining the unitary operator

$$U(b) = e^{iqb/\hbar}$$

(q operator, b a c-number), we can see that also p possesses only a continuous spectrum (from $-\infty$ to ∞), and that the eigenvectors do not have a finite norm and are normalized to a delta function according to

$$\langle p_x|p'_x\rangle = \delta(p_x - p'_x).$$

With the help of the observables p, q any other observable $F(p, q)$ can be constructed representing a dynamical variable.

10.3.1 The translation operator $U(a)$

We saw above, cf. Eq. (10.29), that with

$$U(a) = e^{-ipa/\hbar} \equiv U,$$

we obtain

$$qU|x\rangle = U(q + a)|x\rangle = (x + a)U|x\rangle, \tag{10.30}$$

or

$$qU(a)|x'\rangle = (x' + a)U(a)|x'\rangle. \tag{10.31}$$

[‡]Observe that $\langle\psi|A(p,q)|\psi\rangle = \langle\psi|U^\dagger U A(p,q) U^\dagger U|\psi\rangle$, $U^\dagger U = \mathbb{1}$.

Moreover, since $q|x'\rangle = x'|x'\rangle$, we have

$$q|x' + a\rangle = (x' + a)|x' + a\rangle. \tag{10.32}$$

Comparing Eqs. (10.31) and (10.32) we see that

$$|x' + a\rangle \propto U(a)|x'\rangle, \quad \text{i.e.} \quad U(a)|x'\rangle = c|x' + a\rangle. \tag{10.33}$$

Since U is unitary, it follows that

$$\begin{aligned}
\delta(x'' - x') &= \langle x''|U^\dagger(a)U(a)|x'\rangle = \langle x'' + a|c^*c|x' + a\rangle \\
&= c^*(a, x'')c(a, x')\delta(x'' - x'),
\end{aligned}$$

i.e.

$$|c(a, x')| = 1. \tag{10.34}$$

We choose the phase such that

$$U(a)|0\rangle = |a\rangle, \quad \text{i.e.} \quad |x'\rangle = U(x')|0\rangle. \tag{10.35}$$

It follows that

$$\begin{aligned}
U(a)|x'\rangle &= U(a)U(x')|0\rangle = U(a + x')|0\rangle \\
&= |x' + a\rangle,
\end{aligned} \tag{10.36}$$

i.e. the operator $U(a)$ acts to shift the value x' by the amount a. The operator is therefore called *translation operator*.

It also follows that

$$\langle x'|U(a)|x''\rangle = \langle x'|x'' + a\rangle = \delta(x' - x'' - a). \tag{10.37}$$

With the help of the latter expression we can calculate, for instance, off-diagonal elelemts of the $\{x\}$-representation of the momentum operator, i.e. $\langle x'|p|x''\rangle$. In the expression of Eq. (10.37) we set (with ϵ infinitesimal)

$$U(\epsilon) \simeq \mathbb{1} - \frac{i}{\hbar}p\epsilon. \tag{10.38}$$

Then

$$\langle x'|U(\epsilon)|x''\rangle = \delta(x' - x'' - \epsilon) \simeq \delta(x' - x'') - \frac{i}{\hbar}\epsilon\langle x'|p|x''\rangle,$$

i.e.

$$\langle x'|p|x''\rangle = \frac{\hbar}{i}\lim_{\epsilon \to 0}\frac{\delta(x' - x'') - \delta(x' - x'' - \epsilon)}{\epsilon},$$

or

$$\langle x'|p|x''\rangle = \frac{\hbar}{i}\delta'(x' - x''). \tag{10.39}$$

We mention in passing that, since

$$\delta(x) = \frac{1}{2\pi} \int dp e^{ipx}, \quad \text{i.e.} \quad \delta'(x) = \frac{1}{2\pi} \int dp\, ip\, e^{ipx},$$

it follows that $\delta'(x)$ is an odd function of x.[§] It follows that

$$\langle x''|p|x'\rangle = -\frac{\hbar}{i}\delta'(x'-x'') = \langle x'|p|x''\rangle^*. \tag{10.40}$$

This expresses that p is hermitian.

10.4   Equations of Motion

Let $|\psi(t_0)\rangle$ be the state vector at time t_0 of a completely isolated system (i.e. isolated from any measuring apparatus which would disturb the system and hence its behaviour in the course of time) and let $|\psi(t)\rangle$ be its state vector at a later time $t > t_0$ (with no interference in between). For the way the state of the system develops in the course of time we make the following postulate:

Postulate: A linear operator $U(t, t_0)$ exists which is such that[¶]

$$|\psi(t)\rangle = U(t, t_0)|\psi(t_0)\rangle. \tag{10.41}$$

The operator $U(t, t_0)$ is called the *time development operator*.

We assume first of all, a conservative system.   In this case the classical Hamilton function does not depend explicitly on t. Let $|u_E\rangle$ be an eigenvector of the Hamilton operator H with eigenvalue E, i.e.

$$H|u_E(t_0)\rangle = E|u_E(t_0)\rangle. \tag{10.42}$$

We introduce another postulate:

Postulate: The time dependence of the vector $|u_E(t)\rangle$ is given by the relation

$$|u_E(t)\rangle = e^{-iH(t-t_0)/\hbar}|u_E(t_0)\rangle, \tag{10.43}$$

so that

$$U(t, t_0) = e^{-iH(t-t_0)/\hbar}. \tag{10.44}$$

Differentiating $U(t, t_0)$ with respect to t we obtain

$$i\hbar\frac{d}{dt}U(t, t_0) = HU(t, t_0). \tag{10.45}$$

[§]Observe that $\delta'(-x) = \frac{1}{2\pi}\int_{-\infty}^{\infty} dp\, ip\, e^{-ipx} = \frac{1}{2\pi}\int_{\infty}^{-\infty} d(-p)\,(-ip)\,e^{ipx} = -\delta'(x)$.

[¶]This is in conformity with our earlier postulate in Sec. 5.4. See Eqs. (5.27) and (5.35).

We now demonstrate that Eq. (10.45) has very general validity. In the immediate considerations it is not necessary to restrict ourselves to systems with classical analogues. From Eq. (10.41) we obtain

$$
\begin{aligned}
|\psi(t_2)\rangle &= U(t_2, t_1)|\psi(t_1)\rangle = U(t_2, t_1)U(t_1, t_0)|\psi(t_0)\rangle \\
&= U(t_2, t_0)|\psi(t_0)\rangle,
\end{aligned}
$$

i.e.

$$
U(t_2, t_0) = U(t_2, t_1)U(t_1, t_0). \tag{10.46}
$$

But obviously (cf. Eq. (10.44)) we have

$$
U(t, t) = \mathbb{1}.
$$

Hence

$$
U(t, t_0)U(t_0, t) = U(t, t) = \mathbb{1},
$$

and therefore

$$
[U(t, t_0)]^{-1} = U(t_0, t).
$$

Let us set for small values of ϵ:

$$
U(t + \epsilon, t) = \mathbb{1} - \frac{i}{\hbar}\epsilon\tilde{H}(t). \tag{10.47}
$$

The operator $\tilde{H}$ is d efined by this relation. Then with

$$
U(t + \epsilon, t_0) = U(t + \epsilon, t)U(t, t_0)
$$

we have

$$
\begin{aligned}
\frac{d}{dt}U(t, t_0) &= \lim_{\epsilon \to 0}\frac{U(t + \epsilon, t_0) - U(t, t_0)}{\epsilon} = \lim_{\epsilon \to 0}\frac{[U(t + \epsilon, t) - \mathbb{1}]U(t, t_0)}{\epsilon} \\
&\overset{(10.47)}{=} -\frac{i}{\hbar}\tilde{H}(t)U(t, t_0),
\end{aligned}
$$

or

$$
i\hbar\frac{d}{dt}U(t, t_0) = \tilde{H}(t)U(t, t_0) \tag{10.48}
$$

with $U(t_0, t_0) = \mathbb{1}$. Evidently $U(t, t_0)$ can be expressed as an integral as the solution of Eq. (10.48):

$$
U(t, t_0) = \mathbb{1} - \frac{i}{\hbar}\int_{t_0}^{t} dt'\, \tilde{H}(t')U(t', t_0) \tag{10.49}
$$

(differentiating we obtain immediately Eq. (10.48)).

We obtain the differential equation for the development of states of the system in the course of time by differentiation of Eq. (10.41), i.e.

$$\frac{d}{dt}|\psi(t)\rangle = \left[\frac{d}{dt}U(t, t_0)\right]|\psi(t_0)\rangle,$$

i.e. the *Schrödinger equation*

$$i\hbar\frac{d}{dt}|\psi(t)\rangle = \tilde{H}|\psi(t)\rangle. \tag{10.50}$$

Comparison with the equation postulated earlier, Eq. (5.30), requires the identification

$$\tilde{H}(t) = H \equiv \text{ Hamilton operator}. \tag{10.51}$$

Since

$$|\psi(t + dt)\rangle = U(t + dt, t)|\psi(t)\rangle = \left(\mathbb{1} - \frac{i}{\hbar}Hdt\right)|\psi(t)\rangle$$

and since H is hermitian, it follows that

$$U(t + dt, t) = \mathbb{1} - \frac{i}{\hbar}Hdt \tag{10.52}$$

is the operator of an infinitesimal unitary transformation (recall that $U = \mathbb{1} + i\epsilon F$ satisfies the unitarity condition $UU^\dagger = \mathbb{1}$ provided $F = F^\dagger$). The operator $U(t, t_0)$ can therefore be interpreted as a product of a sequence of infinitesimal unitary transformations.

One describes as *Schrödinger picture* the present description of a system in which the state of the system is represented in terms of a vector $|\psi(t)\rangle$ which changes with time t, whereas the physical quantities are represented by observables in $\mathcal{H}$ which do not depend explicitly on t. The eigenvectors of these observables are also constant in time.

Now let $|\psi(t)\rangle$ be the state of the sytem at time t, the instant at which a measurement is performed on the system. Then the probability to find the system in a state $|\chi\rangle$ is

$$|\langle\chi|\psi\rangle|^2 = \langle\chi|\psi\rangle\langle\psi|\chi\rangle.$$

With unitary transformations we can pass over to equivalent descriptions. Since the only measurable quantities are moduli of scalar products, and these remain unchanged under unitary transformations, the predictions of different descriptions are identical. Another such description is the *Heisenberg picture*. In the following the subscripts S and H indicate Schrödinger and Heisenberg picture quantities. Then

$$|\psi_S(t)\rangle = U(t, t_0)|\psi_S(t_0)\rangle \tag{10.53}$$

and

$$|\psi_H\rangle = U^\dagger(t,t_0)|\psi_S(t)\rangle = |\psi_S(t_0)\rangle = \text{constant in time.} \qquad (10.54)$$

Observables transform correspondingly with a similarity transformation, so that a matrix element (i.e. a scalar product) remains unchanged:

$$A_H(t) = U^\dagger(t,t_0)A_S U(t,t_0). \qquad (10.55)$$

Thus A_H is explicitly time-dependent, i.e. subject to a continuous change, even if A_S is not time-dependent. For $A_S(q,p)$ in the Schrödinger picture $\partial A_S/\partial t = 0$. It does not always have to be the case that A_S does not contain an explicit time dependence; for instance in cases of non-equilibrium or if part of a system disappears through absorption, but here, as a rule, we consider only cases without explicit time dependence of A_S. With explicit time dependence, i.e. $A_S(q,p,t)$, we have to take into account the contribution $\partial A_S/\partial t$. Differentiating Eq. (10.55) and using Eq. (10.48), we obtain

$$
\begin{aligned}
i\hbar\frac{d}{dt}A_H &= -U^\dagger H A_S U + i\hbar U^\dagger \frac{\partial A_S}{\partial t} U + U^\dagger A_S H U \\
&= U^\dagger[A_S, H]U + i\hbar U^\dagger \frac{\partial A_S}{\partial t} U. \qquad (10.56)
\end{aligned}
$$

Here H is the Hamilton operator in the Schrödinger picture. In the Heisenberg picture it is

$$H_H = U^\dagger H U, \qquad (10.57)$$

and

$$
\begin{aligned}
U^\dagger[A_S, H]U &= U^\dagger A_S H U - U^\dagger H A_S U \\
&= (U^\dagger A_S U)(U^\dagger H U) - (U^\dagger H U)(U^\dagger A_S U) \\
&= A_H H_H - H_H A_H \\
&= [A_H, H_H], \qquad (10.58)
\end{aligned}
$$

so that

$$i\hbar\frac{d}{dt}A_H = [A_H, H_H] + i\hbar U^\dagger \frac{\partial A_S}{\partial t} U,$$

or

$$i\hbar\frac{d}{dt}A_H = [A_H, H_H], \quad \text{if} \quad \frac{\partial A_S}{\partial t} = 0. \qquad (10.59)$$

This equation is called *Heisenberg equation of motion*. Thus, in order to obtain the Heisenberg picture (description), the vector space of the Schrödinger picture (description) is transformed in such a way that the state of the system is described by a vector $|\psi_H\rangle$ which is constant in time. However the associated observables are time-dependent.

Wave mechanics is now seen to be the formulation of quantum mechanics in the Schrödinger picture. In many cases the Schrödinger picture is more amenable for explicit calculations. In the Heisenberg picture essential properties of a system can frequently be recognized more easily in relation to their counterparts in classical mechanics — in both cases the time development of a system is given by the time dependence of the dynamical variables. In particular, in the case of the fundamental dynamical variables q_i, p_i we have in the Heisenberg picture (replacing in the above A_H by q_i, p_i):

$$\frac{dq_i}{dt} = \frac{1}{i\hbar}[q_i, H] = \frac{\partial H}{\partial p_i} \tag{10.60a}$$

and

$$\frac{dp_i}{dt} = \frac{1}{i\hbar}[p_i, H] = -\frac{\partial H}{\partial q_i}, \tag{10.60b}$$

where the expressions on the right follow from Eqs. (10.21a) and (10.21b). We see that formally one obtains Hamilton's equations of classical mechanics, provided the Poisson brackets in the latter are replaced by commutators. We also observe that the momentum p_i is conserved when $[p_i, H] = 0$.

Finally we add a comment concerning conserved quantities. An observable C_H is called a conserved quantity, if

$$\frac{d}{dt}C_H = 0, \quad \text{i.e.} \quad [C_H, H_H] = 0. \tag{10.61}$$

Thus conserved quantities commute with the Hamilton operator. This holds also in the case of the Schrödinger picture, since in the transition from one to the other the commutator remains unchanged.

We end with a word of caution. The "pictures" just explained — with regard to the time development of a quantum system — are not to be confused with the representations of the theory. In our treatment of representations, as in the transition from the configuration or position space representation to the momentum space representation with the help of Fourier transforms, we used in essence unitary transformations of matrices. In the treatment of the pictures we used unitary transformations of operators (and vectors).

Example 10.2: Energy and time uncertainty relation[||]
Use the Heisenberg equation of motion and the general relation (10.13) to derive the uncertainty relation for energy and time.

Solution: We have for an observable A

$$\langle A \rangle = \langle \psi_H | A_H | \psi_H \rangle.$$

[||] For an extensive discussion see B. A. Orfanopoulos [224].

Then, since $|\psi_H\rangle$ is time-independent,

$$\frac{d}{dt}\langle A\rangle = \left\langle\frac{dA_H}{dt}\right\rangle = \frac{1}{i\hbar}\langle[A_H, H_H]\rangle + \left\langle\frac{\partial A_S}{\partial t}\right\rangle. \tag{10.62}$$

The Hamiltonian H does not depend on time. Let $|\psi\rangle \equiv |\psi_H\rangle$ represent the state of the system. We construct the following vectors using Eq. (10.7a), i.e.

$$\hat{A}|\psi\rangle \equiv (A - \langle A\rangle)|\psi\rangle, \quad \hat{H}|\psi\rangle \equiv (H - \langle H\rangle)|\psi\rangle, \tag{10.63}$$

so that with Eqs. (10.13) and (10.15):

$$(\triangle A)^2(\triangle H)^2 \geq |\langle\psi|\hat{A}\hat{H}|\psi\rangle|^2 \quad (\equiv |\langle\psi|\hat{H}\hat{A}|\psi\rangle|^2) \geq \frac{\hbar^2}{4}|\langle\psi|[\hat{A},\hat{H}]|\psi\rangle|^2. \tag{10.64}$$

Using Eq. (10.62) and $\partial A_S/\partial t = 0$, we can replace

$$\langle[A_H, H_H]\rangle \quad \text{by} \quad i\hbar\frac{d}{dt}\langle A\rangle.$$

Hence we obtain

$$\frac{\triangle A}{d\langle A\rangle/dt}\triangle H \geq \frac{1}{2}\hbar, \tag{10.65}$$

and, if we set

$$\triangle H = \triangle E \quad \text{and} \quad \tau_A = \frac{\triangle A}{d\langle A\rangle/dt},$$

we obtain

$$\tau_A\triangle E \geq \frac{1}{2}\hbar. \tag{10.66}$$

This means, the "centre of mass" of the statistical distribution of A, i.e. $\langle A\rangle$, is displaced by $\triangle A$ in the interval $\triangle\tau = \tau_A$. Such a characteristic time interval can be defined for any dynamical variable. If τ is the smallest such characteristic interval of time, then if a measurement is made at time t', this is practically the same as at time t for times t with $|t - t'| < \tau$.

We defined earlier stationary states as states with a definite energy E and wave function $\Psi = \psi(\mathbf{r})\exp(-iEt/\hbar)$. The probability density

$$P(\mathbf{r}) = |\psi(\mathbf{r})|^2 = \langle\psi|\mathbf{r}\rangle\langle\mathbf{r}|\psi\rangle$$

is independent of t, i.e. a position or momentum observation is independent of t. This means that for a physical variable $A(q, p)$ we then have

$$0 = \frac{d}{dt}\langle A\rangle = \frac{1}{i\hbar}\langle[A, H]\rangle = 0, \tag{10.67}$$

and

$$\tau_A \propto \frac{1}{d\langle A\rangle/dt} \quad \text{is} \quad \infty \quad \text{and} \quad \triangle E = 0. \tag{10.68}$$

The eigenvalues a of A (i.e. of $A|\phi\rangle = a|\phi\rangle$) are then called "*good quantum numbers*". To put it in a different form: The eigenvalues of an observable are called "good" if $[A, H] = 0$.

10.5   States of Finite Lifetime

When E is real and $P(\mathbf{x})$ is independent of t and $\int d\mathbf{x} P(\mathbf{x}) = 1$, the wave function $\psi_E(\mathbf{x}, t) = \psi(\mathbf{x}) \exp(-iEt/\hbar)$ describes states called *bound states* of unlimited lifetime, i.e. states with sharp energy ($\triangle E = 0$). States with finite lifetime ($\triangle t = \tau \neq \infty$) are those whose probability density falls off after a certain length of time called "*lifetime*". On the basis of the uncertainty relation $\triangle E \triangle t \gtrsim \hbar$ these must have an uncertainty $\triangle E$ in their energy spectrum. This means, the wave function $\psi_{E_0}(\mathbf{x}, t)$ of such a state must be the superposition of states with different energies about E_0 and a corresponding weight function f so that

$$\psi_{E_0}(\mathbf{x}, t) = \int dE' \psi_{E'}(\mathbf{x}) e^{-iE't/\hbar} f_{E_0}(E'). \tag{10.69}$$

(Observe that if ψ consists of only one state with energy E in the domain of integration of E', then $f_E(E') = \delta(E - E')$ and integration with respect to E' yields again $\psi_E(\mathbf{x}, t) = \psi_E(\mathbf{x}) \exp(-iEt/\hbar)$). Recall for instance radioactive decay. Energy and decay length of the radiated α-particles are characteristic properties of naturally radioactive nuclei. But the number of radioactive (thus excited) nuclei decreases exponentially, as is observed. This means that in the case of decaying particles, the particle density diminishes in the course of time ($t > 0$). The wave function of such a state with energy E_0 and lifetime $\tau \equiv \hbar/\gamma > 0$ could, for instance, have around E_0 the form

$$\psi_{E_0}(\mathbf{x}, t) = \psi_{E_0}(\mathbf{x}) e^{-iE_0 t/\hbar} e^{-\gamma t/2\hbar} \theta(t). \tag{10.70}$$

From Eqs. (10.69) and (10.70) we can determine the function $f_{E_0}(E)$ by integration with respect to t. First we obtain from Eq. (10.69):

$$\begin{aligned}
\int_{-\infty}^{\infty} \psi_{E_0}(\mathbf{x}, t) e^{iEt/\hbar} dt &= \int_{-\infty}^{\infty} dt \int dE' \psi_{E'}(\mathbf{x}) f_{E_0}(E') e^{i(E-E')t/\hbar} \\
&= 2\pi\hbar \int dE' \psi_{E'}(\mathbf{x}) f_{E_0}(E') \delta(E - E') \\
&= 2\pi\hbar f_{E_0}(E) \psi_E(\mathbf{x}). \tag{10.71}
\end{aligned}$$

From Eq. (10.70) we obtain:

$$\int_{-\infty}^{\infty} \psi_{E_0}(\mathbf{x}, t) e^{iEt/\hbar} dt = \psi_{E_0}(\mathbf{x}) \int_0^{\infty} dt e^{\{-\gamma/2 + i(E-E_0)\}t/\hbar}$$

$$= \psi_{E_0}(\mathbf{x}) \left[\frac{e^{\{-\gamma/2 + i(E-E_0)\}t/\hbar}}{-\gamma/2\hbar + i(E - E_0)/\hbar} \right]_{t=0}^{\infty} = \frac{\hbar}{(-i)(E - E_0 + i\gamma/2)} \psi_{E_0}(\mathbf{x}). \tag{10.72}$$

From the last two equations we deduce for E close to E_0, i.e. $\psi_E(\mathbf{x}) \simeq \psi_{E_0}(\mathbf{x})$, that[**]

$$f_{E_0}(E) = -\frac{1}{2\pi i}\frac{1}{E - E_0 + i\gamma/2}. \tag{10.73}$$

This result is known as the *Breit–Wigner formula*. Considered as a function of E, the formula says that $f_{E_0}(E)$ possesses a simple pole at $E = E_0 - i\gamma/2$, $\gamma > 0$, whose real part E_0 specifies the energy of the state and and whose imaginary part $\gamma/2$ specifies the lifetime $\tau = \hbar/\gamma$. The state with lifetime $\tau = \hbar/\gamma$ is not a discrete state. A state is discrete if its immediate neighbourhood (in energy) does not contain some other state, which means that a nucleous with discrete energy E_0 does not possesses some other admissable level close to E_0. We obtained the result (10.73) precisely by assuming with

$$f_{E_0}(E) \neq \delta(E - E_0)$$

a "smearing" of states around the energy E_0 (with uncertainty $\triangle E$). We see therefore that the states with lifetime $\tau < \infty$ and $\tau\triangle E \gtrsim \hbar$ belong to the continuum of the spectrum. They are called *"resonance states"*.

10.6 The Interaction Picture

One more motion picture of quantum mechanical systems is in use, and is called *"interaction picture"* or *"Dirac picture"*. This description contains effectively parts of the other two pictures and is particularly useful when Eq. (10.45) can be solved only approximately (in actual fact, of course, every unitary transformation of states and operators defines a possible "picture").

We saw previously that if the state of a system is known at time t_0, i.e. $|\psi(t_0)\rangle$, then $|\psi(t)\rangle$ for $t > t_0$ follows from Eq. (10.41), i.e.

$$|\psi(t)\rangle = U(t, t_0)|\psi(t_0)\rangle, \tag{10.74}$$

where U is the solution of Eq. (10.45), i.e.

$$i\hbar\frac{d}{dt}U(t, t_0) = HU(t, t_0). \tag{10.75}$$

Thus the solution of this equation is the main problem. Let $U^{(0)}(t, t_0)$ be an approximate but unitary solution of this equation, i.e.

$$U = U^{(0)}(t, t_0)U'(t, t_0), \quad (U^{(0)} \text{ unitary}). \tag{10.76}$$

[**]For specific applications see e.g. Eqs. (14.105) and (20.11).

Then

$$i\hbar\frac{d}{dt}(U^{(0)}U') = HU^{(0)}U',$$ (10.77)

i.e.

$$U^{(0)\dagger}\left[i\hbar U^{(0)}\frac{d}{dt}U' = HU^{(0)}U' - i\hbar\frac{dU^{(0)}}{dt}U'\right],$$

or

$$i\hbar\frac{d}{dt}U' = U^{(0)\dagger}\left(HU^{(0)} - i\hbar\frac{dU^{(0)}}{dt}\right)U'$$ (10.78)

with the initial condition

$$U'(t_0, t_0) = 1.$$

For a "good" approximation $U \approx U^{(0)}$ we have

$$HU^{(0)} - i\hbar\frac{d}{dt}U^{(0)} \approx 0,$$ (10.79)

i.e. with Eq. (10.78)

$$\frac{d}{dt}U' \sim 0,$$

i.e. U' has only a weak t-dependence (i.e. varies only slowly with t).

Now let $H^{(0)}(t)$, with Hamiltonian $H = H^{(0)} + H'$, be defined by the relation

$$H^{(0)}U^{(0)} - i\hbar\frac{dU^{(0)}}{dt} = 0 \text{ (exact)}, \quad U^{(0)} = e^{-iH^{(0)}t/\hbar},$$ (10.80)

i.e.

$$H^{(0)}(t) = i\hbar\frac{dU^{(0)}}{dt}U^{(0)\dagger}.$$ (10.81)

In addition we have

$$H^{(0)\dagger}(t) = -i\hbar U^{(0)}\left(\frac{dU^{(0)}}{dt}\right)^{\dagger}.$$

But $U^{(0)}$ is unitary, i.e. $U^{(0)}U^{(0)\dagger} = 1$, i.e.

$$U^{(0)}\frac{dU^{(0)\dagger}}{dt} = -\frac{dU^{(0)}}{dt}U^{(0)\dagger},$$

so that

$$H^{(0)\dagger}(t) = i\hbar\frac{dU^{(0)}}{dt}U^{(0)\dagger} = H^{(0)}(t).$$ (10.82)

It follows that

$$H^{(0)}(t) \text{ is hermitian.}$$ (10.83)

We use now the subdivision of H into unperturbed part and interaction, i.e.

$$H = H^{(0)}(t) + H', \tag{10.84}$$

where H' is also hermitian. We multiply this relation from the left by $U^{(0)\dagger}$ and from the right by $U^{(0)}$ and obtain

$$
\begin{aligned}
U^{(0)\dagger} H U^{(0)} &= U^{(0)\dagger}(H^{(0)}U^{(0)}) + U^{(0)\dagger}H'U^{(0)} \\
&= i\hbar U^{(0)\dagger}\frac{dU^{(0)}}{dt} + U^{(0)\dagger}H'U^{(0)},
\end{aligned} \tag{10.85}
$$

where we used Eq. (10.80). We now multiply this equation from the right by U' and insert the first expression on the right (of Eq. (10.85) multiplied by U') on the right of Eq. (10.78). We also set

$$H_I' = U^{(0)\dagger}H'U^{(0)}, \tag{10.86}$$

and obtain from Eq. (10.78)

$$
\begin{aligned}
i\hbar\frac{dU'}{dt} &= U^{(0)\dagger}HU^{(0)}U' - \left[i\hbar U^{(0)\dagger}\frac{dU^{(0)}}{dt}U' \right] \\
&\overset{(10.85)}{=} U^{(0)\dagger}HU^{(0)}U' - \left[U^{(0)\dagger}HU^{(0)}U' - U^{(0)\dagger}H'U^{(0)}U' \right] \\
&= H_I'U'.
\end{aligned} \tag{10.87}
$$

We define correspondingly as interaction picture state

$$|\psi_I(t)\rangle = U^{(0)\dagger}|\psi_S(t)\rangle, \quad U^{(0)}|\psi_I(t)\rangle = |\psi_S(t)\rangle, \tag{10.88}$$

and as interaction picture version of an observable A the quantity

$$A_I(t) = U^{(0)\dagger}A_S U^{(0)}, \tag{10.89}$$

where $|\psi_S(t)\rangle$ and A_S are state and observable in the Schrödinger picture. Then

$$i\hbar\frac{d}{dt}|\psi_S(t)\rangle = i\hbar\frac{d}{dt}U^{(0)}|\psi_I(t)\rangle = U^{(0)}i\hbar\frac{d}{dt}|\psi_I(t)\rangle + i\hbar\left(\frac{dU^{(0)}}{dt}\right)|\psi_I(t)\rangle \tag{10.90}$$

and

$$H_I'|\psi_I(t)\rangle = U^{(0)\dagger}H'U^{(0)}U^{(0)\dagger}|\psi_S(t)\rangle = U^{(0)\dagger}H'|\psi_S(t)\rangle. \tag{10.91}$$

Since the Schrödinger equation is given by

$$i\hbar\frac{d}{dt}|\psi_S(t)\rangle = H|\psi_S(t)\rangle = (H^{(0)} + H')|\psi_S(t)\rangle, \tag{10.92}$$

we obtain by starting from the right hand side of Eq. (10.90) and replacing
the left hand side of this equation by the right hand side of Eq. (10.92)

$$U^{(0)} i\hbar \frac{d}{dt}|\psi_I(t)\rangle + i\hbar \left(\frac{dU^{(0)}}{dt}\right)|\psi_I(t)\rangle = H^{(0)} U^{(0)}|\psi_I(t)\rangle + H' U^{(0)}|\psi_I(t)\rangle,$$

so that with Eq. (10.80) (and multiplying by $U^{(0)\dagger}$)

$$i\hbar \frac{d}{dt}|\psi_I(t)\rangle = U^{(0)\dagger} H' U^{(0)}|\psi_I(t)\rangle,$$

i.e. (note the difference between H' and H'_I)

$$i\hbar \frac{d}{dt}|\psi_I(t)\rangle = H'_I|\psi_I(t)\rangle, \qquad H'_I = U^{(0)\dagger} H' U^{(0)}. \tag{10.93}$$

Thus the vector $\psi_I(t)\rangle$ satisfies an equation like the Schrödinger equation,
i.e. the vector varies with time (different from the Heisenberg picture). Since
H'_I is assumed to be small (i.e. a perturbation so that $H \simeq H^{(0)}$), we have

$$\frac{d}{dt}|\psi_I(t)\rangle \sim 0,$$

i.e. near zero and $|\psi_I(t)\rangle$ is almost constant in time.

In order to obtain the equation of motion of an interaction picture observable A_I, we differentiate with respect to time the operator defined by
Eq. (10.89), i.e.

$$A_I(t) = U^{(0)\dagger} A_S U^{(0)}. \tag{10.94}$$

Then we have (normally with $\partial A_S/\partial t = 0$, cf. comments after Eq. (10.55))

$$
\begin{aligned}
i\hbar \frac{d}{dt} A_I(t) \quad = \quad & i\hbar \frac{dU^{(0)\dagger}}{dt} A_S U^{(0)} + i\hbar U^{(0)\dagger} \frac{\partial A_S}{\partial t} U^{(0)} + i\hbar U^{(0)\dagger} A_S \frac{dU^{(0)}}{dt} \\
\overset{(10.80)}{=} \quad & -U^{(0)\dagger} H^{(0)} A_S U^{(0)} + i\hbar U^{(0)\dagger} \frac{\partial A_S}{\partial t} U^{(0)} \\
& + U^{(0)\dagger} A_S H^{(0)} U^{(0)} \\
= \quad & -U^{(0)\dagger} H^{(0)} U^{(0)} U^{(0)\dagger} A_S U^{(0)} + i\hbar U^{(0)\dagger} \frac{\partial A_S}{\partial t} U^{(0)} \\
& + U^{(0)\dagger} A_S U^{(0)} U^{(0)\dagger} H^{(0)} U^{(0)} \\
\overset{(10.89),(10.94)}{=} \quad & -H_I^{(0)} A_I + i\hbar \frac{\partial A_I}{\partial t} + A_I H_I^{(0)},
\end{aligned}
$$

i.e. (with $\partial A_S/\partial t = 0$)

$$i\hbar \frac{d}{dt} A_I(t) = [A_I, H_I^{(0)}]. \tag{10.95}$$

Thus in the interaction picture the physical quantities A are represented
by time-dependent observables which satisfy a type of Heisenberg equation
(10.60a) or (10.60b) with $H_I^{(0)}$ instead of H.

10.7   Time-Dependent Perturbation Theory

We consider time-dependent perturbation theory as an application of the interaction picture. We have the Hamiltonian

$$H = H_0 + H', \quad H' = H'(t), \tag{10.96}$$

in which the perturbation part $H'(t)$ depends explicitly on time and H_0 replaces $H^{(0)}$ in the previous discussion. We assume again that the spectrum and the eigenvectors of H_0 are known, i.e.

$$H_0|\varphi_n\rangle = E_n|\varphi_n\rangle, \quad \langle\varphi_n|\varphi_m\rangle = \delta_{nm}, \quad \sum_n |\varphi_n\rangle\langle\varphi_n| = \mathbb{1}. \tag{10.97}$$

The equation to solve is Eq. (10.93), i.e.

$$i\hbar\frac{d}{dt}|\psi_I(t)\rangle = H'_I(t)|\psi_I(t)\rangle, \tag{10.98}$$

where according to (10.87) and (10.94)

$$H'_I = U^{(0)\dagger}(t, t_0)H'U^{(0)}(t, t_0). \tag{10.99}$$

We obtain the operator $U^{(0)}$ from Eq. (10.80) as

$$U^{(0)}(t, t_0) = e^{-iH_0(t-t_0)/\hbar}. \tag{10.100}$$

Instead of the original Schrödinger equation we now solve Eq. (10.98). Integration with respect to t yields

$$|\psi_I(t)\rangle = |\psi_I(t_0)\rangle - \frac{i}{\hbar}\int_{t_0}^t dt' H'_I(t')|\psi_I(t')\rangle. \tag{10.101}$$

Iteration of this inhomogeneous integral equation yields the so-called *Neumann series*, i.e.

$$\begin{aligned}
|\psi_I(t)\rangle &= |\psi_I(t_0)\rangle - \frac{i}{\hbar}\int_{t_0}^t dt' H'_I(t')|\psi_I(t_0)\rangle \\
&\quad + \left(-\frac{i}{\hbar}\right)^2 \int_{t_0}^t dt' \int_{t_0}^{t'} dt'' H'_I(t')H'_I(t'')|\psi_I(t_0)\rangle + \cdots.
\end{aligned}$$

$$\tag{10.102}$$

We assume that before the perturbation H' is switched on at time t_0, the system is in the Schrödinger eigenstate $|\psi_S^{(0)}\rangle_m$ of H_0 with energy E_m. Since

H_0 is independent of time, the time dependence of $|\psi_S^{(0)}\rangle_m$ can be separated as for stationary states, i.e. we can write

$$|\psi_S^{(0)}\rangle_m = e^{-iH_0t/\hbar}|\varphi_m\rangle = e^{-iE_mt/\hbar}|\varphi_m\rangle. \qquad (10.103)$$

The actual state $|\psi_S\rangle$ of the system is solution of

$$i\hbar\frac{\partial}{\partial t}|\psi_S\rangle = H|\psi_S\rangle.$$

We express this state $|\psi_S\rangle$ as a superposition of the states $|\psi_S^{(0)}\rangle_n$ of H_0 with coefficients which as a result of the time-dependent perturbation depend on time. Hence we set

$$|\psi_S\rangle = \sum_n a_n(t)|\psi_S^{(0)}\rangle_n.$$

We wish to know the probability for the system to be in a state $|\psi_S^{(0)}\rangle_n$ at time t after the switching-on of the perturbation H'. This probability is the modulus squared of the projection of the Schrödinger state $|\psi_S\rangle$ onto $|\psi_S^{(0)}\rangle_n$. The corresponding amplitude or appropriate matrix element is

$$_n\langle\psi_S^{(0)}|\psi_S\rangle = \sum_m a_m(t)\,_n\langle\psi_S^{(0)}|\psi_S^{(0)}\rangle_m = a_n(t). \qquad (10.104)$$

Here we insert Eq. (10.103) for $m \to n$ and use Eq. (10.88) and obtain

$$_n\langle\psi_S^{(0)}|\psi_S\rangle = \langle\varphi_n|e^{iH_0t/\hbar}|\psi_S\rangle = \langle\varphi_n|e^{iH_0t/\hbar}U^{(0)}|\psi_I(t)\rangle = \langle\varphi_n|\psi_I(t)\rangle. \qquad (10.105)$$

In the interaction picture the H_0-state m at time t is

$$|\psi_I(t)\rangle = e^{iH_0t/\hbar}|\psi_S^{(0)}\rangle_m = |\varphi_m\rangle. \qquad (10.106)$$

We insert this expression into Eq. (10.102) and truncate the series after the first perturbation contribution, assuming that this supplies a sufficiently good approximation:

$$|\psi_I(t)\rangle = |\varphi_m\rangle - \frac{i}{\hbar}\int_{t_0}^t dt'\,H_I'(t')|\varphi_m\rangle. \qquad (10.107)$$

With this we obtain from Eq. (10.105)

$$\begin{aligned}
_n\langle\psi_S^{(0)}|\psi_S\rangle &= \langle\varphi_n|\varphi_m\rangle - \frac{i}{\hbar}\int_{t_0}^t dt'\,\langle\varphi_n|H_I'(t')|\varphi_m\rangle \\
&= \delta_{nm} - \frac{i}{\hbar}\int_{t_0}^t dt'\,\langle\varphi_n|e^{\frac{i}{\hbar}H_0(t'-t_0)}H'(t')e^{-\frac{i}{\hbar}H_0(t'-t_0)}|\varphi_m\rangle \\
&= \delta_{nm} - \frac{i}{\hbar}\int_{t_0}^t dt'\,\langle\varphi_n|e^{\frac{i}{\hbar}E_n(t'-t_0)}H'(t')e^{-\frac{i}{\hbar}E_m(t'-t_0)}|\varphi_m\rangle.
\end{aligned}$$

$$(10.108)$$

Hence the probability for the transition of the system from state m into the state $n \neq m$ is

$$|a_n(t)|^2 = \left|\frac{1}{\hbar}\int_{t_0}^{t} dt' e^{\frac{i}{\hbar}(E_n - E_m)t'} \langle \varphi_n | H'(t') | \varphi_m \rangle\right|^2 \equiv W_{nm}(t). \qquad (10.109)$$

10.8 Transitions into the Continuum

In many practical applications one is interested in transitions into a continuum. This is the case, for instance, if in the scattering of a particle off some target its momentum $\mathbf{p}$ changes to $\mathbf{p}'$, or if by emission of a photon with continuously variable momentum an atom passes from one state into another. A further example is provided by α-decay, where the momenta of the α-particles form a continuum.

We assume that the perturbation is switched on at some time t_0, but that otherwise the perturbation is time-independent. One reason why we begin with these simplifying assumptions is also to obtain reasonably simple expressions. Thus we set

$$H'(t) = H' \theta(t), \quad t_0 = 0. \qquad (10.110)$$

With this we obtain from Eq. (10.109) the transition probability

$$
\begin{aligned}
W_{mn}(t) &= \frac{1}{\hbar^2}\left|\int_0^{t} dt' e^{\frac{i}{\hbar}(E_n - E_m)t'} \langle \varphi_n | H' | \varphi_m \rangle\right|^2 \\
&= \left|\frac{e^{\frac{i}{\hbar}(E_n - E_m)t} - 1}{(E_n - E_m)}\langle \varphi_n | H' | \varphi_m \rangle\right|^2 \\
&= \left|\frac{\sin\{\frac{1}{2\hbar}(E_n - E_m)t\}}{\hbar\frac{1}{2\hbar}(E_n - E_m)}\right|^2 |\langle \varphi_n | H' | \varphi_m \rangle|^2. \qquad (10.111)
\end{aligned}
$$

Different from Eq. (10.25) we here have the positive quantity (note t in the denominator)

$$\frac{\sin^2\{\frac{1}{2\hbar}(E_n - E_m)t\}}{\{\frac{1}{2\hbar}(E_n - E_m)\}^2 t} = \frac{\sin^2 \alpha t}{\alpha^2 t}.$$

Consider the function

$$
\begin{aligned}
\delta_t(\alpha) &:= \frac{1}{\pi}\frac{\sin^2 \alpha t}{\alpha^2 t} \\
&= \begin{cases} \dfrac{t}{\pi} & \text{for} \quad \alpha \to 0, \\[2mm] \leq \dfrac{1}{\pi\alpha^2 t} & \text{for} \quad \alpha \neq 0. \end{cases} \qquad (10.112)
\end{aligned}
$$

The behaviour of this function or distribution is illustrated in Fig. 10.2.

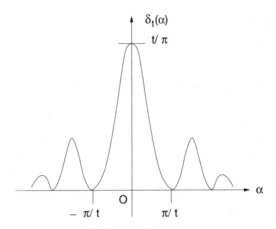

Fig. 10.2 The delta function for $t \to \infty$.

We see that this quantity has the same behaviour (in particular for $\alpha \to 0$) as the function $\delta_\kappa(x)$ we considered previously. Hence

$$\lim_{t\to\infty} \delta_t(\alpha) = \delta(\alpha). \qquad (10.113)$$

Hence also (shifting t to the other side)

$$\frac{\sin^2\{\frac{1}{2\hbar}(E_n - E_m)t\}}{\{\frac{1}{2\hbar}(E_n - E_m)\}^2} = \pi t \delta_t\left(\frac{E_n - E_m}{2\hbar}\right),$$

or

$$\lim_{t\to\infty} \frac{1}{\pi t} \frac{\sin^2\{\frac{1}{2\hbar}(E_n - E_m)t\}}{\{\frac{1}{2\hbar}(E_n - E_m)\}^2} = \delta\left(\frac{E_n - E_m}{2\hbar}\right) = 2\hbar\delta(E_n - E_m). \qquad (10.114)$$

It follows that for large but finite times t

$$W_{mn}(t) \sim \frac{2\pi}{\hbar} t \delta(E_n - E_m)|\langle \varphi_n|H'|\varphi_m\rangle|^2. \qquad (10.115)$$

This expression has a well defined value unequal to zero only if the energies E_n belong to the continuum. Otherwise, if the energies belonged to the discrete spectrum, we would have $E_n - E_m \neq 0$, though sufficiently small, and no other state would be nearby, so that W_{nm} would be zero.

In the case of transitions into the continuum we have to consider the transition probability into the interval dE_n at E_n. We assume that the matrix elements for transitions into this infinitesimal element can be taken as equal. Let $\rho(E_n)dE_n$ be the number of states in the interval dE_n. Then, if $\rho(E_n)$ varies only weakly with E_n, the transition probability into this set of states is

$$\sum_n W_{mn}(t) = \int dE_n \, \rho(E_n)W_{mn}(t) = \rho(E_m)\frac{2\pi|\langle\varphi_n|H'|\varphi_m\rangle|^2}{\hbar}t. \quad (10.116)$$

The *transition rate* Γ is the *transition probability per unit time*, i.e.

$$\Gamma = \frac{1}{t}\sum_n W_{mn}(t) = \rho(E_m)\frac{2\pi}{\hbar}|\langle\varphi_n|H'|\varphi_m\rangle|^2. \quad (10.117)$$

The formulas (10.115) and (10.117) were derived by Pauli in 1928.[*] In view of the usefulness of Eq. (10.117) in many applications, Fermi gave this formula the name *"golden rule"*.[†] The result makes sense as long as the uncertainty $\triangle E$ in the energy satisfies for finite but large times t the relation

$$\triangle E \gg \frac{2\pi\hbar}{t} \gg \delta\epsilon, \quad (10.118)$$

where $\delta\epsilon$ is the separation of states around E_n which in the case of the continuum goes to zero.

If we are concerned with pure scattering, for instance from a potential, so that initial and final states belong to the continuum, then Eq. (10.117) provides the outgoing particle current (see Example 10.4).

Example 10.3: Application of test functions to formula (10.113)
Use the method of test functions to verify in the sense of distribution theory the result (10.113).

Solution: For test functions $\phi(x) \in S(\mathbb{R})$ the delta distribution is defined by the functional

$$\int \delta(x)\phi(x)dx = \phi(0).$$

In the case of the functional (10.113) this is[‡]

$$\lim_{t\to\infty} \frac{1}{\pi}\int_{-\infty}^{\infty} dx\,\frac{\sin^2(xt)}{x^2 t}\phi(x) = \phi(0).$$

We choose a test function which falls off at infinity faster than any inverse power of $|x|$:

$$\phi(x) = e^{-|x|}, \quad \phi(0) = 1.$$

We evaluate the integral

$$I = \int_0^{\infty} dx\,\frac{\sin^2(tx)}{x^2}e^{-x}$$

---

[*] According to F. Schwabl [246], p. 266, in the article of W. Pauli [227].
[†] See E. Fermi [92], pp. 75, 142.
[‡] Recall also the integral $\int_{-\infty}^{\infty} dx \sin^2 x/x^2 = \pi$.

with the help of Tables of Integrals.[§] We have

$$\int_0^\infty e^{-|p|x} \sin ax \sin bx \frac{dx}{x^2} = \frac{p}{4} \ln \frac{p^2 + (a-b)^2}{p^2 + (a+b)^2} + \frac{a}{2} \tan^{-1} \frac{2pb}{p^2 + a^2 - b^2}$$
$$+ \frac{b}{2} \tan^{-1} \frac{2pa}{p^2 + b^2 - a^2},$$

so that for $a = b = t, p = 1$ we obtain $I = -(1/4)\ln(1 + 4t^2) + t \tan^{-1}(2t)$. Then

$$\lim_{t\to\infty} \frac{1}{\pi} \int_{-\infty}^\infty dx \frac{\sin^2(xt)}{x^2 t} e^{-|x|} = 2 \lim_{t\to\infty} \left[-\frac{1}{4t\pi} \ln(1 + 4t^2) + \frac{t}{t\pi} \tan^{-1}(2t) \right] = \frac{2}{\pi} \tan^{-1}(\infty) = 1.$$

Hence the result is $\phi(0)$ as expected.

Example 10.4: Differential cross section from the "golden rule"

Obtain from the "golden rule" (10.117) for $H' = V(\mathbf{x})$ the differential cross section

$$\frac{d\sigma}{d\Omega} = \left(\frac{\mu}{2\pi\hbar} \right)^2 \left| \int d\mathbf{x}\, V(\mathbf{x}) e^{i\boldsymbol{\kappa}\cdot\mathbf{x}} \right|^2,$$

where $\boldsymbol{\kappa} = \mathbf{k}_m - \mathbf{k}_n, E = \hbar^2 k^2/2\mu$. Hint: Use for $\varphi_n(\mathbf{x})$ a plane wave ansatz with box normalization (volume $V = L^3$) and obtain $d\sigma$ from the ratio of outgoing and incoming particle fluxes.

Solution: We set

$$\varphi_n(\mathbf{x}) = \frac{1}{L^{3/2}} e^{i\mathbf{k}\cdot\mathbf{x}}, \qquad \varphi_m(\mathbf{x}) = \frac{1}{L^{3/2}} e^{i\mathbf{k}_0\cdot\mathbf{x}},$$

so that

$$\langle\varphi_n|H'|\varphi_m\rangle = \frac{1}{L^3} \int d\mathbf{x}\, e^{i\boldsymbol{\kappa}\cdot\mathbf{x}} V(\mathbf{x}), \qquad \boldsymbol{\kappa} = \mathbf{k}_0 - \mathbf{k}.$$

The periodic boundary conditions $\varphi_n(x_i) = \varphi_n(x_i \pm L/2)$ imply $k_i L/2 = \pm n_i \pi$, with $n_i = 0, 1, 2, \ldots$. We wish to know the probability of transitions per unit time into the infinitesimal angular region $d\Omega = d(-\cos\theta)d\varphi$ around the direction of the angles (θ, φ). Then

$$\rho(E_k)dE_k = \left(\frac{L}{2\pi} \right)^3 d\mathbf{k} = \left(\frac{L}{2\pi} \right)^3 k^2 dk d\Omega.$$

Since $E_k = \hbar^2 k^2/2\mu, dE_k = \hbar^2 k\, dk/\mu$, and hence

$$\rho(E_k) = \left(\frac{L}{2\pi} \right)^3 \frac{k^2 dk d\Omega\, \mu}{\hbar^2 k dk}.$$

It follows that

$$\Gamma \simeq \frac{2\pi}{\hbar} \rho(E_k)|\langle\varphi_n|H'|\varphi_m\rangle|^2 = \frac{2\pi}{\hbar} \left(\frac{L}{2\pi} \right)^3 \frac{k^2 dk d\Omega\, \mu}{\hbar^2 k dk} \frac{1}{L^6} \left| \int V(\mathbf{x}) e^{i\boldsymbol{\kappa}\cdot\mathbf{x}} d\mathbf{x} \right|^2,$$

which is the number of particles scattered per unit time into the solid angle element $d\Omega$ with only one particle sent into the volume L^3 (since the incoming wave is normalized to 1). The ingoing particle flux is therefore $\mathbf{v}/L^3$, with $\mathbf{v}$ the velocity of the particles, and is obtained as

$$\mathbf{j}_{\text{in}} = \frac{i\hbar}{2\mu} (\varphi_m^* \boldsymbol{\nabla}\varphi_m - \varphi_m \boldsymbol{\nabla}\varphi_m^*) = \frac{i\hbar}{2\mu} \frac{2}{L^3} e^{-i\mathbf{k}_0\cdot\mathbf{x}} (-i\mathbf{k}_0) e^{i\mathbf{k}_0\cdot\mathbf{x}} = \frac{\hbar\mathbf{k}_0}{\mu L^3} = \frac{\mathbf{v}}{L^3}.$$

[§]See I. S. Gradshteyn and I. M. Ryzhik, formula 3.947, p. 491.

The differential cross section is defined as

$$d\sigma = \frac{\text{outgoing particle current} \times r^2 d\Omega}{\text{ingoing particle current}} = \frac{\Gamma}{(\hbar k_0)/(\mu L^3)},$$

i.e.
$$\frac{d\sigma}{d\Omega} = \left(\frac{\mu}{2\pi\hbar}\right)^2 \left|\int d\mathbf{x}\, V(\mathbf{x}) e^{i\boldsymbol{\kappa}\cdot\mathbf{x}}\right|^2,$$

where for purely elastic scattering $|\mathbf{k}| = |\mathbf{k}_0|$. For a possible application of this result, specifically in the case of the Coulomb potential, see the discussion after Eq. (21.69).

10.9 General Time-Dependent Method

For a better understanding of the *"golden rule"* derived above, we present another derivation using the usual method of time-dependent perturbation theory, by which we mean without leaving the Schrödinger picture.[¶] We start from the Hamiltonian

$$H = H_0 + H', \tag{10.119}$$

where $H' = H'(t)$ is a time-dependent perturbation term and $H_0|\varphi_n\rangle = E_n|\varphi_n\rangle$. Thus we assume that the eigenfunctions $\varphi_n = \langle\mathbf{x}|\varphi_n\rangle$ of H_0 are known. The problem is to solve the time-dependent Schrödinger equation

$$i\hbar\frac{\partial}{\partial t}|\psi\rangle = H|\psi\rangle. \tag{10.120}$$

In the case of Eq. (10.119) only the time dependence of H_0 can be separated from the state $|\psi\rangle$ in the form of the exponential factor of a stationary state. We expand the wave function $\psi(\mathbf{x}, t) \equiv \langle\mathbf{x}|\psi\rangle$ therefore in terms of the eigenfunctions of H_0 with time-dependent coefficients. Thus we set

$$\psi(\mathbf{x}, t) = \sum_n a_n(t)\psi_n^{(0)}(\mathbf{x}, t) + \int_{\text{continuum}} a(E, t)\psi^{(0)}(E, \mathbf{x}, t)dE$$

$$\equiv \mathcal{S}a_n(t)\psi_n^{(0)}(\mathbf{x}, t). \tag{10.121}$$

Here $\mathcal{S}$ stands for summation over the discrete part of the spectrum and integration over the continuum with the orthogonality conditions

$$\int d\mathbf{x}\psi_n^{(0)*}(\mathbf{x})\psi_m^{(0)}(\mathbf{x}) = \delta_{mn}, \tag{10.122a}$$

and

$$\int d\mathbf{x}\psi^{(0)*}(E, \mathbf{x})\psi^{(0)}(E', \mathbf{x}) = \delta(E - E'). \tag{10.122b}$$

[¶]See also L. Schiff [243], pp. 148, 193.

The discrete and continuous eigenfunctions of H_0 are

$$\psi_n^{(0)}(\mathbf{x}, t) = e^{-iE_n t/\hbar}\varphi_n(\mathbf{x}), \quad \psi^{(0)}(E, \mathbf{x}, t) = e^{-iEt/\hbar}\varphi(E, \mathbf{x}). \qquad (10.123)$$

We now insert these expressions into Eq. (10.120), i.e.

$$\psi(\mathbf{x}, t) = \mathcal{S}a_n(t)\varphi_n(\mathbf{x})e^{-iE_n t/\hbar}, \qquad (10.124)$$

into

$$i\hbar\frac{\partial\psi}{\partial t} = H\psi,$$

and obtain

$$i\hbar\mathcal{S}\dot{a}_n(t)\varphi_n(\mathbf{x})e^{-iE_n t/\hbar} + \mathcal{S}a_n(t)E_n\varphi_n(\mathbf{x})e^{-iE_n t/\hbar}$$
$$= \mathcal{S}a_n(t)(H_0 + H')\varphi_n(\mathbf{x})e^{-iE_n t/\hbar}$$
$$= \mathcal{S}a_n(t)(E_n + H')\varphi_n(\mathbf{x})e^{-iE_n t/\hbar}, \qquad (10.125)$$

where $\dot{a}_n(t) = da_n(t)/dt$. We multiply this equation from the left by $\varphi_k^*(\mathbf{x})$ and integrate over all space. Then

$$\int d\mathbf{x}\varphi_k^*(\mathbf{x})[i\hbar\mathcal{S}\dot{a}_n(t)\varphi_n(\mathbf{x})e^{-iE_n t/\hbar} + \mathcal{S}a_n(t)E_n\varphi_n(\mathbf{x})e^{-iE_n t/\hbar}]$$
$$= \int d\mathbf{x}\varphi_k^*(\mathbf{x})[\mathcal{S}a_n(t)(E_n + H')\varphi_n(\mathbf{x})e^{-iE_n t/\hbar}]. \qquad (10.126)$$

The second term on the left of this equation and the first term on the right cancel out. Next we use the orthogonality relations

$$\int d\mathbf{x}\varphi_k^*(\mathbf{x})\varphi_n(\mathbf{x}) = \begin{cases} \delta_{kn} & \text{in case of discrete } \varphi's, \\ \delta(E_k - E_n) & \text{in case of continuous } \varphi's. \end{cases} \qquad (10.127)$$

Observe that these conditions do not contradict Eqs. (10.122a) and (10.122b), since

$$\int \psi^{(0)*}(E, \mathbf{x}, t)\psi^{(0)}(E', \mathbf{x}, t)d\mathbf{x} = e^{i(E-E')t/\hbar}\delta(E - E') = \delta(E - E').$$

Thus with Eq. (10.127) we obtain from Eq. (10.126)

$$i\hbar\dot{a}_k(t)e^{-iE_k t/\hbar} = \mathcal{S}a_n(t)e^{-iE_n t/\hbar}H'_{kn}, \qquad (10.128)$$

where

$$H'_{kn} = \int d\mathbf{x}\varphi_k^*(\mathbf{x})H'\varphi_n(\mathbf{x}) \equiv \langle\varphi_k|H'|\varphi_n\rangle. \qquad (10.129)$$

We can rewrite the equation as

$$i\hbar\dot{a}_k(t) = \mathcal{S}H'_{kn}a_n(t)e^{i(E_k-E_n)t/\hbar}. \tag{10.130}$$

This result should be compared with the Schrödinger equation (10.120), since it is equivalent to the latter. We observe that the amplitude a_k of a definite eigenfunction φ_k has effectively replaced the wave function ψ.

In order to solve Eq. (10.130) we develop a perturbation theory. In the first place we replace in Eq. (10.130) H'_{kn} by $\lambda H'_{kn}$ and at the end we allow λ to tend to 1. Thus λ serves as a perturbation parameter which permits us to equate coefficients of the same power of λ on both sides of an equation. Now we set

$$a_k(t) = a_k^{(0)}(t) + \lambda a_k^{(1)}(t) + \lambda^2 a_k^{(2)}(t) + \cdots \tag{10.131}$$

and insert this into Eq. (10.130) along with the parameter λ in front of H'_{kn}. We obtain

$$i\hbar[\dot{a}_k^{(0)}(t) + \lambda\dot{a}_k^{(1)}(t) + \lambda^2\dot{a}_k^{(2)}(t) + \cdots]$$
$$= \mathcal{S}\lambda H'_{kn}(t)e^{i(E_k-E_n)t/\hbar}[a_n^{(0)}(t) + \lambda a_n^{(1)}(t) + \lambda^2 a_n^{(2)}(t) + \cdots].$$

Comparing coefficients of the same powers of λ on both sides, we obtain the equations:

$$i\hbar\dot{a}_k^{(0)}(t) = 0,$$
$$i\hbar\dot{a}_k^{(1)}(t) = \mathcal{S}H'_{kn}(t)e^{i(E_k-E_n)t/\hbar}\dot{a}_n^{(0)}(t),$$
$$\cdots$$
$$i\hbar\dot{a}_k^{(s+1)}(t) = \mathcal{S}H'_{kn}(t)e^{i(E_k-E_n)t/\hbar}\dot{a}_n^{(s)}(t)$$
$$\cdots . \tag{10.132}$$

These equations can be integrated successively. We restrict ourselves here to the two lowest order equations, i.e. to perturbation theory in the lowest order. Thus first we have

$$a_k^{(0)} = \text{const.} \tag{10.133}$$

The numbers $a_k^{(0)} = \text{const.}$ determine the state of the system at time $t = 0$, i.e. they are fixed by the *initial conditions*. Here we assume (our assumption above): At time $t = 0$ all coefficients $a_k^{(0)}$ are zero except one and only one, let's say $a_m^{(0)} \neq 0$, and we set

$$a_k^{(0)} = \delta_{km} \quad \text{or} \quad a_k^{(0)} = \delta(E_k - E_m), \tag{10.134}$$

depending on whether E_m belongs to the discrete part of the spectrum of H_0 or to its continuum. Integrating the second of the equations (10.132), we obtain

$$i\hbar a_k^{(1)} = \int_{-\infty}^{t} S H'_{kn}(t') e^{i(E_k - E_n)t'/\hbar} \delta_{mn} dt',$$

i.e.

$$i\hbar a_k^{(1)} = \int_{-\infty}^{t} H'_{km}(t') e^{i(E_k - E_m)t'/\hbar} dt'. \tag{10.135}$$

We put the additional constant of integration equal to zero, so that

$$i\hbar a_k^{(1)}(t = -\infty) = 0.$$

What is the physical significance of Eq. (10.135)? The quantity $a_k^{(1)}(t)$ is the amplitude which determines the probability that as a result of the perturbation $\lambda H'_{km}$ the system makes a transition from the initial state m (i.e. φ_m) into the state k. Later we shall discuss scattering off a Coulomb potential in terms of the so-called *Born approximation* of the scattering amplitude which we there write $f(\theta, \varphi)$ in the context of time-independent perturbation theory. We shall see that in this approximation the scattering amplitude is effectively the Fourier transform of the potential, i.e. an expression of the following type

$$f_{(\mathbf{k},\mathbf{k}')}^{\text{Born}} = -\frac{\mu}{2\pi\hbar^2} \int e^{i(\mathbf{k}-\mathbf{k}')\cdot\mathbf{x}'} V(\mathbf{x}') d\mathbf{x}'. \tag{10.136}$$

Comparing Eqs. (10.135) and (10.136) we see that both "amplitudes" are Fourier transforms and look similar. The amplitude f^{Born} is a Fourier transform in spatial coordinates, whereas $a_k^{(1)}$ is a Fourier transform in time (a distinction which is easy to remember!).

Equation (10.135) assumes a particularly simple form if — as we considered previously — the perturbation H' is taken to be time-independent except that it is switched on at a particular time $t = 0$ and is again switched off at some later time $t > 0$. In this case we obtain from Eq. (10.135)

$$i\hbar a_k^{(1)}(t) = H'_{km} \int_0^t e^{i(E_k - E_m)t/\hbar} dt' = H'_{km} \left[\frac{e^{i(E_k - E_m)t/\hbar} - 1}{i(E_k - E_m)/\hbar} \right].$$

Hence the probability to find the system at time t in the state $n \neq m$ is (with $\lambda \to 1$)

$$
\begin{aligned}
|a_n^{(1)}(t)|^2 &= \frac{|H'_{nm}|^2}{(E_n - E_m)^2} \left[e^{i(E_n - E_m)t/\hbar} - 1 \right] \left[e^{-i(E_n - E_m)t/\hbar} - 1 \right] \\
&= \frac{4|H'_{nm}|^2 \sin^2\{ \frac{E_n - E_m}{2\hbar} t \}}{(E_n - E_m)^2}
\end{aligned}
\tag{10.137}
$$

in agreement with Eq. (10.111).

Chapter 11

The Coulomb Interaction

11.1 Introductory Remarks

The quasi-particle quantization of the harmonic oscillator is of fundamental significance in view of its role as the prototype for the quantization of fields, such as the electromagnetic field (the quantization in these contexts is often called "second quantization", meaning the inclusion of the creation and annihilation of field quanta). The Coulomb interaction on the other hand is of specific significance for its relevance in atomic and nuclear physics. Both of these important potentials are singled out from a large number of cases by the fact that in their case the Schrödinger equation can be solved explicitly and completely in closed form. Both cases therefore also serve in many applications as the basic unperturbed problem of a perturbation theory.

11.2 Separation of Variables, Angular Momentum

In considering the Coulomb potential in a realistic context we have to consider three space dimensions. Let $\mathbf{r} := (x, y, z)$ be the vector separating two particles which are otherwise characterized by indices 1 and 2 with electric charges $Z_1 e$ and $Z_2 e$. The electrostatic interaction of the particles is the Coulomb potential (later we take $Z_1 = 1, Z_2 \equiv Z$)

$$V(r) = -\frac{Z_1 Z_2 e^2}{r}, \quad r = |\mathbf{r}|. \tag{11.1}$$

We let $\mathbf{p}_i, \mathbf{r}_i, m_i$ be respectively the momentum, the position coordinate and the mass of particle i with $i = 1, 2$. The Hamilton operator is then given by

$$\mathcal{H} = \frac{\mathbf{p}_1^2}{2m_1} + \frac{\mathbf{p}_2^2}{2m_2} - \frac{Z_1 Z_2 e^2}{r}. \tag{11.2}$$

199

We introduce *centre of mass* and *relative coordinates* $\mathbf{R}$ and $\mathbf{r}$ by setting

$$\mathbf{R} = \frac{m_1\mathbf{r}_1 + m_2\mathbf{r}_2}{m_1 + m_2}, \quad \mathbf{r} = \mathbf{r}_1 - \mathbf{r}_2 \tag{11.3}$$

and *centre of mass momentum* $\mathbf{P}$ and the *relative velocity* $\mathbf{p}/m_0$ by setting

$$\mathbf{P} = \mathbf{p}_1 + \mathbf{p}_2, \quad \mathbf{p} = m_0\left(\frac{\mathbf{p}_1}{m_1} - \frac{\mathbf{p}_2}{m_2}\right), \text{ where } m_0 = \frac{m_1 m_2}{m_1 + m_2}. \tag{11.4}$$

The mass m_0 is usually described as the *reduced mass*. Solving Eqs. (11.3) for the individual position coordinates we obtain

$$\mathbf{r}_1 = \mathbf{R} + \frac{m_2}{m_1 + m_2}\mathbf{r}, \quad \mathbf{r}_2 = \mathbf{R} - \frac{m_1}{m_1 + m_2}\mathbf{r}. \tag{11.5}$$

The momenta $\mathbf{P}, \mathbf{p}$ are such that as operators in the position space representation or $\{\mathbf{R}, \mathbf{r}\}$ representation they have the differential operator forms

$$\mathbf{P} = -i\hbar\frac{\partial}{\partial\mathbf{R}}, \quad \mathbf{p} = -i\hbar\frac{\partial}{\partial\mathbf{r}}.$$

Setting $\mathbf{R} = (X, Y, Z)$ and $\mathbf{r} = (x, y, z)$ one verifies that

$$\frac{\partial}{\partial X} = \frac{\partial x_1}{\partial X}\frac{\partial}{\partial x_1} + \frac{\partial x_2}{\partial X}\frac{\partial}{\partial x_2} = \frac{\partial}{\partial x_1} + \frac{\partial}{\partial x_2} \tag{11.6a}$$

etc. so that as expected according to Eq. (11.4) $\mathbf{P} = \mathbf{p}_1 + \mathbf{p}_2$. Similarly

$$\frac{\partial}{\partial x} = \frac{\partial x_1}{\partial x}\frac{\partial}{\partial x_1} + \frac{\partial x_2}{\partial x}\frac{\partial}{\partial x_2} = \frac{m_0}{m_1}\frac{\partial}{\partial x_1} - \frac{m_0}{m_2}\frac{\partial}{\partial x_2} \tag{11.6b}$$

etc. so that as expected we obtain the expression for $\mathbf{p}$ in Eq. (11.4). But now

$$\frac{\mathbf{p}^2}{2m_0} = \frac{m_0}{2}\left(\frac{\mathbf{p}_1}{m_1} - \frac{\mathbf{p}_2}{m_2}\right)^2 = \frac{m_0}{2}\left(\frac{\mathbf{p}_1^2}{m_1^2} + \frac{\mathbf{p}_2^2}{m_2^2} - 2\frac{\mathbf{p}_1 \cdot \mathbf{p}_2}{m_1 m_2}\right)$$

$$= \frac{\mathbf{p}_1^2 m_2}{2m_1(m_1 + m_2)} + \frac{\mathbf{p}_2^2 m_1}{2m_2(m_1 + m_2)} - \frac{\mathbf{p}_1 \cdot \mathbf{p}_2}{m_1 + m_2} \tag{11.7a}$$

and

$$\frac{\mathbf{P}^2}{2(m_1 + m_2)} = \frac{(\mathbf{p}_1 + \mathbf{p}_2)^2}{2(m_1 + m_2)}$$

$$= \frac{\mathbf{p}_1^2 m_1}{2m_1(m_1 + m_2)} + \frac{\mathbf{p}_2^2 m_2}{2m_2(m_1 + m_2)} - \frac{\mathbf{p}_1 \cdot \mathbf{p}_2}{m_1 + m_2}, \tag{11.7b}$$

so that

$$\frac{\mathbf{P}^2}{2(m_1 + m_2)} + \frac{\mathbf{p}^2}{2m_0} = \frac{\mathbf{p}_1^2}{2m_1} + \frac{\mathbf{p}_2^2}{2m_2}. \tag{11.8}$$

It follows that

$$\mathcal{H} = \frac{\mathbf{P}^2}{2(m_1 + m_2)} + \frac{\mathbf{p}^2}{2m_0} + V(\mathbf{r}) \tag{11.9}$$

and so

$$\begin{aligned}
\mathcal{H} &= \mathcal{H}_{\text{cm}} + H \\
&= \text{centre of mass energy } + \text{ energy of relative motion.} \tag{11.10}
\end{aligned}$$

One can convince oneself that the transformation from $(\mathbf{r}_1, \mathbf{r}_2)$ to $(\mathbf{R}, \mathbf{r})$ is a canonical transformation. Hence Hamilton's equations apply and we obtain, with $M := m_1 + m_2$:

$$\dot{\mathbf{R}} = \frac{\partial \mathcal{H}_{\text{cm}}}{\partial \mathbf{P}}, \quad \dot{\mathbf{P}} = -\frac{\partial \mathcal{H}_{\text{cm}}}{\partial \mathbf{R}} \tag{11.11}$$

and

$$\dot{\mathbf{r}} = \frac{\partial H}{\partial \mathbf{p}}, \quad \dot{\mathbf{p}} = -\frac{\partial H}{\partial \mathbf{r}}, \tag{11.12}$$

i.e.

$$\dot{\mathbf{R}} = \frac{\mathbf{P}}{M}, \quad \dot{\mathbf{P}} = 0, \tag{11.13}$$

and

$$\dot{\mathbf{r}} = \frac{\mathbf{p}}{m_0}, \quad \dot{\mathbf{p}} = -\boldsymbol{\nabla}V. \tag{11.14}$$

We observe that the equations of motion in $\mathbf{R}, \mathbf{P}$ on the one hand and those in $\mathbf{r}, \mathbf{p}$ on the other are completely separated. The motion of the centre of mass is that of a particle of mass $M = m_1 + m_2$ in uniform motion along a straight line. The equations of the relative motion have the form of the equations of a single particle of mass m_0 moving in a potential $V(r)$.

Canonical quantization of the classical theory implies that we pass from the Cartesian variables r_i, p_j, R_i, P_j over to operators $\hat{r}_i, \hat{p}_j, \hat{R}_i, \hat{P}_j$ (which in the following we write again r_i, p_j, R_i, P_j), for which we postulate the commutator relations[*]

$$[r_i, p_j] = i\hbar\delta_{ij}, \quad [R_i, P_j] = i\hbar\delta_{ij}, \quad (i, j = x, y, z). \tag{11.15}$$

In accordance with Eq. (11.10) we set

$$\mathcal{H} = \mathcal{H}_{\text{cm}} + H, \tag{11.16}$$

---

[*]Although we write r_i, we mean $r_1 \equiv r_x \equiv x$, etc. The canonical quantization must always be performed in Cartesian coordinates.

$$\mathcal{H}_{\text{cm}} = \frac{\mathbf{P}^2}{2M}, \tag{11.17}$$

$$H = \frac{\mathbf{p}^2}{2m_0} + V(\mathbf{r}). \tag{11.18}$$

In the position space representation the time independent *Schrödinger equation* is therefore

$$\left[\left(-\frac{\hbar^2}{2M}\triangle_R\right) + \left(-\frac{\hbar^2}{2m_0}\triangle_r + V(\mathbf{r})\right)\right]\Psi(\mathbf{R},\mathbf{r}) = E_{\text{total}}\Psi(\mathbf{R},\mathbf{r}). \tag{11.19}$$

For *stationary states* with energy E the equation possesses a complete system of eigensolutions Ψ which can be written

$$\Psi(\mathbf{R},\mathbf{r}) = \Phi(\mathbf{R})\psi(\mathbf{r}), \tag{11.20}$$

where

$$\mathcal{H}_{\text{cm}}\Phi(\mathbf{R}) \equiv \left[-\frac{\hbar^2}{2M}\triangle_R\right]\Phi(\mathbf{R}) = E_R\Phi(\mathbf{R}), \tag{11.21a}$$

and

$$H\psi(\mathbf{r}) \equiv \left[-\frac{\hbar^2}{2m_0}\triangle_r + V(\mathbf{r})\right]\psi(\mathbf{r}) = E\psi(\mathbf{r}) \tag{11.21b}$$

with

$$E_{\text{total}} = E_R + E.$$

Thus the two-particle problem is practically reduced to an effective one-particle problem, i.e. to that of the *Schrödinger wave equation*

$$\left[-\frac{\hbar^2}{2m_0}\triangle_r + V(\mathbf{r})\right]\psi(\mathbf{r}) = E\psi(\mathbf{r}). \tag{11.22}$$

If, as in the case of the Coulomb potential, $V(\mathbf{r}) = V(r)$, it is reasonable to go to *spherical coordinates* r, θ, φ:

$$\begin{aligned} x &= r\sin\theta\,\cos\varphi, \\ y &= r\sin\theta\,\sin\varphi, \\ z &= r\cos\theta, \ \ 0 \leq r \leq \infty, \ 0 \leq \theta \leq \pi, \ -\pi \leq \varphi \leq \pi. \end{aligned} \tag{11.23}$$

In these coordinates we have

$$\triangle_r\psi = \frac{1}{r^2}\frac{\partial}{\partial r}\left(r^2\frac{\partial\psi}{\partial r}\right) + \frac{1}{r^2\sin\theta}\frac{\partial}{\partial\theta}\left(\sin\theta\frac{\partial\psi}{\partial\theta}\right) + \frac{1}{r^2\sin^2\theta}\frac{\partial^2\psi}{\partial\varphi^2}. \tag{11.24}$$

We can rewrite this in a different form by defining the operator

$$p_r := -i\hbar \frac{1}{r} \frac{\partial}{\partial r} r = -i\hbar \left(\frac{\partial}{\partial r} + \frac{1}{r} \right). \tag{11.25}$$

In Example 11.1 it is shown that p_r is hermitian provided

$$\lim_{r \to 0} r\psi(r) = 0. \tag{11.26}$$

However, the operator $-i\hbar\partial/\partial r$ is not hermitian! The operator p_r obviously satisfies the relation

$$[r, p_r] = i\hbar. \tag{11.27}$$

We can now write (see verification below)

$$\mathbf{p}^2 \psi = -\hbar^2 \triangle_r \psi = \left(p_r^2 + \frac{l^2}{r^2} \right) \psi, \quad (r \neq 0), \tag{11.28}$$

where

$$
\begin{aligned}
p_r^2 \psi &= -\hbar^2 \frac{1}{r} \frac{\partial}{\partial r} \left(r \frac{1}{r} \frac{\partial}{\partial r}(r\psi) \right) = -\hbar^2 \frac{1}{r} \frac{\partial}{\partial r} \left(r \frac{\partial \psi}{\partial r} + \psi \right) \\
&= -\hbar^2 \frac{1}{r} \left(2 \frac{\partial \psi}{\partial r} + r \frac{\partial^2 \psi}{\partial r^2} \right) = -\hbar^2 \frac{1}{r^2} \frac{\partial}{\partial r} \left(r^2 \frac{\partial \psi}{\partial r} \right)
\end{aligned}
\tag{11.29}
$$

and (compare with Eq. (11.24))

$$l^2 \psi \equiv -\frac{\hbar^2}{\sin^2 \theta} \left\{ \sin \theta \frac{\partial}{\partial \theta} \left(\sin \theta \frac{\partial \psi}{\partial \theta} \right) + \frac{\partial^2 \psi}{\partial \varphi^2} \right\}. \tag{11.30}$$

We now demonstrate that $\mathbf{l}, \mathbf{l}^2 = l^2$, is the operator of the relative angular momentum, i.e. that the above expression follows from

$$\mathbf{l} := \mathbf{r} \times \mathbf{p} = -i\hbar(\mathbf{r} \times \boldsymbol{\nabla}). \tag{11.31}$$

We have

$$\mathbf{l}^2 = (\mathbf{r} \times \mathbf{p}) \cdot (\mathbf{r} \times \mathbf{p}),$$

where

$$(\mathbf{r} \times \mathbf{p})_i = \sum_{j,k} \epsilon_{ijk} r_j p_k, \quad i, j, k = 1, 2, 3, \tag{11.32}$$

with

$$
\begin{aligned}
\epsilon_{ijk} &= +1, \quad \text{if } i, j, k \text{ a clockwise permutation of } 1, 2, 3, \\
&= -1, \quad \text{if } i, j, k \text{ an anticlockwise permutation of } 1, 2, 3, \\
&= 0, \quad \text{otherwise.}
\end{aligned}
\tag{11.33}
$$

It follows that

$$\sum_i \epsilon_{ijk}\epsilon_{ij'k'} = \delta_{jj'}\delta_{kk'} - \delta_{jk'}\delta_{kj'}. \tag{11.34}$$

With this we obtain — keeping in mind the ordering of r and p —

$$l^2 = \sum_{i,j,k} \epsilon_{ijk}r_j p_k \epsilon_{ij'k'} r_{j'} p_{k'} = \sum_{j,k}(r_j p_k r_j p_k - r_j p_k r_k p_j). \tag{11.35}$$

Using Eq. (11.15) we have

$$r_j p_k r_j p_k = r_j(r_j p_k - i\hbar\delta_{jk})p_k \tag{11.36}$$

and

$$\begin{aligned}
r_j p_k r_k p_j &= r_j p_k(p_j r_k + i\hbar\delta_{jk}) = r_j p_k p_j r_k + i\hbar\delta_{jk} r_j p_k \\
&= r_j p_j(r_k p_k - i\hbar\delta_{kk}) + i\hbar r_j p_j. \tag{11.37}
\end{aligned}$$

It follows that (note the part defined as $-\delta$)

$$\begin{aligned}
l^2 &= (\mathbf{r}^2 \mathbf{p}^2 - i\hbar\mathbf{r}\cdot\mathbf{p}) - (\mathbf{r}\cdot\mathbf{p})^2 + 3i\hbar(\mathbf{r}\cdot\mathbf{p}) - i\hbar\mathbf{r}\cdot\mathbf{p} \\
&= \mathbf{r}^2 \mathbf{p}^2 \underbrace{-(\mathbf{r}\cdot\mathbf{p})^2 + i\hbar(\mathbf{r}\cdot\mathbf{p})}_{-\delta}. \tag{11.38}
\end{aligned}$$

Since

$$r^2 = x^2 + y^2 + z^2, \qquad r\frac{\partial}{\partial r} = \mathbf{r}\cdot\frac{\partial}{\partial \mathbf{r}}$$

we have (cf. Eq. (11.25))

$$\mathbf{r}\cdot\mathbf{p} = -i\hbar\mathbf{r}\cdot\nabla = rp_r + i\hbar \tag{11.39}$$

and hence

$$\delta = (\mathbf{r}\cdot\mathbf{p})(\mathbf{r}\cdot\mathbf{p} - i\hbar) = rp_r(rp_r + i\hbar). \tag{11.40}$$

With Eq. (11.27) we obtain

$$r(p_r r)p_r = r(rp_r - i\hbar)p_r \tag{11.41}$$

and

$$\delta = r^2 p_r^2. \tag{11.42}$$

This means that

$$l^2 = r^2(\mathbf{p}^2 - p_r^2), \tag{11.43}$$

so that

$$\mathbf{p}^2 = p_r^2 + \frac{l^2}{r^2}. \tag{11.44}$$

Comparison of Eqs. (11.28) and (11.30) yields

$$\mathbf{l}^2 = l^2 = -\frac{\hbar^2}{\sin^2\theta}\left\{\sin\theta\frac{\partial}{\partial\theta}\left(\sin\theta\frac{\partial\psi}{\partial\theta}\right) + \frac{\partial^2\psi}{\partial\varphi^2}\right\}. \tag{11.45}$$

Thus in Eq. (11.30) l^2 is to be identified with the square of the angular momentum operator. The Schrödinger equation of the relative motion can therefore be written

$$\left[\frac{p_r^2}{2m_0} + \frac{l^2}{2m_0 r^2} + V(r)\right]\psi(r,\theta,\varphi) = E\psi(r,\theta,\varphi) \tag{11.46}$$

with the condition (11.26), i.e.

$$\lim_{r\to 0} r\psi(\mathbf{r}) = 0. \tag{11.47}$$

Example 11.1: Proof that p_r is hermitian
Show that if

$$\mathcal{D}_{p_r} := \left\{\psi \in \mathcal{H} = \mathcal{L}^2(\mathbb{R}^3), \ \mathbf{r}\in\mathbb{R}^3, \ r:=|\mathbf{r}| : \ \lim_{r\to 0} r\psi(r) = 0\right\}, \quad p_r = \frac{\hbar}{i}\frac{1}{r}\frac{\partial}{\partial r}r,$$

the operator p_r is hermitian, i.e. $(p_r\phi, \psi) = (\phi, p_r\psi), \ \forall\phi, \psi \in \mathcal{D}_{p_r} \subset \mathcal{H}$.

Solution: Set $d\Omega_s = \sin\theta d\theta d\varphi, \ \theta \in [0,\pi], \varphi \in [0, 2\pi]$. Then

$$
\begin{aligned}
(\phi, p_r\psi) &= \int \phi(r,\theta,\varphi)^* p_r\psi(r,\theta,\varphi)r^2 d\Omega_s = \frac{\hbar}{i}\int\phi(r,\theta,\varphi)^*\frac{1}{r}\frac{\partial}{\partial r}(r\psi(r,\theta,\varphi))r^2 drd\Omega_s \\
&= \frac{\hbar}{i}\int r\phi(r,\theta,\varphi)^*\frac{\partial}{\partial r}(r\psi(r,\theta,\varphi))drd\Omega_s.
\end{aligned}
$$

With partial integration with respect to r this becomes

$$
\begin{aligned}
(\phi, p_r\psi) &= \int d\Omega_s\left[r\phi(r,\theta,\varphi)^* r\psi(r,\theta,\varphi)\right]_{r=0}^{r=\infty} - \frac{\hbar}{i}\int\frac{\partial}{\partial r}(r\phi^*(r,\theta,\varphi))r\psi(r,\theta,\varphi)drd\Omega_s \\
&= -\frac{\hbar}{i}\int\left\{\frac{1}{r}\frac{\partial}{\partial r}(r\phi(r,\theta,\varphi))\right\}^*\psi(r,\theta,\varphi)r^2 drd\Omega_s \\
&= \int\left\{\frac{\hbar}{i}\frac{1}{r}\frac{\partial}{\partial r}(r\phi(r,\theta,\varphi))\right\}^*\psi(r,\theta,\varphi)r^2 drd\Omega_s = (p_r\phi, \psi).
\end{aligned}
$$

11.2.1 Separation of variables

We deduce from the expressions of H and $\mathbf{l}^2$ that

$$[H, \mathbf{l}^2] = 0, \quad H = \frac{1}{2m_0}\left(p_r^2 + \frac{\mathbf{l}^2}{r^2}\right) + V(r), \tag{11.48}$$

i.e. H and $\mathbf{l}^2$ possess a common system of basis eigenvectors. We therefore first search for a complete system of eigenfunctions of $\mathbf{l}^2$. Since $\mathbf{l} = \mathbf{r} \times \mathbf{p}$, we have

$$
\begin{aligned}
[l_x, l_y] &= [yp_z - zp_y, zp_x - xp_z] = [yp_z, zp_x] + [zp_y, xp_z] \\
&= y[p_z, z]p_x + p_y[z, p_z]x = i\hbar(xp_y - yp_x) \\
&= i\hbar l_z.
\end{aligned}
\tag{11.49}
$$

By cyclic permutation of x, y, z we have altogether

$$
[l_x, l_y] = i\hbar l_z, \quad [l_y, l_z] = i\hbar l_x, \quad [l_z, l_x] = i\hbar l_y.
\tag{11.50}
$$

Thus the components of $\mathbf{l}$ do not commute pairwise, i.e. for any two of the components there is no common complete system of eigenfunctions, and this implies in the terminology we used earlier that their determination cannot be "*sharp*" simultaneously, i.e. they are incompatible variables. However,

$$
[l_z, l_x^2] = i\hbar(l_y l_x + l_x l_y), \quad [l_z, l_y^2] = -i\hbar(l_y l_x + l_x l_y), \quad [l_z, l_z^2] = 0,
$$

so that

$$
[l_z, \mathbf{l}^2] = 0,
\tag{11.51}
$$

and correspondingly

$$
[l_x, \mathbf{l}^2] = 0, \quad [l_y, \mathbf{l}^2] = 0.
\tag{11.52}
$$

This means that $\mathbf{l}^2, l_z$ (or $\mathbf{l}^2, l_x$ or $\mathbf{l}^2, l_y$) are simultaneously "sharp" determinable variables and hence have a common system of basis functions. We can now proceed as in the case of the harmonic oscillator and determine their eigenfunctions.

Example 11.2: Verify that $[l_i, q_j] = i\hbar\epsilon_{ijk}q_k$

Solution: This can be verified in analogy to the preceding equations ($i = 1, 2, 3; q_1 = x, q_2 = y, q_3 = z$).

11.3 Representation of Rotation Group

Our objective now is to develop a representation of the rotation group in analogy to the energy representation of the simple harmonic oscillator. In order to determine the eigenvectors of the operators $\mathbf{l}^2$ and l_z one defines

$$
l_\pm = l_x \pm il_y.
\tag{11.53}
$$

Either of these operators is the hermitian conjugate of the other. These operators here play a role analogous to that of the quasi-particle operators

$A, A^\dagger$ in the case of the linear harmonic oscillator. In the present context the operators are called *shift operators* for reasons which will be seen below. With the help of Eq. (11.49) we obtain the following relations:

$$[l_z, l_+] = \hbar l_+, \quad [l_z, l_-] = -\hbar l_-, \quad [l_+, l_-] = 2\hbar l_z. \tag{11.54}$$

On the other hand we obtain from Eqs. (11.51), (11.52):

$$0 = [\mathbf{l}^2, l_+] = [\mathbf{l}^2, l_-] = [\mathbf{l}^2, l_z], \tag{11.55}$$

and with Eq. (11.54)

$$\mathbf{l}^2 = \frac{1}{2}(l_+ l_- + l_- l_+) + l_z^2 = l_+ l_- + l_z^2 - \hbar l_z. \tag{11.56}$$

One now defines certain ket-vectors as eigenvectors of l_z which span the space $\mathcal{H}_l$ of the appropriate states; in particular one defines $|l\rangle \in \mathcal{H}_l$ as eigenvector of l_z with the largest eigenvalue l_0, assuming a finite dimensional representation space.[†] Thus we have (**l** being a bounded operator)

$$l_z|l\rangle = l_0 \hbar |l\rangle \quad (l_0 : \text{max.}). \tag{11.57}$$

Then from Eq. (11.54):

$$l_z l_- |l\rangle = (l_- l_z - l_- \hbar)|l\rangle = (l_0 - 1)\hbar l_- |l\rangle, \tag{11.58}$$

i.e. $l_-|l\rangle$ is eigenvector of l_z with eigenvalue $(l_0 - 1)\hbar$. Similarly we have

$$l_z(l_-)^2|l\rangle = (l_0 - 2)\hbar(l_-)^2|l\rangle$$

and more generally

$$l_z(l_-)^r|l\rangle = (l_0 - r)\hbar(l_-)^r|l\rangle. \tag{11.59}$$

We write or define:

$$|l - r\rangle := (l_-)^r|l\rangle \quad \text{so that} \quad l_z|l - r\rangle \overset{(11.59)}{=} \hbar(l_0 - r)|l - r\rangle. \tag{11.60}$$

Similarly for l_+ with the help of Eq. (11.54):

$$l_z l_+ |l\rangle = (l_+ l_z + l_+ \hbar)|l\rangle = (l_0 + 1)\hbar l_+ |l\rangle. \tag{11.61}$$

But l_0 is by definition the largest eigenvalue of l_z. Therefore $(l_0 + 1)\hbar$ cannot be an eigenvalue of l_z. Hence we must have

$$l_+|l\rangle = 0. \tag{11.62}$$

[†]At this stage it is not yet decided whether l_0 is an integer or not! But we know that l_0 is real, since l_z is hermitian.

Thus for $\mathbf{l}^2$ we obtain:

$$\mathbf{l}^2|l\rangle \stackrel{(11.56)}{=} \left[\frac{1}{2}(l_+l_- + l_-l_+) + l_z^2\right]|l\rangle = \left[\frac{1}{2}l_+l_- + l_z^2\right]|l\rangle$$

$$\stackrel{(11.54)}{=} \left[\frac{1}{2}(l_-l_+ + 2l_z\hbar) + l_z^2\right]|l\rangle = (l_z\hbar + l_z^2)|l\rangle$$

$$\stackrel{(11.57)}{=} l_0(l_0 + 1)\hbar^2|l\rangle. \tag{11.63}$$

Hence $|l\rangle$ is eigenvector of $\mathbf{l}^2$ with eigenvalue $l_0(l_0+1)\hbar^2$. Since $\mathbf{l}^2$ commutes with l_-, i.e.

$$[\mathbf{l}^2, l_-] = 0,$$

(cf. Eq. (11.55)) we have

$$\mathbf{l}^2 l_-|l\rangle = l_-\mathbf{l}^2|l\rangle = l_0(l_0 + 1)\hbar^2 l_-|l\rangle, \tag{11.64}$$

i.e. $l_-|l\rangle$ is also eigenvector of $\mathbf{l}^2$ with eigenvalue $l_0(l_0+1)\hbar^2$. From Eq. (11.64) we obtain on multiplication by l_-:

$$l_-\mathbf{l}^2 l_-|l\rangle = \mathbf{l}^2 l_-^2|l\rangle = l_0(l_0 + 1)\hbar^2 l_-^2|l\rangle,$$

and more generally

$$\mathbf{l}^2(l_-)^r|l\rangle = l_0(l_0 + 1)\hbar^2(l_-)^r|l\rangle$$

$$\stackrel{(11.60)}{=} l_0(l_0 + 1)\hbar^2|l - r\rangle. \tag{11.65}$$

We assumed that $\mathcal{H}_l$ is finite dimensional so that the sequence of eigenvectors of l_z, i.e.

$$|l\rangle, |l - 1\rangle, \ldots, |l - r\rangle$$

terminates at some number (let us say) $r = n$ (n a positive integer), i.e.

$$l_-|l - n\rangle \stackrel{(11.60)}{=} (l_-)^{n+1}|l\rangle = 0. \tag{11.66}$$

It then follows from the second of relations (11.56) that

$$\mathbf{l}^2|l - n\rangle = (l_+l_- + l_z^2 - l_z\hbar)|l - n\rangle$$

$$\stackrel{(11.66)}{=} (l_z^2 - l_z\hbar)|l - n\rangle$$

$$\stackrel{(11.60)}{=} \{(l_0 - n)^2 - (l_0 - n)\}\hbar^2|l - n\rangle. \tag{11.67}$$

But according to Eqs. (11.60) and (11.65) for $r = n$:

$$\mathbf{l}^2|l - n\rangle = \mathbf{l}^2(l_-)^n|l\rangle = l_0(l_0 + 1)\hbar^2|l - n\rangle, \tag{11.68}$$

so that with Eq. (11.67) we obtain (observe $l_0(l_0 + 1) = -l_0(-l_0 - 1)$):

$$l_0(l_0 + 1) = \{(l_0 - n)^2 - (l_0 - n)\} = (l_0 - n)(l_0 - n - 1),$$

i.e.

$$-l_0 = l_0 - n, \quad l_0 = \frac{n}{2}. \tag{11.69}$$

From this it follows that l_0 must be either half integral or integral. In the following we consider primarily the case of integral values of l_0. The operators l_+, l_-, l_z transform the vectors $|l\rangle, \ldots, |l - n\rangle$ into one other. The dimension of the representation space is therefore $n + 1 = 2l_0 + 1$. The basis vectors $|l\rangle, \ldots, |l - n\rangle$ are eigenvectors of l_z with eigenvalues $l_0\hbar, (l_0 - 1)\hbar, \ldots, -l_0\hbar$.

As in the case of the harmonic oscillator we introduce normalized basis vectors. We write these[‡] for integral values of l_0:

$$|l, m\rangle,$$

where

$$l_z|l, m\rangle = m\hbar|l, m\rangle \quad \text{with} \quad -l_0 \leq m \leq l_0. \tag{11.70}$$

According to Eq. (11.65) these are eigenvectors of $\mathbf{l}^2$ with eigenvalues $l_0(l_0 + 1)\hbar^2$. The nonvanishing matrix elements of $\mathbf{l} = (l_+, l_-, l_z)$ are then:

$$\begin{aligned} \langle l, m|l_z|l, m\rangle &= m\hbar, \\ \langle l, m \pm 1|l_\pm|l, m\rangle &= \sqrt{(l_0 \pm m + 1)(l_0 \mp m)}\hbar. \end{aligned} \tag{11.71}$$

These expressions should be compared with the corresponding ones in the case of the harmonic oscillator. The normalization factor in Eq. (11.71) can be established in analogy to the method used in the case of the harmonic oscillator; we skip this here. We therefore write the $2l_0 + 1$ orthonormal vectors:

$$|l, m\rangle : \quad |l, l_0\rangle, |l, l_0 - 1\rangle, \ldots, |l, -l_0\rangle.$$

According to Eqs. (11.63) and (11.70) they satisfy the eigenvalue equations (with $m = l_0, l_0 - 1, \ldots, -l_0$):

$$\mathbf{l}^2|l, m\rangle = l_0(l_0 + 1)\hbar^2|l, m\rangle, \quad l_z|l, m\rangle = m\hbar|l, m\rangle.$$

The phases are chosen such that[§]

$$\begin{aligned} l_+|l, m\rangle &= \sqrt{l_0(l_0 + 1) - m(m + 1)}|l, m + 1\rangle\hbar \\ &= \sqrt{(l_0 + m + 1)(l_0 - m)}|l, m + 1\rangle\hbar, \\ l_-|l, m\rangle &= \sqrt{l_0(l_0 + 1) - (m - 1)m}|l, m - 1\rangle\hbar \\ &= \sqrt{(l_0 - m + 1)(l_0 + m)}|l, m - 1\rangle\hbar. \end{aligned}$$

[‡]D. M. Brink and G. R. Satchler [38], p. 17.
[§]See e.g. A. Messiah [195], Vol. II, p. 24.

Then, as above,

$$l_+|l, l_0\rangle = 0, \quad l_-|l, -l_0\rangle = 0.$$

11.4   Angular Momentum:Angular Representation

In the case of the harmonic oscillator we arrived at the position or config-uration space dependent wave functions by considering the position space representation of states. We now perform the analogous procedure for the angular momentum or, as this is sometimes described, for the simple "*rotor*", for which the eigenfunctions $|\psi\rangle$ are given by

$$\mathbf{l}^2|\psi\rangle = l_0(l_0 + 1)\hbar^2|\psi\rangle. \tag{11.72}$$

In view of the spherical symmetry it is reasonable to use spherical polar coordinates r, θ, φ, as remarked earlier. In these coordinates we obtain:

$$l_z = xp_y - yp_x = -i\hbar\frac{\partial}{\partial\varphi},$$

$$l_\pm = l_x \pm il_y = \pm\hbar e^{\pm i\varphi}\left(\frac{\partial}{\partial\theta} \pm i\cot\theta\frac{\partial}{\partial\varphi}\right), \tag{11.73}$$

so that

$$\mathbf{l}^2 = -\hbar^2\left[\frac{1}{\sin\theta}\frac{\partial}{\partial\theta}\left(\sin\theta\frac{\partial}{\partial\theta}\right) + \frac{1}{\sin^2\theta}\frac{\partial^2}{\partial\varphi^2}\right]. \tag{11.74}$$

We choose the eigenfunctions of $\mathbf{l}^2$ and l_z as basis vectors of the ("irre-ducible") representation, i.e. with

$$\mathbf{l}^2 Y_{l_0}^m(\theta, \varphi) = l_0(l_0 + 1)\hbar^2 Y_{l_0}^m(\theta, \varphi),$$
$$l_z Y_{l_0}^m(\theta, \varphi) = m\hbar Y_{l_0}^m(\theta, \varphi),$$

where $l_0 = 0, 1, 2, \ldots$ and $m = -l_0, -l_0 + 1, \ldots, l_0$. With the ansatz

$$Y(\theta, \varphi) = \Theta(\theta)\Phi(\varphi), \quad \Phi \propto e^{im\varphi}, \tag{11.75}$$

the variables contained in the expression (11.74) can be separated:

$$\left[-\frac{1}{\sin\theta}\frac{\partial}{\partial\theta}\left(\sin\theta\frac{\partial}{\partial\theta}\right) - \frac{m^2}{\sin^2\theta}\right]\Theta(\theta) = l_0(l_0 + 1)\Theta(\theta), \tag{11.76}$$

$$-\frac{\partial^2}{\partial\varphi^2}\Phi(\varphi) = m^2\Phi(\varphi). \tag{11.77}$$

Without prior knowledge about the integral nature of l_0 and m we can show that indeed these have to assume the integral values derived above. We observe that l_0 and m define the eigenvalues of Eqs. (11.76) and (11.77) and these are determinable by the requirement of uniqueness of the original wave function and its square integrability. Uniqueness implies immediately that m has to be integral, so that

$$e^{im\varphi} = e^{im(\varphi \pm 2\pi)}.$$

The functions $Y_l^m(\theta, \varphi)$ are called *spherical harmonics*. These functions are defined in such a way that they are normalized to unity on the unit sphere, and that their phases are such that $Y_l^0(0,0)$ is real and positive. The functions Y_l^m satisfy the orthonormality condition

$$\int d\Omega Y_l^{m*} Y_{l'}^{m'} \equiv \int_0^{2\pi} d\varphi \int_0^\pi \sin\theta d\theta Y_l^{m*}(\theta, \varphi) Y_{l'}^{m'}(\theta, \varphi) = \delta_{mm'}\delta_{ll'}, \quad (11.78)$$

and the completeness relation

$$\sum_{l=0}^{\infty} \sum_{m=-l}^{l} Y_l^{m*}(\theta, \varphi) Y_l^m(\theta', \varphi') = \frac{\delta(\theta - \theta')\delta(\varphi - \varphi')}{\sin\theta} \equiv \delta(\Omega - \Omega'). \quad (11.79)$$

The connection of these hyperspherical functions with the *associated Legendre functions* $P_l^m, m \geq 0$, is given by

$$Y_l^m(\theta, \varphi) = (-1)^m \left[\frac{(2l+1)(l-m)!}{4\pi(l+m)!} \right]^{1/2} P_l^m(\cos\theta) e^{im\varphi}. \quad (11.80)$$

In particular for $m = 0$:

$$Y_l^0 = \sqrt{\frac{2l+1}{4\pi}} P_l(\cos\theta).$$

Furthermore

$$Y_0^0 = \frac{1}{\sqrt{4\pi}}, \quad Y_1^0 = \sqrt{\frac{3}{4\pi}} \cos\theta, \quad Y_2^0 = \sqrt{\frac{5}{16\pi}} (3\cos^2\theta - 1),$$

$$Y_3^0 = \sqrt{\frac{7}{16\pi}} (5\cos^3\theta - 3\cos\theta). \quad (11.81)$$

One should note that the functions Y_l^m are square integrable only for the given integral values of m and l. This is the case if one demands that $\mathbf{l}^2$ and l_z possess a common complete, orthonormalized system of eigenfunctions. We mention in passing that in spherical coordinates

$$l_z \equiv xp_y - yp_x = -i\hbar \frac{\partial}{\partial\varphi}. \quad (11.82)$$

The number l is referred to as the *"orbital quantum number"* and m as *"magnetic quantum number"*. Instead of the values $l = 0, 1, 2, \ldots$ respectively, the designation $s, p, d, f, g, h, \ldots$ is also in use. The associated Legendre functions or spherical functions $P_l^m(x)$ satisfy the differential equation

$$(1 - x^2)P_l^{m\prime\prime} - 2(m + 1)xP_l^{m\prime} + [l(l + 1) - m(m + 1)]P_l^m = 0. \qquad (11.83)$$

Their connection with the Legendre functions $P_l(x)$ is given by the relation

$$P_l^m(x) = -\frac{d^m}{dx^m}P_l(x) \qquad (11.84)$$

with the differential equation

$$(1 - x^2)P_l'' - 2xP_l' + l(l + 1)P_l = 0. \qquad (11.85)$$

Setting

$$y = (1 - x^2)^{m/2}P_l^m(x) \qquad (11.86)$$

we obtain the equation

$$(1 - x^2)y'' - 2xy' + \left[l(l + 1) - \frac{m^2}{1 - x^2}\right]y = 0. \qquad (11.87)$$

The number m must be an integer already for reasons of uniqueness of the wave function, i.e. of Y_l^m, i.e. $\exp(im\varphi)$, in the replacement $\varphi \to \varphi + 2\pi$. If we did not know yet that $l = l_0 =$ an integer ≥ 0, we would now have to search for those values of l for which the solutions are square integrable. It is instructive to pursue the appropriate arguments. Consider the equation for $P_l(x), m = 0$, i.e.

$$(1 - x^2)y'' - 2xy' + l(l + 1)y = 0. \qquad (11.88)$$

For the solution we use the ansatz of a power series, i.e.

$$y = \sum_{i=0}^{\infty} a_i x^{i+\kappa}, \quad (\kappa \text{ lowest power } \geq 0). \qquad (11.89)$$

Inserting this expansion into Eq. (11.88) we obtain for every value of x

$$\sum_i a_i(\kappa + i)(\kappa + i - 1)x^{\kappa+i-2}$$

$$- \sum_i a_i[(\kappa + i)(\kappa + i - 1) + 2(\kappa + i) - l(l + 1)]x^{\kappa+i} = 0. \qquad (11.90)$$

For $i = 0, 1$ we obtain by comparing coefficients

$$a_0 \kappa(\kappa - 1) = 0, \quad a_1(\kappa + 1)\kappa = 0, \tag{11.91}$$

from which we deduce that $\kappa = 0, 1$. Comparing the coefficients of $x^{\kappa+j}$ we obtain

$$a_{j+2}(\kappa + j + 2)(\kappa + j + 1) = a_j[(\kappa + j)(\kappa + j + 1) - l(l + 1)]. \tag{11.92}$$

This is a recurrence relation. For $|x| < 1$ we can obtain from this the coefficients of a convergent series, but also those for a different series for $|x| > 1$. However, these are of little interest to us here. For the solution in accordance with (11.78) to be square integrable, they have to be polynomials (recall the orthogonalization procedure of E. Schmidt). This is precisely the case, when the series terminates after a finite number of terms. We see that this happens precisely when l is a positive or negative integer or zero.¶

11.5   Radial Equation for Hydrogen-like Atoms

The common set of eigenfunctions of the operators $H, \mathbf{l}^2$ and l_z are the solutions of the Schrödinger equation of the form

$$\psi_l^m(r, \theta, \varphi) = Y_l^m(\theta, \varphi)\chi_l(r), \tag{11.93}$$

where $\chi_l(r)$ is the solution of the *radial equation* (cf. Eq. (11.46))

$$\left[\frac{p_r^2}{2m_0} + \frac{l(l + 1)\hbar^2}{2m_0 r^2} + V(r) - E \right] \chi_l(r) = 0 \tag{11.94}$$

with (cf. first line of Eq. (11.29))

$$p_r^2 \equiv -\hbar^2 \frac{1}{r} \frac{\partial^2}{\partial r^2} r.$$

We set

$$y_l = r\chi_l. \tag{11.95}$$

The hermiticity condition for p_r then requires that (cf. Eq. (11.26))

$$y_l(r = 0) = 0.$$

We have therefore (multiplying Eq. (11.94) from the left by r)

$$\left[-\frac{\hbar^2}{2m_0} \frac{d^2}{dr^2} + l(l + 1)\frac{\hbar^2}{2m_0 r^2} + V(r) - E \right] y_l(r) = 0 \tag{11.96}$$

¶For further details see H. Margenau and G. M. Murphy [190], p. 93.

with

$$\langle \psi_l^m | \psi_l^m \rangle = \int_0^\infty r^2 dr |\chi_l(r)|^2 = \int_0^\infty |y_l(r)|^2 dr. \qquad (11.97)$$

We thus have a problem similar to that of a particle of mass m in the domain $(0, \infty)$ of a one-dimensional space.

In the first place we investigate the behaviour of y_l in the neighbourhood of the origin. We assume that $V(r)$ has at $r = 0$ at most a singularity of the form of $1/r$. This applies to the Coulomb potential but also to Coulomb-like potentials like screened Coulomb potentials or the Yukawa potential. In these cases the behaviour of y_l near the origin is dominated by the centrifugal term and thus Eq. (11.96) can be shown to possess a regular solution there which behaves like $r^{l+1}(1 + O(r))$ in approaching $r = 0$. The equation also has an irregular solution, which behaves like $1/r^l$ in approaching $r = 0$. In order to verify this behaviour we set

$$y_l(r) = r^s(1 + a_1 r + a_2 r^2 + \cdots),$$

and substitute this into Eq. (11.96). Equating the coefficient of $1/r^2$ to zero, we obtain the so-called "indicial equation"

$$-s(s - 1) + l(l + 1) = 0, \quad \text{i.e.} \quad s = l + 1, -l.$$

Obviously the irregular solution has to be rejected, since it satisfies neither the condition $y_l(0) = 0$, nor the condition of normalizability for $l \neq 0$.

We now consider the hydrogen atom with $V(r) = -e^2/r$ (i.e. $Z_1 = Z_2 = 1$ in Eq. (11.1)); an analogous consideration applies to hydrogen-like atoms like, for instance, He^+, Li^{++}, and others. The wave function of the relative motion is the solution $\psi(\mathbf{r})$ of the equation

$$\left[-\frac{\hbar^2}{2m_0} \triangle - \frac{e^2}{r} \right] \psi(\mathbf{r}) = E\psi(\mathbf{r}), \quad m_0 = \frac{m_e M_p}{m_e + M_p}. \qquad (11.98)$$

Here

$$\psi(\mathbf{r}) = Y_l^m(\theta, \varphi) \frac{y_l(r)}{r} \qquad (11.99)$$

and

$$y_l'' + \left[\epsilon + \frac{2m_0}{\hbar^2} \frac{e^2}{r} - \frac{l(l + 1)}{r^2} \right] y_l = 0 \qquad (11.100)$$

with

$$\epsilon = \frac{2m_0 E}{\hbar^2}. \qquad (11.101)$$

We set

$$\kappa = (-\epsilon)^{1/2} = \frac{(-2m_0 E)^{1/2}}{\hbar}. \qquad (11.102)$$

For $E > 0$ the solutions oscillate in the region $r \to \infty$, for $E < 0$ they decrease exponentially. We set

$$x = 2\kappa r. \tag{11.103}$$

We set as *Bohr radius**

$$a = \frac{(4\pi\epsilon_0)\hbar^2}{m_0 e^2} \quad (= 0.529 \times 10^{-10} \text{ meters})$$

and

$$\nu = \frac{1}{\kappa a} = \frac{m_0 e^2}{\hbar} \left(-\frac{1}{2m_0 E} \right)^{1/2} = \frac{e^2}{\hbar c} \left(-\frac{m_0 c^2}{2E} \right)^{1/2}. \tag{11.104}$$

With these substitutions we can rewrite Eq. (11.100) as[†]

$$\left[\frac{d^2}{dx^2} - \frac{l(l+1)}{x^2} + \frac{\nu}{x} - \frac{1}{4} \right] y_l = 0. \tag{11.105}$$

11.6 Discrete Spectrum of the Coulomb Potential

11.6.1 The eigenvalues

Here we are interested in the solution which is regular at the origin (i.e. behaves there like x^{l+1}) extended to an exponentially decreasing branch at infinity. We can see from Eq. (11.105) that for $x \to \infty$ there is a solution behaving like $\exp(-x/2)$. We set therefore

$$y_l = x^{l+1} e^{-x/2} v_l(x). \tag{11.106}$$

With this ansatz we separate from the solution the behaviour around the origin as well as that at infinity. It remains to determine the function $v_l(x)$ for a complete determination of the solution. Inserting the ansatz into Eq. (11.105) we obtain the equation for $v_l(x)$, i.e.

$$\left[x \frac{d^2}{dx^2} + (2l + 2 - x) \frac{d}{dx} - (l + 1 - \nu) \right] v_l(x) = 0. \tag{11.107}$$

This equation has the form of a *confluent hypergeometric equation*, i.e.

$$x\Phi'' + (b - x)\Phi' - a\Phi = 0. \tag{11.108}$$

---

*The factor $4\pi\epsilon_0$ appears in SI units.

[†]This equation is known as Whittaker's equation; see e.g. W. Magnus and F. Oberhettinger [181], p.88.

The regular solution of this equation is the *confluent hypergeometric series* (also called *Kummer series*)

$$\Phi(a, b; x) = 1 + \frac{a}{b} \frac{x}{1!} + \frac{a(a+1)}{b(b+1)} \frac{x^2}{2!} + \cdots = \sum_{n=0}^{\infty} \frac{(a+n-1)!(b-1)!}{(a-1)!(b+n-1)!} \frac{x^n}{n!}.$$
(11.109)

In our case we can write the series

$$v_l(x) = \Phi(l+1-\nu, 2l+2; x) = \sum_{p=0}^{\infty} \frac{\Gamma(l+1+p-\nu)}{\Gamma(l+1-\nu)} \frac{(2l+1)!}{(2l+1+p)!} \frac{x^p}{p!}. \quad (11.110)$$

The expression[*]

$$\Gamma(x+1) = x!$$

is the *gamma function*, i.e. the factorial for an arbitrary argument. The function $\Gamma(z)$ is defined by the integral

$$\Gamma(z) := \int_0^{\infty} e^{-t} t^{z-1} dt. \quad (11.111)$$

Important properties of the function $\Gamma(z)$ are:
(1) $z\Gamma(z) = \Gamma(z+1), z(z-1)! = z!$,
(2) $\Gamma(z)\Gamma(1-z) = \frac{\pi}{\sin \pi z}, (-z)!(z-1)! = \frac{\pi}{\sin \pi z}$, called *inversion* or *reflection* formula,
(3) $\Gamma(2z) = \frac{2^{2z-1}}{\sqrt{\pi}} \Gamma(z)\Gamma\left(z + \frac{1}{2}\right), \sqrt{\pi}(2z)! = 2^{2z} z!(z-1/2)!$, called *duplication* formula,
(4) $\Gamma(\frac{1}{2}) = \sqrt{\pi}$,
(5) $\Gamma(n+1) = n!$. (11.112)

The gamma function is a nonvanishing analytic, more precisely, meromorphic function, of its argument z with simple poles at $z = -n$ and residues there given by $(-1)^n/n!$ as may be deduced from the property (2) above.

In general the function Φ is an infinite series and behaves at $x \to \infty$ like

$$x^{-(l+1+\nu)} e^x.$$

Obviously such a series is useless in our case, since the exponential function (generated by the infinite series) would destroy the asymptotic behaviour of the wave function which we separated off with Eq. (11.106), and hence the

---

[*]Some writers, particularly pure mathematicians, are very strict in reserving the factorial exclusively for positive integers and the gamma function for all other cases. However, with many of these around in some calculation, and for both cases, it is convenient to use only the factorial notation.

latter would not be square integrable. Thus the infinite series must break off somewhere, i.e. must be a polynomial. This is achieved, if the series terminates for certain values of ν, i.e. for

$$\nu = n = l + 1 + n', \quad n' = 0, 1, 2, \ldots. \tag{11.113}$$

This is a quantization condition, which is thus seen to be a consequence of demanding the square integrability of the wave function. The quantum number n' is called *radial quantum number*. This number is equal to the number of finite zeros of the radial part of the wave function (excepting the origin), which are also described as its *nodes*. One defines as the *principal quantum number* the number n. With (cf. Eqs. (11.101) to (11.104))

$$n = \nu = \frac{1}{\kappa a} = \frac{m_0 e^2}{\hbar} \left(-\frac{1}{2 m_0 E} \right)^{1/2},$$

we obtain the energy spectrum of the *discrete states* with angular momentum l, i.e.[†]

$$E_{l,n'} = -\left(\frac{e^2}{\hbar c} \right)^2 \frac{m_0 c^2}{2(l + 1 + n')^2}, \tag{11.114a}$$

$$E_n = -\left(\frac{e^2}{\hbar c} \right)^2 \frac{m_0 c^2}{2n^2}, \quad n = 1, 2, 3, \ldots. \tag{11.114b}$$

It should be noted that the magnetic quantum number m does not appear in this expression. For every value E_n, i.e. every n, the *orbital* or *azimuthal quantum number* l can assume the values $0, 1, 2, \ldots$. The *degeneracy* of the levels E_n is therefore of degree[‡]

$$\sum_{l=0}^{n-1} (2l + 1) = 2 \sum_{l=0}^{n-1} l + n = 2 \frac{n}{2}(n - 1) + n = n(n - 1) + n = n^2. \tag{11.114c}$$

(Recall that for an arithmetic series $a + (a + d) + \cdots + (a + (n - 1)d) = na + n(n - 1)d/2$; in our case: $a = 0$, $d = 1$.)

[†] Observe that with property (2) of the gamma function we can re-express the ratio of gamma functions in Eq. (11.110) as follows:

$$\frac{\Gamma(l + 1 + p - \nu)}{\Gamma(l + 1 - \nu)} = \frac{\pi}{\Gamma(\nu - l - p) \sin\{\pi(\nu - l - p)\}} \frac{\sin\{\pi(\nu - l)\}\Gamma(\nu - l)}{\pi}$$

$$= \frac{(-1)^p \Gamma(\nu - l)}{\Gamma(\nu - l - p)} = (-1)^p \frac{n'!}{(n' - p)!}.$$

[‡] Note that $2l + 1$ is the number of possible orientations of the angular momentum vector $\mathbf{l}$ with respect to a preferred direction, and so to the number of different values of m.

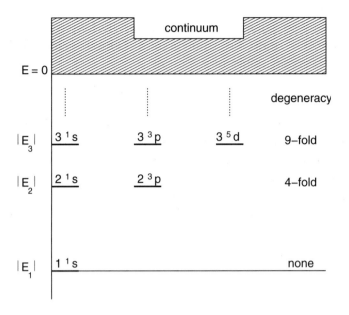

Fig. 11.1 Hydrogen atom energy levels.

The following *spectroscopic description of states* is customary:

$$nl \longrightarrow ns, \ np, \ nd, \ ne, \ nf \ldots,$$

where n denotes the *principal quantum number* and $s, p, d, \ldots$ stand for $l = 0, 1, 2, \ldots$ respectively.[§] One also writes

$$n^{2l+1}l,$$

where the superscript $2l+1$ at the top left of l gives the degree of degeneracy. The spectrum of the hydrogen atom thus has the form shown schematically in Fig. 11.1.

The above treatment of the hydrogen atom has obvious shortcomings:
(a) The hydrogen atom was treated nonrelativistically; relativistic corrections are of the order of $v^2/c^2 \sim O(E_n/m_0c^2)$ as we illustrate in Example 11.4.
(b) The spin of the electron was not taken into account, and hence the fine structure of the energy levels is excluded.
(c) The nucleus was treated as pointlike and not as something with structure.

[§]One describes as *Lyman series* the final state with $n = 1, \therefore l = 0$, from an initial state $n \geq 2, \therefore l = 1$. One describes as *Balmer series* the final state $n = 2, \therefore l = 0$ or 1, from an initial state $n > 2, l = 1$ to $n = 2, l = 0$ (called principal series, hence p), from $n, l = 2$ to $n = 2, l = 1$ (called diffuse series, hence d), and from $n, l = 0$ to $n = 2, l = 0$ (called sharp series, hence s).

The associated radial wave function is

$$y_l(x) = x^{l+1} e^{-x/2} \Phi(-n', 2l+2; x),$$

$$\Phi(-n', 2l+2; x) = \sum_{p=0}^{n'} \frac{(-x)^p}{p!} \frac{n'!(2l+1)!}{(n'-p)!(2l+1+p)!}. \qquad (11.115)$$

The spectrum (11.114b) yields the binding energy of the states. The mass of the hydrogen atom is given by

$$\text{mass}_H = M_p + m_e + E < M_p + m_e. \qquad (11.116)$$

11.6.2 Laguerre polynomials: Various definitions in use!

The polynomial obtained by terminating the hypergeometric series (11.110) is usually re-expressed in terms of orthogonal polynomials known as *associated Laguerre polynomials*. Unfortunately, however, several different definitions are in use, so that one always has to check the definition. We mention here three different definitions used in the literature. Each can be recognized by reference to the generating function.

Apart from factors, the polynomial contained in y_l of Eq. (11.115) is effectively the *associated Laguerre polynomial* normally written $L_r^k(z)$. We distinguish between different definitions in use. Thus Messiah [195] defines the original or stem *Laguerre polynomial* as :

$$\left. \begin{aligned} {}^*L_r(z) &\equiv {}^*L_r^0(z) = e^z \frac{d^r}{dz^r} (e^{-z} z^r), \\ {}^*L_r^k(z) &= \left(-\frac{d}{dz} \right)^k L_{r+k}(z), \end{aligned} \right\} \quad k, r = 0, 1, 2, 3, \dots . \qquad (11.117)$$

The function ${}^*L_r^k(z)$ is a polynomial of degree r, in fact, it is given by[¶]

$$^*L_r^k(z) = \frac{[(r+k)!]^2}{r!k!} \Phi(-r, k+1; z) \equiv (r+k)! \, L_r^k(z), \qquad (11.118a)$$

where $L_r^k(z)$ is the associated Laguerre polynomial as normally defined and mostly written $L_r^{(k)}(z)$. With the help of Eq. (11.115) we can re-express this function as

$$^*L_r^k(z) = \frac{[(r+k)!]^2}{r!k!} \sum_{p=0}^{r} \frac{(-x)^p}{p!} \frac{r!k!}{(r-p)!(k+p)!}. \qquad (11.118b)$$

[¶]This definition of ${}^*L_r^k$ is used by A. Messiah [195], Vol. I, Appendix B.1.2. Unfortunately the definition mostly used in mathematical literature differs. For clarity and later use we denote the function used there by its normal form L_r^k; see, for instance, W. Magnus and F. Oberhettinger [181], p. 84, and I. S. Gradshteyn and I. M. Ryzhik [122], formula 7.414(3), p. 844.

The *differential equation* of the *associated Laguerre polynomials* is (same for $^*L_r^k(z)$)

$$\left[z\frac{d^2}{dz^2} + (k+1-z)\frac{d}{dz} + r \right] L_r^k(z) = 0. \tag{11.119}$$

In particular one has (we cite these for later comparison)

$$L_0(z) = 1, \quad L_1(z) = 1-z, \quad L_2(z) = 1 - 2z + \frac{1}{2}z^2, \dots,$$

and

$$L_0^1(z) = 1, \quad L_1^1(z) = 2-z, \quad L_2^1(z) = 3 - 3z + \frac{1}{2}z^2, \dots .$$

The *generating function* of the associated Laguerre polynomials — obtained by searching for an integral representation of the solution of this differential equation which is of the type of a Laplace transform — is for $|t| < 1$ and expanded in ascending powers of t (cf. Example 11.3):

$$\begin{aligned}
\frac{e^{-zt/(1-t)}}{(1-t)^{k+1}} &= \sum_{r=0}^{\infty}\left[\sum_{i=0}^{\infty}\frac{(-z)^i}{i!}\frac{(k+r)!}{(i+k)!(r-i)!} \right]t^r \\
&= \sum_{r=0}^{\infty}\frac{t^r}{(k+r)!}\,{}^*L_r^k(z) = \sum_{r=0}^{\infty}L_r^k(z)t^r. \tag{11.120}
\end{aligned}$$

It will be seen that the *orthonormality condition* is$^{\parallel}$

$$\int_0^\infty e^{-z}z^{k\,*}L_p^k(z)\,{}^*L_q^k(z)dz = \frac{[(p+k)!]^3}{p!}\delta_{pq}. \tag{11.121}$$

In Example 11.3 we show in a typical calculation how this normalization condition can be obtained with the help of the generating function.

For our purposes here it is convenient to refer to yet another definition of the associated Laguerre polynomials. It is said that this is the definition preferred in applied mathematics.** The associated Laguerre polynomial here, which we write $^\dagger L_r^k$, is defined by††

$$^\dagger L_r^k(z) = \frac{d^k}{dz^k}L_r^0(z), \quad ^\dagger L_r^k(z) = \frac{(-1)^k(r!)^2}{k!(r-k)!}\Phi(k-r, k+1; z), \quad (r \le k). \tag{11.122}$$

$^{\parallel}$See A. Messiah [195], Vol. I, Appendix B.1.2.

**I. N. Sneddon [255], p. 163. The reader is there warned that "*care must be taken in reading the literature to ensure that the particular convention being followed is understood*".

††For this definition and the following equations we refer to I. N. Sneddon [255], pp. 162 - 164. There at some points $n+l$ is misprinted $n+1$.

The differential equation of these associated Laguerre polynomials is[‡‡]

$$\left[z\frac{d^2}{dz^2} + (k+1-z)\frac{d}{dz} + r - k \right]{}^\dagger L_r^k(z) = 0. \tag{11.123}$$

For the purpose of clarity and comparison with the polynomials of the other definition it may help to see that here in particular

$${}^\dagger L_0(z) = 1, \quad {}^\dagger L_1(z) = 1 - z, \quad {}^\dagger L_2(z) = 2 - 4z + z^2, \ldots,$$

and

$${}^\dagger L_1^1(z) = -1, \quad {}^\dagger L_2^1(z) = -4 + 2z, \quad {}^\dagger L_3^1(z) = -18 + 18z - 3z^2, \ldots .$$

The generating function of the associated Laguerre polynomials so defined is (observe $(-t)^k$ as compared with (11.120))

$$(-t)^k \frac{e^{-zt/(1-t)}}{(1-t)^{k+1}} = \sum_{p=k}^\infty \frac{t^p}{p!} {}^\dagger L_p^k(z). \tag{11.124a}$$

The orthonormality condition is (cf. Example 11.8):

$$\int_0^\infty e^{-z} z^{k+1} {}^\dagger L_p^k(z) {}^\dagger L_q^k(z) dz = \frac{(2p-k+1)(p!)^3}{(p-k)!} \delta_{pq}. \tag{11.124b}$$

Comparing Eqs. (11.118a) and (11.122) we see that

$${}^\dagger L_{r+k}^k(z) = (-1)^{k*} L_r^k(z) = (-1)^k (k+r)! L_r^k(z). \tag{11.124c}$$

Example 11.3: Generating function of Laguerre polynomials
With a Laplace transform ansatz for the solution of a second order differential equation like Eq. (11.119) derive the generating function of associated Laguerre polynomials (11.120), and obtain with this the normalization condition (11.121).

Solution: We consider the following equation of the form of Eq. (11.119):

$$ty''(t) + (a+1-t)y'(t) + by(t) = 0 \quad \text{with ansatz} \quad y(t) = \int s(p)e^{-pt} dp.$$

The limits will be determined later. Then, with $y'(t) = -\int ps(p)e^{-pt} dp$ and partial integration,

$$
\begin{aligned}
ty'(t) &= -t\int ps(p)e^{-pt} dp = \left[ps(p)e^{-pt}\right] - \int \frac{d}{dp}(ps(p))e^{-pt} dp, \\
t\frac{d^2y}{dt^2} &= -\left[p^2 s(p)e^{-pt}\right] + \int \frac{d}{dp}(p^2 s(p))e^{-pt} dp,
\end{aligned}
$$

[‡‡]Comparison with Eq. (11.107) implies $k = 2l+1$ and $r = n+l$.

since $d(p^2 s)/dp = ps + pd(ps)/dp$. We demand (to be considered later for integration contours) the following condition, which removes [. . .] from the above equation:

$$C(p) := p(p+1)s(p)e^{-pt} = 0.$$

Then the preceding two equations (inserted into the original differential equation) leave us with

$$\int e^{-pt} dp \left[\frac{d}{dp} \{ p(p+1)s(p) \} - (a+1)ps(p) + bs(p) \right] = 0.$$

Hence the expression in square brackets vanishes. It follows that

$$\frac{1}{p(p+1)s(p)} \frac{d}{dp} \{ p(p+1)s(p) \} = \frac{a+1}{p+1} - \frac{b}{p(p+1)} = \frac{a+b+1}{p+1} - \frac{b}{p}.$$

Integrating the equation we obtain $s(p)$ and hence $y(t)$:

$$\ln[p(p+1)s(p)] = \ln \left[(p+1)^{a+b+1} p^{-b} \right], \quad \text{i.e.} \quad s(p) \propto \frac{(p+1)^{a+b}}{p^{b+1}}, \quad y(t) \propto \int \frac{(p+1)^{a+b}}{p^{b+1}} e^{-pt} dp.$$

The condition $C(p) = 0$ now implies for possible contours of integration in the plane of complex p:

$$C(p) = p(p+1)s(p)e^{-pt} = \frac{(p+1)^{a+b+1}}{p^b} e^{-pt} = 0, \quad t > 0.$$

Possible paths satisfying this condition are: (1) If $b < 0$ from 0 to ∞, (2) from -1 to ∞, (3) around $p = -1$ if $a+b$ is a negative integer, and (4) around $p = 0$ if $b > 0$ and integral. Case (1) with $p \to \infty$ implies $y(t) \sim \int dp p^{a-1} e^{-pt} \sim t^{-a}, a = 2l + 2$; this is the rejected irregular solution. Case (2) implies the exponentially increasing solution $\sim e^{+t}$; case (3) similarly. Thus in the case of Eq. (11.119) with a and b as there, with requirement of a finite solution, the only possibility is the last. Hence one defines now as *associated Laguerre polynomial* L_b^a and as the *generating function* $g(p)$ of these (Cauchy's residue theorem requires the coefficient of $1/p$):

$$L_b^a(t) := \frac{1}{2\pi i} \oint_{p=0} \frac{(p+1)^{a+b}}{p^{b+1}} e^{-pt} dp = \text{coefficient of } p^b \text{ in } g(p) := (p+1)^{a+b} e^{-pt},$$

and one has (as polynomial of degree b, since for $n > b : (b-n)! \to \infty$, agreeing with Eq. (11.118b))

$$
\begin{aligned}
L_b^a(t) &= \text{coefficient of } p^b \text{ in } \sum_{n=0}^{\infty} (p+1)^{a+b} \frac{(-pt)^n}{n!} \\
&= \sum_{n=0}^{\infty} \text{coefficient of } p^{b-n} \text{ in } (p+1)^{a+b} \frac{(-t)^n}{n!} = \sum_{n=0}^{\infty} \frac{(-t)^n}{n!} \frac{(a+b)!}{(b-n)!(a+n)!}.
\end{aligned}
$$

An alternative integral representation is obtained by setting $p = q/(1-q)$ or $q = p/(p+1)$. Then $dp = dq/(1-q)^2, p+1 = 1/(1-q)$ and

$$L_b^a(t) = \frac{1}{2\pi i} \oint_{q=0} \frac{dq}{q^{b+1}(1-q)^{a+1}} e^{-tq/(1-q)} = \text{coefficient of } q^b \text{ in } \frac{e^{-tq/(1-q)}}{(1-q)^{a+1}} \equiv G(t).$$
(11.124d)

In order to obtain the *orthogonality and normalization of the associated Laguerre polynomials*, we insert this contour integral representation of the polynomial in the following integral and obtain

$$
\begin{aligned}
I &= \int_0^{\infty} L_b^a(t) L_c^a(t) t^a e^{-t} dt \\
&= \frac{1}{(2\pi i)^2} \oint \oint \frac{dp\,dq}{p^{b+1}(1-p)^{a+1} q^{c+1}(1-p)^{a+1}} \int_{t=0}^{\infty} t^a e^{-t[p/(1-p)+q/(1-q)+1]} dt.
\end{aligned}
$$

We evaluate the integral on the right with the help of the integral representation of the *gamma function*, Eq. (11.111), and obtain

$$\int_{t=0}^{\infty} t^a e^{-t[p/(1-p)+q/(1-q)+1]} dt = a! \left\{ \frac{(1-p)(1-q)}{(1-pq)} \right\}^{a+1}.$$

With this the integral I becomes

$$I = \frac{a!}{(2\pi i)^2} \oint \oint \frac{dpdq}{p^{b+1}q^{c+1}} (1-pq)^{-a-1} = \frac{a!}{2\pi i} \oint \frac{dp}{p^{b+1}} \times \text{coefficient of } q^c \text{ in } (1-pq)^{-a-1}.$$

The coefficient here required is obtained from the expansion as

$$(-p)^c \frac{(-a-1)!}{c!(-c-a-1)!} \quad \text{or} \quad p^c \frac{(c+a)!}{c!a!}$$

with use of the inversion formula (11.112). Clearly only the second form is sensible, so that

$$I = \frac{(c+a)!}{c!} \frac{1}{2\pi i} \oint \frac{dp}{p^{b-c+1}} = \frac{(a+c)!}{c!} \quad \text{if} \quad c = b, \quad \text{and} \quad 0 \text{ if } c \neq b,$$

$$\text{or} \quad \int_0^{\infty} L_b^a(t) L_c^a(t) t^a e^{-t} dt = \frac{(a+c)!}{c!} \delta_{bc}.$$

11.6.3 The eigenfunctions

According to the previous discussion we obtain n^2 orthogonal eigenfunctions in association with the eigenvalue E_n. The wave function of the state (n, l, m) is

$$\psi_{nlm} = a^{-3/2} N_{nl} F_{nl}\left(\frac{2r}{na}\right) Y_l^m(\theta, \varphi), \quad \text{where} \quad F_{nl}(x) = \frac{1}{x} y_l(x) \quad (11.125a)$$

and the factor $a^{-3/2}$ with $a =$ Bohr radius, is introduced for dimensional reasons. Then N_{nl} is a dimensionless normalization constant and (note that we use the associated Laguerre polynomials as in the book of Sneddon [255])

$$F_{nl}(x) = x^l e^{-x/2 \dagger} L_{n+l}^{2l+1}(x). \quad (11.125b)$$

The normalization condition ($\int |\psi|^2 d\mathbf{r} = 1$) determines N_{nl} (which we leave as an exercise!):

$$N_{nl} = \frac{2}{n^2} \left\{ \frac{(n-l-1)!}{[(n+l)!]^3} \right\}^{1/2}. \quad (11.126)$$

It is useful to know the following integral for certain cases:

$$I_{nm}^k(p) := \int_0^{\infty} e^{-x} x^{p+k-1\dagger} L_n^k(x)^{\dagger} L_m^k(x) dx. \quad (11.127)$$

One finds in particular (see also Example 11.7, and for the explicit derivation of the second case below Example 11.8):*

$$I_{nn}^k(1) = \frac{(n!)^3}{(n-k)!},$$

$$I_{nn}^k(2) = \frac{(n!)^3}{(n-k)!}(2n-k+1),$$

$$I_{nn}^k(3) = \frac{(n!)^3}{(n-k)!}(6n^2 - 6nk + k^2 + 6n - 3k + 2). \qquad (11.128)$$

These expressions permit us to obtain the mean or expectation values $\langle 1/r \rangle$ and $\langle r \rangle$, from which we can deduce information on the behaviour of the electron:

$$
\left\langle \frac{1}{r} \right\rangle = \frac{\int_0^\infty \frac{1}{r} F_{nl}^2 \left(\frac{2r}{na} \right) r^2 dr}{\int_0^\infty F_{nl}^2 \left(\frac{2r}{na} \right) r^2 dr} = \frac{2}{na} \frac{\int_0^\infty \frac{1}{x} F_{nl}^2(x) x^2 dx}{\int_0^\infty F_{nl}^2(x) x^2 dx}
$$

$$
= \frac{2}{na} \frac{I_{n+l,n+l}^{2l+1}(1)}{I_{n+l,n+l}^{2l+1}(2)} = \frac{2}{na} \frac{1}{(2n + 2l - 2l - 1 + 1)}
$$

$$
= \frac{1}{n^2 a}. \qquad (11.129)
$$

Analogously we have

$$
\langle r \rangle = \frac{\int_0^\infty r F_{nl}^2 \left(\frac{2r}{na} \right) r^2 dr}{\int_0^\infty F_{nl}^2 \left(\frac{2r}{na} \right) r^2 dr} = \frac{na}{2} \frac{\int_0^\infty x F_{nl}^2(x) x^2 dx}{\int_0^\infty F_{nl}^2(x) x^2 dx} = \frac{na}{2} \left[\frac{I_{n+l,n+l}^{2l+1}(3)}{I_{n+l,n+l}^{2l+1}(2)} \right]
$$

$$
= \frac{na}{2} \frac{\{6(n+l)^2 - 6(n+l)[(2l+1) - 1] + (2l+1)^2 - 3(2l+1) + 2\}}{\{2(n+l) - (2l+1) + 1\}}
$$

$$
= \frac{a}{2} [3n^2 - l(l+1)]. \qquad (11.130)
$$

We see that the mean value of r is the larger the larger n is; thus on the average the electron is the farther away from the proton or nucleus the larger n is. This behaviour is indicated in Fig. 11.2. For the ground state ($n = 1, l = 0$) we have

$$
\left\langle \frac{1}{r} \right\rangle_{1s} = \frac{1}{a}, \qquad \langle r \rangle_{1s} = \frac{3}{2} a. \qquad (11.131)
$$

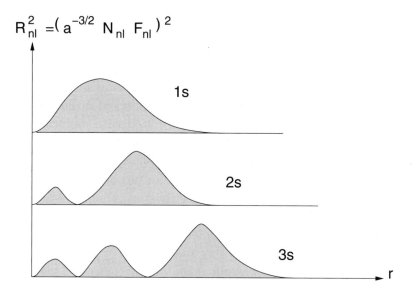

$$R_{nl}^2 = (a^{-3/2} N_{nl} F_{nl})^2$$

Fig. 11.2 Hydrogen atom wave functions.

In the case in which l assumes its maximal value $n-1$, we have

$$L_{n+l}^{2l+1}(x) = L_{2n-1}^{2n-1}(x) = 1,$$

and ψ becomes particularly simple. In this case

$$
\begin{aligned}
\langle r \rangle &= \frac{a}{2}[3n^2 - (n-1)n] = \frac{a}{2}[2n^2 + n] \\
&= an\left(n + \frac{1}{2}\right).
\end{aligned}
\tag{11.132}
$$

For $\langle r^2 \rangle$ one obtains

$$\langle r^2 \rangle = n^2 \left(n + \frac{1}{2}\right)(n+1)a^2, \tag{11.133}$$

or

$$\langle r^2 \rangle \simeq n^4 a^2 \quad \text{for large } r,$$

i.e. $\sqrt{\langle r^2 \rangle} \sim n^2 a$. This corresponds to $E \simeq -e^2/2n^2 a$ like classically with circular radius $n^2 a$.[†] We obtain therefore for the quadratic deviation

$$\triangle r = \sqrt{\langle r^2 \rangle - \langle r \rangle^2} = \frac{1}{2}na\sqrt{2n+1} = \frac{\langle r \rangle}{\sqrt{2n+1}}, \tag{11.134}$$

---

[*]These are taken from I. S. Sneddon [255], Problem 5.20, pp. 173/174, with $n + l$ replaced by n and $2l + 1$ by k.

[†]In the Kepler problem of a particle of mass m_0 in Classical Mechanics with the Newton

or

$$\frac{\triangle r}{\langle r \rangle} = \frac{1}{\sqrt{2n+1}}. \tag{11.135}$$

This expression becomes very small for large values of n. This means, the electron remains practically localized within a spherical surface of radius $\langle r \rangle$. However, for these values of l it is not distributed with uniform probability over the spherical surface, but rather like classically in a plane (i.e. the states Y_l^l are predominantly concentrated in the xy-plane).

The above treatment of the hydrogen atom has obvious shortcomings:
(a) The hydrogen atom was treated nonrelativistically; relativistic corrections are of the order of $v^2/c^2 \sim O(E_n/m_0c^2)$. In order to obtain an impression of relativistic corrections in a simple context, we consider in Example 11.4 the relativistic equation in cylindrical coordinates.
(b) The spin of the electron was not taken into account and hence the spin fine structure of the energy levels is excluded.
(c) The nucleus was treated as pointlike and not as something with structure.

Example 11.4: The relativistic eigenenergy using cylindrical coordinates
Using the formula for the relativistic total energy E of a particle of mass m_0 and charge e moving in the Coulomb potential $V = -Ze/r$ (with E numerically equal to the corresponding Hamiltonian H for a conservative system), and separating this in cylindrical coordinates r, θ, determine the energy E by evaluating the Bohr–Sommerfeld–Wilson integral of "old quantum theory".

Solution: We have

$$H = \sqrt{m_0^2 c^4 + \mathbf{p}^2 c^2} + eV, \quad \text{i.e.} \quad \mathbf{p}^2 = \frac{(E-eV)^2}{c^2} - m_0^2 c^2.$$

In cylindrical coordinates, r, θ,

$$\mathbf{p}^2 = p_r^2 + \frac{1}{r^2} p_\theta^2 \quad \text{and} \quad \oint p_i dq_i = n_i h, \quad i = \theta, r.$$

Since θ is a cyclic coordinate, we have $p_\theta = \text{const.} = n_\theta \hbar$. Thus we have to evaluate the following

gravitational potential written $V(r) = -m_0 \mu / r$, the Kepler period is given by

$$T_{\text{Kepler}} = \frac{2\pi a_{\text{Kepler}}^{3/2}}{\sqrt{\mu}},$$

where a_{Kepler} is the length of the semi-major axis of the elliptic orbit. Thus, identifying $\sqrt{\langle r^2 \rangle}$ with a_{Kepler}, we see that

$$T_{\text{Kepler}} = \frac{2\pi}{\sqrt{\mu}} \langle r^2 \rangle^{3/4} = \frac{2\pi}{\sqrt{\mu}} (n^2 a)^{3/2} = \frac{2\pi}{\sqrt{\mu}} a^{3/2} n^3.$$

See also Example 14.3, in which periods are obtained from the Bohr–Sommerfeld–Wilson quantization condition.

integral which we achieve with the *"method of poles at infinity"* explained in Example 16.2. Thus

$$
\begin{aligned}
n_r h = \oint p_r dq_r &= \oint \sqrt{\frac{E^2}{c^2} - \frac{2EeV(r)}{c^2} + \frac{e^2 V^2(r)}{c^2} - m_0^2 c^2 - \frac{n_\theta^2 \hbar^2}{r^2}} \, dr \\
&= \oint \sqrt{\frac{E^2}{c^2} - m_0^2 c^2 + \frac{1}{r}\left(\frac{2EZe^2}{c^2}\right) - \frac{1}{r^2}\left(n_\theta^2 \hbar^2 - \frac{Z^2 e^4}{c^2}\right)} \, dr \\
&\equiv i \oint \sqrt{A + \frac{2B}{r} + \frac{C}{r^2}} \, dr = -2\pi\left(\sqrt{C} + \frac{B}{\sqrt{A}}\right),
\end{aligned}
$$

where the constants are identified by comparison. It follows that (with $h = 2\pi\hbar$)

$$
2\pi n_r \hbar = -2\pi\left[n_\theta \hbar \sqrt{1 - \frac{Z^2 e^4}{c^2 \hbar^2 n_\theta^2}} - \frac{Ze^2 E}{c^2 \sqrt{m_0^2 c^2 - E^2/c^2}}\right].
$$

We then obtain with $\alpha = e^2/\hbar c \simeq 1/137$, the *"fine structure constant"*, (and subtract $m_0 c^2$ to obtain the ordinary nonrelativistic energy)

$$
\begin{aligned}
E &= m_0 c^2 \left[1 + \frac{Z^2 e^4}{c^2 \hbar^2 [n_r + n_\theta \sqrt{1 - Z^2 e^4/c^2 \hbar^2 n_\theta^2}]}\right]^{-1/2} \\
&= m_0 c^2 \left[1 + \frac{\alpha^2 Z^2}{n_r + n_\theta \sqrt{1 - \alpha^2 Z^2/n_\theta^2}}\right]^{-1/2}.
\end{aligned}
$$

Therefore the energy is no longer a function of $n = n_r + n_\theta$, but depends very slightly on n_θ alone which gives rise to the "fine structure" of hydrogen lines.

11.6.4 Hydrogen-like atoms in parabolic coordinates

The Schrödinger equation of a hydrogen-like atom can also be separated in *parabolic coordinates* ξ, η, φ , which are related to Cartesian coordinates in the following way, as also indicated in Fig. 11.3, where $\varphi = 0, y = 0$:

$$
x = \sqrt{\xi\eta}\cos\varphi, \quad y = \sqrt{\xi\eta}\sin\varphi, \quad z = \frac{1}{2}(\xi - \eta); \tag{11.136a}
$$

$$
\xi = r(1 + \cos\theta) = r + z, \quad \eta = r(1 - \cos\theta) = r - z, \quad \varphi = \varphi, \tag{11.136b}
$$

where $r = \sqrt{x^2 + y^2 + z^2}, \theta, \varphi$ are the spherical coordinates. Whenever there is degeneracy, it is possible to separate the Schrödinger equation in other coordinates. The deeper reason for this are symmetry properties of the problem. In the case of degeneracy with respect to the magnetic quantum number m (as in the present case) one can find linear combinations of hyperspherical functions $Y_m^l(\theta, \varphi)$, which correspond to a new choice of polar axis.[‡]

[‡]See L. Schiff [243], p. 86.

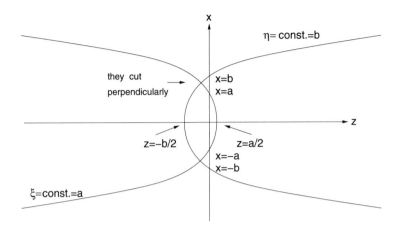

Fig. 11.3 Parabolas $\xi = $ const. and $\eta = $ const. with z-axis as axis of rotation.

In terms of the coordinates (11.136b) the Schrödinger equation of relative motion (11.22) becomes — as we show in Example 11.3 —

$$-\frac{\hbar^2}{2m_0}\left\{\frac{4}{\xi+\eta}\left[\frac{\partial}{\partial\xi}\left(\xi\frac{\partial\psi}{\partial\xi}\right)+\frac{\partial}{\partial\eta}\left(\eta\frac{\partial\psi}{\partial\eta}\right)\right]+\frac{1}{\xi\eta}\frac{\partial^2\psi}{\partial\varphi^2}\right\}-\frac{2Z_1Z_2e^2}{\xi+\eta}\psi = E\psi.$$

$$(11.137)$$

One now sets the total wave function $\psi = \mathcal{E}(\xi)\mathcal{N}(\eta)\Phi(\varphi)$ and divides the equation by ψ, so that the variables can be separated. Then

$$\frac{1}{\Phi}\frac{\partial^2\Phi}{\partial\varphi^2} = -m^2, \quad \Phi \propto e^{\pm im\varphi},$$

where the separation constant m^2 must be such that $m = 0, 1, 2, \ldots$ for uniqueness of the wave function. Multiplying the remaining equation by $(\xi+\eta)/4$ and setting the ξ-part equal to $-\lambda$, a separation constant, one obtains the following two equations:

$$\frac{\partial}{\partial\xi}(\xi\mathcal{E}') + \left[\frac{m_0E}{2\hbar^2} + \frac{\lambda}{\xi} - \frac{m^2}{4\xi^2}\right](\xi\mathcal{E}) = 0,$$

$$\frac{\partial}{\partial\eta}(\eta\mathcal{N}') + \left[\frac{m_0E}{2\hbar^2} + \left(\frac{m_0Z_1Z_2e^2}{\hbar^2} - \lambda\right)\frac{1}{\eta} - \frac{m^2}{4\eta^2}\right](\eta\mathcal{N}) = 0.$$

Since these equations are alike, there is really only one basic equation to solve here. We rewrite the first of these equations with the help of the relation

$$\frac{1}{\xi^{1/2}}\frac{\partial}{\partial\xi}(\xi\mathcal{E}') = \left(\frac{\partial^2}{\partial\xi^2} + \frac{1}{4\xi^2}\right)(\xi^{1/2}\mathcal{E}).$$

Then

$$\frac{\partial^2}{\partial \xi^2}(\xi^{1/2}\mathcal{E}) + \left[\frac{m_0 E}{2\hbar^2} + \frac{\lambda}{\xi} - \frac{m^2 - 1}{4\xi^2}\right](\xi^{1/2}\mathcal{E}) = 0.$$

To obtain a closer analogy with the separation in the spherical polar case, we set (since E is negative for the discrete spectrum we consider)

$$\frac{m_0 E}{2\hbar^2} = -\frac{1}{4}\beta^2, \quad \frac{\lambda}{\beta} = n', \quad \frac{1}{\beta}\left[\frac{m_0 Z_1 Z_2 e^2}{\hbar^2} - \lambda\right] = n'', \quad \rho = \beta\xi \quad \text{and} \quad \sigma = \beta\eta.$$

Then the first of the two equations becomes

$$\frac{\partial^2}{\partial \rho^2}(\rho^{1/2}\mathcal{E}) - \left[\frac{1}{4} - \frac{n'}{\rho} + \frac{m^2 - 1}{4\rho^2}\right](\rho^{1/2}\mathcal{E}) = 0.$$

In the spherical polar case we had (cf. Eq. (11.105))

$$l(l+1) \equiv \left(\frac{m}{2} - \frac{1}{2}\right)\left(\frac{m}{2} + \frac{1}{2}\right).$$

Hence we write $l = (m-1)/2$. Then by comparison with the spherical polar separation the answer is as in Eqs. (11.106) and (11.125b) (dividing by $\rho^{1/2}$)

$$\mathcal{E} \propto e^{-\rho/2}\rho^{m/2}L_{n_1}^{2l+1=m}(\rho), \quad n_1 \equiv n' - \frac{1}{2}(m+1) = 0, 1, 2, \ldots.$$

Analogously we have

$$\mathcal{N} \propto e^{-\sigma/2}\sigma^{m/2}L_{n_2}^m(\sigma), \quad n_2 = n'' - \frac{1}{2}(m+1) = 0, 1, 2, \ldots.$$

Adding now

$$n' + n'' = n_1 + n_2 + m + 1 = \frac{m_0 Z_1 Z_2 e^2}{\beta\hbar^2} \equiv n \quad \text{(say)},$$

we obtain the energy eigenvalues given by

$$-E = \frac{\hbar^2\beta^2}{2m_0} = \frac{\hbar^2}{2m_0}\frac{m_0^2 Z_1^2 Z_2^2 e^4}{\hbar^4 n^2} = \frac{m_0 Z_1^2 Z_2^2 e^4}{2\hbar^2 n^2}.$$

This agrees with our earlier formulas like (11.114b). We leave the calculation of the degree of degeneracy to Example 11.6.

Example 11.5: The Schrödinger equation in parabolic coordinates
Perform the transformation of the metric, i.e. $ds^2 = dx^2 + dy^2 + dz^2$, as well as the Laplacian $\triangle \equiv \boldsymbol{\nabla}^2$ and the Schrödinger equation, from Cartesian to parabolic coordinates.

Solution: We begin with z of Eq. (11.136a), i.e. $z = (\xi - \eta)/2$, and so $dz = (d\xi - d\eta)/2$. Thus

$$dz^2 = \frac{1}{4}[d\xi^2 + d\eta^2 - 2d\xi\,d\eta].$$

Similarly we proceed with dx and dy. Thus, for instance, see Eq. (11.136a),

$$dx = d(\sqrt{\xi\eta}\cos\varphi) = \frac{1}{2}\sqrt{\frac{\eta}{\xi}}\cos\varphi d\xi + \frac{1}{2}\sqrt{\frac{\xi}{\eta}}\cos\varphi d\eta - \sqrt{\xi\eta}\sin\varphi d\varphi.$$

One now performs the square of this and obtains similarly dy^2. Since the lines of constant ξ are orthogonal to lines of constant η (they cut perpendicularly), all cross terms cancel in the sum of the squares, and one obtains

$$ds^2 = \frac{1}{4}\left(\frac{\xi+\eta}{\xi}\right)d\xi^2 + \frac{1}{4}\left(\frac{\xi+\eta}{\eta}\right)d\eta^2 + \xi\eta d\varphi^2 \equiv (g_\xi d\xi)^2 + (g_\eta d\eta)^2 + (g_\phi d\varphi)^2,$$

where $g_\xi d\xi$, etc. are elements of length. The general formula for the Laplacian is[§]

$$\Delta = \frac{1}{g_\xi g_\eta g_\varphi}\left[\frac{\partial}{\partial\xi}\left\{\frac{g_\eta g_\varphi}{g_\xi}\frac{\partial}{\partial\xi}\right\} + \frac{\partial}{\partial\eta}\left\{\frac{g_\xi g_\varphi}{g_\eta}\frac{\partial}{\partial\eta}\right\} + \frac{\partial}{\partial\varphi}\left\{\frac{g_\xi g_\eta}{g_\varphi}\frac{\partial}{\partial\varphi}\right\}\right].$$

It follows that

$$\Delta = \frac{4}{\xi+\eta}\left[\frac{\partial}{\partial\xi}\left(\xi\frac{\partial}{\partial\xi}\right) + \frac{\partial}{\partial\eta}\left(\eta\frac{\partial}{\partial\eta}\right)\right] + \frac{1}{\xi\eta}\frac{\partial^2}{\partial\varphi^2}.$$

Example 11.6: Calculation of degree of degeneracy

Show that the degree of degeneracy of the wave functions in terms of parabolic coordinates agrees with that obtained earlier for spherical polar coordinates, Eq. (11.114c).

Solution: For the wave functions $\psi = \mathcal{E}(\xi)\mathcal{N}(\eta)\Phi(\varphi)$, i.e.

$$\psi = e^{\pm im\varphi}e^{-(\rho+\sigma)/2}(\rho\sigma)^{m/2}L_{n_1}^m(\rho)L_{n_2}^m(\sigma),$$

we have to find the number of values of the quantum numbers n_1, n_2 and m which give n. Beware! For $m = 0$ there is only one m-state, for $m \neq 0$ there are two, as the two signs of the exponential indicate. We consider first the case $m = 0$. Here we have (see preceding text) $n_1 + n_2 = n - 1$. We use the method of selector variables, and attach a factor ω to each term introduced in the wave function. We then sum all possibilities and select the coefficient of ω^{n-1} in the result. Thus we want to obtain the coefficient of ω^{n-1} in

$$\Sigma_2 \equiv \sum_{n_1=0}^{\infty}\sum_{n_2=0}^{\infty}\omega^{n_1}\omega^{n_2} = \frac{1}{(1-\omega)^2} = 1 + 2\omega + 3\omega^2 + \cdots.$$

The coefficient of ω^{n-1} in this expansion is seen to be n. In the case $m \neq 0$ we have $n_1 + n_2 = n - m - 1$ and therefore the number of the coefficient of ω^{n-m-1} in the double sum Σ_2 is $n - m$. Observing that the highest possible value of m is $n - 1$ (i.e. when $n_1 = 0 = n_2$) and summing over m, the total number of wave functions we have is the number n with $m = 0$ plus twice the number with $m \neq 0$ starting from $m = 1$, i.e.[*]

$$n + 2\sum_{m=1}^{n-1}(n-m) = n + 2\sum_{m=1}^{n-1}n - 2\sum_{m=1}^{n-1}m = n + 2n(n-1) - \frac{2}{2}(n-1)n = n^2.$$

[§]See e.g. W. Magnus and F. Oberhettinger [181], last chapter.

[*]Recall that the sum of an arithmetic progression is given by $a + (a+d) + (a+2d) + \cdots + \{a + (n-2)d\} = (n-1)a + (n-1)(n-2)d/2$.

Example 11.7: Stark effect in hydrogen-like atoms

Consider a perturbation $v \equiv eFz$, where F is the electric field, and determine the average value of v for a given state, i.e. $eF\langle z \rangle$, which will then give the change in energy of that state to the first order (i.e. with Rayleigh–Schrödinger perturbation theory).

Solution: Using formulas and results of above we have to determine

$$\langle v \rangle = eF\langle z \rangle = \frac{1}{2}eF\langle \xi - \eta \rangle = \frac{1}{2\beta}eF\langle \rho - \sigma \rangle.$$

We obtain the volume element dV in parabolic coordinates from the above results as

$$dV = g_\xi g_\eta g_\varphi d\xi d\eta d\varphi = \frac{1}{4}(\xi + \eta)d\xi d\eta d\varphi \propto (\rho + \sigma)d\rho d\sigma.$$

With the help of the wave functions we derived above, we now have

$$
\begin{aligned}
\langle \rho - \sigma \rangle &= \frac{\int d\rho d\sigma \psi^*(\rho - \sigma)(\rho + \sigma)\psi}{\int d\rho d\sigma \psi^*(\rho + \sigma)\psi} \\
&= \frac{\int_0^\infty \int_0^\infty e^{-(\rho+\sigma)}(\rho\sigma)^m [L_{n_1}^m(\rho)L_{n_2}^m(\sigma)]^2 (\rho^2 - \sigma^2)d\rho d\sigma}{\int_0^\infty \int_0^\infty e^{-(\rho+\sigma)}(\rho\sigma)^m [L_{n_1}^m(\rho)L_{n_2}^m(\sigma)]^2 (\rho + \sigma)d\rho d\sigma} \\
&= \frac{I_{n_1}^{m+2}I_{n_2}^m - I_{n_2}^{m+2}I_{n_1}^m}{I_{n_1}^{m+1}I_{n_2}^m + I_{n_2}^{m+1}I_{n_1}^m}, \quad \text{where} \quad I_q^p = \int_0^\infty d\rho e^{-\rho}\rho^p [L_q^m(\rho)]^2.
\end{aligned}
$$

By contour integration (see Examples 11.8 and 11.9) one finds that

$$
I_{n_2}^m = \frac{(n_2 + m)!}{n_2!}, \qquad I_{n_1}^{m+1} = \frac{(n_1 + m)!}{n_1!}(2n_1 + m + 1),
$$

$$
I_{n_1}^{m+2} = \frac{(n_1 + m)!}{n_1!}\{6n_1^2 + 6n_1(m + 1) + (m + 1)(m + 2)\}.
$$

Apart from the power of the leading factorial (which results from the definition of the associated Laguerre polynomial in the generating function), these expressions agree with those of Eq. (11.128) if one makes there the substitutions $n \to n_1$ (or n_2) $+ m, k \to m$. In fact, with the connection (11.124c), we can relate the expressions I_q^p to those of Eq. (11.128). Thus e.g., with Eq. (11.127) and setting $n = r + k$,

$$
I_{nn}^k(1) = \int_0^\infty dx e^{-x}x^k [{}^\dagger L_n^k(x)]^2, \quad I_{r+k,r+k}^k(1) = \int_0^\infty e^{-x}x^k [(k + r)! L_r^k(x)]^2 = [(k + r)!]^2 I_r^k.
$$

It follows that

$$
\begin{aligned}
\langle \rho - \sigma \rangle &= \frac{\{6n_1^2 + 6n_1(m + 1) + (m + 1)(m + 2)\} - \{n_1 \rightleftharpoons n_2\}}{\{2n_1 + m + 1\} + \{n_1 \rightleftharpoons n_2\}} \\
&= \frac{6(n_1 - n_2)(n_1 + n_2 + m + 1)}{2(n_1 + n_2 + m + 1)} = 3(n_1 - n_2).
\end{aligned}
$$

Therefore the change in energy of the state to first order is given by, with $Z_1 \to Ze, Z_2 \to e$,

$$
\langle v \rangle = 3(n_1 - n_2)\frac{eF}{2}\left(\frac{n\hbar^2}{m_0 Z e^2}\right).
$$

Thus in this case nondegenerate perturbation theory could be used as discussed in Example 8.1 (cf. Sec. 8.6). However, if we expressed the perturbation in spherical polar coordinates, i.e. with $z = r\cos\theta$, we would have to use degenerate perturbation theory and obtain the same result.

Example 11.8: Laguerre linear expectation value integral

Evaluate the following integral (which is different from the normalization integral of Example 11.3 by one power of t, and is a special case of Example 11.9) with the help of the generating function $G(t)$ of associated Laguerre polynomials obtained in Example 11.3:

$$I := \int_0^\infty e^{-t} [L_n^m(t)]^2 t^{m+1} dt, \quad L_n^m(t) = \text{coefficient of } q^n \text{ in } G(t) = \frac{e^{-tq/(1-q)}}{(1-q)^{m+1}}.$$

Solution: Using the generating function for both associated Laguerre polynomials, the integral I is the coefficient of $p^n q^n$ in the expression

$$\frac{1}{[(1-p)(1-q)]^{m+1}} \int_0^\infty e^{-t[1+p/(1-p)+q/(1-q)]} t^{m+1} dt, \quad 1 + \frac{p}{1-p} + \frac{q}{1-q} = \frac{1-pq}{(1-q)(1-p)} \equiv \alpha$$

(see Eq. (11.124d) of the alternative integral representation obtained in Example 11.3). With $s = \alpha t, dt = ds/\alpha$, the integral arising here becomes with Eq. (11.111)

$$\int_0^\infty e^{-\alpha t} t^{m+1} dt = \int_0^\infty \frac{e^{-s} s^{m+1}}{\alpha^{m+2}} ds = \frac{(m+1)!}{\alpha^{m+2}}.$$

For use in the following recall that

$$(1+x)^{-n} = \sum_{r=0}^\infty (-1)^r \frac{(n+r-1)!}{(n-1)! r!} x^r.$$

We now evaluate the integral I as the coefficient of $p^n q^n$ in the expansion of

$$\frac{(m+1)![(1-q)(1-p)]^{m+2}}{[(1-p)(1-q)]^{m+1}(1-pq)^{m+2}} = \text{coefficient of } p^n q^n \text{ in } \frac{(m+1)!(1-p)(1-q)}{(1-pq)^{(m+2)}}.$$

Now the coefficient of q^n in $(1-q)(1-pq)^{-(m+2)}$ is equal to the coefficient of $(q^n - q^{n-1})$ in $(1-pq)^{-(m+2)}$. The coefficient of q^n in the expansion of $(1-pq)^{-\mu}$ is $p^n(n+\mu-1)!/n!(\mu-1)!$. It follows therefore that the coefficient of $(q^n - q^{n-1})$ in $(1-pq)^{-(m+2)}$ is

$$= \frac{p^n(n+m+1)!}{n!(m+1)!} - \frac{p^{n-1}(n+m)!}{(n-1)!(m+1)!}.$$

Finally I is the coefficient of p^n in the expansion of

$$(1-p)\left\{ \frac{p^n(n+m+1)!}{n!} - \frac{p^{n-1}(n+m)!}{(n-1)!} \right\} = \text{coefficient of } (p^n - p^{n-1}) \text{ in}$$

$$\left\{ \frac{p^n(n+m+1)!}{n!} - \frac{p^{n-1}(n+m)!}{(n-1)!} \right\} = \frac{(n+m+1)!}{n!} + \frac{(n+m)!}{(n-1)!} = \frac{(n+m)!}{n!}(2n+m+1).$$

This verifies the expression $I_{n_1}^{m+1}$ in Example 11.7 and (remembering the definition of the associated Laguerre polynomila there!) the second of relations (11.128). In Example 11.9 we consider integrals with an arbitrary power of t.

Example 11.9: Laguerre expectation value integrals

Evaluate the following integral with the help of the generating function of Laguerre polynomials:

$$I_{bc}^a(i) := \int_0^\infty dt\, t^i [L_b^a(t) L_c^a(t) t^a e^{-t}], \quad i, a, b, c \text{ positive integers}.$$

Solution: One considers the following integral and obtains $I^a_{bc}(i)$ from the coefficient of α^i:

$$I(\alpha) = \int_0^\infty dt\, e^{-\alpha t}[L^a_b(t)L^a_c(t)t^a e^{-t}], \quad I^a_{bc}(i) = i!(-1)^i \times \text{ coefficient of } \alpha^i \text{ in the expansion of } I(\alpha).$$

Using the contour integral representation (11.124d) of an associated Laguerre polynomial, we can re-express $I(\alpha)$ in the form:

$$I(\alpha) = \frac{1}{(2\pi i)^2} \oint \oint \frac{dp\,dq}{p^{b+1}(1-p)^{a+1}q^{c+1}(1-q)^{a+1}} j(\alpha),$$

where $j(\alpha)$ contains the integration with respect to t which may be evaluated with the help of the integral representation (11.111) of the gamma or factorial function, i.e.

$$j(\alpha) = \int_0^\infty dt\, t^a e^{-t[p/(1-p)+q/(1-q)+1+\alpha]} = a!\left[\frac{(1-p)(1-q)}{1-pq+\alpha(1-q)(1-p)}\right]^{a+1}.$$

It follows that (using Cauchy's residue theorem in integrating around $q = 0$)

$$I(\alpha) = \frac{a!}{(2\pi i)^2} \oint \oint \frac{dp\,dq}{p^{b+1}q^{c+1}}[1 - pq + \alpha(1-q)(1-p)]^{-a-1} = \frac{a!}{2\pi i} \oint \frac{dp}{p^{b+1}} \times \text{ coefficient of } q^c \text{ in}$$

$$[1 - pq + \alpha(1-q)(1-p)]^{-a-1} \equiv [1 + \alpha(1-p) - q(p + \alpha - \alpha p)]^{-a-1}.$$

The coefficient of q^c in this expression is

$$\frac{1}{[1+\alpha(1-p)]^{a+1}}\left(\frac{p+\alpha-\alpha p}{1+\alpha-\alpha p}\right)^c \frac{(c+a)!}{c!a!} = \frac{\alpha^c(c+a)!g(p)}{c!a!(1+\alpha)^{a+c+1}},$$

where

$$g(p) = \frac{[1+\frac{1-\alpha}{\alpha}p]^c}{[1-\frac{\alpha}{1+\alpha}p]^{a+c+1}} = \sum_{r=0}^\infty \sum_{v=r}^\infty \frac{c!(a+c+v-r)!\alpha^{v-2r}(1-\alpha)^r}{(c-r)!r!(a+c)!(v-r)!(1+\alpha)^{v-r}}p^v.$$

$$\therefore I(\alpha) = \frac{a!}{2\pi i} \oint \frac{dp}{p^{b+1}} \frac{\alpha^c(c+a)!}{c!a!(1+\alpha)^{a+c+1}}g(p) = \sum_{r=0}^\infty J_{c+b-2r}f(\alpha),$$

where with $\beta = a+c+1+b$ (observe that for $\alpha = 0$ a term with $c+b-2r = 0$ remains)

$$J_{c+b-2r} = \frac{\alpha^{c+b-2r}(a+c+b-r)!}{(c-r)!r!(b-r)!} \quad \text{and} \quad f(\alpha) = \frac{(1-\alpha)^r}{(1+\alpha)^{\beta-r}} = \sum_{w=0}^\infty \sum_{y=w}^\infty K_{y,w}(r),$$

$$K_{y,w}(r) = \frac{r!(\beta-r-1+y-w)!(-\alpha)^y}{(r-w)!w!(\beta-r-1)!(y-w)!}.$$

Special cases: (1) In the case $\alpha = 0$, we have $2r = b + c = $ even, so that

$$J_0 = \frac{(a+\frac{b+c}{2})!}{(\frac{c-b}{2})!(\frac{b-c}{2})!(\frac{b+c}{2})!} \xrightarrow{b=c} \frac{(a+b)!}{b!}.$$

(2) In the case of one power of α, we have to have $y + (c+b-2r) = 1$, i.e. (a) $y = 1, c+b-2r = 0$ (i.e. $c + b$ even) and/or $y = 0, c + b - 2r = 1$ (i.e. $c + b$ odd). In case (a) we have

$$I(\alpha) = J_0 \sum_{w=0}^\infty K_{1,w}\left(\frac{b+c}{2}\right), \quad \text{where} \quad J_0 = \frac{(a+\frac{b+c}{2})!}{(\frac{c-b}{2})!(\frac{b-c}{2})!(\frac{b+c}{2})!},$$

and (only $w = 0$ and 1 arise in the sum)

$$
K_{1,w}\left(\frac{b+c}{2}\right) = \frac{(\frac{b+c}{2})!(\beta - \frac{b+c}{2} - w)!(-\alpha)}{(\frac{b+c}{2} - w)!w!(\beta - \frac{b+c}{2} - 1)!(1 - w)!},
$$

$$
\sum_{w=0}^{\infty} K_{1,w}\left(\frac{b+c}{2}\right) = \left(\beta - \frac{b+c}{2}\right)(-\alpha) + \left(\frac{b+c}{2}\right)(-\alpha) = -\beta\alpha.
$$

Hence (recall that $I_{bc}^{a}(i) = i!(-1)^{i} \times$ coefficient of α^{i} in $I(\alpha)$)

$$
I = \frac{(a + \frac{b+c}{2})!(-\beta\alpha)}{(\frac{c-b}{2})!(\frac{b-c}{2})!(\frac{b+c}{2})!} \xrightarrow{b=c} \frac{(a+b)!(-\beta\alpha)}{b!} = -\frac{(a+b)!}{b!}(a + 2b + 1)\alpha.
$$

These results are seen to agree with the expressions $I_{n_2}^{m}, I_{n_1}^{m+1}$ cited in Example 11.7.

11.7 Continuous Spectrum of Coulomb Potential

In this and the following sections we consider the scattering of a charged particle in the presence of a Coulomb potential. In view of the slow fall-off of the potential at infinity, the standard principles of nonrelativistic scattering theory, like the asymptotic condition and definitions of the scattering amplitude and the phase shift, cannot really — and this means in a rigorous sense — be applied in this case. One should actually consider a screened Coulomb potential* and consider quantities like the partial wave expansion (to be defined in a later section) as distributions. Since such a rigorous treatment is beyond the scope of this text, we consider the traditional treatment here and refer the interested reader to literature in which it is demonstrated that the derivation given below finds its rigorous justification on a basis of distribution theory.[†]

We have the Schrödinger equation of relative motion of particles with charges $Z_1 e, Z_2 e$ given by

$$
\left[-\frac{\hbar^2}{2m_0}\triangle - \frac{Z_1 Z_2 e^2}{r}\right]\psi(\mathbf{r}) = E\psi(\mathbf{r}). \tag{11.138}
$$

Here we set (this defines the velocity v_0)

$$
E \equiv \frac{\hbar^2 k^2}{2m_0} \equiv \frac{1}{2}m_0 v_0^2, \qquad \gamma = -\frac{Z_1 Z_2 e^2}{\hbar v_0}. \tag{11.139}
$$

---

*Screened Coulomb potentials are considered in Chapter 16, however parallel to the Coulomb case.

[†]A very readable source to consult, also with regard to earlier literature, is the work of J. R. Taylor [267].

Then Eq. (11.138) becomes

$$\left(\triangle + k^2 - \frac{2\gamma k}{r}\right)\psi(\mathbf{r}) = 0. \tag{11.140}$$

The simple Coulomb potential permits particularly simple solutions. In particular the equation possesses a *regular solution* of the form (also $\psi \sim \exp(ikr)$, for $r \to \infty$)

$$\psi_r = Ae^{ikz}f(r - z), \quad z = r\cos\theta, \quad A = \text{const.} \tag{11.141}$$

Inserting this expression into Eq. (11.140) and using (11.136b), i.e. writing

$$\psi_r = Ae^{ik(\xi - \eta)/2}f(\eta) \text{ and } v = ik\eta, \tag{11.142}$$

we obtain the equation

$$\left[v\frac{d^2}{dv^2} + (1 - v)\frac{d}{dv} + i\gamma\right]f(v) = 0. \tag{11.143}$$

This equation is of the form of a *hypergeometric equation*, that we encountered previously. According to Eqs. (11.108) and (11.109) we can therefore write the solution

$$\psi_r = Ae^{ikz}\Phi(-i\gamma, 1; v = ik(r - z)), \tag{11.144}$$

where

$$\Phi(a, b; z) = 1 + \frac{a}{b}\frac{z}{1!} + \frac{a(a + 1)}{b(b + 1)}\frac{z^2}{2!} + \cdots.$$

Our question is now: How does ψ_r behave for $r \to \infty$, i.e. what is the behaviour of $\Phi(a, b; z)$ for $|z| \to \infty$? We can find the appropriate formula in books on Special Functions:[‡]

$$\Phi(a, b; z) = W_1(a, b; z) + W_2(a, b; z), \tag{11.145}$$

with the asymptotic expansions

$$W_1(a, b; z) \simeq \frac{\Gamma(b)}{\Gamma(b - a)}(-z)^{-a}\sum_{n=0}^{\infty}\frac{\Gamma(n + a)\Gamma(n + a - b + 1)}{\Gamma(a)\Gamma(a - b + 1)}\frac{(-z)^{-n}}{n!},$$
$$(-\pi < \arg(-z) < \pi), \tag{11.146}$$

[‡]See, for instance, W. Magnus and F. Oberhettinger [181], pp. 86 - 87 under Kummer functions.

$$W_2(a, b; z) \simeq \frac{\Gamma(b)}{\Gamma(a)} e^z z^{a-b} \sum_{n=0}^{\infty} \frac{\Gamma(n+1-a)}{\Gamma(1-a)} \frac{\Gamma(n+b-a)}{\Gamma(b-a)} \frac{(z)^{-n}}{n!},$$

$$(-\pi < \arg(z) < \pi) \qquad (11.147)$$

It follows that for $r \to \infty$, with $z = r\cos\theta$,

$$\psi_r = Ae^{ikz}[W_1(-i\gamma, 1; ik(r-z)) + W_2(-i\gamma, 1; ik(r-z))] \equiv \psi_i + \psi_d, \quad (11.148)$$

where

$$\psi_i \simeq Ae^{ikz} \frac{\Gamma(1)}{\Gamma(1+i\gamma)} [ik(z-r)]^{i\gamma} \left[1 + O\left(\frac{1}{r}\right)\right], \qquad (11.149)$$

$$\psi_d \simeq Ae^{ikz} \frac{e^{ik(r-z)}}{\Gamma(-i\gamma)} [ik(r-z)]^{-1-i\gamma} \left[1 + O\left(\frac{1}{r}\right)\right], \qquad (11.150)$$

i.e.

$$\psi_r \simeq \frac{A(-i)^{i\gamma}}{\Gamma(1+i\gamma)} \left\{ e^{i(kz+\gamma\ln k(r-z))} - \frac{\gamma\Gamma(1+i\gamma)}{k(r-z)\Gamma(1-i\gamma)} e^{i(kr-\gamma\ln k(r-z))} \right\},$$

$$(11.151)$$

where

$$\exp[-i\gamma\ln k(r-z)] = \exp[-i\gamma\ln kr(1-\cos\theta)]$$
$$= \exp[-i\gamma\ln 2kr - i\gamma\ln\sin^2(\theta/2)].$$

We now define a quantity $f(\theta)$ by the asymptotic relation

$$\psi_r \propto \left\{ e^{i(kz+\gamma\ln k(r-z))} + \frac{f(\theta)}{r} e^{i(kr-\gamma\ln 2kr)} \right\}. \qquad (11.152)$$

Since $z = r\cos\theta$, a comparison of Eq. (11.151) with Eq. (11.152) determines the quantity $f(\theta)$ as

$$f(\theta) = -\frac{\gamma}{\underbrace{k(1-\cos\theta)}_{2\sin^2(\frac{\theta}{2})}} \frac{\Gamma(1+i\gamma)}{\Gamma(1-i\gamma)} \exp\left[-i\gamma\ln\sin^2\left(\frac{\theta}{2}\right) \right]$$

$$= -\frac{\gamma}{2k\sin^2(\frac{\theta}{2})} \exp\left[2i\arg\Gamma(1+i\gamma) - i\gamma\ln\sin^2\left(\frac{\theta}{2}\right) \right]. (11.153)$$

The asymptotic solution represents the *stationary scattering state* ($E =$ const.) of a particle with incident momentum $\hbar k$ in the direction of z for large distances away from the scattering centre. Without the logarithmic phase factors, i.e. if ψ_r were

$$\psi_r \propto \left[e^{ikz} + f(\theta)\frac{e^{ikr}}{r} \right], \qquad (11.154)$$

the wave could be looked at as the sum of an incoming plane wave $\exp(ikz)$ and an outgoing scattered wave. The quantity $|f(\theta)|^2$ would then be a measure for the scattering in the direction θ. The problem of the *logarithmic phase*, also called *phase anomaly*, that we encounter here is a characteristic of the Coulomb potential. As a consequence of the masslessness of the photon, the effective range of the Coulomb potential is infinite. In other words, the Coulomb potential has an influence on the incoming wave even as far as the asymptotic domain.

11.7.1 The Rutherford formula

The above considerations permit us to calculate the *differential cross section* which is a measurable quantity. This cross section is defined by the ratio of the outgoing particle current (in direction θ) to the current of the ingoing particles.

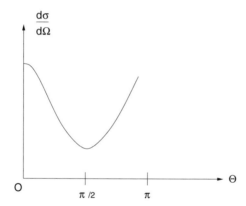

Fig. 11.4 Variation of the observed differential cross section with θ.

We define as *current density* the quantity

$$\mathbf{j} := \frac{\hbar}{2im_0}[\psi^*(\boldsymbol{\nabla}\psi) - \psi(\boldsymbol{\nabla}\psi)^*]. \tag{11.155}$$

Since the gradient is proportional to the configuration space representation of the momentum, we see that this expression is the exact analogue of a current density in classical considerations. For the incoming wave $\psi_i = \exp(ikz)$, $z \to -\infty$, the current density is

$$j_i = \frac{\hbar}{2im_0}(2ik) = \frac{\hbar k}{m_0} \equiv v_0. \tag{11.156}$$

In the case of the Coulomb potential we now ignore the logarithmic part of the phase (see discussion later); in the current this would in any case contribute only something of order $1/r$. In Eq. (11.154) we chose the prefactor of the incoming wave in (11.152) as 1. Correspondingly the other part of (11.152) yields as density of the outgoing current

$$j_r = \frac{1}{r^2}|f(\theta)|^2 v_0. \tag{11.157}$$

The *differential scattering cross section* $d\sigma$ into the solid angle element $d\Omega$ is defined by the ratio

$$\frac{d\sigma}{r^2 d\Omega} = \frac{j_r}{j_i}, \tag{11.158}$$

which in the present case is with $|f(\theta)|$ obtained from Eq. (11.153):

$$\frac{d\sigma}{d\Omega} = |f(\theta)|^2 = \frac{\gamma^2}{4k^2 \sin^4\left(\dfrac{\theta}{2}\right)}. \tag{11.159}$$

The expression $f(\theta)$, which yields the scattering cross section, is called *scattering amplitude*. We recognize the result (11.159) as the *Rutherford formula* which can also be derived purely classically. That this formula retains its validity here in quantum mechanics is something of a coincidence. We make a few more important observations. We see that σ depends only on the modulus of the potential (i.e. the square γ^2). This implies that scattering takes place for the attractive as well as the repulsive potential. We obtain the *total scattering cross section* by integration:

$$\sigma_{\text{tot}} = \int \frac{d\sigma}{d\Omega} d\Omega = \frac{\gamma^2}{4k^2} \int \frac{d\Omega}{\sin^4\left(\dfrac{\theta}{2}\right)}. \tag{11.160}$$

We observe that this expression diverges in the forward direction, i.e. for $\theta \to 0$. In reality this divergence does not occur, since in actual fact the Coulomb potential occurs only in a screened form. In the context of quantum electrodynamics one refers to "*vacuum polarization*" and relates this to the idea that a charged particle like the electron polarizes the otherwise neutral vacuum by attracting virtual charges of opposite polarity and repelling virtual charges of the same polarity, thereby screening itself off from the surroundings. This implies effectively that the Coulomb potential becomes (virtually) a screened potential or Yukawa-type potential of the form

$$V_{\text{screened}} = -\frac{Z_1 Z_2 e^2}{r} e^{-r/r_0}.$$

This screening of the Coulomb potential insures that the scattering cross section in forward direction is actually finite as observed experimentally and indicated in Fig.11.4.

11.8 Scattering of a Wave Packet

We saw earlier that in quantum mechanics a particle is described by a wave packet, whose maximum moves with the particle velocity, and which provides the *uncertainties* arising in quantum mechanics. The wave packet is a superposition of waves $\psi(\mathbf{x}, t)$, i.e. of continuum wave functions, which in the case of free motion (zero potential) are simply plane waves. A wave packet with momentum maximum at $\mathbf{k} = \mathbf{k}_0$ (with momentum $\mathbf{p} = \hbar\mathbf{k}$) is therefore given by the following expression in which $\mathbf{x}_0$ can be considered to be an impact parameter, i.e.

$$\psi_{\mathbf{k}_0}(\mathbf{x} - \mathbf{x}_0, t) = \int \frac{d\mathbf{k}}{(2\pi)^3} A_{\mathbf{k}_0}(\mathbf{k})\psi_{\mathbf{k}}(\mathbf{x} - \mathbf{x}_0, t), \qquad (11.161)$$

where $A_{\mathbf{k}_0}(\mathbf{k}) \simeq A(\mathbf{k} - \mathbf{k}_0)$ and

$$\psi_{\mathbf{k}}(\mathbf{x}, t) = U(t, t_0)\psi_{\mathbf{k}}(\mathbf{x}, t_0) = e^{-iH(t-t_0)/\hbar}\psi_{\mathbf{k}}(\mathbf{x}, t_0), \qquad (11.162)$$

and if H is time-independent one has the stationary wave function

$$\psi_{\mathbf{k}}(\mathbf{x}, t) = e^{-iE_k(t-t_0)/\hbar}\psi_{\mathbf{k}}(\mathbf{x}), \quad \text{where} \quad \psi_{\mathbf{k}}(\mathbf{x}) \equiv \psi_{\mathbf{k}}(\mathbf{x}, t_0). \qquad (11.163)$$

The particle velocity $\mathbf{v}$ is

$$\mathbf{v} = \frac{\hbar\mathbf{k}}{m_0}, \qquad \mathbf{v}_0 = \frac{\hbar\mathbf{k}_0}{m_0}, \qquad (11.164)$$

and $\psi_{\mathbf{k}}(\mathbf{x})$ is a solution of the continuum with energy

$$E \equiv E_k = \frac{\hbar^2 k^2}{2m_0} \geq 0, \quad E_k \simeq E_{k_0} + (\mathbf{k} - \mathbf{k}_0) \cdot \mathbf{v}_0\hbar, \qquad (11.165)$$

of the equation

$$\left[-\frac{\hbar^2}{2m_0}\nabla^2 + V(\mathbf{x})\right]\psi_{\mathbf{k}}(\mathbf{x}) = E_k\psi_{\mathbf{k}}(\mathbf{x}). \qquad (11.166)$$

Let the incoming free wave packet at time $t \geq t_0$, i.e. the wave packet in the absence of a scattering potential, be given by

$$\psi_{\mathbf{k}_0}^{(0)}(\mathbf{x} - \mathbf{x}_0, t) = \int \frac{d\mathbf{k}}{(2\pi)^3} A_{\mathbf{k}_0}(\mathbf{k})\frac{e^{i\mathbf{k}\cdot(\mathbf{x}-\mathbf{x}_0)}}{(2\pi)^{3/2}}e^{-iE_k(t-t_0)/\hbar}$$

$$\overset{(11.165)}{\simeq} e^{-\frac{i}{\hbar}[E_{k_0} - \hbar\mathbf{k}_0\cdot\mathbf{v}_0](t-t_0)}\psi_{\mathbf{k}_0}^{(0)}(\mathbf{x} - \mathbf{x}_0 - \mathbf{v}_0(t - t_0), t_0). \qquad (11.167)$$

In Example 11.10 we show that the solution $\psi_{\mathbf{k}}(\mathbf{x})$ of Eq. (11.166), i.e. the stationary scattering wave, can be written

$$\psi_{\mathbf{k}}(\mathbf{x}) = \frac{e^{i\mathbf{k}\cdot\mathbf{x}}}{(2\pi)^{3/2}} + \frac{2m_0}{\hbar^2} \int d\mathbf{x}' \frac{e^{ik|\mathbf{x}-\mathbf{x}'|}}{|\mathbf{x}-\mathbf{x}'|} V(\mathbf{x}')\psi_{\mathbf{k}}(\mathbf{x}'). \qquad (11.168)$$

We saw earlier (cf. Eq. (11.154)) that for $|\mathbf{x}| \to \infty$, $\mathbf{k} = k\mathbf{e}_z$, this solution has the following asymptotic behaviour in which $f(\theta,\varphi)$ is the scattering amplitude with respect to the incoming plane wave, i.e.

$$\psi_{\mathbf{k}}(\mathbf{x} - \mathbf{x}_0) \simeq e^{-i\mathbf{k}\cdot\mathbf{x}_0}\psi_{\mathbf{k}}(\mathbf{x}),$$

$$\psi_{\mathbf{k}}(\mathbf{x}) = \frac{1}{(2\pi)^{3/2}}\left[e^{i\mathbf{k}\cdot\mathbf{x}} + \frac{e^{ikr}}{r}f_{\mathbf{k}}(\theta,\varphi) + O\left(\frac{1}{r^2}\right)\right]. \quad (11.169)$$

The wave packet in the presence of the potential is obtained from the free form (11.167) by replacing the plane wave $\exp(i\mathbf{k}\cdot\mathbf{x})$ by this scattered wave. From Eqs. (11.161) and (11.163) we obtain

$$\psi_{\mathbf{k}_0}(\mathbf{x} - \mathbf{x}_0, t) = \int \frac{d\mathbf{k}}{(2\pi)^3} A_{\mathbf{k}_0}(\mathbf{k})e^{-iE_k(t-t_0)/\hbar}\psi_{\mathbf{k}}(\mathbf{x} - \mathbf{x}_0). \qquad (11.170)$$

With Eqs. (11.169) and (11.167) we have

$$\psi_{\mathbf{k}_0}(\mathbf{x} - \mathbf{x}_0, t)$$
$$\simeq \int \frac{d\mathbf{k}}{(2\pi)^3} A_{\mathbf{k}_0}(\mathbf{k})e^{-iE_k(t-t_0)/\hbar}\frac{1}{(2\pi)^{3/2}}\left[e^{i\mathbf{k}\cdot\mathbf{x}} + \frac{e^{ikr}}{r}f_{\mathbf{k}}(\theta,\varphi)\right]e^{-i\mathbf{k}\cdot\mathbf{x}_0}$$
$$= \psi_{\mathbf{k}_0}^{(0)}(\mathbf{x} - \mathbf{x}_0, t) + \frac{1}{r}\int \frac{d\mathbf{k}}{(2\pi)^3}\frac{A_{\mathbf{k}_0}(\mathbf{k})}{(2\pi)^{3/2}}e^{-iE_k(t-t_0)/\hbar+ikr}f_{\mathbf{k}}(\theta,\varphi)e^{-i\mathbf{k}\cdot\mathbf{x}_0}$$
$$\simeq \psi_{\mathbf{k}_0}^{(0)}(\mathbf{x} - \mathbf{x}_0, t) + \frac{1}{r}f_{\mathbf{k}_0}(\theta,\varphi)\int \frac{d\mathbf{k}}{(2\pi)^3}\frac{A_{\mathbf{k}_0}(\mathbf{k})}{(2\pi)^{3/2}}e^{i[kr-E_k(t-t_0)/\hbar]}e^{-i\mathbf{k}\cdot\mathbf{x}_0}.$$
$$(11.171)$$

The phase of $f_{\mathbf{k}}(\theta,\varphi)$ could be carried along but we ignore it. Then with

$$k = \sqrt{(\mathbf{k}_0 + \mathbf{k} - \mathbf{k}_0)^2} \simeq \sqrt{k_0^2 + 2\mathbf{k}_0\cdot(\mathbf{k}-\mathbf{k}_0)} \simeq k_0 + \frac{\mathbf{k}_0\cdot(\mathbf{k}-\mathbf{k}_0)}{k_0},$$
$$(11.172)$$

and the expansion of E_k in Eq. (11.165)

$$kr - \frac{E_k(t-t_0)}{\hbar} = k_0 r - \frac{E_{k_0}(t-t_0)}{\hbar} + (\mathbf{k}-\mathbf{k}_0)\cdot\underbrace{\left[\frac{\mathbf{k}_0}{k_0}\{r - v_0(t-t_0)\}\right]}_{\mathbf{x}'},$$
$$(11.173)$$

so that with Eqs. (11.167), (11.171),

$$\psi_{\mathbf{k}_0}(\mathbf{x} - \mathbf{x}_0, t) - \psi^{(0)}_{\mathbf{k}_0}(\mathbf{x} - \mathbf{x}_0, t)$$

$$= \frac{1}{r} f_{\mathbf{k}_0}(\theta, \varphi) \int \frac{d\mathbf{k}}{(2\pi)^3} \frac{A_{k_0}(\mathbf{k})}{(2\pi)^{3/2}} e^{-i\mathbf{k}\cdot\mathbf{x}_0} e^{i[k_0 r - E_{k_0}(t-t_0)/\hbar]} e^{i(\mathbf{k}-\mathbf{k}_0)\cdot\mathbf{x}'}$$

$$= \frac{1}{r} f_{\mathbf{k}_0}(\theta, \varphi) e^{-i\mathbf{k}_0\cdot\mathbf{x}_0} \int \frac{d\mathbf{k}}{(2\pi)^3} \frac{A_{k_0}(\mathbf{k})}{(2\pi)^{3/2}} e^{i(\mathbf{k}-\mathbf{k}_0)\cdot(\mathbf{x}'-\mathbf{x}_0)} e^{i[k_0 r - E_{k_0}(t-t_0)/\hbar]}$$

$$= \frac{1}{r} f_{\mathbf{k}_0}(\theta, \varphi) e^{-i\mathbf{k}_0\cdot\mathbf{x}_0} e^{i[k_0 r - E_{k_0}(t-t_0)/\hbar]} \psi^{(0)}_{\mathbf{k}_0}(\mathbf{x}' - \mathbf{x}_0, t_0). \tag{11.174}$$

Here $\psi^{(0)}_{\mathbf{k}_0}(\mathbf{x}' - \mathbf{x}_0, t_0)$ has the form of the wave packet in radial direction as indicated in Fig. 11.5. The scattering amplitude is the same as that given by Eq. (11.154), however, with the energy of the maximum momentum of the wave packet.

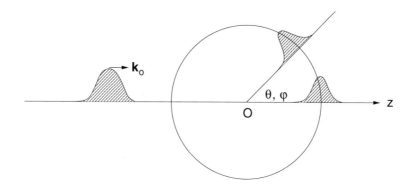

Fig. 11.5 Scattering of a wave packet with scattering centre O.

Example 11.10: Schrödinger equation as integral equation

Demonstrate the conversion of the Schrödinger equation into an integral equation, i.e.

$$(\triangle + k^2)\psi(\mathbf{r}) = \frac{2m_0}{\hbar^2} V(\mathbf{r})\psi(\mathbf{r}) \longrightarrow \psi(\mathbf{r}) = \frac{1}{(2\pi)^{3/2}} e^{i\mathbf{k}\cdot\mathbf{r}} + \frac{2m_0}{\hbar^2} \int d\mathbf{r}' G(\mathbf{r}, \mathbf{r}') V(\mathbf{r}')\psi(\mathbf{r}'),$$

and determine the Green's function $G(\mathbf{r}, \mathbf{r}')$.

Solution: We define the Green's function $G(\mathbf{r}, \mathbf{r}')$ by the equation

$$(\triangle + k^2)G(\mathbf{r}, \mathbf{r}') = -4\pi\delta(\mathbf{r} - \mathbf{r}').$$

A particular solution of the Schrödinger equation is $-\int G(\mathbf{r}, \mathbf{r}')U(\mathbf{r}')\psi(\mathbf{r}')d\mathbf{r}'/4\pi$ (as we saw earlier). Adding a suitably normalized solution of the free equation, the solution can be written

$$\psi_k(\mathbf{r}) = \frac{1}{(2\pi)^{3/2}} e^{i\mathbf{k}\cdot\mathbf{r}} - \frac{1}{4\pi} \int d\mathbf{r}' G(\mathbf{r}, \mathbf{r}')U(\mathbf{r}')\psi_k(\mathbf{r}'), \quad U(\mathbf{r}) \equiv \frac{2m_0}{\hbar^2} V(\mathbf{r}).$$

We find a set of Green's functions $G(\mathbf{r}, \mathbf{r}') \equiv G(\mathbf{r} - \mathbf{r}')$ with the Fourier transforms

$$G(\mathbf{r}) = \int d\mathbf{k}' g(\mathbf{k}') e^{i\mathbf{k}' \cdot \mathbf{r}}, \qquad \delta(\mathbf{r}) = \frac{1}{(2\pi)^3} \int d\mathbf{k}' e^{i\mathbf{k}' \cdot \mathbf{r}},$$

$$g(k') = \frac{1}{2\pi^2} \frac{1}{k'^2 - k^2}, \qquad G(\mathbf{r}) = \frac{1}{2\pi^2} \int \frac{d\mathbf{k}'}{k'^2 - k^2} e^{i\mathbf{k}' \cdot \mathbf{r}}.$$

Integrating out the angles we obtain

$$
\begin{aligned}
G(\mathbf{r}) &= \frac{1}{2\pi^2} \int_{k'=0}^{\infty} \int_{\theta=0}^{\pi} \int_{\varphi=0}^{2\pi} \frac{k'^2 dk' d(-\cos\theta) d\varphi}{k'^2 - k^2} e^{ik'r\cos\theta} \\
&= \frac{1}{\pi r} \int_0^{\infty} \frac{k' dk'}{i(k'^2 - k^2)} (e^{ik'r} - e^{-ik'r}) = \frac{1}{\pi r} \int_{-\infty}^{\infty} \frac{k' dk'}{i(k'^2 - k^2)} e^{ik'r} \\
&= -\frac{1}{\pi r} \frac{d}{dr} \int_{-\infty}^{\infty} \frac{dk'}{k'^2 - k^2} e^{ik'r}.
\end{aligned}
$$

Since this integral does not exist, we relate it to a properly defined contour integral with the two simple poles at $k' = \pm\sqrt{k^2 + i\epsilon} = \pm(k + i\epsilon/2k)$, $\epsilon > 0$ small, displaced slightly away from the real axis as indicated in Fig. 11.6. Thus we consider the contour integral taken around the contour C_+, which is such that the contribution along the infinite semi-circle vanishes. Hence we write $G(\mathbf{r}) \to G_+(\mathbf{r})$,

$$
\begin{aligned}
G_+(\mathbf{r}) &= \lim_{\epsilon \to 0} -\frac{1}{\pi r} \frac{d}{dr} \oint_{C_+} \frac{dk'}{k'^2 - (k^2 + i\epsilon)} e^{ik'r} = -\frac{1}{\pi r} \frac{d}{dr} \left[2\pi i \sum \text{residues} \right] \\
&= -\frac{1}{\pi r} \frac{d}{dr} 2\pi i \left(\frac{e^{ik'r}}{k' + k + i\epsilon/2k} \right)_{k' = k + i\epsilon/2k} \\
&= -\frac{1}{r} \frac{d}{dr} \left(i\frac{e^{ikr}}{k} \right) = \frac{1}{r} e^{ikr}.
\end{aligned}
$$

With the choice of an analogous contour C_- in the lower half plane, one obtains another Green's function given by

$$G_-(\mathbf{r}) = \frac{1}{r} e^{-ikr}.$$

These are the so-called outgoing and ingoing Green's functions.

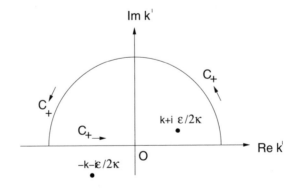

Fig. 11.6 The contour C_+ in the upper half of the k'-plane.

11.9 Scattering Phase and Partial Waves

We observed above that the Coulomb potential is associated with some special effects which one does not expect in general, e.g. the logarithmic phase as a consequence of the slow decrease at infinity and a special symmetry which permits separation in other coordinate systems. Before we introduce the general form of the so-called partial wave expansion, we define first the scattering phase for the case of the Coulomb problem proceeding from our previous considerations.

If we proceed from the radial Schrödinger equation, i.e. Eq. (11.100) — and we saw that in the particular case of the Coulomb potential this is not essential — we have to solve the following equation for the scattering case obtained from Eq. (11.140) (previously we solved this only for the bound state problem)[§]

$$y_l'' + \left[k^2 - \frac{2\gamma k}{r} - \frac{l(l+1)}{r^2} \right] y_l = 0. \tag{11.175}$$

Here we set (as in the case of the hydrogen atom)

$$y_l = e^{ikr}(kr)^{l+1}\phi_l(\xi), \qquad \xi = -2ikr. \tag{11.176}$$

Then

$$\xi \frac{d^2}{d\xi^2}\phi_l + (2l+2-\xi)\frac{d}{d\xi}\phi_l - (l+1+i\gamma)\phi_l = 0. \tag{11.177}$$

Similar to above we obtain a regular solution $y_l^{(r)}$ with

$$\phi_l = F(l+1+i\gamma, 2l+2; \xi), \tag{11.178}$$

whose continuation to infinity is again given by (cf. Eq. (11.145))

$$F(a, b; z) = W_1(a, b; z) + W_2(a, b; z). \tag{11.179}$$

Thus we proceed as above. Then

$$
\begin{aligned}
y_l^{(r)} &= e^{ikr}(kr)^{l+1}F(l+1+i\gamma, 2l+2; \xi) \\
&= e^{ikr}(kr)^{l+1}[W_1(l+1+i\gamma, 2l+2; -2ikr) \\
&\qquad + W_2(l+1+i\gamma, 2l+2; -2ikr)]
\end{aligned}
$$

[§]For comparison with calculations in Chapter 16 note that for $\hbar = c = 1$ together with $2m_0 = 1$ and hence $E = k^2$ the quantity γ here contains a factor k, i.e. under the stated conditions $2k\gamma = Z_1 Z_2 e^2 = -M_0$, where M_0 is a parameter in Chapter 16.

Using Eqs. (11.146) and (11.147) this becomes asymptotically

$$
\begin{aligned}
y_l^{(r)} \;\simeq\; & e^{ikr}(kr)^{l+1}\left[\frac{\Gamma(2l+2)}{\Gamma(l+1-i\gamma)}(2ikr)^{-l-1-i\gamma}\right.\\
& \left.+\frac{\Gamma(2l+2)}{\Gamma(l+1+i\gamma)}e^{-2ikr}(-2ikr)^{i\gamma-l-1}\right]\\
=\; & \frac{\Gamma(2l+2)}{2^{l+1}}\left[\frac{(2kr)^{-i\gamma}(i)^{-l-1-i\gamma}}{\Gamma(l+1-i\gamma)}e^{ikr}+\frac{(2kr)^{i\gamma}(-i)^{i\gamma-l-1}}{\Gamma(l+1+i\gamma)}e^{-ikr}\right]\\
=\; & \frac{\Gamma(2l+2)}{2^{l+1}}\left[\frac{e^{ikr-i\gamma\ln 2kr}}{\Gamma(l+1-i\gamma)}e^{i\pi(-l-1-i\gamma)/2}\right.\\
& \left.+\frac{e^{-ikr+i\gamma\ln 2kr}}{\Gamma(l+1+i\gamma)}e^{-i\pi(i\gamma-l-1)/2}\right]\\
=\; & \frac{\Gamma(2l+2)}{2^{l+1}}e^{\pi\gamma/2}\left[\frac{e^{ikr-i\gamma\ln 2kr-i\pi(l+1)/2}}{\Gamma(l+1-i\gamma)}\right.\\
& \left.+\frac{e^{-ikr+i\gamma\ln 2kr+i\pi(l+1)/2}}{\Gamma(l+1+i\gamma)}\right].
\end{aligned}
\tag{11.180}
$$

We set

$$
e^{2i\tilde{\delta}_l}:=\frac{\Gamma(l+1+i\gamma)}{\Gamma(l+1-i\gamma)},
\tag{11.181}
$$

so that

$$
e^{i\tilde{\delta}_l}\Gamma(l+1-i\gamma)=e^{-i\tilde{\delta}_l}\Gamma(l+1+i\gamma).
$$

Then

$$
\begin{aligned}
y_l^{(r)} \;\overset{r\to\infty}{\simeq}\; & \frac{\Gamma(2l+2)e^{\pi\gamma/2}e^{i\tilde{\delta}_l}}{2^{l+1}\Gamma(l+1+i\gamma)}\left[e^{i\tilde{\delta}_l+ikr-i\gamma\ln 2kr-i\pi(l+1)/2}\right.\\
& \left.+e^{-i\tilde{\delta}_l-ikr+i\gamma\ln 2kr+i\pi(l+1)/2}\right]\\
=\; & \frac{\Gamma(2l+2)e^{\pi\gamma/2}e^{i\tilde{\delta}_l}}{2^{l}\Gamma(l+1+i\gamma)}\cos\left(kr+\tilde{\delta}_l-\gamma\ln 2kr-\frac{\pi}{2}(l+1)\right)\\
=\; & \frac{\Gamma(2l+2)e^{\pi\gamma/2}e^{i\tilde{\delta}_l}}{2^{l}\Gamma(l+1+i\gamma)}\sin\left(kr-\gamma\ln 2kr-\frac{1}{2}\pi l+\tilde{\delta}_l\right).
\end{aligned}
\tag{11.182}
$$

We set

$$
\Gamma(l+1+i\gamma):=Re^{i\theta},\quad \theta=\arg\Gamma(l+1+i\gamma)
\tag{11.183}
$$

Then

$$
e^{2i\tilde{\delta}_l}=\frac{Re^{i\theta}}{Re^{-i\theta}}=e^{2i\theta}\;\longrightarrow\;\theta=\tilde{\delta}_l.
$$

Hence
$$\tilde{\delta}_l = \arg \Gamma(l + 1 + i\gamma). \tag{11.184}$$

The phase $\delta_l = \tilde{\delta}_l - l\pi/2$ is called *Coulomb scattering phase*. The expression

$$e^{2i\delta_l} = e^{2i\tilde{\delta}_l - i\pi l}$$

is called *S-matrix element* or *scattering matrix element*. One defines as *regular Coulomb wave* $F_l(\gamma, kr)$ the regular solution with the following asymptotic behaviour:

$$F_l \overset{r \to \infty}{\simeq} \sin(kr - \gamma \ln 2kr + \delta_l) \tag{11.185}$$

(also called regular spherical Coulomb function).

Analogously one defines as nonregular spherical Coulomb function or *Jost solution* the outgoing or incoming wave $u_l^{(+)}, u_l^{(-)}$ respectively with the following asymptotic behaviour:

$$u_l^{(\pm)} \overset{r \to \infty}{\simeq} e^{\pm i(kr - \gamma \ln 2kr - l\pi/2)}. \tag{11.186}$$

The factor $e^{2i\delta_l}$ between the Jost solutions (in the regular solution continued to infinity), i.e.

$$S \equiv e^{2i\delta_l} = \frac{\Gamma(l + 1 + i\gamma)}{\Gamma(l + 1 - i\gamma)} e^{-i\pi l} \tag{11.187}$$

contains almost the entire physical information with regard to the Coulomb potential. For instance the poles of S are given by

$$l + 1 + i\gamma = -n', \quad n' = 0, 1, 2, \ldots,$$

i.e. with Eq. (11.139)

$$l + 1 + n' = -i\gamma = i \frac{Z_1 Z_2 e^2 m_0^{1/2}}{\hbar \sqrt{E} \sqrt{2}}. \tag{11.188}$$

Squaring this expression we obtain the wellknown formula for the binding energy of bound states (as a comparison with Eq. (11.114a) reveals). That the energy E appears in Eq. (11.188) in the form $\sqrt{E}$, indicates that — looked at analytically — the scattering amplitude $f(\theta)$ possesses at $E = 0$ a branch point of the square root type, i.e. where the continuum of the spectrum starts.

In the present quantum mechanical considerations l is, as we saw, a positive integer. However, considering l as a function of E extended into the complex plane, i.e. considering

$$l \longrightarrow \alpha_n(E),$$

and plotting these for different values of n, we obtain trajectories which are known as *Regge trajectories*. For the Coulomb potential a typical such trajectory is indicated in Fig. 11.7.[¶] These Regge trajectories can be shown to determine the asymptotic high energy behaviour of scattering amplitudes and hence cross sections. We shall return to these in Chapter 16 in the consideration of screened Coulomb potentials or Yukawa potentials and indicate other important aspects of these.

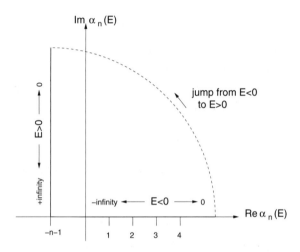

Fig. 11.7 A Regge trajectory of the Coulomb potential.

We mentioned above that the logarithmic phase does not arise in the case of short-range potentials. In these cases, and thus in general, we obtain the *partial wave expansion of the scattering amplitude* as follows. The large r asymptotic behaviour of the solution which is regular at the origin in the case of the scattering problem can be written

$$\psi_{\text{reg}} = \psi_{\text{in}} + \psi_{\text{scatt}}, \qquad (11.189)$$

where as before the incoming wave in the direction of z is

$$\psi_{\text{in}} = e^{ikz} \qquad (11.190)$$

and the scattering solution

$$\psi_{\text{scatt}} = f(\theta)\frac{e^{ikr}}{r} + O\left(\frac{1}{r^2}\right). \qquad (11.191)$$

[¶]V. Singh [252].

For a spherically symmetric potential the amplitude f depends only on θ. The incoming plane wave can be re-expressed as a superposition of incoming and outgoing spherical waves with appropriate components of l and with a definite relative phase. The mathematical expression can be found in Tables of Special Functions (there cf. *Legendre functions*) and its asymptotic expression for $r \to \infty$ and $z = r \cos \theta$ is given by[‖]

$$e^{ikz} \simeq \frac{i}{2kr} \sum_l (2l+1) \left\{ e^{il\pi} e^{-ikr} - e^{ikr} \right\} P_l(\cos\theta). \tag{11.192}$$

Thus the incoming and outgoing spherical waves have a definite relative phase. The potential disturbs this phase relationship (by delay or absorption or redirection). When we solve the Schrödinger equation and calculate the asymptotic behaviour of the regular solution we obtain correspondingly

$$\psi_{\text{reg}} \simeq \frac{i}{2kr} \sum_l (2l+1) \left\{ e^{il\pi} e^{-ikr} - \eta_l e^{2i\delta_l} e^{ikr} \right\} P_l(\cos\theta), \tag{11.193}$$

where $\eta_l e^{2i\delta_l}$ determines the change in amplitude and phase of the outgoing spherical wave. It follows therefore that

$$
\begin{aligned}
\psi_{\text{scatt}} \simeq\ & f(\theta) \frac{e^{ikr}}{r} = \psi_{\text{reg}} - e^{ikz} \\
=\ & \frac{i}{2kr} \sum_l (2l+1) \left\{ e^{il\pi} e^{-ikr} - \eta_l e^{2i\delta_l} e^{ikr} \right\} P_l(\cos\theta) \\
& - \frac{i}{2kr} \sum_l (2l+1) \left\{ e^{il\pi} e^{-ikr} - e^{ikr} \right\} P_l(\cos\theta) \\
=\ & \frac{1}{k} \sum_l (2l+1) \left(\frac{\eta_l e^{2i\delta_l} - 1}{2i} \right) P_l(\cos\theta) \frac{e^{ikr}}{r}.
\end{aligned}
\tag{11.194}
$$

This therefore yields the following *partial wave expansion* of the scattering amplitude

$$f(\theta) = \frac{1}{2ik} \sum_l (2l+1)[S(l,k) - 1] P_l(\cos\theta) \tag{11.195}$$

with the scattering matrix element

$$S(l,k) = \eta_l e^{2i\delta_l(k)}, \tag{11.196}$$

in which $\eta_l \neq 1$ indicates absorption.

[‖] See e.g. W. Magnus and F. Oberhettinger [181], pp. 21 - 22. There the expansion is given in terms of Bessel functions J_ν which have to be re-expressed in terms of the asymptotic behaviour of Hankel functions $H_\nu^{(1,2)}$.

Chapter 12

Quantum Mechanical Tunneling

12.1 Introductory Remarks

One of the main objectives of this text is the presentation of methods of calculation of typically quantum mechanical effects which are generally not so easily derivable in nontrivial contexts but are important as basic phenomena in standard examples. Such a vital quantum mechanical phenomenon is so-called *tunneling*, i.e. that quantum mechanically a particle may have a small probability of being in a spatial domain, which a particle in classical mechanics would never be able to intrude. Similarly difficult is, in general, the computation of finite *lifetimes* of a quantum mechanical state. It is not surprising, therefore, that the explicit derivation of such quantities in nontrivial contexts, as — for instance — for trigonometric or double well potentials, in texts on quantum mechanics is rare. In fact, one of the main and now well-known methods to perform such calculations, namely the instanton (or more generally pseudoparticle) method, was developed only in the last two to three decades. These newer methods will be presented in detail in later chapters. It is very instructive, however, to explore first much simpler models in order to acquire an impression of the type of results to be expected. Thus in this chapter we consider traditional "square well" potentials in one-dimensional contexts, which in one form or another, can be found in any traditional text on quantum mechanics. In particular we consider cases which illustrate the occurrence of tunneling and that of *resonances*, i.e. states of a finite lifetime. For various general aspects and applications of tunneling we refer to other literature.[*]

---

[*]See e.g. M. Razavy [236].

12.2 Continuity Equation and Conditions

The statistical interpretation of the Schrödinger wave function $\psi(\mathbf{x}, t)$ was developed by Born in 1926. We assumed this already in the preceding since we interpreted the normalization integral

$$\int_{\text{all space}} d\mathbf{x}\, |\psi(\mathbf{x}, t)|^2 = 1 \tag{12.1}$$

as the probability described by $\psi(\mathbf{x}, t)$ that the one particle concerned is somewhere in space at time t, so that $\rho = |\psi|^2$ is the probability density for a spatial measurement at time t. Correspondingly one interprets as *probability current density* the expression

$$\mathbf{j}(\mathbf{x}, t) = \frac{\hbar}{2m_0} \left\{ \psi^* \frac{1}{i} \boldsymbol{\nabla} \psi - \psi \frac{1}{i} \boldsymbol{\nabla} \psi^* \right\}. \tag{12.2}$$

With the help of the Schrödinger equation

$$i\hbar \frac{\partial}{\partial t} \psi(\mathbf{x}, t) = H\psi(\mathbf{x}, t), \quad -i\hbar \frac{\partial}{\partial t} \psi^*(\mathbf{x}, t) = [H\psi(\mathbf{x}, t)]^* \tag{12.3}$$

we can obtain the *continuity equation* which describes the conservation of probability in analogy to the case of the conservation of charge in electrodynamics. According to Eq. (12.2) we have

$$
\begin{aligned}
i\hbar \frac{\partial}{\partial t} |\psi|^2 &= i\hbar \left(\frac{\partial}{\partial t} \psi^* \right) \psi + \psi^* \left(i\hbar \frac{\partial}{\partial t} \psi \right) \\
&= -(H\psi)^* \psi + \psi^* (H\psi) \\
&= -(V\psi)^* \psi + \psi^* (V\psi) + \frac{\hbar^2}{2m_0} (\triangle \psi)^* \psi - \psi^* \frac{\hbar^2}{2m_0} (\triangle \psi) \\
&= 0 + \frac{\hbar^2}{2m_0} \boldsymbol{\nabla} \cdot [(\boldsymbol{\nabla} \psi^*) \psi - \psi^* (\boldsymbol{\nabla} \psi)] \\
&= \frac{\hbar}{i} \boldsymbol{\nabla} \cdot \mathbf{j},
\end{aligned}
$$

i.e.

$$i\hbar \left(\frac{\partial \rho}{\partial t} + \boldsymbol{\nabla} \cdot \mathbf{j} \right) = 0. \tag{12.4}$$

Then

$$\int_{F_0} \mathbf{j} \cdot d\mathbf{F} \tag{12.5}$$

is the probability that per unit time a particle passes through the surface F_0.

The time-independent Schrödinger equation is a second order differential equation in **x**. If $\psi(\mathbf{x})$ and $\boldsymbol{\nabla}\psi(\mathbf{x})$ are known at every point **x**, the equation can be integrated at every point, even for discontinuous potentials. It is therefore sensible to demand that $\psi(\mathbf{x})$ and $\boldsymbol{\nabla}\psi(\mathbf{x})$ be continuous, finite and single-valued at every point **x** in order to insure that $\psi(\mathbf{x})$ is a unique representative of the state of a system. From this follows that both ρ and **j** have to be finite and continuous everywhere.[†] For the singular delta function potential a modification of these demands is necessary, since this potential implies a discontinuity of the derivative at its singularity (as will be seen below).

An important consequence of the wave-like nature of the wave function $\psi(\mathbf{x})$ is that it can differ from zero also in domains which are classically not accessible to the particle. Such a region is for instance the domain where V is larger than the total energy E. Thus it is possible that the probability for the particle to be in such a domain is nonzero, although in general small. It is therefore quantum mechanically possible for a "particle" to pass through a classically forbidden region. This effect is known as *"tunneling"*. The wave phenomenon which permits transmission and reflection is familiar from electrodynamics where the laws of reflection and refraction of optics are derived from continuity conditions at the interface between dielectric media applied to the fields **E** and **B** of Maxwell's equations. In much the same way that we define there reflection and transmission coefficients, we can do this also here. If there is no absorption, the problem is analogous to that of the scattering process from a potential in a one-dimensional case, which is the case we consider here.

12.3 The Short-Range Delta Potential

It is very instructive to consider in detail a potential $V(x) = V_0\delta(x)$ with the "shape" of a one-dimensional delta function. This case plays a special role, as we mentioned above. The Schrödinger equation for this case is[*]

$$\psi'' + [k^2 - 2\alpha\delta(x)]\psi = 0, \quad V(x) = +V_0\delta(x), \quad V_0 > 0, \tag{12.6}$$

where

$$\alpha = \frac{m_0 V_0}{\hbar^2}, \quad k^2 = \frac{2m_0 E}{\hbar^2}. \tag{12.7}$$

[†]See L. Schiff [243], pp. 29 - 33.

[*]Maybe the reader dislikes being confronted again with the one-dimensional case. However, higher dimensional delta potentials do not have properties like that in one dimension. Delta potentials in more than one dimension do not permit bound states and scattering. Nonetheless, with regularization their study is very instructive for illustrating basic concepts of modern quantum field theory, as shown by P. Gosdzinsky and R. Tarrach [119].

Integrating the equation over a small interval of x around 0 we see, that the derivative of ψ is discontinuous there:

$$[\psi']^{\epsilon}_{-\epsilon} + k^2 \int_{-\epsilon}^{\epsilon} \psi dx - 2\alpha \int_{-\epsilon}^{\epsilon} \delta(x)\psi(x)dx = 0, \tag{12.8}$$

i.e.

$$[\psi']_{0+} - [\psi']_{0-} = 2\alpha\psi(0). \tag{12.9}$$

States with $k^2 > 0$ correspond to those of free particles that move with constant velocity $v = \hbar k/m_0$, except in domains of the potential V. Consider the case of a purely outgoing wave in the region $x > 0$ (no reflection back from infinity). In the region $x < 0$ we then have both types of waves: the incoming or incident wave to the right as well as the wave reflected from the potential, i.e.

$$\psi = \begin{cases} e^{ikx} + Be^{-ikx} & \text{for} \quad x < 0, \\ Ce^{ikx} & \text{for} \quad x > 0. \end{cases} \tag{12.10}$$

The continuity conditions at $x = 0$ are therefore

$$\text{for } \psi: \quad 1 + B = C, \tag{12.11a}$$

and according to Eq. (12.9)

$$\text{for } \psi': \quad ikC - (ik - ikB) = 2\alpha C \quad \text{or} \quad 2\alpha(1 + B). \tag{12.11b}$$

It follows that

$$ik - ik(C - 1) - ikC = -2\alpha C,$$

i.e.

$$C = \frac{ik}{-\alpha + ik} \tag{12.12a}$$

and hence

$$B = C - 1 = -\frac{\alpha}{\alpha - ik}. \tag{12.12b}$$

Thus for ψ we obtain

$$\psi = \begin{cases} e^{ikx} - \frac{\alpha}{\alpha-ik}e^{-ikx} & \text{for} \quad x < 0, \\ -\frac{ik}{\alpha-ik}e^{ikx} & \text{for} \quad x > 0. \end{cases} \tag{12.13}$$

One defines respectively as *reflection* and *transmission coefficients* R and T the squares of the respective amplitudes, i.e.

$$R = \left| -\frac{\alpha}{\alpha - ik} \right|^2 = \frac{\alpha^2}{\alpha^2 + k^2}, \tag{12.14}$$

and

$$T = \left| \frac{ik}{\alpha - ik} \right|^2 = 1 - R. \tag{12.15}$$

Furthermore we see that similar to the S-matrix which we encountered in the case of the Coulomb potential, the quantity T, as coefficient of the outgoing wave, possesses simple poles at $\alpha \pm ik = 0$, i.e. at $k^2 = -\alpha^2$, or

$$E = -\frac{\hbar^2}{2m_0}\alpha^2. \tag{12.16}$$

The corresponding wave function proportional to e^{ikx} for $x < 0$ associated with the pole for $k = -i\alpha$ is[†]

$$\psi(x) = \sqrt{\alpha}e^{-\alpha|x|}, \tag{12.17}$$

when normalized to 1, i.e.

$$\int_{-\infty}^{\infty} dx |\psi(x)|^2 = 1.$$

In the *bound state problem* the Schrödinger equation is

$$\psi'' + [k^2 + 2\alpha\delta(x)]\psi = 0, \quad k^2 = -\kappa^2 \leq 0.$$

We write the solution

$$\psi = Ce^{-\kappa|x|} = \begin{cases} Ce^{-\kappa x} & \text{for} \quad x > 0, \\ Ce^{\kappa x} & \text{for} \quad x < 0, \end{cases} \tag{12.18}$$

so that

$$\psi(0+) = \psi(0-)$$

and

$$\psi'(0+) - \psi'(0-) = -2\alpha\psi(0), \quad -\kappa C - (\kappa)C = -2\alpha C, \quad \kappa = \alpha.$$

Thus

$$\psi = Ce^{-\alpha|x|},$$

and C follows from the normalization:

$$1 = \int_{-\infty}^{\infty} dx C^2 e^{-2\alpha|x|} = 2C^2 \int_0^{\infty} dx e^{-2\alpha|x|} = \frac{C^2}{\alpha}.$$

Hence $C = \sqrt{\alpha}$ and therefore

$$\psi = \sqrt{\alpha}e^{-\alpha|x|}. \tag{12.19}$$

Thus the delta function potential supports exactly one bound state.

[†]Note that the physical pole has negative imaginary part. See e.g. R. Omnès and M. Froissart [223] or other books on scattering theory like R. Newton [219].

12.4   Scattering from a Potential Well

We can immediately transfer the above considerations in a similar way to scattering from a potential well[‡] as indicated in Fig. 12.1.

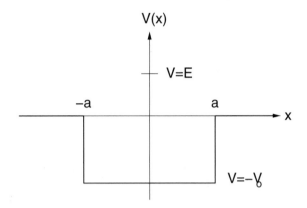

Fig. 12.1 The square well potential.

The Schrödinger equation is

$$\psi''(x) + \frac{2m_0}{\hbar^2}[E - V(x)]\psi(x) = 0.$$

We want to consider the case of stationary states for $E > 0$. We set

$$k = \sqrt{\frac{2m_0 E}{\hbar^2}}, \quad \kappa = \sqrt{\frac{2m_0(E + V_0)}{\hbar^2}}, \tag{12.20}$$

so that

$$\begin{aligned}
\psi'' + k^2\psi = 0 \quad & \text{for} \quad |x| > a, \\
\psi'' + \kappa^2\psi = 0 \quad & \text{for} \quad |x| < a.
\end{aligned} \tag{12.21}$$

The solution of the Schrödinger equation then consists of the following parts:

$$\psi(x) = \begin{cases}
Ae^{ikx} + Be^{-ikx} & \text{with} \quad x \in [-\infty, -a], \\
Ce^{-i\kappa x} + De^{i\kappa x} & \text{with} \quad x \in [-a, a], \\
Fe^{ikx} + Ge^{-ikx} & \text{with} \quad x \in [a, \infty].
\end{cases} \tag{12.22}$$

[‡]We follow here largely F. Schwabl [246], p. 58. Similar treatments of square well potentials or barriers can be found in most books on quantum mechanics, or in the lectures of E. Fermi [92], p. 55. The latter contains also specific applications in nuclear physics.

The connection relations obtained by equating $\psi(x)$ and its derivative from left and right at $x = -a$ and a yield the equations

$$
\begin{aligned}
Ae^{-ika} + Be^{ika} &= Ce^{i\kappa a} + De^{-i\kappa a}, \\
ikAe^{-ika} - ikBe^{ika} &= -i\kappa Ce^{i\kappa a} + Di\kappa e^{-i\kappa a},
\end{aligned}
\tag{12.23}
$$

and

$$
\begin{aligned}
Ce^{-i\kappa a} + De^{i\kappa a} &= Fe^{ika} + Ge^{-ika}, \\
-i\kappa Ce^{-i\kappa a} + Di\kappa e^{i\kappa a} &= Fike^{ika} - ikGe^{-ika}.
\end{aligned}
\tag{12.24}
$$

The first set of equations can be rewritten as

$$
\begin{pmatrix} e^{-ika} & e^{ika} \\ e^{-ika} & -e^{ika} \end{pmatrix} \begin{pmatrix} A \\ B \end{pmatrix} = \begin{pmatrix} e^{i\kappa a} & e^{-i\kappa a} \\ -\frac{\kappa}{k}e^{i\kappa a} & \frac{\kappa}{k}e^{-i\kappa a} \end{pmatrix} \begin{pmatrix} C \\ D \end{pmatrix},
$$

or

$$
\begin{pmatrix} A \\ B \end{pmatrix} = \frac{1}{2} \begin{pmatrix} e^{ika} & e^{ika} \\ e^{-ika} & -e^{-ika} \end{pmatrix} \begin{pmatrix} e^{i\kappa a} & e^{-i\kappa a} \\ -\frac{\kappa}{k}e^{i\kappa a} & \frac{\kappa}{k}e^{-i\kappa a} \end{pmatrix} \begin{pmatrix} C \\ D \end{pmatrix},
\tag{12.25}
$$

i.e.

$$
\begin{pmatrix} A \\ B \end{pmatrix} = M(a) \begin{pmatrix} C \\ D \end{pmatrix},
\tag{12.26}
$$

where

$$
M(a) = \frac{1}{2} \begin{pmatrix} (1 - \frac{\kappa}{k})e^{i\kappa a + ika} & (1 + \frac{\kappa}{k})e^{-i\kappa a - ika} \\ (1 + \frac{\kappa}{k})e^{i\kappa a - ika} & (1 - \frac{\kappa}{k})e^{-i\kappa a - ika} \end{pmatrix}.
\tag{12.27}
$$

Similarly Eq. (12.24) yields

$$
\begin{pmatrix} F \\ G \end{pmatrix} = M(-a) \begin{pmatrix} C \\ D \end{pmatrix}.
\tag{12.28}
$$

From these matrix equations we obtain

$$
\begin{pmatrix} A \\ B \end{pmatrix} = M(a)M(-a)^{-1} \begin{pmatrix} F \\ G \end{pmatrix},
\tag{12.29}
$$

where

$$
M(-a)^{-1} = \frac{1}{2} \begin{pmatrix} (1 + \frac{\kappa}{k})e^{i\kappa a + ika} & (1 - \frac{\kappa}{k})e^{i\kappa a - ika} \\ (1 - \frac{\kappa}{k})e^{-i\kappa a + ika} & (1 + \frac{\kappa}{k})e^{-i\kappa a - ika} \end{pmatrix}.
\tag{12.30}
$$

Hence with the complex quantities

$$\epsilon : = \frac{i\kappa}{k} - \frac{k}{i\kappa}, \quad \eta := \frac{i\kappa}{k} + \frac{k}{i\kappa},$$

$$1 + \frac{1}{4}\epsilon^2 = \frac{1}{4}\eta^2, \quad \eta^2 = -\eta\eta^*, \quad \epsilon^2 = -\epsilon\epsilon^*. \qquad (12.31)$$

we have

$$\begin{pmatrix} A \\ B \end{pmatrix} = N(a) \begin{pmatrix} F \\ G \end{pmatrix}, \qquad (12.32)$$

$$N(a) = \begin{pmatrix} [\cos 2\kappa a - \frac{1}{2}\epsilon \sin 2\kappa a]e^{2ika} & -\frac{1}{2}\eta \sin 2\kappa a \\ \frac{1}{2}\eta \sin 2\kappa a & [\cos 2\kappa a + \frac{1}{2}\epsilon \sin 2\kappa a]e^{-2ika} \end{pmatrix}.$$

Thus if we now set $G = 0$ in order to have only an outgoing wave proportional to $\exp(ikx)$ to the right of the well, we obtain

$$A = F\left(\cos 2\kappa a - \frac{1}{2}\epsilon \sin 2\kappa a\right)e^{2ika}, \quad B = \frac{1}{2}\eta F \sin 2\kappa a. \qquad (12.33)$$

One defines as *transmission amplitude* the ratio of outgoing to incoming amplitudes, i.e. the quantity

$$T(E) := \frac{F}{A} = \frac{e^{-2ika}}{\cos 2\kappa a - \frac{1}{2}\epsilon \sin 2\kappa a} \qquad (12.34)$$

and as *transmission coefficient* (note ϵ is complex!)

$$\begin{aligned} |T(E)|^2 &= \frac{1}{1 - (1 + \frac{1}{4}\epsilon^2)\sin^2 2\kappa a} \frac{1}{1 - \frac{1}{4}\eta^2 \sin^2 2\kappa a} \\ &= \frac{1}{1 + \frac{1}{4}\eta\eta^* \sin^2 2\kappa a}. \end{aligned} \qquad (12.35)$$

Inserting Eq. (12.34) into Eq. (12.22) we obtain

$$\psi(x) = \begin{cases} Ae^{ikx} + AT(E)\frac{\eta}{2}\sin 2\kappa a e^{-ikx} & \text{with} \quad x \in [-\infty, -a], \\ Ce^{-i\kappa x} + De^{i\kappa x} & \text{with} \quad x \in [-a, a], \\ AT(E)e^{ikx} & \text{with} \quad x \in [a, \infty]. \end{cases} \qquad (12.36)$$

One defines as *reflection coefficient* $|R|^2$ the modulus squared of the reflected amplitude divided by the amplitude of the incoming wave, i.e.

$$|R|^2 = \left| T(E)\frac{\eta}{2}\sin 2\kappa a \right|^2. \qquad (12.37)$$

On the other hand according to Eq. (12.35) and using Eq. (12.31):

$$1 - |T(E)|^2 = \frac{-\frac{\eta^2}{4}\sin^2 2\kappa a}{1 - \frac{\eta^2}{4}\sin^2 2\kappa a} = -\frac{\eta^2}{4}\sin^2 2\kappa a |T(E)|^2$$

$$= \left|\frac{\eta}{2}\sin 2\kappa a T(E)\right|^2 = |R|^2, \tag{12.38}$$

as expected. We see in particular from Eq. (12.34) that the amplitude $F = AT(E)$ of the outgoing wave has a simple pole where

$$\cos 2\kappa a - \frac{1}{2}\epsilon\sin 2\kappa a = 0, \tag{12.39}$$

i.e. where

$$\cos 2\kappa a - \frac{i}{2}\left(\frac{\kappa}{k} + \frac{k}{\kappa}\right)\sin 2\kappa a = 0. \tag{12.40}$$

With the relation[§]

$$\cot(2\kappa a) = \frac{1}{2}[\cot(\kappa a) - \tan(\kappa a)],$$

we obtain

$$\cot(\kappa a) - \tan(\kappa a) = i\frac{k}{\kappa} - \frac{\kappa}{ik}, \tag{12.41}$$

i.e. an equation of the type

$$f^{-1} - f = g^{-1} - g,$$

which is satisfied if either $f = g$ or $f = -g^{-1}$, i.e.

$$\cot(\kappa a) = i\frac{k}{\kappa} \quad \text{or} \quad \tan(\kappa a) = -\frac{ik}{\kappa}. \tag{12.42}$$

These equations have no solution for κ and k real, nor when both parameters are purely imaginary. However, in the domain of k imaginary and κ real, they have solutions, i.e. for

$$-V_0 < E < 0. \tag{12.43}$$

Thus we set

$$k = -i\sqrt{\frac{2m_0|E|}{\hbar^2}}, \tag{12.44}$$

[§]H. B. Dwight [81], formula 406.12, p. 83.

and obtain the equations

$$\kappa \tan(\kappa a) = -\sqrt{\frac{2m_0|E|}{\hbar^2}} \quad \text{and} \quad \kappa \cot(\kappa a) = \sqrt{\frac{2m_0|E|}{\hbar^2}}. \tag{12.45}$$

These are the equations that one obtains from the connection relations for the case of Eq. (12.42) which then leads to *bound states* with associated even and odd wave functions.[¶] We observe that $T(E)$ has poles at those values of E which supply binding energies, and is itself infinite there. This can be understood. The infinity of the amplitude of the outgoing wave corresponds to a zero of the amplitude of the incoming wave, i.e. $A = 0$. The vanishing of A implies that because k is imaginary the factor $\exp(ikx)$ does not diverge exponentially for $x \to -\infty$. The remaining wave part becomes exponentially decreasing.

We now return to the *scattering states*. The amplitude of the transmitted wave assumes its maximum value where (cf. Eq. (12.35)) for $E > 0$

$$\sin 2\kappa a = 0, \quad \text{i.e.} \quad 2\kappa a = n\pi, \quad n = 0, \pm 1, \pm 2, \ldots. \tag{12.46}$$

This condition implies the eigenvalues

$$E \to E_R = n^2 \frac{\hbar^2 \pi^2}{8m_0 a^2} - V_0 > 0. \tag{12.47}$$

These *states in the continuum* are called *resonances*. At their positions

$$\cos(2\kappa a)|_{E_R} = (-1)^n, \quad \tan(2\kappa a)_{E_R} = 0. \tag{12.48}$$

We expand $T(E)$ around these energy values, i.e. around the values given by Eq. (12.46). Then according to Eq. (12.34),

$$T(E)e^{2ika} = \frac{1}{\cos(2\kappa a)[1 - \frac{i}{2}(\frac{\kappa}{k} + \frac{k}{\kappa})\tan(2\kappa a)]}. \tag{12.49}$$

Taylor expansion of the denominator in powers of $(E - E_R)$ yields

$$\frac{1}{2}\left(\frac{\kappa}{k} + \frac{k}{\kappa}\right)\tan(2\kappa a) = \frac{2}{\Gamma}(E - E_R),$$

where Γ is given by[‖]

$$\frac{2}{\Gamma} = \left[\frac{1}{2}\left(\frac{\kappa}{k} + \frac{k}{\kappa}\right)\frac{d(2\kappa a)}{dE}\right]_{E_R} = \frac{1}{2}\frac{\sqrt{2m_0}a}{\hbar}\frac{2E_R + V_0}{\sqrt{E_R(E_R + V_0)}}. \tag{12.50}$$

[¶]See e.g. F. Schwabl [246], pp. 64 - 65.
[‖]Note that $d\tan(2\kappa a)/d(2\kappa a)$ evaluated at E_R is 1. Also, from Eq. (12.20),

$$\frac{d(2\kappa a)}{dE} = \frac{a}{\hbar}\frac{2m_0}{\sqrt{2m_0(E_R + V_0)}}, \quad \frac{\kappa}{k} + \frac{k}{\kappa} = \sqrt{\frac{E + V_0}{E}} + \sqrt{\frac{E}{E + V_0}} = \frac{2E + V_0}{\sqrt{E}\sqrt{E + V_0}}.$$

We obtain therefore

$$T(E)e^{2ika} = \frac{1}{(-1)^n[1 - i\frac{2}{\Gamma}(E - E_R)]} = (-1)^n \frac{i\Gamma/2}{E - E_R + i\Gamma/2}, \quad (12.51)$$

i.e. $T(E)$ has poles at

$$E = E_R - i\frac{\Gamma}{2}. \quad (12.52)$$

Here E_R is the energy of the resonance and $2/\Gamma$ the *lifetime* of the state. This result is physically very plausible. Since we have reflection as well as transmission there are states with repeated reflection and finally transmission. These are precisely the resonance states.

12.5 Degenerate Potentials and Tunneling

In the following we consider a potential which is very similar to a periodic potential and can serve to illustrate a number of aspects of the process of tunneling in quantum mechanics. We consider a potential consisting of a chain of rectangular barriers as depicted in Fig. 12.2. A potential of this type is known as a *Penney–Kronig potential*. We restrict ourselves to the period $-2a \leq x \leq 2a$.

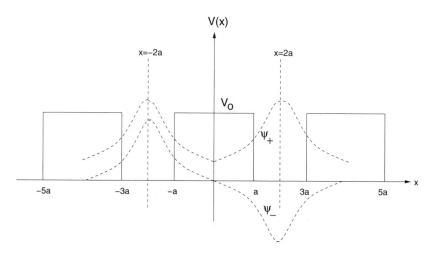

Fig. 12.2 The Penney–Kronig potential.

The Schrödinger equation to be solved is

$$\psi''(x) + \frac{2m_0}{\hbar^2}(E - V(x))\psi = 0, \quad V(x) = \theta(x+a)V_0\theta(a-x), \quad -2a \leq x \leq 2a.$$

(a)   We consider first the case of $V_0 = \infty$. The particle of mass m_0 whose quantum mechanics is described by the Schrödinger equation is free to move in the domains under consideration to the left and to the right of the central barrier. In view of the periodicity we have

$$\psi(x) = \psi(x + 4a), \qquad (12.53)$$

and we choose

$$\psi(x) \propto \sin \sqrt{\frac{2m_0 E}{\hbar^2}} \begin{pmatrix} x + 3a \\ \text{or} \\ 3a - x \end{pmatrix}, \quad x \in \begin{cases} [-3a, -a] \\ \text{and } [a, 3a] \end{cases}. \qquad (12.54)$$

There is the additional boundary condition (for $V_0 = \infty$)

$$\psi(-a) = 0 = \psi(a),$$

since the particle cannot penetrate into the infinitely high wall. This boundary condition implies the quantization of the energy with

$$\sqrt{\frac{2m_0 E}{\hbar^2}}(2a) = n\pi, \quad E \to E_n = \frac{n^2 \pi^2 \hbar^2}{8a^2 m_0}, \quad n = 0, 1, 2, \ldots. \qquad (12.55)$$

The Hamilton operator of the problem, i.e. the differential operator

$$H = -\frac{\hbar^2}{2m_0} \frac{d^2}{dx^2} + V(x),$$

is even in x, i.e., with $\mathcal{P}$ as *parity operator*, $\mathcal{P}H\mathcal{P}^{-1} = H$, and hence for $H\psi(x) = E\psi(x)$:

$$\mathcal{P}H\mathcal{P}^{-1}\mathcal{P}\psi(x) = E\mathcal{P}\psi(x), \quad H\mathcal{P}\psi(x) = E\mathcal{P}\psi(x), \quad H\psi(-x) = E\psi(-x),$$

so that with $H\psi(x) = E\psi(x)$:

$$H[\psi(x) \pm \psi(-x)] = E[\psi(x) \pm \psi(-x)]. \qquad (12.56)$$

Thus the problem has a two-fold *degeneracy*: Associated with every eigenvalue E there is an even and an odd eigenfunction and

$$E \to E_n = \frac{\hbar^2}{2m_0} \frac{n^2 \pi^2}{4a^2}.$$

(b)   Next we consider the case $V_0 > E$. We set

$$k^2 = \frac{2m_0 E}{\hbar^2}, \quad -\kappa^2 = \frac{2m_0(E - V_0)}{\hbar^2} < 0. \qquad (12.57)$$

We restrict ourselves again to the period $-2a \leq x \leq 2a$. We can write the solution as consisting of the following pieces:

$$\psi(x) = \begin{cases} A \sin k(x+3a) & \text{with} \quad x \in [-2a, -a], \\ Be^{\kappa x} + Ce^{-\kappa x} & \text{with} \quad x \in [-a, a], \\ D \sin k(3a-x) & \text{with} \quad x \in [a, 2a]. \end{cases} \tag{12.58}$$

This wave function is no longer restricted to the interior of a box. We therefore have different boundary conditions. We demand these for even and odd wave functions $\psi_{\pm}(x)$ which we can clearly construct from $\psi(x)$ (as examples see the dotted lines in Fig. 12.2). We expect, of course, that these are such that for $V_0 \to \infty$ they approach those of the previous case. We obtain even and odd functions $\psi_{\pm}$ by setting

$$\frac{A}{D} = \pm 1, \quad \frac{B}{C} = \pm 1. \tag{12.59}$$

This means that

$$\psi_{\pm}(x) = \begin{cases} A \sin k(x+3a) & \text{with} \quad x \in [-2a, -a], \\ Be^{\kappa x} \pm Be^{-\kappa x} & \text{with} \quad x \in [-a, a], \\ \pm A \sin k(3a-x) & \text{with} \quad x \in [a, 2a]. \end{cases} \tag{12.60}$$

The boundary conditions that $\psi_{\pm}$ have to satisfy, which, however, we shall not exploit here, are

$$\psi_{-}(0) = 0, \quad \psi_{+}(0) = \text{const.}, \quad \psi'_{+}(0) = 0, \quad \psi'_{-}(0) = \text{const.}, \tag{12.61}$$

so that (W meaning Wronskian)

$$W[\psi_{+}, \psi_{-}] \neq 0.$$

The central region of the solution is the classically forbidden domain.

At $x = \pm a$ we now have to connect ψ and ψ' of neighbouring domains. At $x = -a$ the connecting relations are

$$\begin{aligned} A \sin(k2a) &= Be^{-\kappa a} \pm Be^{\kappa a} = \pm[Be^{\kappa a} \pm Be^{-\kappa a}], \\ kA \cos(k2a) &= \kappa Be^{-\kappa a} \mp \kappa Be^{\kappa a} = \mp \kappa[Be^{\kappa a} \mp Be^{-\kappa a}], \end{aligned} \tag{12.62}$$

and at $x = a$:

$$\begin{aligned} Be^{\kappa a} \pm Be^{-\kappa a} &= \pm A \sin(k2a), \\ \kappa Be^{\kappa a} \mp \kappa Be^{-\kappa a} &= \mp kA \cos(k2a). \end{aligned} \tag{12.63}$$

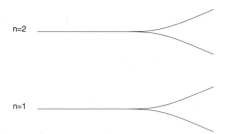

Fig. 12.3 Level degeneracy removed by tunneling.

In general one now rewrites the equations in matrix form. However, in the present case it is easier to continue with the above equations. Taking the ratio of the equations in both cases, we obtain

$$-\frac{\kappa}{k}\tan(2ak) = \frac{e^{\kappa a} \pm e^{-\kappa a}}{e^{\kappa a} \mp e^{-\kappa a}} = \begin{cases} \coth(\kappa a) \\ \tanh(\kappa a) \end{cases}. \qquad (12.64)$$

These are transcendental equations for the determination of the eigenvalues of the even and odd eigenfunctions. For large values of V_0, i.e. for $V_0 \to \infty$, we expect the eigenvalues to approach asymptotically those of the case $V_0 = \infty$. It is plausible therefore to use appropriate expansions, i.e. we set

$$\tanh x = 1 - 2e^{-2x} + O(e^{-4x}) \quad \text{and} \quad \coth x = 1 + 2e^{-2x} + O(e^{-4x}), \quad (12.65)$$

and we expand $\tan(2ak)$ about $n\pi$, the limiting value for $\kappa = \infty$:

$$\begin{aligned} \tan(2ak) &= \tan(n\pi) + \frac{(2ak - n\pi)}{1!}\frac{1}{\cos^2(n\pi)} + \cdots \\ &= 2ak - n\pi + O[(2ak - n\pi)^2]. \end{aligned} \qquad (12.66)$$

Inserting these expansions into Eq. (12.64) we obtain

$$-\frac{\kappa}{k}[(2ak - n\pi) + O[(2ak - n\pi)^2]] = 1 \pm 2e^{-2\kappa a},$$

i.e. (with re-insertion of $n\pi/2a$ for k)

$$2ak \simeq n\pi - \frac{n\pi}{2a\kappa}\left(1 \pm 2e^{-2\kappa a}\right) + O\left(\frac{n\pi}{2a\kappa}\right). \qquad (12.67)$$

The value of k belonging to the upper sign, say k_{even}, is the one belonging to the even solution, and is — as we see — less than that belonging to the lower sign case, i.e. the odd solution with, say, $k = k_{\text{odd}}$. Thus we observe a splitting of the asymptotically degenerate eigenvalues:

$$\frac{k_{\text{even}}}{k_{\text{odd}}} = k_{\text{asympt.}} \mp \triangle k, \quad \text{where} \quad \triangle k \propto e^{-2\kappa a}. \qquad (12.68)$$

Thus the effect of tunneling — here an exponentially small contribution — removes the degeneracy of the asymptotically degenerate states, as indicated in Fig. 12.3.

The following example is an attempt to transcribe the quantum mechanical effect into a macroscopic situation, and is a reformulated version of a problem set in lectures of Fermi.**

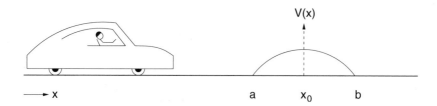

Fig. 12.4 A car meeting a bump in the road.

Example 12.1: **A car's quantum mechanical probability to pass a bump**
A very slowly moving car of mass m_0 (kinetic energy almost zero) encounters on the road between $x = a$ and $x = b, b > a$, a sinusoidal bump in the road with a height of one meter at its peak as indicated in Fig. 12.4. Classically the car is unable to overcome the bump, but quantum mechanically there is a finite probability for this to succeed. Compute this probability.

Solution: The wave functions ψ_0, ψ_V in regions $V = 0, V \neq 0$ are given by the equations

$$\psi_0'' + \frac{2m_0 E}{\hbar^2}\psi_0 = 0, \quad \psi_V'' + \frac{2m_0}{\hbar^2}[E - V(x)]\psi_V = 0.$$

Hence in dominant order

$$\psi_0 \simeq \exp\left[\pm i \sqrt{\frac{2m_0 E}{\hbar^2}} x\right], \quad \psi_V \simeq \exp\left[-\int^x \sqrt{\frac{2m_0}{\hbar^2}(V(x) - E)} dx\right].$$

Thus with $E \sim 0$ the probability P is given by the square of the *transmission amplitude* divided by the square of the incident amplitude, i.e.

$$P = \left|\frac{\psi_V}{\psi_0}\right|^2 \simeq \exp\left[-2\sqrt{\frac{2m_0}{\hbar^2}} \int_a^b dx \sqrt{V(x)}\right], \tag{12.69}$$

where the potential $V(x)$ is (with g the acceleration due to gravity)

$$V(x) = m_0 g y(x), \quad y(x) = \sin\left[\pi \frac{x - a}{b - a}\right],$$

$$y'(x_0) = 0, \quad x_0 = \frac{1}{2}(a + b) \quad \text{and} \quad y(x_0) = 1. \tag{12.70}$$

Setting

$$w(x) = \pi \frac{x - a}{b - a}, \quad w(x_0) = \frac{\pi}{2},$$

**E. Fermi [92], p. 57.

we have

$$\int_a^b dx \sqrt{V(x)} = \sqrt{m_0 g} \frac{2(b-a)}{\pi} \int_0^{\pi/2} dw \sin^{1/2} w. \tag{12.71}$$

The integral can be looked up in Tables of Integrals. We find[tt] (with $(-1/2)! = \sqrt{\pi}$)

$$\int_0^{\pi/2} dx \sin^{1/2} x = \sqrt{\frac{2}{\pi}} \left[\left(-\frac{1}{4} \right)! \right]^2.$$

Looking up the value of the factorial in Tables,[tt], we obtain $(-1/4)! = (4/3)(3/4)! = (4/3) \times 0.9191 \simeq 1.2$. We can now evaluate the probability. We have

$$P \simeq \exp\left[-\frac{8m_0}{\hbar} \sqrt{g} \frac{(b-a)}{\pi^{3/2}} \left(\frac{4}{3} \times 0.9191 \right)^2 \right]. \tag{12.72}$$

Assuming a mass of 1 ton of the car (i.e. 1000 kg) and using the following values of the natural constants: $g = 9.807 \text{ m s}^{-2}$, $\hbar = 1.055 \times 10^{-34} \text{ J s}$, and assuming a length of $b - a = 100$ m of the base of the bump, one evaluates an approximate probability of

$$\exp[-6.4 \times 10^{39}].$$

[tt]I. S. Gradshteyn and I. M. Ryzhik [122], formulas 3.621.1, p. 369 and 8.384.1, p. 950, which give

$$\int_0^{\pi/2} dx \sin^{\mu-1} x = 2^{\mu-2} B\left(\frac{\mu}{2}, \frac{\mu}{2} \right) = 2^{\mu-2} \frac{[(\mu/2 - 1)!]^2}{(\mu - 1)!},$$

where $B(x, y)$ is the *beta function* (see Eq. (15.17)).
[tt]For instance E. Jahnke and F. Emde [143], p. 14.

Chapter 13

Linear Potentials

13.1 Introductory Remarks

We distinguish here between three different types of linear potentials. In the present chapter we consider in some detail the first of these which is the potential of a freely falling particle in one space dimension. This is the *problem of Galilei in quantum mechanics.*[*] In Chapter 14, Example 14.3, we consider the corresponding case of a particle above the flat surface of the Earth. Finally, in Chapter 15, we consider the linear potential in three space dimensions that forbids the particle to escape. Thus, whereas the first case permits only a continuous spectrum, the last case allows only a discrete spectrum. An additional reason to consider the linear potential at this stage is its appearance in the following chapter in connection with the matching of WKB solutions across a turning point. Our treatment in the present chapter aims predominantly at an exploration of the quantum mechanics of the freely falling particle in a domain close to its classical behaviour.

13.2 The Freely Falling Particle: Quantization

Under quantization of the freely falling particle we understand here the solution of the Schrödinger equation for the linear potential. This is the quantum mechanical version of the problem of Galilei in one space dimension. We shall see that in this particular case the wave function can be written as a superposition of de Broglie waves, each of which propagates with the classical momentum and energy of a Galileian particle. Then considering the probability distribution determined by the squared modulus of the wave function,

---

[*]A highly abridged treatment of this problem is given, for instance, in the book of G. Süssmann [266], p. 144.

we shall see that at large times t its behaviour is determined entirely by the parameters describing the classical motion of the Galileian particle.

13.2.1   Superposition of de Broglie waves

We consider a particle of mass m_0 falling freely in the time-independent homogeneous field of the gravitational force

$$F(x,t) = m_0 g$$

with potential

$$V(x,t) = -m_0 g x + V_0, \tag{13.1}$$

where g is the acceleration due to gravity and we can put $V_0 = 0$. We see that the case $g < 0$ can be related to the reflection $x \to -x$. A different "*linear potential*" is the so-called "*confinement potential*"

$$V(x,t) = c|x| + V_0, \quad c > 0, \tag{13.2}$$

which is the case with a discrete spectrum. We consider here the case of Eq. (13.1). Recall the classical treatment of the freely falling particle. The equation of motion follows for instance from the derivative of the constant energy, i.e. from

$$\frac{d}{dx}\left[\frac{1}{2}m_0\dot{x}^2 + V(x)\right] = 0, \quad V(x) = -m_0 g x, \quad x > 0, \quad g > 0,$$

and yields $m_0 \ddot{x} = m_0 g$. In the present case of quantum mechanics we have

$$i\hbar\frac{\partial}{\partial t}\psi(x,t) = \left[-\frac{\hbar^2}{2m_0}\frac{\partial^2}{\partial x^2} - m_0 g x\right]\psi(x,t). \tag{13.3}$$

In general the momentum representation is of little importance. The present case is an exception and a good example of its applicability. Therefore we use the Fourier transforms

$$\psi(x,t) = \frac{1}{\sqrt{2\pi}}\int_{-\infty}^{\infty} dk\, e^{ikx}\tilde{\psi}(k,t), \tag{13.4a}$$

$$\tilde{\psi}(k,t) = \frac{1}{\sqrt{2\pi}}\int_{-\infty}^{\infty} dx\, e^{-ikx}\psi(x,t), \tag{13.4b}$$

and consider the following partial integration in order to re-express the last term in Eq. (13.3) as a partial derivative $\partial/\partial k$:

$$\frac{1}{\sqrt{2\pi}}\int dk e^{ikx}(im_0 g)\frac{\partial}{\partial k}\tilde{\psi}(k,t)$$

$$= \frac{e^{ikx}}{\sqrt{2\pi}}(im_0 g)\tilde{\psi}(k,t)\Big|_{k=-\infty}^{k=\infty} - \int dk\frac{d}{dk}\left[\frac{1}{\sqrt{2\pi}}e^{ikx}(im_0 g)\right]\tilde{\psi}(k,t)$$

$$= \frac{m_0 g x}{\sqrt{2\pi}}\int dk e^{ikx}\tilde{\psi}(k,t)$$

$$= m_0 g x\psi(x,t).$$

The vanishing of the boundary contributions is to be understood in the sense of distribution theory (i.e. multiplication by a test function and subsequent integration). Equation (13.3) yields us now by inserting for $\psi(x,t)$ the expression (13.4a)

$$i\hbar\frac{\partial}{\partial t}\tilde{\psi}(k,t) = \frac{\hbar^2 k^2}{2m_0}\tilde{\psi}(k,t) - im_0 g\frac{\partial}{\partial k}\tilde{\psi}(k,t). \tag{13.5}$$

We rewrite this equation in the following form

$$\left(i\hbar\frac{\partial}{\partial t} + im_0 g\frac{\partial}{\partial k}\right)\tilde{\psi}(k,t) = \frac{\hbar^2 k^2}{2m_0}\tilde{\psi}(k,t), \tag{13.6}$$

and wish to solve it. This can be achieved by first converting this partial differential equation with two partial derivatives into an ordinary differential equation with the one total derivative d/dt. To this end we consider the following substitution in Eq. (13.6) (and to avoid confusion we do not introduce new functions and variables):

$$k \to k + \frac{m_0 g t}{\hbar}. \tag{13.7}$$

In the wave function on the left hand side of Eq. (13.6) the argument k is then replaced by this. Applying the total derivative d/dt to this function, we obtain the differential operator on the left of Eq. (13.6), i.e.

$$i\hbar\frac{d}{dt}\tilde{\psi}\left(k + \frac{m_0 g t}{\hbar}, t\right) = \left(i\hbar\frac{\partial}{\partial t} + im_0 g\frac{\partial}{\partial k}\right)\tilde{\psi}\left(k + \frac{m_0 g t}{\hbar}, t\right). \tag{13.8}$$

With this result and the substitution (13.7) applied to every quantity in Eq. (13.6) the latter equation becomes

$$i\hbar\frac{d}{dt}\tilde{\psi}\left(k + \frac{m_0 g t}{\hbar}, t\right) = \frac{\hbar^2}{2m_0}\left(k + \frac{m_0 g t}{\hbar}\right)^2\tilde{\psi}\left(k + \frac{m_0 g t}{\hbar}, t\right). \tag{13.9}$$

This is now an ordinary differential equation of the first order and can immediately be integrated to give an exponential, i.e. the following result

$$\tilde{\psi}\left(k+\frac{m_0 g t}{\hbar},t\right) = \tilde{\psi}_0(k) \exp\left[\frac{\hbar}{2m_0 i}\int_0^t dt'\left(k+\frac{m_0 g t'}{\hbar}\right)^2\right]. \qquad (13.10)$$

Note the amplitude function $\tilde{\psi}_0(k)$ in front. We have thus obtained the solution of the transformed equation. To obtain the solution of the original equation (13.6), we have to reverse the substitutions and replace in (13.10) k by

$$\text{back}: \quad k \to k - \frac{m_0 g t}{\hbar}.$$

The solution of Eq. (13.6) is then

$$\begin{aligned}
\tilde{\psi}(k,t) &= \tilde{\psi}_0\left(k-\frac{m_0 g t}{\hbar}\right)\exp\left[\frac{\hbar}{2m_0 i}\int_0^t dt'\left\{k+\frac{m_0 g}{\hbar}(t'-t)\right\}^2\right] \\
&= \tilde{\psi}\left(k-\frac{m_0 g t}{\hbar},t\right).
\end{aligned} \qquad (13.11)$$

In Example 13.1 we verify that this solution indeed solves Eq. (13.6). We observe that the solution is actually a function of $k_t = k - m_0 g t/\hbar$.

Example 13.1: **Verification of momentum representation solution**
Verify by direct differentiation that the solution (13.11) solves the partial differential equation (13.6).

Solution: We apply the two operator expressions on the left of Eq. (13.6) to the solution (13.11). Thus we have first (the second term arising from the upper integration limit, the third term from the integrand)

$$i\hbar\frac{\partial}{\partial t}\tilde{\psi}(k,t) = i\hbar\frac{\partial\tilde{\psi}_0}{\partial k}\left(\frac{-m_0 g}{\hbar}\right)\exp\left[\cdots\right] + \frac{\hbar^2}{2m_0}k^2\tilde{\psi}(k,t) - g\hbar\int_0^t dt'\left\{k+\frac{m_0 g}{\hbar}(t'-t)\right\}\tilde{\psi}(k,t).$$

On the other hand the second operator expression implies

$$im_0 g\frac{\partial}{\partial k}\tilde{\psi}(k,t) = im_0 g\frac{\partial\tilde{\psi}_0}{\partial k}\exp\left[\cdots\right] + g\hbar\int_0^t dt'\left\{k+\frac{m_0 g}{\hbar}(t'-t)\right\}\tilde{\psi}(k,t).$$

Thus addition of both indeed leaves the expression on the right hand side of Eq. (13.6).

Our next step is to re-express the solution (13.11) in a neater form. We define the wave number k_t as

$$k_t := k - \frac{m_0 g}{\hbar}t. \qquad (13.12)$$

Then the integral in the exponential of the solution (13.11) can be written

$$I := \int_0^t dt' \left\{ k + \frac{m_0 g}{\hbar}(t' - t) \right\}^2 = \int_0^t dt' \left\{ k_t + \frac{m_0 g}{\hbar}t' \right\}^2$$

$$= t k_t^2 - \frac{m_0 k_t}{\hbar}t^2 + \frac{m_0^2 g^2}{\hbar^2}\frac{t^3}{3}. \tag{13.13}$$

Inserting here the explicit expression for k_t, we can write the integral

$$I = -\frac{1}{3}\frac{\hbar}{m_0 g}(k_t)^3 + \frac{1}{3}\frac{\hbar}{m_0 g}\left(k_t - \frac{m_0 g}{\hbar}t \right)^3$$

$$= -\frac{1}{3}\frac{\hbar}{m_0 g}\left(k - \frac{m_0 g}{\hbar}t \right)^3 + \frac{1}{3}\frac{\hbar}{m_0 g}k^3. \tag{13.14}$$

With the definition

$$I_t(k) := \frac{1}{3}\left(k - \frac{m_0 g}{\hbar}t \right)^3 \tag{13.15}$$

we obtain

$$I = -\frac{\hbar}{m_0 g}[I_t(k) - I_0(k)]. \tag{13.16}$$

Next we evaluate the Fourier transform of $\tilde{\psi}(k, t)$ defined by Eq. (13.4a). Inserting $\tilde{\psi}(k, t)$ into this integral and recalling the property (13.11), we obtain

$$\psi(x, t) = \frac{1}{\sqrt{2\pi}} \int_{-\infty}^{\infty} dk e^{ik_t x} \tilde{\psi}(k_t, t)$$

$$= \frac{1}{\sqrt{2\pi}} \int_{-\infty}^{\infty} dk\, \tilde{\psi}_0 \left(k - \frac{m_0 g}{\hbar}t \right) \exp\left[ik_t x \right.$$

$$\left. - \frac{\hbar^2}{2m_0^2 ig}[I_t(k) - I_0(k)] \right]. \tag{13.17}$$

Inserting the explicit expression for the argument of the exponential, we have

$$\psi(x, t) = \frac{1}{\sqrt{2\pi}} \int_{-\infty}^{\infty} dk \tilde{\psi}_0 \left(k - \frac{m_0 g t}{\hbar} \right) \exp\left[ik_t x \right.$$

$$\left. - \frac{\hbar^2}{2m_0^2 ig}\left\{ -\frac{m_0 g t}{\hbar}k^2 + \frac{m_0^2 g^2 t^2}{\hbar^2}k - \frac{m_0^3 g^3 t^3}{3\hbar^3} \right\} \right], \tag{13.18}$$

With

$$k_t = k - \frac{m_0 g t}{\hbar}, \quad \omega_t = \frac{\hbar}{2m_0}k_t^2 = \frac{\hbar^2}{2m_0}k^2 - gkt + \frac{g^2 m_0}{2\hbar}t^2, \tag{13.19a}$$

and

$$
\begin{aligned}
\int_0^t dt'\omega_{t'} &= \frac{\hbar}{2m_0}k^2t - gk\frac{t^2}{2} + \frac{g^2m_0}{2\hbar}\frac{t^3}{3} \\
&= -\frac{\hbar^2}{2m_0^2g}\left[-\frac{m_0gk^2}{\hbar}t + \frac{m_0^2g^2}{\hbar^2}kt^2 - \frac{g^3m_0^3}{\hbar^3}\frac{t^3}{3}\right], \text{(13.19b)}
\end{aligned}
$$

we can rewrite $\psi(x,t)$ as

$$
\psi(x,t) = \frac{1}{\sqrt{2\pi}}\int_{-\infty}^{\infty} dk\tilde{\psi}_0(k_t))\exp\left[i\left\{ k_t x - \int_0^t \omega_{t'}dt'\right\}\right]. \tag{13.20}
$$

This result can be interpreted as follows. $\tilde{\psi}_0(k)$ is the probability amplitude with wave vector k at time $t = 0$ and varies in the course of time with the time-dependent wave vector $k_t = k - m_0gt/\hbar$. These wave vectors vary in accordance with Eq. (13.19a) exactly as for a falling particle. With the help of the de Broglie relation this wave vector corresponds to the particle momentum $p_t = p - m_0gt$ or to its velocity $v_t = v - gt, v = k\hbar/m_0$. The associated kinetic energy of the particle is

$$
\frac{1}{2}mv_t^2 = \hbar\omega_t. \tag{13.21}
$$

In other words, the relation (13.20) expresses the wave function $\psi(x,t)$ as a decomposition into de Broglie waves. Such a decomposition is possible only for homogeneous force fields.

13.2.2 Probability distribution at large times

Next we inquire about the behaviour of the probability density $|\psi(x,t)|^2$ for $t \to \infty$ (distant future or past). To this end we convert Eq. (13.18) into a different form in order to be able to perform the integration with respect to k. The k-dependent terms in the argument of the exponential are

$$
T := ikx - \frac{\hbar^2}{2m_0^2ig}\left\{ -\frac{m_0gt}{\hbar}k^2 + \frac{m_0^2g^2t^2}{\hbar^2}k\right\}. \tag{13.22}
$$

We can rearrange these terms in the following way with the k-dependent contributions contained in a quadratic form:

$$
\begin{aligned}
T &= \frac{\hbar t}{2m_0i}\left(k - \frac{m_0x}{\hbar t} - \frac{m_0gt}{2\hbar}\right)^2 + ix\left(\frac{m_0gt}{2\hbar}\right) \\
&\quad - \frac{\hbar t}{2m_0i}\left\{ \left(\frac{m_0gt}{2\hbar}\right)^2 + \left(\frac{m_0x}{\hbar t}\right)\right\}.
\end{aligned} \tag{13.23}
$$

The integral (13.18) thus becomes

$$
\psi(x,t) = \frac{1}{\sqrt{2\pi}} \int_{-\infty}^{\infty} dk \tilde{\psi}_0(k_t) \exp\left[-ix\left(\frac{m_0 g t}{\hbar}\right) + \frac{m_0 g^2 t^3}{6i\hbar} + T \right]
$$

$$
= \frac{1}{\sqrt{2\pi}} \exp\left[-ix\left(\frac{m_0 g t}{2\hbar}\right) + \frac{m_0 g^2 t^3}{24i\hbar} - \frac{m_0 x^2}{2\hbar t i} \right]
$$

$$
\times \int_{-\infty}^{\infty} dk \tilde{\psi}_0(k_t) \exp\left[\frac{\hbar t}{2m_0 i}\left(k - \frac{m_0 x}{\hbar t} - \frac{m_0 g t}{2\hbar} \right)^2 \right]. \quad (13.24)
$$

We now set

$$
\xi := \sqrt{\frac{\hbar t}{2m_0}}\left(k - \frac{m_0 x}{\hbar t} - \frac{m_0 g t}{2\hbar} \right), \quad k = \frac{m_0 x}{\hbar t} + \frac{m_0 g t}{2\hbar} + \sqrt{\frac{2m_0}{\hbar t}}\xi. \quad (13.25)
$$

The amplitude $\tilde{\psi}_0(k_t)$ then becomes for $|t| \to \infty$:

$$
\tilde{\psi}_0(k_t)\Big|_{|t|\to\infty} = \tilde{\psi}_0\left(k - \frac{m_0 g t}{\hbar} \right)\Big|_{|t|\to\infty} \simeq \tilde{\psi}_0\left(\frac{m_0 x}{\hbar t} - \frac{m_0 g t}{2\hbar} \right),
$$

and, being now independent of ξ, may be put in front of the integral. Then

$$
\psi(x,t) \simeq \frac{1}{\sqrt{2\pi}} \exp\left[-ix\frac{m_0 g t}{2\hbar} + \frac{m_0 g^2 t^3}{24i\hbar} - \frac{m_0 x^2}{2\hbar t i} \right]
$$

$$
\times \sqrt{\frac{2m_0}{\hbar t}}\tilde{\psi}_0\left(\frac{m_0 x}{\hbar t} - \frac{m_0 g t}{2\hbar} \right) \int_{-\infty}^{\infty} d\xi e^{-i\xi^2}. \quad (13.26)
$$

For $|t| \to \infty$ the definition of ξ defines a distance $x_{\rm cl}$ given by

$$
k - \frac{m_0 x_{\rm cl}}{\hbar t} - \frac{m_0 g t}{2\hbar} = O\left(\frac{\xi}{\sqrt{|t|}} \right) \to 0,
$$

i.e.
$$
x_{\rm cl} = vt - \frac{1}{2}gt^2, \quad v = \frac{k\hbar}{m_0}. \quad (13.27)
$$

Thus $x_{\rm cl}$ is the distance travelled by the particle as determined in classical mechanics. We evaluate the *Fresnel integral* in Eq. (13.26) with the formula

$$
\int_{-\infty}^{\infty} d\xi e^{i\alpha\xi^2} = \left(\frac{i\pi}{\alpha} \right)^{1/2} \quad (13.28)
$$

(for a verification replace α by $\alpha + i\mu, \mu > 0$, integrate and then take the limit $\mu \to 0$). It follows that

$$
\psi(x,t) \simeq \sqrt{\frac{m_0}{i\hbar t}} \exp\left[i\left(-\frac{m_0 g x t}{2\hbar} + \frac{m_0 x^2}{2\hbar t} - \frac{m_0 g^2 t^3}{24\hbar} \right) \right] \tilde{\psi}_0\left(\frac{m_0 x}{\hbar t} - \frac{m_0 g t}{2\hbar} \right).
$$

$$
(13.29)
$$

Replacing here x by the large time value $x_{\rm cl}$, this expression can be brought into the following form:

$$\psi(x,t) \simeq \sqrt{\frac{m_0}{i\hbar t}} \exp\left[i\left(k_t x_{\rm cl} - \int_0^t \omega_{t'} dt'\right)\right] \tilde{\psi}_0(k_t). \tag{13.30}$$

We see therefore: The large time asymptotic behaviour of the wave function is determined entirely by the classical values of x (i.e. $x_{\rm cl}$), k_t and ω_t.

13.3 Stationary States

In this section we show that the Schrödinger equation with the potential (13.1) has only a continuous spectrum. For *stationary states* we set

$$\psi(x,t) = e^{-iEt/\hbar}\phi(x) \tag{13.31}$$

and obtain from Eq. (13.3) the time-independent Schrödinger equation

$$\frac{d^2\phi(x)}{dx^2} + \frac{2m_0}{\hbar^2}[E + m_0 g x]\phi(x) = 0. \tag{13.32}$$

For $x \to \infty$ the function $\phi(x)$ behaves like

$$\phi(x) \quad \propto \quad \exp\left[\pm i \int^x dx\left\{\frac{2m_0}{\hbar^2}(E + m_0 g x)\right\}^{1/2}\right]$$
$$\sim \quad \exp\left[\pm \frac{2i}{3}(2m_0^2 g/\hbar^2)^{1/2} x^{3/2}\right] \tag{13.33}$$

for $E \ll |m_0 g x|$. These are incoming and outgoing waves. For $x \to -\infty$ exponentially increasing and decreasing solutions are possible.

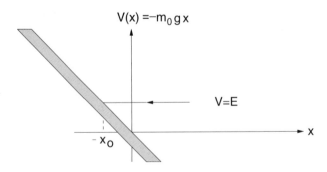

Fig. 13.1 A turning point at $x = -x_0$.

Classically the particle coming in from the right with energy E cannot penetrate into the potential barrier since (where $V > E$) with conservation of energy $E - V = p^2/2m$, its momentum p would be imaginary there, which is not possible for real particles. However quantum mechanically this is possible with a small, i.e. exponentially decreasing, probability. The point $-x_0$, cf. Fig. 13.1, at which $E - V = 0$, i.e. at which the classical particle bounces back, i.e. is reflected, is therefore called a *turning point*. Quantum mechanically this is a point at which periodic solutions of the Schrödinger equation go over into exponential solutions or vice versa. Thus the spectrum does not contain states which are normalizable over the entire domain $x \in [-\infty, \infty]$, and hence is continuous and extends from $-\infty$ to $+\infty$. In view of the simple form of the potential in the present case the equation can be integrated in the momentum or k representation, and we can establish for the resulting continuum solutions the orthonormality and completeness relations. The *Fourier transform* is given by the relations (13.4a) and (13.4b), i.e.

$$\psi(x,t) = \frac{1}{\sqrt{2\pi}} \int_{-\infty}^{\infty} dk e^{ikx} \tilde{\psi}(k,t). \qquad (13.34a)$$

Then

$$i\hbar \frac{\partial}{\partial t} \psi(x,t) = E\psi(x,t) = \frac{1}{\sqrt{2\pi}} \int_{-\infty}^{\infty} dk e^{ikx} E\tilde{\psi}(k,t),$$

but also

$$i\hbar \frac{\partial}{\partial t} \psi(x,t) = \frac{1}{\sqrt{2\pi}} \int_{-\infty}^{\infty} dk e^{ikx} i\hbar \frac{\partial \tilde{\psi}(k,t)}{\partial t},$$

so that

$$i\hbar \frac{\partial \tilde{\psi}(k,t)}{\partial t} = E\tilde{\psi}(k,t). \qquad (13.34b)$$

It therefore follows from Eq. (13.6) that also for $\tilde{\psi}(k,t) \to \tilde{\phi}(k)$

$$\left(E + im_0 g \frac{\partial}{\partial k} \right) \tilde{\phi}(k) = \frac{\hbar^2 k^2}{2m_0} \tilde{\phi}(k). \qquad (13.35)$$

It follows that

$$\int \frac{d\tilde{\phi}}{\tilde{\phi}} = \int \frac{1}{im_0 g} \left(\frac{\hbar^2 k^2}{2m_0} - E \right) dk,$$

i.e.

$$\tilde{\phi}(k) = \tilde{c} \exp\left[\frac{i}{m_0 g} \left(Ek - \frac{\hbar^2 k^3}{6m_0} \right) \right] \qquad (13.36)$$

with $\tilde{c} = $ const. We determine the constant from the condition that the continuum solutions are to be orthonormalized to a delta function, i.e. we have

$$\int_{-\infty}^{\infty} dx \phi_E^*(x) \phi_{E'}(x) = \delta(E - E').$$ (13.37)

Using

$$\delta(x) = \frac{1}{2\pi} \int e^{ikx} dk,$$

one verifies that with

$$\phi_E(x) = \frac{1}{\sqrt{2\pi}} \int dk e^{ikx} \tilde{\phi}_E(k), \quad \tilde{\phi}_E(k) = \frac{1}{\sqrt{2\pi}} \int dx e^{-ikx} \phi_E(x),$$ (13.38)

we have for $\tilde{\phi}_E(k)$:

$$\int_{-\infty}^{\infty} dk \tilde{\phi}_E^*(k) \tilde{\phi}_{E'}(k) = \int_{-\infty}^{\infty} dx \phi_E^*(x) \phi_{E'}(x) = \delta(E - E').$$ (13.39)

Inserting (13.36) into (13.39) we obtain

$$\int dk \tilde{c} \tilde{c}^* \exp\left[\frac{i}{m_0 g} k(E - E')\right] = \delta(E - E').$$

It follows that

$$\tilde{c} \tilde{c}^* = \frac{1}{2\pi m_0 g}, \quad \tilde{c} = \frac{1}{\sqrt{2\pi m_0 g}},$$ (13.40)

i.e. apart from a constant phase which we choose to be zero. In a similar way we can verify the validity of the completeness relation, i.e. the relations

$$\int_{-\infty}^{\infty} dE \tilde{\phi}_E(k) \tilde{\phi}_E^*(k') = \delta(k - k'), \quad \int_{-\infty}^{\infty} dE \phi_E(x) \phi_E^*(x') = \delta(x - x').$$ (13.41)

Next we insert (13.36) (with $\tilde{c}$ replaced by (13.40)) into (13.38), so that

$$\phi_E(x) = \frac{1}{\sqrt{2\pi}} \int dk e^{ikx} \frac{1}{\sqrt{2\pi m_0 g}} \exp\left[\frac{i}{m_0 g}\left(Ek - \frac{\hbar^2 k^3}{6m_0}\right)\right].$$ (13.42)

This integral can be rewritten in terms of the *Airy function Ai(s)*. The Airy function $Ai(s)$ can be defined by the following integral:

$$Ai(s) = \frac{1}{2\pi} \int_{-\infty}^{\infty} dt \exp\left[i\left(st - \frac{1}{3}t^3\right)\right] = \frac{1}{\pi} \int_0^{\infty} dt \cos\left(\frac{1}{3}t^3 - st\right).$$ (13.43)

Setting

$$y = \frac{2m_0^2 g}{\hbar^2} x + \frac{2m_0}{\hbar^2} E,$$

we can write $\phi_E(x)$:

$$\phi_E(x) = \frac{1}{2\pi\sqrt{m_0 g}} \int dk \exp\left[i\frac{\hbar^2}{2m_0^2 g}\left(yk - \frac{1}{3}k^3\right)\right].$$

With the substitution $k = \mu t$ we change the variable to t, so that the integral becomes

$$\phi_E(x) = \frac{1}{2\pi}\frac{\mu}{\sqrt{m_0 g}} \int_{-\infty}^{\infty} dt \exp\left[i\frac{\hbar^2}{2m_0^2 g}\mu^3\left\{\frac{y}{\mu^2}t - \frac{1}{3}t^3\right\}\right].$$

Choosing

$$\mu = \left(\frac{2m_0^2 g}{\hbar^2}\right)^{1/3}, \quad \text{and setting} \quad s = \frac{y}{\mu^2} = \frac{1}{\mu^2}\left[\mu^3 x + \frac{2m_0}{\hbar^2}E\right], \quad (13.44)$$

the function $\phi_E(x)$ is then expressed in terms of the Airy function:

$$\phi_E(x) = \frac{\mu}{\sqrt{m_0 g}} Ai\left(\mu x + \frac{2m_0}{\hbar^2}\frac{E}{\mu^2}\right). \quad (13.45)$$

One can now derive normalization and completeness integrals for the Airy functions. For instance

$$\int dx\, Ai(y + x)Ai^*(x + z) = \delta(y - z). \quad (13.46)$$

This follows since with Eq. (13.43)

$$\int dx\, Ai(y + x)Ai^*(x + z)$$

$$= \frac{1}{(2\pi)^2} \int ds \int dt \exp\left[i\left\{s(y + x) - \frac{1}{3}s^3\right\}\right]\exp\left[-i\left\{t(x + z) - \frac{1}{3}t^3\right\}\right]$$

$$= \int \frac{ds}{2\pi} \int \frac{dt}{2\pi} \int dx\, e^{ix(s-t)}e^{i(sy-tz)}e^{-\frac{i}{3}(s^3-t^3)}$$

$$= \int \frac{ds}{2\pi} \int dt\, \delta(s - t)e^{i(sy-tz)}e^{-\frac{i}{3}(s^3-t^3)}$$

$$= \int \frac{dt}{2\pi}e^{it(y-z)} = \delta(y - z).$$

We will encounter the Airy function $Ai(s)$ again in Chapter 14 in the matching of exponential WKB solutions to periodic WKB solutions, and in Chapter 15 in the computation of energy eigenvalues for the three-dimensional linear potential. In the next section we derive the important asymptotic expansions of $Ai(s)$ for both positive and negative values of s, and we shall see that one branch has periodic behaviour, whereas the other is of exponential type. For a further solution Bi of Airy functions we refer in Chapter 14 to Tables.

13.4 The Saddle Point or Stationary Phase Method

The integrand of Eq. (13.43) contains the phase (with $x \to z$):

$$\Phi(z) := sz - \frac{1}{3}z^3. \tag{13.47}$$

Consider the following integral along some as yet unspecified contour C in the plane of complex z:

$$I := \frac{1}{2\pi} \int_C dz \, e^{i\Phi(z)} \quad \text{with} \quad z = re^{i\theta}. \tag{13.48}$$

We have

$$\Phi(z) = sre^{i\theta} - \frac{1}{3}r^3 e^{3i\theta}$$
$$= \left(sr\cos\theta - \frac{1}{3}r^3\cos 3\theta \right) + i\left(sr\sin\theta - \frac{1}{3}r^3\sin 3\theta \right). \tag{13.49}$$

We want the integral to exist and so desire that

$$\lim_{r \to \infty} e^{i\Phi(z)} = 0, \tag{13.50}$$

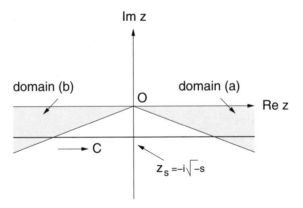

Fig. 13.2 The contour C for $s < 0$ with ends in the angular domains (a) and (b).

Thus we must have

$$\sin 3\theta < 0, \text{ i.e. } -\pi \le 3\theta \le 0, \text{ i.e. } -\frac{1}{3}\pi \le \theta \le 0 \text{ for } r \to \infty, \tag{13.51a}$$

and since with $\sin 3\theta < 0$ also $\sin(3\theta + 2\pi) < 0$:

$$-\pi \le 3\theta + 2\pi \le 0, \text{ or } -\pi \le \theta \le -\frac{2}{3}\pi \text{ for } r \to \infty. \tag{13.51b}$$

The integral (13.45) exists for the contour C, provided its ends at infinity satisfy these conditions. It is reasonable to choose the contour in such a way that only a short piece of it contributes substantially to the integral. Such a piece can be found in the neighbourhood of the so-called *saddle point* z_s, at which the phase becomes stationary, i.e. where the derivative of the phase function vanishes, i.e.

$$0 = \Phi'(z_s) = s - z_s^2, \quad s = z_s^2 \equiv r_s^2 e^{2i\theta_s}. \tag{13.52}$$

Let s be real and $s < 0$. Then $\theta_s = \pm \pi/2$, so that $\exp(\pm 2i\theta_s) = \exp(\pm i\pi) = -1$ and $-s = r_s^2$. Then

$$z_s = r_s e^{i\theta_s} = \sqrt{-s} e^{\pm i\pi/2} = \pm i\sqrt{-s}. \tag{13.53}$$

Since we shall have the contour C in the lower half plane, we choose

$$z_s = -i\sqrt{-s}, \quad s < 0, \tag{13.54}$$

and we set with σ real:

$$z := z_s + \frac{\sigma}{(-s)^{1/4}}, \tag{13.55}$$

where σ is the new variable with $-\infty < \sigma < \infty$. Inserting (13.55) into (13.48), we obtain

$$
\begin{aligned}
I &= \frac{1}{2\pi} \int_C \frac{d\sigma}{(-s)^{1/4}} \exp\left[i\left\{ \Phi(z_s) + \frac{1}{2}(z - z_s)^2 \Phi''(z_s) + \cdots \right\} \right] \\
&\simeq \frac{1}{2\pi} \frac{e^{i\Phi(z_s)}}{(-s)^{1/4}} \int_C d\sigma e^{\frac{i}{2}(z-z_s)^2 \Phi''(z_s)}.
\end{aligned}
\tag{13.56}
$$

But

$$\Phi(z_s) = s z_s - \frac{1}{3} z_s^3 = -i\left[\sqrt{-s} s + \frac{1}{3}(\sqrt{-s})^3 \right] = i\frac{2}{3}(-s)^{3/2},$$

and

$$\Phi''(z_s) = -2z_s \overset{(13.54)}{=} 2i\sqrt{-s}. \tag{13.57}$$

We now choose the path of integration C at fixed $\Im z = -i\sqrt{-s}$ with $\sigma \in [-\infty, \infty]$ as indicated in Fig. 13.2. Then the ends of the contour lie in the domains specified by Eqs. (13.51a) and (13.51b), and the main contribution to the integral comes from the σ region around the saddle point. Hence

$$I = \frac{e^{-\frac{2}{3}(-s)^{3/2}}}{2\pi(-s)^{1/4}} \int_{-\infty}^{\infty} d\sigma e^{-\sigma^2 - \frac{i\sigma^3}{3(-s)^{1/2}} + \cdots}, \tag{13.58}$$

where the contribution $\propto \sigma^3$ is obtained from $\Phi'''(z_s)$ whose evaluation we do not reproduce here. For $(-s) \to \infty$ the integral converges towards $\sqrt{\pi}$, so that

$$Ai(s) \overset{-s \to \infty}{\simeq} \frac{1}{2\sqrt{\pi}(-s)^{1/4}} \exp\left[-\frac{2}{3}(-s)^{3/2}\right]. \qquad (13.59)$$

In Example 13.2 we show that for $s \to \infty$:

$$Ai(s) \overset{s \to \infty}{\simeq} \frac{1}{\sqrt{\pi}s^{1/4}} \cos\left(\frac{2}{3}s^{3/2} - \frac{\pi}{4}\right). \qquad (13.60)$$

The reason for describing the point z_s as a saddle point is that the integral in Eq. (13.56) decreases exponentially around $\sigma = 0$ for real values of σ, but would increase exponentially for imaginary values of σ — thus describing a surface around that point very analogous to that of a saddle. We observe that the Airy function has a periodic behaviour in one direction but an exponential behaviour in the opposite direction. This is an important aspect in the WKB method to be considered in Chapter 14.

With the help of the substitutions (13.44) we can convert Eq. (13.32) with the solution (13.42) into an equation with a more appealing form. According to Eqs. (13.44) and (13.45)

$$\phi_E(x) \sim Ai(z),$$

where — with $\epsilon = (2/\hbar^2 m_0 g^2)^{1/3}$ — we have

$$z = \mu x + \frac{2m_0}{\mu^2} \frac{E}{\hbar^2} = \left(\frac{2}{\hbar^2 m_0 g^2}\right)^{1/3} \left[E + \left(\frac{2\hbar^2 m_0^3 g^3}{2\hbar^2}\right)^{1/3} x\right] = \epsilon[E + m_0 g x]. \qquad (13.61)$$

Equation (13.32) therefore becomes

$$\frac{d^2\phi}{dz^2} + z\phi = 0. \qquad (13.62)$$

This is the *Airy differential equation* with one solution

$$\phi(z) = Ai(z). \qquad (13.63)$$

The Airy function can be re-expressed in terms of cylinder or Bessel functions $Z_\nu(z)$, as one can see from the following equation[†]

$$u'' + \beta^2 \gamma^2 z^{2\beta-2} u = 0 \qquad (13.64)$$

[†]See I. S. Gradshteyn and I. M. Ryzhik [122], Sec. 8.491.

with solution

$$u = z^{1/2} Z_{1/2\beta}(\gamma z^\beta). \tag{13.65}$$

With expansions (13.59) and (13.60) we have thus obtained (with the help of the saddle point method) the asymptotic expansions of the Airy function $Ai(z)$ which play an important role in Chapter 14.

Example 13.2: **Periodic asymptotic behaviour of the Airy function**
Obtain with the saddle point method the asymptotic large-s behaviour of the Airy function $Ai(s)$.

Solution: We have (cf. Eq. (13.43))

$$Ai(s) = \frac{1}{2\pi} \int_C dz\, e^{i\Phi(z)}, \quad \Phi(z) = sz - \frac{1}{3}z^3.$$

This time s is real and positive. Hence the saddle points are given by

$$0 = \Phi'(z_s) = s - z_s^2, \quad z_s = \pm\sqrt{s} \equiv z_\pm$$

with

$$\Phi''(z) = -2z, \quad \Phi''(z_\pm) = \mp 2\sqrt{s}.$$

Thus there are two symmetrically placed real saddle points on the real axis. Choosing the contour C to lie along the real axis, the conditions (13.51a) and (13.51b) are satisfied. We have therefore

$$\int_C dz\, e^{i\Phi(z)} = \int_{-\infty}^0 dz\, e^{i\Phi(z)} + \int_0^\infty dz\, e^{i\Phi(z)}.$$

In these integrals we set respectively

$$z = z_- + \frac{\rho}{s^{1/4}}, \quad z = z_+ + \frac{\rho}{s^{1/4}}$$

with

$$\Phi(z) = \Phi(z_\pm) + \frac{1}{2}(z - z_\pm)^2 \Phi''(z_\pm).$$

Then, changing to the variable ρ,

$$Ai(s) = \frac{1}{2\pi s^{1/4}} e^{i\phi(z_-)} \int_{-\infty}^\infty d\rho\, e^{i\rho^2} + \frac{1}{2\pi s^{1/4}} e^{i\phi(z_+)} \int_{-\infty}^\infty d\rho\, e^{-i\rho^2}.$$

Introducing a new variable σ by setting

$$\rho = \sigma - i|\sigma|, \quad d\rho = -\frac{i}{|\sigma|}[\sigma + i|\sigma|]d\sigma,$$

we obtain

$$\int_{-\infty}^0 d\rho\, e^{i\rho^2} = \frac{1}{\sqrt{2}} e^{i\pi/4} \int_{-\infty}^0 d\sigma\, e^{-2|\sigma|^2} \quad \text{and} \quad \int_0^\infty d\rho\, e^{-i\rho^2} = \frac{1}{\sqrt{2}} e^{-i\pi/4} \int_0^\infty d\sigma\, e^{-2|\sigma|^2}.$$

It follows that for $s \to \infty$:

$$Ai(s) \simeq \frac{1}{\sqrt{\pi} s^{1/4}} \cos\left(\frac{2}{3}s^{3/2} - \frac{\pi}{4}\right).$$

Example 13.3: **Derivation of the Stirling approximation of a factorial**
Obtain with the saddle point method from the integral defining the gamma or factorial function
the Stirling approximation.

Solution: Briefly, the method of steepest descent evaluates the integral

$$\oint e^{f(z)}dz$$

by expanding $f(z)$ about its extremum at z_0 with $z - z_0 = ix, dz = idx$ (observe that this implies
an integration parallel to the axis of imaginary z through the saddle point). Then, since $f'(z_0) = 0$
and $(z - z_0)^2 = -x^2$,

$$\oint e^{f(z)}dz \simeq e^{f(z_0)} i \int_{-\infty}^{\infty} e^{-x^2 f''(z_0)/2} dx = i\sqrt{\frac{2\pi}{f''(z_0)}} e^{f(z_0)}.$$

The gamma function $\Gamma(n+1)$ or factorial function $n!$ is defined as in Eq. (11.111), i.e.

$$n! = \int_0^{\infty} e^{-z} z^n dz = \int_0^{\infty} e^{-z + n \ln z} dz.$$

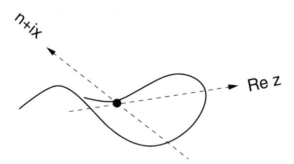

Fig. 13.3 The saddle point at $z = n + ix$.

Thus here $f(z) = -z + n \ln z, z_0 = n, f''(z_0) = -n/z_0^2 = -1/n$, and hence integrating as described
above through the saddle point indicated in Fig. 13.3, one obtains

$$n! \simeq \sqrt{2\pi} n^{n+1/2} e^{-n},$$

where we used $\int_{-\infty}^{\infty} dx \exp(-w^2 x^2/2) = \sqrt{2\pi/w^2}$.

Chapter 14

Classical Limit and WKB Method

14.1 Introductory Remarks

One of the most successful methods of solving the Schrödinger equation is the WKB method, named after its first promulgators in quantum mechanics: Wentzel, Kramers and Brillouin.[*] The method had, however, been developed previously by others in different contexts.[†] The central problem of the method is the topic of *"connection formulas"*, i.e. the linkage of solutions above a turning point to those below. A critical overview of the history of the method, of misunderstandings and other aspects may be found in the monograph of Dingle.[‡] Descriptions of the method can be found in most books on quantum mechanics and in some monographs.[§] Our main interest here focusses on the use of the method as an alternative to perturbation theory (supplemented by boundary conditions) and to the path integral method so that the results can be compared. It goes without saying that such comparisons are very instructive. The WKB method we consider in this chapter is a precursor to our uses of the path integral method in later chapters in the sense that the expansions are around the classical limit, and the role played by $\hbar$ in our considerations here, will there be that of a coupling parameter. The WKB result is therefore frequently described as the *semiclassical approximation*. The central issue of WKB solutions is their continuation in the sense of matched asymptotic expansions across a turning point (i.e. where

---

[*]G. Wentzel [282], H. A. Kramers [153] and L. Brillouin [37].

[†]The most prominent earlier expounder is H. Jeffreys [144].

[‡]R. B. Dingle [70]. For the history see pp. 316 - 317, for misunderstandings pp. 292 – 295.

[§]See e.g. N. Fröman and P. O. Fröman [99].

$E - V$ changes sign). It is then possible to derive in a general form the relation known as Bohr–Sommerfeld–Wilson quantization condition.* We illustrate the use of this condition by application to several examples.

14.2   Classical Limit and Hydrodynamics Analogy

In this section we consider the limit $\hbar \to 0$. We shall see that in this limit the Schrödinger equation describes a steady stream of noninteracting Newtonian particles.

We start from the Schrödinger equation

$$i\hbar \frac{\partial \Psi}{\partial t} = H\Psi \quad \text{with} \quad H = \frac{\mathbf{p}^2}{2m_0} + V(\mathbf{r}).$$

For stationary states Ψ, for which the observation of position and momentum is independent of the time at which the observation takes place, the energy has a definite value E with

$$\Psi(\mathbf{r}, t) = e^{-iEt/\hbar}\psi(\mathbf{r}) \quad \text{and} \quad |\Psi(\mathbf{r}, t)|^2 = |\psi(\mathbf{r})|^2.$$

It is for these states that one obtains the time-independent Schrödinger equation

$$\triangle\psi(\mathbf{r}) + \frac{2m_0}{\hbar^2}[E - V(\mathbf{r})]\psi(\mathbf{r}) = 0. \tag{14.1}$$

In this equation we set, the function S being called *phase function*,

$$\psi(\mathbf{r}) = A(\mathbf{r})e^{iS(\mathbf{r})/\hbar}, \tag{14.2}$$

$A(\mathbf{r})$ and $S(\mathbf{r})$ being real functions of $\mathbf{r}$, so that

$$\nabla\psi \equiv \frac{\partial \psi(\mathbf{r})}{\partial \mathbf{r}} = \left(\nabla A(\mathbf{r}) + \frac{i}{\hbar}A(\mathbf{r})\nabla S\right)e^{iS(\mathbf{r})/\hbar},$$

$$\triangle\psi = \nabla \cdot \nabla\psi$$

$$= \left[\triangle A + \frac{i}{\hbar}\nabla \cdot (A\nabla S) + \frac{i}{\hbar}\nabla S \cdot \left\{\nabla A + \frac{i}{\hbar}A\nabla S\right\}\right]e^{iS(\mathbf{r})/\hbar}$$

$$= \left[\triangle A + \frac{i}{\hbar}(\nabla A \cdot \nabla S + A\triangle S)\right.$$

$$\left. + \frac{i}{\hbar}\left\{\nabla S \cdot \nabla A + \frac{i}{\hbar}A(\nabla S)^2\right\}\right]e^{iS(\mathbf{r})/\hbar}. \tag{14.3}$$

---

*See e.g. the derivation in Example 18.7.

Inserting this into Eq. (14.1) and separating real and imaginary parts, we obtain the following two equations:

$$E - V(\mathbf{r}) + \frac{\hbar^2}{2m_0}\frac{\triangle A}{A} - \frac{1}{2m_0}(\boldsymbol{\nabla} S)^2 = 0 \tag{14.4}$$

and

$$2\boldsymbol{\nabla} A \cdot \boldsymbol{\nabla} S + A\triangle S = 0. \tag{14.5}$$

Apart from these we also wish to consider the equations that follow in a similar way from the time-dependent Schrödinger equation. We set in this equation,

$$i\hbar\frac{\partial\Psi}{\partial t} = H\Psi = \left[-\frac{\hbar^2}{2m_0}\triangle + V(\mathbf{r})\right]\Psi, \tag{14.6}$$

now with A also a function of t,

$$\Psi(\mathbf{r},t) = A(\mathbf{r},t)e^{iS(\mathbf{r},t)/\hbar}, \tag{14.7}$$

and obtain

$$
i\hbar\left[\frac{\partial A}{\partial t} + \frac{i}{\hbar}\frac{\partial S}{\partial t}A\right]
$$
$$
= -\frac{\hbar^2}{2m_0}\left[\triangle A - \frac{A(\boldsymbol{\nabla} S)^2}{\hbar^2} + \frac{2i}{\hbar}\boldsymbol{\nabla} A\cdot\boldsymbol{\nabla} S + \frac{i}{\hbar}A\triangle S\right] + V(\mathbf{r})A. \tag{14.8}
$$

Taking real and imaginary parts of these equations we obtain the equations:

$$\frac{\partial S}{\partial t} + \frac{(\boldsymbol{\nabla} S)^2}{2m_0} = \frac{\hbar^2}{2m_0}\frac{\triangle A}{A} - V(\mathbf{r}), \tag{14.9}$$

and

$$m_0\frac{\partial A}{\partial t} + \boldsymbol{\nabla} A\cdot\boldsymbol{\nabla} S + \frac{1}{2}A\triangle S = 0. \tag{14.10}$$

These equations are equivalent to the Schrödinger equation. We can identify Eq. (14.10) with a *continuity equation* by defining as *probability density*

$$\rho := \Psi^*\Psi = [A(\mathbf{r},t)]^2. \tag{14.11}$$

We define as *probability current density* the vector quantity

$$\mathbf{J} := \Re\left[\Psi^*\frac{\hbar}{im_0}\boldsymbol{\nabla}\Psi\right], \tag{14.12}$$

where

$$\boldsymbol{\nabla}\Psi = \left[\frac{\boldsymbol{\nabla} A}{A} + \frac{i}{\hbar}\boldsymbol{\nabla} S\right]\Psi. \tag{14.13}$$

This means

$$\Psi^* \frac{\hbar}{im_0}\boldsymbol{\nabla}\Psi = \frac{\hbar}{im_0}A\boldsymbol{\nabla}A + \frac{1}{m_0}A^2\boldsymbol{\nabla}S,$$

and so

$$\mathbf{J} = \frac{1}{m_0}A^2\boldsymbol{\nabla}S. \qquad (14.14)$$

But now

$$\frac{\partial\rho}{\partial t} \overset{(14.11)}{=} \frac{\partial}{\partial t}(A^2) \qquad (14.15)$$

and

$$\boldsymbol{\nabla}\cdot\mathbf{J} = \frac{1}{m_0}\boldsymbol{\nabla}\cdot(A^2\boldsymbol{\nabla}S) = \frac{1}{m_0}[\boldsymbol{\nabla}(A^2)\cdot\boldsymbol{\nabla}S + A^2\triangle S]. \qquad (14.16)$$

From Eq. (14.10) we obtain

$$m_0A\frac{\partial A}{\partial t} + A\boldsymbol{\nabla}A\cdot\boldsymbol{\nabla}S + \frac{1}{2}A^2\triangle S = 0.$$

Multiplying this equation by 2 and recalling that the function A was introduced as being real, we have

$$m_0\frac{\partial}{\partial t}(A^2) + \boldsymbol{\nabla}(A^2)\cdot\boldsymbol{\nabla}S + A^2\triangle S = 0. \qquad (14.17)$$

With Eqs. (14.15) and (14.16) this becomes

$$m_0\frac{\partial\rho}{\partial t} + m_0\boldsymbol{\nabla}\cdot\mathbf{J} = 0, \quad \text{i.e.} \quad \frac{\partial\rho}{\partial t} + \boldsymbol{\nabla}\cdot\mathbf{J} = 0. \qquad (14.18)$$

This is the *equation of continuity* that we encountered already earlier. Since we obtained this equation from Eq. (14.10), the implications of this equation are also those of the equation of continuity.

Next we investigate the significance of Eq. (14.9) for the phase function S. Since

$$\mathbf{J} = \frac{1}{m_0}A^2\boldsymbol{\nabla}S = \frac{\rho}{m_0}\boldsymbol{\nabla}S, \qquad (14.19)$$

it is suggestive to define the following vector quantities

$$\frac{d\mathbf{r}}{dt} \equiv \mathbf{v} := \frac{\mathbf{J}}{\rho} = \frac{1}{m_0}\boldsymbol{\nabla}S. \qquad (14.20)$$

With this we can rewrite Eq. (14.9) as

$$\frac{\partial S}{\partial t} + \frac{1}{2}m_0\mathbf{v}^2 + V = \frac{\hbar^2}{2m_0}\frac{\triangle A}{A}. \qquad (14.21)$$

In the limit $\hbar^2 \to 0$ this becomes

$$\frac{\partial S}{\partial t} + \frac{1}{2}m_0 \mathbf{v}^2 + V = 0. \tag{14.22}$$

Constructing the gradient of this equation, we obtain

$$\frac{\partial}{\partial t}\boldsymbol{\nabla} S + \frac{m_0}{2}\boldsymbol{\nabla}(\mathbf{v} \cdot \mathbf{v}) + \boldsymbol{\nabla} V = 0, \tag{14.23}$$

and hence[†]

$$\frac{\partial}{\partial t}(m_0 \mathbf{v}) + m_0(\mathbf{v} \cdot \boldsymbol{\nabla})\mathbf{v} + \boldsymbol{\nabla} V = 0. \tag{14.24}$$

However,

$$\frac{d}{dt} = \frac{\partial}{\partial t} + \frac{d\mathbf{r}}{dt} \cdot \frac{\partial}{\partial \mathbf{r}} = \frac{\partial}{\partial t} + \mathbf{v} \cdot \boldsymbol{\nabla},$$

so that the equation can be written

$$\frac{d}{dt}(m_0 \mathbf{v}) + \boldsymbol{\nabla} V = 0. \tag{14.25}$$

We recognize this equation as the classical equation of motion of the particle of mass m_0. We see therefore that in the limit $\hbar \to 0$ the particle behaves like one moving according to Newton's law in the force field of the potential V. We deduced this equation solely from the phase S in the limit $\hbar \to 0$.

The motion of the particle or rather its probability is altogether given by the above equation of continuity. This means, the wave function Ψ describes effectively a fluid of particles of mass m_0. The particles have no interaction with each other. This result is very general. This is the analogy of the Schrödinger equation for $\hbar^2 \to 0$ with hydrodynamics.

In the case of *stationary states* we have $\partial \rho / \partial t = 0$ (which is analogous to the condition of stationary currents, i.e. $\boldsymbol{\nabla} \cdot \mathbf{J} = 0$, or that of equilibrium in electrodynamics) and $\partial S / \partial t = -E$ (from $\Psi \propto \exp(-iEt/\hbar)$). The equation of continuity then describes the stationary flow of a fluid.

One should note that in no way do we simply obtain the classical equation of motion from the Schrödinger equation in the limit $\hbar^2 \to 0$. In the first place we require the ansatz (14.2) for the probability amplitude. This expression with $\hbar$ in the denominator shows, that the quantum theory implies an *essential singularity* at $\hbar = 0$ (i.e. a singularity different from that of a pole). An expansion in ascending powers of $\hbar$ therefore starts with contributions of the order of $1/\hbar^2, 1/\hbar$, which for very small values of $\hbar$ yield the dominant contributions. Such series are typical for expansions in the neighbourhood of an *essential singularity*; they are asymptotic expansions.

[†]Recall the formula $\boldsymbol{\nabla}(\mathbf{u} \cdot \mathbf{v}) = (\mathbf{v} \cdot \boldsymbol{\nabla})\mathbf{u} + (\mathbf{u} \cdot \boldsymbol{\nabla})\mathbf{v} + \mathbf{v} \times \mathrm{rot}\,\mathbf{u} + \mathbf{u} \times \mathrm{rot}\,\mathbf{v}$, so that $\boldsymbol{\nabla}(\mathbf{v} \cdot \mathbf{v}) = 2(\mathbf{v} \cdot \boldsymbol{\nabla})\mathbf{v} + 2\mathbf{v} \times \mathrm{rot}\,\mathbf{v}$, where, with the help of Eq. (14.20), $\mathrm{rot}\,\mathbf{v} = (1/m_0)\mathrm{rot}\,\mathrm{grad}\,S = 0$.

14.3   The WKB Method

14.3.1   The approximate WKB solutions

Without exaggeration one can say that (apart from numerical methods) the method we now describe is the only one that permits us to solve a differential equation of the second order with coefficients which are nontrivial functions of the independent variable over a considerable interval of the variable. As remarked at the beginning, the method is widely familar under the abbreviations of the names of Wentzel, Kramers and Brillouin. There are also other more specific descriptions, such as *"Liouville–Green method"* and *"phase integral method".*[‡]

The basic idea of the WKB method is to use the series expansion in ascending powers of $\hbar$ and then to neglect contributions of higher order. We consider the one-dimensional time-independent Schrödinger equation

$$y'' + \frac{2m_0}{\hbar^2}[E - V(x)]y = 0. \tag{14.26}$$

We observe already here that an expansion in rising powers of $\hbar$, i.e.

$$\frac{\hbar^2}{2m_0}\frac{1}{(E - V)},$$

corresponds to an expansion in decreasing powers of $(E - V)$. This means that the WKB method can make sense not only where one is interested in the classical limit but also for $E - V(x)$ and hence E large. The expansion in powers of $\hbar^2$, or $\hbar^2/2m_0(E - V)$, is an asymptotic expansion.

We set, similar to our earlier procedure with A and S real functions,[§]

$$y = A(x)e^{iS(x)/\hbar} \equiv e^{iW(x)/\hbar}, \tag{14.27}$$

where

$$W(x) = S(x) + \frac{\hbar}{i}\ln A(x). \tag{14.28}$$

The calculations we performed at the beginning imply the following equations (compare with Eqs. (14.4) and (14.5))

$$\left(\frac{dS}{dx}\right)^2 - 2m_0(E - V) = \frac{\hbar^2}{A}\frac{d^2A}{dx^2} \tag{14.29}$$

and

$$2\frac{dA}{dx}\frac{dS}{dx} + A\frac{d^2S}{dx^2} = 0. \tag{14.30}$$

[‡]For details see R. B. Dingle [70], p. 318.
[§]We follow here partially A. Messiah [195], Sec. 6.2.

Equation (14.30) corresponds to the equation of continuity (i.e. Eq. (14.10)). Integrating this second equation we obtain

$$\int \frac{A'}{A} dx = -\frac{1}{2} \int \frac{S''}{S'} dx,$$

so that

$$\ln A = \text{const.} - \frac{1}{2} \ln S', \quad A = c(S')^{-1/2}. \tag{14.31}$$

This expression can be substituted into Eq. (14.28). But first we note that:

$$A = c(S')^{-1/2}, \quad \frac{dA}{dx} = A\left[-\frac{1}{2}\frac{S''}{S'}\right],$$

$$\frac{d^2 A}{dx^2} = A\left[-\frac{1}{2}\frac{S'''}{S'} + \frac{3}{4}\frac{(S'')^2}{(S')^2}\right].$$

With this Eq. (14.29) becomes (note that this is still exact!)

$$(S')^2 - 2m_0(E - V) = \hbar^2\left[\frac{3}{4}\frac{(S'')^2}{(S')^2} - \frac{1}{2}\frac{S'''}{S'}\right]. \tag{14.32}$$

In the WKB method one now puts

$$S := S_0(x) + \hbar^2 S_1(x) + (\hbar^2)^2 S_2(x) + \cdots,$$
$$S' = S_0' + \hbar^2 S_1' + \cdots,$$
$$S'' = S_0'' + \hbar^2 S_1'' + \cdots. \tag{14.33}$$

Up to and including contributions of the order of $\hbar^2$ Eq. (14.32) is therefore

$$(S_0')^2 + 2\hbar^2 S_0' S_1' - 2m_0(E - V) = \hbar^2\left[\frac{3}{4}\left(\frac{S_0''}{S_0'}\right)^2 - \frac{1}{2}\frac{S_0'''}{S_0'}\right] \tag{14.34}$$

Equating coefficients of the same power of $\hbar$ we obtain

$$(S_0')^2 = 2m_0(E - V) \tag{14.35}$$

and

$$2S_0' S_1' = \left[\frac{3}{4}\left(\frac{S_0''}{S_0'}\right)^2 - \frac{1}{2}\frac{S_0'''}{S_0'}\right]. \tag{14.36}$$

Equation (14.35) can be integrated. Since the expression

$$S_0' = \pm\sqrt{2m_0(E - V)}$$

has the form of a momentum, it is reasonable to set (cf. Eq. (1.19), $p = h/\lambda$)

$$\lambdabar(x) := \frac{\hbar}{\sqrt{2m_0(E-V)}}. \qquad (14.37)$$

Then

$$S_0' = \pm \frac{\hbar}{\lambdabar} \quad \text{and} \quad S_0 = \pm \hbar \int^x \frac{dx}{\lambdabar(x)}. \qquad (14.38)$$

For Eq. (14.27) we obtain now with Eq. (14.31):

$$\begin{aligned}
y &= \exp\left[\frac{iW(x)}{\hbar}\right] = \exp\left[\frac{i}{\hbar}\left(S + \frac{\hbar}{i}\ln A\right)\right] \\
&= \exp\left[\frac{i}{\hbar}S - \frac{1}{2}\ln S' + \ln c\right] \\
&\simeq \exp\left[\frac{i}{\hbar}(S_0 + \hbar^2 S_1) - \frac{1}{2}\ln(S_0' + \hbar^2 S_1') + \ln c\right] \\
&\simeq \frac{c}{(S_0')^{1/2}}\exp\left[\frac{i}{\hbar}S_0\right] + O(\hbar^2),
\end{aligned}$$

and hence

$$y = c'(\lambdabar)^{1/2}\exp\left[\pm i \int \frac{dx}{\lambdabar(x)}\right]$$

or

$$y(x) = \alpha\sqrt{\lambdabar(x)}\cos\left(\int^x \frac{dx}{\lambdabar(x)} + \beta\right), \qquad (14.39)$$

where α and β are constants. The solution is seen to be periodic provided $\lambdabar(x)$ is real, i.e. $E > V(x)$. For $V(x) > E$ (this is the classically forbidden domain) we set

$$l(x) = \frac{\hbar}{\sqrt{2m_0(V(x)-E)}}. \qquad (14.40)$$

Then

$$y(x) = \sqrt{l(x)}\left[\gamma\exp\left\{\int^x \frac{dx}{l(x)}\right\} + \delta\exp\left\{-\int^x \frac{dx}{l(x)}\right\}\right]. \qquad (14.41)$$

We now ask: What are the conditions of validity of Eqs. (14.39) and (14.41)? We note first that the expansion in powers of $\hbar^2$ does not converge, and instead is asymptotic which we shall not establish in detail here.[¶] For the expansion (14.33) to make sense as an asymptotic series, we expect that in any case

$$S_0 \gg \hbar^2 S_1.$$

[¶]See R. B. Dingle [70], Chapter XIII.

We have

$$S_0 = \pm\hbar \int^x \frac{dx}{\lambdabar(x)}.$$

(14.42)

and S_1 follows from (14.36). Since

$$\frac{[(S_0')^{-1/2}]''}{(S_0')^{-1/2}} = \frac{[-\frac{1}{2}(S_0')^{-3/2}S_0'']'}{(S_0')^{-1/2}} = -\frac{1}{2}\left(\frac{S_0'''}{S_0'}\right) + \frac{3}{4}\left(\frac{S_0''}{S_0'}\right)^2$$

we have (cf. Eq. (14.36))

$$2S_0'S_1' = [(S_0')^{-1/2}]''(S_0')^{1/2}.$$

With $S_0' = \pm\hbar/\lambdabar, E > V(x)$, we obtain

$$2S_1' = \left[\left(\pm\frac{\hbar}{\lambdabar}\right)^{-1/2}\right]''\left(\pm\frac{\hbar}{\lambdabar}\right)^{-1/2}$$

and hence

$$\pm 2\hbar S_1' = [\lambdabar^{1/2}]''\lambdabar^{1/2} = \left[\frac{1}{2}\lambdabar^{-1/2}\lambdabar'\right]'\lambdabar^{1/2}$$

$$= \left[\frac{1}{2}\lambdabar^{-1/2}\lambdabar'' - \frac{1}{4}\lambdabar^{-3/2}(\lambdabar')^2\right]\lambdabar^{1/2}$$

so that

$$\pm\hbar S_1' = \left(\frac{1}{4}\lambdabar'' - \frac{1}{8}\frac{(\lambdabar')^2}{\lambdabar}\right).$$

It follows that

$$\hbar S_1 = \pm\left(\frac{1}{4}\lambdabar' - \frac{1}{8}\int \frac{(\lambdabar')^2}{\lambdabar}dx\right).$$

(14.43)

The condition $S_0 \gg \hbar^2 S_1$ therefore implies that

$$\pm\hbar\int\frac{dx}{\lambdabar} \gg \pm\hbar\left(\frac{1}{4}\lambdabar' - \frac{1}{8}\int dx\frac{(\lambdabar')^2}{\lambdabar}\right).$$

(14.44)

This condition is satisfied provided E or $E - V(x)$ is sufficiently large or

$$\lambdabar = \hbar/\sqrt{2m_0(E - V(x))}$$

is sufficiently small. Setting

$$R := |\lambdabar'| = \frac{|m_0\hbar V'(x)|}{|2m_0(E - V(x)|^{3/2}},$$

(14.45a)

we can rewrite the inequality (14.44) as

$$\int dx \sqrt{2m_0(E - V(x))} \left\{ 1 + \frac{1}{8}R^2 \right\} \gg \frac{1}{4}R\hbar, \qquad (14.45b)$$

so that the condition $S_0 \gg \hbar^2 S_1$ is also satisfied by demanding that $R \ll 1$. In a similar way we have for $E < V(x)$: $l'(x) \ll 1$. Roughly speaking one therefore requires E to be large in both cases. In the neighbourhood of $E = V(x)$, i.e. around the classical turning point, the WKB approximation is in general invalid (see below).

We can derive the equation whose exact solutions are the WKB approximations of the Schrödinger equation. In terms of λ the original Schrödinger equation is

$$y'' + \frac{1}{\lambda^2}y = 0, \quad \lambda(x) = \frac{\hbar}{\sqrt{2m_0(E - V(x))}}. \qquad (14.46)$$

The desired equation has the solutions

$$y(x) = a\lambda^{1/2} \exp\left[\pm i \int \frac{dx}{\lambda}\right]. \qquad (14.47)$$

Differentiating this twice — we omit the details — we obtain the equation

$$y'' + \left[\frac{1}{\lambda^2} - \frac{(\lambda^{1/2})''}{\lambda^{1/2}}\right]y = 0. \qquad (14.48)$$

The (dominant) WKB approximations are exact solutions of this equation.

14.3.2   Turning points and matching of WKB solutions

Let $x = a$ be the value of x at which $E = V(a)$. How do the solutions of Eq. (14.48) behave in the neighbourhood of such a point called "*turning point*", at which $E - V(x)$ changes sign? We note first that in this case near $x = a$:

$$E - V(x) = V(a) - V(x) \simeq -(x - a)V'(a),$$

and hence (with $m_0 = 1/2, \hbar = 1$)

$$\lambda(x) \simeq \frac{1}{\sqrt{E - V(x)}} \sim \frac{1}{(x - a)^{1/2}},$$

$$\sqrt{\lambda} \sim \frac{1}{(x - a)^{1/4}},$$

with the derivative relations

$$(\sqrt{\lambda})' \sim -\frac{1}{4}\frac{1}{(x-a)^{5/4}},$$

$$(\sqrt{\lambda})'' \sim \frac{5}{16}\frac{1}{(x-a)^{9/4}},$$

$$\frac{(\sqrt{\lambda})''}{\sqrt{\lambda}} \sim \frac{1}{(x-a)^2}.$$

These relations imply that Eq. (14.48) possesses a singularity of the form of $1/(x-a)^2$ at every turning point $x = a$. In the immediate neighbourhood of such a point the Schrödinger equation can therefore not be replaced by Eq. (14.48), since this requires different approximations there. It is therefore necessary to find other solutions in the neighbourhood of $x = a$ and then to match these to the WKB solutions in adjoining domains as indicated in Fig. 14.1.

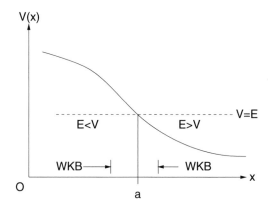

Fig. 14.1 The regions around the turning point at $x = a$.

As indicated in Fig. 14.1 there are WKB solutions on either side of the turning point and some distance away from it. On one side, i.e. in the domain $E > V$ the WKB solutions are oscillatory, and on the other side, in the domain $E < V$ the WKB solutions are exponentially increasing or decreasing. The transition is provided by *matching relations*, which connect one domain with the other. We shall not derive these conditions here in detail but rather make them plausible in Sec. 14.3.3 after stating them first here. We assume we have a turning point as indicated in Fig. 14.1. We define solutions y_1 and y_2 with branches to the left and to the right as follows (for definiteness note the + signs):

$$y_1 \sim \begin{cases} \sqrt{l}\exp\left(+\int_x^a \dfrac{dx}{l}\right) & \text{for} \quad x \ll a, \\[4mm] -\sqrt{\lambdabar}\sin\left(\int_a^x \dfrac{dx}{\lambdabar} - \dfrac{\pi}{4}\right) & \text{for} \quad x \gg a \end{cases} \tag{14.49}$$

and

$$y_2 \sim \begin{cases} \dfrac{\sqrt{l}}{2}\exp\left(-\int_x^a \dfrac{dx}{l}\right) & \text{for} \quad x \ll a, \\[4mm] \sqrt{\lambdabar}\cos\left(\int_a^x \dfrac{dx}{\lambdabar} - \dfrac{\pi}{4}\right) & \text{for} \quad x \gg a. \end{cases} \tag{14.50}$$

One should note that the solution $Ay_1 + By_2$ for $A \neq 0$, has the same asymptotic behaviour in the domain $x \ll a$ as Ay_1. Put differently: We could add a lot of contributions to Ay_1 without affecting the asymptotic form of Ay_1. This means the asymptotic part suffices only in the exponentially decreasing case. Note also the extra factor 2 in y_2.

The *first* WKB *matching condition* in the direction indicated by the arrow (and for the potential decreasing from left to right, as in Fig. 14.1) is then:

$$\frac{\sqrt{l}}{2}\exp\left(-\int_x^a \frac{dx}{l}\right) \Longrightarrow \sqrt{\lambdabar}\cos\left(\int_a^x \frac{dx}{\lambdabar} - \frac{\pi}{4}\right). \tag{14.51}$$

It can be shown that the *second* WKB *matching condition* in the opposite direction (and for the potential rising from right to left as in Fig. 14.1) is given by the following relation with the + sign in the argument of the exponential on the right and no factor of 2:

$$\sqrt{\lambdabar}\cos\left(\int_a^x \frac{dx}{\lambdabar} + \frac{\pi}{4}\right) \Longrightarrow \sqrt{l}\exp\left(+\int_x^a \frac{dx}{l}\right). \tag{14.52}$$

For a potential rising from left to right,

and $E \leqq V$ for $a \leqq x$ the same formulas apply with the same direction of the arrow, however with the limits of integration x, a interchanged. Thus, in order to make the formulas independent of whether the potential is rising or falling, one could simply replace everywhere

$$\int_x^a \cdots \quad \text{by} \quad \left|\int_x^a \cdots\right|.$$

In that case Eq. (14.51) — always with the decreasing exponential on the left — is valid in both directions. One should remember that the matching relations above are those for the dominant terms in the WKB expansion. If higher order contributions are to be taken into account, the matching relations have to be altered accordingly.

14.3.3 Linear approximation and matching

How do we arrive at the matching relations given above? We observed above that Eq. (14.48), whose exact solutions are the dominant WKB approximations, is singular at a turning point $x = a$. Thus around that point we can not use the WKB solutions and hence have to find others of the original equation (14.46). Therefore we consider now this original equation in the domain around $x = a$, i.e. the equation

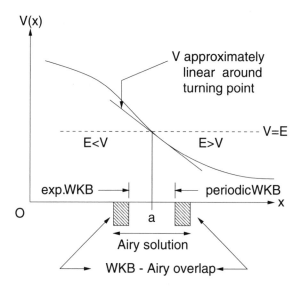

Fig. 14.2 The overlap regions around the turning point at $x = a$.

$$y'' + \frac{1}{\lambda^2}y = 0. \qquad (14.53)$$

Since at $x \simeq a$, as we saw,

$$\frac{1}{\lambda^2} \sim (x - a),$$

the equation becomes approximately

$$y'' = (x - a)\chi_1 y, \qquad (14.54)$$

where χ_1 is a proportionality factor which we choose to be constant. With the substitution

$$z = (x - a)\chi_1^{1/3},$$ (14.55)

the equation becomes the *Airy differential equation*

$$\frac{d^2y}{dz^2} = zy.$$ (14.56)

We have discussed some aspects of solutions of this equation in Chapter 13. In particular we encountered a solution written Ai. A second solution is written Bi. In the following we require the asymptotic behaviour of the solutions $Ai(-z), Bi(-z)$ for $|z| \to \infty$. This is given by the following expressions, as may be seen by referring back to Eqs. (13.59), (13.60), or to the literature:[||]

$$Ai(-z) \simeq \begin{cases} \dfrac{1}{2\sqrt{\pi}} \dfrac{1}{(-z)^{1/4}} e^{-\frac{2}{3}(-z)^{3/2}} & \text{for} \quad x - a \ll 0, \\[2ex] \left. \begin{aligned} & \dfrac{1}{\sqrt{\pi}} \dfrac{1}{z^{1/4}} \sin\left(\dfrac{2}{3}z^{2/3} + \dfrac{1}{4}\pi\right) \\ & = \dfrac{1}{\sqrt{\pi}} \dfrac{1}{z^{1/4}} \cos\left(\dfrac{2}{3}z^{2/3} - \dfrac{1}{4}\pi\right) \end{aligned} \right\} & \text{for} \quad x - a \gg 0, \end{cases}$$

(14.57a)

$$Bi(-z) \simeq \begin{cases} \dfrac{1}{\sqrt{\pi}} \dfrac{1}{(-z)^{1/4}} e^{\frac{2}{3}(-z)^{3/2}} & \text{for} \quad x - a \ll 0, \\[2ex] \left. \begin{aligned} & \dfrac{1}{\sqrt{\pi}} \dfrac{1}{z^{1/4}} \cos\left(\dfrac{2}{3}z^{2/3} + \dfrac{1}{4}\pi\right) \\ & = -\dfrac{1}{\sqrt{\pi}} \dfrac{1}{z^{1/4}} \sin\left(\dfrac{2}{3}z^{2/3} - \dfrac{1}{4}\pi\right) \end{aligned} \right\} & \text{for} \quad x - a \gg 0. \end{cases}$$

(14.57b)

To the same degree of approximation as the equation $y'' = (x - a)\chi_1 y$, we have

$$\int_a^x \frac{dx}{\lambdabar} \simeq \int_a^x \chi_1^{1/2}(x-a)^{1/2}dx = \frac{2}{3}(x-a)^{3/2}\chi_1^{1/2} = \frac{2}{3}z^{3/2}.$$ (14.58)

Replacing in Eqs. (14.57a) and (14.57b)

$$\frac{2}{3}z^{3/2} \quad \text{by} \quad \int_a^x \frac{dx}{\lambdabar}, \qquad z^{-1/4} \quad \text{by} \quad \sqrt{\lambdabar},$$

[||]Cf. R. B. Dingle [70], p.291, or M. Abramowitz and I. A. Stegun [1], formulas 10.4.59-60, p. 448.

and

$$\frac{2}{3}(-z)^{3/2} \quad \text{by} \quad \int_x^a \frac{dx}{l}, \quad (-z)^{-1/4} \quad \text{by} \quad \sqrt{l},$$

we obtain the matching relations (14.51) and (14.52). In this sense we have now verified these relations.

We summarize what we have achieved. In the neighbourhood of a turning point the Schrödinger equation becomes an Airy equation (in the leading approximation). At and around the turning point the solution of the Schrödinger equation is therefore given by Airy functions, e.g. by $Ai(z)$. These solutions, however, are only valid in a small interval around the turning point at $x = a$ as indicated in Fig. 14.2. As we approach a limit of this interval in going away from the turning point, we enter the domain of validity of a WKB solution. The latter domain is a small region where the solutions from either direction overlap, and hence must be proportional (in view of the uniqueness of the solution there). In the direction to the left as indicated in Fig. 14.1 the Airy solution becomes proportional to the exponential WKB solution and to the right to the trigonometric WKB solution. In this way different asymptotic branches of one and the same solution are matched across a turning point.

As an illustration of the use of the WKB solutions we apply these in Example 14.1 to the quartic oscillator.

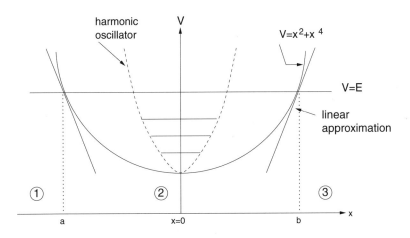

Fig. 14.3 The anharmonic potential well.

Example 14.1: Quartic oscillator and quantization condition
With the WKB matching relations derive for the case of the quartic oscillator with potential $V(x) = x^2 + x^4$ the Bohr–Sommerfeld–Wilson quantization condition given by

$$\int_a^b \frac{dx}{\lambda} \equiv \frac{1}{\hbar} \int_a^b \sqrt{2m_0(E - V(x))}\,dx = (2n+1)\frac{\pi}{2}, \quad n = 0, 1, 2, \ldots, \quad (14.59a)$$

where $x = a, b$ are the two turning points. Expressed as an integral over a complete cycle from one turning point back to it the relation is:

$$\oint dx \sqrt{2m_0 (E - V(x))} = \left(n + \frac{1}{2} \right) h. \tag{14.59b}$$

Solution: We consider an anharmonic potential well as depicted in Fig. 14.3. In the neighbourhood of the minimum at $x = 0$ we could approximate the potential by the harmonic oscillator. The corresponding eigenfunction would be approximations of the proper eigenfunctions around the origin. Consequently only the lowest eigenvalues would be reasonably well approximated by those of the harmonic oscillator. Let $E = V(x)$ at $x = a$ and $x = b$. Then around these points the potential would be well approximated by a linear potential (as we saw above). Here we are interested in the discrete states (the only ones here). Hence we search for solutions which are exponentially decreasing in the far regions, i.e. regions 1 and 3 in Fig. 14.3. In these domains the dominant WKB approximations are

$$y_1(x) = \frac{1}{2} c \sqrt{l} \exp\left(-\int_x^a \frac{dx}{l} \right) \text{ for } x \ll a, \quad y_3(x) = \frac{1}{2} c' \sqrt{l} \exp\left(-\int_b^x \frac{dx}{l} \right) \text{ for } x \gg b. \tag{14.60}$$

Using the matching relation (14.51), the continuations of these functions into the central region 2 are:

$$y_a(x) = c \sqrt{\lambdabar} \cos\left(\int_a^x \frac{dx}{\lambdabar} - \frac{\pi}{4} \right), \quad y_b(x) = c' \sqrt{\lambdabar} \cos\left(\int_x^b \frac{dx}{\lambdabar} - \frac{\pi}{4} \right) \text{ for } a \lesssim x \lesssim b. \tag{14.61}$$

Evidently these functions have to continue themselves into each other (since the wave function has to be unique). The condition for this is

$$y_a(x) = y_b(x) \quad \text{for} \quad x \in (a, b). \tag{14.62}$$

The condition for this to be exactly satisfied is — as we show now — that the Bohr–Sommerfeld–Wilson quantization condition holds and $c' = (-1)^n c, n = 0, 1, 2, \ldots$. In order to see this we consider

$$
\begin{aligned}
\cos\left(\int_x^b \frac{dx}{\lambdabar} - \frac{\pi}{4} \right) &= \cos\left(\int_a^b \frac{dx}{\lambdabar} - \int_a^x \frac{dx}{\lambdabar} - \frac{\pi}{4} \right) = \cos\left(\int_a^b \frac{dx}{\lambdabar} + \frac{\pi}{4} - \int_a^x \frac{dx}{\lambdabar} - \frac{\pi}{2} \right) \\
&= \sin\left(\int_a^b \frac{dx}{\lambdabar} \right) \cos\left(\frac{\pi}{4} - \int_a^x \frac{dx}{\lambdabar} \right) + \cos\left(\int_a^b \frac{dx}{\lambdabar} \right) \sin\left(\frac{\pi}{4} - \int_a^x \frac{dx}{\lambdabar} \right) \\
&= (-1)^n \cos\left(\int_a^x \frac{dx}{\lambdabar} - \frac{\pi}{4} \right)
\end{aligned}
$$

provided

$$\int_a^b \frac{dx}{\lambdabar} = (2n + 1) \frac{\pi}{2}, \quad n = 0, 1, 2, \ldots.$$

The potential now has to be inserted into this condition and the discrete eigenvalues E_n of the problem are obtained. We argued earlier that the WKB method is suitable in the case of large values of E. According to Eq. (14.44) this implies large values of n. One can also express this by saying that the linear approximation of the potential around the turning points must extend over a length of several wave-lengths and is therefore valid for large values of n, whereas the approximation of the eigenvalues by comparison with the harmonic oscillator in the domain of the minimum, extends only over a length of very few wave-lengths and is therefore valid for n small. Sometimes this distinction becomes imprecise, as in the case of the harmonic oscillator itself, for which the WKB approximation surprisingly yields already the exact energy eigenvalue as one can verify by evaluating in its case the Bohr–Sommerfeld–Wilson rule.

14.4 Bohr–Sommerfeld–Wilson Quantization

A quantization condition which is very useful in practice and in a wide spectrum of applications is — and remains in spite of its old fashioned reputation — the quantization condition established by N. Bohr, A. Sommerfeld and W. Wilson by supplementing classical mechanics by Planck's discretization.

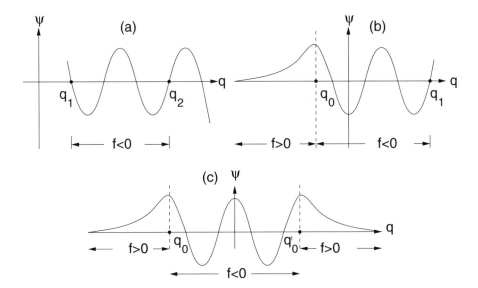

Fig. 14.4 Three cases with different pairs of zeros.

In this so-called *"old quantum theory"* (i.e. before Heisenberg's discovery of the canonical algebra and the formulation of the Schrödinger equation) this condition was always given as $\oint p\,dq = (n+1)h, n = 0,1,2,\ldots$. This relation is sometimes wrong. The *corrected form** is, with $n = 0,1,2,\ldots$,

$$\int_{q_1}^{q_2} p\,dq = \left[n+1-\frac{1}{4}\times \text{ number of turning points}\right]\pi\hbar, \qquad (14.63a)$$

or, equivalently,

$$\oint p\,dq = \left[n+1-\frac{1}{4}\times \text{ number of turning points}\right]h. \qquad (14.63b)$$

Consider the Schrödinger equation in the abbreviated form

$$\frac{d^2\psi(q)}{dq^2} - f(q)\psi(q) = 0. \qquad (14.64)$$

---

*The author learned this in lectures (1956) of R. B. Dingle, and knows of no published form or script.

One pair of solutions is (as we know from the earlier sections of this chapter)

$$\psi_{\mp}(q) \propto \frac{1}{f^{1/4}} \exp\left[\mp \int dq \sqrt{f(q)}\right], \quad \text{if} \quad f(q) \text{ is positive,}$$

$$\tilde{\psi}(q) \propto \frac{1}{(-f)^{1/4}} \frac{\sin}{\cos}\left[\int dq \sqrt{-f(q)}\right], \quad \text{if} \quad f(q) \text{ is negative.}$$

$$(14.65)$$

We use the knowledge of these solutions to find the eigenvalues of the equation. There are three cases.

Case (a): The case of *two trigonometrical zeros*, i.e. the wave function is to vanish at points $q = q_1$ and q_2 with the function $f(q)$ remaining negative in between as illustrated in Fig. 14.4(a). We have the case of trigonometric solutions. Since sine and cosine have zeros spaced at intervals of π, this means

$$\int dq \sqrt{-f} = (n+1)\pi, \quad n \text{ an integer} \geq 0.$$

In quantum mechanics we have

$$\frac{d^2\psi(q)}{dq^2} + \kappa^2 \psi(q) = 0, \quad \text{where} \quad \kappa = \frac{p}{\hbar}, \quad -f = \kappa^2, \quad \therefore \quad \sqrt{-f} = \frac{p}{\hbar}.$$

It follows that

$$\int_{q_1}^{q_2} p\, dq = (n+1)\pi\hbar = \frac{1}{2}(n+1)h, \quad \oint p\, dq = 2\int_{q_1}^{q_2} p\, dq = (n+1)h. \quad (14.66)$$

Thus in this case the so-called old quantum theory is correct.

Before we continue we recall from Eq. (14.51) the linkage between the trigonometrical and the exponential solutions across the position q_0 in space at which $f(q) = 0$, i.e the relation (with a shift of $\pi/2$)

$$\frac{1}{2f^{1/4}} \exp\left[-\left|\int_{q_0}^{q} \sqrt{f(q)}dq\right|\right] \leftrightarrow \frac{1}{(-f)^{1/4}} \sin\left[\left|\int_{q_0}^{q} \sqrt{-f(q)}dq\right| + \frac{1}{4}\pi\right].$$

$$(14.67)$$

Case (b): Next we consider, as illustrated in Fig. 14.4(b), the case of $\psi(q) = 0$ at $q = q_1$, where $f(q)$ is negative, i.e. this is *a trigonometrical zero*, together with $q \to -\infty$, where $f(q)$ is positive, i.e. this is *an exponential zero*. In this case the wave function for $q > q_0$ is

$$\frac{1}{(-f)^{1/4}} \sin\left[\int_{q_0}^{q} \sqrt{-f(q)}dq + \frac{1}{4}\pi\right].$$

This will vanish at $q = q_1$ if

$$\int_{q_0}^{q_1} \sqrt{-f(q)}dq + \frac{1}{4}\pi = (n+1)\pi, \quad \text{i.e.} \quad \int_{q_0}^{q_1} \sqrt{-f(q)}dq = \left[(n+1) - \frac{1}{4}\right]\pi.$$
(14.68)

It follows that

$$\oint p\, dq = \left[(n+1) - \frac{1}{4}\right]h.$$
(14.69)

Thus in this case the so-called old quantum theory is wrong.

Case (c): Finally we consider the case of $\psi(q) = 0$ at $q = \pm\infty$ exponentially as illustrated in Fig. 14.4(c). For the wave function to be exponentially decreasing to the left of q_0 and using Eq. (14.67):

- the wave function for $q > q_0$ must be proportional to

$$\frac{1}{(-f)^{1/4}} \sin\left[\int_{q_0}^{q} \sqrt{-f(q)}dq + \frac{1}{4}\pi\right], \text{ and}$$

- for the wave function to be exponentially decreasing to the right of q_0', the wave function for $q < q_0'$ must be proportional to

$$\frac{1}{(-f)^{1/4}} \sin\left[\int_{q}^{q_0'} \sqrt{-f(q)}dq + \frac{1}{4}\pi\right] \propto \frac{1}{(-f)^{1/4}} \sin\left[\int_{q_0'}^{q} \sqrt{-f(q)}dq - \frac{1}{4}\pi\right]$$

(where we reversed the order of integration to obtain a positive integral and multiplied through by minus 1). Since both cases refer to the same region $q_0 < q < q_0'$, the sines must be proportional. Therefore their arguments differ only by $(n+1)\pi$, i.e.

$$\left[\int_{q_0}^{q} \sqrt{-f(q)}dq + \frac{1}{4}\pi\right] - \left[\int_{q_0'}^{q} \sqrt{-f(q)}dq - \frac{1}{4}\pi\right] = (n+1)\pi.$$

Hence we obtain the condition

$$\int_{q_0}^{q_0'} \sqrt{-f(q)}dq = \left[(n+1) - \frac{1}{2}\right]\pi, \quad \text{or} \quad \oint p\, dq = \left[(n+1) - \frac{1}{2}\right]h. \quad (14.70)$$

We can therefore summarize the results in the form of Eq. (14.63a) or (14.63b). As illustrations we consider Examples 14.2 and 14.3.

Example 14.2: The harmonic oscillator
Use the relation (14.63b) to obtain the eigenenergies of the quantized harmonic oscillator.

Solution: As in the case of the quartic potential, this is the case of two turning points. Hence we have

$$\oint p\, dq = \left(n + \frac{1}{2}\right).$$

With potential $V(q) = 2\pi^2 m_0 \nu^2 q^2$, this requires evaluation of the integral

$$4 \int_0^{q_1} \sqrt{2m_0(E - V(q))} dq \quad \text{with} \quad q_1 = \sqrt{E/2\pi^2 m_0 \nu^2}.$$

The result is E/ν, so that $E = h\nu(n + 1/2)$.

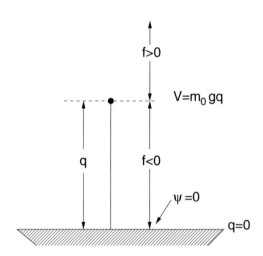

Fig. 14.5 A particle above a flat Earth.

Example 14.3: A particle in the gravitational field
A particle of mass m_0 is to be considered at a height q above the flat surface of the Earth. Calculate its quantized energy.

Solution: This is the case of one trigonometrical zero of the wave function ψ at $q = 0$ and an exponential zero at $q \to \infty$, as illustrated in Fig. 14.5. Thus there is one turning point at $E = m_0 g q$, where g is the acceleration due to gravity.[†] Hence we have the case of the integral

$$\oint p \, dq = \left[(n+1) - \frac{1}{4}\right] h.$$

Inserting the potential this becomes

$$2 \int_{q=0}^{q_0 = E/m_0 g} dq \sqrt{2m_0(E - m_0 g q)} = \left[(n+1) - \frac{1}{4}\right] h.$$

Evaluation of the integral yields the energy

$$E_n = \left(\frac{9}{32} m_0 g^2 h^2\right)^{1/3} \left[n + \frac{3}{4}\right]^{2/3}. \tag{14.71}$$

For $m_0 = 1 \times 10^{-3}$ kg, $g = 9.807$ ms^{-2}, and $h = 6.63 \times 10^{-34}$ J s, one obtains $E \simeq 2.28 \times 10^{-23}$ J $\times (n + 3/4)^{2/3}$.

[†]Thus a trigonometrical zero is the condition of absolutely no penetrability beyond it, whereas a turning point does, in principle, permit this although with rapidly diminishing probability.

14.5 Further Examples

Example 14.4: WKB level splitting formula

Derive from the WKB solutions of the Schrödinger equation with symmetric double well potential the WKB level splitting formula

$$\triangle_n^{\text{WKB}} E \equiv \frac{1}{2}\triangle E(q_0 = 2n + 1) = \frac{\omega\hbar}{\pi}\exp\left[-\int_{-z_0}^{z_0} dz\sqrt{-E + V(z)}/\hbar\right],$$

where $\mp z_0$ are the left and right barrier turning points, and ω is the oscillator frequency in either well.

Solution: The proof is contained in Example 18.7.

Example 14.5: Period of oscillation between two turning points

Use the Bohr–Sommerfeld–Wilson quantization rule to obtain for the period T of oscillation of a particle of mass m_0 between two turning points the relation

$$\frac{T}{2\pi} = \frac{1}{\omega} = \frac{m_0}{2\pi}\oint\frac{dx}{p} = \frac{\partial}{\partial E}\hbar\left(n + \frac{1}{2}\right), \quad n = 0, 1, 2, \ldots, \tag{14.72}$$

where $p = \sqrt{2m_0(E - V)}$.

Solution: We have

$$\frac{1}{\hbar}\int_{x_1}^{x_2} dx\, p = \int_{x_1}^{x_2} dx\sqrt{\frac{2m_0}{\hbar^2}(E - V)} = \left(n + \frac{1}{2}\right)\pi. \tag{14.73}$$

Hence, with $p = m_0 dx/dt$,

$$\frac{\partial}{\partial E}\int_{x_1}^{x_2} dx\sqrt{\frac{2m_0}{\hbar^2}(E - V)} = \frac{m_0}{\hbar}\int_{x_1}^{x_2}\frac{dx}{p} = \frac{1}{\hbar}\int_{t_1}^{t_2} dt, \tag{14.74}$$

or

$$\frac{t_2 - t_1}{2\pi} = \frac{m_0}{2\pi}\int_{x_1}^{x_2}\frac{dx}{p} = \frac{\hbar}{2\pi}\frac{\partial}{\partial E}\left(n + \frac{1}{2}\right)\pi.$$

For a period T from x_1 to x_2 and back to x_1 this implies

$$\frac{T}{2\pi} = \frac{1}{\omega} = \frac{m_0}{2\pi}\oint\frac{dx}{p} = \frac{\partial}{\partial E}\hbar\left(n + \frac{1}{2}\right). \tag{14.75}$$

In the case of the harmonic oscillator we have

$$\hbar\left(n + \frac{1}{2}\right) = \frac{E}{\omega},$$

which verifies the formula immediately.

In the case of screened Coulomb potentials (cf. Chapter 16) one obtains the quantization relation (for $M_0 = $ const.)

$$N = l + n + 1 = \left|\frac{M_0}{2\sqrt{E}} + \cdots\right|.$$

Hence

$$\frac{\partial}{\partial E}\hbar\left(N + \frac{1}{2}\right) \simeq \left|\frac{\partial}{\partial E}\left[\hbar\frac{M_0}{2\sqrt{E}}\right]\right| = \frac{\hbar}{4}\frac{M_0}{E^{3/2}} = \frac{\hbar}{4}M_0\left(\frac{2N}{M_0}\right)^3 = \frac{2N^3}{M_0^2}\hbar,$$

and

$$\frac{T}{2\pi} \simeq \frac{2N^3}{M_0^2}\hbar \qquad (14.76)$$

in agreement with the result obtained from the classical *Kepler period* for the Coulomb potential (cf. Eq. (11.133)).

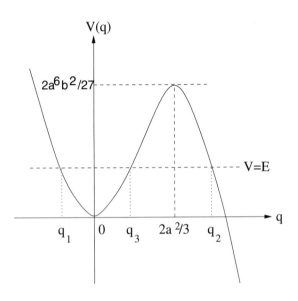

Fig. 14.6 The cubic potential.

Example 14.6: WKB method applied to the cubic potential

Consider the cubic potential

$$V(q) = \frac{1}{2}b^2q^2(a^2 - q). \qquad (14.77)$$

Determine (a) the turning points, and (b) evaluate the WKB exponential $\exp[-2I_{\text{barrier}}]$ and the WKB prefactor $2I_{\text{well}}$, which determine the imaginary part of the energy.

Solution: The potential has a finite minimum at $q = 0$ and a finite maximum at $q = 2a^2/3$ as indicated in Fig. 14.6. It follows that the energy between the minimum and this maximum lies in the range

$$0 \le E \le E_{\text{max}} = \frac{2}{27}a^6b^2. \qquad (14.78)$$

(a) We determine first the three turning points at $q_i = q_1, q_2, q_3$ for $E > 0$ given by $E - V = 0$. The q^2-term in this equation can be removed by transforming the equation to a cubic in y, where

$$y = q - \frac{1}{3}a^2, \quad \text{i.e.} \quad q = y + \frac{1}{3}a^2.$$

The equation $E - V = 0$ then becomes after a few steps of algebra

$$y^3 - Py + Q \equiv (y - y_1)(y - y_2)(y - y_3) = 0, \quad \text{where} \quad P = \frac{1}{3}a^4 \quad \text{and} \quad Q = 2\left(\frac{E}{b^2} - \frac{a^6}{27}\right). \qquad (14.79)$$

The roots of this equation are known[‡] and for $4P^3 > 27Q^2$ are real and expressed in terms of an angle θ given by (observe that with the choice of the minus sign, $\theta = 0$ implies $E_{\max}$ of Eq. (14.78))

$$\cos 3\theta = \sqrt{\frac{27\,Q^2}{4P^3}} \stackrel{-}{=} (+) \left(1 - \frac{27E}{a^6 b^2}\right) \leq 1, \ \theta = \pm 60° \text{ when } E = 0. \tag{14.80}$$

Inserting the roots y_1, y_2, y_3 into $q = y + a^2/3$, we have

$$q_1 = \frac{1}{3}a^2[1 - 2\cos\theta], \ q_2 = \frac{1}{3}a^2[1 - 2\cos(\theta + 120°)], \ q_3 = \frac{1}{3}a^2[1 - 2\cos(\theta - 120°)] \tag{14.81}$$

with $0 \leq \theta \leq 60°$ for $0 \leq E \leq E_{\max}$. Here $\cos(\theta \pm 120°) = -\frac{1}{2}(\cos\theta \pm \sqrt{3}\sin\theta)$ and[§]

$$
\begin{aligned}
q_1 - q_3 &= \frac{2a^2}{3}[-\cos\theta + \cos(\theta - 120°)] = a^2\left[-\cos\theta + \frac{\sin\theta}{\sqrt{3}}\right], \\
q_1 - q_2 &= \frac{2a^2}{3}[-\cos\theta + \cos(\theta + 120°)] = a^2\left[-\cos\theta - \frac{\sin\theta}{\sqrt{3}}\right], \\
q_2 - q_3 &= \frac{2a^2}{3}[-\cos(\theta + 120°) + \cos(\theta - 120°)] = \frac{2a^2}{\sqrt{3}}\sin\theta.
\end{aligned}
\tag{14.82}
$$

In order to determine the relative positions of the roots q_1, q_2, q_3, we set $\theta = 60° - \epsilon, \epsilon > 0$, and expand the roots in powers of ϵ. One obtains the ordering in Fig. 14.6, i.e.

$$q_1 \simeq \frac{a^2}{3}\{-\sqrt{3}\epsilon\} < 0, \quad q_2 \simeq \frac{a^2}{3}\{3\} > 0, \quad q_3 \simeq \frac{a^2}{3}\{\sqrt{3}\epsilon\} > 0.$$

(b)   We see that q_1 and q_3 merge to 0 with $\epsilon \to 0$. Thus we can now write the WKB integral from one barrier turning point to the other as (recalling the factor $\sqrt{2}/b$ in front of E in Eq. (14.79) and then setting $m_0 = 1, \hbar = 1$)

$$I_{\text{barrier}} := \int_{q_3}^{q_2} dq \sqrt{\frac{2m_0}{\hbar^2}(E - V)} = \sqrt{2}\frac{b}{\sqrt{2}} \int_{q_3}^{q_2} dq \sqrt{(q - q_1)(q_2 - q)(q - q_3)}, \ q_1 < q_3 \leq q < q_2. \tag{14.83}$$

The integral may now be evaluated with the help of Tables of Integrals. We obtain[¶]

$$I_{\text{barrier}} := b(q_2 - q_3)^2(q_2 - q_1)g \int_0^{u_1} du \, \text{sn}^2 u \, \text{cn}^2 \, \text{dn}^2 u, \ u_1 = \text{sn}^{-1}\sqrt{\frac{q_2 - q_3}{q_2 - q_3}} = 1, \ u_1 = K(k), \tag{14.84}$$

and the elliptic modulus k and the parameter g are given by

$$
\begin{aligned}
k^2 &= \frac{q_2 - q_3}{q_2 - q_1} = \frac{2\tan\theta}{\sqrt{3} + \tan\theta} \leq 1, \quad \theta \leq 60°, \quad \tan 60° = \sqrt{3}, \\
k'^2 &= 1 - k^2 = \frac{\sqrt{3} - \tan\theta}{\sqrt{3} + \tan\theta} \leq 1, \quad \therefore \tan\theta \neq \sqrt{3}, \theta \neq -60°, \\
g &= \frac{2}{\sqrt{q_2 - q_1}} = \frac{2}{a\sqrt{\cos\theta + \frac{\sin\theta}{\sqrt{3}}}} = \frac{2}{a}(1 - k'^2 + k'^4)^{1/4}.
\end{aligned}
\tag{14.85}
$$

The angle $\theta = 60°$ corresponds to $E = 0$, and the angle $\theta = 0°$ to $E = E_{\max}$. Therefore $E = 0$ corresponds to $k^2 = 1$. With the other integral from Tables we obtain ($K(k)$ being the complete

---

[‡]L. M. Milne–Thomson [196], p. 37.
[§]To help we cite H. B. Dwight [81], p. 78: $\sin 120° = \sqrt{3}/2, \cos 120° = -1/2$.
[¶]P. F. Byrd and M. D. Friedman [40], formulas 236.08, p. 80 and 361.04, p. 212.

elliptic integral of the first kind, $E(u)$ the incomplete elliptic integral of the second kind, with $K(k=1) = \infty, E(u=0) = 0, E(u=K(k)) \equiv E(k), E(k=1) = 1$)

$$G(k) \equiv \int_0^{u_1 = K(k)} du\, \text{sn}^2 u\, \text{cn}^2\, \text{dn}^2 u = \frac{1}{15k^4}[k'^2(k^2-2)K(k) + 2(k^4+k'^2)E(k)] \xrightarrow{k\to 1} \frac{2}{15}. \quad (14.86)$$

Next we set

$$\gamma^2 \equiv \frac{\tan\theta}{\sqrt{3}}, \quad \cos\theta + \frac{\sin\theta}{\sqrt{3}} = \frac{1+\gamma^2}{\sqrt{1+3\gamma^4}}, \quad (\cos\theta)^{-2} = 1 + 3\gamma^4 = 4\frac{(1-k'^2+k'^4)}{(1+k'^2)^2},$$

$$1 + \gamma^2 = \frac{2}{1+k'^2} \quad \text{and} \quad \sqrt{\cos\theta + \frac{\sin\theta}{\sqrt{3}}} = (1-k'^2+k'^4)^{-1/4}. \quad (14.87)$$

Inserting $G(k)$ together with the expressions for the prefactors into (14.84) we obtain

$$\begin{aligned}
I_{\text{barrier}} &= b\left(\frac{2a^2\sin\theta}{\sqrt{3}}\right)^2 a^2 \left(\cos\theta + \frac{\sin\theta}{\sqrt{3}}\right) \frac{2}{a\sqrt{\cos\theta + \frac{\sin\theta}{\sqrt{3}}}} G(k) \\
&= 8ba^5 \frac{\gamma^4\sqrt{1+\gamma^2}}{(1+3\gamma^4)^{5/4}} \underbrace{\frac{1}{15k^4}[k'^2(k^2-2)K(k) + 2(k^4+k'^2)E(k)]}_{G(k)} \\
&= 2ba^5 \frac{(1+k'^2)^2}{(1-k'^2+k'^4)^{5/4}} G(k) \xrightarrow{k^2\to 1} \frac{4ba^5}{15}. \quad (14.88)
\end{aligned}$$

For one complete round from one turning point back to it, we obtain in the limit of $E = 0$:

$$\oint dq\sqrt{\frac{2m_0}{\hbar^2}(E-V)} = 2I_{\text{barrier}} = \frac{8ba^5}{15}. \quad (14.89)$$

This expression will again be obtained later with the help of configurations called "*bounces*" (cf. Eq. (24.32)).

Next we evaluate the required integral across the well from q_1 to q_3 at energy E. Whereas the integrand of the above barrier integral is effectively a momentum, the integrand of the integral across the well is effectively the inverse of this momentum. Thus

$$\begin{aligned}
I_{\text{well}} : &= \int_{q_1}^{q_3} dq \frac{1}{\sqrt{\frac{2m_0}{\hbar^2}(E-V)}} = \frac{1}{\sqrt{2}}\frac{\sqrt{2}}{b}\int_{q_1}^{q_3} dq \frac{1}{\sqrt{(q-q_1)(q_2-q)(q-q_3)}}, \\
&= \frac{i}{b}\int_{q_1}^{q_3} dq \frac{1}{\sqrt{(q-q_1)(q_2-q)(q_3-q)}}, \quad q_1 < q \le q_3 < q_2. \quad (14.90)
\end{aligned}$$

We can evaluate this integral again with the use of Tables of Integrals. Thus[||]

$$I_{\text{well}} = \frac{i}{b}gF(\psi, \tilde{k}), \quad (14.91)$$

where

$$\tilde{k}^2 = \frac{q_3-q_1}{q_2-q_1} = \frac{a^2[\cos\theta - \frac{\sin\theta}{\sqrt{3}}]}{a^2[\cos\theta + \frac{\sin\theta}{\sqrt{3}}]} = \frac{\sqrt{3}-\tan\theta}{\sqrt{3}+\tan\theta} = k'^2, \quad g = \frac{2}{\sqrt{q_2-q_1}} \quad \text{as in (14.85)},$$

$$\sin\psi = \sqrt{\frac{q_3-q_1}{q_3-q_1}} = 1, \quad \psi = \frac{\pi}{2}, \quad F(\pi/2, k') = K(k'). \quad (14.92)$$

[||]P. F. Byrd and M. D. Friedman [40], formula 233.00, p. 72.

Hence (using $K(0) = \pi/2$)

$$I_{\text{well}} = i\frac{2K(k')}{ab}(1 - k'^2 + k'^4)^{1/4} \xrightarrow{k \to 1} i\frac{\pi}{ab} = i\frac{\pi}{w}, \tag{14.93}$$

where w is the harmonic oscillator frequency in the well.** However, we cannot expect the ratio $\exp(-2I_{\text{barrier}})/2I_{\text{well}} \propto w\exp(-2I_{\text{barrier}})$ evaluated at $k = 1$, and hence at $E = 0$, to represent a physical decay rate. To obtain the latter we have to use the quantum mechanical expression approximated by $E = \hbar w(n + 1/2)$, so that even in the case of the ground state the zero point energy will contribute to the prefactor, in fact through logarithmic contributions contained in the argument of the exponential for $k \neq 1$.

With algebra — which it is impossible to reproduce here in detail — one can derive expansions in ascending powers of k'^2 (which is small) of all relevant quantities. Thus one obtains from (14.85)

$$\cos 3\theta = \frac{1}{2}\frac{(1 - k^2 - 2k^4)(1 + k'^2)}{(1 - k^2 + k^4)^{3/2}} = -1 + \frac{27}{8}k'^4(1 + k'^2) + \cdots. \tag{14.94}$$

Comparing this equation with Eq. (14.80), we obtain

$$E = \frac{1}{8}a^6b^2k'^4(1 + k'^2). \tag{14.95}$$

Comparing this with the harmonic oscillator approximation $E \to E_n = ab(n + \frac{1}{2})$, we obtain

$$k'^4 = 4\frac{2n + 1}{a^5 b}, \quad n = 0, 1, 2, \ldots. \tag{14.96}$$

Expanding the coefficients of the elliptic integrals in Eq. (14.88) in rising powers of k'^2, we obtain

$$\frac{15}{2a^5 b}I_{\text{barrier}} = -k'^2\left[1 + \frac{25}{4}k'^2 + \frac{589}{32}k'^4 + \cdots\right]K(k) + 2\left[1 + \frac{17}{4}k'^2 + \frac{285}{32}k'^4 + \cdots\right]E(k). \tag{14.97}$$

Here we insert the corresponding expansions of the complete elliptic integrals†† (note the argument of K and E is k):

$$K(k) = \ln\left(\frac{4}{k'}\right) + \frac{k'^2}{4}\left[\ln\left(\frac{4}{k'}\right) - 1\right] + \cdots,$$

$$E(k) = 1 + \frac{k'^2}{2}\left[\ln\left(\frac{4}{k'}\right) - \frac{1}{2}\right] + \frac{3k'^4}{16}\left[\ln\left(\frac{4}{k'}\right) - \frac{13}{12}\right] + \cdots. \tag{14.98}$$

With these expansions one obtains

$$I_{\text{barrier}} = \frac{2}{15}a^5 b\left[2 + 8k'^2 - \frac{15}{8}k'^4\ln\left(\frac{4}{k'}\right) + \frac{497}{32}k'^4\cdots\right] \tag{14.99}$$

**For the potential approximated around the origin as $V \simeq a^2b^2q^2/2$ the eigenvalues are $E = \hbar w(n+1/2)$ with $w = ab$. The same expression for E is obtained by applying the "*method of poles at infinity*" to the Bohr–Sommerfeld–Wilson condition (14.63b). With $q = 1/z$ the integral (14.89) can be expanded to exhibit a simple pole at $z = 0$ allowing evaluation with Cauchy's residue theorem:

$$\left(n + \frac{1}{2}\right)h = \oint\frac{b\sqrt{-a^2}dz}{-z^3}\sqrt{1 - \frac{1}{a^2 z} - \frac{2Ez^2}{a^2 b^2}} \simeq \oint\frac{b\sqrt{-a^2}dz}{-z^3}\left[1 - \frac{Ez^2}{a^2 b^2}\right] = \frac{2\pi ib\sqrt{-a^2}E}{a^2 b^2} = \frac{2\pi E}{ab}.$$

††I. S. Gradshteyn and I. M. Ryzhik [122], formulas 8.113.3 and 8.114.3, pp. 905, 906.

For $k'^2 \to 0$ the first and the third terms dominate, so that we obtain

$$I_{\text{barrier}} = \frac{4}{15}a^5 b - \frac{2n+1}{4}\ln\left(\frac{2^6 a^5 b}{2n+1}\right) + O(\sqrt{a^5 b}). \tag{14.100}$$

It follows that (for one complete orbit back to the original turning point)

$$\exp[-2I_{\text{barrier}}] \simeq \left(\frac{2^5 a^5 b}{n+\frac{1}{2}}\right)^{n+\frac{1}{2}} e^{-\frac{8}{15}a^5 b}. \tag{14.101}$$

Correspondingly we obtain for the full period $2I_{\text{well}}$:

$$2I_{\text{well}} = \frac{4i}{ba}K(k')(1 - k'^2 + k'^4)^{1/4} \simeq \pm i\frac{2\pi}{ba}. \tag{14.102}$$

Thus with the WKB method we obtain

$$\frac{\exp[-2I_{\text{barrier}}]}{2I_{\text{well}}} = \pm i\frac{ba}{2\pi}\left(\frac{2^5 a^5 b}{n+\frac{1}{2}}\right)^{n+\frac{1}{2}} e^{-\frac{8}{15}a^5 b}. \tag{14.103}$$

With *Stirling's formula* in the form of what we later (with Eq. (18.178)) call the *Furry factor* f_n set equal to one, i.e.

$$\left(\frac{1}{n+\frac{1}{2}}\right)^{n+\frac{1}{2}} \simeq \sqrt{2\pi}\frac{e^{-(n+\frac{1}{2})}}{n!},$$

the result becomes

$$\frac{\exp[-2I_{\text{barrier}}]}{2I_{\text{well}}} = \pm i\frac{ba}{\sqrt{2\pi}}\frac{(2^5 a^5 b)^{n+\frac{1}{2}}}{n!}e^{-(n+\frac{1}{2})}e^{-\frac{8}{15}a^5 b}, \tag{14.104}$$

and we obtain for the ground state ($n = 0$)

$$i\gamma \equiv \frac{\exp[-2I_{\text{barrier}}]}{2I_{\text{well}}} = \pm i\sqrt{8a^5 b}\frac{\sqrt{2ab}}{\sqrt{e\pi}}e^{-\frac{8}{15}a^5 b}. \tag{14.105}$$

As argued in Chapters 24 and 26 one expects this WKB result to agree with the one loop path integral result using bounces. We observe that with $\sqrt{2/e} \approx 1$, the result agrees with the ground state path integral result (24.96) for the imaginary part of the energy as in the *Breit–Wigner formula* (10.73), i.e.

$$E_n = E_n^{(0)} - \frac{i}{2}\gamma, \quad \gamma \simeq \frac{\hbar\omega}{2\pi}\exp[-2I_{\text{barrier}}],$$

where (cf. also Eq. (20.11))

$$\frac{\hbar\omega}{2\pi} = \left[\oint \frac{dq}{\sqrt{\frac{2m_0}{\hbar^2}(E - V)}}\right]^{-1}.$$

It would be interesting to derive the same quantity with the perturbation method and to compare the results; this has not yet been done.

Chapter 15

Power Potentials

15.1 Introductory Remarks

The particular linear potential we considered in Chapter 13, describing free fall under gravity, led to a spectrum which contains only scattering states. In the present chapter we consider briefly the three-dimensional potential

$$V(\mathbf{r}) \propto |\mathbf{r}| = r,$$

which permits only bound states. This potential became widely popular in the spectroscopy of elementary particles, i.e. the classification of the various states of nucleons and mesons and other particles, after the realization that quarks as their constituents might not exist as free particles and, in fact, that it may not even be possible to extract individual quarks experimentally with any finite amount of energy. This would mean that the force binding the quarks together would not decrease with increasing separation as in the case of e.g. the Coulomb potential. There are numerous, at least indirect, indications from the nonabelian generalization of quantized Maxwell theory, now known as *quantum chromodynamics*, that this is indeed the case. The study of this potential, along with inclusion of angular momentum and spin effects and their interactions, led to a classification of quark bound states which is in surprisingly good agreement with a large amount of experimental data particularly in the case of heavy quarks.* In the present chapter we consider various aspects of this potential and extend this consideration to

---

*These investigations became very popular after the discovery of the heavy charmonium bound state Ψ and were naturally extended to sufficient complexity to permit comparison with experimental measurements. See particularly E. Eichten, K. Gottfried, T. Kinoshita, K. D. Lane and T.–M. Yan [82], C. Quigg and J. L. Rosner [230], [231], [232], [233] and H. B. Thacker, C. Quigg and J. L. Rosner [269].

general power potentials. We leave consideration of the logarithmic potential to some remarks and Example 15.1.

15.2 The Power Potential

The force

$$\mathbf{F} = -\boldsymbol{\nabla}V = -\text{const.}\frac{\mathbf{r}}{r}$$

acting between two particles is constant and directed towards the origin (of the relative coordinate). This force maintains this value irrespective of how far apart the particles are separated. These particles therefore cannot be separated by any finite amount of energy.

In order to have a clear starting point, we return to the Schrödinger equation in three dimensions, i.e.

$$-\frac{\hbar^2}{2\mu}\boldsymbol{\nabla}^2\Psi(\mathbf{r}) + [V(r) - E]\Psi(\mathbf{r}) = 0, \tag{15.1}$$

where μ is the reduced mass of the two particles of masses m_1, m_2, i.e.

$$\mu = \frac{m_1 m_2}{m_1 + m_2},$$

and $\mathbf{r}$ is the relative coordinate. For a central potential $V(r)$ we write

$$\Psi(\mathbf{r}) = Y_{lm}(\theta, \phi)\frac{1}{r}u(r) \equiv Y_{lm}(\theta, \phi)R(r), \quad R(r) = \frac{u(r)}{r}, \tag{15.2}$$

and obtain

$$u''(r) + \frac{2\mu}{\hbar^2}\left[E - V(r) - \frac{l(l+1)\hbar^2}{2\mu r^2}\right]u(r) = 0 \tag{15.3}$$

with the boundary conditions[†]

$$u(0) = 0, \quad u'(0) = \left[\frac{u(r)}{r}\right]_{r=0} = R(0), \tag{15.4}$$

and the normalization

$$\int d\mathbf{r}|\Psi(\mathbf{r})|^2 = 1, \quad \int_0^\infty dr\{u(r)\}^2 = 1. \tag{15.5}$$

Instead of considering the linear potential, we consider immediately the more general power potential

$$V(r) = \lambda r^\nu, \quad \nu \geq 1, \; \lambda > 0. \tag{15.6}$$

[†]We have $u(r) = rR(r), u \sim r^{l+1} \rightarrow R \sim r^l$, and so $u'(r) \sim (l+1)r^l = (l+1)u(r)/r = (l+1)R(r)$, and $u'(0) \rightarrow [u(r)/r]_0$.

In Chapter 14 we obtained the Bohr–Sommerfeld–Wilson quantization condition (14.63a), (14.63b) for the one-dimensional Schrödinger equation. We now want to find a quantization condition for the three-dimensional problem, which has been reduced to an effective one-dimensional problem in the polar coordinate r. We therefore proceed as follows.

One-dimensional case: We consider first briefly a particle of mass m_0 in a symmetric one-dimensional potential with the symmetry

$$V(x) = V(-x) \quad \text{and} \quad V(0) = 0 \tag{15.7}$$

without loss of generality. Then

$$\lambdabar(x) = \lambdabar(-x), \quad \lambdabar(x) = \hbar/\sqrt{2m_0(E - V(x))},$$

and the turning points are at

$$a = -b := -x_c. \tag{15.8a}$$

We also assume that

$$V'(x) \geq 0, \quad x > 0. \tag{15.8b}$$

This condition implies that we have only one pair of turning points, so that with Eq. (15.8a) we can write the quantization condition (14.63b):

$$2 \int_0^{x_c} \frac{dx}{\lambdabar} = (2N + 1)\frac{\pi}{2}, \quad N = 0, 1, 2 \ldots$$

or

$$\int_0^{x_c} \frac{dx}{\lambdabar} = \left(N + \frac{1}{2}\right)\frac{\pi}{2}, \quad N = 0, 1, 2, \ldots. \tag{15.9}$$

Three-dimensional case: We transcribe the above considerations of the one-dimensional case into that of the three-dimensional case by restricting ourselves in the latter case to S waves (no centrifugal potential!), and there identify the S-wave states with those states of the one-dimensional case whose wave functions vanish at the origin — these are the nonsymmetric ones with N odd — as required by Eq. (15.4) in the three-dimensional case. This means that we transcribe Eq. (15.9) into the three-dimensional case in the following form:

$$\int_0^{r_c} dr \sqrt{2\mu(E - V(r))} = \left[\underbrace{(2n - 1)}_{N, \text{ odd integer}} + \frac{1}{2}\right]\frac{\pi\hbar}{2} = \left(n - \frac{1}{4}\right)\pi\hbar. \tag{15.10}$$

Here r_c is the turning point and $n = 1, 2, 3, \ldots$ is to be interpreted as the principal quantum number in three dimensions. In order to to obtain the eigenvalues $E \to E_n$ for the radial potential (15.6) we have to evaluate the integral

$$I := \int_0^{r_c} dr \sqrt{2\mu(E - \lambda r^\nu)} = \left(n - \frac{1}{4}\right)\pi\hbar. \qquad (15.11)$$

Here the turning point r_c is given by

$$E - \lambda r_c^\nu = 0, \quad r_c = \left(\frac{E}{\lambda}\right)^{1/\nu}. \qquad (15.12)$$

Now,

$$I = \sqrt{2\mu E} \int_0^{r_c} dr \left[1 - \frac{\lambda}{E} r^\nu\right]^{1/2}.$$

We set

$$t = \frac{\lambda}{E} r^\nu, \quad r = \left(\frac{Et}{\lambda}\right)^{1/\nu}, \qquad (15.13)$$

so that

$$dt = \frac{\lambda}{E}\nu r^{\nu-1} dr, \quad dr = \frac{E}{\lambda\nu}\frac{dt}{(Et/\lambda)^{(\nu-1)/\nu}}$$

or

$$dr = \frac{E}{\lambda\nu}\left(\frac{\lambda}{E}\right)^{(\nu-1)/\nu} t^{(1-\nu)/\nu} dt.$$

It follows that

$$I = \sqrt{2\mu E}\frac{E}{\lambda\nu}\left(\frac{\lambda}{E}\right)^{(\nu-1)/\nu}\int_0^{t_c} dt\, t^{(1-\nu)/\nu}(1-t)^{1/2}. \qquad (15.14)$$

From Eqs. (15.12) and (15.13) we obtain

$$t_c = \frac{\lambda}{E}r_c^\nu = \frac{\lambda}{E}\frac{E}{\lambda} = 1, \qquad (15.15)$$

so that

$$I = \sqrt{2\mu E}\frac{E}{\lambda\nu}\left(\frac{\lambda}{E}\right)^{(\nu-1)/\nu}\int_0^1 dt\, t^{(1-\nu)/\nu}(1-t)^{1/2}. \qquad (15.16)$$

The integral on the right is the integral representation of the *beta function* $B(x,y)$ defined by

$$B(x,y) = \frac{\Gamma(x)\Gamma(y)}{\Gamma(x+y)} \equiv \frac{(x-1)!(y-1)!}{(x+y-1)!} = \int_0^1 t^{x-1}(1-t)^{y-1} dt, \qquad (15.17)$$

so that by comparison $y = 3/2$ and $x = 1 + (1 - \nu)/\nu$, and

$$I = \sqrt{2\mu E} \, \frac{E}{\lambda\nu} \left(\frac{\lambda}{E}\right)^{(\nu-1)/\nu} \frac{\Gamma(\frac{3}{2})\Gamma(1 + \frac{1-\nu}{\nu})}{\Gamma(\frac{3}{2} + 1 + \frac{1-\nu}{\nu})} = \left(n - \frac{1}{4}\right)\pi\hbar, \qquad (15.18)$$

or

$$E^{\frac{3}{2} - (\frac{\nu-1}{\nu})} = \frac{(n - \frac{1}{4})\pi\hbar\nu\Gamma(\frac{3}{2} + \frac{1}{\nu})\lambda^{1/\nu}}{\Gamma(\frac{3}{2})\Gamma(\frac{1}{\nu})\sqrt{2\mu}},$$

or

$$E_n = \left[\frac{(n - \frac{1}{4})\pi\Gamma(\frac{3}{2} + \frac{1}{\nu})\lambda^{1/\nu}}{\Gamma(\frac{3}{2})\frac{1}{\nu}\Gamma(\frac{1}{\nu})\sqrt{2\mu/\hbar^2}}\right]^{2\nu/(\nu+2)}. \qquad (15.19)$$

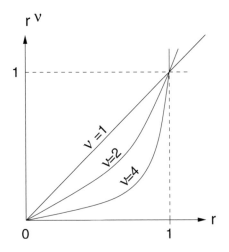

Fig. 15.1 Approach to square well with $\nu \to \infty$.

One can convince oneself now, for instance simply graphically as indicated in Fig. 15.1, that in the limit $\nu \to \infty$ the potential approaches the shape of an infinitely high square well.[‡] In this limit the eigenvalues become:[§]

$$E_n = \left[\frac{(n - \frac{1}{4})\pi\Gamma(\frac{3}{2})}{\Gamma(\frac{3}{2})\Gamma(1)\sqrt{2\mu/\hbar^2}}\right]^2 = \frac{n^2\pi^2}{2\mu}\hbar^2. \qquad (15.20)$$

These are precisely the eigenvalues which one obtains from the vanishing of the (periodic) eigenfunction at the wall of the square well (of course $n - 1/4$ has to be replaced by n since the WKB approximation is only accidentally correct for small values of n).[¶]

[‡] Compare with Eq. (12.55).
[§] $\Gamma(z) = (z-1)!, \Gamma(z+1) = z!, \frac{1}{\nu}\Gamma(\frac{1}{\nu}) = \Gamma(1 + \frac{1}{\nu}) = (\frac{1}{\nu})!$.
[¶] The case $\nu = 4$ is that of the *pure anharmonic oscillator* also discussed by M. Weinstein [281].

In the case of $\nu = 2$, the harmonic oscillator, we obtain

$$
\begin{aligned}
E_n &= \left[\frac{(n - \frac{1}{4})\pi\Gamma(2)\lambda^{1/2}}{\Gamma(\frac{3}{2})\Gamma(\frac{3}{2})\sqrt{2\mu/\hbar^2}} \right] = \frac{(n - \frac{1}{4})\pi\lambda^{1/2}}{\frac{1}{4}\pi\sqrt{2\mu/\hbar^2}} \\
&= (4n - 1)\sqrt{\frac{\lambda}{2\mu}}\hbar.
\end{aligned}
\tag{15.21}
$$

This result agrees with that for the one-dimensional harmonic oscillator r^2 in the form[‖]

$$
\phi''(r) + \frac{2\mu}{\hbar^2}(E - \lambda r^2)\phi(r) = 0
\tag{15.22}
$$

and with the boundary condition $\phi(0) = 0$ which selects from the usual eigenfunctions the odd ones.

In the case $\nu = 1$, the case of the linear potential, we obtain

$$
E_n = \left[\frac{(n - \frac{1}{4})\pi\Gamma(\frac{5}{2})\lambda}{\Gamma(\frac{3}{2})\sqrt{2\mu/\hbar^2}} \right]^{2/3} = \left[\frac{3\pi}{2}\left(n - \frac{1}{4} \right)\frac{\lambda\hbar}{\sqrt{2\mu}} \right]^{2/3}.
\tag{15.23}
$$

The eigenvalues for the linear potential can also be obtained from the zeros of the Airy function. The potential

$$
V(x) = \lambda|x|
\tag{15.24}
$$

is discontinuous at $x = 0$ (this means the derivative there jumps from positive to negative). Physically it does not make sense for the probability amplitude to have a discontinuity there. It is therefore necessary to demand its continuity there in the sense of equality of the first derivatives from either direction (apart from the equality of the values of the functions there from either direction). But for our present purposes this applies only in as far as the solutions $\phi(x)$ which are exponentially decreasing at infinity vanish at $x = 0$, since these are the ones we are interested in (cf. discussion at the beginning of this section). In the one-dimensional consideration these are precisely the odd wave functions as illustrated by an example in Fig. 15.2. For the following we set in Eq. (13.61) $\hbar^2 = 1, m_0 = \frac{1}{2}, g = -2$, so that the appropriate Schrödinger equation becomes

$$
\frac{d^2\phi}{dx^2} + (E - x)\phi(x) = 0, \quad x \geq 0.
\tag{15.25}
$$

[‖]Comparison with Eq. (6.2) implies $\lambda \equiv \mu\omega^2/2, \omega \equiv \sqrt{2\lambda/\mu}$, and with Eq. (6.41) for the eigenvalues $(N + \frac{1}{2})\hbar\omega \rightarrow (2n - 1 + \frac{1}{2})\hbar\omega = (4n - 1)\frac{1}{2}\hbar\omega$.

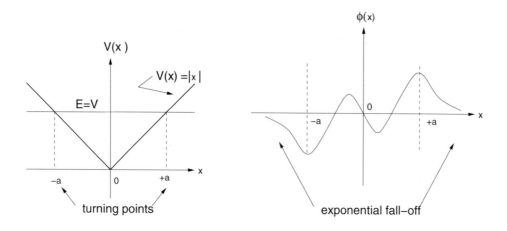

Fig. 15.2 Behaviour of an odd wave function at the origin.

With $z := E - x$ this is

$$\frac{d^2\phi(z)}{dz^2} + z\phi(z) = 0.$$

According to Eqs. (14.56) and (14.57a) one solution of this equation is the *Airy function*

$$\phi(z) \propto Ai(z) = Ai(E - x). \tag{15.26}$$

For

$$z \equiv s := E - x \to -\infty, \quad x \to +\infty,$$

Eq. (13.59) applies, i.e.

$$Ai(s) \simeq \frac{1}{2\sqrt{\pi}(-s)^{1/4}} \exp\left[-\frac{2}{3}(-s)^{3/2}\right], \tag{15.27}$$

i.e. this wave function has the required exponentially decreasing behaviour at infinity. In the domain

$$-a < x < +a, \quad \text{i.e. } E > V \text{ or } s > 0,$$

and for $s \to \infty$ the trigonometric behaviour of Eq. (13.60) applies, i.e.

$$Ai(s) \simeq \frac{1}{\sqrt{\pi}s^{1/4}} \cos\left(\frac{2}{3}s^{3/2} - \frac{\pi}{4}\right). \tag{15.28}$$

In particular at $x = 0$, i.e. $s = E$, we must have as a result of the boundary condition $\phi(x) = 0$:

$$\phi(x)|_{x=0} \propto Ai(E - x)|_{x=0} = 0, \tag{15.29}$$

Table 15.1: *S*-**Wave Energy Eigenvalues** [*]
for $V(r) = r$ (with $\hbar = 2\mu = 1$)

n	E_n from $[\frac{3\pi}{2}(n - \frac{1}{4})]^{2/3}$	E_n from $\mathrm{Ai}(E_n) = 0$
1	2.32025	2.33811
2	4.08181	4.08795
3	5.51716	5.52056
4	6.78445	6.78671
5	7.94249	7.94413
6	9.02137	9.02265
7	10.03914	10.04017
8	11.00767	11.00852
9	11.93528	11.93602
10	12.82814	12.82878

i.e. the eigenvalues E_n are determined by the zeros of the Airy function. The expression (15.28) is really only valid for s large and $s \to \infty$, but for large values of E Eq. (15.29) implies that

$$\cos\left(\frac{2}{3}E^{3/2} - \frac{\pi}{4}\right) = 0, \tag{15.30}$$

i.e.

$$\frac{2}{3}E^{3/2} - \frac{\pi}{4} = \frac{\pi}{2}(2n - 1), \quad n = 1, 2, 3, \ldots$$

(an odd function has an odd number of zeros!), i.e. $(2/3)E^{3/2} = \pi n - (\pi/4)$, or

$$E \to E_n = \left[\frac{3}{2}\pi\left(n - \frac{1}{4}\right)\right]^{2/3}. \tag{15.31}$$

Actually this expression is only valid for E and hence n large, so that $n - 1/4 \simeq n$. It is interesting to note that in the dominant approximation both methods agree. It is possible, of course, to obtain the zeros of the Airy function numerically. Table 15.1 demonstrates the quality of the WKB approximation by comparison with the exact values.

---

[*]This Table is reprinted from C. Quigg and J. L. Rosner, *Quantum Mechanics with Applications to Quarkonium*, Table 2, p. 201, [231], copyright of North–Holland Publ. Co., with permission from Elsevier.

15.3 The Three-Dimensional Wave Function

The linear potential has in particular been used in the investigation of the spectrum of heavy quark-antiquark pairs. In order to be comparable with experimental data these investigations, of course, had to be supplemented by inclusion of relativistic corrections as well as other contributions arising for instance from spin and angular momentum interactions. The resulting masses agree in most cases very well with those extracted from experimental observations. In general these quark-antiquark pairs, e.g. the charmanticharm meson Ψ, have only a finite lifetime and decay into other particles such as electron-positron pairs, $e\bar{e}$. The decay widths Γ, which are inversely proportional to the lifetimes, can be shown to be proportional to the modulus-squared of the particle wave function at the origin, i.e.[**]

$$\Gamma(\Psi \to e\bar{e}) \propto |\Psi(0)|^2.$$

We indicate briefly here — without entering into further details — how these quantities can be calculated.

We saw previously (see for instance Eq. (14.50)) that in the domain $V < E$ the leading WKB approximation of the wave function $u(r)$ is given by the periodic function

$$u(r) = N\sqrt{\lambdabar(r)}\cos\left(\int_r^{r_c}\frac{dr'}{\lambdabar(r')} - \frac{\pi}{4}\right), \quad \lambdabar = \frac{\hbar}{\sqrt{2\mu(E-V)}}, \qquad (15.32)$$

where N is the normalization constant.[*] We can determine the constant N in the leading approximation by demanding that

$$1 = 2\int_0^{r_c} dr\,[u(r)]^2,$$

i.e.

$$N^2\int_0^{r_c} dr\lambdabar(r)\cos^2\left(\int_0^r\frac{dr'}{\lambdabar(r')} - \frac{\pi}{4}\right) = \frac{1}{2}. \qquad (15.33)$$

Here we replace the oscillatory part $\cos^2(\cdots)$ by a mean value, i.e.

$$\cos^2(\cdots) \to \frac{1}{r_c}\int_0^{r_c} dr\,\cos^2(\cdots). \qquad (15.34a)$$

[**]The most frequently quoted reference for this result is R. P. van Royen and V. F. Weisskopf [239]. See also J. D. Jackson [140].

[*]Note that this makes N a WKB normalization constant. We shall encounter WKB normalization constants at numerous points in later chapters, e.g. in Chapters 24 and 26.

In accordance with the relation

$$\frac{1}{T} \int_0^{T=2\pi/\omega} dt \, \cos^2 \omega t = \frac{1}{2}, \tag{15.34b}$$

this averaging implies a numerical factor like $1/2$, so that approximately

$$N^2 \approx \left[\int_0^{r_c} \lambda(r) dr \right]^{-1}, \quad \lambda = \frac{\hbar}{\sqrt{2\mu(E-V)}}. \tag{15.35}$$

Variation or differentiation of Eq. (15.10) with respect to n implies (which is permissible for small separations of neighbouring levels)

$$2\mu \frac{1}{2} \frac{dE_n}{dn} \int_0^{r_c} \frac{dr}{\sqrt{2\mu(E_n - V)}} = \pi\hbar,$$

i.e.

$$\mu \frac{dE_n}{dn} \int_0^{r_c} \lambda(r) dr = \pi\hbar^2, \tag{15.36}$$

so that with Eq. (15.35)

$$N^2 \approx \frac{\mu}{\pi\hbar^2} \frac{dE_n}{dn}. \tag{15.37}$$

According to Eqs. (15.2) and (15.4) we have for $l = m = 0$:

$$|\Psi(0)|^2 = |u'(0)|^2 |Y_{00}(\theta, \phi)|^2 = \frac{1}{4\pi} |u'(0)|^2. \tag{15.38}$$

From Eq. (15.32) we obtain to leading order

$$u'(r) \simeq \frac{N}{\sqrt{\lambda(r)}} \sin \left[\int_r^{r_c} \frac{dr'}{\lambda(r')} - \frac{\pi}{4} \right], \tag{15.39}$$

and therefore

$$
\begin{aligned}
u'(0) &= \frac{N}{\sqrt{\lambda(0)}} \sin \left[\int_0^{r_c} \frac{dr'}{\lambda(r')} - \frac{\pi}{4} \right] \\
&= \frac{N}{\sqrt{\lambda(0)}} \sin \left[\int_0^{r_c} \frac{1}{\hbar} dr' \sqrt{2\mu(E-V)} - \frac{\pi}{4} \right] \\
&\overset{(15.10)}{=} \frac{N}{\sqrt{\lambda(0)}} \sin \left[\left(n - \frac{1}{4} \right) \pi - \frac{\pi}{4} \right] \\
&= \frac{N}{\sqrt{\lambda(0)}} \sin \left[\left(n - \frac{1}{2} \right) \pi \right], \quad n = 1, 2, \ldots \\
&= (-1)^{n-1} \frac{N}{\sqrt{\lambda(0)}}. \tag{15.40}
\end{aligned}
$$

Since we are considering potentials with $V(0) = 0$, we have

$$u'(0) = (-1)^{n-1} \frac{N}{\hbar^{1/2}} (2\mu E_n)^{1/4}. \tag{15.41}$$

Hence Eq. (15.38) implies

$$|\Psi(0)|^2 = \frac{1}{4\pi} \frac{N^2 (2\mu E_n)^{1/2}}{\hbar}, \tag{15.42}$$

and hence with Eq. (15.37) we obtain:

$$\begin{aligned}
|\Psi(0)|^2 &= \frac{1}{4\pi} \frac{(2\mu E_n)^{1/2}}{\hbar} \frac{\mu}{\pi \hbar^2} \frac{dE_n}{dn} \\
&= \frac{(2\mu)^{3/2}}{8\pi^2 \hbar^3} E_n^{1/2} \frac{dE_n}{dn}.
\end{aligned} \tag{15.43}$$

Inserting here, for instance, the expression (15.23) for the energy levels of the linear potential, we can investigate the dependence of the decay rates on n (note that these would be decays into particles different from those bound by the linear potential — a consideration we cannot enter into more detail here). An enormous literature exists on the subject in view of its relevance to the spectroscopy of mesons and baryons made up of quark constituents and the necessity of checks with results derived from experiments. Thus only theoretical investigations performed immediately after the discovery of the particles Ψ and Υ are basically of an analytical nature[†] and therefore have not been performed largely with numerical methods and the fitting of as few parameters as possible.[‡] Using the result (15.43) or the results of corresponding calculations one finds the following dependence of $|\Psi(0)|^2$ on the quantum numbers $n = 1, 2, 3, \ldots$ for various potentials:
(a) For a confinement potential of power ν:

$$|\Psi(0)|^2 \propto n^{2(\nu-1)/(\nu+2)}, \tag{15.44}$$

(b) for the logarithmic potential[§]

$$|\Psi(0)|^2 \propto \frac{1}{2n+1}, \tag{15.45}$$

[†]See the references of Quigg and Rosner cited at the beginning of this chapter.

[‡]For one such investigation which includes the Coulomb potential in addition to a confining potential see e.g. S. K. Bose, G. E. Hite and H. J. W. Müller-Kirsten [28]. The case of arbitrary positive power potentials as above has been considered by R. S. Kaushal and H. J. W. Müller–Kirsten [147].

[§]S. K. Bose and H. J. W. Müller–Kirsten [29].

(c) and for the Coulomb potential[¶]

$$|\Psi(0)|^2 \propto \frac{1}{n^3}. \tag{15.46}$$

We can now link the value of the wave function at the origin to the oscillation period T using the arguments of Example 14.5, and there Eq. (14.75). We observed there in the case of the Coulomb potential that the period is proportional to n^3, in agreement with Eq. (15.46) and the classical Kepler period of Eq. (11.133).

Example 15.1: Regge trajectories of the logarithmic potential
Consider the radial Schrödinger equation for the potential $V(r) = g\ln(r/r_0), g > 0$, i.e.

$$\frac{d^2\psi}{dr^2} + \left(\alpha - \beta\ln r - \frac{\gamma}{r^2}\right)\psi = 0, \quad \alpha = \frac{2\mu E'}{\hbar^2}, \quad E' = E + g\ln(r_0), \quad \beta = \frac{2\mu g}{\hbar^2}, \quad \gamma = l(l+1).$$

With the substitutions $r = \exp(z-c), -\infty < z < \infty, c = -\alpha/\beta, L^2 = (l+1/2)^2$, as well as $h^4 = 4\beta\exp[(2\alpha - \beta)/\beta], \psi = \exp[(z-c)/2]$, rewrite the equation as

$$\frac{d^2\phi}{dz^2} + [-L^2 + U(z)]\phi, \quad U(z) = -z\beta e^{2\alpha/\beta}e^{2z}.$$

Expanding $U(z)$ about its extremum at $z_0 = -1/2$ and using the perturbation method of Dingle and Müller derive the expansions

$$\left(l + \frac{1}{2}\right)^2 = \frac{1}{8}h^4 - \frac{1}{2}qh^2 - \frac{1}{2^3 3^2}(33q^2 + 1) - \frac{1}{2^5 3^3 h^2}q(35q^2 - 1) + \cdots,$$

$$E_q = \frac{1}{2}q - \frac{1}{2}\ln\left(\frac{8\mu g r_0^2}{\hbar^2}\right) + \cdots, \quad q = 2n + 1, \quad n = 0, 1, 2, \ldots.$$

Solution: Details can be found in the literature.[‖]

[¶]L. D. Landau and E. M. Lifshitz [157].
[‖]S. K. Bose and H. J. W. Müller–Kirsten [29].

Chapter 16

Screened Coulomb Potentials

16.1 Introductory Remarks

We observed previously that the infinite range of the Coulomb potential —
the concept of range being defined more precisely below — leads to a scattering phase with a logarithm-r contribution. This implies, that no matter how
far the particle is scattered away from the source of the Coulomb potential,
it will always notice the latter's presence. This effect sounds unphysical, and
in fact, it is. In reality, in microscopic and hence quantum physics with the
possibility of real and virtual creation of particles and hence a vacuum state,
a charge as source of a Coulomb force leads to an effective polarization of the
vacuum like that of a dielectric, so that like charges are repelled and unlike
charges are attracted by each other. The result is a *screening* of the Coulomb
potential which thus attributes it a finite range r_0. Phenomenologically the
screened attractive Coulomb potential may be written

$$V(r) = -g^2 \frac{e^{-r/r_0}}{r},$$

where r is the distance from the source charge and the charges and other
constants are collected in the coupling g^2. It is clear that for $r_0 \to \infty$ the
potential becomes of Coulomb type. The exponential insures a rapid fall-off
of the potential at large distances and for this reason the parameter r_0 can be
looked at as a measure of the *range* of the potential. A very similar potential
is the *Yukawa potential*

$$V(r) = -g^2 \frac{e^{-\mu r}}{r},$$

in which μ has the dimension of an inverse length or equivalently that of a
mass in natural units. Historically this potential arose with the realization

that the strong nuclear force is mediated by the exchange of mesons, and that, in fact the parameter μ represents the mass of such a spinless meson.* The relativistic equation of motion of such a meson when free is the Klein–Gordon equation which results from the quantization of the classical relativistic energy momentum relation of a particle of mass μ, i.e. from the relation

$$-p_0^2 + \mathbf{p}^2 + \mu^2 c^2 = 0. \tag{16.1}$$

We encountered Green's functions earlier. A Green's function is effectively the inverse of a quantity called *propagator* which is intimately related to the expression in Eq. (16.1) and is written in spacetime-dimensional Minkowskian notation $1/(p^\nu p_\nu + \mu^2)$. In the literature reference is sometimes made to the so-called *static limit of a relativistic propagator*. What is meant is that the four-dimensional Minkowskian Fourier transform of the propagator is given by the relation

$$\frac{1}{(2\pi)^4} \int d^4k \frac{e^{ikx}}{\mathbf{k}^2 + \mu^2} = \frac{1}{4\pi} \delta(x_0) \frac{e^{-\mu|\mathbf{x}|}}{|\mathbf{x}|}, \tag{16.2}$$

or $(\mathbf{p} = \hbar\mathbf{k})$

$$\frac{1}{(2\pi)^4} \frac{1}{\mathbf{k}^2 + \mu^2} = \frac{1}{(2\pi)^4} \int e^{-i\mathbf{k}\cdot\mathbf{x}} d^3x \frac{e^{-\mu|\mathbf{x}|}}{4\pi|\mathbf{x}|}.$$

We observe that the Fourier transform of the propagator is effectively the Yukawa potential.

Yukawa potentials and superpositions of such potentials play an important role in nuclear physics, and in view of their relation to the exchange of virtual elementary particles have therefore been objects of intense study in elementary particle theory. The realization that mesons and baryons are made up of quarks does not really change that picture at lower energies. Since, however, the mathematical expression with an exponential is not so easy to handle analytically, one frequently resorts in calculations to the expansion of the potential in rising powers of r. Thus in the following we consider a generalized Yukawa potential which can be expanded as a power series in r and is written

$$V(r) = \sum_{i=-1}^{\infty} M_{i+1}(-r)^i, \tag{16.3}$$

where for the real potentials we consider here, all coefficients M_i are real and independent of the energy $E = k^2$. Regge trajectories — which we encounter

---

*The S-matrix theory of strong interactions was actually initiated by W. Heisenberg [133], who later — as is wellknown — deviated from this idea and invested his efforts into the study of nonlinear spinor field theory.

in this context — are functions which interpolate integral (i.e. physical) values of angular momentum as functions of energy E. They are usually written $l = \alpha_n(E)$, where n is an integer and we referred to these already in Chapter 11 in the simple case of the Coulomb potential. Regge trajectories were realized to play an important role in the high energy behaviour of hadronic scattering amplitudes.[†] Regge trajectories arise as poles of the S-matrix in the plane of complex angular momentum. Naturally it is easiest to familiarize oneself with these by studying solvable potential models.[‡]

In the following we consider screened Coulomb or Yukawa potentials of the type of Eq. (16.3) in the radial Schrödinger equation for the partial wave $\psi(l, k; r)$, i.e.

$$\left[\frac{d^2}{dr^2} + k^2 - \frac{l(l+1)}{r^2} - V(r) \right] \psi(l, k; r) = 0, \qquad (16.4)$$

where $E = k^2$ and $\hbar = c = 1 = 2m_0$, m_0 being the reduced mass of the system. Potentials expandable as in Eq. (16.3) have been considered by various authors.[§] Our intention here is to consider these potentials as a generalization of the Coulomb potential and hence to proceed along similar lines in the derivation of Regge trajectories, bound state eigenenergies and the S-matrix. As discussed in detail by Bethe and Kinoshita[¶] one can start by arguing that a countably infinite number of Regge poles may be defined in the region of large negative energies $E = k^2$ of the radial Schrödinger equation by requiring $l + n + 1$ for $n = 0, 1, 2, \ldots$ to be of the order of $1/k$, i.e.

$$l + n + 1 = -\frac{\triangle_n(K)}{2K}, \quad |K| \to \infty, \qquad (16.5)$$

where $K = ik$ and $\triangle_n$ is an expansion in descending powers of K. Basically this argument amounts to an argument similar to that used in the case of the Coulomb potential where the integer n arose from the requirement that the wave function be normalizable, and hence that the confluent hypergeometric series there obtained has to break off after a finite number of terms in order not to destroy this behaviour. Thus in the present case it turns out that it is easiest to calculate first the expansion for the expression $l + n + 1$, i.e. for the Regge trajectories or Regge or l-plane poles of the S-matrix[‖] $l \equiv l_n(K)$

[†]Standard references are the monographs of R. Omnès and M. Froissart [223], S. C. Frautschi [97] and E. J. Squires [258].

[‡]The approximate behaviour of Regge trajectories for the Yukawa potential has also been calculated by H. Cheng and T. T. Wu [47].

[§]S. Mandelstam [185]. See also the other references below.

[¶]H. A. Bethe and T. Kinoshita [21].

[‖]These were first considered by C. Lovelace and D. Masson [180], however, only for the cases of $n = 0, 1$ in the above.

as a function of the energy.** The energy is later obtained by reversion of the resulting series.††Thus it will be seen that with perturbation expansions the problem of the screened Coulomb potential can be solved practically as completely as the Coulomb problem.

16.2 Regge Trajectories

Since our treatment here aims at obtaining an S-matrix as a generalization of that of the Coulomb potential, we naturally assume conditions on the potential $V(r)$ which are such that the S-matrix is meromorphic in the entire plane of complex angular mommentum. The conditions for this have been investigated in the literature* and may be summarized as follows:

$$V(r) = \int_{\mu_0}^{\infty} d\mu\, \sigma(\mu) e^{-\mu r}/r,$$

$$\int_{0}^{\infty} d\rho\, \rho |V(\rho e^{i\theta})| < \infty \text{ for all } |\theta| < \pi/2,$$

$$rV(r) \text{ regular at } r = 0.$$

Under these conditions the S-matrix is meromorphic (i.e. has only simple poles) in the plane of complex angular momentum and in the complex k-plane $(E = k^2)$ cut along the imaginary axis.† Considering such superpositions of Yukawa potentials with

$$\int_{\mu_0}^{\infty} \sigma(\mu) \mu^n d\mu < \text{const. for all } n,$$

one can assume an expansion of the potential $V(r)$ in ascending powers of r, starting with the power r^{-1} of the Coulomb potential.

Proceeding now as in the case of the Coulomb potential we change the variable of Eq. (16.4) to $z = -2Kr$ and set

$$\psi(l, k; z) = e^{-z^2/2} z^{l+1} \chi(l, k; z) \tag{16.6}$$

Then χ is a solution of the equation

$$\mathcal{D}_a \chi = \frac{1}{2K}(M_0 - \triangle_n(K))\chi + \frac{1}{2K} \sum_{i=1}^{\infty} \left(\frac{z}{2K}\right)^i M_i \chi, \tag{16.7}$$

**H. J. W. Müller [201] and [202]; H. J. W. Müller and K. Schilcher [203]. In the following we follow mainly the last two of these references.

††H. J. W. Müller [204].

*See in particular A. Bottino, A. M. Longoni and T. Regge [31] and E. J. Squires [258].

†In the case of the Coulomb potential (cf. Section 11.9) the cut starts at $E = k^2 = 0$, in the present case at $k^2 = 0$ or $k^2 - M_1 = 0$, as may be seen from Eqs. (16.15) and (16.19).

where

$$\mathcal{D}_a = z\frac{d^2}{dz^2} + (b-z)\frac{d}{dz} - a$$

and (cf. Eq. (16.5))

$$a = l + 1 + \frac{\triangle_n(K)}{2K} = -n \quad \text{and} \quad b = 2l + 2 = -2n - \frac{\triangle_n(K)}{K}. \qquad (16.8)$$

The right hand side of Eq. (16.7) is seen to be of order $1/K$, so that to leading order we have

$$\mathcal{D}_a\chi^{(0)} = 0, \quad \chi^{(0)} = \Phi(a,b;z) \equiv \Phi(a),$$

where $\Phi(a,b;z)$ is seen to be a confluent hypergeometric function which for reasons of convenience we abreviate in the following as $\Phi(a)$. We also observe that the ansatz (16.5) is equivalent to $a = -n$, which means that the hypergeometric series breaks off after a finite number of terms, as in the case of the Coulomb problem. The function $\Phi(a,b;z)$ is known to satisfy a recurrence relation which we write here for convenience in the form

$$z\Phi(a) = (a,a+1)\Phi(a+1) + (a,a)\Phi(a) + (a,a-1)\Phi(a-1), \qquad (16.9)$$

where

$$(a,a+1) = a-b+1, \quad (a,a) = b-2a, \quad (a,a-1) = a-1. \qquad (16.10)$$

By a repeated application of the recurrence relation (16.9) we obtain

$$z^m\Phi(a) = \sum_{j=-m}^{m} S_m(a,j)\Phi(a+j). \qquad (16.11a)$$

The coefficients $S_m(a,j)$ may be computed from a recurrence relation which follows from the coefficients (16.10):

$$\begin{aligned} S_m(a,j) &= (a+j-1, a+j)S_{m-1}(a,j-1) + (a+j, a+j)S_{m-1}(a,j) \\ &\quad + (a+j+1, a+j)S_{m-1}(a,j+1). \end{aligned} \qquad (16.11b)$$

The associated boundary conditions are: $S_0(a,0) = 1$, all other $S_0(a,i \neq 0) = 0$, and all $S_m(a,j)$ for $|j| > m$ are zero.[‡]

[‡]Recurrence relations for coefficients of perturbation expansions for Yukawa, anharmonic and cosine potentials have been derived in L. K. Sharma and H. J. W. Müller–Kirsten [249].

Substituting the first approximation $\chi^{(0)} = \Phi(a)$ into the right hand side of Eq. (16.7), the latter can be written

$$R_a^{(0)} = \frac{1}{2K}[a,a]_1\Phi(a) + \sum_{i=1}^{\infty}\frac{1}{(2K)^{i+1}}\sum_{j=-i}^{i}[a,a+j]_{i+1}\Phi(a+j), \quad (16.12a)$$

where

$$[a,a]_1 = M_0 - \triangle_n(K), \quad [a,a+j]_{i+1} = M_i S_i(a,j), \ 0 \le |j| \le i. \quad (16.12b)$$

The usefulness of this notation can now be seen in the ease with which it permits the calculation of any number of higher-order perturbation terms. Any term $\mu\Phi(a+n)$ on the right hand side of Eq. (16.7) and hence in $R_a^{(0)}$ may be cancelled out by adding to $\chi^{(0)}$ a contribution $\mu\Phi(a+n)/n$ except, of course, when $n = 0$. This follows from the fact that

$$\mathcal{D}_a\Phi(a) = 0, \quad \mathcal{D}_{a+n}\Phi(a+n) = 0, \quad \mathcal{D}_{a+n} = \mathcal{D}_a - n,$$

and hence

$$\mathcal{D}_a\Phi(a+n) = n\Phi(a+n),$$

so that

$$\mathcal{D}_a\left[\mu\frac{\Phi(a+n)}{n}\right] = \mu\Phi(a+n).$$

The coefficient of the sum of all the remaining terms in $\Phi(a)$ is then set equal to zero and determines to that order of approximation the quantity $\triangle_n(K)$. This equation is seen to be

$$\begin{aligned}
0 &= \frac{1}{2K}[a,a]_1 + \frac{1}{(2K)^2}[a,a]_2 + \frac{1}{(2K)^3}[a,a]_3 \\
&\quad + \frac{1}{(2K)^4}\left[[a,a]_4 + \frac{[a,a+1]_2}{1}[a+1,a]_2 + \frac{[a,a-1]_2}{-1}[a-1,a]_2\right] \\
&\quad + \cdots.
\end{aligned} \quad (16.13)$$

One can now construct coefficients with their recurrence relations for the individual terms of this expansion, i.e. for the coefficients $M^{(i)}$ of the expansion

$$\triangle_n(K) = M_0 + \sum_{i=1}^{\infty}\frac{1}{(2K)^i}M^{(i)}.$$

Details can be found in the references cited above. Evaluating the first few terms of the expansion (16.13) one obtains the quantity $\triangle_n(K)$ and hence with Eq. (16.5) the Regge trajectories $l = l_n(K)$. We first cite the final result

and then demonstrate the calculation in Example 16.1 by evaluating the first
two terms. The result is

$$
\begin{aligned}
\triangle_n(K) \;=\; & M_0 - \frac{1}{2K^2}[n(n+1)M_2 + M_1 M_0] - \frac{(2n+1)M_0 M_2}{4K^3} \\
& + \frac{1}{8K^4}[3M_4(n-1)n(n+1)(n+2) + 2M_3 M_0(3n^2 + 3n - 1) \\
& + 6M_2 M_1 n(n+1) + 2M_2 M_0^2 + 3M_1^2 M_0] \\
& + \frac{(2n+1)}{8K^5}[3M_4 M_0(n^2 + n - 1) + 3M_3 M_0^2 + M_2^2 n(n+1) \\
& + 4M_2 M_1 M_0] + O(K^{-7}).
\end{aligned}
\tag{16.14}
$$

Inserting this expansion into Eq. (16.5), we obtain the expansion for the
Regge poles, i.e.

$$
\begin{aligned}
l_n(K) \;=\; & -n - 1 - \frac{M_0}{2K} + \frac{1}{4K^3}[n(n+1)M_2 + M_1 M_0] + \frac{(2n+1)M_0 M_2}{8K^4} \\
& - \frac{1}{16K^5}[3M_4(n-1)n(n+1)(n+2) + 2M_3 M_0(3n^2 + 3n - 1) \\
& + 6M_2 M_1 n(n+1) + 2M_2 M_0^2 + 3M_1^2 M_0] \\
& - \frac{(2n+1)}{16K^6}[3M_4 M_0(n^2 + n - 1) + 3M_3 M_0^2 + M_2^2 n(n+1) \\
& + 4M_2 M_1 M_0] + O(K^{-8}).
\end{aligned}
\tag{16.15}
$$

The same expansion may be derived by the WKB method from the Bohr–
Sommerfeld–Wilson integral for three space dimensions as explained in Example 16.2.[§] This is an important observation which indicates the equivalence of the methods. Obviously the expansion — which, of course, is an
asymptotic expansion for large values of the energy — is not very useful in
the very interesting domain around energy zero. However, it is clear that one
expects the trajectories to be finitely closed curves with asymptotes given by
those of the Coulomb potential. Thus the expected appearance is that shown
in Fig. 16.1. With numerical methods one can achieve more, including an
exploration of the domain of small energies E. Such plots of numerically computed Regge trajectories for specific values of the overall coupling constant
and energies varying from minus infinity to plus infinity have been given by
various authors.[¶] The plots confirm the expected behaviour for strongly attractive potentials but exhibit also a superficially unexpected departure into
the lower half of the complex l-plane in the case of the first few trajectories
(counting in terms of the quantum number n) for weak coupling. Analytical

---

[§]Observe there the quadratic form of the centrifugal potential!

[¶]See in particular A. Ahmadzadeh, P. G. Burke and C. Tate [6].

expressions of the behaviour of a Regge trajectory in the immediate neigh-bourhood of $E = 0$ are practically unknown. However, this is a particularly interesting domain since at the position soon after this point at integral l (as at $l = 2$ in Fig. 16.1), where the imaginary part of l is very small, one can expect a resonance with the lifetime determined by this imaginary part. In fact, just as a decay width Γ represents the width of the resonance in energy, so the imaginary part α_I of $l = \alpha_n$ represents its width in angular momen-tum. The conjugate variable to energy is time, and the lifetime Δt of the resonance satisfies the relation $\Gamma \Delta t \sim \hbar$. Similarly the conjugate variable to angular momentum is angle, and the angle $\Delta \theta$, through which the particle orbits during the course of the resonance, satisfies the relation $\alpha_I \Delta \theta \sim \hbar$. For a resonance with a long lifetime, α_I is small and $\Delta \theta$ is large. For a bound state $\alpha_I = 0$ and the orbit becomes permanent.[||]

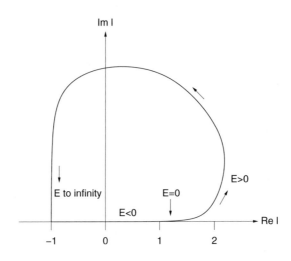

Fig. 16.1 Typical Regge trajectory for a strongly attractive Yukawa potential.

Example 16.1: Evaluation of perturbation terms

Use the above formulae to verify the first two terms in the expansion of $\triangle_n$ for the Yukawa Regge trajectories.

Solution: We evaluate the first three terms of expansion (16.13). From Eq. (16.12b) we obtain

$$[a, a]_2 = M_1 S_1(a, 0) \quad \text{and} \quad [a, a]_3 = M_2 S_2(a, 0).$$

From Eq. (16.11b) we obtain
$$S_1(a.0) = (a, a) S_0(a, 0).$$

Other terms vanish since $S_0(a, \pm 1) = 0$. Hence

$$S_1(a, 0) = (a, a) = b - 2a = -2n - \frac{\triangle_n}{K} + 2n = -\frac{\triangle_n}{K}.$$

[||]See S. C. Frautschi [97], p. 114.

Analogously we evaluate

$$
\begin{aligned}
S_2(a,0) &= (a-1,a)S_1(a,-1) + (a,a)S_1(a,0) + (a+1,a)S_1(a,1) \\
&= (a-b)S_1(a,-1) + (b-2a)S_1(a,0) + aS_1(a,1).
\end{aligned}
$$

We obtain the quantities $S_1(a,i)$ again from Eq. (16.11b). Thus (with terms which are 0):

$$
\begin{aligned}
S_1(a,-1) &= (a,a-1)S_0(a,0) + 0 + 0 = (a-1), \\
S_1(a,0) &= (a,a)S_0(a,0) + 0 + 0 = (b-2a), \\
S_1(a,1) &= (a,a+1)S_0(a,0) + 0 + 0 = (a-b+1).
\end{aligned}
$$

Hence

$$
\begin{aligned}
S_2(a,0) &= (a-b)(a-1) + (a-b+1)a + \left(\frac{\triangle_n}{K}\right)^2 \\
&= -(n+1)\left(n + \frac{\triangle_n}{K}\right) - n\left(n+1+\frac{\triangle_n}{K}\right) + \left(\frac{\triangle_n}{K}\right)^2.
\end{aligned}
$$

Inserting these expressions into expansion (16.13) we obtain

$$
\begin{aligned}
0 &= \frac{1}{2K}(M_0 - \triangle_n) + \frac{1}{(2K)^2}M_1\left(-\frac{2\triangle_n}{2K}\right) \\
&\quad + \frac{1}{(2K)^3}M_2\left[-(n+1)\left(n+\frac{\triangle_n}{K}\right) - n\left(n+1+\frac{\triangle_n}{K}\right) + \left(\frac{\triangle_n}{K}\right)^2\right] \\
&\simeq \frac{1}{2K}(M_0 - \triangle_n) - \frac{2}{(2K)^3}[M_1 M_0 + n(n+1)M_2].
\end{aligned}
$$

Hence

$$
\triangle_n = M_0 - \frac{1}{2K^2}[n(n+1)M_2 + M_1 M_0] + \cdots.
$$

Example 16.2: Eigenvalue approximation by "poles at infinity"

Verify the dominant behaviour of the Regge trajectories of Yukawa potentials by contour integration of the Bohr–Sommerfeld–Wilson integral in which $l(l+1)$ is replaced by $(l+1/2)^2$.

Solution: Ignoring higher order terms in r, we have to evaluate (cf. Sec. 14.4) with $m_0 = 1/2$ and around the classical orbit the integral in the following relation

$$
\left(n + \frac{1}{2}\right)h = \oint_{\text{classical orbit}} pdr \simeq \oint dr\left[-K^2 + \frac{M_0}{r} - \frac{(l+\frac{1}{2})^2\hbar^2}{r^2}\right]^{1/2}.
$$

The integral is most easily evaluated by the method of "poles at infinity". One sets $z = 1/r$ so that, using the *Cauchy formula* (6.65) and observing that in z there is one pole at the origin and one in approaching infinity in view of the expansions

$$
\frac{\sqrt{A + 2Bz + Cz^2}}{z^2} = \frac{\sqrt{A}}{z^2} + \frac{B}{\sqrt{A}z} + \cdots \quad \text{and} \quad = \frac{\sqrt{C}}{z} + O\left(\frac{1}{z^2}\right),
$$

$$
\begin{aligned}
\oint dr\sqrt{A + \frac{2B}{r} + \frac{C}{r^2}} &= -\oint \frac{dz}{z^2}\sqrt{A + 2Bz + Cz^2} \stackrel{(6.65)}{=} -2\pi i\left[\frac{d}{dz}\sqrt{A + 2Bz + Cz^2}\right]_{z=0,\infty} \\
&= 2\pi i\left[\frac{B + Cz}{\sqrt{A + 2Bz + Cz^2}}\right]_{z=0,\infty} \\
&= 2\pi i\left(\frac{B}{\sqrt{A}} + \sqrt{C}\ \text{with ambiguous signs of square roots}\right).
\end{aligned}
$$

Thus here

$$\left(n + \frac{1}{2}\right)h = i \oint dr \left[K^2 - \frac{M_0}{r} + \frac{\hbar^2(l+\frac{1}{2})^2}{r^2}\right]^{1/2} = -2\pi\left[\left(l + \frac{1}{2}\right)\hbar - \frac{M_0}{2\sqrt{K^2}}\right],$$

$$(n + l + 1)h = 2\pi\frac{M_0}{\pm 2K} = \frac{M_0}{\pm 2\hbar K}h.$$

in agreement with our expressions above. This shows that $l(l+1)$ has to appear in the Bohr–Sommerfeld–Wilson integral as $(l+1/2)^2$.

Example 16.3: Calculation of Regge trajectories by the WKB method

Use the WKB method to verify the first two terms in the expansion of $\triangle_n$ for the Yukawa Regge trajectories, i.e. evaluate — with $l(l+1)$ replaced by $(l + \frac{1}{2})^2$ and $V(r) \equiv -M_0/r + v(r)$ — the following equation

$$n + \frac{1}{2} = \oint dr \left[-K^2 + \frac{M_0}{r} - \frac{(l+\frac{1}{2})^2}{r^2} - v\right]^{1/2}, \quad n = 0, 1, 2, \ldots,$$

by expanding the right hand side in rising powers of v and evaluating the individual integrals. The turning points in this case are at $r = 0$ and $r = \infty$.

Solution: For details of the solution we refer to papers of Boukema.[*]

16.3   The S-Matrix

It is clear that if we now work through the usual procedure for the derivation of the S-matrix we pick up a logarithmic phase as in the case of the Coulomb potential. Here we do not perform this procedure. We know that corresponding to every Coulomb Regge trajectory we have a corresponding one in the Yukawa case. Thus we can write down the S-matrix for the present case by exploiting the limiting case of the Coulomb potential.[†] First, however, we can use Eq. (16.5) together with the expansion (16.14) for $\triangle_n(K)$ in order to re-express the latter in terms of l so that

$$l + 1 + \frac{\triangle_l(K)}{2K} = -n, \quad n = 0, 1, 2, \ldots.$$

With this inversion we obtain

$$\triangle_l(K) = M_0 - \frac{1}{2K^2}[l(l+1)M_2 + M_0 M_1]$$

$$+ \frac{1}{8K^4}[3(l-1)l(l+1)(l+2)M_4 + 2M_3 M_0(3l^2 + 3l - 1)$$

$$+ 6l(l+1)M_2 M_1 + 3M_2 M_0^2 + 3M_1^2 M_0] + O(K^{-6}). \quad (16.16)$$

---

[*]J. I. Boukema [34] and [35]. In the second paper the second order WKB approximation is used and shown to yield complete agreement with the terms given in Eq. (16.15).

[†]See Chapter 11 and V. Singh [252].

We observe that $\triangle_l(K)$ has the property

$$\triangle_l(K) = \triangle_l(-K) \tag{16.17}$$

and is therefore real for real values of the potential coefficients M_i.

Paralleling the case of the Coulomb potential in Chapter 11, we can now write down the S-matrix in terms of the scattering phase δ_l as the expression

$$S(l,K) \equiv e^{2i\delta_l} = \frac{\Gamma(l+1+\frac{\triangle_l(K)}{2K})}{\Gamma(l+1-\frac{\triangle_l(K)}{2K})}e^{-i\pi l}. \tag{16.18}$$

We observe that the poles of the S-matrix are given by

$$l+1+\frac{\triangle_l(K)}{2K} = -n, \quad n = 0,1,2\ldots,$$

which are precisely the expressions yielding the Regge trajectories of above. We observe also that the S-matrix is unitary as a consequence of the result (16.17). One can use the explicit expression (16.18) of the S-matrix now to explore further aspects, such as, for instance, the behaviour of the scattering phase in the domain of high energies.[‡]

16.4 The Energy Expansion

It may have been noticed that in the above considerations we could have combined the constant M_1 with the energy into a combination $K^2 + M_1$, and could then have carried out the perturbation procedure not in inverse powers of K but in inverse powers of $\sqrt{K^2 + M_1}$. This has been done[§] and one obtains

$$l = -n - 1 + \frac{M_0}{2\sqrt{K^2+M_1}} - \frac{n(n+1)M_2}{4(K^2+M_1)^{3/2}} + \frac{(2n+1)M_0M_2}{8(K^2+M_1)^2}$$

$$+ \frac{1}{16(K^2+M_1)^{5/2}}\bigg[3M_4(n-1)n(n+1)(n+2) + 2M_2M_0^2$$

$$+2M_3M_0(3n^2+3n-1)\bigg] - O[(K^2+M_1)^{-3}] \tag{16.19}$$

(two additional terms are given in the literature). Expanding the square root $\sqrt{K^2+M_1}$ for $|M_1/K^2| < 1$ one regains $l_n(K)$ with the expansion (16.14). Now reversing the expansion (16.19) we obtain the energy, i.e. the following

---

[‡]See e.g. H. J. W. Müller [211].
[§]See H. J. W. Müller [204].

expansion, in which the leading term is the usual expression of the Balmer formula for the Coulomb potential:

$$
\begin{aligned}
K^2 = {} & -M_1 + \frac{M_0^2}{4(l+n+1)^2}\left[1 - 4n(n+1)\frac{M_2}{M_0}(l+n+1)^2\right. \\
& + 4(2n+1)\frac{M_2}{M_0^3}(l+n+1)^3 \\
& + \frac{4(l+n+1)^4}{M_0^6}\left\{3M_4M_0(n-1)n(n+1)(n+2) - 3M_2^2n^2(n+1)^2\right. \\
& \left. + 2M_3M_0^2(3n^2+3n-1) + 2M_2M_0^3\right\} \\
& - \frac{24(2n+1)(l+n+1)^5}{M_0^6}\left\{M_4M_0(n^2+n-1) + M_3M_0^3\right. \\
& \left. - M_2^2n(n+1)\right\} \\
& - \frac{4(l+n+1)^6}{M_0^9}\left\{10M_6M_0^2(n-2)(n-1)n(n+1)(n+2)(n+3)\right. \\
& + 2M_5M_0^3\{5n(n+1)(3n^2+3n-10)+12\} + 4M_3M_0^5 \\
& + 2M_4M_0^4(6n^2+6n-11) + 2M_2^2M_0^3(9n^2+9n-1) \\
& - 10M_3M_2M_0^2n(n+1)(3n^2+3n+2) + 20M_2^3n^3(n+1)^3 \\
& \left.\left. - 30M_4M_2M_0(n-1)n^2(n+1)^2(n+2)\right\} + \cdots\right].
\end{aligned}
\tag{16.20}
$$

Extensive investigations of the energy eigenvalues for the Yukawa potential can be found in the literature.[¶]

16.5 The Sommerfeld–Watson Transform

The basic theoretical tool for the exploration of Regge poles is a representation of the scattering amplitude given by the so-called Sommerfeld-Watson transform.[*] We recall from Chapter 11 the definition of the scattering amplitude $F(\theta, k)$ as coefficient of the outgoing spherical wave in the asymptotic behaviour of the wave function $\psi(r)$, i.e. in the following expression with an

---

[¶]See in particular G. J. Iafrate and L. B. Mendelsohn [135] and E. Zauderer [286]. Large coupling expansions derived in H. J. W. Müller–Kirsten and N. Vahedi–Faridi [212] contain as special case the expansions of the first pair of authors. Analogous expansions for specific energy-dependent Yukawa potentials have been investigated by A. E. A. Warburton [279].

[*]According to S. C. Frautschi [97], p. 107, the transformation was introduced in the form given below by G. N. Watson [280] in 1918 and later resurrected by A. Sommerfeld [256], p. 282.

ingoing plane wave in the direction of z (here with $\hbar = 1, c = 1, m_0 = 1/2$):

$$\psi(r) \overset{r \to \infty}{\longrightarrow} e^{ikz} + F(\theta, k) \frac{e^{ikr}}{r}. \tag{16.21}$$

The scattering amplitude determines the experimentally measurable cross section σ given by

$$\frac{d\sigma}{d\Omega} = |F(\theta, k)|^2. \tag{16.22}$$

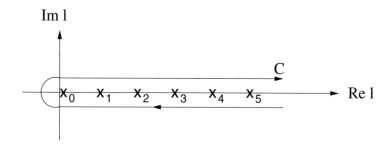

Fig. 16.2 The integration contour C.

The scattering amplitude possesses the partial wave expansion

$$F(\theta, k) = \sum_{l=0}^{\infty} (2l + 1) f(l, k) P_l(\cos \theta), \tag{16.23}$$

in which (cf. Eq. (11.187))

$$f(l, k) = \frac{1}{2ik}[S(l, k) - 1] = \frac{1}{k} e^{i\delta_l(k)} \sin \delta_l(k) \quad \text{and} \quad S(l, k) = e^{2i\delta_l(k)} e^{-i\pi l}, \tag{16.24}$$

$\delta_l(k)$ being the phase shift.[†] The idea is now to consider each term of the expansion (16.23) as the contribution of the residue of a pole in the plane of complex l of some suitably constructed contour integral. For this purpose we consider the following integral taken along the contour C shown in Fig. 16.2 in the complex l-plane:

$$F(\theta, k) = \frac{1}{2i} \int_C dl \, \frac{(2l + 1)}{\sin \pi l} f(l, k) P_l(- \cos \theta). \tag{16.25}$$

Note the minus sign in the argument of the Legendre function. Setting $l = n + x, |x| \ll 1, n = 0, 1, 2, \ldots$, so that

$$\sin \pi l = \sin \pi(n + x) \simeq (-1)^n \pi x = (-1)^l \pi(l - n).$$

[†]Actually $\delta_l = \bar{\delta}_l - l\pi/2$ as in Eqs. (11.184) to (11.187).

Then integrating with the help of Cauchy's residue theorem, we obtain

$$F(\theta, k) = \frac{1}{2i} \int_C dl \, \frac{(2l+1)}{\pi(l-n)} f(l,k) \underbrace{(-1)^l P_l(-\cos\theta)}_{P_l(+\cos\theta)}$$

$$= \sum_{n=0}^{\infty} (2n+1) f(n,k) P_n(\cos\theta), \tag{16.26}$$

which is the usual partial wave expansion.

In order to see the relevance of Regge trajectories in a reaction of particles, we have to digress a little and introduce a few simple ideas which played an important role in the development of particle physics. Consider the reaction of a particle $a(p)$ with four-momentum p_μ colliding with a particle $b(q)$ having four-momentum q_μ and producing a particle $c(p')$ and a particle $d(q')$, as indicated in Fig. 16.3. Thus the direct reaction may, for instance, be elastic scattering of a off b, in which case $c = a$ and $d = b$. Such a direct reaction and its "crossed channel reactions" are shown schematically in Fig. 16.3. The reaction therefore describes the following processes which are also described as (reaction) "*channels*" and in which an overline symbol stands for the appropriate antiparticle (like π meson with positive charge and that with negative charge):

$$\begin{aligned} s : & \quad a + b \quad \longrightarrow \quad c + d, \\ t : & \quad a + \bar{c} \quad \longrightarrow \quad \bar{b} + d, \\ u : & \quad a + \bar{d} \quad \longrightarrow \quad c + \bar{b}. \end{aligned} \tag{16.27}$$

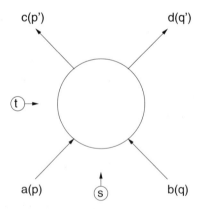

Fig. 16.3 Reaction channels and their respective Mandelstam variables.

For the momenta indicated we set (with metric $+, -, -, -$)

$$s = (p+q)^2, \quad t = (q-q')^2, \quad u = (p-q')^2, \tag{16.28}$$

and for simplicity we assume here that all particles are spinless and have the same mass m_0. The variables s, t, u are known as *Mandelstam variables*. Thus, if s describes the square of the total energy in the s channel, the variables t and u would describe momentum transfers in the crossed channels. One now makes two hypotheses. (a) Mandelstam's hypothesis: All three processes (described by the s, t, u channels) are determined by one and the same relativistically invariant scattering amplitude $A(s, t, u)$ (if any of the external particles has nonzero spin, there will be several such amplitudes which together describe all three processes), which is an analytic function of the variables s, t, u, except for poles and cuts which characterize the reactions in the three channels.[‡] (b) Chew's hypothesis: All (composite) particles lie on Regge trajectories.[§]

Since the three variables s, t, u are not independent, one can write the amplitude as depending only on two, e.g. $A(s, t)$. For a Lorentz-invariant normalization of the initial and final states, one then has

$$A(s,t) = \sqrt{s}F_s(s, \theta_s), \quad A(s,t) = \sqrt{t}F_t(t, \theta_t) \tag{16.29}$$

and with self-explanatory meaning for the cross section of the reactions

$$\frac{d\sigma}{d\Omega_s} = |F_s(s, \theta_s)|^2, \quad \text{etc.,} \tag{16.30}$$

with partial wave expansions (observe the non-invariant factor k has been removed)

$$F_s = \sum_l (2l + 1)F_l(q_s)P_l(\cos \theta_s), \quad F_t = \sum_l (2l + 1)F_l(q_t)P_l(\cos \theta_t). \tag{16.31}$$

Taking all particles to have the same mass m_0, the only kinematics we require in the following is given by the relations (e.g. if we had been considering the s reaction originally, the momentum k there would correspond to q_s)

$$s = -4(q_s^2 + m_0^2), \quad t = -2q_s^2(1 - \cos \theta_s), \quad u = -2q_s^2(1 + \cos \theta_s),$$
$$t = 4(q_t^2 + m_0^2), \quad s = -2q_t^2(1 - \cos \theta_t). \tag{16.32}$$

In the Yukawa picture of a reaction the dynamics is described in terms of the exchange of mesons π (called pions). In a Regge theory on the other hand, the dynamics is described by the exchange of quantum numbers, the quantities carrying these quantum numbers are the Regge trajectories. Symbolically we then have the situation shown in Fig. 16.4.

[‡]See e.g. S. Mandelstam [184] and R. Blankenbecler, M. L. Goldberger, N. N. Khuri and S. B. Treiman [25].

[§]See G. F. Chew [48].

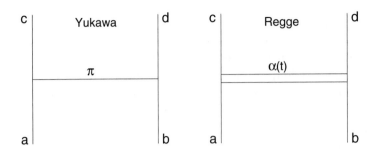

Fig. 16.4 Yukawa versus Regge theory.

Thus one is interested in establishing a representation of the s-channel amplitude in terms of t-channel Regge trajectories. However, the t-channel partial wave expansion is *not* valid in the physical region of the s-channel (i.e. in the domain of s-channel physical values of the kinematical variables of the t-channel). For $s \to \infty$: $\cos \theta_t = 1 + s/2q_t^2 \to \infty$. But the partial wave expansion

$$\sum_l (2l + 1) F_l(q_t) P_l(\cos \theta_t)$$

converges only within the so-called *Lehmann ellipse*[¶] which is shown in Fig. 16.5 for Yukawa potentials

$$V(r) = -\frac{1}{r} \int_{s=m_0^2}^{\infty} dr \, \sigma(\mu) e^{-\mu r}.$$

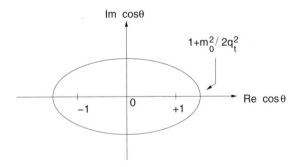

Fig. 16.5 The Lehmann ellipse.

It is known from the Mandelstam representation that the amplitude possesses a branch point at $\cos \theta = 1 + m_0^2/2q_t^2$. But for integral values of l, the Legendre polynomial $P_l(\cos \theta)$ does not possess a cut. So what can one do?

[¶]H. Lehmann [162].

The answer is, to use the Sommerfeld–Watson transform, i.e. the contour representation (16.26) but for a different choice of the contour. Thus consider now the same integral but taken along the closed contour C' shown in Fig. 16.6. Again we have

$$\oint_{C'} \cdots = 2\pi i \sum_n \text{residues } \beta_n.$$

For Yukawa potentials one can show that the integral along the curved portions A and A' of the circle at infinity tends to zero. Moreover, all poles of $f(l,k)$ have $\Im l > 0$, and there is only a finite number $N+1$ in the domain to the right of $\Re l = -1/2$.

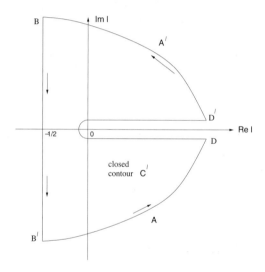

Fig. 16.6 The integration contour C'.

Then

$$\int_{DD'} \cdots + \int_{BB'} \cdots = 2\pi i \sum_n \text{residues } \beta_n,$$

i.e.

$$
\begin{aligned}
F(\theta,k) &= \pi \sum_{n=0}^{N} \frac{[2\alpha_n(k)+1]\beta_n(k)}{\sin \pi \alpha_n(k)} P_{\alpha_n(k)}(-\cos\theta) \\
&\quad + \frac{1}{2i} \int_{-\frac{1}{2}-i\infty}^{-\frac{1}{2}+i\infty} \frac{(2l+1)dl}{\sin \pi l} F(l,k) P_l(-\cos\theta) \\
&\equiv F(s,t).
\end{aligned}
\tag{16.33}
$$

Since
$$\text{for } s \to \infty : \quad \cos \theta_t = 1 + \frac{s}{2q_t^2} \to \infty$$

and (from Tables of Special Functions)

$$\text{for } |z| \to \infty : \quad P_\alpha(z) \simeq \frac{\Gamma(\alpha + \frac{1}{2})}{\sqrt{\pi}\Gamma(\alpha + 1)} (2z)^\alpha,$$

we have $P_{-1/2+iR}(z) \sim O(1/\sqrt{z}) \to 0$ for $s \to \infty$ (i.e. for high energies in the s-channel). It follows that under these conditions, i.e. at high s-channel energies, the amplitude can be represented as a sum over t-channel Regge poles, i.e.

$$F(s,t) = \pi \sum_n \frac{[2\alpha_n(t) + 1]\beta_n(t)}{\sin \pi\alpha_n(t)} P_{\alpha_n(t)}(-\cos \theta_t). \tag{16.34}$$

This is the asymptotic Regge expansion of the invariant amplitude for $s \to \infty$. In particular, if $\alpha_0(t)$ is the Regge pole with largest real part, we have

$$F(s,t) \approx B(t)s^{\alpha_0(t)}. \tag{16.35}$$

This result is valid for $s \to \infty$ and t negative (cf. Eq. (16.32)) and finite. The result demonstrates that the high energy behaviour of an s-channel reaction is determined by the leading Regge trajectory in the crossed t-channel. Apart from more refined details, this behaviour has been confirmed in high energy hadronic reactions.

16.6 Concluding Remarks

The potentials we considered above have wide application under the name of Yukawa potentials in nuclear physics and as screened Coulomb potentials in atomic physics. In subnuclear physics they played an important role before the advent of the quark idea and thus of quantum chromodynamics. Regge trajectories have, of course, also been investigated in the case of other potentials.[||]

The significance of Regge trajectories is evident from the fact that they determine the high energy behaviour of scattering amplitudes, as we discussed briefly in the above. The slope of the Regge function $\alpha_n(k^2)$ is an

---

[||] Simple cases are the harmonic potential and the square-well potential; for plots and discussion see R. Omnès and M. Foissart [223], p. 42. Potentials with a short-range repulsion more singular than $1/r^2$ have been considered by N. Limić [177] and E. Predazzi and T. Regge [229]. The logarithmic potential has been investigated in a way similar to the Yukawa potential above by S. K. Bose and H. J. W. Müller–Kirsten [29].

important parameter in string theory. Regge poles are not the only possible singularities of a scattering amplitude in the plane of complex angular momentum. Singularities appear also in the form of cuts in the plane of complex angular momentum, called *"Regge cuts"*, and also as so-called *"fixed poles"*.**
Whereas the former are related to absorptive properties of a reaction, the latter are related to the distinction between *"elementary"* and *"composite particles"*, a composite particle being one with structure, like a meson or a baryon, and is composed of quarks.

The above treatment of screened Coulomb potentials is incomplete. The small imaginary parts of eigenvalues or lifetimes of resonances have not yet been calculated. The way to do this is similar to calculations in Chapters 18 and 20. Related aspects are discussed in Sec. 20.1.

** For an overview see e.g. H. J. W. Müller–Kirsten [209].

Chapter 17

Periodic Potentials

17.1 Introductory Remarks

From the beginning of applications of quantum mechanics periodic potentials of various types were immediately considered, motivated mainly by the regularities of the crystal structure of matter. Apart from the discontinuous Kronig–Penney potential consisting of a periodic repetition of rectangular barriers, the most immediate candidate is a trigonometric form like that of the the cosine function. The Schrödinger equation with this potential, however, is effectively the Mathieu equation whose solution has for a long time been considered as being very difficult. The reason for these difficulties is, that the Mathieu equation lies outside the scope of equations which can be reduced to hypergeometric type. The Mathieu equation can be obtained as limiting cases of spheroidal wave equations and the elliptic Lamé and ellipsoidal equations which represent a further level of complication.* These elliptic equations involve Jacobian elliptic functions like $\mathrm{sn}(x)$ which are functions that interpolate between the π- or 2π-periodic trigonometric functions on the one hand and the non-periodic hyperbolic functions on the other, and are themselves periodic with period e.g. $2K$ or $4K$, K being the complete elliptic integral of the first kind. In essence these periodic Jacobian elliptic functions, depending on the parameter $k, 0 \leq k^2 \leq 1$, called *elliptic modulus*, are not much harder to handle than trigonometric functions, since many analogous relations hold. Thus Jacobian elliptic functions also pro-

---

*To be precise: Separation of the wave equation $\triangle\psi + \chi^2\psi = 0$ in ellipsoidal coordinates leads to the ellipsoidal wave equation with $\mathrm{sn}^2 z$ and $\mathrm{sn}^4 z$ terms, separation of Laplace's equation $\triangle\psi = 0$ in ellipsoidal coordinates leads to the Lamé equation (no $\mathrm{sn}^4 z$ term). Taking in the ellipsoidal wave equation the limit $k \to 1$ (k elliptic modulus) and putting $\tanh z = \sin\theta$, the ellipsoidal wave equation reduces to the trigonometric form of the spheroidal wave equation. See F. M. Arscott [11], p. 19 and p. 25, example 14.

339

vide *periodic potentials* like trigonometric functions. The cosine potential of
the Mathieu equation can therefore more precisely be described as *Mathieu*
or *trigonometric potential* and the potential $\text{sn}^2(x)$ of the Lamé equation as
Lamé or *elliptic potential*. For large coupling, these periodic potentials may
be approximated by series of degenerate harmonic oscillator potentials. The
finite heights of the periodic functions of the potentials permit tunneling
from one well to another and thereby produce a splitting of the asymptotically degenerate harmonic oscillator levels. A main objective of this chapter
is the calculation of these level splittings following the original calculations
of Dingle and Müller[†] with the perturbation method described in Sec. 8.7,
and to relate these to the weak coupling bands or regions of stability. Both
of these domains are important for a host of other considerations.

The Mathieu equation, or *S*-wave Schrödinger equation with the cosine
potential is conveniently written

$$\frac{d^2y}{dz^2} + [\lambda - 2h^2 \cos 2z]y = 0, \tag{17.1}$$

where h^2 is a parameter and $-\pi \le z \le \pi$. We might mention already here the
much less familiar but more general Lamé equation, or *S*-wave Schrödinger
equation with elliptic potential, which is conveniently written

$$\frac{d^2y}{dz^2} + [\lambda - 2\kappa^2 \text{sn}^2 z]y = 0, \tag{17.2}$$

where $\kappa^2 = n(n + 1)k^2$ and n real and $> -1/2$ and $0 < z < 2K$. Here the
Jacobian elliptic function $\text{sn}\, z$ is a periodic function analogous to a sine function and has real period $2K$.[‡] For $n = 0, 1, 2 \ldots$ the Lamé equation possesses
$2n+1$ polynomial solutions, each with its own characteristic eigenvalue. This
is a significant difference compared with the Mathieu equation, which does
not possess any polynomial solutions. In fact, it is this difference which
implies in the case of the Mathieu equation the existence of the parameter
function ν called *Floquet exponent*.[§] The Mathieu equation can be obtained
from the Lamé equation in the limit of $n \to \infty, k^2 \to 0$ which means that in
this limit the Lamé polynomials degenerate into the periodic Mathieu functions, i.e. those of integral order. The first of these conditions is already

---

[†]Mathieu equation: R. B. Dingle and H. J. W. Müller [73]; Lamé equation: H. J. W. Müller,
[205].

[‡]The reader who encounters these Jacobian elliptic functions here for the first time, need not
be afraid of them. Like the trigonometric functions sine, cosine and tangent, the three Jacobian
elliptic functions $\text{sn}\, z, \text{cn}\, z$ and $\text{dn}\, z$ are handled with analogous formulas, such as $\text{sn}^2 z + \text{cn}^2 z = 1$
and double and half argument formulas etc. which one looks up in books like that of Milne–
Thomson [196] or Tables when required. The most important formulas for our purposes here are
collected in Appendix A.

[§]See also the discussion in Sec. 17.4.1.

seen to rule out polynomials, and one can imagine that a nontrivial parity factor $\exp(in\pi)$ then turns into a complicated phase factor $\exp(i\nu\pi)$. It is this additional parameter function appearing in the solution of the Mathieu equation which attributes the equation its reputation as being particularly hard to handle.

Although the mathematics literature on the subject uses the physical concepts of *stability* and *instability*, these concepts are rarely explained as such there. In classical mechanics stability is treated for instance in connection with planetary motion. The orbits of planets are stable in the sense that deviations from the recurring elliptic orbits are small. Thus the solution of the appropriate Newton equation is essentially a periodic function like $\cos\theta$ (in the case of planetary motion this yields the polar equation of an ellipse), i.e. a bounded function. The instability of an orbit would be evident either from an unbounded spiralling away to infinity or from a collapse into the centre. Correspondingly quantum mechanics — which requires the electrons of an atom to move around the nucleus — explains the stability of atoms. Thus a bounded trigonometric solution is indicative of stability and an unbounded exponential solution of instability. Looking at the Mathieu equation for $h^2 \to 0$, i.e.

$$y'' + \lambda y = O(h^2),$$

we see that in a plot of h^2 versus λ, as in Fig. 17.1, the semi-axis $\lambda > 0, h^2 = 0$ belongs to the domain of stability, since the solution there is of the form $\cos\sqrt{\lambda}z$. Adding the negative term $-2h^2\cos 2z$ to λ, we see that for sufficiently large values of h^2 the periodicity or boundedness of the solution is destroyed and hence becomes one of instability.¶

17.2 Cosine Potential: Weak Coupling Solutions

We consider first the weak coupling case, i.e. the case of h^2 small. Here an important aspect is the determination of the domains of stability of the solutions and the boundaries of these domains. Our considerations below are deliberately made simple and detailed because their treatment in purely mathematical texts requires more time to become accustomed to.

17.2.1 The Floquet exponent

We consider the given Mathieu equation with argument z and compare this with the same equation but with z replaced by $z + \pi$, i.e. we compare the

---

¶For interesting related discussions see M. Salem and T. Vachaspati [242].

equations

$$\frac{d^2y(z)}{dz^2} + [\lambda - 2h^2\cos 2z]y(z) = 0,$$

$$\frac{d^2y(z+\pi)}{d(z+\pi)^2} + [\lambda - 2h^2\underbrace{\cos 2(z+\pi)}_{\cos 2z}]y(z+\pi) = 0.$$

We see that there are solutions which are proportional, i.e. (with T_π as translation operator)

$$T_\pi y(z) \equiv y(z+\pi) = \sigma y(z), \quad \sigma = \text{const.} \tag{17.3}$$

The parameter σ is therefore the eigenvalue of the operator T_π. Our first objective is its determination which amounts to the determination of the parameter called *Floquet exponent* below. Setting $z = 0, \pi$, we obtain

$$\sigma = \frac{y(\pi)}{y(0)}, \quad \sigma = \frac{y(2\pi)}{y(\pi)}, \quad \text{i.e.} \quad \sigma^2 = \frac{y(2\pi)}{y(0)} = 1.$$

Now suppose we write the equation for h^2 small

$$y'' + \lambda y = O(h^2).$$

Then solutions are

$$y(z) \propto e^{\pm i\sqrt{\lambda}z}[1 + O(h^2)] \quad \text{or} \quad \begin{matrix}\cos\sqrt{\lambda}z\\\sin\sqrt{\lambda}z\end{matrix}[1 + O(h^2)],$$

or linear combinations. A solution of the second order differential equation is determined completely only with specification of boundary conditions which determine the two integration constants. We set (here and in the following frequently apart from contributions of $O(h^2)$) with constants A, B and a, b:

$$y(z) = A\cos\sqrt{\lambda}z + B\sin\sqrt{\lambda}z$$

and

$$y_+(z) = a\cos\sqrt{\lambda}z, \quad y_-(z) = b\sin\sqrt{\lambda}z, \tag{17.4}$$

with (for convenience in connection with later equations)

$$y_+'(z) = a\sqrt{\lambda}\sin\sqrt{\lambda}z = -\frac{a\sqrt{\lambda}}{b}y_-(z),$$

$$y_-'(z) = b\sqrt{\lambda}\cos\sqrt{\lambda}z = \frac{b\sqrt{\lambda}}{a}y_+(z).$$

The solution $y(z)$ is an arbitrary solution, whereas the solutions y_+ and y_- have here been chosen specifically as even and odd around $z = 0$ respectively. This means, we choose these with the following set of boundary conditions:

$$y_+(0) = a, \quad y'_+(0) = 0, \quad y_-(0) = 0, \quad y'_-(0) = b\sqrt{\lambda}$$

with *Wronskian* $W[y_+, y_-] = y_+(0)y'_-(0) - y_-(0)y'_+(0) = ab\sqrt{\lambda}$. We can write therefore, with constants α and β,

$$y(z) = \alpha y_+(z) + \beta y_-(z), \quad y'(z) = \alpha y'_+(z) + \beta y'_-(z).$$

Using Eq. (17.3) we obtain

$$\begin{aligned} y(z + \pi) &= \sigma[\alpha y_+(z) + \beta y_-(z)], \\ y'(z + \pi) &= \sigma[\alpha y'_+(z) + \beta y'_-(z)]. \end{aligned}$$

Setting in these last equations $z = 0$ and using the previous pair of equations and the boundary conditions, we obtain

$$\begin{aligned} y(\pi) &= \alpha y_+(\pi) + \beta y_-(\pi) &= \sigma[\alpha a], \\ y'(\pi) &= \alpha y'_+(\pi) + \beta y'_-(\pi) &= \sigma[\beta b\sqrt{\lambda}] \end{aligned}$$

or

$$\begin{pmatrix} y_+(\pi) - a\sigma & y_-(\pi) \\ y'_+(\pi) & y'_-(\pi) - \sigma b\sqrt{\lambda} \end{pmatrix} \begin{pmatrix} \alpha \\ \beta \end{pmatrix} = 0.$$

For linear independence of α and β, the determinant of the matrix must vanish, i.e.

$$[y_+(\pi) - a\sigma][y'_-(\pi) - \sigma b\sqrt{\lambda}] - y'_+(\pi)y_-(\pi) = 0$$

or

$$ab\sqrt{\lambda}\sigma^2 - \{ay'_-(\pi) + b\sqrt{\lambda}y_+(\pi)\}\sigma + \{y_+(\pi)y'_-(\pi) - y'_+(\pi)y_-(\pi)\} = 0.$$

Now from Eq. (17.4),

$$ay'_-(\pi) = b\sqrt{\lambda}y_+(\pi) \quad \text{and} \quad by'_+(\pi) = -a\sqrt{\lambda}y_-(\pi),$$

and the Wronskian

$$W[y_+, y_-] = y_+(\pi)y'_-(\pi) - y_-(\pi)y'_+(\pi) = ab\sqrt{\lambda}$$

in agreement with its value at $z = 0$ above. This expression is equal to the coefficient of σ^2 in the quadratic equation for σ. The roots $\sigma_\pm$ are therefore obtained as

$$\sigma_\pm = \frac{2b\sqrt{\lambda}y_+(\pi) \pm \sqrt{4b^2\lambda y_+^2(\pi) - 4a^2b^2\lambda}}{2ab\sqrt{\lambda}}.$$

Setting
$$f = 2b\sqrt{\lambda}y_+(\pi), \qquad g = 2ab\sqrt{\lambda},$$

we have
$$\sigma_\pm = \frac{f \pm \sqrt{f^2 - g^2}}{g}, \qquad \sigma_+ + \sigma_- = \frac{2y_+(\pi)}{a}, \qquad \sigma_+\sigma_- = 1.$$

From this we see that (as one can also verify explicitly with some manipulations) one root is the inverse of the other, i.e. $\sigma_+ = 1/\sigma_-$. Setting
$$\sigma_+ = e^{i\pi\nu},$$

we have for the sum of the roots
$$\sigma_+ + \sigma_- = 2\cos\pi\nu$$

or$^{\|}$
$$\cos\pi\nu = \frac{y_+(\pi)}{a} \overset{(17.4)}{=} \cos\sqrt{\lambda}\pi. \tag{17.5}$$

The parameter ν is known as *Floquet exponent*. Thus this Floquet exponent is determined by the value at $z = \pi$ of the solution which is even around $z = 0$ and is independent of the normalization constant a of the even solution.**

An important consequence of Eq. (17.5) is that for $h^2 = 0$ we have $\nu = \sqrt{\lambda}$ and hence more generally
$$\nu^2 = \lambda + O(h^2). \tag{17.6}$$

We can derive various terms of this expansion perturbatively with the method of Sec. 8.7. This is the easiest application of that method since only simple trigonometric expressions are involved. This calculation is demonstrated in Example 17.1. The result for λ is the expansion

$$\begin{aligned}\lambda &= \nu^2 + \frac{h^4}{2(\nu^2 - 1)} + \frac{(5\nu^2 + 7)h^8}{32(\nu^2 - 1)^3(\nu^2 - 4)} \\ &+ \frac{(9\nu^4 + 58\nu^2 + 29)h^{12}}{64(\nu^2 - 1)^5(\nu^2 - 4)(\nu^2 - 9)} + O(h^{16}).\end{aligned} \tag{17.7}$$

This series may be reversed to yield the Floquet exponent, i.e.

$$\begin{aligned}\nu^2 &= \lambda - \frac{h^4}{2(\lambda - 1)} - \frac{(13\lambda - 25)h^8}{32(\lambda - 1)^3(\lambda - 4)} \\ &- \frac{(45\lambda^3 - 455\lambda^2 + 1291\lambda - 1169)h^{12}}{64(\lambda - 1)^5(\lambda - 4)^2(\lambda - 9)} + O(h^{16})\end{aligned} \tag{17.8}$$

$^{\|}$Observe that for integral values of ν (in lowest order of h^2), this implies $y_+(\pi)/a = \pm 1$. From this condition we determine later (cf. Eq. (17.22)) the boundary conditions of periodic solutions.

**This point is not immediately clear from mathematics literature, where the constants are usually taken as unity from the beginning.

and so

$$\nu = \sqrt{\lambda} + \frac{h^4}{4(1-\lambda)\sqrt{\lambda}} - \frac{(8 - 35\lambda + 15\lambda^2)h^8}{64(\lambda-4)(\lambda-1)^3\lambda\sqrt{\lambda}} + O(h^{12}). \qquad (17.9a)$$

One observes immediately that these expansions cannot hold for integral values of ν or λ. These cases therefore have to be dealt with separately, as will be done later. With a perturbation theory ansatz for y_+ around $h^2 = 0$ as in Example 17.1 below one obtains the following expansion for $\cos\pi\nu$ which is fairly obvious from Eqs. (17.5) and (17.9a) and is therefore not derived here in detail[tt]

$$\cos\pi\nu = \cos\pi\sqrt{\lambda} + h^4 \frac{\pi\sin\pi\sqrt{\lambda}}{4\sqrt{\lambda}(\lambda-1)}$$

$$+ h^8 \left[\frac{15\lambda^2 - 35\lambda + 8}{64(\lambda-1)^3(\lambda-4)\lambda\sqrt{\lambda}} \pi\sin\pi\sqrt{\lambda} - \frac{\pi^2\cos\pi\sqrt{\lambda}}{32\lambda(\lambda-1)^2} \right]$$

$$+ O(h^{12}). \qquad (17.9b)$$

Next we observe that if we demand that the solutions $y_+(z)$ and $y_-(z)$ satisfy the condition (17.3), we must have $\sigma^2 = 1$ and hence

$$e^{2i\sqrt{\lambda}\pi} = 1,$$

i.e.

$$\sqrt{\lambda} = \pm n, \quad \lambda = n^2, \quad n \text{ an integer.}$$

It follows that in these cases

$$\nu^2 = n^2[1 + O(h^2)]. \qquad (17.10)$$

For these integral cases of ν we demonstrate in Example 17.2 how the expansions for small values of h^2 may be obtained perturbatively.

Example 17.1: The eigenvalue λ for non-integral ν and h^2 small
Use the perturbation method of Sec. 8.7 to obtain the eigenvalue λ as a perturbation expansion for nonintegral values of ν and h^2 small.

Solution: We write the eigenvalue equation $\lambda = \nu^2 - 2h^2\triangle$ and insert this into the Mathieu equation $y'' + [\lambda - 2h^2\cos 2z]y = 0$, which can then be rewritten as

$$D_\nu y = 2h^2(\triangle + \cos 2z)y, \quad \text{where} \quad D_\nu := \frac{d^2}{dz^2} + \nu^2. \qquad (17.11)$$

To lowest order the solution y is

$$y^{(0)} = y_\nu = \cos\nu z \quad \text{or} \quad \sin\nu z \quad \text{or} \quad e^{\pm i\nu z}. \qquad (17.12)$$

[tt]See J. Meixner and F. W. Schäfke [193], p. 124.

The complete solutions of these cases are written respectively in self-evident notation

$$ce_\nu(z, h^2), \quad se_\nu(z, h^2), \quad me_{\pm\nu}(z, h^2).$$

We observe that in these cases

$$
\begin{aligned}
2\cos 2z \, \cos\nu z &= \cos(\nu + 2)z + \cos(\nu - 2)z, \\
2\cos 2z \, \sin\nu z &= \sin(\nu + 2)z + \sin(\nu - 2)z, \\
2\cos 2z \, e^{\pm i\nu z} &= e^{\pm i(\nu+2)z} + e^{\pm i(\nu-2)z},
\end{aligned}
$$

i.e. in each case

$$2\cos 2z \, y_\nu = y_{\nu+2} + y_{\nu-2}. \tag{17.13}$$

The first approximation $y^{(0)} = y_\nu$ leaves unaccounted on the right hand side of Eq. (17.11) terms amounting to

$$
\begin{aligned}
R_\nu^{(0)} &= 2h^2(\triangle + \cos 2z)y_\nu = 2h^2 \triangle y_\nu + h^2(y_{\nu+2} + y_{\nu-2}) \\
&\equiv h^2[(\nu, \nu - 2)y_{\nu-2} + (\nu, \nu)y_\nu + (\nu, \nu + 2)y_{\nu+2}],
\end{aligned}
$$

where

$$(\nu, \nu) = 2\triangle \quad \text{and} \quad (\nu, \nu \pm 2) = 1. \tag{17.14}$$

We now observe that

$$D_\nu y_\nu = 0, \quad \text{so that also} \quad D_{\nu+\alpha} y_{\nu+\alpha} = 0,$$

and hence

$$D_{\nu+\alpha} = D_\nu + \alpha(2\nu + \alpha) \quad \text{and} \quad D_\nu y_{\nu+\alpha} = -\alpha(2\nu + \alpha)y_{\nu+\alpha}. \tag{17.15}$$

Hence a term $\mu y_{\nu+\alpha}$ on the right hand side of Eq. (17.11) and so in $R_\nu^{(0)}$ may be cancelled out by adding to $y^{(0)}$ the new contribution

$$\frac{\mu y_{\nu+\alpha}}{-\alpha(2\nu + \alpha)},$$

except, of course, when $\alpha = 0$ or $2\nu + \alpha = 0$. We assume for the time being that $\alpha \neq 0$ and $2\nu + \alpha \neq 0$ (the latter case requires separate consideration, see Example 17.2) which applies when ν is nonintegral. The right hand side of Eq. (17.14) therefore leads to the following next-to leading order contribution

$$y^{(1)} = h^2\left[\frac{(\nu, \nu - 2)}{2(2\nu - 2)}y_{\nu-2} + \frac{(\nu, \nu + 2)}{-2(2\nu + 2)}y_{\nu+2}\right].$$

Then up to $O(h^2)$ the sum $y^{(0)} + y^{(1)}$ is the solution provided the remaining term in $R_\nu^{(0)}$ vanishes, i.e. $h^2(\nu, \nu) = 0$, i.e. up to $O(h^2)$ we have $\triangle = 0$, or

$$\triangle = O(h^4).$$

Next we treat terms $y_{\nu+\alpha}$ in $y^{(1)}$ in a similar way, i.e. since $y^{(0)} = y_\nu$ leaves unaccounted $R_\nu^{(0)}$, so similarly $y^{(1)}$ leaves uncompensated

$$R_\nu^{(1)} = h^2\left[\frac{(\nu, \nu - 2)}{2(2\nu - 2)}R_{\nu-2}^{(0)} + \frac{(\nu, \nu + 2)}{-2(2\nu + 2)}R_{\nu+2}^{(0)}\right].$$

Proceeding in this manner we obtain the solutions ce, se and me as[*]

$$y = y^{(0)} + y^{(1)} + y^{(2)} + \cdots = y_\nu + \sum_{i=1}^{\infty} h^{2i} \sum_{j=-i, j\neq 0}^{i} p_{2i}(2j)y_{\nu+j} \tag{17.16}$$

---

[*]These small-h^2 expansions are convergent, as shown, for instance, by J. Meixner and F. W. Schäfke [193]. With our perturbation formalism — as demonstrated in cases considered in Secs. 17.3.3 and 19.2.2 — we can write down recurrence relations for the perturbation coefficients $p_{2i}(2j)$ together with boundary conditions. The recurrence relation can also be obtained by substituting the right hand side of Eq. (17.16) directly into the original equation in the form of Eq. (17.11).

together with the equation determining $\triangle$, i.e.

$$0 = h^2(\nu,\nu) + h^4 \left[\frac{(\nu,\nu-2)}{2(2\nu-2)}(\nu-2,\nu) + \frac{(\nu,\nu+2)}{-2(2\nu+2)}(\nu+2,\nu) \right] + \cdots . \qquad (17.17)$$

Inserting 1 for the step-coefficients of Eq. (17.14), we obtain immediately

$$0 = 2\triangle h^2 + \frac{h^4}{2(\nu^2-1)} + \cdots ,$$

which verifies the term of order h^4 in Eq. (17.7). The higher order terms naturally require a little more algebra. This completes the determination of the solutions for nonintegral values of ν in the domain of small values of h^2. These expansions are actually convergent with a definite radius of convergence as explained in the mathematical literature.[†] Note, however, the solutions are not yet normalized.

In the case of integral values of ν, one has to deal with each integral case separately as in the following example.

Example 17.2: The eigenvalue λ for $\nu=1$ and h^2 small
Use the perturbation method of Chapter 8 to obtain the perturbation expansion of λ for $\nu = 1$ and h^2 small.

Solution: We start as in Example 17.1 with $\lambda = \nu^2 - 2h^2\triangle$ but with $\nu = 1$. We consider the specific unperturbed solution $y_\nu = \cos\nu z$ which is the dominant contribution $y^{(0)}$ of our solution y. For $\nu = 1$ we then have $y_1 = \cos z = y_{-1}$. Following the first few arguments of Example 17.1 we now have the situation that the contribution y_1 to the entire solution leaves uncompensated on the right hand side of the equation the terms amounting to

$$\begin{aligned} R^{(0)}_{\nu=1} &= 2h^2(\triangle + \cos 2z)y_{\nu=1} = 2h^2\triangle y_1 + h^2(y_3 + y_{-1}) \\ &= h^2[(2\triangle + 1)y_1 + y_3]. \end{aligned} \qquad (17.18)$$

Again applying Eq. (17.15) (but with $\nu = 1$ and $\alpha = 2$) we obtain the next to leading contribution to the solution as

$$y^{(1)} = \frac{h^2}{\{-2(2\nu+2)\}_{\nu=1}}y_3 = -\frac{h^2}{8}y_3.$$

This contribution leaves uncompensated the terms amounting to

$$R^{(1)}_1 = -\frac{h^2}{8}2h^2(\triangle + \cos 2z)y_3 \overset{(17.13)}{=} -\frac{h^4}{4}\left(\triangle y_3 + \frac{y_5 + y_1}{2}\right).$$

Hence the next contribution to the solution is

$$y^{(2)} = -\frac{h^4}{4}\left(\triangle\frac{y_3}{-8} + \frac{y_5}{2(-4)(2+4)}\right) = -\frac{h^4}{4}\left(\triangle\frac{y_3}{-8} + \frac{y_5}{-48}\right).$$

Proceeding in this manner we obtain as the complete solution the sum

$$y = y^{(0)} + y^{(1)} + y^{(2)} + \cdots$$

provided the coefficient of the sum of the contributions in y_1 contained in $R^{(0)}_1, R^{(1)}_1, \ldots$ is set equal to zero, thus determining the quantity $\triangle$. Hence we obtain

$$0 = h^2(2\triangle + 1) - \frac{h^4}{8} + \cdots .$$

[†]See e.g. J. Meixner and F. W. Schäfke [193], p. 121.

From this we obtain

$$-2h^2\triangle = h^2 - \frac{h^4}{8} + \cdots.$$

Inserting this into $\lambda = \nu^2 - 2h^2\triangle$ for $\nu = 1$ we obtain (cf. Eq. (17.24c) below)

$$\lambda = 1 + h^2 - \frac{h^4}{8} + \cdots. \tag{17.19a}$$

This expression agrees with the result given in the literature (there the eigenvalue is called a_1).[‡]
The associated solution is

$$
\begin{aligned}
y &= y^{(0)} + y^{(1)} + \cdots = \cos z - \frac{h^2}{8}\cos 3z - \frac{h^4}{4}\left(\triangle\frac{\cos 3z}{-8} + \frac{\cos 5z}{-48}\right) + \cdots \\
&= \cos z - \frac{h^2}{8}\cos 3z - \frac{h^4}{4}\left(\frac{\cos 3z}{16} + \frac{\cos 5z}{-48}\right) + \cdots.
\end{aligned} \tag{17.19b}
$$

This expansion agrees with that given in by Meixner and Schäfke [193] (p. 123) (there the solution is called $ce_1(z, h^2)$) except for the overall normalization which implies that in our (still unnormalized) case above there are no contributions $\cos z$ in the higher order contributions. With normalization (not to 1, see below!) one obtains the additional terms given by Meixner and Schäfke [193].

We introduce a normalization constant c and rename the normalized solution y then y_c. The specific normalization here is taken as

$$1 = \frac{1}{\pi}\int_{-\pi}^{\pi} dz\, ce_1^2(z, h^2), \quad \text{and uses} \quad \int_0^{\pi} dz \cos mz \cos nz = \frac{\pi}{2}\delta_{mn}, \quad m, n \text{ integers.}$$

Thus we have

$$
\begin{aligned}
1 &= \frac{1}{\pi}\int_{-\pi}^{\pi} dz\, y_c^2 = \frac{1}{\pi}c^2\int_{-\pi}^{\pi} dz\left[\cos z - \frac{h^2}{8}\cos 3z - \cdots\right]^2, \\
&= \frac{1}{\pi}c^2\int_{-\pi}^{\pi} dz\left[\cos^2 z + \frac{h^4}{64}\cos^2 3z + \cdots\right] = c^2\left[1 + \frac{h^4}{64} + \cdots\right].
\end{aligned}
$$

It follows that

$$c = \left[1 - \frac{h^4}{128} + \cdots\right],$$

and the normalized solution is therefore

$$
\begin{aligned}
y_c &= \left[1 - \frac{h^4}{128} + \cdots\right]\left[\cos z - \frac{h^2}{8}\cos 3z + \cdots\right] \\
&= \cos z - \frac{h^2}{8}\cos 3z - \frac{h^4}{4}\left(\frac{\cos 3z}{16} + \frac{\cos 5z}{-48} + \frac{\cos z}{32}\right) - \cdots
\end{aligned}
$$

in agreement with Meixner and Schäfke [193].

In the above we considered the case of ν an integer and calculated the eigenvalue λ. The reverse situation is later of importance, i.e. the case of λ an integer, and ν is to be found in ascending powers of h^2. We consider this case again in an example.

[‡]J. Meixner and F. W. Schäfke [193], p. 120.

Example 17.3: The Floquet exponent ν for $\lambda=4$ and h^2 small

Show that the expansion of the Floquet exponent ν around $\sqrt{\lambda} = 2$ is given by the following *complex* expansion:

$$\nu = 2 - \frac{i\sqrt{5}}{3}\left(\frac{h}{2}\right)^4 + \frac{7i}{108\sqrt{5}}\left(\frac{h}{2}\right)^8 + \frac{11851i}{31104\sqrt{5}}\left(\frac{h}{2}\right)^{12} + \cdots .$$

Solution: We proceed as follows which demonstrates explicitly how the singular factors cancel out systematically. We set $\lambda - 4 \approx 4\epsilon$ and expand the cosine and sine expressions appearing in Eq. (17.9b) about $\lambda = 4$. We thus obtain the expansions

$$\cos\sqrt{\lambda}\pi = \cos 2\pi + (\lambda - 4)(-\sin\sqrt{\lambda}\pi)_{\lambda=4}\frac{\pi}{2\sqrt{\lambda}} + \cdots = 1 - \frac{(\lambda - 4)^2\pi^2}{8} + \cdots,$$

$$\sin\sqrt{\lambda}\pi = \sin 2\pi + (\lambda - 4)(\cos\sqrt{\lambda}\pi)_{\lambda=4}\frac{\pi}{2\sqrt{\lambda}} + \cdots = \frac{\pi(\lambda - 4)}{2\sqrt{\lambda}} + \cdots.$$

Substituting these expressions into Eq. (17.9b) and considering the approach $\lambda \to 4$ gives

$$\cos\pi\nu = \left[1 - \frac{\pi^2}{8}(\lambda - 4)^2 + \cdots\right] + \frac{h^4\pi^2(\lambda - 4)}{4\sqrt{\lambda}(\lambda - 1)2\sqrt{\lambda}}$$

$$+ h^8\left[\frac{(15\lambda^2 - 35\lambda + 8)\pi^2(\lambda - 4)}{64(\lambda - 1)^3(\lambda - 4)\lambda\sqrt{\lambda}2\sqrt{\lambda}} - \frac{\pi^2}{32\lambda(\lambda - 1)^2}\right] + \cdots$$

$$= \left[1 - \frac{(\lambda - 4)^2\pi^2}{8} + \cdots\right] + \frac{h^4\pi^2}{8\lambda(\lambda - 1)}\left[(\lambda - 4) + \cdots\right]$$

$$+ \frac{h^8\pi^2}{128\lambda^2(\lambda - 1)^3}\left[(11\lambda^2 - 31\lambda + 8) + \cdots\right] + O(h^{12}).$$

Hence — observe the cancellation of factors $(\lambda - 4)$ in the term of $O(h^8)$ in the limit $\epsilon \to 0$ —

$$\cos\pi\nu = 1 + \frac{h^8\pi^2(11 \times 16 - 31 \times 4 + 8)}{2^7 4^2 3^3} + \cdots = 1 + \frac{5\pi^2h^8}{2^9 3^2} + \cdots.$$

Setting in $\cos\pi\nu$ on the left hand side of Eq. (17.9b) $\nu = 2 + \delta$, so that

$$\cos\pi\nu = \cos\pi(2 + \delta) = \cos 2\pi\cos\pi\delta = 1 - \frac{\pi^2\delta^2}{2} + \cdots,$$

and comparing with the above, we obtain[§]

$$\delta = \pm i\frac{\sqrt{5}}{3}\left(\frac{h}{2}\right)^4,$$

and therefore

$$\nu = 2 - \frac{i\sqrt{5}}{3}\left(\frac{h}{2}\right)^4 + \frac{7i}{108\sqrt{5}}\left(\frac{h}{2}\right)^8 + \cdots.$$

Expansions of ν for integral values of $\sqrt{\lambda}$ have recently been given with a larger number of terms:[¶]

$$\sqrt{\lambda} = 3: \quad \nu = 3 - \frac{1}{6}\left(\frac{h}{2}\right)^4 + \frac{133}{4320}\left(\frac{h}{2}\right)^8 + \frac{311}{1555200}\left(\frac{h}{2}\right)^{12} + \cdots,$$

$$\sqrt{\lambda} = 4: \quad \nu = 4 - \frac{1}{15}\left(\frac{h}{2}\right)^4 - \frac{137}{27000}\left(\frac{h}{2}\right)^8 + \frac{305843}{680400000}\left(\frac{h}{2}\right)^{12} + \cdots.$$

[§]R. Manvelyan, H. J. W. Müller–Kirsten, J.–Q. Liang and Yunbo Zhang [189].
[¶]S. S. Gubser and A. Hashimoto [124].

Our next immediate aim is to specify the solutions for which Eq. (17.10) applies, i.e. solutions for integral values of ν.

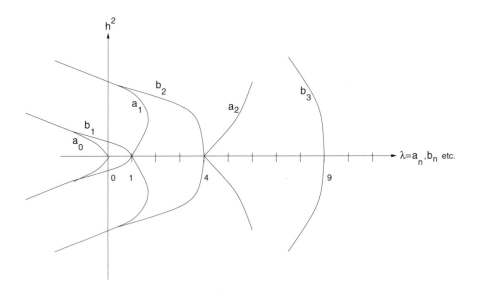

Fig. 17.1 Boundaries of domains of stability.

17.2.2   Four types of periodic solutions

The Mathieu equation allows four types of different solutions — briefly: Even and odd solutions of periods π and 2π. These are defined by specific boundary conditions. We now derive these boundary conditions, concentrating again on the leading term in the even and odd solutions $y_\pm(z)$ for ease of understanding. Taking in particular Eqs. (17.4) into account, we have[||]

$$y_+(z \pm \pi) = a\cos\sqrt{\lambda}(z \pm \pi) = y_+(z)\cos\sqrt{\lambda}\pi \mp \frac{a}{b}y_-(z)\sin\sqrt{\lambda}\pi,$$

$$= y_+(z)\frac{y_+(\pi)}{a} \pm y_-(z)\frac{y_+'(\pi)}{b\sqrt{\lambda}} = y_+(z)\frac{y_+(\pi)}{y_+(0)} \pm y_-(z)\frac{y_+'(\pi)}{y_-'(0)}, \quad (17.20a)$$

and

$$y_-(z \pm \pi) = b\sin\sqrt{\lambda}(z \pm \pi) = y_-(z)\cos\sqrt{\lambda}\pi \pm b\cos\sqrt{\lambda}z\sin\sqrt{\lambda}\pi,$$

$$= y_-(z)\frac{y_-'(\pi)}{b\sqrt{\lambda}} \pm y_+(z)\frac{y_-(\pi)}{a} = y_-(z)\frac{y_-'(\pi)}{y_-'(0)} \pm y_+(z)\frac{y_-(\pi)}{y_+(0)}. \quad (17.20b)$$

[||]These conditions, as also conditions (17.21a) and (17.21b), may also be found, for instance, in F. M. Arscott [11], pp. 28 - 29. We verify these for the large-h^2 solutions in Example 17.6.

We select the equations with lower signs and put $z = \pi/2$. Then

$$y_+\left(\frac{\pi}{2}\right) = y_+\left(\frac{\pi}{2}\right)\frac{y_+(\pi)}{a} \pm y_-\left(\frac{\pi}{2}\right)\frac{y'_+(\pi)}{b\sqrt{\lambda}},$$

$$-y_-\left(\frac{\pi}{2}\right) = y_-\left(\frac{\pi}{2}\right)\frac{y'_-(\pi)}{b\sqrt{\lambda}} \pm y_+\left(\frac{\pi}{2}\right)\frac{y_-(\pi)}{a},$$

and for the derivatives

$$-y'_+\left(\frac{\pi}{2}\right) = y'_+\left(\frac{\pi}{2}\right)\frac{y_+(\pi)}{a} \pm y'_-\left(\frac{\pi}{2}\right)\frac{y'_+(\pi)}{b\sqrt{\lambda}},$$

$$y'_-\left(\frac{\pi}{2}\right) = y'_-\left(\frac{\pi}{2}\right)\frac{y'_-(\pi)}{b\sqrt{\lambda}} \pm y'_+\left(\frac{\pi}{2}\right)\frac{y_-(\pi)}{a},$$

From these equations we obtain $y_+(\pi)$ in terms of functions at $\pi/2$ by eliminating $y'_+(\pi)$ from the first and the third equations. Then

$$y_+\left(\frac{\pi}{2}\right)\left[1 - \frac{y_+(\pi)}{a}\right] = y_-\left(\frac{\pi}{2}\right)\left\{\frac{-y'_+(\frac{\pi}{2})[1 + \frac{y_+(\pi)}{a}]}{y'_-(\frac{\pi}{2})}\right\},$$

which after some rearrangement can be written

$$\frac{y_+(\pi)}{a} = 1 + \frac{2y_-(\frac{\pi}{2})y'_+(\frac{\pi}{2})}{y_+(\frac{\pi}{2})y'_-(\frac{\pi}{2}) - y_-(\frac{\pi}{2})y'_+(\frac{\pi}{2})}.$$

Using the Wronskian (which we actually had above!)

$$W[y_+, y_-] = y_+\left(\frac{\pi}{2}\right)y'_-\left(\frac{\pi}{2}\right) - y_-\left(\frac{\pi}{2}\right)y'_+\left(\frac{\pi}{2}\right)$$

$$= a\left(\cos\sqrt{\lambda}\frac{\pi}{2}\right)^2 b\sqrt{\lambda} + ab\left(\sin\sqrt{\lambda}\frac{\pi}{2}\right)^2\sqrt{\lambda}$$

$$= ab\sqrt{\lambda},$$

we obtain

$$\frac{y_+(\pi)}{a} = 1 + \frac{2y_-(\frac{\pi}{2})y'_+(\frac{\pi}{2})}{ab\sqrt{\lambda}}. \tag{17.21a}$$

Since (with the help of the Wronskian)

$$y_-\left(\frac{\pi}{2}\right)y'_+\left(\frac{\pi}{2}\right) = y_+\left(\frac{\pi}{2}\right)y'_-\left(\frac{\pi}{2}\right) - ab\sqrt{\lambda},$$

we also have

$$\frac{y_+(\pi)}{a} = 1 + 2\frac{y_+(\frac{\pi}{2})y'_-(\frac{\pi}{2}) - ab\sqrt{\lambda}}{ab\sqrt{\lambda}},$$

and hence**

$$\frac{y_+(\pi)}{a} = -1 + \frac{2y_+(\frac{\pi}{2})y'_-(\frac{\pi}{2})}{ab\sqrt{\lambda}}. \tag{17.21b}$$

For integral values of ν (in lowest order of h^2) we know from Eq. (17.5) that the left hand sides of Eqs. (17.21a) and (17.21b) must be ± 1. Thus for these solutions the right hand sides of these equations imply the vanishing of the functions $y_\pm$ or their derivatives at $z = \pi/2$. We see that there are four possibilities and hence four different types of functions. These may be subdivided into classes depending on whether the integral value $n = 0, 1, 2, \ldots$ of ν is even or odd, and whether the function is even or odd. These are defined by the following boundary conditions, where the notation for the corresponding eigenvalue is put alongside on the right:

$$\begin{aligned} y'_+(\tfrac{\pi}{2}) &= 0 \quad \text{with} \quad \lambda \to a_{2n}(h^2), \; y \to \mathrm{ce}_{2n \equiv q_0 - 1}, \\ y_+(\tfrac{\pi}{2}) &= 0 \quad \text{with} \quad \lambda \to a_{2n+1}(h^2), \; y \to \mathrm{ce}_{2n+1 \equiv q_0}, \\ y'_-(\tfrac{\pi}{2}) &= 0 \quad \text{with} \quad \lambda \to b_{2n+1}(h^2), \; y \to \mathrm{se}_{2n+1 \equiv q_0}, \\ y_-(\tfrac{\pi}{2}) &= 0 \quad \text{with} \quad \lambda \to b_{2n+2}(h^2), \; y \to \mathrm{se}_{2n+2 \equiv q_0 + 1} \end{aligned} \tag{17.22}$$

The functions so defined have respectively period $\pi, 2\pi, 2\pi, \pi$, in this order. The ordering of the eigenvalues which then results is

$$a_0 < b_1 < a_1 < b_2 < \cdots < b_n < a_n < \cdots \quad \text{for } h^2 > 0. \tag{17.23}$$

We can now see how the domains of stability and instability arise. This is shown schematically in Fig. 17.1 for the first few eigenvalues which are boundaries of domains of stability (as discussed in Sec. 17.1).

Finally we cite here some expansions of the eigenvalues along with those of the associated solutions (an explicit example, the case of a_1 with solution ce_1 (of period 2π), was treated in detail in Example 17.2):

$$a_0 = -\frac{h^4}{2} + \frac{7h^8}{128} - \frac{29h^{12}}{2304} + \cdots, \quad \sqrt{2}\mathrm{ce}_0 = 1 - \frac{h^2}{2} + \cdots, \tag{17.24a}$$

$$b_1 = 1 - h^2 - \frac{h^4}{8} + \frac{h^6}{64} - \cdots, \quad \mathrm{se}_1 = \sin z - \frac{h^2}{8} \sin 3z + \cdots, \tag{17.24b}$$

$$a_1 = 1 + h^2 - \frac{h^4}{8} - \frac{h^6}{64} - \cdots, \quad \mathrm{ce}_1 = \cos z - \frac{h^2}{8} \cos 3z + \cdots, \tag{17.24c}$$

$$b_2 = 4 - \frac{h^4}{12} + \frac{5h^8}{13824} - \cdots, \quad \mathrm{se}_2 = \sin 2z - \frac{h^2}{12} \sin 4z + \cdots, \tag{17.24d}$$

**To avoid confusion we emphasize: In Sec. 19.3.2 we require this relation for nonintegral values of ν. There the second term on the right is nonzero.

$$a_2 = 4 + \frac{5h^4}{12} - \frac{763h^8}{13824} + \cdots, \quad ce_2 = \cos 2z - h^2 \frac{\cos 4z - 3}{12} + \cdots. \quad (17.24e)$$

17.3 Cosine Potential: Strong Coupling Solutions

17.3.1 Preliminary remarks

In the case of large couplings h^2 we observe from Fig. 17.1 that the boundaries of regions of stability merge to lines. The cosine potential $\cos 2z$ depicted in Fig. 17.2 suggests that for very large fluctuations the potential can be approximated by a number of independent and infinitely high degenerate harmonic oscillator potentials. Consequently one expects the eigenvalues in the large h^2 domain to be given approximately by those of the harmonic oscillator. In the case of the cosine potential, of course, the walls are not infinitely high and hence tunneling occurs from one well into another. The main objective of the present section is therefore the calculation of the tunneling effect in the form of a splitting of the otherwise asymptotically degenerate harmonic oscillator levels.[††]

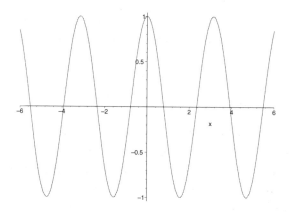

Fig. 17.2 The cosine potential.

Since we have already a definite notation for the levels (boundaries of regions of stability) in the domain of small values of h^2 we naturally want to be able to relate these correctly to those in the asymptotic large-h^2 domain. Thus in Fig. 17.3 we show schematically how the low energy solutions for integral values of the Floquet exponent are related to the asymptotically degenerate harmonic oscillator eigenvalues. We can see from the inverse of Fig. 17.2 or by calculation that the potential $2h^2 \cos 2z$ has (harmonic oscillator-like)

[††]We follow here R. B. Dingle and H. J. W. Müller [73].

minima at $z = \pm \pi/2$. We therefore wish to expand it about these points. Thus we rewrite the Mathieu equation as

$$y'' + \left[\lambda + 2h^2 \cos 2 \left(z \pm \frac{1}{2}\pi \right) \right] y = 0$$

and take

$$\cos 2 \left(z \pm \frac{1}{2}\pi \right) \simeq 1 - 2 \left(z \pm \frac{1}{2}\pi \right)^2.$$

Changing the independent variable now to

$$\chi = 2h^{1/2} \left(z \pm \frac{1}{2}\pi \right),$$

the equation is approximated by

$$\frac{d^2 y}{d\chi^2} + \left[\frac{\lambda + 2h^2}{4h} - \frac{\chi^2}{4} \right] y = 0. \tag{17.25}$$

This is a Weber (or parabolic cylinder) equation which we encountered earlier (cf. Secs. 8.2 and 8.3) and is, of course, simply the one-dimensional Schrödinger equation for the harmonic oscillator. This equation has normalizable solutions for

$$\frac{\lambda + 2h^2}{2h} = q_0 \equiv 2n + 1, \quad n = 0, 1, 2, \ldots.$$

In the case of finite heights of the potential barriers, as in the case of the cosine potential, there is tunneling from one barrier to the next. Thus in this case the quantity on the left is not an exact odd integer, but only approximately so, and we set this equal to q. Taking the remaining terms of the cosine expansion into account, we therefore set

$$\lambda = -2h^2 + 2hq + \frac{\triangle}{8}, \tag{17.26}$$

where $\triangle/8$ is a remainder.

17.3.2   The solutions

We insert Eq. (17.26) into the original Mathieu equation and obtain

$$y'' + 2h \left[q + \frac{\triangle}{16h} - 2h \cos^2 z \right] y = 0. \tag{17.27}$$

For $h \to \infty$ the solutions of this equation are

$$y \sim \exp\left(\pm 2h \int^z \cos z \, dz \right) = \exp(\pm 2h \sin z).$$

Writing $y = A \exp(\pm 2h \sin z)$, and substituting this into Eq. (17.27) we obtain

$$A'' \pm 4h \cos z \, A' + 2h\left(q \mp \sin z + \frac{\triangle}{16h} \right) A = 0. \tag{17.28}$$

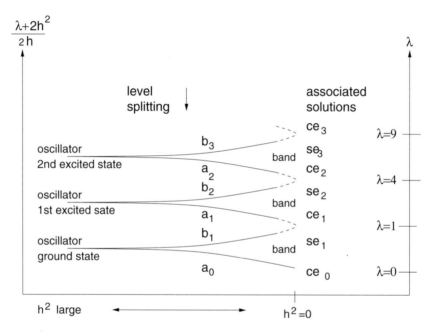

Fig. 17.3 Schematic picture of Mathieu equation eigenvalues.

Choosing

$$y(z) = A(z)e^{2h \sin z}, \tag{17.29}$$

the equation can be written in the form

$$D_q^{(A)} A = -\frac{1}{2^6 h}(16A'' + 2\triangle A), \tag{17.30}$$

where

$$D_q^{(A)} := \cos z \frac{d}{dz} + \frac{1}{2}(q - \sin z). \tag{17.31}$$

We make the important observation here that if one solution of the type (17.29) is known, another solution, the linearly independent one, can be obtained by changing the sign of z or alternatively the signs of both q and h throughout.

The solution $A_q(z)$ of the first order equation $D_q^{(A)} A_q(z) = 0$ is

$$
A_q(z) = \frac{\cos^{\frac{1}{2}(q-1)}\left(\frac{1}{4}\pi + \frac{1}{2}z\right)}{\sin^{\frac{1}{2}(q+1)}\left(\frac{1}{4}\pi + \frac{1}{2}z\right)} = \frac{\cos^{\frac{1}{2}(q-1)}\left(\frac{1}{4}\pi + \frac{1}{2}z\right)}{\cos^{\frac{1}{2}(q+1)}\left(\frac{1}{4}\pi - \frac{1}{2}z\right)}. \tag{17.32}
$$

To lowest order the solution A is $A^{(0)} = A_q$. With some algebra one finds that

$$
16 A_q'' + 2\triangle A_q = (q, q+4) A_{q+4} + (q, q) A_q + (q, q-4) A_{q-4}, \tag{17.33}
$$

where

$$
\begin{aligned}
(q, q+4) &= (q+1)(q+3), \\
(q, q) &= 2[(q^2+1) + \triangle], \\
(q, q-4) &= (q-1)(q-3).
\end{aligned} \tag{17.34}
$$

The leading approximation $A^{(0)} = A_q$ therefore leaves uncompensated terms on the right hand side of Eq. (17.30) amounting to

$$
R_q^{(0)} = -\frac{1}{2^6 h}[(q, q+4) A_{q+4} + (q, q) A_q + (q, q-4) A_{q-4}]. \tag{17.35}
$$

But since

$$
D_q^{(A)} A_q = 0, \quad D_{q+4i}^{(A)} A_{q+4i} = 0,
$$

and so

$$
D_q^{(A)}\left(\frac{\mu A_{q+4i}}{-2i}\right) = \mu A_{q+4i},
$$

a term μA_{q+4i} on the right hand side of Eq. (17.30) can be cancelled out by adding to $A^{(0)}$ a new contribution $-\mu A_{q+4i}/2i$ — except, of course, when $i = 0$ — the terms not involving A_q in (17.35) can be cancelled out by adding to $A^{(0)}$ the next order contribution

$$
A^{(1)} = \frac{1}{2^7 h}\left[\frac{(q, q+4)}{1} A_{q+4} + \frac{(q, q-4)}{-1} A_{q-4}\right]. \tag{17.36}
$$

Then $A^{(0)} + A^{(1)}$ is the approximation of A to that order provided the co-efficient of the term in (17.35) not yet taken care of is set equal to zero, i.e.

$$
(q, q) = 0.
$$

This equation determines the quantity $\triangle$ in the eigenvalue equation to the same order of approximation. In its turn the contribution $A^{(1)}$ leaves uncompensated terms on the right hand side of Eq. (17.30) amounting to

$$
\begin{aligned}
R_q^{(1)} &= \frac{1}{2^7 h}\left[\frac{(q, q+4)}{1}R_{q+4}^{(0)} + \frac{(q, q-4)}{-1}R_{q-4}^{(0)}\right] \\
&= -\frac{1}{2^{13}h^2}\left[\frac{(q, q+4)}{1}(q+4, q+8)A_{q+8} + \frac{(q, q+4)}{1}(q+4, q+4)A_{q+4}\right. \\
&\quad + \left\{\frac{(q, q+4)}{1}(q+4, q) + \frac{(q, q-4)}{-1}(q-4, q)\right\}A_q \\
&\quad \left. + \frac{(q, q-4)}{-1}(q-4, q-4)A_{q-4} + \frac{(q, q-4)}{-1}(q-4, q-8)A_{q-8}\right].
\end{aligned}
$$

$$(17.37)$$

Here all the terms, except those involving A_q, can be taken care of by adding to $A^{(0)} + A^{(1)}$ the next order contribution $A^{(2)}$, where

$$
\begin{aligned}
A^{(2)} &= \frac{1}{2^{14}h^2}\left[\frac{(q, q+4)}{1}\frac{(q+4, q+8)}{2}A_{q+8} + \frac{(q, q+4)}{1}\frac{(q+4, q+4)}{1}A_{q+4}\right. \\
&\quad + \frac{(q, q-4)}{-1}\frac{(q-4, q-4)}{-1}A_{q-4} \\
&\quad \left. + \frac{(q, q-4)}{-1}\frac{(q-4, q-8)}{-2}A_{q-8}\right].
\end{aligned}
$$

$$(17.38)$$

Clearly we can continue in this way. Then for

$$A = A^{(0)} + A^{(1)} + A^{(2)} + \cdots$$

to satisfy Eq. (17.30) and hence $y = Ae^{2h\sin z}$ to be a solution of the Mathieu equation, the sum of the terms in A_q in Eqs. (17.35) and (17.37) and so on — left uncompensated — must vanish. Thus

$$
0 = -\frac{(q, q)}{2^6 h} - \frac{1}{2^{13}h^2}\left[\frac{(q, q+4)}{1}(q+4, q) + \frac{(q, q-4)}{-1}(q-4, q)\right] - \cdots,
$$ (17.39)

which is the equation from which the quantity $\triangle$ and hence the eigenvalue λ is determined. We note here again that if one solution $y = A(z)e^{2h\sin z}$ is known, an associated solution $y = \overline{A}(z)e^{-2h\sin z}$ with the same eigenvalue and thus the same expansion (17.39) is obtained by changing the sign of z or alternatively the signs of both q and h throughout. Thus in particular

$$\overline{A}(z, q, h) = A(-z, q, h), \quad \overline{A}(z, q, h) = A(z, -q, -h).$$ (17.40)

Finally we note that the above trigonometric solutions of the Mathieu equation are valid in the domains

$$\left| \cos\left(\frac{1}{4}\pi \pm \frac{1}{2}z\right) \right| \gg \frac{1}{\sqrt{h}}, \tag{17.41}$$

so that the early successive contributions decrease in magnitude. We observe that for the large values of h we are considering here, this means the solutions are valid in particular around $z = 0$ since then $1/\sqrt{2} \gg 1/\sqrt{h} \sim 0$.

Before we study the eigenvalue equation in more detail, we show that two other pairs of solutions with a similar coefficient structure and the same eigenvalue expansion can be found which are valid in adjoining domains of the variable z.

We return to Eqs. (17.28) and write the upper equation for a solution $y = B\exp(2h\sin z)$ in new notation

$$B'' + 4h\cos z\, B' + 2h\left(q - \sin z + \frac{\triangle}{16h}\right)B = 0. \tag{17.42}$$

Changing now the independent variable to

$$w(z) = 4\sqrt{h}\cos\left(\frac{1}{4}\pi + \frac{1}{2}z\right), \tag{17.43}$$

it is found that

$$D_q^{(B)} = \frac{1}{2^6 h}4\left[w^2\frac{d^2 B}{dw^2} + w(1 - w^2)\frac{dB}{dw} - \left(w^2 + \frac{1}{2}\triangle\right)B\right], \tag{17.44}$$

where

$$D_q^{(B)} = \frac{d^2}{dw^2} - w\frac{d}{dw} + \frac{1}{2}(q - 1). \tag{17.45}$$

We recognize $D_q^{(B)}$ as the differential operator of the *Hermite equation*, i.e. a solution $B_q(w)$ of the equation $D_q^{(B)}B_q(w) = 0$ is $B_q(w) \propto He_{(q-1)/2}(w)$, where $He_n(w)$ is a *Hermite polynomial* when n is an integer.[‡‡] Hence in dominant order the solution B is $B^{(0)} = B_q(w)$, and it is convenient to set

$$B_q(w) = \frac{He_{\frac{1}{2}(q-1)}(w)}{2^{\frac{1}{4}(q-1)}\left[\frac{1}{4}(q - 1)\right]!}, \tag{17.46}$$

[‡‡]We encountered Hermite polynomials earlier in Chapter 6. The differential equation there is Eq. (6.67). Note different definitions in use, and discussed there.

because then as shown in Example 17.4 (using the known recurrence relations of $He_m(w)$) the perturbation remainder can be linearized, i.e.

$$-4\left\{ w^2 \frac{d^2 B_q}{dw^2} + w(1 - w^2)\frac{dB_q}{dw} - \left(w^2 + \frac{1}{2}\Delta\right)B_q \right\}$$
$$= (q, q+4)B_{q+4} + (q, q)B_q + (q, q-4)B_{q-4}, \qquad (17.47)$$

where *the coefficients are the same as before*, i.e. as in Eq. (17.34), which is *an amazing and unique property of the Mathieu solutions* — cf. also the same coefficients in the case of functions C below. Comparison with the case of solution A now shows that the form of that solution can be taken over here except that everywhere $A_q(z)$ has to be replaced by $B_q(w)$, the expansion for the quantity Δ being identical with that of the previous case. The solution with contributions $B^{(0)}, B^{(1)}, \ldots$ forms a rapidly decreasing expansion provided that

$$|w(z)| \ll \sqrt{h}, \quad \text{i.e.} \quad \left| \cos\left(\frac{1}{4}\pi + \frac{1}{2}z\right) \right| \ll 1. \qquad (17.48)$$

This shows that this solution is valid in particular around $z = \pi/2$. Again a change in the sign of z throughout yields the associated Mathieu function $y = \overline{B}\exp(-2h\sin z)$ with $\overline{B}[w(z)] = B[w(-z)]$ and domain of validity in particular around $z = -\pi/2$.

Finally we obtain a third pair of solutions, i.e. $y = C\exp(2h\sin z)$ and $\overline{C}\exp(-2h\sin z)$, with the equation

$$C'' + 4h\cos z\, C' + 2h\left(q - \sin z + \frac{\Delta}{16h}\right)C = 0, \qquad (17.49)$$

by changing the independent variable to

$$w(-z) = 4\sqrt{h}\cos\left(\frac{1}{4}\pi - \frac{1}{2}z\right). \qquad (17.50)$$

Then the dominant order solution is $C^{(0)}(w) = C_q(w)$ with

$$D_q^{(C)}C_q(w) = 0, \quad D_q^{(C)} = \frac{d^2}{dw^2} + w\frac{d}{dw} + \frac{1}{2}(q+1),$$

and — again choosing the multiplicative factor suitably —

$$C_q(w) = 2^{\frac{1}{4}(q+1)}\left[\frac{1}{4}(q-3)\right]!He^*_{-\frac{1}{2}(q+1)}(w), \qquad (17.51)$$

since then (using the known recurrence relations of He_m^*)

$$-4\left\{w^2\frac{d^2C_q}{dw^2} + w(1+w^2)\frac{dC_q}{dw} + \left(w^2 + \frac{1}{2}\triangle\right)C_q\right\}$$
$$= (q,q+4)C_{q+4} + (q,q)C_q + (q,q-4)C_{q-4}, \qquad (17.52)$$

and the coefficients are again the same as before. Again the solutions have the same structure as in the previous two cases and the eigenvalue expansion remains unchanged. The solution $y = C\exp(2h\sin z)$ is a decreasing asymptotic expansion provided that

$$|w(-z)| \ll \sqrt{h}, \quad \text{i.e.} \quad \left|\cos\left(\frac{1}{4}\pi - \frac{1}{2}z\right)\right| \ll 1. \qquad (17.53)$$

This solution is valid in particular around $z = -\pi/2$. Again a change in sign of z throughout yields the associated solution $y = \overline{C}\exp(-2h\sin z)$ with $\overline{C}[w(z)] = B[w(-z)]$ and domain of validity in particular around $z = \pi/2$. These domains of validity are indicated in Fig. 17.4. We note that

$$\overline{C}(z,q,h) \propto B(z,-q,-h), \qquad (17.54)$$

where the proportionality factor is complex.

We have thus determined three pairs of solutions of the Mathieu equation, each of them associated with one and the same expansion for the eigenvalue. A closer look at that expansion is our next objective.

Example 17.4: Hermite function linearization of perturbation
Using the recurrence relations (6.59d) of Hermite functions $He_n(w)$, show that the perturbation remainder of Eq. (17.44) can be linearized with the same coefficients as in the case of the trigonometric solutions.

Solution: Inserting the dominant Hermite function solution into the right hand sid of Eq. (17.44), we have a remainder

$$\mathcal{R} := w^2 He_n'' + w(1-w^2)He_n' - \left(w^2 + \frac{1}{2}\triangle\right)He_n, \quad \text{where} \quad n = \frac{1}{2}(q-1).$$

Using the recurrence relations

$$wHe_n(w) = He_{n+1}(w) + nHe_{n-1}(w), \qquad He_n'(w) = wHe_n(w) - He_{n+1}(w),$$

the remainder $\mathcal{R}$ can be re-expressed as

$$\mathcal{R} = -(n+1)He_{n+2} - \left[(2n^2+2n+1) + \frac{1}{2}\triangle\right]He_n - n^2(n-1)He_{n-2}$$
$$= -\frac{1}{2}(q+1)He_{\frac{1}{2}(q+3)} - \left[\frac{1}{2}(q^2+1) + \frac{1}{2}\triangle\right]He_{\frac{1}{2}(q-1)} - \frac{1}{8}(q-1)^2(q-3)He_{\frac{1}{2}(q-5)}.$$

It follows that with the definition of $B_q(w)$ as

$$B_q(w) = \frac{He_{\frac{1}{2}(q-1)}(w)}{2^{\frac{1}{4}(q-1)}[\frac{1}{4}(q-1)]!},$$

that

$$-\frac{4\mathcal{R}}{2^{\frac{1}{4}(q-1)}[\frac{1}{4}(q-1)]!} = (q,q+4)B_{q+4}(w) + (q,q)B_q(w) + (q,q-4)B_{q-4}(w),$$

with $(q,q+4) = (q+1)(q+3), (q,q) = 2[(q^2+1) + \triangle], (q,q-4) = (q-1)(q-3)$.

17.3.3 The eigenvalues

We saw above that along with each solution the following expansion results:

$$-2\triangle = M_1 + \frac{1}{2^7 h}M_2 + \frac{1}{2^{14}h^2}M_3 + \cdots, \qquad (17.55)$$

where

$$
\begin{aligned}
M_1 &= 2(q^2+1), \\
M_2 &= \frac{(q,q+4)}{1}(q+4,q) + \frac{(q,q-4)}{-1}(q-4,q), \\
M_3 &= \frac{(q,q+4)}{1}\frac{(q+4,q+4)}{1}(q+4,q) \\
&\quad + \frac{(q,q-4)}{-1}\frac{(q-4,q-4)}{-1}(q-4,q) \qquad (17.56)
\end{aligned}
$$

and so on. The coefficients have thus been constructed in a way which now allows the formulation of a recurrence relation. We see that (a) every coefficient M_r results from a sequence of r moves from q back to q, no intermediate move to or from q being allowed. (b) The allowed moves are $+4, 0$ and -4. In addition (c) each move except the last has to be divided by the final displacement from q divided by 4.

We can write the solutions obtained previously, e.g. the solution $y = \exp(2h\sin z)A(z)$, in the following compact form with coefficients $P_r(q,j)$:

$$
\begin{aligned}
y &= e^{2h\sin z}A(z) \\
&= e^{2h\sin z}\sum_{r=0}^{\infty}\frac{1}{(2^7 h)^r}\sum_{j=-r, j\neq 0}^{r}P_r(q,j)A_{q+4j} \\
&= e^{2h\sin z}\Big[A_q + \frac{1}{2^7 h}\{P_1(q,1)A_{q+4} + P_1(q,-1)A_{q-4}\} \\
&\quad + \frac{1}{(2^7 h)^2}\{P_2(q,2)A_{q+8} + P_2(q,1)A_{q+4} \\
&\quad + P_2(q,-1)A_{q-4} + P_2(q,-2)A_{q-8}\} + \cdots\Big]. \qquad (17.57)
\end{aligned}
$$

Clearly

$$P_1(q,1) = \frac{(q,q+4)}{1}, \quad P_1(q,-1) = \frac{(q,q-4)}{-1},$$

$$P_2(q,2) = \frac{(q,q+4)}{1}\frac{(q+4,q+8)}{2}, \quad P_2(q,1) = \frac{(q,q+4)}{1}\frac{(q+4,q+4)}{1},$$

$$P_3(q,3) = \frac{(q,q+4)}{1}\frac{(q+4,q+8)}{2}\frac{(q+8,q+12)}{3},$$

$$P_3(q,2) = \frac{(q,q+4)}{1}\frac{(q+4,q+8)}{2}\frac{(q+8,q+4)}{1}$$
$$+\frac{(q,q+4)}{1}\frac{(q+4,q+4)}{1}\frac{(q+4,q+4)}{1},$$

and so on. By the above rules we can write down the recurrence relation for the evolution of a coefficient P_r by steps from the coefficients P_{r-1}:

$$jP_r(q,j) = (q+4j-4,q+4j)P_{r-1}(q,j-1) + (q+4j,q+4j)P_{r-1}(q,j)$$
$$+(q+4j+4,q+4j)P_{r-1}(q,j+1) \qquad (17.58)$$

with $P_0(q,0) = 1$ and all other $P_0(q,j \neq 0) = 0$. From the coefficients (17.34) we deduce that

$$P_r(q,-j) = (-1)^r P_r(-q,j). \qquad (17.59)$$

One can now write down formulas for M_{2r} and M_{2r+1} in terms of the coefficients $P_r(q,j)$. Each term in M_{2r} can be considered as the product of the contribution from a sequence of r moves starting from q and ending at $q + 4j$, and the contribution from a sequence of r moves from $q + 4j$ back to q, i.e. as $jP_r^2(q,j)$, the factor j removing a duplication of factors in the denominator at the junction. Thus we have, for instance,

$$M_2 = P_1^2(q,1) - P_1^2(q,-1),$$

and in general*

$$M_{2r} = \sum_{j=1}^{r} j[P_r^2(q,j) - P_r^2(q,-j)] = 2\sum_{j=1}^{r} jP_r^2(q,j),$$

$$M_{2r+1} = \sum_{j=1}^{r} j[P_r(q,j)P_{r+1}(q,j) - P_r(q,-j)P_{r+1}(q,-j)]$$

$$= 2\sum_{j=1}^{r} jP_r(q,j)P_{r+1}(q,j). \qquad (17.60)$$

---

*R. B. Dingle and H. J. W. Müller [73].

In this way one obtains the eigenvalue expansion

$$
\begin{aligned}
\lambda \ = \ & -2h^2 + 2hq - \frac{1}{2^3}(q^2+1) - \frac{1}{2^7 h}q(q^2+3) \\
& -\frac{1}{2^{12}h^2}(5q^4 + 34q^2 + 9) - \frac{1}{2^{17}h^3}q(33q^4 + 410q^2 + 405) \\
& -\frac{1}{2^{20}h^4}(63q^6 + 1260q^4 + 2943q^2 + 486) \\
& -\frac{1}{2^{25}h^5}q(527q^6 + 15617q^4 + 69001q^2 + 41607) \\
& -\frac{1}{2^{31}h^6}(9387q^8 + 388780q^6 + 2845898q^4 + 4021884q^2 + 506979) \\
& -\frac{1}{2^{37}h^7}q(175045q^8 + 9702612q^6 + 107798166q^4 \\
& \qquad\qquad + 288161796q^2 + 130610637) - \cdots
\end{aligned}
\tag{17.61}
$$

These are the terms given explicitly in the reference cited above.[†] Terms up to and including those of order $1/h^5$ had been obtained by others and by different methods.[‡] We observe that the expansion remains unchanged under the replacements $q \to -q, h \to -h$, as pointed out earlier. Up to this point, of course, q is only known to be approximately an odd integer $q_0 = 2n + 1$. Its precise deviation from q_0, as a result of tunneling between wells, remains to be determined next. It will be seen in Chapter 18 that in the case of anharmonic oscillators, the solutions and eigenvalue expansions again have the same type of symmetries.

17.3.4 The level splitting

Above we obtained three pairs of solutions along with their domains of validity as decreasing asymptotic expansions. These domains in the interval $-\pi/2 \le z \le \pi/2$ are indicated in Fig. 17.4. We observe that there are regions, where solutions overlap. This means that the in approaching such a domain the appropriate solutions must become proportional in view of their uniqueness. In fact we demonstrate in Example 17.5 the following proportionalities there:

$$
\begin{aligned}
B[w(z)] = \alpha A(z), \qquad \alpha = & \ \frac{(8h)^{\frac{1}{4}(q-1)}}{[\frac{1}{4}(q-1)]!}\left[1 - \frac{1}{2^7 h}(3q^2 - 2q + 3)\right. \\
& \left. + \frac{1}{2^{15}h^2}(9q^4 - 92q^3 + 70q^2 - 284q + 57) + \cdots\right]
\end{aligned}
\tag{17.62a}
$$

[†]R. B. Dingle and H. J. W. Müller [73].
[‡]E. L. Ince [136], [137]; S. Goldstein [115], [116].

and

$$\overline{C}[w(z)] = \overline{\alpha}\overline{A}(z), \qquad \overline{\alpha} = \frac{[\frac{1}{4}(q-3)]!}{(8h)^{\frac{1}{4}(q+1)}}\left[1 + \frac{1}{2^7 h}(3q^2 + 2q + 3)\right.$$

$$\left. + \frac{1}{2^{15}h^2}(9q^4 + 92q^3 + 70q^2 + 284q + 57) + \cdots\right].$$

$$(17.62b)$$

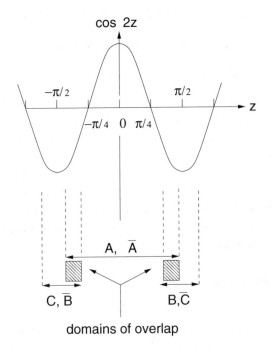

Fig. 17.4 Domains of solutions and their domains of overlap.

Example 17.5: Proportionality of solutions in domains of overlap

Show that in their common domain of validity $B = \alpha A$ and $\overline{C} = \overline{\alpha}\overline{A}$, and determine the constants α and $\overline{\alpha}$.

Solution: We begin with the solution containing the function $A(z)$. The first term of the expansion of $A(z)$ is the function $A_q(z)$ given by Eq. (17.32). We write and expand this function as follows:

$$
\begin{aligned}
A_q(z) &= \cos^{\frac{1}{2}(q-1)}\left(\frac{1}{4}\pi + \frac{1}{2}z\right)\left[1 - \cos^2\left(\frac{1}{4}\pi + \frac{1}{2}z\right)\right]^{-\frac{1}{4}(q+1)} \\
&= \cos^{\frac{1}{2}(q-1)}\left(\frac{1}{4}\pi + \frac{1}{2}z\right)\left[1 + \frac{(q+1)}{1!4}\cos^2\left(\frac{1}{4}\pi + \frac{1}{2}z\right)\right. \\
&\quad \left. + \frac{(q+1)(q+5)}{2!4^2}\cos^4\left(\frac{1}{4}\pi + \frac{1}{2}z\right) + \cdots\right].
\end{aligned}
$$

Inserting this into $A = A^{(0)} + A^{(1)} + \cdots$, with e.g. $A^{(1)}$ given by Eq. (17.36), the coefficient of the dominant factor $\cos^{(q-1)/2}(\pi/4 + z/2)$ in the result is seen to be

$$a = 1 + \frac{1}{2^7 h} \frac{(q, q-4)}{-1} \frac{(q-3)}{1!4} + O\left(\frac{1}{h^2}\right) = 1 - \frac{(q-1)(q-3)^2}{2^9 h} + \cdots .$$

Similarly substituting the asymptotic expansion of the Hermite function[§] into the expression (17.46), we obtain

$$B_q[w(z)] = \frac{H_{(q-1)/2}(w)}{2^{\frac{1}{4}(q-1)} [\frac{1}{4}(q-1)]!} = \frac{w^{\frac{1}{2}(q-1)}}{2^{\frac{1}{4}(q-1)} [\frac{1}{4}(q-1)]!} \left[1 - \frac{(q-1)(q-3)}{1!(8w^2)} + \cdots \right].$$

Inserting this into the expansion of $B[w(z)] = B^{(0)} + B^{(1)} + \cdots$ where

$$w(z) = 4h^{1/2} \cos\left(\frac{1}{4}\pi + \frac{1}{2}z\right),$$

the coefficient of the dominant factor $\cos^{(q-1)/2}(\pi/4 + z/2)$ in the result is seen to be

$$b = \frac{(8h)^{\frac{1}{4}(q-1)}}{[\frac{1}{4}(q-1)]!} \left[1 + \frac{1}{2^7 h} \frac{(q, q+4)}{1} \frac{(q+4, q)}{-1} \frac{1}{4(q+3)} + \cdots \right]$$

$$= \frac{(8h)^{\frac{1}{4}(q-1)}}{[\frac{1}{4}(q-1)]!} \left[1 - \frac{1}{2^9 h} (q+1)^2 (q+3) + \cdots \right].$$

It follows therefore that over their common range of validity $B = \alpha A$ with $\alpha = b/a$, i.e. with

$$\alpha = \frac{(8h)^{\frac{1}{4}(q-1)}}{[\frac{1}{4}(q-1)]!} \left[1 - \frac{1}{2^7 h}(3q^2 - 2q + 3) + \frac{1}{2^{15} h^2}(9q^4 - 92q^3 + 70q^2 - 284q + 57) + \cdots \right].$$

The other relation is obtained similarly. For further details and explicit expressions of higher order terms we refer to Dingle and Müller [73].

In order to link our asymptotic solutions with the Mathieu functions of integral order defined earlier in the consideration of small values of h^2, we have to impose the appropriate boundary conditions. These are the conditions of Eqs. (17.22). As we saw there, these are conditions imposed on even and odd solutions at the point $z = \pi/2$. Hence we have to construct even and odd solutions and impose the boundary conditions at $z = \pi/2$. Since the solution with $\overline{A}$ is obtained from that with A by a change of sign of z, we have as even and odd solutions

$$y_{\pm} \propto A(z)e^{2h\sin z} \pm \overline{A}(z)e^{-2h\sin z}. \tag{17.63}$$

In Example 17.6 at the end of this section we derive for these large-h^2 solutions the conditions similar to those given in various equations in Sec. 17.2.2 for the case of small-h^2 solutions and thus insure that in both cases the same boundary conditions are obeyed.

[§] This is essentially the expansion of the parabolic cylinder function with the exponential factor removed — see Eq. (6.61) for the connection and Eq. (8.53) for the expansion.

We can extend the solutions (17.63) to the domain around $z = \pi/2$ by using the proportionalities (17.62a) and (17.62b). Then

$$y_{\pm} \propto \frac{B[w(z)]}{\alpha} e^{2h \sin z} \pm \frac{\overline{C}[w(z)]}{\overline{\alpha}} e^{-2h \sin z}. \tag{17.64}$$

These are now the solutions around $z = \pi/2$ which permit us to apply the boundary conditions (17.22) and to correlate the solutions to those for small values of h^2. Thus $y_+ \equiv ce$ and $y_- \equiv se$. The functions defined as in Eq. (17.22) are therefore given by the following boundary conditions:

$$y'_+\left(\frac{\pi}{2}\right) = 0 \quad : \quad \left[\frac{1}{\alpha}\left(\frac{\partial B}{\partial w}\right)_{w=0} e^{2h} + \frac{1}{\overline{\alpha}}\left(\frac{\partial \overline{C}}{\partial w}\right)_{w=0} e^{-2h}\right]\left(\frac{\partial w}{\partial z}\right)_{z=\pi/2} = 0,$$

$$y_+\left(\frac{\pi}{2}\right) = 0 \quad : \quad \frac{B[w=0]}{\alpha} e^{2h} + \frac{\overline{C}[w=0]}{\overline{\alpha}} e^{-2h} = 0,$$

$$y'_-\left(\frac{\pi}{2}\right) = 0 \quad : \quad \left[\frac{1}{\alpha}\left(\frac{\partial B}{\partial w}\right)_{w=0} e^{2h} - \frac{1}{\overline{\alpha}}\left(\frac{\partial \overline{C}}{\partial w}\right)_{w=0} e^{-2h}\right]\left(\frac{\partial w}{\partial z}\right)_{z=\pi/2} = 0,$$

$$y_-\left(\frac{\pi}{2}\right) = 0 \quad : \quad \frac{B[w=0]}{\alpha} e^{2h} - \frac{\overline{C}[w=0]}{\overline{\alpha}} e^{-2h} = 0.$$

$$\tag{17.65}$$

For the evaluation of these conditions we require the following expressions involving Hermite functions which we obtain from Tables of Special Functions:[¶]

$$H_\nu(w=0) = \frac{\sqrt{\pi}2^{\nu/2}}{[-\frac{1}{2}(\nu+1)]!}, \quad \left(\frac{\partial H_\nu}{\partial w}\right)_{w=0} = -\frac{\sqrt{2\pi}2^{\nu/2}}{[-\frac{1}{2}(\nu+2)]!}. \tag{17.66}$$

[¶]E.g. W. Magnus and F. Oberhettinger [181], p. 80, give the following expressions for polynomials with argument zero:

$$He_{2n}(0) = \frac{(-1)^n(2n)!}{2^n n!}, \quad He_{2n+1}(0) = 0.$$

With the help of the duplication and inversion formulas of factorials, and then replacing $2n$ by ν, one obtains the first of relations (17.66). Here ν is the not necessarily integral index of the Hermite function $He_\nu(w)$ with Hermite equation

$$\left(\frac{d^2}{dw^2} - w\frac{d}{dw} + \nu\right)He_\nu(w) = 0.$$

The second of relations (17.66) then follows from the recurrence relation (6.59d), i.e.

$$He'_\nu(w) = wHe_\nu(w) - He_{\nu+1}(w), \quad \text{i.e.} \quad He'_\nu(0) = -He_{\nu+1}(0).$$

With these expressions we obtain by insertion into B and its derivative, and similarly for $\overline{C}$ and its derivative,

$$B[w=0] = \frac{\sqrt{\pi}}{[\frac{1}{4}(q-1)]![-\frac{1}{4}(q+1)]!} \left[1 - \frac{q}{2^5 h} + \frac{q^4 - 82q^2 - 39}{2^{14}h^2} - \cdots \right],$$

(17.67a)

$$\left(\frac{\partial B}{\partial w}\right)_{w=0} = -\frac{\sqrt{2\pi}}{[\frac{1}{4}(q-1)]![-\frac{1}{4}(q+3)]!} \left[1 - \frac{q}{2^4 h} + \frac{q^4 - 106q^2 - 87}{2^{14}h^2} - \cdots \right],$$

(17.67b)

$$\overline{C}[w=0] = -\frac{\sqrt{\pi}[\frac{1}{4}(q-3)]! \sin\{\frac{\pi}{4}(q-1)\}}{[\frac{1}{4}(q-1)]!} \left[1 - \frac{q}{2^5 h} + \frac{q^4 - 82q^2 - 39}{2^{14}h^2} - \cdots \right],$$

(17.67c)

$$\Re\left(\frac{\partial \overline{C}}{\partial w}\right)_{w=0} = \sqrt{2\pi} \cos\left\{\frac{\pi}{4}(q-1)\right\} \left[1 - \frac{q}{2^4 h} + \frac{q^4 - 106q^2 - 87}{2^{14}h^2} - \cdots \right].$$

(17.67d)

Considering now the second of conditions (17.65) we have

$$\frac{B[w=0]}{\overline{C}[w=0]} = -\frac{\alpha}{\overline{\alpha}} e^{-4h}.$$

(17.68)

Inserting here expansions (17.67a) and (17.67c), and using the reflection formula (8.14) from which one obtains

$$\left[-\frac{1}{4}(q+1)\right]! = \frac{\pi}{\cos\{\frac{\pi}{4}(q-1)\}[\frac{1}{4}(q-3)]!},$$

and the duplication formula $(2z)!\sqrt{\pi} = z!(z-1/2)!2^{2z}$, from which one obtains

$$\left[\frac{1}{4}(q-1)\right]! \left[\frac{1}{4}(q-3)\right]! = \frac{\sqrt{\pi}}{2^{(q-1)/2}} \left[\frac{1}{2}(q-1)\right]!,$$

the condition reduces to the equation

$$\cot\left\{\frac{\pi}{4}(q-1)\right\} = \sqrt{\frac{\pi}{2}} \frac{(16h)^{q/2} e^{-4h}}{[\frac{1}{2}(q-1)]!} \left[1 - \frac{3(q^2+1)}{2^6 h}\right.$$
$$\left. + \frac{1}{2^{13}h^2}(9q^4 - 40q^3 + 18q^2 - 136q + 9) + \cdots \right]. \quad (17.69)$$

If the right hand side of this equation were exactly equal to zero, the solutions would be given by $q = q_0 = 3, 7, 11, \ldots$. Hence expanding the cotangent about these points, we have

$$\cot\left\{\frac{\pi}{4}(q-1)\right\} = -\frac{1}{4}\pi(q-q_0) + O[(q-q_0)^3].$$

Hence we obtain

$$q - q_0 \simeq -2\sqrt{\frac{2}{\pi}} \frac{(16h)^{q_0/2}e^{-4h}}{[\frac{1}{2}(q_0 - 1)]!} \left[1 - \frac{3(q_0^2 + 1)}{2^6 h} \right.$$
$$\left. + \frac{1}{2^{13}h^2}(9q_0^4 - 40q_0^3 + 18q_0^2 - 136q_0 + 9) + \cdots \right]. \quad (17.70)$$

Clearly the last of conditions (17.65) leads to the same result except for a change in sign.

Considering now the first of conditions (17.65) we have

$$\frac{(\partial B/\partial w)_{w=0}}{(\partial \overline{C}/\partial z)_{w=0}} = -\frac{\alpha}{\bar{\alpha}} e^{-4h}. \quad (17.71)$$

Proceeding as above one obtains

$$\tan\left\{ \frac{\pi}{4}(q - 1) \right\} = -\sqrt{\frac{\pi}{2}} \frac{(16h)^{q/2}e^{-4h}}{[\frac{1}{2}(q - 1)]!} \left[1 - \frac{3(q^2 + 1)}{2^6 h} \right.$$
$$\left. + \frac{1}{2^{13}h^2}(9q^4 - 40q^3 + 18q^2 - 136q + 9) + \cdots \right]. (17.72)$$

This leads again to the result (17.70) except that now $q_0 = 1, 5, 9, \ldots$. The remaining condition (17.65) yields the same result except for a change in sign. Thus we have

$$q - q_0 = \mp 2\sqrt{\frac{2}{\pi}} \frac{(16h)^{q_0/2}e^{-4h}}{[\frac{1}{2}(q_0 - 1)]!} \left[1 - \frac{3(q_0^2 + 1)}{2^6 h} \right.$$
$$+ \frac{1}{2^{13}h^2}(9q_0^4 - 40q_0^3 + 18q_0^2 - 136q_0 + 9)$$
$$- \frac{1}{2^{19}h^3}(9q_0^6 - 120q_0^5 + 467q_0^4 - 528q_0^3 + 3307q_0^2 - 408q_0 + 1089)$$
$$\left. + \cdots \right]. \quad (17.73)$$

We now return to the eigenvalue $\lambda(q)$ and expand this around an odd integer q_0, i.e. we use

$$\lambda(q) \simeq \lambda(q_0) + (q - q_0)\left(\frac{\partial \lambda}{\partial q} \right)_{q_0}.$$

Differentiating $\lambda(q)$ of Eq. (17.61), this becomes

$$
\begin{aligned}
\lambda(q) &= \lambda(q_0) + (q - q_0)2h\left\{1 - \frac{q_0}{2^3 h} - \frac{3(q_0^2 + 1)}{2^8 h^2} - \cdots\right\}\\[2mm]
&= \lambda(q_0) \mp \frac{(16h)^{\frac{1}{2}q_0 + 1}e^{-4h}}{(8\pi)^{1/2}[\frac{1}{2}(q_0 - 1)]!}\left[1 - \frac{1}{2^6 h}(3q_0^2 + 8q_0 + 3)\right.\\[2mm]
&\qquad \left. + \frac{1}{2^{13}h^2}(9q_0^4 + 8q_0^3 - 78q_0^2 - 88q_0 - 87) - \cdots\right]\\[2mm]
&\equiv \lambda_{\mp}(q_0),\quad q_0 = 2n + 1,\quad n = 0, 1, 2, \dots .
\end{aligned}
\tag{17.74a}
$$

The difference between the value of λ (earlier called a_n and b_{n+1}) for the even Mathieu functions ce_{q_0} or ce_{q_0-1} (upper, i.e. minus sign), and that for the odd Mathieu functions se_{q_0+1} and se_{q_0} (lower, i.e. plus sign) give the so-called *level splitting* $\triangle\lambda(q_0)$ as a consequence of tunneling indicated schematically in Fig. 17.3. In the literature that we follow here[*] in all cases several more higher order terms have been given.[†] From Eq. (17.74a) we obtain for the level splitting in dominant order[‡]

$$
\begin{aligned}
\underbrace{\triangle\lambda(q_0)}_{\equiv \triangle E(q_0)} &= \underbrace{\lambda_+(q_0)}_{E_{odd}(q_0)} - \underbrace{\lambda_-(q_0)}_{E_{even}(q_0)} \simeq \frac{2(16h)^{\frac{1}{2}q_0 + 1}}{(8\pi)^{1/2}[\frac{1}{2}(q_0 - 1)]!}e^{-4h}\\[3mm]
&= \frac{(16h)^{n+\frac{3}{2}}}{n!\sqrt{2\pi}}e^{-4h}.
\end{aligned}
\tag{17.74b}
$$

The above results for the cosine potential are in many respects important, quite apart from applications. The complete, though approximate, solution of the Schrödinger equation for the basic and nontrivial periodic potential is important for a thorough understanding of its quantum mechanics, just as any solvable problem is important for the insights it offers in a concrete and transparent form. In particular we saw that the perturbation expansion yielded only the degenerate eigenvalue expansion, and not the separation of these as a result of tunneling. Effectively the degenerate eigenvalue expansion results from the anharmonic terms contained in the power expansion of the cosine potential. The tunneling effect, made evident by the splitting of the asymptotically degenerate (harmonic) oscillator levels, results from the boundary conditions. In the literature one finds frequently the statement that the splitting is a *nonperturbative effect*. We see here that

---

[*]R. B. Dingle and H. J. W. Müller [73].

[†]The dominant contributions were also found by S. Goldstein [116]. See also F. M. Arscott [11], p. 120, example 3.

[‡]In Sec. 20.3.3 we call this difference $2\delta E_{q_0}$.

this nonperturbative effect, evident through the exponential factor $\exp(-4h)$ in Eq. (17.74b), results from the imposition of boundary conditions.[§] The boundary conditions we imposed here are those of a self-adjoint problem which has real eigenvalues.

The explicit derivation of the level splitting (17.74b) is also important for various other reasons. The splitting can also be derived by the path-integral method as will be shown in Chapter 26, and one can compare the methods. The splitting can also be obtained with some methods of large order of perturbation theory. In fact the perturbation theory developed in this context and tested by application to the cosine potential is of such generality that the recurrence relation of the coefficients of the eigenvalue expansion can be looked at as a difference equation, which one can then try to solve in order to obtain the behaviour of the coefficients of the late terms of the expansion.

As mentioned at the beginning, the Schrödinger equation with the cosine potential is a special case of several other more general cases. We shall not consider these in detail here but do consider briefly the elliptic potential in the next section without going into extensive calculations which can be looked up in the literature. The *associated* or *modified Mathieu equation* with $\cosh z$ instead of $\cos z$ is another important equation which we shall study in detail in Chapter 19 in connection with a singular potential. Naturally one expects the solutions there to be related to those of the periodic case considered here.

Example 17.6: Translation of Solutions

For the large-h^2 even and odd solutions (17.63) — and using only the leading terms — derive the following set of equations with shifted arguments similar to those of the small-h^2 case of Sec. 17.2.2:

$$
y_+(z \pm \pi) = \frac{y_+(\pm\pi)}{y_+(0)} y_+(z) \pm \frac{y'_+(\pm\pi)}{y'_-(0)} y_-(z),
$$

$$
y_-(z \pm \pi) = \pm \frac{y_-(\pm\pi)}{y_+(0)} y_+(z) + \frac{y'_-(\pm\pi)}{y'_-(0)} y_-(z).
$$

Solution: The derivation requires a cumbersome tracking of minus signs. Thus we do not reproduce every step. We begin with the solution containing the function $A(z)$. From Eq. (17.32) we obtain

$$
A_q(z \pm \pi) = (-1)^{\pm\frac{1}{2}(q\mp1)} \overline{A}_q(z), \qquad \overline{A}_q(z \pm \pi) = -(-1)^{\mp\frac{1}{2}(q\mp1)} A_q(z)
$$

(the extra minus sign in the second equation coming from $A_q(z \pm 2\pi) = -A_q(z)$). We consider for simplicity only the dominant contributions and replace throughout -1 to some power by $\exp(i\pi)$ to that power and combine such terms into cosines and sines. Then we obtain first

$$
y_\pm(z) = y_A(z) \pm y_{\overline{A}}(z), \quad y_A(z) = A_q(z)e^{2h\sin z}, \quad y_{\overline{A}}(z) = \overline{A}_q(z)e^{-2h\sin z}.
$$

[§]The exponential factor represents, in effect, the exponential factor we encountered in the semiclassical method where it has the structure of the factor $\exp(\text{classical action}/\hbar)$.

Replacing $y_A, y_{\overline{A}}$ by $y_\pm$, we obtain — with $A_q(z + \pi) = (-1)^{(q-1)/2} A_q(-z)$, etc. —

$$y_+(z \pm \pi) = \pm i \sin\left\{\frac{\pi}{2}(q \mp 1)\right\} y_+(z) - \cos\left\{\frac{\pi}{2}(q \mp 1)\right\} y_-(z).$$

From this we deduce for $z = 0$, since $A_q(0) = \sqrt{2} = \overline{A}_q(0)$:

$$y_+(\pm\pi) = \pm i \sin\left\{\frac{\pi}{2}(q \mp 1)\right\} y_+(0), \quad y_+(0) \simeq 2\sqrt{2}.$$

The derivatives can be handled with the help of Eq. (17.31), the operator having solution $A_q(z)$. Thus $A'_q(0) = -(q/2)A_q(0) = -\overline{A}'_q(0)$ and

$$A'_q(z = \pm\pi) = \frac{1}{2}qA_q(z = \pm\pi), \quad \overline{A}'_q(z = \pm\pi) = -\frac{1}{2}q\overline{A}_q(z = \pm\pi).$$

Then $y'_+(0) = 0, y'_-(0) = \sqrt{2}(4h - q)$, and

$$y'_+(z = \pm\pi) = \sqrt{2}(q - 4h)\cos\left\{\frac{1}{2}\pi(q - 1)\right\} = -\sqrt{2}(q - 4h)\cos\left\{\frac{1}{2}\pi(q + 1)\right\}.$$

Similarly we find

$$
\begin{aligned}
y_-(z \pm \pi) &= (-1)^{\pm\frac{1}{2}(q\mp 1)} y_{\overline{A}}(z) + (-1)^{\mp\frac{1}{2}(q\mp 1)} y_a(z) \\
&= \cos\left\{\frac{1}{2}\pi(q \mp 1)\right\} y_+(z) \mp i\sin\left\{\frac{1}{2}\pi(q \mp 1)\right\} y_-(z),
\end{aligned}
$$

and by putting $z = 0$:

$$y_-(\pm\pi) = \cos\left\{\frac{1}{2}\pi(q \mp 1)\right\} y_+(0).$$

Finally for the derivative of the odd solution we obtain

$$y'_-(\pi) = [a'_q(\pi) - 2h\, A_q(\pi)] - [\overline{A}'_q(\pi) + 2h\overline{A}_q(\pi)] = i\sqrt{2}(q - 4h)\sin\left\{\frac{1}{2}\pi(q - 1)\right\},$$

where $A_q(0) = \overline{A}_q(0) = \sqrt{2}$. Replacing sines and cosines by the functional expressions, we obtain the conditions stated at the beginning. From the equations for the sine and cosine we can deduce the value of the Wronskian at $z = \pi$ in the leading approximation for large h:

$$W[y_+, y_-]_{z=\pi} = [y_+ y'_- - y_- y'_+]_{z=\pi} = -[y_+(0)]^2(q - 4h) \simeq -8(q - 4h) \simeq 32h.$$

17.4 Elliptic and Ellipsoidal Potentials

17.4.1 Introduction

Our intention here is partly to deal with other periodic potentials, but also to enable a brief familiarization with the *Lamé equation*, which is easiest by comparison with the foregoing treatment of the Mathieu equation. The Lamé equation has not been a widely known equation of mathematical physics so far. But more recently it has been observed to arise in various contexts. In particular the Lamé equation was recognized to arise as the equation of small fluctuations about instanton solutions for practically all basic potentials.[*] As Arscott [11] (p. 194) remarks, the new stage of the development of the investigation of the Lamé equation was really initiated by Ince[†] around 1940, and was continued by Erdélyi,[‡] and the work of both prepared the ground for present-day investigations. The most conspicuous difference compared with the Mathieu equation is, as alluded to at the beginning, the non-occurrence of the Floquet exponent. Arscott [11] (p. 194) remarks to parallels with the Mathieu equation:"... *This line of investigation, however, soon encounters grave difficulties and there has not been developed up till now any general theory of Lamé's equation at all comparable with that of the Mathieu equation ... for general n the solution of Lamé's equation is not single-valued owing to the singularities in the finite part of the u-plane, so that Floquet's theorem cannot be immediately applied ....*"

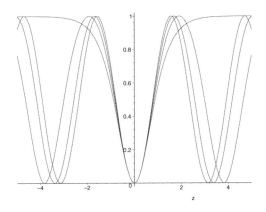

Fig. 17.5 The potential $\text{sn}^2(z, k)$ for $k = 0, 0.5, 0.75, 0.98, 1.0$ and $-5 \leq z \leq 5$.

If one separates the wave equation[§]

$$\nabla^2 \Psi + l^2 \Psi = 0$$

in ellipsoidal coordinates (which we do not need to consider here), one arrives at three equations of which one is the *ellipsoidal wave equation*

$$\frac{d^2y}{du^2} + [\Lambda - \kappa^2 \text{sn}^2 u - \Omega^2 k^4 \text{sn}^4 u] y = 0. \tag{17.75}$$

---

[*]This means double well, inverted double well and cosine potentials (J.–Q. Liang, H. J. W. Müller–Kirsten and D. H. Tchrakian [165]), but also cubic potentials.

[†]E. L. Ince [138].

[‡]A. Erdélyi [85].

[§]F. M. Arscott [11], p. 19.

Here κ^2 and Λ are separation constants and $\Omega^2 = l^2(a^2 - c^2), a > b > c$ being related to the lengths of the three axes of the ellipsoid in a Cartesian coordinate system. If we put $\Omega = 0$, Eq. (17.75) reduces to Lamé's equation and one writes $\kappa^2 = n(n+1)k^2$, where n is real and $\geq -1/2$ (and n is an integer in the case of solutions called Lamé polynomials) where $k, |k| \leq 1$, is the elliptic modulus of the Jacobian elliptic functions $\operatorname{sn}u$, $\operatorname{cn}u$ and $\operatorname{dn}u$. The range of the independent variable u is $0 < u < 2K$, K being the complete elliptic integral of the first kind. The function sn^2u is plotted in Fig. 17.5 for several values of k. In order to distinguish the above equation from that with the *Lamé potential*, we refer to the potential consisting of the two terms with sn^2u and sn^4u as the *ellipsoidal potential*.

17.4.2 Solutions and eigenvalues

In the following we sketch the main points of the method of deriving asymptotic expansions of the eigenvalues and eigenfunctions, and of the derivation of the level splitting[¶]. Although the results have been calculated for the ellipsoidal wave equation, we consider mainly the Lamé equation, i.e. the equation

$$\frac{d^2y}{du^2} + [\Lambda - \kappa^2\operatorname{sn}^2u]y = 0, \qquad (17.76)$$

which can be looked at as a Schrödinger equation with periodic potential $\kappa^2\operatorname{sn}^2u$ where $\operatorname{sn}u$ is one of the Jacobian elliptic functions of period $2K$. In comparisons with the Schrödinger equation, the usual factor $-\hbar^2/2m_0$, in front of the second derivative has to be kept in mind (m_0 being the mass).

The first step is to write the eigenvalue Λ as

$$\Lambda(q, \kappa) = q\kappa + \frac{\triangle(q, \kappa)}{8}, \qquad (17.77)$$

where $q \to q_0 = 2N + 1, N = 0, 1, 2, \ldots$ in the case $\kappa^2 \to \infty$, i.e. for very high barriers (harmonic oscillator approximation around a minimum of the potential). For barriers of finite height the parameter q is only approximately an odd integer q_0 in view of tunneling effects.

The second step is to insert (17.77) into Eq. (17.76) and to write the solution

$$y = A(u)\exp\left\{-\int \kappa\operatorname{sn}u\, du\right\} = A(u)[f(u)]^{\kappa/2k}, \qquad (17.78)$$

[¶] We follow in this brief recapitulation the description in H. J. W. Müller–Kirsten, Jian-zu Zhang and Yunbo Zhang [206].

where

$$f(u) = \left(\frac{dnu + kcnu}{dnu - kcnu}\right).$$

For large values of κ the equation for $A(u)$ can be solved iteratively resulting in an asymptotic expansion for $A(u)$ and concurrently one for the remainder in Eq. (17.77), i.e. $\triangle$. A second solution is written

$$y = \overline{A}(u) \exp\left\{\int \kappa snu \, du\right\}. \tag{17.79}$$

The very useful property of these solutions is that for the same value of $\triangle$ (which remains unchanged under the combined replacements $q \rightarrow -q, \kappa \rightarrow -\kappa$)

$$\overline{A}(u) = A(u + 2K), \quad \overline{A}(u, q, \kappa) = A(u, -q, -\kappa). \tag{17.80}$$

The domain of validity of these solutions is that away from an extremum of the potential, more precisely for

$$\left|\frac{dnu \mp cnu}{dnu \pm cnu}\right| \gg \frac{1}{\kappa}.$$

Thus one can construct solutions $Ec(u), Es(u)$, which are respectively even in u (or snu) or odd, i.e.

$$\begin{matrix} Ec(u) \\ Es(u) \end{matrix} \propto A(u)[f(u)]^{\kappa/2k} \pm \overline{A}(u)[f(u)]^{-\kappa/2k}. \tag{17.81}$$

Since these expansions are not valid at the extrema of the potential (where the boundary conditions are to be imposed), one has to derive new sets of solutions there and match these to the former (i.e. determine their proportionality factors) in domains of overlap (their extreme regions of validity).

Thus in the third step two more pairs of solutions $B, \overline{B}$ and $C, \overline{C}$ replacing $A, \overline{A}$ are derived, one pair in terms of Hermite functions of a real variable, the other in terms of those of an imaginary variable, by transforming the equations for $A, \overline{A}$ into equations in terms of the variables

$$z(u) = \frac{\sqrt{8\kappa}}{k'}\left(\frac{dnu \mp cnu}{dnu \pm cnu}\right)^{1/2}, \quad k' = \sqrt{1 - k^2}. \tag{17.82}$$

Solving the resulting equations iteratively as before, one obtains again the same expansion for Λ, but solutions B, C which are valid for

$$\left|\frac{dnu \pm cnu}{dnu \mp cnu}\right| \ll 1.$$

In their regions of overlap one can determine the proportionality factors $\alpha, \overline{\alpha}$ of

$$B = \alpha A, \quad \overline{C} = \overline{\alpha}\overline{A}. \tag{17.83}$$

Then

$$\frac{\mathrm{Ec}(u)}{\mathrm{Es}(u)} \propto \frac{B[z(u)]}{\alpha}[f(u)]^{\kappa/2k} \pm \frac{\overline{C}[z(u)]}{\overline{\alpha}}[f(u)]^{-\kappa/2k}. \tag{17.84}$$

Each of the solutions thus derived is associated with one and the same expansion of the eigenvalue Λ. This expansion is found to be in the more general case of Eq. (17.75), the ellipsoidal equation,[||]

$$
\begin{aligned}
\Lambda \;=\; & q\kappa - \frac{1}{2^3}(1+k^2)(q^2+1) - \frac{q}{2^6\kappa}\{(1+k^2)^2(q^2+3) - 4k^2(q^2+5)\} \\
& -\frac{1}{2^{10}\kappa^2}\{(1+k^2)^3(5q^4+34q^2+9) \\
& -4k^2(1+k^2)(5q^4+34q^2+9) - 384\Omega^2 k^4(q^2+1)\} \\
& -\frac{q}{2^{14}\kappa^3}\{(1+k^2)^4(33q^4+410q^2+405) \\
& -24k^2(1+k^2)^2(7q^4+90q^2+95) + 16k^4(9q^4+130q^2+173) \\
& +512\Omega^2 k^4(1+k^2)(q^2+11)\} - \frac{1}{2^{16}\kappa^4}\{(1+k^2)^5(63q^6+1260q^4 \\
& +2943q^2+486) - 8k^2(1+k^2)^3(49q^6+1010q^4+1493q^2+432) \\
& +16k^4(1+k^2)(35q^6+760q^4+2043q^2+378) \\
& -64\Omega^2 k^4(1+k^2)^2(5q^4+34q^2+9) + 256\Omega^2 k^6(5q^4-38q^2-63)\} \\
& +\cdots .
\end{aligned}
\tag{17.85}
$$

The first three terms of this expansion were first given by Ince [138], who obtained this expansion for the eigenvalues of Lamé's equation (i.e. $\Omega = 0$).

17.4.3 The level splitting

In the fourth and final step one applies the appropriate boundary conditions on these solutions, i.e. one sets at $u = 0$ and $u = 2K$ altogether[**]

$$\mathrm{Ec}(2K) = \mathrm{Ec}(0) = 0, \quad \mathrm{Es}(2K) = \mathrm{Es}(0) = 0, \tag{17.86}$$

as well as

$$\left(\frac{\partial \mathrm{Ec}}{\partial u}\right)_{2K} = \left(\frac{\partial \mathrm{Ec}}{\partial u}\right)_{0} = 0, \quad \left(\frac{\partial \mathrm{Es}}{\partial u}\right)_{2K} = \left(\frac{\partial \mathrm{Es}}{\partial u}\right)_{0} = 0. \tag{17.87}$$

[||]H. J. W. Müller [205].
[**]A. Erdélyi, W. Magnus, F. Oberhettinger and F. G. Tricomi [87], p. 64.

These conditions define respectively functions $Ec_n^{q_0}, Es_n^{q_0+1}, Ec_n^{q_0-1}$ and $Es_n^{q_0}$ of periods $4K, 2K, 2K$ and $4K$ respectively. Evaluating these one obtains (from factors of factorials in q and $-q$) expressions $\cot\{\pi(q-1)/4\} = \cdots$ and $\tan\{\pi(q-1)/4\} = \cdots$ (in much the same way as in the case of Mathieu functions), from which the difference $q - q_0$ is obtained by expansion around zeros. One obtains with q_0 an odd integer

$$
\begin{aligned}
q - q_0 &= \mp 2\sqrt{\frac{2}{\pi}}\left(\frac{1+k}{1-k}\right)^{-\kappa/k}\left(\frac{8\kappa}{1-k^2}\right)^{q_0/2}\frac{1}{[\frac{1}{2}(q_0-1)]!}\Bigg[1 \\
&\quad -\frac{3(q_0^2+1)(1+k^2)}{2^5\kappa} \\
&\quad \frac{1}{3.2^{11}\kappa^2}\{3(1+k^2)^2(9q_0^4 - 40q_0^3 + 18q_0^2 - 136q_0 + 9) \\
&\quad +256k^2 q_0(q_0^2+5)\} - \cdots\Bigg].
\end{aligned}
\tag{17.88}
$$

Here the upper sign refers to $Ec_n^{q_0-1}$ or $Ec_n^{q_0}$, and the lower to $Es_n^{q_0+1}$ or $Es_n^{q_0}$, q_0 being an odd integer. Finally expanding

$$
\begin{aligned}
\Lambda(q) &\simeq \Lambda(q_0) + (q-q_0)\left(\frac{\partial\Lambda}{\partial q}\right)_{q_0} \overset{(17.85)}{=} \Lambda(q_0) + (q-q_0)\kappa\Bigg[1 - \frac{q_0(1+k^2)}{2^2\kappa} \\
&\quad -\frac{1}{2^6\kappa^2}\{3(1+k^2)^2(q_0^2+1) - 4k^2(q_0^2 + 2q_0 + 5)\} + \cdots\Bigg],
\end{aligned}
\tag{17.89}
$$

one obtains the eigenvalues from which the level splitting can be deduced. One finds[††]

$$
\begin{aligned}
\Lambda(q) &\simeq \Lambda(q_0) \mp 2\kappa\sqrt{\frac{2}{\pi}}\left(\frac{1+k}{1-k}\right)^{-\kappa/k}\left(\frac{8\kappa}{1-k^2}\right)^{q_0/2}\frac{1}{[\frac{1}{2}(q_0-1)]!}\Bigg[1 \\
&\quad -\frac{1}{2^5\kappa}(1+k^2)(3q_0^2 + 8q_0 + 3) \\
&\quad +\frac{1}{3.2^{11}\kappa^2}\{3(1+k^2)^2(9q_0^4 + 8q_0^3 - 78q_0^2 - 88q_0 - 87) \\
&\quad +128k^2(2q_0^3 + 9q_0^2 + 10q_0 + 15)\} - \cdots\Bigg].
\end{aligned}
\tag{17.90}
$$

For the two lowest levels $q_0 = 1$ and one obtains for their separation

$$
\triangle\Lambda(1) \simeq \frac{2(4\kappa)^{3/2}(1-k)^{\kappa-1/2}}{(2\pi)^{1/2}2^\kappa}\left[1 + (1-k)\left\{\kappa\left(\frac{1}{2} - \ln 2\right) + \frac{1}{4}\right\} + O(\kappa^{-1})\right].
\tag{17.91}
$$

[††]H. J. W. Müller [205].

This result agrees with a result of Dunne and Rao[‡‡] who calculated this expression using instanton methods, in which they dubbed the classical configurations *Lamé instantons*. One can see, however, that calculations with the Schrödinger equation are not only easier, but also more easily generalizable.

17.4.4 Reduction to Mathieu functions

Under certain limiting conditions the ellipsoidal wave equation reduces to the Mathieu equation. Thus if $k \to 0$ and $n \to \infty$ in such a way that

$$\kappa^2 = n(n+1)k^2 \sim \quad \text{finite},$$

i.e. $\kappa^2 \equiv h^2$ (say), then $\operatorname{sn} u \to \sin u$ and Lamé's equation ($\Omega = 0$) becomes

$$y'' + \{\Lambda - 4h^2 \sin^2 u\}y = 0. \tag{17.92}$$

Replacing u by $x \pm \pi/2$, this equation becomes

$$y'' + \{\Lambda - 2h^2 - 2h^2 \cos 2x\}y = 0.$$

Hence the conditions

$$\Omega = 0, \quad k = 0, \quad \kappa = 2h, \quad \Lambda - 2h^2 = \lambda, \quad u = x \pm \frac{\pi}{2} \tag{17.93}$$

reduce the periodic ellipsoidal wave functions and their eigenvalues to corresponding Mathieu functions and their eigenvalues.[*]

Apart from the choice of notation, the conditions (17.86), (17.87) agree with those of Ince.[†] One can verify that under the conditions stated the results of this case of the ellipsoidal wave equation reduce to the corresponding results of the Mathieu equation. Without going into details we mention again (as at the beginning) that there is also a specialization from the ellipsoidal wave equation and its solutions to spheroidal wave equations and their solutions.[‡]

[‡‡]G. V. Dunne and K. Rao [79].

[*]Thus in the case of the dominant contribution of Eq. (17.90) one has in this limit:

$$\left(\frac{1+k}{1-k}\right)^{-\kappa/k} = e^{-\frac{\kappa}{k}\ln[(1+k)/(1-k)]} \to e^{-4h}, \quad \left(\frac{8\kappa}{1-k^2}\right)^{q_0/2} \to (16h)^{q_0/2},$$

and hence

$$\Lambda(q) \to \Lambda(q_0) \mp 4h\sqrt{\frac{2}{\pi}}e^{-4h}\frac{(16h)^{q_0/2}}{[\frac{1}{2}(q_0-1)]!}\left[1 + O\left(\frac{1}{h}\right)\right]$$

in agreement with Eq. (17.74a).

[†]E. L. Ince [138].

[‡]H. J. W. Müller [206].

17.5 Concluding Remarks

The potentials we considered above will re-appear in later chapters. In particular we shall encounter the *Lamé equation* as the *equation of small fluctuations* around classical configurations associated with cosine, double well, inverted double well and cubic potentials, i.e. all basic potentials, thus revealing an unexpected significance of this not so wellknown equation of mathematical physics.[§] This is therefore a very important equation which in the limit of infinite period becomes a *Pöschl–Teller equation.*[¶] The equations considered above also appear in diverse new problems of physics. Thus, for instance, in the problem of two parallel solenoids the lines of constant electromagnetic vector potential $|\mathbf{A}|$ are elliptic with the Hamiltonian separating into a Mathieu equation and an associated Mathieu equation. This is an interesting problem, also because the role played by the Floquet exponent in this problem is not yet well understood.[‖] Another recent appearance of the Mathieu equation is in the study of the mass spectrum of a scalar field in a world with latticized and circular continuum space.[**] The associated Mathieu equation appears also in string theory in connection with fluctuations about a $D3$-brane (see Chapter 19).

[§]J.–Q. Liang, H. J. W. Müller–Kirsten and D. H. Tchrakian [165]. See Chapter 25.

[¶]A host of related elliptic equations has recently been discovered and studied. See e.g. A. Ganguly [103]. A generalized associated Lamé potential has been considered by A. Khare and U. Sukhatme [148].

[‖]See Z.–Y. Gu and S.–W. Quian [123], A. D. de Veigy and S. Ouvry [276] and Jian–zu Zhang, H. J. W. Müller–Kirsten and J. M. S. Rana [287].

[**]Y. Cho, N. Kan and K. Shiraishi [49].

Chapter 18

Anharmonic Oscillator Potentials

18.1 Introductory Remarks

The anharmonic (quartic) oscillator* has repeatedly been the subject of detailed investigations related to perturbation theory. The recent work of R. Friedberg and T. D. Lee [98] referred to it as a *"long standing difficult problem of a quartic potential with symmetric minima"*. In particular the investigations of Bender and Wu,[†] which related analyticity considerations to perturbation theory and hence to the large order behaviour of the eigenvalue expansion attracted widespread interest.[‡] The fact that a main part of their work was concerned with the calculation of the imaginary part of the eigenenergy in the non-selfadjoint case which permits tunneling, demonstrates that derivations of such a quantity are much less familiar than calculations of discrete bound state eigenenergies in quantum mechanical problems. This lack of popularity of the calculation of complex eigenvalues even in texts on quantum mechanics may be attributed to the necessity of matching of various branches of eigenfunctions in domains of overlap and to the necessary imposition of suitable boundary conditions, both of which make the calculation more difficult.

In the cases treated most frequently in the literature the anharmonic

---

*We follow here largely P. Achuthan, H. J. W. Müller–Kirsten and A. Wiedemann [4] and a revised version of parts of this reference by J.–Q. Liang and H. J. W. Müller–Kirsten [163].

[†]C. M. Bender and T. T. Wu [18], [19].

[‡]Thus A. Turbiner [273] remarks:*"It can not be an exaggeration to say that after the seminal papers by C. Bender and T. T. Wu, in nearly a thousand of physics articles the problem of the anharmonic oscillator was touched in one way or another. This seemingly simple problem revealed extremely rich internal structure ..."*. There is no end to this: An entirely new approach to anharmonic oscillators was recently developed by M. Weinstein [281].

oscillator potential is defined by the sum of an harmonic oscillator potential and a quartic contribution. These contributions may be given different signs, and thus lead to very different physical situations, which are nonetheless linked as a consequence of their common origin which is for all one and the same basic differential equation. To avoid confusion we specify first the potential $V(z)$ in the Schrödinger equation

$$\frac{d^2 y(z)}{dz^2} + [E - V(z)]y(z) = 0$$

for the different cases which are possible and illustrate these in Fig. 18.1. The three different cases are:

(1) *Discrete eigenvalues with no tunneling*: In this case

$$V(z) = \frac{1}{4}|h^4|z^2 + \frac{1}{2}|c^2|z^4,$$

(2) *Discrete eigenvalues with tunneling*: In this case, described as the case of the double well potential,

$$V(z) = -\frac{1}{4}|h^4|z^2 + \frac{1}{2}|c^2|z^4,$$

(3) *Complex eigenvalues with tunneling*: In this case, with the potential described as an inverted double well potential,

$$V(z) = \frac{1}{4}|h^4|z^2 - \frac{1}{2}|c^2|z^4.$$

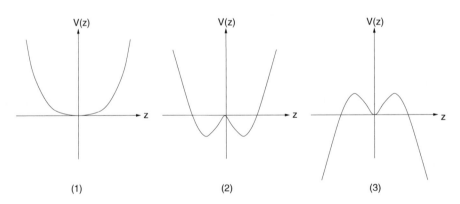

Fig. 18.1 The three different types of anharmonic potentials.

Case (1) is obviously the simplest with the anharmonic term implying simply a shift of the discrete harmonic oscillator eigenvalues with similarly normalisable wave functions. The shift of the eigenvalues is best calculated with straightforward perturbation theory. The result is an expansion in descending powers of h^2. It is this expansion which led to a large number of investigations culminating (so to speak) in the work of Bender and Wu who established the asymptotic nature of the expansion.

Case (2) is also seen to allow only discrete eigenvalues (the potential rising to infinity on either side), however the central hump with troughs on either side permits tunneling and hence (if the hump is sufficiently high) a splitting of the asymptotically degenerate eigenvalues in the wells on either side which vanishes in the limit of an infinitely high central hump. The eigenvalues of this case are given by Eq. (18.175) below. The level splitting will be rederived from path integrals in Chapter 26.

Case (3) is seen to be very different from the first two cases, since the potential decreases without limit on either side of the centre. The boundary conditions are non-selfadjoint and hence the eigenvalues are complex. This type of potential allows tunneling through the barriers and hence a passage out to infinity so that a current can be defined. If the barriers are sufficiently high we expect the states in the trough to approximate those of an harmonic oscillator, however with decay as a consequence of tunneling. The eigenvalues of this case are given by Eq. (18.86) below. The imaginary part will be rederived from path integrals in Chapter 26.

The question is therefore: How does one calculate the eigenvalues in these cases from the differential equation? This is the question we address in this chapter, and we present a fairly complete treatment of the case of large values of h^2 along lines parallel to those in our treatment of the cosine potential in Chapter 17. We do not dwell on Case (1) since this is effectively included in the first part of Case (3), except for a change of sign of $|c^2|$. Thus we are mainly concerned with the double well potential and its inverted form. We begin with the latter. In this case our aim is to obtain the aforementioned complex eigenvalue. In the case of the double well potential our aim is to obtain the separation of harmonic oscillator eigenvalues as a result of tunneling between the two wells. Since these exponentially small contributions are related to the behaviour of the late terms of the eigenvalue expansions (as we shall see in Chapter 20), and this behaviour is that of asymptotic expansions our treatment largely terminates this much-discussed topic, i.e. it will not be possible to obtain the exponentially small contributions with convergent expansions, as is sometimes hoped. Calculations of complex eigenvalues (imaginary parts of eigenenergies) are rare in texts on quantum mechanics. We therefore consider in this chapter and in Chapter 20 in detail some prominent examples and in such a way, that the general applicability of the method becomes evident.

18.2   The Inverted Double Well Potential

18.2.1   Defining the problem

We consider the case of the inverted double well potential depicted as Case
(3) in Fig. 18.1 and more specifically in Fig. 18.2. The potential in this case
is given by

$$V(z) = -v(z), \quad v(z) = -\frac{1}{4}h^4 z^2 + \frac{1}{2}c^2 z^4, \tag{18.1}$$

for h^4 and c^2 real and positive, and the Schrödinger equation to be considered
is

$$\frac{d^2 y}{dz^2} + [E + v(z)]y = 0. \tag{18.2}$$

We adopt the following conventions which it is essential to state in order to
assist comparison with other literature. We take here $\hbar = 1$ and the mass
m_0 of the particle $= 1/2$. This implies that results for $m_0 = 1$ (a frequent
convention in field theory considerations) differ from those obtained here by
factors of $2^{1/2}$, a point which has to be kept in mind in comparisons. If
suffixes $1/2, 1$ refer to the two cases, we can pass from one case to the other
by making the replacements:

$$E_{1/2} = 2E_1, \quad h^4_{1/2} = 2h^4_1, \quad c^2_{1/2} = 2c^2_1. \tag{18.3}$$

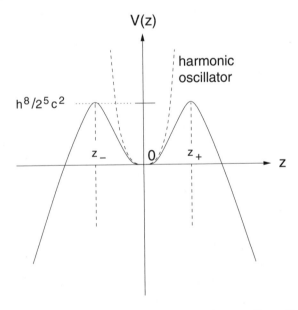

Fig. 18.2 The inverted double well potential with (hatched) oscillator potential.

Introducing a parameter q and a quantity $\triangle \equiv \triangle(q, h)$, and a variable w defined by setting

$$E = \frac{1}{2}qh^2 + \frac{\triangle}{2h^4} \quad \text{and} \quad w = hz, \tag{18.4}$$

we can rewrite Eq. (18.2) as

$$\mathcal{D}_q(w)y(w) = -\frac{1}{h^6}(\triangle + c^2 w^4)y(w) \tag{18.5}$$

with

$$\mathcal{D}_q(w) = 2\frac{d^2}{dw^2} + q - \frac{w^2}{2}. \tag{18.6}$$

In the domain of w finite, $|h^2| \to \infty$ and c^2 finite, the harmonic part of the potential dominates over the quartic contribution and Eq. (18.5) becomes

$$\mathcal{D}_q(w)y(w) = O\left(\frac{1}{h^6}\right). \tag{18.7a}$$

The problem then reduces to that of the pure harmonic oscillator with $y(w)$ a *parabolic cylinder function* $D_n(w)$, i.e.

$$y(w) \propto D_{\frac{1}{2}(q-1)}(w) \quad \text{and} \quad q = q_0 = 2n + 1, \quad n = 0, 1, 2, \ldots. \tag{18.7b}$$

The perturbation expansion in descending powers of h suggested by the above considerations is therefore an expansion around the central minimum of $V(z)$ at $z = 0$. The positions $z_\pm$ of the maxima of $V(z)$ on either side of $z = 0$ in the case $c^2 > 0$ are obtained from

$$v'(z_\pm) = 0 \quad \text{as} \quad z_\pm = \pm\frac{h^2}{2c} \tag{18.8}$$

with

$$v''(z_\pm) = h^4 \quad \text{and} \quad V(z_\pm) = \frac{h^8}{2^5 c^2}.$$

Thus for $c^2 > 0$ and relatively small, and h^2 large the eigenvalues are essentially perturbatively shifted eigenvalues of the harmonic oscillator as is evident from Fig. 18.2.

The problem here is to obtain the solutions in various domains of the variable, to match these in domains of overlap, then to specify the necessary boundary conditions and finally to exploit the latter for the derivation of the complex eigenvalue. The result will be that derived originally by Bender and Wu, although our method of matched asymptotic expansions here (which parallels that used in the case of the cosine potential) is different.

18.2.2   Three pairs of solutions

We are concerned with the equation

$$\frac{d^2y(z)}{dz^2} + \left[E - \frac{h^4z^2}{4} + \frac{c^2z^4}{2} \right] y(z) = 0 \tag{18.9}$$

where

$$E = \frac{1}{2}qh^2 + \frac{\triangle}{2h^4}. \tag{18.10}$$

Here again q is a parameter still to be determined from boundary conditions, and $\triangle = \triangle(q,h)$ is obtained from the perturbation expansion of the eigenvalue, as encountered and explained earlier. Inserting (18.10) into (18.9) we obtain

$$\frac{d^2y}{dz^2} + \left[\frac{1}{2}qh^2 + \frac{\triangle}{2h^4} - \frac{h^4z^2}{4} + \frac{c^2z^4}{2} \right] y = 0. \tag{18.11}$$

The solutions in terms of parabolic cylinder functions are valid around $z = 0$ and extend up to $z \simeq O(1/h^2)$, as we shall see. Before we return to these solutions we derive a new pair which is valid in the adjoining domains. Thus these solutions are not valid around $z = 0$. In order to arrive at these solutions we set in Eq. (18.11)

$$y(z) = A(z) \exp\left[\pm i \int^z dz \left\{ -\frac{h^4z^2}{4} + \frac{c^2z^4}{2} \right\}^{1/2} \right]. \tag{18.12}$$

Then $A(z)$ is found to satisfy the following equation

$$A''(z) \pm 2i \left\{ -\frac{h^4z^2}{4} + \frac{c^2z^4}{2} \right\}^{1/2} A'(z) \pm iA(z)\frac{d}{dz}\left\{ -\frac{h^4z^2}{4} + \frac{c^2z^4}{2} \right\}^{1/2}$$

$$+ \left[\frac{1}{2}qh^2 + \frac{\triangle}{2h^4} \right] A(z) = 0. \tag{18.13}$$

Later we will be interested in the construction of wave functions which are even or odd around $z = 0$. This construction is simplified by the consideration of symmetry properties of our solutions which arise at this point. We observe — before touching the square roots in Eq. (18.13) — that one equation (of the two alternatives) follows from the other by changing the sign of z throughout. This observation allows us to define the pair of solutions

$$y_A(z) = A(z) \exp\left[+i \int^z dz \left\{ -\frac{h^4z^2}{4} + \frac{c^2z^4}{2} \right\}^{1/2} \right], \tag{18.14a}$$

$$\overline{y}_A(z) = \overline{A}(z) \exp\left[-i \int^z dz \left\{ -\frac{h^4z^2}{4} + \frac{c^2z^4}{2} \right\}^{1/2} \right], \tag{18.14b}$$

with

$$\overline{A}(z) = A(-z) \quad \text{and} \quad \overline{y}_A(z) = y_A(-z), \tag{18.15}$$

where $A(z)$ is the solution of the upper of Eqs. (18.13) and $\overline{A}(z)$ that of the lower of these equations. We take the square root by setting

$$\left\{ -\frac{z^2 h^4}{4} \right\}^{1/2} = (\overset{+}{-}) i \frac{zh^2}{2}. \tag{18.16}$$

For large h^2 we can write the Eqs. (18.13)

$$\mp z A'(z) \mp \frac{1}{2} A(z) + \frac{1}{2} q A(z) = O\left(\frac{1}{h^2}\right).$$

We define $A_q(z)$ as the solution of the equation

$$z A_q'(z) - \frac{1}{2}(q-1)A_q(z) = 0, \tag{18.17}$$

i.e.

$$A_q(z) = z^{\frac{1}{2}(q-1)} \equiv \frac{1}{(z^2)^{1/4}} \exp\left[\frac{1}{2} q \int^z \frac{dz}{(z^2)^{1/2}}\right]. \tag{18.18}$$

We define correspondingly

$$\overline{A}_q(z) = z^{-\frac{1}{2}(q+1)} = A_{-q}(z) \equiv \frac{1}{(z^2)^{1/4}} \exp\left[-\frac{1}{2} q \int^z \frac{dz}{(z^2)^{1/2}}\right]. \tag{18.19}$$

We see that one solution follows from the other by replacing z by $-z$. Clearly $A_q(z), \overline{A}_q(z)$ approximate the solutions of Eqs. (18.13) and we can develop a perturbation theory along the lines of our method as employed in the case of periodic potentials. One finds that these solutions are associated with the same asymptotic expansion for $\triangle$ and hence E (given by Eq. (18.34) below) — to be derived in detail in Example 18.1 and as a verification again in connection with the solution y_B — as the other solutions. Since these higher order contributions are of little interest for our present considerations, we do not pursue their calculation. Thus we now have the pair of solutions

$$y_A(z) = \exp\left[i \int^z dz \left\{ -\frac{1}{4} z^2 h^4 + \frac{1}{2} c^2 z^4 \right\}^{1/2}\right] \left[A_q(z) + O\left(\frac{1}{h^2}\right)\right], \tag{18.20a}$$

$$\overline{y}_A(z) = \exp\left[-i \int^z dz \left\{ -\frac{1}{4} z^2 h^4 + \frac{1}{2} c^2 z^4 \right\}^{1/2}\right] \left[\overline{A}_q(z) + O\left(\frac{1}{h^2}\right)\right]. \tag{18.20b}$$

These expansions are valid as decreasing asymptotic expansions in the domain

$$|z| > O\left(\frac{1}{h}\right),$$

i.e. away from the central minimum. With proper care in selecting signs of square roots we can use the solutions (18.20a) and (18.20b) to construct solutions $y_\pm(z)$ which are respectively even and odd under the parity transformation $z \to -z$ (or equivalently $q \to -q, h^2 \to -h^2$), i.e. we write

$$y_\pm(z) = \frac{1}{2}[y_A(q, h^2; z) \pm \bar{y}_A(q, h^2; z)]. \tag{18.21}$$

Example 18.1: Calculation of eigenvalues along with solutions of type A
Use the solutions of type A, i.e. (18.20a), (18.20b), to obtain in leading order the eigenvalues E, i.e.

$$E(q, h^2) = \frac{1}{2}qh^2 + \frac{\triangle}{2h^4}, \quad \triangle = -\frac{3}{2}c^2(q^2 + 1) + O\left(\frac{1}{h^6}\right).$$

Solution: We rewrite the upper of Eqs. (18.13) in the following form with power expansion of the square root quantities and division by h^2:

$$-zA'(z) + \frac{1}{2}(q-1)A(z) = -\frac{1}{h^2}A''(z) - \frac{\triangle}{2h^6}A(z) + \sum_{i=1}^{\infty}\left(\frac{2c^2z^2}{h^4}\right)^i[\alpha_i zA'(z) + \frac{1}{2}\beta_i A(z)],$$

where the expansion coefficients are given by

$$\alpha_0 = 1, \quad \alpha_1 = -\frac{1}{2}, \quad \alpha_2 = -\frac{1}{8}, \quad \alpha_3 = -\frac{1}{16}, \quad \alpha_4 = -\frac{5}{128}, \dots,$$

$$\beta_0 = 1, \quad \beta_1 = -\frac{3}{2}, \quad \beta_2 = -\frac{5}{8}, \quad \beta_3 = -\frac{7}{16}, \dots.$$

Using Eqs. (18.17) and (18.18) we obtain

$$A_q'(z) = \frac{1}{2}\frac{(q-1)}{z}A_q(z) = \frac{1}{2}(q-1)A_{q-2}(z)$$

and from this or separately

$$A_q''(z) = \left[\frac{1}{2}\frac{(q-1)}{z}\right]^2 A_q(z) - \frac{1}{2}\frac{(q-1)}{z^2}A_q(z) = \frac{1}{4}(q-1)(q-3)A_{q-4}(z).$$

The lowest order solution $A^{(0)} = A_q(z)$ therefore leaves uncompensated on the right hand side of the equation for A the terms amounting to

$$R_q^{(0)} = -\frac{1}{4h^2}(q-1)(q-3)A_{q-4} - \frac{\triangle}{2h^6}A_q - \frac{c^2}{2h^4}(q+2)A_{q+4} - \left(\frac{c^2}{2h^4}\right)^2(q+4)A_{q+8}$$

$$+ \frac{1}{2}\sum_{i=3}^{\infty}\left(\frac{2c^2z^2}{h^4}\right)^i[(q-1)\alpha_i + \beta_i]A_q.$$

Clearly one now uses the relation
$$z^{2i}A_q(z) = A_{q+4i}(z).$$

In this way $R_q^{(0)}$ is expressed as a linear combination of functions $A_{q+4i}(z)$. As always in the procedure, one observes that with

$$\mathcal{D}_q := -z\frac{d}{dz} + \frac{1}{2}(q-1), \quad \mathcal{D}_{q+4i} = \mathcal{D}_q + 2i, \quad \mathcal{D}_q\frac{\mu A_{q+4i}}{-2i} = \mu A_{q+4i}.$$

Thus a term μA_{q+4i} in $R_q^{(0)}$ can be taken care of by adding to $A^{(0)}$ the contribution $\frac{\mu A_{q+4i}}{-2i}$ except, of course, when $i = 0$. In this way we obtain the next order contribution $A^{(1)}$ to $A^{(0)}$, and the coefficient of terms with $i = 0$, i.e. those in $A_q(z)$, give an equation from which $\triangle$ is determined. Hence in the present case

$$A^{(1)} = -\frac{1}{4h^2}(q-1)(q-3)\frac{A_{q-4}}{2} - \frac{c^2}{2h^4}(q+2)\frac{A_{q+4}}{-2} - \left(\frac{c^2}{2h^4}\right)^2(q+4)\frac{A_{q+8}}{-4} + \cdots.$$

In its turn $A^{(1)}$ leaves uncompensated terms amounting to

$$\begin{aligned} R_q^{(1)} &= -\frac{1}{4h^2}(q-1)(q-3)\frac{1}{2}\bigg\{ -\frac{1}{4h^2}(q-5)(q-7)A_{q-8} - \frac{c^2}{2h^4}(q-2)A_q \\ &\quad -\frac{\triangle}{2h^6}A_{q-4} - \left(\frac{2c^2}{h^4}\right)^2\frac{q}{16}A_{q+4} + \cdots \bigg\} \\ &\quad -\frac{c^2}{2h^4}\frac{(q+2)}{-2}\bigg\{ -\frac{1}{4h^2}(q+3)(q+1)A_q - \frac{c^2}{2h^4}(q+6)A_{q+8} - \frac{\triangle}{2h^6}A_{q+4} + \cdots \bigg\} + \cdots. \end{aligned}$$

The sum of terms with A_q in $R_q^{(0)}, R_q^{(1)}, \ldots$ must then be set equal to zero. Hence to the order we are calculating here

$$0 = \frac{c^2}{(4h^2)(4h^4)}[(q-1)(q-2)(q-3) - (q+1)(q+2)(q+3)] + O\left(\frac{1}{h^6}\right) - \frac{\triangle}{2h^6},$$

i.e.

$$0 = -\frac{c^2}{2^4 h^6}12(q^2+1) - \frac{\triangle}{2h^6} + O\left(\frac{1}{h^6}\right).$$

It follows that

$$\triangle = -\frac{3}{2}c^2(q^2+1) + O\left(\frac{1}{h^6}\right).$$

The same result is obtained below in connection with the solution of type B.

In the next two pairs of solutions the exponential factor of the above solutions of type A is contained in the *parabolic cylinder functions* (which are effectively exponentials times Hermite functions).

We return to Eq. (18.5). The solutions $y_q(z)$ of the equation

$$\mathcal{D}_q(w)y_q(w) = 0, \quad w = hz, \tag{18.22}$$

are parabolic cylinder functions $D_{\frac{1}{2}(q-1)}(\pm w)$ and $D_{-\frac{1}{2}(q+1)}(\pm iw)$ (observe that Eq. (18.22) is invariant under the combined substitutions $q \to -q, w \to \pm iw$) or functions

$$B_q(w) = \frac{D_{\frac{1}{2}(q-1)}(\pm w)}{[\frac{1}{4}(q-1)]!2^{\frac{1}{4}(q-1)}} \quad \text{and} \quad C_q(w) = \frac{D_{-\frac{1}{2}(q+1)}(\pm iw)2^{\frac{1}{4}(q+1)}}{[-\frac{1}{4}(q+1)]!}.$$

$$\tag{18.23}$$

The solutions B_q satisfy the following recurrence relation (obtained from the basic recurrence relation for parabolic cylinder functions given in the literature[§])

$$w^2 y_q = \frac{1}{2}(q+3)y_{q+4} + q y_q + \frac{1}{2}(q-3)y_{q-4}. \tag{18.24}$$

The extra factors in Eq. (18.23) have been inserted to make this recurrence relation assume this particularly symmetric and appealing form.[¶] For higher even powers of w we write

$$w^{2i} y_q = \sum_{j=-i}^{i} S_{2i}(q, 4j) y_{q+4j}, \tag{18.25}$$

where in the case $i = 2$:

$$
\begin{aligned}
S_4(q, \pm 8) &= \frac{1}{4}(q \pm 3)(q \pm 7), \\
S_4(q, \pm 4) &= (q \pm 2)(q \pm 3), \\
S_4(q, 0) &= \frac{3}{2}(q^2 + 1).
\end{aligned}
\tag{18.26}
$$

The first approximation $y(w) = y^{(0)}(w) = y_q(w) = B_q(w)$ therefore leaves uncompensated terms amounting to

$$R_q^{(0)} = -\frac{1}{h^6}(\triangle + c^2 w^4)y_q(w) \equiv -\frac{1}{h^6}\sum_{j=-2}^{2}[q, q+4j]y_{q+4j}, \tag{18.27}$$

where

$$[q, q] = \triangle + c^2 S_4(q, 0), \quad \text{and for } j \neq 0: \quad [q, q+4j] = c^2 S_4(q, 4j). \tag{18.28}$$

Now, since $\mathcal{D}_q y_q = 0$, we also have $\mathcal{D}_{q+4j} y_{q+4j} = 0$, $\mathcal{D}_{q+4j} = \mathcal{D}_q + 4j$, and so

$$\mathcal{D}_q y_{q+4j}(w) = -4j y_{q+4j}(w). \tag{18.29}$$

[§]A. Erdélyi, W. Magnus, F. Oberhettinger and F. G. Tricomi [86], pp. 115 - 123. Comparison with our notation is easier if this reference is used.

[¶]As an alternative to $B_q(w)$ in Eq. (18.23) one can choose the solutions as $\tilde{B}_q(w)$ with

$$\tilde{B}_q(w) = \frac{D_{\frac{1}{2}(q-1)}(w)}{[\frac{1}{4}(q-3)]! 2^{\frac{1}{4}(q-1)}}.$$

These satisfy the recurrence relation

$$w^2 y_q(w) = \frac{1}{2}(q+1)y_{q+4} + q y_q + \frac{1}{2}(q-1)y_{q-4}.$$

Actually these factors can also be extracted from WKB solutions for large values of q. See Example 18.3.

Hence a term μy_{q+4j} on the right hand side of Eq. (18.27) can be removed by adding to $y^{(0)}$ the contribution $(-\mu/4j)y_{q+4j}$. Thus the next order contribution to y_q is

$$y^{(1)}(w) = \frac{1}{h^6}\sum_{j=-2,j\neq 0}^{2}\frac{[q,q+4j]}{4j}y_{q+4j}. \tag{18.30}$$

For the sum $y(w) = y^{(0)}(w) + y^{(1)}(w)$ to be a solution to that order we must also have to that order

$$[q,q] = 0, \quad \text{i.e.} \quad \triangle = -\frac{3}{2}(q^2+1)c^2 + O\left(\frac{1}{h^6}\right). \tag{18.31}$$

Proceeding in this way we obtain the solution

$$y = y^{(0)}(w) + y^{(1)}(w) + y^{(2)}(w) + \cdots \tag{18.32}$$

with the corresponding equation from which $\triangle$ can be obtained, i.e.

$$0 = \frac{1}{h^6}[q,q] + \left(\frac{1}{h^6}\right)^2\sum_{j\neq 0}\frac{[q,q+4j]}{4j}[q+4j,q] + \cdots. \tag{18.33}$$

Evaluating this expansion and inserting the result for $\triangle$ into Eq. (18.10) we obtain

$$E(q,h^2) = \frac{1}{2}qh^2 - \frac{3c^2}{4h^4}(q^2+1) - \frac{c^4}{h^{10}}q(4q^2+29) + O\left(\frac{1}{h^{16}}\right). \tag{18.34}$$

We observe that odd powers of q arise in combination with odd powers of $1/h^2$, and even powers of q in combination with even powers of $1/h^2$, so that the entire expansion is invariant under the interchanges

$$q \to -q, \quad h^2 \to -h^2.$$

This type of invariance is a property of a very large class of eigenvalue problems. Equation (18.34) is the expansion of the eigenenergies E of Case (1) with $q = q_0 = 2n+1, n = 0,1,2,\ldots$ and $|c^2|$ replaced by $-|c^2|$. In Case (3) the parameter q is only approximately an odd integer in view of tunneling.

We can now write the solution $y(w)$ in the form

$$y(w) = y_q(w) + \sum_{i=1}^{\infty}\left(\frac{1}{h^6}\right)^i\sum_{j=-2i,j\neq 0}^{2i}P_i(q,q+4j)y_{q+4j}(w), \tag{18.35}$$

where for instance

$$P_1(q, q \pm 4) = \frac{[q, q \pm 4]}{\pm 4} = c^2 \frac{(q \pm 2)(q \pm 3)}{\pm 4},$$

$$P_2(q, q \pm 4) = \frac{[q, q \pm 4]}{\pm 4} \frac{[q \pm 4, q \pm 4]}{\pm 4} + \frac{[q, q \pm 8]}{\pm 8} \frac{[q \pm 8, q \pm 4]}{\pm 4}$$
$$+ \frac{[q, q \mp 4]}{\mp 4} \frac{[q \mp 4, q \pm 4]}{\pm 4},$$

and so on. Again we can write down a recurrence relation for the coefficients $P_i(q, q + 4j)$ in complete analogy to other applications of the method, i.e.

$$4t P_i(q, q + 4t) = \sum_{j=-2}^{2} P_{i-1}(q, q + 4j + 4t)[q + 4j + 4t, q + 4t] \qquad (18.36)$$

with the boundary conditions

$$
\begin{aligned}
P_0(q, q) &= 1, \quad \text{and for } j \neq 0 \text{ all other } P_0(q, q + 4j) = 0, \\
P_{i \neq 0}(q, q) &= 0, \\
P_i(q, q + 4j) &= 0 \quad \text{for} \quad |j| > 2i \text{ or } |j| \geq 2i + 1. \qquad (18.37)
\end{aligned}
$$

For further details concerning these coefficients, their recurrence relations and the solutions of the latter we refer to the literature.[||] Since our starting equation (18.11) is invariant under a change of sign of z, we may infer that given one solution $y(z)$, there is another solution $y(-z)$. We thus have the following pair of solutions

$$y_B(z) = \left[B_q(w) + \sum_{i=1}^{\infty} \left(\frac{1}{h^6} \right)^i \sum_{j=-2i, j \neq 0}^{2i} P_i(q, q + 4j) B_{q+4j}(w) \right]_{w=hz, \arg z = 0},$$

$$\bar{y}_B(z) = [y_B(z)]_{\arg z = \pi} = [y_B(-z)]_{\arg z = 0}. \qquad (18.38)$$

These solutions are suitable in the sense of decreasing asymptotic expansions in the domains

$$|z| \lesssim O\left(\frac{1}{h^2} \right), \quad \arg z \sim 0, \pi.$$

They are linearly independent there as long as q is not an integer.

Our third pair of solutions is obtained from the *parabolic cylinder functions of complex argument*. We observed earlier that these are obtained by making the replacements

$$q \to -q, \quad w \to \pm i w.$$

[||]P. Achuthan, H. J. W. Müller–Kirsten and A. Wiedemann [4]

These solutions are therefore defined by the following substitutions:

$$y_C(z) = [y_B(z)]_{q \to -q, h \to ih}, \quad \overline{y}_C(z) = [\overline{y}_B(z)]_{q \to -q, h \to ih} \tag{18.39}$$

with the same coefficients $P_i(q, q + 4j)$ as in y_B. The solutions $y_C, \overline{y}_C$ are suitable asymptotically decreasing expansions in one of the domains

$$|z| \lesssim O\left(\frac{1}{h^2}\right), \quad \arg z \sim \mp \frac{\pi}{2}.$$

We emphasize again that all three pairs of solutions are associated with the same expansion of the eigenvalue $E(q, h^2)$ in which odd powers of q are associated with odd powers of h^2, so that the eigenvalue expansion remains unaffected by the interchanges $q \to -q, h^2 \to -h^2$ as long as corrections resulting from boundary conditions are ignored. We add parenthetically that all our solutions here are unnormalized as is clear from the fact that the function in the dominant term (e.g. $A_q(z)$) does not appear in any of the higher order terms. Such contributions arise, in fact, with normalization since the normalization constants are also asymptotic expansions.

18.2.3 Matching of solutions

We saw that the solutions of types B and C are valid around the central minimum at $|z| = 0$, the solutions of type A being valid away from the minimum. Thus in the transition region some become proportional. In order to be able to extract the proportionality factor between two solutions, one has to stretch each by appropriate expansion to the limit of its domain of validity. In this bordering domain the adjoining branches of the overall solution then differ by a proportionality constant.[**]

First we deal with the exponential factor

$$\exp\left[i \int^z dz \left\{ -\frac{1}{4}z^2 h^4 + \frac{1}{2}c^2 z^4 \right\}^{1/2}\right]$$

$$= \exp\left[-\frac{h^6}{8c^2} \int^z d\left(\frac{2c^2 z^2}{h^4}\right)\left\{1 - \frac{2c^2 z^2}{h^4}\right\}^{1/2}\right]$$

in the solutions of type A which are not valid around $z = 0$. Integrating and expanding as follows since h^2 is assumed to be large, we obtain:

$$\exp\left[\frac{h^6}{8c^2}\frac{2}{3}\left\{1 - \frac{2c^2 z^2}{h^4}\right\}^{3/2}\right] = \exp\left[\frac{h^6}{12c^2} - \frac{h^2 z^2}{4} + O\left(\frac{z^4}{h^2}\right)\right]. \tag{18.40}$$

[**]Variables like those we use here for expansion about the minimum of a potential (e.g. like w of Eqs. (18.4)), (18.5) are known in some mathematical literature as *"stretching variables"* and are there discussed in connection with matching principles, see e.g. J. Mauss [192].

Considering the pair of solutions $y_A(z), \bar{y}_A(z)$ we see that in the direction of $z = 0$ (of course, not around that point)

$$
\begin{aligned}
y_A(z) &= e^{h^6/12c^2} e^{-\frac{1}{4}z^2 h^2} \left[z^{\frac{1}{2}(q-1)} + O\left(\frac{1}{h^2}\right) \right], \\
\bar{y}_A(z) &= e^{-h^6/12c^2} e^{\frac{1}{4}z^2 h^2} \left[z^{-\frac{1}{2}(q+1)} + O\left(\frac{1}{h^2}\right) \right].
\end{aligned} \tag{18.41}
$$

The cases of the solutions of types B and C require a careful look at the parabolic cylinder functions since these differ in different regions of the argument of the variable z. Thus from the literature [86] we obtain

$$
D_{\frac{1}{2}(q-1)}(w) = w^{\frac{1}{2}(q-1)} e^{-\frac{1}{4}w^2} \sum_{i=0}^{\infty} \frac{[\frac{1}{2}(q-1)]!}{i! [\frac{1}{2}(q - 4i - 1)]!} \frac{1}{(-2w^2)^i}, \quad |\arg w| < \frac{3}{4}\pi,
\tag{18.42}
$$

but

$$
\begin{aligned}
D_{\frac{1}{2}(q-1)}(w) = {}& w^{\frac{1}{2}(q-1)} e^{-\frac{1}{4}w^2} \sum_{i=0}^{\infty} \frac{[\frac{1}{2}(q-1)]!}{i![\frac{1}{2}(q - 4i - 1)]!} \frac{1}{(-2w^2)^i} \\
& - \frac{(2\pi)^{1/2} e^{-i\frac{\pi}{2}(q-1)}}{[-\frac{1}{2}(q+1)]!} \frac{e^{\frac{1}{4}w^2}}{w^{\frac{1}{2}(q+1)}} \sum_{i=0}^{\infty} \frac{[-\frac{1}{2}(q+1)]!}{i![-\frac{1}{2}(q + 4i + 1)]!} \frac{1}{(2w^2)^i} \\
& \text{with } \frac{5}{4}\pi > \arg w > \frac{1}{4}\pi.
\end{aligned}
\tag{18.43}
$$

The function $D_{\frac{1}{2}(q-1)}(w)$ has a similarly complicated expansion for

$$
-\frac{1}{4\pi} > \arg w > -\frac{5}{4}\pi.
$$

From (18.42) we obtain for the solution $y_B(z), w = hz$:

$$
y_B(z) \simeq B_q(w) = \frac{(h^2 z^2)^{\frac{1}{4}(q-1)}}{[\frac{1}{4}(q-1)]! 2^{\frac{1}{4}(q-1)}} e^{-\frac{1}{4}h^2 z^2} \left[1 + O\left(\frac{1}{h^2}\right) \right].
\tag{18.44}
$$

In the solution $\bar{y}_B(z)$, with z in $y_B(z)$ replaced by $-z$, we would have to substitute correspondingly the expression (18.43) (since $z \to -z$ implies $\arg z = \pm \pi$). We do not require this at present. Comparing the solution $y_A(z)$ of Eq. (18.41) with the solution $y_B(z)$ of Eq. (18.44) we see that in their common domain of validity

$$
y_A(z) = \frac{1}{\alpha} y_B(z)
\tag{18.45}
$$

with

$$\alpha = \frac{(h^2)^{\frac{1}{4}(q-1)}e^{-\frac{h^6}{12c^2}}}{[\frac{1}{4}(q-1)]!2^{\frac{1}{4}(q-1)}}\left[1+O\left(\frac{1}{h^2}\right)\right]. \tag{18.46}$$

However, the ratio of $\bar{y}_A(z), \bar{y}_B(z)$ is not a constant.

We proceed similarly with the solutions $y_C(z), \bar{y}_C(z)$. Inserting the expansion (18.42) into $\bar{y}_C(z)$ we obtain

$$\bar{y}_C(z) = \frac{(-h^2z^2)^{-\frac{1}{4}(q+1)}e^{\frac{1}{4}h^2z^2}}{[-\frac{1}{4}(q+1)]!2^{-\frac{1}{4}(q+1)}}\left[1+O\left(\frac{1}{h^2}\right)\right]. \tag{18.47}$$

Comparing this behaviour of the solution $\bar{y}_C(z)$ with that of solution $\bar{y}_A(z)$ of Eq. (18.41), we see that in their common domain of validity

$$\bar{y}_A(z) = \frac{1}{\bar{\alpha}}\bar{y}_C(z), \tag{18.48}$$

where

$$\bar{\alpha} = \frac{(-h^2)^{-\frac{1}{4}(q+1)}e^{\frac{h^6}{12c^2}}}{[-\frac{1}{4}(q+1)]!2^{-\frac{1}{4}(q+1)}}\left[1+O\left(\frac{1}{h^2}\right)\right]. \tag{18.49}$$

Again there is no such simple relation between $y_A(z)$ and $y_C(z)$.

18.2.4 Boundary conditions at the origin
(A) Formulation of the boundary conditions

The more difficult part of the problem is to recognize the boundary conditions we have to impose. Looking at the potential we are considering here — as depicted in Fig. 18.2 — we see that near the origin the potential behaves like that of the harmonic oscillator in fact — our large-h^2 solutions require this for large h^2. Thus the boundary conditions to be imposed there are the same as in the case of the harmonic oscillator for alternately even and odd wave functions. Recalling the solutions $y_\pm(z)$ which we defined with Eq. (18.21) as even and odd about $z = 0$, we see that at the origin we have to demand the conditions

$$y'_+(0) = 0 \quad \text{and} \quad y_-(0) = 0 \tag{18.50}$$

and $y_+(0) \neq 0, y'_-(0) \neq 0$. The first of the conditions (18.50) will be seen to imply $q_0 \equiv 2n+1 = 1, 5, 9, \ldots$ and the second $q_0 = 3, 7, 11, \ldots$. For instance $q_0 = 1$ (or $n = 0$) implies a ground state wave function with the shape of a Gauss curve above $z = 0$, i.e. large probability for the particle to be found thereabouts. At $z = 0$ the solutions of type A are invalid; hence we have to use the proportionalities just derived in order to match these to the solutions valid around the origin. Then imposing the above boundary conditions we obtain

$$0 = y'_+(0) = \lim_{z \to 0} \frac{1}{2}[y'_A(z) + \overline{y}'_A(z)] = \frac{1}{2}\left[\frac{1}{\alpha}y'_B(0) + \frac{1}{\alpha}\overline{y}'_C(0)\right] \qquad (18.51)$$

and

$$0 = y_-(0) = \lim_{z \to 0} \frac{1}{2}[y_A(z) - \overline{y}_A(z)] = \frac{1}{2}\left[\frac{1}{\alpha}y_B(0) - \frac{1}{\alpha}\overline{y}_C(0)\right]. \qquad (18.52)$$

Thus we obtain the equations

$$\frac{y'_B(0)}{\overline{y}'_C(0)} = -\frac{\alpha}{\alpha} \quad \text{and} \quad \frac{y_B(0)}{\overline{y}_C(0)} = \frac{\alpha}{\alpha}. \qquad (18.53)$$

Clearly we now have to evaluate the solutions involved and their derivatives at the origin. We leave the detailed calculations to Example 18.2.

Example 18.2: Evaluation of $y_B(0)$, $\overline{y}_C(0)$, $y'_B(0)$, $\overline{y}'_C(0)$
Show that — with $w = hz$ — the leading terms of the quantities listed are given by

$$B_q(0) = \frac{1}{\sqrt{\pi}} \frac{[\frac{1}{4}(q-3)]!}{[\frac{1}{4}(q-1)]!} \sin\{\frac{\pi}{4}(q+1)\}, \qquad \overline{C}_q(0) = \frac{\sqrt{\pi}}{[-\frac{1}{4}(q+1)]![\frac{1}{4}(q-1)]!},$$

$$\left[\frac{d}{dw}B_q(w)\right]_0 = -\sqrt{\frac{2}{\pi}} \sin\{\frac{\pi}{4}(q+3)\}, \qquad \left[\frac{d}{dw}\overline{C}_q(w)\right]_0 = i\sqrt{\frac{2}{\pi}} \sin\{\frac{\pi}{4}(q-3)\}. \qquad (18.54)$$

In fact, with the *reflection formula* $(-z)!(z-1)! = \pi/\sin \pi z$, one finds that

$$[-\frac{1}{4}(q+1)]! = \frac{\pi}{[\frac{1}{4}(q-3)]! \sin\{\frac{\pi}{4}(q+1)\}} \quad \text{and hence} \quad \frac{B_q(0)}{\overline{C}_q(0)} = 1. \qquad (18.55)$$

Solution: From the literature, e.g. M. Abramowitz and I. A. Stegun [1], we obtain

$$D_{\frac{1}{2}(q-1)}(0) = \frac{\sqrt{\pi} 2^{\frac{1}{4}(q-1)}}{[-\frac{1}{4}(q+1)]!}, \qquad D'_{\frac{1}{2}(q-1)}(0) = -\frac{\sqrt{\pi} 2^{\frac{1}{4}(q+1)}}{[-\frac{1}{4}(q+3)]!}. \qquad (18.56)$$

Thus with the help of the reflection formula cited above:

$$B_q(0) = \frac{D_{\frac{1}{2}(q-1)}(0)}{[\frac{1}{4}(q-1)]!2^{\frac{1}{4}(q-1)}} = \frac{\sqrt{\pi}}{[\frac{1}{4}(q-1)]![-\frac{1}{4}(q+1)]!} = \frac{1}{\sqrt{\pi}} \frac{[\frac{1}{4}(q-3)]!}{[\frac{1}{4}(q-1)]!} \sin\{\frac{\pi}{4}(q+1)\} \qquad (18.57)$$

and

$$\left[\frac{d}{dw}B_q(w)\right]_0 = -\frac{\sqrt{2\pi}}{[\frac{1}{4}(q-1)]![-\frac{1}{4}(q+3)]!} = -\sqrt{\frac{2}{\pi}} \sin\{\frac{\pi}{4}(q+3)\}. \qquad (18.58)$$

Expressions for $\overline{C}_q(0)$, $[\overline{C}'_q(w)]_{w=0}$ follow with the help of the *"circuit relation"* of parabolic cylinder functions given in the literature as

$$D_{\frac{1}{2}(q-1)}(w) = e^{-i\frac{\pi}{2}(q-1)}D_{\frac{1}{2}(q-1)}(-w) + \frac{\sqrt{2\pi}e^{-i\frac{\pi}{4}(q+1)}}{[-\frac{1}{2}(q+1)]!}D_{-\frac{1}{2}(q+1)}(-iw).$$

From this relation we obtain

$$D_{-\frac{1}{2}(q+1)}(0) = \sqrt{\frac{\pi}{2}} \frac{D_{\frac{1}{2}(q-1)}(0)}{[\frac{1}{2}(q-1)]! \cos\{\frac{\pi}{4}(q-1)\}}. \tag{18.59}$$

Inserting from (18.56), using the above reflection formula and the *duplication formula* $\sqrt{\pi}(2z)! = 2^{2z} z!(z-1/2)!$, we obtain

$$D_{-\frac{1}{2}(q+1)}(0) = \frac{\sqrt{\pi}}{2^{\frac{1}{4}(q+1)}[-\frac{1}{4}(q-1)]!}.$$

From this we derive

$$\begin{aligned}
\overline{C}_q(0) &= \frac{D_{-\frac{1}{2}(q+1)}(0)}{2^{-\frac{1}{4}(q+1)}[-\frac{1}{4}(q+1)]!} \\
&= \frac{\sqrt{\pi}}{[-\frac{1}{4}(q+1)]![\frac{1}{4}(q-1)]!} = \frac{[\frac{1}{4}(q-3)]!}{\sqrt{\pi}[\frac{1}{4}(q-1)]!} \sin\{\frac{\pi}{4}(q+1)\}. \tag{18.60}
\end{aligned}$$

Similarly we obtain

$$[D'_{-\frac{1}{2}(q+1)}(iw)]_{w=0} = \frac{-i\sqrt{\pi}}{2^{\frac{1}{4}(q-1)}[\frac{1}{4}(q-3)]!} \tag{18.61}$$

and

$$\begin{aligned}
[\overline{C}'_q(w)]_0 &= \frac{D'_{-\frac{1}{2}(q+1)}(0)}{[-\frac{1}{4}(q+1)]! 2^{-\frac{1}{4}(q+1)}} = \frac{-i\sqrt{2\pi}}{[-\frac{1}{4}(q+1)]![\frac{1}{4}(q-3)]!} \\
&= -i\sqrt{\frac{2}{\pi}} \sin\{\frac{\pi}{4}(q+1)\} = i\sqrt{\frac{2}{\pi}} \sin\{\frac{\pi}{4}(q-3)\}. \tag{18.62}
\end{aligned}$$

(B) Evaluation of the boundary conditions

We now evaluate Eqs. (18.53) in dominant order and insert the appropriate expressions for α and $\overline{\alpha}$ from Eqs. (18.46), (18.49). Starting with the derivative expression we obtain (apart from contributions of order $1/h^2$)

$$-\frac{1}{i}\frac{\sin\{\frac{\pi}{4}(q+3)\}}{\sin\{\frac{\pi}{4}(q-3)\}} = -\frac{(h^2)^{\frac{1}{4}(q-1)}[-\frac{1}{4}(q+1)]! 2^{-\frac{1}{4}(q+1)}}{[\frac{1}{4}(q-1)]! 2^{\frac{1}{4}(q-1)}(-h^2)^{-\frac{1}{4}(q+1)}} e^{-\frac{h^6}{6c^2}}.$$

We rewrite the left hand side as

$$-\frac{1}{i}\frac{\sin\{\frac{\pi}{4}(q+3)\}}{\sin\{\frac{\pi}{4}(q+3-6)\}} = i\tan\{\frac{\pi}{4}(q+3)\}.$$

We rewrite the right hand side of the derivative equation again with the help of the inversion and duplication formulas and obtain

$$-\pi \frac{(\frac{h^4}{4})^{q/4}(-1)^{\frac{1}{4}(q+1)} e^{-h^6/6c^2}}{[\frac{1}{4}(q-1)]![\frac{1}{4}(q-3)]! \sin\{\frac{\pi}{4}(q+1)\}} = \sqrt{\frac{\pi}{2}} \frac{(h^4)^{q/4}(-1)^{\frac{1}{4}(q+1)}}{[\frac{1}{2}(q-1)]! \cos\{\frac{\pi}{4}(q+3)\}} e^{-\frac{h^6}{6c^2}}.$$

Then the derivative relation of Eqs. (18.53) becomes

$$\sin\left\{\frac{\pi}{4}(q+3)\right\} = -i\sqrt{\frac{\pi}{2}}\frac{(h^4)^{q/4}(-1)^{\frac{1}{4}(q+1)}}{[\frac{1}{2}(q-1)]!}e^{-\frac{h^6}{6c^2}}. \tag{18.63}$$

Proceeding similarly with the second of relations (18.53), we obtain

$$\cos\left\{\frac{\pi}{4}(q+3)\right\} = -\sqrt{\frac{\pi}{2}}\frac{(h^4)^{q/4}(-1)^{\frac{1}{4}(q+1)}}{[\frac{1}{2}(q-1)]!}e^{-\frac{h^6}{6c^2}}. \tag{18.64}$$

In each of Eqs. (18.63) and (18.64) the right hand side is an exponentially small quantity. In fact the left hand side of (18.63) vanishes for $q = q_0 = 1, 5, 9, \ldots$ and the left hand side of (18.64) for $q = q_0 = 3, 7, 11, \ldots$. With a Taylor expansion about q_0 the left hand side of (18.63) becomes

$$(q-q_0)\frac{\pi}{4}\cos\left\{\frac{\pi}{4}(q_0+3)\right\} + \cdots \simeq (q-q_0)\frac{\pi}{4}(-1)^{-\frac{1}{4}(q_0+3)}.$$

It follows that we obtain for the even function with $q = q_0 = 1, 5, 9, \ldots$

$$(q-q_0) \simeq \pm\frac{2\sqrt{2}}{\sqrt{\pi}}\frac{(h^2)^{q_0/2}}{[\frac{1}{2}(q_0-1)]!}e^{-\frac{h^6}{6c^2}}. \tag{18.65}$$

Expanding similarly the left hand side of Eq. (18.64) about $q_0 = 3, 7, 11, \ldots$, we again obtain (18.65) but now for the odd function with these values of q_0. We have thus obtained the conditions resulting from the boundary conditions at $z = 0$. Our next task is to extend the solution all the way to the region beyond the shoulders of the inverted double well potential and to impose the necessary boundary conditions there. Thus we have to determine these conditions first.

18.2.5 Boundary conditions at infinity
(A) Formulation of the boundary conditions

We explore first the conditions we have to impose at $|z| \to \infty$. Recall the original Schrödinger equation (18.2) with potential (18.1). For $c^2 < 0$ and the solution $y(z)$ square integrable in $-\infty < \Re z < \infty$, the energy E is real; this is the case of the purely discrete spectrum (the differential operator being selfadjoint for the appropriate boundary conditions, i.e. the vanishing of the wave functions at infinity). This is Case (1) of Fig. 18.1. The analytic continuation of one case to the other is accomplished by replacing $\pm c^2$ by $\mp c^2$ or, equivalently, by the rotations

$$E \to e^{i\pi}E = -E, \quad z \to e^{i\pi/2}z, \quad z^2 \to -z^2.$$

One can therefore retain c^2 as it is and perform these rotations. It is then necessary to insure that when one rotates to the case of the purely discrete spectrum without tunneling, the resulting wave functions vanish at infinity and thus are square integrable. Thus in our case here the behaviour of the solutions at infinity has to be chosen such that this condition is satisfied. Now, for $\Re z \to \pm\infty$ we have

$$y(z) \sim \exp\left\{ \pm i \int^z dz \left[\frac{c^2}{2} z^4 \right]^{1/2} \right\} = \exp\left\{ \pm i \left(\frac{c^2}{2} \right)^{1/2} \frac{z^3}{3} \right\}. \tag{18.66}$$

In order to decide which solution or combination of solutions is compatible with the square integrability in the rotated (c^2 reversed) case, we set

$$z = |z| e^{i\theta}, \quad z^3 = |z|^3 e^{3i\theta}.$$

Then

$$\exp\left\{ - i \left(\frac{c^2}{2} \right)^{1/2} \frac{z^3}{3} \right\} = \exp\left\{ - i \left(\frac{c^2}{2} \right)^{1/2} \frac{|z|^3}{3} (\cos 3\theta + i \sin 3\theta) \right\}$$

$$\propto \exp\left\{ \left(\frac{c^2}{2} \right)^{1/2} \frac{|z|^3}{3} \sin 3\theta \right\}.$$

This expression vanishes for $|z| \to +\infty$ if the angle θ lies in the range $-\pi < 3\theta < 0$, i.e. if $-\pi/3 < \theta < 0$. Thus

$$\exp\left\{ - i \left(\frac{c^2}{2} \right)^{1/2} \frac{z^3}{3} \right\} \to 0 \quad \text{for } \Re z \to +\infty \text{ in } \arg z \in \left(-\frac{\pi}{3}, 0 \right),$$

$$\exp\left\{ + i \left(\frac{c^2}{2} \right)^{1/2} \frac{z^3}{3} \right\} \to 0 \quad \text{for } \Re z \to -\infty \text{ in } \arg z \in \left(0, \frac{\pi}{3} \right). \tag{18.67}$$

Rotating z by $\pi/2$, i.e. replacing $\sin 3\theta$ by

$$\sin 3\left(\theta + \frac{\pi}{2} \right) = - \cos 3\theta,$$

we see that the solution with the exponential factor is exponentially decreasing for $|z| \to \infty$ provided that $\cos 3\theta > 0$, i.e. in the domain $-\pi/2 < 3\theta < \pi/2$, or

$$-\frac{\pi}{6} < \theta < \frac{\pi}{6}.$$

In the case of the inverted double well potential under consideration here (i.e. the case of complex E), we therefore demand that for $\Re z \to +\infty$ and $-\pi/3 < \arg z < 0$ the wave functions have decreasing phase, i.e.

$$y(z) \sim \exp\left\{ - i \left(\frac{c^2}{2} \right)^{1/2} \frac{z^3}{3} \right\}, \quad c^2 > 0. \tag{18.68}$$

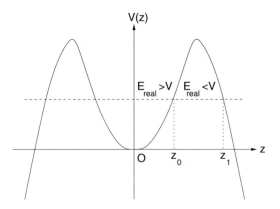

Fig. 18.3 The inverted double well potential with turning points z_0, z_1.

This is the boundary condition also used by Bender and Wu [18]. For $c^2 < 0$ we have correspondingly

$$y(z) \sim \exp\left\{ \pm \left(\frac{|c^2|}{2}\right)^{1/2} \frac{z^3}{3} \right\}, \quad c^2 < 0, \quad \text{for } z \to \mp\infty.$$

This is not the asymptotic behaviour of a wave function of the simple harmonic oscillator. We have to remember that we have various branches of the solutions $y(z)$ in different domains of z.

Our procedure now is to continue the even and odd solutions (18.21) to $+$ infinity and to demand that they satisfy the condition (18.68) for $c^2 > 0$. Equating to zero the coefficient of the term with sign opposite to that in the exponential of Eq. (18.68) will lead to our second condition which together with the first obtained from boundary conditions at the origin determines the imaginary part of the eigenvalue E.

(B) Evaluation of the boundary conditions

The following considerations (usually for real z) require some algebraic steps which could obscure the basic procedure. We therefore explain our procedure first. Our even and odd solutions $y_\pm(z)$ (cf. Eq. (18.21)) were defined in terms of solutions of type A which have a wide domain of validity. Looking at Fig. 18.3 we see that at a given energy E and to the right of $z = 0$ (which is the only region we consider for reasons of symmetry) there are two turning points z_0, z_1. Thus we have to match the solutions of type A first to solutions to the left of z_1 and then extend these to solutions to the right, and there impose the boundary condition (18.68) on $y_\pm(z)$ (by demanding that the coefficient of the solutions with other behaviour be zero). We do this extension with the help of WKB solutions, i.e. we match the WKB solutions to the left of z_1 to the solutions of type A, and then use the WKB procedure (called "linear matching" across the turning point) to obtain the dominant WKB solutions beyond z_1. The distant turning point at z_1 as indicated in Fig. 18.3 is given by (using Eq. (18.4) for E and ignoring nondominant terms)

$$
-\frac{1}{4}z^2 h^4 + \frac{1}{2}c^2 z^4 \simeq -\frac{1}{2}qh^2, \quad \text{i.e.} \quad z_1 \simeq \frac{h^2}{\sqrt{2c^2}}\left(1 - \frac{2qc^2}{h^6}\right) \simeq \frac{h^2}{\sqrt{2c^2}}. \quad (18.69)
$$

The WKB solutions have been discussed in Chapter 14. From there or the literature* we obtain in the domain $V > \Re E$ to the left of z_1 as the dominant terms of the WKB solutions

$$
y_{\mathrm{WKB}}^{(l,z_1)}(z) = \left[-\frac{1}{2}qh^2 + \frac{1}{4}z^2 h^4 - \frac{1}{2}c^2 z^4\right]^{-1/4}
$$

$$
\times \exp\left\{\int_z^{z_1} dz\left[-\frac{1}{2}qh^2 + \frac{1}{4}z^2 h^4 - \frac{1}{2}c^2 z^4\right]^{1/2}\right\},
$$

$$
\overline{y}_{\mathrm{WKB}}^{(l,z_1)}(z) = \left[-\frac{1}{2}qh^2 + \frac{1}{4}z^2 h^4 - \frac{1}{2}c^2 z^4\right]^{-1/4}
$$

$$
\times \exp\left\{-\int_z^{z_1} dz\left[-\frac{1}{2}qh^2 + \frac{1}{4}z^2 h^4 - \frac{1}{2}c^2 z^4\right]^{1/2}\right\}, \quad (18.70)
$$

where $z < z_1$, i.e. $z^2 h^4/4 > c^2 z^4/2$. In using these expressions it has to be remembered that the moduli of the integrals have to be taken.† To the right of the turning point at z_1 these solutions match on to

$$
y_{\mathrm{WKB}}^{(r,z_1)}(z) = \left[\frac{1}{2}qh^2 - \frac{1}{4}z^2 h^4 + \frac{1}{2}c^2 z^4\right]^{-1/4}
$$

$$
\times \cos\left\{\int_z^{z_1} dz\left[\frac{1}{2}qh^2 - \frac{1}{4}z^2 h^4 + \frac{1}{2}c^2 z^4\right]^{1/2} + \frac{\pi}{4}\right\},
$$

$$
\overline{y}_{\mathrm{WKB}}^{(r,z_1)}(z) = 2\left[\frac{1}{2}qh^2 - \frac{1}{4}z^2 h^4 + \frac{1}{2}c^2 z^4\right]^{-1/4}
$$

$$
\times \sin\left\{-\int_z^{z_1} dz\left[\frac{1}{2}qh^2 - \frac{1}{4}z^2 h^4 + \frac{1}{2}c^2 z^4\right]^{1/2} + \frac{\pi}{4}\right\}.
$$

$$
(18.71)
$$

We now come to the algebra of evaluating the integrals in the above solutions. We begin with the exponential factors occurring in Eqs. (18.70),

---

*R.B. Dingle [70], p. 291, equations (21), (22) or A. Messiah, Vol. I [195], Sec. 6.2.4.
†R. B. Dingle [70], p. 291.

i.e.

$$
\begin{aligned}
& E_\pm \\
&= \exp\left\{\pm\int_z^{z_1} dz\left[-\frac{1}{2}qh^2 + \frac{1}{4}z^2h^4 - \frac{1}{2}c^2z^4\right]^{1/2}\right\} \\
&= \exp\left[\pm\frac{h^6}{8c^2}\int_z^{z_1} d\left(\frac{2c^2z^2}{h^4}\right)\left\{1 - \frac{2c^2z^2}{h^4}\right\}^{1/2}\left\{1 - \frac{2q}{z^2h^2}\frac{1}{[1-\frac{2c^2z^2}{h^4}]}\right\}^{1/2}\right] \\
&\simeq \exp\left[\mp\frac{h^6}{8c^2}\left\{\frac{2}{3}\left[1 - \frac{2c^2z^2}{h^4}\right]^{3/2}\right\}_z^{z_1} \mp\frac{qh^4}{8c^2}\int_z^{z_1}\frac{\frac{1}{z^2}d\left(\frac{2c^2z^2}{h^4}\right)}{[1-\frac{2c^2z^2}{h^4}]^{1/2}}\right],
\end{aligned}
$$

and so

$$
E_\pm = \exp\left[\pm\frac{h^6}{8c^2}\frac{2}{3}\left\{1 - \frac{2c^2z^2}{h^4}\right\}^{3/2} \mp\frac{q}{2}\int_z^{z_1}\frac{dz}{z[\frac{h^4}{2c^2}-z^2]^{1/2}}\left(\frac{h^4}{2c^2}\right)^{1/2}\right].
\tag{18.72}
$$

Here the first part is the exponential factor contained in $y_A(z), \bar{y}_A(z)$ respectively (cf. Eq. (18.40)). In the remaining factor we have (looking up Tables of Integrals)

$$
\int_z^{z_1}\frac{dz}{z[\frac{h^4}{2c^2}-z^2]^{1/2}} = \left\{-\left(\frac{2c^2}{h^4}\right)^{1/2}\ln\left|\frac{1}{z}\left\{\left(\frac{h^4}{2c^2}\right)^{1/2} + \left(\frac{h^4}{2c^2}-z^2\right)^{1/2}\right\}\right|\right\}_z^{z_1},
$$

and hence (with use of Eq. (18.69))

$$
\int_z^{z_1}\frac{dz}{z[\frac{h^4}{2c^2}-z^2]^{1/2}} = +\left(\frac{2c^2}{h^4}\right)^{1/2}\ln\left|\frac{1}{z}\left\{\left(\frac{h^4}{2c^2}\right)^{1/2} + \left(\frac{h^4}{2c^2}-z^2\right)^{1/2}\right\}\right|.
$$

Since we are interested in determining the proportionality of two solutions in their common domain of validity we require only the dominant z-dependence contained in this expression. We obtain this factor by expanding the expression in powers of z/z_1 (since in the integral $z < z_1$). Thus the above factor yields

$$
\left(\frac{2c^2}{h^4}\right)^{1/2}\ln\left|\frac{2}{z}\left(\frac{h^4}{2c^2}\right)^{1/2}\right|,
$$

so that (cf. Eq. (18.40))

$$
\begin{aligned}
E_\pm &= \exp\left[\pm\frac{h^6}{8c^2}\frac{2}{3}\left\{1 - \frac{2c^2z^2}{h^4}\right\}^{3/2} \mp\frac{q}{2}\ln\left|2\left(\frac{h^4}{2c^2}\right)^{1/2}\right|\right]z^{\pm q/2} \\
&= z^{\pm q/2}\left[2\left(\frac{h^4}{2c^2}\right)^{1/2}\right]^{\mp q/2}\exp\left[\pm i\int^z dz\left\{-\frac{1}{4}z^2h^4 + \frac{1}{2}c^2z^4\right\}^{1/2}\right].
\end{aligned}
\tag{18.73}
$$

Thus at the left end of the domain of validity of the WKB solutions we have

$$y_{\text{WKB}}^{(l,z_1)}(z) \simeq \frac{E_+}{\left[\frac{1}{4}z^2h^4\right]^{1/4}} \quad \text{and} \quad \bar{y}_{\text{WKB}}^{(l,z_1)}(z) \simeq \frac{E_-}{\left[\frac{1}{4}z^2h^4\right]^{1/4}}. \tag{18.74}$$

Comparing these solutions now with the solutions (18.20a) and (18.20b), we see that in their common domain of validity

$$y_A(z) = \beta y_{\text{WKB}}^{(l,z_1)}(z), \quad \bar{y}_A(z) = \bar{\beta}\bar{y}_{\text{WKB}}^{(l,z_1)}(z), \tag{18.75}$$

where

$$\beta = \left[\frac{h^2}{2}\right]^{1/2}\left[2\frac{(h^2)}{(2c^2)^{1/2}}\right]^{q/2} \quad \text{and} \quad \bar{\beta} = \left[-\frac{h^2}{2}\right]^{1/2}\left[2\frac{(-h^2)}{(2c^2)^{1/2}}\right]^{-q/2} \tag{18.76a}$$

or

$$\frac{\beta}{\bar{\beta}} = \left[\frac{2h^2}{(2c^2)^{1/2}}\right]^q \frac{(-1)^{q/2}}{\sqrt{-1}} \tag{18.76b}$$

apart from factors $[1+O(1/h^2)]$. In these expressions we have chosen the signs of square roots of h^4 so that the conversion symmetry under replacements $q \to -q, h^2 \to -h^2$ is maintained.

Returning to the even and odd solutions defined by Eqs. (18.21) we now have

$$\begin{aligned} y_\pm(z) &= \frac{1}{2}[y_A(z) \pm \bar{y}_A(z)] = \frac{1}{2}[\beta y_{\text{WKB}}^{(l,z_1)}(z) \pm \bar{\beta}\bar{y}_{\text{WKB}}^{(l,z_1)}(z)] \\ &= \frac{1}{2}[\beta y_{\text{WKB}}^{(r,z_1)}(z) \pm \bar{\beta}\bar{y}_{\text{WKB}}^{(r,z_1)}(z)]. \end{aligned} \tag{18.77}$$

Now in the domain $z \to \infty$ we have

$$\begin{aligned} \int_{z_1}^z \left[\frac{1}{2}qh^2 - \frac{1}{4}z^2h^4 + \frac{1}{2}c^2z^4\right]^{1/2} &\simeq \int_{z_1}^z dz \frac{cz^2}{\sqrt{2}}\left[1 - \frac{h^4}{4c^2z^2}\right] \simeq \left[\frac{cz^3}{3\sqrt{2}}\right]_{z_1}^z \\ &= \frac{cz^3}{3\sqrt{2}} - \frac{h^6}{12c^2}. \end{aligned} \tag{18.78}$$

Inserting this into the solutions (18.71) and these into (18.77) we can rewrite the even and odd solutions for $\Re z \to \infty$ as (by separating cosine and sine into their exponential components)

$$\begin{aligned} y_\pm(z) &\simeq \frac{1}{2}\left[\frac{1}{2}c^2z^4\right]^{-1/4}\left[S_+(\pm)\exp\left\{i\left(\frac{cz^3}{3\sqrt{2}} - \frac{h^6}{12c^2}\right)\right\}\right. \\ &\quad \left. + S_-(\pm)\exp\left\{-i\left(\frac{cz^3}{3\sqrt{2}} - \frac{h^6}{12c^2}\right)\right\}\right], \end{aligned} \tag{18.79}$$

where

$$S_+(\pm) = \left(\frac{1}{2}\beta \pm \frac{1}{i}\overline{\beta}\right)\exp\left(i\frac{\pi}{4}\right),$$

$$S_-(\pm) = \left(\frac{1}{2}\beta \mp \frac{1}{i}\overline{\beta}\right)\exp\left(-i\frac{\pi}{4}\right). \tag{18.80}$$

Imposing the boundary condition that the even and odd solutions have the asymptotic behaviour given by Eq. (18.68), we see that we have to demand that

$$S_+(\pm) = 0, \quad \text{i.e.} \quad \frac{1}{2}\beta \pm \frac{1}{i}\overline{\beta} = 0. \tag{18.81}$$

Inserting expressions (18.76a) for β and $\overline{\beta}$, this equation can be rewritten as

$$(-h^2)^{q/2} = (-h^2)^{q/2}\frac{i\beta}{2\overline{\beta}} = i\frac{2^{q-1}(-1)^q}{\sqrt{-1}}\left(\frac{h^6}{2c^2}\right)^{q/2} \tag{18.82a}$$

or as the replacement

$$(-h^2)^{q_0/2} \Rightarrow \overset{+}{(-)} 2^{q_0-1}\left(\frac{h^6}{2c^2}\right)^{q_0/2}. \tag{18.82b}$$

This is our second condition along with Eq. (18.65). Inserting this into the latter equation we obtain (the factor "i" arising from the minus sign on the left of Eq. (18.82b))

$$(q - q_0) = \pm i\sqrt{\frac{2}{\pi}}\frac{2^{q_0}\left(\frac{h^6}{2c^2}\right)^{q_0/2}}{[\frac{1}{2}(q_0 - 1)]!}e^{-\frac{h^6}{6c^2}}. \tag{18.83}$$

with $q_0 = 1, 3, 5, \dots$.

18.2.6   The complex eigenvalues

We now return to the expansion of the eigenvalues, i.e. Eq. (18.34),

$$E(q, h^2) = \frac{1}{2}qh^2 - \frac{3c^2}{4h^4}(q^2 + 1) - \frac{c^4q}{h^{10}}(4q^2 + 29) + O\left(\frac{1}{h^{16}}\right). \tag{18.84}$$

Expanding about $q = q_0$ we obtain

$$\begin{aligned}
E(q, h^2) &= E(q_0, h^2) + (q - q_0)\left(\frac{dE}{dq}\right)_{q_0} + \cdots \\
&= E(q_0, h^2) + (q - q_0)\frac{h^2}{2} + \cdots.
\end{aligned} \tag{18.85}$$

Clearly the expression for $(q - q_0)$ has to be inserted here giving in the dominant approximation

$$
E = E(q_0, h^2) \overset{+}{(-)} i \frac{2^{q_0} h^2 \left(\dfrac{h^6}{2c^2}\right)^{q_0/2}}{(2\pi)^{1/2}[\frac{1}{2}(q_0 - 1)]!} e^{-h^6/6c^2}.
$$

The final result is therefore

$$
E = \frac{1}{2}q_0 h^2 - \frac{3c^2}{4h^4}(q_0^2 + 1) - \frac{c^4 q_0}{h^{10}}(4q_0^2 + 29) + O\left(\frac{1}{h^{16}}\right)
$$

$$
\overset{+}{(-)} i \frac{2^{q_0} h^2 \left(\dfrac{h^6}{2c^2}\right)^{q_0/2}}{(2\pi)^{1/2}[\frac{1}{2}(q_0 - 1)]!} e^{-h^6/6c^2}. \tag{18.86}
$$

The imaginary part of this expression agrees with the result of Bender and Wu (see formula (3.36) of their Ref. [19]) for $\hbar = 1$ and in their notation

$$
q_0 = 2K + 1, \quad \frac{h^6}{2c^2} = \epsilon.
$$

In the above we did not require the matching of WKB solutions at the lower turning point z_0 to the solutions $y_A(z), \bar{y}_A(z)$. Since this is not without interest (e.g. in comparison with the work of Bender and Wu [19]) we deal with this in Example 18.3. The large order behaviour of the eigenvalue expansion (of Case (1)) which Bender and Wu derived from their result (18.86) will be dealt with in Chapter 20.

Example 18.3: Matching of WKB solutions to others at z_0
Determine the proportionality constants $\rho(q, h^2), \bar{\rho}(q, h^2)$ of the relations

$$
y_{\text{WKB}}^{(r,z_0)}(z) = \rho \bar{y}_A(z), \quad \bar{y}_{\text{WKB}}^{(r,z_0)}(z) = \bar{\rho} y_A(z),
$$

where the superscript $\{r, z_0\}$ means "to the right of the turning point z_0".

Solution: The turning point at z_0 close to the local minimum is given by

$$
\frac{1}{2}qh^2 - \frac{1}{4}z^2 h^4 + \frac{1}{2}c^2 z^4 \simeq 0, \quad \text{i.e.} \quad z_0 = \frac{(2q)^{1/2}}{h}\left(1 + \frac{2qc^2}{h^6} + \cdots\right) \simeq \frac{(2q)^{1/2}}{h}.
$$

In the domain $\Re E > V$ to the left of z_0 the dominant terms of the WKB solutions are (cf. above)

$$
y_{\text{WKB}}^{(l,z_0)}(z) = \left[\frac{1}{2}qh^2 - \frac{1}{4}z^2 h^4 + \frac{1}{2}c^2 z^4\right]^{-1/4}
$$

$$
\times \cos\left\{\int_z^{z_0} dz\left[\frac{1}{2}qh^2 - \frac{1}{4}z^2 h^4 + \frac{1}{2}c^2 z^4\right]^{1/2} + \frac{\pi}{4}\right\},
$$

$$
\bar{y}_{\text{WKB}}^{(l,z_0)}(z) = 2\left[\frac{1}{2}qh^2 - \frac{1}{4}z^2 h^4 + \frac{1}{2}c^2 z^4\right]^{-1/4}
$$

$$
\times \sin\left\{-\int_z^{z_0} dz\left[\frac{1}{2}qh^2 - \frac{1}{4}z^2 h^4 + \frac{1}{2}c^2 z^4\right]^{1/2} + \frac{\pi}{4}\right\},
$$

where $z_0 > z$. In the domain $V > \Re E$ to the right of z_0 the dominant terms of the WKB solutions which match on to $y_{\mathrm{WKB}}^{(l,z_0)}(z), \overline{y}_{\mathrm{WKB}}^{(l,z_0)}(z)$ are respectively

$$
y_{\mathrm{WKB}}^{(r,z_0)}(z) = \left[-\frac{1}{2}qh^2 + \frac{1}{4}z^2h^4 - \frac{1}{2}c^2z^4 \right]^{-1/4}
$$
$$
\times \exp\left\{ \int_z^{z_0} dz \left[-\frac{1}{2}qh^2 + \frac{1}{4}z^2h^4 - \frac{1}{2}c^2z^4 \right]^{1/2} \right\},
$$

$$
\overline{y}_{\mathrm{WKB}}^{(r,z_0)}(z) = \left[-\frac{1}{2}qh^2 + \frac{1}{4}z^2h^4 - \frac{1}{2}c^2z^4 \right]^{-1/4}
$$
$$
\times \exp\left\{ -\int_z^{z_0} dz \left[-\frac{1}{2}qh^2 + \frac{1}{4}z^2h^4 - \frac{1}{2}c^2z^4 \right]^{1/2} \right\},
$$

where $z > z_0$. Expanding the square root occurring in $y_{\mathrm{WKB}}^{(l,z_0)}(z), \overline{y}_{\mathrm{WKB}}^{(l,z_0)}(z)$, we have

$$
\left[\frac{1}{2}qh^2 - \frac{1}{4}z^2h^4 + \frac{1}{2}c^2z^4 \right]^{1/2} = \left(\frac{1}{2}qh^2 \right)^{1/2} \left\{ 1 - \frac{z^2}{2z_0^2} + \cdots \right\},
$$

and we see that

$$
y_{\mathrm{WKB}}^{(l,z_0)}(z) \simeq \frac{1}{[\frac{1}{2}qh^2]^{1/4}} \cos\left\{ \left(q - \sqrt{\frac{q}{2}}hz \right) + \frac{\pi}{4} \right\}, \quad \overline{y}_{\mathrm{WKB}}^{(l,z_0)}(z) \simeq \frac{2}{[\frac{1}{2}qh^2]^{1/4}} \sin\left\{ \left(q - \sqrt{\frac{q}{2}}hz \right) + \frac{\pi}{4} \right\}.
$$

In other words, the (approximate) solutions[‡]

$$
\exp\left\{ \pm i \left(\frac{q}{2} \right)^{1/2} hz \right\}
$$

combined with the particular constants contained in $y_{\mathrm{WKB}}^{(l,z_0)}, \overline{y}_{\mathrm{WKB}}^{(l,z_0)}$ match on to the WKB solutions with exponential behaviour to the right of z_0.

On the other hand in the exponentially behaving WKB solutions the quartic interaction term acts as a correction to the harmonic term close to z_0, and we have

$$
\int_{z_0}^{z} dz \left[-\frac{1}{2}qh^2 + \frac{1}{4}z^2h^4 - \frac{1}{2}c^2z^4 \right]^{1/2} = \int_{z_0}^{z} dz \left[-\frac{1}{2}qh^2 + \frac{1}{4}z^2h^4 \right]^{1/2} \left(1 + O\left(\frac{1}{h^2} \right) \right)
$$
$$
\simeq \frac{h^2}{2} \int_{z_0}^{z} (z^2 - z_0^2)^{1/2} = \frac{h^2}{2} \left[\frac{1}{2}z(z^2 - z_0^2)^{1/2} - \frac{1}{2}z_0^2 \ln|z + (z^2 - z_0^2)^{1/2}| \right]_{z_0}^{z}
$$
$$
= \frac{h^2}{4} \left[z^2 - \frac{q}{h^2} - \frac{2q}{h^2} \ln\left(\frac{\sqrt{2}zh}{\sqrt{q}} \right) \right].
$$

Then to the same order of approximation

$$
y_{\mathrm{WKB}}^{(r,z_0)}(z) \simeq \left[\frac{z^2h^4}{4} - \frac{c^2z^4}{2} \right]^{-1/4} \exp\left(\frac{h^2z^2}{4} \right) e^{-q/4} \left(\frac{2^{1/2}hz}{q^{1/2}} \right)^{-q/2}
$$
$$
\simeq \left(\frac{2}{h} \right)^{1/2} \frac{\exp(\frac{1}{4}h^2z^2)}{(h^2z^2)^{(q+1)/4}} \left(\frac{2e}{q} \right)^{-q/4},
$$

$$
\overline{y}_{\mathrm{WKB}}^{(r,z_0)}(z) \simeq \left(\frac{2}{h} \right)^{1/2} \frac{(h^2z^2)^{(q-1)/4}}{\exp(\frac{1}{4}h^2z^2)} \left(\frac{2e}{q} \right)^{q/4}.
$$

[‡]Solutions of this type (which are asymptotic forms for $q \to \infty$) have been investigated in the literature; cf. N. Schwind [247].

It may be observed that $\overline{y}_{\mathrm{WKB}}^{(r,z_0)}(z)$ corresponds to solution (3.13) of Bender and Wu [19]. We can reexpress these relations with the help of *Stirling's formula*, i.e.

$$(j-1)! = (2\pi)^{1/2} e^{-j} j^{j-\frac{1}{2}} \left[1 - O\left(\frac{1}{j}\right)\right], \quad \text{so that} \quad \left[\frac{1}{4}q\right]! 2^{q/2} \simeq (2\pi)^{1/2} e^{-q/4} q^{q/4}.$$

Hence

$$y_{\mathrm{WKB}}^{(r,z_0)}(z) \simeq \left(\frac{1}{\pi h}\right)^{1/2} \frac{[\frac{1}{4}(q-1)]!}{(h^2 z^2)^{\frac{1}{4}(q+1)}} 2^{\frac{1}{4}(q-1)} e^{\frac{1}{4}h^2 z^2}$$

and

$$\overline{y}_{\mathrm{WKB}}^{(r,z_0)}(z) \simeq \left(\frac{1}{\pi h}\right)^{1/2} \frac{[-\frac{1}{4}(q+1)]!}{2^{\frac{1}{4}(q+1)}} (h^2 z^2)^{\frac{1}{4}(q-1)} e^{-\frac{1}{4}h^2 z^2}$$

for $q \neq$ odd integers. In this form the WKB solutions reveal their similarity with the solutions $y_B, \overline{y}_B, y_C, \overline{y}_C$ and demonstrate that the factors which we inserted (e.g. in Eq. (18.23)) appear quite naturally. The last expression can be made acceptable for $q =$ odd integers by applying Stirling's formula in the denominator of the second of the previous pair of expressions. Then

$$\overline{y}_{\mathrm{WKB}}^{(r,z_0)}(z) \simeq \left(\frac{4\pi}{h}\right)^{1/2} \frac{1}{[\frac{1}{4}(q-3)]! 2^{\frac{1}{4}(q+1)}} (h^2 z^2)^{\frac{1}{4}(q-1)} e^{-\frac{1}{4}h^2 z^2}$$

Comparing now the expressions (18.41) for $y_A(z), \overline{y}_A(z)$ with expressions $\overline{y}_{\mathrm{WKB}}^{(r,z_0)}, y_{\mathrm{WKB}}^{(r,z_0)}$, we see that the factors $\rho, \overline{\rho}$ requested at the beginning are given by

$$\rho = \left(\frac{1}{\pi h}\right)^{1/2} \frac{[\frac{1}{4}(q-1)]! 2^{\frac{1}{4}(q-1)}}{(h^2)^{\frac{1}{4}(q+1)}} e^{h^6/12c^2}, \quad \overline{\rho} = \left(\frac{1}{\pi h}\right)^{1/2} \frac{[-\frac{1}{4}(q+1)]!(h^2)^{\frac{1}{4}(q-1)}}{2^{\frac{1}{4}(q+1)}} e^{-h^6/12c^2}.$$

18.3 The Double Well Potential

18.3.1 Defining the problem

In dealing with the case of the symmetric double well potential, we shall employ basically the same procedure as above or, in fact, as in the case of the Mathieu equation. But there are significant differences.

We consider the following equation

$$\frac{d^2 y(z)}{dz^2} + [E - V(z)] y(z) = 0 \tag{18.87}$$

with double well potential

$$V(z) = v(z) = -\frac{1}{4} z^2 h^4 + \frac{1}{2} c^2 z^4 \quad \text{for} \quad c^2 > 0, \ h^4 > 0. \tag{18.88}$$

The minima of $V(z)$ on either side of the central maximum at $z = 0$ are located at

$$z_{\pm} = \pm \frac{h^2}{2c} \quad \text{with} \quad V(z_{\pm}) = -\frac{h^8}{2^5 c^2}, \tag{18.89}$$

and

$$V^{(2)}(z_\pm) = h^4, \quad V^{(3)}(z_\pm) = \pm 6ch^2,$$
$$V^{(4)}(z_\pm) = 12c^2, \quad V^{(i)}(z_\pm) = 0, \; i \geq 5. \tag{18.90}$$

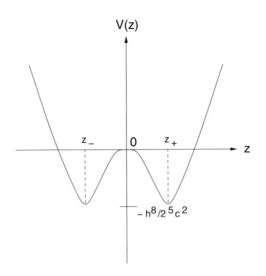

Fig. 18.4 The double well potential.

In order to obtain a rough approximation of the eigenvalues we expand the potential about the minima at $z_\pm$ and obtain

$$\frac{d^2 y}{dz^2} + \left[E - V(z_\pm) - \frac{1}{2}(z - z_\pm)^2 h^4 + O[(z - z_\pm)^3] \right] y = 0. \tag{18.91}$$

We set

$$\frac{1}{4}h_\pm^4 \equiv \frac{1}{2}h^4, \quad h_\pm^2 = \sqrt{2}h^2 \tag{18.92}$$

(thus we sometimes use h^4 and sometimes $h_\pm^4$) and

$$E - V(z_\pm) = \frac{1}{2}q_\pm h_\pm^2 + \frac{\triangle}{h_\pm^4}. \tag{18.93}$$

With the further substitution

$$\omega_\pm = h_\pm(z - z_\pm), \tag{18.94}$$

the previous equation (18.91) becomes

$$\mathcal{D}_{q_\pm}(\omega_\pm)y(\omega_\pm) = O\left(\frac{1}{h_\pm^3}\right)y, \tag{18.95a}$$

where

$$\mathcal{D}_{q_{\pm}}(\omega_{\pm}) = 2\frac{d^2}{d\omega^2} + q_{\pm} - \frac{1}{2}\omega^2. \tag{18.95b}$$

By comparison of Eqs. (18.95a) and (18.95b) with the equation of *parabolic cylinder functions* $u(z) \equiv D_\nu(z)$, i.e.

$$\frac{d^2 u(z)}{dz^2} + \left[\nu + \frac{1}{2} - \frac{1}{4}z^2\right]u(z) = 0,$$

we conclude that in the dominant approximation $q_{\pm}$ is an odd integer, $q_0 = 2n+1, n = 0, 1, 2, \ldots$. Inserting the expression (18.93) for E into Eq. (18.87), we obtain

$$\frac{d^2 y}{dz^2} + \left[\frac{1}{2}q_{\pm}h_{\pm}^2 + \frac{\triangle}{h_{\pm}^4} - \frac{1}{4}h_{\pm}^4 U(z)\right]y = 0, \tag{18.96}$$

where

$$U(z) = \frac{4}{h_{\pm}^4}[V(z) - V(z_{\pm})], \tag{18.97a}$$

and near a minimum at $z_{\pm}$

$$U(z) = (z - z_{\pm})^2 + O[(z - z_{\pm})^3]. \tag{18.97b}$$

Our basic equation, Eq. (18.96), is again seen to be invariant under a change of sign of z. Thus again, given one solution, we obtain another by replacing z by $-z$.

18.3.2 Three pairs of solutions

We define our first pair of solutions $y(z)$ as solutions with the proportionality

$$y(z) = \exp\left[\pm\frac{1}{2}h_{\pm}^2\int^z U^{1/2}(z)dz\right]. \tag{18.98}$$

Evaluating the exponential we can define these as the pair

$$\begin{aligned}
y_A(z) &= A(z)\exp\left[-\frac{1}{\sqrt{2}}\left\{\frac{c}{3}z^3 - \frac{h^4}{4c}z\right\}\right], \\
\overline{y}_A(z) &= \overline{A}(z)\exp\left[+\frac{1}{\sqrt{2}}\left\{\frac{c}{3}z^3 - \frac{h^4}{4c}z\right\}\right].
\end{aligned} \tag{18.99}$$

The equation for $A(z)$ is given by the following equation with upper signs and the equation for $\overline{A}(z)$ by the following equation with lower signs:

$$A''(z)\mp\sqrt{2}\left\{cz^2 - \frac{h^4}{4c}\right\}A'(z)\mp\sqrt{2}czA(z) + \left(\frac{1}{2}q_{\pm}h_{\pm}^2 + \frac{\triangle}{h_{\pm}^4}\right)A(z) = 0. \tag{18.100}$$

Since

$$z_+ = \frac{h^2}{2c}, \quad h_\pm^2 = \sqrt{2}h^2,$$

and selecting z_+ with $q_+ = q$, these equations can be rewritten as

$$(z_+^2 - z^2)A'(z) + (qz_+ - z)A(z) = -\frac{\sqrt{2}}{2c}\left[A''(z) + \frac{\triangle}{h_+^4}A(z)\right], \quad (18.101a)$$

$$(z_+^2 - z^2)\overline{A}'(z) - (qz_+ + z)\overline{A}(z) = \frac{\sqrt{2}}{2c}\left[\overline{A}''(z) + \frac{\triangle}{h_+^4}\overline{A}(z)\right]. \quad (18.101b)$$

To a first approximation for large h^2 we can neglect the right hand sides. The dominant approximation to A is then the function A_q given by the solution of the first order differential equation

$$\mathcal{D}_q A_q(z) = 0, \quad \mathcal{D}_q = (z_+^2 - z^2)\frac{d}{dz} + (qz_+ - z). \quad (18.102)$$

We observe that a change of sign of z in this equation is equivalent to a change of sign of q, but the solution is a different one, i.e. $\overline{A}_q(z) = A_q(-z) = A_{-q}(z)$. Integration of Eq. (18.102) yields the following expression

$$A_q(z) = \frac{1}{|z^2 - z_+^2|^{1/2}}\left|\frac{z - z_+}{z + z_+}\right|^{q/2} = \frac{|z - z_+|^{\frac{1}{2}(q-1)}}{|z + z_+|^{\frac{1}{2}(q+1)}}. \quad (18.103)$$

Looking at Eqs. (18.101a), (18.101b), we observe that the solution $\overline{y}_A(q, h^2; z)$ may be obtained from the solution $y_A(q, h^2; z)$ by either changing the sign of z throughout or — alternatively — the signs of both q and h^2 (and/or c), i.e.

$$\overline{y}_A(q, h^2; z) = y_A(q, h^2; -z) = y_A(-q, -h^2; z). \quad (18.104)$$

Both solutions $y_A(z), \overline{y}_A(z)$ are associated with one and the same expansion for $\triangle$ and hence E. We leave the calculation to Example 18.4. The result is given by

$$\triangle = -c^2(3q^2 + 1) - \frac{\sqrt{2}c^4}{4h^6}q(17q^2 + 19) + \cdots, \quad (18.105)$$

and

$$E(q, h^2) = -\frac{h^8}{2^5c^2} + \frac{1}{\sqrt{2}}qh^2 - \frac{c^2(3q^2 + 1)}{2h^4}$$
$$-\frac{\sqrt{2}c^4}{8h^{10}}q(17q^2 + 19) + O(h^{-16}). \quad (18.106)$$

Example 18.4: Calculation of eigenvalues along with solutions of type A

Show in conjunction with the derivation of solution y_A that the leading terms of $\triangle$ and hence E are given by Eqs. (18.105) and (18.106).

Solution: The structure of the solution (18.103) for $A_q(z)$ is very similar to that of the corresponding solution in considerations of other potentials, such as periodic potentials. Thus it is natural to explore analogous steps. The first such step would be to re-express the right hand side of Eq. (18.101a) with A replaced by A_q as a linear combination of terms $A_{q+2i}, i = 0, \pm 1, \pm 2, \dots$. First, however, we re-express A_q'' in terms of functions of z multiplied by A_q. We know the first derivative of A_q from Eq. (18.102), i.e.

$$A_q'(z) = \frac{qz_+ - z}{z^2 - z_+^2} A_q(z). \tag{18.107}$$

Differentiation yields

$$\begin{aligned}
A_q''(z) &= \frac{A_q(z)}{(z^2 - z_+^2)^2}\left[(z - qz_+)^2 - (z^2 - z_+^2) + 2z(z - qz_+)\right] \\
&= \frac{A_q(z)}{(z^2 - z_+^2)^2}[2z^2 - 4zz_+q + z_+^2(q^2 + 1)]. \tag{18.108}
\end{aligned}$$

We wish to rewrite this expression as a sum

$$\sum_i \text{coefficient}_i A_{q+2i}(z).$$

We also note at this stage the derivative of the entire solution $y_A(z)$ taking into account only the dominant contribution:

$$y_A'(z) \simeq \left[\frac{c}{\sqrt{2}}(z^2 - z_+^2) + \frac{(qz_+ - z)}{(z^2 - z_+^2)}\right]A_q(z)\exp\left[\frac{1}{\sqrt{2}}\left\{\frac{c}{3}z^3 - \frac{h^4}{4c}z\right\}\right]. \tag{18.109}$$

We observe some properties of the function $A_q(z)$ given by Eq. (18.103):

$$A_{q+2i}A_{q+2j} = A_{q+2i+2j}A_q, \qquad \frac{A_{q+2}}{A_q} = \frac{z - z_+}{z + z_+}, \tag{18.110}$$

$$\frac{A_{q+2} + A_{q-2}}{A_q} = 2\frac{z^2 + z_+^2}{z^2 - z_+^2}, \qquad \frac{A_{q+2} - A_{q-2}}{A_q} = -4\frac{zz_+}{z^2 - z_+^2}, \tag{18.111}$$

$$\frac{(A_{q+2} - A_{q-2})^2}{A_q} = (4zz_+)^2\frac{A_q}{(z^2 - z_+^2)^2} = A_{q+4} - 2A_q + A_{q-4}, \tag{18.112}$$

From these we obtain, for instance by *componendo et dividendo*,

$$\begin{aligned}
-\frac{z}{z_+} &= \frac{A_{q+2} + A_q}{A_{q+2} - A_q} = -\left(1 + \frac{A_{q+2}}{A_q}\right)\left[1 + \left(\frac{A_{q+2}}{A_q}\right) + \left(\frac{A_{q+2}}{A_q}\right)^2 + \cdots\right] \\
&= -\left[1 + 2\frac{A_{q+2}}{A_q} + 2\frac{A_{q+4}}{A_q} + 2\frac{A_{q+6}}{A_q} + \cdots\right] \tag{18.113}
\end{aligned}$$

Similarly we obtain

$$-\frac{z_+}{z} = \frac{A_{q+2} - A_q}{A_{q+2} + A_q} = -\left[1 - 2\frac{A_{q+2}}{A_q} + 2\frac{A_{q+4}}{A_q} - 2\frac{A_{q+6}}{A_q} + \cdots\right] \tag{18.114a}$$

and from this

$$\frac{z_+^2}{z^2} = 1 - 4\frac{A_{q+2}}{A_q} + 8\frac{A_{q+4}}{A_q} - 12\frac{A_{q+6}}{A_q} + 16\frac{A_{q+8}}{A_q} - \cdots. \tag{18.114b}$$

Hence with Eq. (18.112)

$$
\begin{aligned}
\frac{z}{(z^2 - z_+^2)^2}A_q &= \frac{z_+}{z_+z}\left(\frac{1}{4z_+}\right)^2[A_{q+4} - 2A_q + A_{q-4}]\\
&= \frac{1}{z_+}\left(\frac{1}{4z_+}\right)^2[A_{q+4} - 2A_q + A_{q-4}]\left[1 - 2\frac{A_{q+2}}{A_q} + 2\frac{A_{q+4}}{A_q} + \cdots\right]\\
&= \frac{1}{z_+}\left(\frac{1}{4z_+}\right)^2[A_{q-4} - 2A_{q-2} + 2A_{q+2} - A_{q+4} + \cdots].
\end{aligned} \tag{18.115}
$$

Finally we have also

$$
\begin{aligned}
\frac{A_q}{(z^2 - z_+^2)^2} &= \frac{z^2 A_q}{(z^2 - z_+^2)^2}\frac{1}{z^2}\\
&= \left(\frac{1}{4z_+}\right)^2[A_{q+4} - 2A_q + A_{q-4}]\frac{1}{z_+^2}\left[1 - 4\frac{A_{q+2}}{A_q} + 8\frac{A_{q+4}}{A_q} - 12\frac{A_{q+6}}{A_q} + \cdots\right]\\
&= \left(\frac{1}{2z_+}\right)^4[A_{q-4} - 4A_{q-2} + 6A_q - 4A_{q+2} + A_{q+4} + \cdots].
\end{aligned} \tag{18.116}
$$

Inserting (18.112), (18.115) and (18.116) into Eq. (18.108), we obtain[§]

$$
\begin{aligned}
A_q'' &= \left(\frac{1}{4z_+}\right)^2[(q-1)(q-3)A_{q-4} - 4(q-1)^2 A_{q-2} + 6(q^2+1)A_q\\
&\quad -4(q+1)^2 A_{q+2} + (q+1)(q+3)A_{q+4}\cdots].
\end{aligned} \tag{18.117}
$$

Here the first approximation of A, $A^{(0)} = A_q$, leaves uncompensated on the right hand side of Eq. (18.101a) the contribution

$$
\begin{aligned}
R_q^{(0)} &= -\frac{\sqrt{2}}{2c}\left[A_q'' + \frac{\triangle}{h_+^4}A_q\right]\\
&= -\frac{2\sqrt{2}c}{2^4 h^4}[(q, q-4)A_{q-4} + (q, q-2)A_{q-2} + (q, q)A_q + (q, q+2)A_{q+2}\\
&\quad +(q, q+4)A_{q+4} + (q, q+6)A_{q+6} + \cdots],
\end{aligned} \tag{18.118}
$$

where the lowest coefficients have been determined above as

$$(q, q\mp 4) = (q\pm 1)(q\mp 3), \quad (q, q\mp 2) = -4(q\mp 1)^2, \quad (q, q) = 6(q^2+1) + z_+^2\frac{2\triangle}{c^2}. \tag{18.119}$$

It is now clear how the calculation of higher order contributions proceeds in our standard way. In particular the dominant approximation of $\triangle$ is obtained by setting $(q, q) = 0$, i.e.

$$\triangle = -c^2(3q^2 + 1), \quad E(q, h^2) = -\frac{h^8}{2^5 c^2} + \frac{1}{\sqrt{2}}qh^2 - \frac{c^2(3q^2+1)}{2h^4} + O\left(\frac{1}{h^6}\right).$$

Since

$$\mathcal{D}_q A_q = 0, \quad \mathcal{D}_{q+2i}A_{q+2i} = 0 \quad \text{and} \quad \mathcal{D}_{q+2i} = \mathcal{D}_q + 2iz_+,$$

[§]The reader may observe the similarity with the corresponding coefficients in the simpler case of the cosine potential, cf. Eqs. (17.33) and (17.34).

we have

$$
\mathcal{D}_q \left(\frac{A_{q+2i}}{-2iz_+} \right) = A_{q+2i},
$$

except, of course, for $i = 0$. The first approximation $A^{(0)} = A_q$ leaves uncompensated on the right hand side of Eq. (18.101a) the contribution $R_q^{(0)}$. Terms μA_{q+2i} in this may therefore be eliminated by adding to $A^{(0)}$ the contribution $A^{(1)}$ given by

$$
A^{(1)} = -\frac{\sqrt{2}c}{2^3 h^4} \left[\frac{(q,q-4)}{8z_+} A_{q-4} + \frac{(q,q-2)}{4z_+} A_{q-2} + \frac{(q,q+2)}{-4z_+} A_{q+2} + \frac{(q,q+4)}{-8z_+} A_{q+4} + \cdots \right].
$$
(18.120)

The sum $A = A^{(0)} + A^{(1)}$ then represents a solution to that order provided the sum of terms in A_q in $R_q^{(0)}$ and $R_q^{(1)}$ is set equal to zero, where $R_q^{(1)}$ is the sum of terms left uncompensated by $A^{(1)}$, i.e.

$$
R_q^{(1)} = \left(-\frac{\sqrt{2}c}{2^3 h^4} \right) \left[\frac{(q,q-4)}{8z_+} R_{q-4}^{(0)} + \frac{(q,q-2)}{4z_+} R_{q-2}^{(0)} + \cdots \right].
$$
(18.121)

This coefficient of A_q set equal to zero yields to that order the following equation

$$
0 = \left(-\frac{\sqrt{2}c}{2^3 h^4} \right)(q,q) + \left(-\frac{\sqrt{2}c}{2^3 h^4} \right)^2 \left[\frac{(q,q-4)(q-4,q)}{8z_+} + \frac{(q,q-2)(q-2,q)}{4z_+} \right.
$$
$$
\left. + \frac{(q,q+2)(q+2,q)}{-4z_+} + \frac{(q,q+4)(q+4,q)}{-8z_+} \right] + \cdots,
$$

which reduces to

$$
0 = 2(3q^2 + 1) + \frac{2}{c^2}\triangle + \frac{\sqrt{2}c^2}{2h^6} q(17q^2 + 19),
$$
(18.122)

thus yielding the next approximation of $\triangle$ as given in Eq. (18.105).

The solutions $y_A(z), \overline{y}_A(z)$ derived above are valid around $z = 0$, i.e. in the domains away from the minima,

$$
|z - z_\pm| > O\left(\frac{1}{h_+^2} \right).
$$

We can define solutions which are even or odd about $z = 0$ as

$$
\begin{aligned}
y_\pm(z) &= \frac{1}{2}[y_A(q,h^2;z) \pm \overline{y}_A(q,h^2;z)] = \frac{1}{2}[y_A(q,h^2;z) \pm y_A(q,h^2;-z)] \\
&= \frac{1}{2}[y_A(q,h^2;z) \pm y_A(-q,-h^2;z)].
\end{aligned}
$$
(18.123)

Considering only the leading approximations considered explicitly above we have (since $A_q(0) = 1/z_+ = \overline{A}_q(0)$, $A_q'(0) = -q/z_+^2$)

$$
y_+(q,h^2;0) = \frac{2c}{h^2}, \quad y_-(q,h^2;0) = 0,
$$

$$
y_+'(q,h^2;0) = 0, \quad y_-'(q,h^2;0) = \frac{h^2}{2\sqrt{2}} - \frac{4qc^2}{h^4}.
$$
(18.124)

Our second pair of solutions, $y_B(z), \overline{y}_B(z)$ is obtained around a minimum of the potential. We see already from Eqs. (18.95a) and (18.95b) that the solution there is of parabolic cylinder type. This means, in this case we use the Schrödinger equation with the potential $V(z)$ expanded about $z_\pm$ as in Eq. (18.91). Inserting (18.93) and setting $w_\pm = h_\pm(z - z_\pm)$, the equation is — with differential operator $\mathcal{D}_q$ as defined by Eq. (18.95b) —

$$\mathcal{D}_q(w_\pm)y(w_\pm) = \frac{1}{h_\pm^6}\left[\pm 2^{5/4}ch^3w_\pm^3 + c^2w_\pm^4 - 2\Delta\right]y(w_\pm).$$

Thus we write the first solution

$$y_B(z) = B_q[w_\pm(z)] + O(h_\pm^{-2}), \quad B_q[w_\pm(z)] = \frac{D_{\frac{1}{2}(q-1)}(w_\pm(z))}{[\frac{1}{4}(q-1)]!2^{\frac{1}{4}(q-1)}}, \quad (18.125a)$$

and another

$$\overline{y}_B(z) = y_B(-z) = \overline{B}_q[w_\pm(z)] + O(h_\pm^{-2}) = B_q[w_\pm(-z)] + O(h_\pm^{-2}) \quad (18.125b)$$

It is clear that correspondingly we have solutions $y_C(z), \overline{y}_C(z)$ with complex variables and $C_q(w)$ given by Eq. (18.23) with appropriate change of parameters to those of the present case. Thus

$$y_C(z) = C_q[w_\pm(-z)] + O(h_\pm^{-2}),$$

$$C_q[w_\pm(-z)] = \frac{D_{-\frac{1}{2}(q+1)}(iw_\pm(-z))2^{\frac{1}{4}(q+1)}}{[-\frac{1}{4}(q+1)]!}, \quad (18.126a)$$

$$\overline{y}_C(z) = y_C(-z) = \overline{C}_q[w_\pm(-z)] + O(h_\pm^{-2}). \quad (18.126b)$$

These are solutions again around a minimum and with $y_B(z), \overline{y}_C(z)$ providing a pair of decreasing asymptotic solutions there (or correspondingly $\overline{y}_B(z), y_C(z)$). We draw attention to two additional points. Since $w_\pm(-z) = -h_\pm(z + z_\pm)$, we have $w_\pm(-z_\pm) = -2h_\pm z_\pm$, but $w_\pm(z_\pm) = 0$. Moreover, in view of the factor "i" in the argument of C_q the solutions $\overline{y}_A, \overline{y}_C$ have the same exponential behaviour near a minimum.

18.3.3 Matching of solutions

Next we consider the proportionality of solutions $y_A(z)$ and $y_B(z)$. Evaluating the exponential factor contained in $y_A(z)$ of Eq. (18.99) for $z \to z_\pm$, we have (cf. (18.97b))

$$
\exp\left[-\frac{1}{2}h_\pm^2 \int_0^z U^{1/2}(z)dz\right] \overset{(18.97b)}{\simeq} \exp\left[-\frac{1}{2}h_\pm^2 \int_0^z (\pm)(z-z_\pm)dz\right]
$$

$$
= \exp\left[-\frac{1}{2}h_\pm^2(\pm)\left(\frac{1}{2}z^2 - zz_\pm\right)\right]
$$

$$
= \exp\left[\mp\frac{1}{4}h_\pm^2(z-z_\pm)^2\right]\exp\left[\pm\frac{1}{4}h_\pm^2 z_\pm^2\right].
$$

For later reference (after Eq. (18.172)) we add here comments on this approximation. We observe that for $z = z_\pm$ the approximation yields $\exp[\pm h_\pm^2 z_\pm^2/4]$. Here $z_\pm = \pm h^2/2c$ are the positions of the minima of the potential. Later we calculate the coordinates of turning points z_0, z_1 and find that these are given by $z_+ + O(1/h)$ for finite q (cf. Eqs. (18.146), (18.147)). Thus for h very large these turning points are very close to the minimum at z_+. Consequently the above integral from $z = 0$ to $z_\pm$ differs from that from $z = 0$ to a turning point (as in expression $I_2(0)$ of Eq. (18.164)) only in a nonleading contribution and hence implies the equivalence of the exponentials in the relation (18.172) obtained later.

Allowing z to approach z_+ in the solution $y_A(z)$, we have

$$
y_A(z) \simeq \frac{|z-z_+|^{\frac{1}{2}(q-1)}}{|2z_+|^{\frac{1}{2}(q+1)}} e^{\frac{1}{4}h_+^2 z_+^2} e^{-\frac{1}{4}h_+^2(z-z_+)^2} = \frac{\left(\frac{w_+}{h_+}\right)^{\frac{1}{2}(q-1)}}{(2z_+)^{\frac{1}{2}(q+1)}} e^{\frac{1}{4}h_+^2 z_+^2} e^{-\frac{1}{4}w_+^2}.
$$
(18.127)

Recalling that around $\arg w_\pm \sim 0$, the dominant term in the power expansion of the parabolic cylinder function is given by

$$
D_\nu(w_\pm) \simeq w_\pm^\nu e^{-w_\pm^2/4},
$$

and comparing with Eq. (18.125a) we see that (considering only dominant contributions) in their common domain of validity

$$
y_A(z) = \frac{1}{\alpha} y_B(z), \quad \alpha = \frac{(h_+)^{\frac{1}{2}(q-1)}(2z_+)^{\frac{1}{2}(q+1)}}{2^{\frac{1}{4}(q-1)}[\frac{1}{4}(q-1)]!} e^{-\frac{1}{4}h_+^2 z_+^2}\left[1 + O\left(\frac{1}{h_+^2}\right)\right].
$$
(18.128)

Similarly we obtain in approaching z_+:

$$
\bar{y}_A(z) \simeq \frac{(2z_+)^{\frac{1}{2}(q-1)}}{(z-z_+)^{\frac{1}{2}(q+1)}} e^{-\frac{1}{4}h_+^2 z_+^2} e^{\frac{1}{4}h_+^2(z-z_+)^2}
$$
(18.129)

and

$$\overline{y}_C(z) \simeq \frac{D_{-\frac{1}{2}(q+1)}(iw_+(z))2^{\frac{1}{4}(q+1)}}{[-\frac{1}{4}(q+1)]!} \simeq \frac{2^{\frac{1}{4}(q+1)}e^{\frac{1}{4}h_+^2(z-z_+)^2}}{[-\frac{1}{4}(q+1)]![ih_+(z-z_+)]^{\frac{1}{2}(q+1)}}.$$
(18.130)

Therefore in their common domain of validity

$$\overline{y}_A(z) = \frac{1}{\alpha}\overline{y}_C(z),$$

$$\overline{\alpha} = \frac{2^{\frac{1}{4}(q+1)}e^{\frac{1}{4}h_+^2 z_+^2}}{(2z_+)^{\frac{1}{2}(q-1)}(-h_+^2)^{\frac{1}{4}(q+1)}[-\frac{1}{4}(q+1)]!}\left[1 + O\left(\frac{1}{h_+^2}\right)\right].$$
(18.131)

We have thus found three pairs of solutions: The two solutions of type A are valid in regions away from the minima, and are both in their parameter dependence asymptotically decreasing there and permit us therefore to define the extensions of the solutions $y_\pm$ which are respectively even and odd about $z = 0$ to the minima. The pair of solutions of type B is defined around $\arg z = 0, \pi$ and the solutions of type C around $\arg z = \pm\pi/2$. The next aspect to be considered is that of boundary conditions. We have to impose boundary conditions at the minima and at the origin. The solutions in terms of parabolic cylinder functions have a wide range of validity, even above the turning points, but it is clear that none of the above solutions can be used at the top of the central barrier. Thus it is unavoidable to appeal to other methods such as the WKB method to apply the necessary boundary conditions at that point. The involvement of these WKB solutions leads to problems, since, basically, they assume large quantum numbers. Various investigations[¶] therefore struggle to overcome this to a good approximation. We achieve the same goal here by demanding our basic perturbation solutions to be interconvertible on the basis of the parameter symmetries of the original equation.

18.3.4   Boundary conditions at the minima
(A) Formulation of the boundary conditions

The present case of the double well potential differs from that of the simple harmonic oscillator potential in having two minima instead of one. Since it is more probable to find a particle in the region of a minimum than elsewhere, we naturally expect the wave function there to be similar to that of the harmonic oscillator, and this means at both minima. Thus

---

[¶]E.g. S. K. Bhattacharya and A. R. P. Rau [22], S. K. Bhattacharya [23], J. D. de Deus [66] and W. H. Furry [102]

the most basic solution would be even with maxima at $z_\pm$, as indicated in Fig. 18.5. However, an even wave function can also pass through zero at these points, as indicated in Fig. 18.5. The odd wave function then exhibits a correspondingly opposite behaviour, as indicated there. At $\Re z \to \pm\infty$ we require the wave functions to vanish so that they are square integrable. We have therefore the following two sets of boundary conditions at the local minima of the double-well potential:

$$y'_+(z_\pm) = 0, \quad y_+(z_\pm) \neq 0, \quad y_-(z_\pm) = 0, \quad y'_-(z_\pm) \neq 0 \qquad (18.132)$$

and

$$y_+(z_\pm) = 0, \quad y'_+(z_\pm) \neq 0, \quad y'_-(z_\pm) = 0, \quad y_-(z_\pm) \neq 0. \qquad (18.133)$$

We have

$$y_\pm(z_\pm) = \frac{1}{2}[y_A(z) \pm \overline{y}_A(z)]_{z \to z_\pm} = \frac{1}{2}\left[\frac{1}{\alpha}y_B(z_\pm) \pm \frac{1}{\alpha}\overline{y}_C(z_\pm)\right] \qquad (18.134)$$

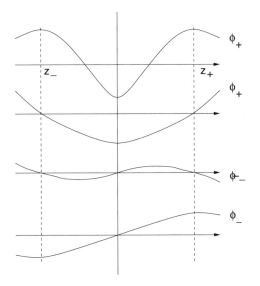

Fig. 18.5 Behaviour of fundamental wave functions.

and

$$y'_\pm(z_\pm) = \frac{1}{2}\left[\frac{1}{\alpha}y'_B(z_\pm) \pm \frac{1}{\alpha}\overline{y}'_C(z_\pm)\right]. \qquad (18.135)$$

Hence the conditions (18.132), (18.133) imply

$$\frac{y_B(z_\pm)}{\overline{y}_C(z_\pm)} = \mp\frac{\alpha}{\alpha}, \quad \text{and} \quad \frac{y'_B(z_\pm)}{\overline{y}'_C(z_\pm)} = \mp\frac{\alpha}{\alpha}. \qquad (18.136)$$

(B) Evaluation of the boundary conditions

Inserting into the first of Eqs. (18.136) the dominant approximations we obtain (cf. also (18.54)

$$
\begin{aligned}
1 &= \mp\frac{\alpha}{\bar{\alpha}} = \mp(-1)^{\frac{1}{4}(q+1)}\frac{(h_+^2)^{q/2}(2z_+)^q[-\frac{1}{4}(q+1)]!}{2^{q/2}[\frac{1}{4}(q-1)]!}e^{-\frac{1}{2}h_+^2 z_+^2} \\
&= \mp(-1)^{\frac{1}{4}(q+1)}\frac{\pi(h_+^2)^{q/2}(2z_+)^q}{2^{q/2}[\frac{1}{4}(q-1)]![\frac{1}{4}(q-3)]!\sin\{\frac{\pi}{4}(q+1)\}}e^{-\frac{1}{2}h_+^2 z_+^2} \\
&= \mp(-1)^{\frac{1}{4}(q+1)}\sqrt{\frac{\pi}{2}}\frac{(h_+^2)^{q/2}(2z_+)^q}{[\frac{1}{2}(q-1)]!\sin\{\frac{\pi}{4}(q+1)\}}e^{-\frac{1}{2}h_+^2 z_+^2}, \quad (18.137)
\end{aligned}
$$

where we used first the reflection formula and then the duplication formula. Thus

$$
\sin\left\{\frac{\pi}{4}(q+1)\right\} = \mp(-1)^{\frac{1}{4}(q+1)}\sqrt{\frac{\pi}{2}}\frac{(h_+^2)^{q/2}(2z_+)^q}{[\frac{1}{2}(q-1)]!}e^{-\frac{1}{2}h_+^2 z_+^2}, \quad (18.138)
$$

Using formulae derived in Example 18.2 we can rewrite the second of Eqs. (18.136) as

$$
\frac{y_B'(z_\pm)}{y_C'(z_\pm)} = -i\cot\left\{\frac{\pi}{4}(q-3)\right\} \equiv -i\cot\left\{\frac{\pi}{4}(q+1)\right\} = \mp\frac{\alpha}{\bar{\alpha}}. \quad (18.139)
$$

Using Eq. (18.134) this equation can be written

$$
-i\cos\left\{\frac{\pi}{4}(q+1)\right\} = \mp(-1)^{\frac{1}{4}(q+1)}\sqrt{\frac{\pi}{2}}\frac{(h_+^2)^{q/2}(2z_+)^q}{[\frac{1}{2}(q-1)]!}e^{-\frac{1}{2}h_+^2 z_+^2}. \quad (18.140)
$$

Now

$$
\begin{aligned}
\sin\left\{\frac{\pi}{4}(q+1)\right\} &\simeq \sin\left\{\frac{\pi}{4}(q_0+1)\right\} + \frac{\pi}{4}(q-q_0)\cos\left\{\frac{\pi}{4}(q_0+1)\right\} + \cdots \\
&\simeq (-1)^{\frac{1}{4}(q_0+1)}(q-q_0)\frac{\pi}{4} \quad \text{for} \quad q_0 = 3,7,11,\ldots
\end{aligned}
$$

$$(18.141)$$

and

$$
\begin{aligned}
\cos\left\{\frac{\pi}{4}(q+1)\right\} &\simeq \cos\left\{\frac{\pi}{4}(q_0+1)\right\} - \frac{\pi}{4}(q-q_0)\sin\left\{\frac{\pi}{4}(q_0+1)\right\} + \cdots \\
&\simeq -(-1)^{\frac{1}{4}(q_0-1)}(q-q_0)\frac{\pi}{4} \quad \text{for} \quad q_0 = 1,5,9,\ldots \\
&= -(q-q_0)\frac{\pi}{4}(-1)^{\frac{1}{4}(q_0+1)}(-1)^{1/2}.
\end{aligned}
$$

$$(18.142)$$

Thus altogether we obtain

$$(q - q_0) \simeq \mp 4 \sqrt{\frac{1}{2\pi}} \frac{(h_+^2)^{q_0/2}(2z_+)^{q_0}}{[\frac{1}{2}(q_0 - 1)]!} e^{-\frac{1}{2}h_+^2 z_+^2}, \quad q_0 = 1, 3, 5, \ldots. \quad (18.143)$$

In Example 18.5 (after determination of the turning points) we rederive this relation using the WKB solutions from above the turning points matched (linearly) to their counterparts below the turning points and then evaluated at the minimum.

We emphasize: We needed the solutions of type A for the definition of even and odd solutions. Since the type A solutions are not valid at the minima, we matched them to the solutions of types B and C which are valid there and hence permit the imposition of boundary conditions at the minima.

18.3.5 Boundary conditions at the origin
(A) Formulation of the boundary conditions

Since our even and odd solutions are defined to be even or odd with respect to the origin, we must also demand this behaviour here along with a nonvanishing Wronskian.

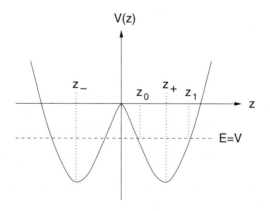

Fig. 18.6 Turning points z_0, z_1.

Hence we have to impose at $z = 0$ the conditions

$$y_-(0) = 0, \quad y'_-(0) \neq 0, \quad y_+(0) \neq 0, \quad y'_+(0) = 0. \quad (18.144)$$

Thus we require the extension of our solutions to the region around the local maximum at $z = 0$. We do this with the help of WKB solutions. We deduce from Eq. (18.96) that the two turning points at z_0 and z_1 to the right of the origin are given by

$$\frac{1}{2}qh_+^2 + \frac{\triangle}{h_+} - \frac{1}{4}h_+^4 U(z) = 0, \qquad (18.145)$$

i.e.

$$\frac{1}{2}qh_+^2 + \frac{1}{4}z^2 h^4 - \frac{1}{2}c^2 z^4 - \frac{h^8}{2^5 c^2} = O\left(\frac{\triangle}{h_+}\right).$$

Using $z_+ = h^2/2c$, one finds that

$$z_0 = \left[\frac{h^4}{4c^2} - \sqrt{\frac{qh_+^2}{c^2} + O\left(\frac{\triangle}{h_+}\right)}\right]^{1/2} = \frac{h^2}{2c} - \left(\frac{2q}{h_+^2}\right)^{1/2} + \frac{2^{1/2}cq}{h^4} + O(h^{-5})$$

$$(18.146)$$

and

$$z_1 = \left[\frac{h^4}{4c^2} + \sqrt{\frac{qh_+^2}{c^2} + O\left(\frac{\triangle}{h_+}\right)}\right]^{1/2} \simeq \frac{h^2}{2c} + \frac{(2q^2)^{1/4}}{h}. \qquad (18.147)$$

Since the minima are at $z_\pm = h^2/2c$ with h^2 very large, we see that the turning points z_0 and z_1 are very close to a minimum for reasonable values of $q \simeq q_0$, the latter being an odd integer. As a consequence, in leading order the integrals to be studied below from z_0 to $z = 0$ are equal to those from z_+ to $z = 0$. We also note that the height of the potential at the turning points is

$$V(z)|_{z_0,z_1} \simeq -\frac{h^8}{2^5 c^2} + \frac{qh}{2^{1/2}}.$$

Thus again we see that for large values of h^2 the turning points are very close to the minima of the potential for nonasymptotically large values of q.

(B) Evaluation of the boundary conditions

We now proceed to evaluate the boundary conditions (18.144). In the domain $0 < z < z_0$, i.e. to the left of z_0 where $V > E$, the dominant terms of the WKB solutions are*

$$y_{\text{WKB}}^{(l,z_0)}(z) = \left[-\frac{1}{2}qh_+^2 + \frac{1}{4}h_+^4 U(z)\right]^{-1/4}$$

$$\times \exp\left(-\int_z^{z_0} dz\left[-\frac{1}{2}qh_+^2 + \frac{1}{4}h_+^4 U(z)\right]^{1/2}\right) \qquad (18.148)$$

---

*The superscript (l, z_0) means "to the left of z_0".

and

$$\overline{y}_{\text{WKB}}^{(l,z_0)}(z) = \left[-\frac{1}{2}qh_+^2 + \frac{1}{4}h_+^4 U(z) \right]^{-1/4}$$

$$\times \exp\left(\int_z^{z_0} dz \left[-\frac{1}{2}qh_+^2 + \frac{1}{4}h_+^4 U(z) \right]^{1/2} \right). \quad (18.149)$$

In order to be able to extend the even and odd solutions to $z = 0$, we have to match $y_A(z), \overline{y}_A(z)$ to $y_{\text{WKB}}^{(l,z_0)}(z)$ and $\overline{y}_{\text{WKB}}^{(l,z_0)}(z)$. We therefore have to consider the exponential factors occurring in (18.148) and (18.149) and consider both types of solutions in a domain approaching but not reaching the minimum of the potential at z_+. Thus we consider in the domain $|z - z_+| > (2q/h_+^2)^{1/2}$:

$$I_1 = \int_z^{z_0} dz \left[-\frac{q}{2}h_+^2 + \frac{h_+^4}{4}U(z) \right]^{1/2} \simeq \pm\frac{h_+^2}{2} \int_z^{z_0} dz \left[(z - z_+)^2 - \frac{2q}{h_+^2} \right]^{1/2}$$

$$= \pm\frac{1}{2}h_+^2 \int_{z-z_+}^{z_0-z_+} dz' \left(z'^2 - \frac{2q}{h_+^2} \right)^{1/2} = \pm\frac{1}{2}h_+^2 \left[\frac{1}{2}z' \left(z'^2 - \frac{2q}{h_+^2} \right)^{1/2} \right.$$

$$\left. - \frac{q}{h_+^2} \ln\left| z' + \left(z'^2 - \frac{2q}{h_+^2} \right)^{1/2} \right| \right]_{z-z_+}^{z_0-z_+}. \quad (18.150)$$

Evaluating this we have

$$\pm I_1$$

$$\simeq \frac{1}{2}h_+^2 \left[\frac{1}{2}\left\{ z'^2\left(1 - \frac{q}{h_+^2 z'^2} \right) \right\}_{z'=-(2q/h_+^2)^{1/2}} \right.$$

$$- \frac{q}{h_+^2} \ln\left| -\left(\frac{2q}{h_+^2} \right)^{1/2} + O\left(\frac{1}{h^5} \right) \right| - \frac{1}{2}(z - z_+)\left\{ (z - z_+)^2 - \frac{2q}{h_+^2} \right\}^{1/2}$$

$$\left. + \frac{q}{h_+^2} \ln\left| (z - z_+) + \left\{ (z - z_+)^2 - \frac{2q}{h_+^2} \right\}^{1/2} \right| \right]$$

$$\simeq \frac{1}{4}q + \frac{1}{4}q\ln\left| \frac{h_+^2}{2q} \right| - \frac{1}{4}h_+^2(z - z_+)^2 + \frac{1}{2}q\ln|2(z - z_+)|. \quad (18.151)$$

In identifying the WKB exponentials we recall that $\overline{y}_{\text{WKB}}$ is exponentially increasing and y_{WKB} is exponentially decreasing. Hence we have for y_{WKB} the exponential factor

$$\exp\left(-\int_z^{z_0} dz \left[-\frac{1}{2}qh_+^2 + \frac{1}{4}h_+^4 U(z) \right]^{1/2} \right)$$

$$\simeq |z - z_+|^{q/2} e^{\frac{1}{4}q} \left(\frac{1}{2}h_+^2 \right)^{q/4} \left(\frac{q}{4} \right)^{-q/4} e^{-\frac{1}{4}(z-z_+)^2 h_+^2}. \quad (18.152a)$$

We observe here that the WKB solution involves unavoidably the quantum number dependent factors $\exp(q/4)$ and $(q/4)^{q/4}$ which do not appear as such in the perturbation solutions. The only way to relate these solutions is with the help of the Stirling formula which converts the product or ratio of such factors into factorials such as those inserted from the beginning into the unperturbed solutions (18.125a) and (18.126a). However, Stirling's formula is the dominant term of the asymptotic expansion of a factorial or gamma function and thus assumes the argument ($\propto q \sim 2n + 1$) to be large (it is known, of course, that the Stirling approximation is amazingly good even for small values of the argument). Thus using the *Stirling formula* $z! \simeq e^{-(z+1)}(z+1)^{z+\frac{1}{2}}\sqrt{2\pi}$, we can write the exponential as

$$
\exp\left(-\int_z^{z_0} dz\left[-\frac{1}{2}qh_+^2 + \frac{1}{4}h_+^4 U(z)\right]^{1/2}\right)
$$
$$
\simeq (2\pi)^{1/2}\frac{|z - z_+|^{q/2}}{[\frac{1}{4}(q-1)!]}\left(\frac{1}{2}h_+^2\right)^{q/4} e^{-\frac{1}{4}(z-z_+)^2 h_+^2}. \qquad (18.152b)
$$

Here $q/4$ was assumed to be large but we write $[\frac{1}{4}(q-1)]!$ since this is the factor appearing in the solution (18.125a). We see therefore, since there is no way to obtain an exact leading order approximation with Stirling's formula for small values of q, the results necessarily require adjustment or normalization there in the q-dependence. This is the aspect investigated by Furry [102]. Since correspondingly

$$
\left[-\frac{1}{2}qh_+^2 + \frac{1}{4}h_+^4 U(z)\right]^{-1/4} \simeq \frac{2^{1/2}}{(z - z_+)^{1/2}(h_+^4)^{1/4}},
$$

we obtain

$$
y_{WKB}^{(l,z_0)}(z) \simeq 2\pi^{1/2}\frac{|z - z_+|^{\frac{1}{2}(q-1)}}{(h_+^4)^{1/4}[\frac{1}{4}(q-1)]!}\left(\frac{1}{2}h_+^2\right)^{q/4} e^{-\frac{1}{4}(z-z_+)^2 h_+^2}
$$
$$
\text{for} \quad |z - z_+| > \left(\frac{2q}{h_+^2}\right)^{1/2}. \qquad (18.153)
$$

This expression is valid to the left of the turning point at z_0 above the minimum at z_+.

In a corresponding manner — i.e. using Stirling's formula (and not the inversion relation) — we have

$$
\overline{y}_{WKB}^{(l,z_0)}(z) = \frac{1}{\sqrt{\pi}(h_+^4)^{1/4}}\frac{[\frac{1}{4}(q-3)]!}{|z - z_+|^{\frac{1}{2}(q+1)}}\left(\frac{1}{2}h_+^2\right)^{-q/4} e^{\frac{1}{4}(z-z_+)^2 h_+^2}, \qquad (18.154)
$$

where $[\frac{1}{4}q]!$ was written as $[\frac{1}{4}(q-3)]!$ for q large. Comparing for $z \to z_+$ the WKB solutions (18.153), (18.154) with the type-A solutions (18.99), (18.103) and (18.104), we obtain in leading order (i.e. multiplied by $(1 + O(1/h_+^2))$) the proportionality constants $\gamma, \overline{\gamma}$ of the matching relations

$$y_{\mathrm{WKB}}^{(l,z_0)}(z) = \gamma y_A(z), \quad \overline{y}_{\mathrm{WKB}}^{(l,z_0)}(z) = \overline{\gamma}\,\overline{y}_A(z), \tag{18.155}$$

i.e.

$$\gamma = \frac{2\pi^{1/2}}{(h_+^4)^{1/4}[\frac{1}{4}(q-1)]!}\left(\frac{1}{2}h_+^2\right)^{q/4}(2z_+)^{\frac{1}{2}(q+1)}e^{-\frac{1}{4}h_+^2 z_+^2} \tag{18.156a}$$

and

$$\overline{\gamma} = \frac{[\frac{1}{4}(q-3)]!}{\pi^{1/2}(h_+^4)^{1/4}}\left(\frac{1}{2}h_+^2\right)^{-q/4}(2z_+)^{-\frac{1}{2}(q-1)}e^{\frac{1}{4}h_+^2 z_+^2}. \tag{18.156b}$$

Using again the duplication formula[†] the ratio of these constants becomes

$$\frac{\gamma}{\overline{\gamma}} = \sqrt{2\pi}\frac{(h_+^2)^{q/2}(2z_+)^q}{[\frac{1}{2}(q-1)]!}e^{-\frac{1}{2}h_+^2 z_+^2} \overset{(18.143)}{\simeq} \mp\frac{2\pi}{4}(q-q_0). \tag{18.157}$$

Since the factorials $[\frac{1}{4}(q-1)]!, [\frac{1}{4}(q-3)]!$ are really correct replacements of $[\frac{1}{4}q]!$ only for q large, this result is somewhat imprecise. However, it is our philosophy here that the factorials with factors occurring in the perturbation solutions are the more natural and hence correct expressions, as the results also seem to support. The relation (18.157) is used in Example 18.5 for the calculation of the *tunneling deviation* $q - q_0$ by using the usual (i.e. linearly matched) WKB solutions, and is shown to reproduce correctly the result (18.143) which was obtained with our perturbation solutions from the boundary conditions at the minimum.

Returning to the even and odd solutions (18.123) we have

$$y_\pm(z) = \frac{1}{2}[y_A(z) \pm \overline{y}_A(z)] = \frac{1}{2\gamma}y_{\mathrm{WKB}}^{(l,z_0)}(z) \pm \frac{1}{2\overline{\gamma}}\overline{y}_{\mathrm{WKB}}^{(l,z_0)}(z). \tag{18.158}$$

Applying the boundary conditions (18.144) we obtain

$$\frac{y_{\mathrm{WKB}}^{(l,z_0)}(0)}{\overline{y}_{\mathrm{WKB}}^{(l,z_0)}(0)} = \frac{\gamma}{\overline{\gamma}}, \quad \frac{y_{\mathrm{WKB}}^{(l,z_0)\prime}(0)}{\overline{y}_{\mathrm{WKB}}^{(l,z_0)\prime}(0)} = -\frac{\gamma}{\overline{\gamma}}. \tag{18.159}$$

[†]In the present case as

$$\left[\frac{1}{4}(q-1)\right]!\left[\frac{1}{4}(q-3)\right]! = \sqrt{\pi}2^{-\frac{1}{2}(q-1)}\left[\frac{1}{2}(q-1)\right]!.$$

Thus we have to consider the behaviour of the integrals occurring in the solutions (18.148), (18.149) near $z = 0$. We have

$$
\begin{aligned}
I_2 &\equiv \int_z^{z_0} dz \left[-\frac{1}{2}qh_+^2 + \frac{1}{4}h_+^4 U(z) \right]^{1/2} \\
&= \int_z^{z_0} dz \left[-\frac{1}{2}qh_+^2 - \frac{1}{4}z^2 h^4 + \frac{1}{2}c^2 z^4 + \frac{h^8}{2^5 c^2} \right]^{1/2} \\
&= \int_z^{z_0} dz \left[\frac{1}{2}\left(\frac{h^4}{2^2 c} - cz^2 \right)^2 - \frac{1}{2}qh_+^2 \right]^{1/2} \\
&\simeq \pm \int_z^{z_0} dz \frac{1}{2^{1/2}}\left(\frac{h^4}{2^2 c} - cz^2 \right) \mp \int_z^{z_0} dz \frac{qh_+^2}{2^{3/2}\left(\frac{h^4}{2^2 c} - cz^2 \right)} \\
&= \frac{c}{\sqrt{2}} \int_z^{z_0} dz \left[\left(\frac{h^4}{4c^2} - \sqrt{q}\frac{h_+}{c} - z^2 \right)\left(\frac{h^4}{4c^2} + \sqrt{q}\frac{h_+}{c} - z^2 \right) \right]^{1/2} (18.160)
\end{aligned}
$$

Setting

$$
b^2 = \frac{h^4}{4c^2} - \sqrt{q}\frac{h_+}{c} \simeq z_0^2, \quad a^2 = \frac{h^4}{4c^2} + \sqrt{q}\frac{h_+}{c}, \quad b^2 < a^2, \qquad (18.161)
$$

we can rewrite the integral as

$$
I_2(z) = \frac{c}{\sqrt{2}} \int_z^b dz \sqrt{(a^2 - z^2)(b^2 - z^2)}. \qquad (18.162)
$$

The integral appearing here is an elliptic integral which can be looked up in Tables.[‡] The elliptic modulus k (with $k'^2 = 1 - k^2$) and an expression u appearing in the integral are defined by

$$
k^2 = \frac{b^2}{a^2} \equiv \frac{1-u}{1+u}, \quad u = \frac{8c}{h_+^3}\sqrt{q} \equiv G\sqrt{2q}, \quad G^2 = \frac{8\sqrt{2}c^2}{h^6}, \qquad (18.163a)
$$

so that

$$
a = \frac{h^2}{2c}[1 + u]^{1/2}. \qquad (18.163b)
$$

The integral $I_2(z)$ evaluated at $z = 0$ is then given by

$$
\begin{aligned}
I_2(0) &= \frac{ca^3}{3\sqrt{2}}[(1 + k^2)E(k) - k'^2 K(k)] = \frac{h^6}{12\sqrt{2}c^2}\sqrt{1+u}[E(k) - uK(k)] \\
&= \frac{2}{3G^2}\sqrt{1+u}[E(k) - uK(k)], \qquad (18.164)
\end{aligned}
$$

[‡]See P. F. Byrd and M. D. Friedman [40], formulae 220.05, p. 60 and 361.19, p. 213.

where $K(k)$ and $E(k)$ are the complete elliptic integrals of the first and second kinds respectively. Here we are interested in the behaviour of the integral in the domain of large h^6/c^2, which implies small G^2. The nontrivial expansions are derived in Example 5, where the result is shown to be (with $q \simeq q_0 = 2n + 1, n = 0, 1, 2, \dots$)

$$
\begin{aligned}
I_2(0) &= \pm \frac{2}{3G^2} \pm \frac{q}{2} \ln\left(\frac{G}{4}\right) \pm \frac{q}{4} \ln\left(\frac{2}{q}\right) \mp \frac{q}{4} \\
&= \pm \frac{2}{3G^2} \pm \frac{1}{2}(2n+1) \ln\left(\frac{G}{4}\right) \pm \frac{1}{4}(2n+1) \ln\left(\frac{2n+1}{2}\right) \\
&\quad \mp \left(\frac{2n+1}{4}\right)
\end{aligned}
\tag{18.165}
$$

in agreement with Ref. [167].[§] Since the integral is to be positive (as required in the WKB solutions) we have to take the upper signs. It then follows that (again in each case in leading order)

$$
y_{\text{WKB}}^{(l,z_0)}(z)|_{z=0} \simeq \frac{\left(\frac{2h^6}{c^2}\right)^{q/4}}{2^{3q/8}\left(\frac{h^8}{2^5 c^2}\right)^{1/4}\left(\frac{q}{4}\right)^{q/4} e^{-q/4}} e^{-\frac{h^6}{12\sqrt{2}c^2}}.
\tag{18.166}
$$

Using *Stirling's formula* we can write this

$$
y_{\text{WKB}}^{(l,z_0)}(z)|_{z=0} \simeq \frac{(2\pi)^{1/2}\left(\frac{2h^6}{c^2}\frac{1}{2^{3/2}}\right)^{q/4}}{\left(\frac{h^8}{2^5 c^2}\right)^{1/4}\left[\frac{1}{4}(q-1)\right]!} e^{-\frac{h^6}{12\sqrt{2}c^2}}.
\tag{18.167}
$$

Correspondingly we find

$$
\overline{y}_{\text{WKB}}^{(l,z_0)}(z)|_{z=0} \simeq \frac{\left[\frac{1}{4}(q-3)\right]!}{(2\pi)^{1/2}\left(\frac{2h^6}{c^2}\frac{1}{2^{3/2}}\right)^{q/4}\left(\frac{h^8}{2^5 c^2}\right)^{1/4}} e^{+\frac{h^6}{12\sqrt{2}c^2}}
\tag{18.168}
$$

and

$$
\frac{d}{dz} y_{\text{WKB}}^{(l,z_0)}(z)\bigg|_{z=0} \simeq \left(\frac{h^8}{2^5 c^2}\right)^{1/4} \frac{(2\pi)^{1/2}\left(\frac{2h^6}{c^2}\frac{1}{2^{3/2}}\right)^{q/4}}{\left[\frac{1}{4}(q-1)\right]!} e^{-\frac{h^6}{12\sqrt{2}c^2}}
\tag{18.169}
$$

and

$$
\frac{d}{dz} \overline{y}_{\text{WKB}}^{(l,z_0)}(z)\bigg|_{z=0} \simeq -\left(\frac{h^8}{2^5 c^2}\right)^{1/4} \frac{\left[\frac{1}{4}(q-3)\right]!}{(2\pi)^{1/2}\left(\frac{2h^6}{c^2}\frac{1}{2^{3/2}}\right)^{q/4}} e^{\frac{h^6}{12\sqrt{2}c^2}}.
\tag{18.170}
$$

[§]The expansion of I_2 used in Ref. [4] misses an n−dependent power of 2 in the result.

With these expressions we obtain now from Eqs. (18.159) — on using once again the duplication formula in the same form as above —

$$\sqrt{2\pi}\,\frac{2^{q/2}}{[\frac{1}{2}(q-1)]!}\left(\frac{2h^6}{c^2}\frac{1}{2^{3/2}}\right)^{q/2}e^{-\frac{h^6}{6\sqrt{2}c^2}}=\frac{\gamma}{\tilde{\gamma}}. \tag{18.171}$$

Comparing this result with that of Eq. (18.157), we can therefore impose the boundary conditions at $z = 0$ by making in Eq. (18.143) the replacement:

$$(h_+^2)^{q/2}(2z_+)^q e^{-\frac{1}{2}h_+^2 z_+^2} \to 2^q\left(\frac{h^6}{2^{3/2}c^2}\right)^{q/2}e^{-\frac{h^6}{6\sqrt{2}c^2}}. \tag{18.172}$$

A corrresponding relation holds for h_+ and z_+ replaced by h_- and z_-. Inserting the expressions for h_+ and z_+ in terms of h, we see that the relation is really an identity (the pre-exponential factors on the left and on the right being equal) with the exponential on the left being an approximation of the exponential on the right (which contains the full action of the instanton). Thus our second condition in the present case of the double well potential turns out to be an identity which confirms our discussion at the beginning of Sec. 18.3.3 concerning the exponentials.

18.3.6 Eigenvalues and level splitting

We now insert the replacement (18.172) into Eq. (18.143) with $q_0 = 2n + 1, n = 0, 1, 2, \ldots$, and obtain

$$(q-q_0) \simeq \mp 4\sqrt{\frac{1}{2\pi}\frac{2^{q_0}\left(\frac{h^6}{2c^2}\right)^{q_0/2}}{2^{q_0/4}[\frac{1}{2}(q_0-1)]!}}e^{-\frac{h^6}{6\sqrt{2}c^2}}, \quad q_0 = 1, 3, 5, \ldots. \tag{18.173}$$

We obtain the energy $E(q, h^2)$ and hence the splitting of asymptotically degenerate energy levels, with the help of Eq. (18.106). Expanding this around an odd integer q_0 we have

$$E(q,h^2) \simeq E_0(q_0,h^2)+(q-q_0)\left(\frac{\partial E}{\partial q}\right)_{q_0} \simeq E_0(q_0,h^2)+(q-q_0)\frac{h^2}{\sqrt{2}}. \tag{18.174}$$

Inserting here the result (18.173), we obtain for $E(q_0, h^2)$ the split expressions

$$E_\pm(q_0,h^2) \simeq E_0(q_0,h^2) \mp \frac{2^{q_0+1}h^2\left(\frac{h^6}{2c^2}\right)^{q_0/2}}{\sqrt{\pi}2^{q_0/4}[\frac{1}{2}(q_0-1)]!}e^{-\frac{h^6}{6\sqrt{2}c^2}}, \quad q_0 = 1, 3, 5, \ldots,$$

$$\tag{18.175}$$

where $E_0(q_0, h^2)$ is given by Eq. (18.106), i.e.[¶]

$$E_0(q_0, h^2) = -\frac{h^8}{2^5 c^2} + \frac{1}{\sqrt{2}}q_0 h^2 - \frac{c^2(3q_0^2 + 1)}{2h^4} - \frac{\sqrt{2}c^4 q_0}{8h^{10}}(17q_0^2 + 19) + O\left(\frac{1}{h^{16}}\right).$$

Thus

$$
\begin{aligned}
\triangle E(q_0, h^2) &= E_-(q_0, h^2) - E_+(q_0, h^2) \\
&\simeq \frac{2^{q_0+2} h^2}{\sqrt{\pi} 2^{q_0/4} [\frac{1}{2}(q_0 - 1)]!} \left(\frac{h^6}{2c^2}\right)^{q_0/2} e^{-\frac{h^6}{2^{1/2} 6c^2}}, \quad \left(\text{mass } m_0 = \frac{1}{2}\right) \\
&= 2^{(2n+9)/4} \frac{h^2}{\sqrt{\pi} n!} \left(\frac{h^6}{c^2}\right)^{n+1/2} e^{-\frac{h^6}{2^{1/2} 6c^2}}. \quad (18.176)
\end{aligned}
$$

Combining Eqs. (18.174) and (18.143) with (18.157), the level splitting, i.e. the difference between the eigenenergies of even and odd states with (here) $q_0 = 2n + 1, n = 0, 1, 2, \ldots$, for finite h^2, can be given by

$$\triangle E(q_0, h^2) \simeq \frac{4}{\pi} \left(\frac{\partial E}{\partial q}\right)_{q_0} \frac{y_{\text{WKB}}^{(l,z_0)}(0)}{\bar{y}_{\text{WKB}}^{(l,z_0)}(0)} = \frac{2}{\pi} \left(\frac{\partial E}{\partial(n + \frac{1}{2})}\right) \frac{y_{\text{WKB}}^{(l,z_0)}(0)}{\bar{y}_{\text{WKB}}^{(l,z_0)}(0)}. \quad (18.177)$$

In the work of Bhattacharya [23] the WKB level splitting is effectively (i.e. in the WKB restricted sense) defined by

$$\triangle_n^{\text{WKB}} E = \frac{1}{\pi} \frac{\partial E}{\partial(n + \frac{1}{2})} \frac{y_{\text{WKB}}^{(l,z_0)}(0)}{\bar{y}_{\text{WKB}}^{(l,z_0)}(0)} = \frac{1}{2} \triangle E(q_0, h^2).$$

Here the derivative $\partial E/\partial n$ corresponds to the usual oscillator frequency. The result (18.176) is described by Bhattacharya [23] as a "*modified well and barrier*" result $\triangle_n^{\text{MWB}}$, the pure WKB result of Banerjee and Bhatnagar [14] (i.e. that without the use of Stirling's formula and so left in terms of e and n^n) being this divided by the *Furry factor*

$$f_n := \left[\frac{1}{\sqrt{2\pi}}\left(\frac{e}{n + \frac{1}{2}}\right)^{n+\frac{1}{2}} n!\right]^{-1}, \quad (18.178)$$

[¶]In S. K. Bhattacharya [23] the "usual WKB approximation" of this expression is — with replacements $h^4/4 \leftrightarrow k, c^2/2 \leftrightarrow \lambda$ — given as

$$E_0 = -\frac{k^2}{4\lambda} + (2k)^{1/2} q - \frac{3}{4}\frac{\lambda}{k}q^2 \longrightarrow -\frac{h^8}{2^5 c^2} + q\frac{h^2}{\sqrt{2}} - q^2\frac{3c^2}{2h^4},$$

i.e. the following expression for large $q \sim 2n + 1$, which evidently supplies some correction terms.

which is unity for $n \to \infty$ with Stirling's formula.[||] Of course, these derivations do not exploit the symmetry of the original equation under the interchanges $q \leftrightarrow -q, h^2 \leftrightarrow -h^2$, as we do here. We see therefore, that if this symmetry is taken into account from the very beginning, the Furry factor corrected WKB result follows automatically. The Furry factor represents effectively a correction factor to the normalization constants of WKB wave functions (which are normally for $m_0 = 1/2, \hbar = 1$, and harmonic oscillator frequency equal to 1 given by $1/\sqrt{2\pi}$ and independent of n as explained by Furry [102] — see also Example 18.6) to yield an improvement of WKB results for small values of n, as is also explained by Bhattacharya [23].[**]

Had we taken the mass m_0 of the particle in the symmetric double well potential equal to 1 (instead of 1/2), we would have obtained the result with E, h^4 and c^2 replaced by $2E, 2h^4$ and $2c^2$ respectively (see Eq. (18.3)). Then

$$E(q_0, h^2) = E_0(q_0, h^2) \mp \frac{2^{q_0 + \frac{1}{2}} h^2 (\frac{h^6}{\sqrt{2}c^2})^{q_0/2}}{\sqrt{\pi} 2^{q_0/4} [\frac{1}{2}(q_0 - 1)]!} e^{-\frac{h^6}{6c^2}}. \qquad (18.179)$$

and $\triangle E$ becomes

$$\triangle_1 E(q_0, h^2) \simeq 2^{q_0} \sqrt{\frac{2}{\pi}} \frac{2h^2}{2^{q_0/4}[\frac{1}{2}(q_0 - 1)]!} \left(\frac{h^6}{2^{1/2}c^2} \right)^{q_0/2} e^{-h^6/6c^2}, \quad (m_0 = 1) \qquad (18.180)$$

with

$$E_0(q_0, h^2)|_{m_0=1} = -\frac{h^8}{2^4 c^2} + q_0 h^2 - \cdots .$$

If in addition the potential is written in the form

$$V(z) = \frac{\lambda}{4} \left(z^2 - \frac{\mu^2}{\lambda} \right)^2, \quad \lambda > 0, \qquad (18.181)$$

[||]W. H. Furry [102], Eq. (66). The factor set equal to 1 represents a somewhat "mutilated" Stirling approximation. This, however, supplies the bridge to the perturbation theory results and removes the unnatural appearance of factors e and n^n. In Chapter 26 we obtain in each case complete agreement of the path integral result with the perturbation theory result with the help of this factor.

[**]Comparing the present work with that of S. K. Bhattacharya [23], we identity the parameter k there with $h^4/4$ here, and λ there with $c^2/2$ here. Then the splitting $\triangle_n^{MWB}$ of Eq. (24) there is

$$\triangle_n^{MWB} = \sqrt{\frac{k}{\pi}} \frac{2^{(10n+13)/4}}{n!} \left(\frac{k^{3/2}}{\lambda} \right)^{n+1/2} \exp\left[-\frac{\sqrt{2}}{3} \frac{k^{3/2}}{\lambda} \right]$$

$$= \frac{h^2}{\sqrt{\pi}n!} 2^{(2n+5)/4} \left(\frac{h^6}{c^2} \right)^{n+1/2} \exp\left[-\frac{h^6}{6\sqrt{2}c^2} \right]$$

in agreement with $\triangle E(q_0, h^2)/2$ of Eq. (18.176) together with Eq. (18.178).

a form frequently used in field theoretic applications, so that by comparison with Eq. (18.88) $h^4 \equiv 2\mu^2, c^2 = \lambda/2$, the level splitting is (with $\hbar = 1$)

$$\triangle_1 E(q_0, h^2) \simeq \frac{2^{q_0+2}\mu}{\pi^{1/2} 2^{q_0/4}[\frac{1}{2}(q_0 - 1)]!} \left(\frac{4\mu^3}{\lambda}\right)^{q_0/2} e^{-8^{1/2}\mu^3/3\lambda}. \qquad (18.182)$$

This result agrees with the ground state ($q_0 = 1$) result of Gildener and Patrascioiu [110] using the path integral method for the evaluation of pseudoparticle (instanton) contributions, provided their result is multiplied by a factor of 2 resulting from a corresponding inclusion of anti-pseudoparticle (anti-instanton) contributions.* A similar correspondence is also found in the case of the cosine potential.† Thus for the case of the ground state Eq. (18.182) implies

$$\triangle_1 E(1, h^2) \simeq 2\sqrt{2}\mu \left(\frac{16\sqrt{2}\mu^3}{\pi\lambda}\right)^{1/2} e^{-\sqrt{8}\mu^3/3\lambda}.$$

Equation (18.180) agrees similarly also with the result for arbitrary levels obtained with the use of periodic instantons‡ and with the results of multi-instanton methods.§

18.3.7 General Remarks

In the above we have attempted a fairly complete treatment of the large-h^2 case of the quartic anharmonic oscillator, carried out along the lines of the corresponding calculations for the cosine potential and thus of the well-established Mathieu equation. In principle one could also consider the case of small values of h^2 and obtain convergent instead of asymptotic expansions; however, these are presumably not of much interest in physics (for reasons explained in Chapter 20). We considered above only the symmetric two-minima potential. The asymmetric case can presumably also be dealt with in a similar way since various references point out that the asymmetric case can be transformed into a symmetric one.¶

---

*See E. Gildener and A. Patrasciociu [110], formula (4.11).

†This will become evident when the perturbation theory result of Chapter 17 is compared with the path integral result in Chapter 26.

‡To help the comparison note that in J.–Q. Liang and H. J.W. Müller–Kirsten [167] the potential is written as

$$V(z) = \frac{\eta^2}{2}\left[z^2 - \frac{m^2}{\eta^2}\right]^2.$$

for $m_0 = 1$. The comparison with Eq. (18.181) therefore implies the correspondence $\eta^2 \leftrightarrow \lambda/2, m^2 \leftrightarrow \mu^2/2$, and g^2 is given as $g^2 = \eta^2/m^3$.

§J. Zinn–Justin and U. D. Jentschura [293], first paper, Eqs. (2.34) and (E.15).

¶See R. Friedberg and T. D. Lee [98], K. Banerjee and S. P. Bhatnagar [14], J. Zinn–Justin and U. D. Jentschura [293], H. M. M. Mansour and H. J. W. Müller–Kirsten [186].

Every now and then literature appears which purports to overcome the allegedly ill-natured "divergent perturbation series" of the anharmonic oscillator problem, and numerical studies are presented to support this claim.‖ It will become clear in Chapter 20 — as is also demonstrated by the work of Bender and Wu [18], [19] — that the expansions considered above are asymptotic. Tables of properties of Special Functions are filled with such expansions derived from differential equations for all the wellknown and less wellknown Special Functions. There is no reason to view the anharmonic oscillator differently. In fact, in principle its Schrödinger equation is an equation akin to equations like the Mathieu or modified Mathieu equations which lie outside the range of hypergeometric types. There is no way to obtain the exponentially small (real or imaginary) nonperturbative contributions derived above with some convergent expansion, since — as we shall see in Chapter 20 — these nonperturbative contributions are directly related to the large-order behaviour of the perturbation expansion and determine this as asymptotic. The immense amount of literature meanwhile accumulated for instance in the case of the Mathieu equation can indicate what else can be achieved along parallel lines in the case of special types of Schrödinger equations, like those for anharmonic oscillators. Sophisticated mathematics — like that of the extensive investigations of Turbiner [273] over several decades — seems to approach the problem from a different angle, as he discusses, for instance, in his recent paper on simple uniform approximation of the logarithmic derivative of the ground state wave function of anharmonic oscillator potentials.

The ground state splitting of the symmetric double well potential has been considered in a countless number of investigations. A reasonable, though incomplete list of references in this direction has been given by Garg [105] beginning with the wellknown though nonexplicit (and hence not really useful) ground state formula in the book of Landau and Lifshitz [156]. Very illuminating discussions of double wells and periodic potentials, mostly in connection with instantons, can be found in an article by Coleman [54]. The wide publicity given to the work of Bender and Wu made pure mathematicians aware of the subject; as some relevant references with their view we cite papers of Harrell and Simon [129], Ashbaugh and Harrell [12] and Harrell [130],[131], [132] and Loeffel, Martin, Simon and Wightman [178]. The double well potential, in both symmetric and asymmetric form, has also been the subject of numerous numerical studies. As references in this direction, though not exclusively numerical, we cite papers of the Uppsala group of Fröman and Fröman [100]. Perturbation theoretical aspects are

---

‖See e.g. B. P. Mahapatra, N. Santi and N. B. Pradhan [182].

employed in papers of Banerjee and Bhatnagar [14], Biswas, Datta, Saxena, Srivastava and Varma [24] and Bhatacharya and Rau [23]. Wave functions of symmetric and asymmetric double well potentials have been considered in the following reference in which it is demonstrated that actual physical tunneling takes place only into those states which have significant overlap with the false vacuum eigenfunction: Nieto, Gutschick, Bender, Cooper and Strottman [221].

Example 18.5: Evaluation of WKB exponential with elliptic integrals
Show that

$$\frac{2}{3G^2}(1+u)^{1/2}[E(k) - uK(k)] \simeq \frac{2}{3G^2} + \frac{q}{2}\ln\left(\frac{G}{4}\right) - \frac{q}{4}\ln\left(\frac{2}{q}\right) - \frac{q}{4}$$
$$= \frac{2}{3G^2} + \frac{1}{2}(2n+1)\ln\left(\frac{G}{4}\right) + \frac{1}{4}(2n+1)\ln\left(\frac{2n+1}{2}\right) - \left(\frac{2n+1}{4}\right). \quad (18.183)$$

Solution: We have

$$k^2 = \frac{1-u}{1+u}, \quad u = G\sqrt{2q}, \quad q \simeq 2n+1. \quad (18.184)$$

Hence we obtain for G close to zero:

$$k^2 = \frac{1 - G\sqrt{2q}}{1 + G\sqrt{2q}} = 1 - 2G\sqrt{2q} + 4G^2q - \cdots, \quad k^2 \sim 1, \quad (18.185)$$

and

$$(1+u)^{1/2} = (1 + G\sqrt{2q})^{1/2} = 1 + G\sqrt{\frac{q}{2}} - \frac{1}{4}G^2q + \cdots, \quad (18.186)$$

and

$$k'^2 = 1 - k^2 = 2G\sqrt{2q} - 4G^2q + \cdots, \quad k'^2 \sim 0. \quad (18.187)$$

We now reexpress various quantities in terms of u. Thus

$$k'^2 = 2u - 2u^2 + \cdots, \quad (18.188)$$

and hence

$$k' = \sqrt{2u}(1-u)^{1/2} = \sqrt{2u}\left(1 - \frac{u}{2} + \cdots\right). \quad (18.189)$$

The following expression appears frequently in the expansions of elliptic integrals. Therefore it is convenient to deal with this here. We have

$$\frac{4}{k'} = \frac{4}{\sqrt{2u}(1 - \frac{u}{2} + \cdots)} = \frac{4}{\sqrt{2u}}\left(1 + \frac{u}{2} + \cdots\right). \quad (18.190)$$

Hence

$$\ln\left(\frac{4}{k'}\right) = \ln\left[\frac{4}{\sqrt{2u}}\left(1 + \frac{u}{2} - \cdots\right)\right] \simeq \ln\left(\frac{4}{\sqrt{2u}}\right) + \frac{u}{2}. \quad (18.191)$$

Our next objective is the evaluation of the elliptic integrals $E(k)$ and $K(k)$ by expanding these in ascending powers of k'^2, which is assumed to be small. We obtain the expansions from Ref. [122] as

$$E(k) = 1 + \frac{1}{2}\left[\ln\left(\frac{4}{k'}\right) - \frac{1}{2}\right]k'^2 + \frac{3}{16}\left[\ln\left(\frac{4}{k'}\right) - \frac{13}{12}\right]k'^4 + \cdots \quad (18.192)$$

and

$$K(k) = \ln\left(\frac{4}{k'}\right) + \frac{1}{4}\left[\ln\left(\frac{4}{k'}\right) - 1\right]k'^2 + \cdots .$$ (18.193)

Consider first $E(k)$:

$$
\begin{aligned}
E(k) &= 1 + \frac{1}{2}\left[\ln\left(\frac{4}{\sqrt{2u}}\right) + \frac{u}{2} - \frac{1}{2}\right]2u(1-u) + \frac{3}{16}\left[\ln\left(\frac{4}{\sqrt{2u}}\right) + \frac{u}{2} - \frac{13}{12}\right]4u^2(1-u)^2 + \cdots \\
&= 1 + \ln\left(\frac{4}{\sqrt{2u}}\right)\left\{u(1-u) + \frac{3}{4}u^2(1-u)^2\right\} + \frac{1}{2}(u-1)u(1-u) \\
&\quad + \frac{3}{4}\frac{1}{2}\left(u - \frac{13}{6}\right)u^2(1-u)^2 + \cdots
\end{aligned}
$$

We can rewrite this as

$$E(k) = 1 + \ln\left(\frac{4}{\sqrt{2u}}\right)\frac{u(1-u)}{4}\{4 + 3u(1-u)\} + \frac{1}{2^4 3}u(u-1)(1-u)\{24 - 3u(6u-13)\}.$$ (18.194)

Analogously we have

$$
\begin{aligned}
uK(k) &= u\left[\ln\left(\frac{4}{\sqrt{2u}}\right) + \frac{u}{2}\right] + \frac{u}{4}\left[\ln\left(\frac{4}{\sqrt{2u}}\right) + \frac{u}{2} - 1\right]2u(1-u) \\
&= \ln\left(\frac{4}{\sqrt{2u}}\right)\left\{u + \frac{1}{2}u^2(1-u)\right\} + \frac{1}{2}u^2\left[1 + (1-u)\left(\frac{u}{2} - 1\right)\right] \\
&= \ln\left(\frac{4}{\sqrt{2u}}\right)\frac{1}{2}\{u^2(1-u) + 2u\} + \frac{1}{4}u^2[2 + (1-u)(u-2)]
\end{aligned}
$$

or

$$
\begin{aligned}
uK(k) &= \ln\left(\frac{4}{\sqrt{2u}}\right)\frac{u}{2}\{u(1-u) + 2\} + \frac{1}{4}u^2[-u^2 + 3u] \\
&= \ln\left(\frac{4}{\sqrt{2u}}\right)\frac{u}{2}\{u(1-u) + 2\} - \frac{1}{4}u^3(u-3).
\end{aligned}
$$ (18.195)

With (18.194) and (18.195) we obtain

$$
\begin{aligned}
&E(k) - uK(k) \\
&= 1 + \ln\left(\frac{4}{\sqrt{2u}}\right)\left[\frac{1}{4}u(1-u)\{4 + 3u(1-u)\} - \frac{1}{2}u\{u(1-u) + 2\}\right] \\
&\quad + \frac{1}{2}u(u-1)(1-u)\frac{1}{2^3 3}\{24 - 3u(6u-13)\} + \frac{1}{4}u^3(u-3) \\
&= 1 + \ln\left(\frac{4}{\sqrt{2u}}\right)\frac{u}{4}[(1-u)\{4 + 3u(1-u)\} - 2\{u(1-u) + 2\}] \\
&\quad + \frac{1}{2.4.6}u(u-1)(1-u)\{24 - 3u(6u-13)\} + \frac{1}{4}u^3(u-3).
\end{aligned}
$$ (18.196)

Now consider the last line here without the factor 2.4.6 in the denominator and in the last step pick out the lowest order terms in u (i.e. up to and including u^2):

$$
\begin{aligned}
&u(u-1)(1-u)\{24 - 3u(6u-13)\} + 12u^3(u-3) \\
&= -u(1-u)^2\{24 - 18u^2 + 39u\} + 12u^3(u-3) \\
&= -3u(1-u)^2\{8 - 6u^2 + 13u\} + 12u^3(u-3) \\
&\simeq -3u.8 + 6u^2.8 - 3.13u^2 = -3u.8 + 9u^2.
\end{aligned}
$$ (18.197)

Now consider the bracket [...] in (18.196), i.e.

$$[(1-u)\{4+3u(1-u)\} - 2\{u(1-u)+2\}] = u(3u^2 - 4u - 3). \tag{18.198}$$

From (18.195) with (18.196) and (18.197) we now obtain

$$E(k) - uK(k) \simeq 1 + \ln\left(\frac{4}{\sqrt{2u}}\right)\frac{1}{4}u^2(3u^2 - 4u - 3) - \frac{1}{2}u + \frac{9u^2}{2.4.6}. \tag{18.199}$$

We now return to the expression on the left of Eq. (18.173) to be evaluated, i.e.

$$I_2 = \frac{2}{3G^2}(1+u)^{1/2}[E(k) - uK(k)].$$

Inserting here from the above expansions the contributions up to and including those of order u^2, we obtain:

$$I_2 \simeq \frac{2}{3G^2}\left(1 + \frac{u}{2} - \frac{1}{8}u^2 + \cdots\right)\left[1 - \ln\left(\frac{4}{\sqrt{2u}}\right)\frac{u^2}{4}(3 + 4u - 3u^2) - \frac{1}{2}u + \frac{3u^2}{16} + O(u^3)\right]. \tag{18.200}$$

Remembering that $u = G\sqrt{2q}$, this becomes

$$
\begin{aligned}
I_2 &\simeq \frac{2}{3G^2} - \frac{u^2}{2G^2}\ln\left(\frac{4}{\sqrt{2u}}\right) + \frac{2}{3G^2}\left(-\frac{3u^2}{8}\right) + \frac{u^2}{8G^2} \\
&= \frac{2}{3G^2} - q\ln\left(\frac{4}{2^{1/2}G^{1/2}2^{1/4}q^{1/4}}\right) - \frac{q}{4} = \frac{2}{3G^2} - \frac{q}{2}\ln\left(\frac{2^{5/2}}{Gq^{1/2}}\right) - \frac{q}{4}. \tag{18.201}
\end{aligned}
$$

Thus

$$
\begin{aligned}
I_2 &\simeq \frac{2}{3G^2} + \frac{q}{2}\ln\left(\frac{G}{4}\right) - \frac{q}{4}\ln\left(\frac{2}{q}\right) - \frac{q}{4} \\
&= \frac{2}{3G^2} + \frac{1}{2}(2n+1)\ln\left(\frac{G}{4}\right) + \frac{1}{4}(2n+1)\ln\left(\frac{2n+1}{2}\right) - \left(\frac{2n+1}{4}\right). \tag{18.202}
\end{aligned}
$$

Example 18.6: Normalization of WKB solutions

Show that the normalized form of the WKB solution (18.148) (for particle mass 1/2 and in dominant order) is given by

$$y_{WKB \text{ normalized}}^{(l,z_0)} = \left(\frac{h^4}{8\pi^2}\right)^{1/4}\frac{\exp[-\int_z^{z_0} dz\sqrt{|E - V(z)|}}{|E - V(z)|^{1/4}}.$$

Solution: From Eqs. (18.128) and (18.156a) we obtain α and γ and hence the ratio (always in the dominant order)

$$\frac{\gamma}{\alpha} = \left(\frac{8\pi^2}{h_+^2}\right)^{1/4} \quad \text{with} \quad h_+^2 = \sqrt{2}h^2.$$

Since (cf. Eqs. (18.128) and (18.155))

$$\alpha y_A(z) = y_B(z) \quad \text{and} \quad y_{WKB}^{(l,z_0)}(z) = \gamma y_A(z),$$

we have, where $w_\pm(z) = h_\pm(z - z_\pm)$,

$$y_B(z) = \frac{\alpha}{\gamma}y_{WKB}^{(l,z_0)}(z) \quad \text{with} \quad y_B(z) \simeq B_q[w_\pm(z)] = \frac{D_{\frac{1}{2}(q-1)}[w_\pm(z)]}{[\frac{1}{4}(q-1)]!2^{\frac{1}{4}(q-1)}}.$$

With the standard normalization of parabolic cylinder functions, i.e.**

$$\int_{-\infty}^{\infty} dw \, D^2_{\frac{1}{2}(q-1)}(w) = \sqrt{2\pi} \left[\frac{1}{2}(q-1)\right]! \quad \text{and} \quad \frac{[\frac{1}{4}(q-1)]![\frac{1}{4}(q-3)]! 2^{\frac{1}{2}(q-1)}}{[\frac{1}{2}(q-1)]!} = \sqrt{\pi},$$

we obtain (for q reasonably large)

$$\left[\frac{1}{2}(q-1)\right]! \sqrt{2\pi} \simeq \int_{-\infty}^{\infty} dz \, h_+ \frac{\{[\frac{1}{4}(q-1)]! 2^{\frac{1}{4}(q-1)}\}^2}{2\pi} y^2_{WKB} \left(\frac{h_+^2}{2}\right)^{1/2},$$

and hence

$$1 \simeq \int_{-\infty}^{\infty} dz \left(\frac{h_+}{\sqrt{4\pi}} y_{WKB}\right)^2.$$

Hence with $h_+/\sqrt{4\pi} = (h^4/8\pi^2)^{1/4}, h^4/2 = h_+^4/4$ this implies

$$y_{WKB,\,\text{normalized}} = \left(\frac{h^4}{8\pi^2}\right)^{1/4} \frac{\exp[-\int_z^{z_0} dz \sqrt{|E - V(z)|}]}{|E - V(z)|^{1/4}}.$$

With $h^4/4 \equiv m^2$ and mass 1 (instead of $1/2$) the result is

$$y_{WKB,\,\text{normalized}} = \left(\frac{m^2}{\pi^2}\right)^{1/4} \frac{\exp[-\int_z^{z_0} dz \sqrt{|E - V(z)|}]}{|E - V(z)|^{1/4}}.$$

We see that the normalization constant is independent of a quantum number. The constant obtained here will arise in Chapter 26 in the evaluation of the normalization constant of the WKB wave function (cf. Eq. (26.35)).

Example 18.7: Recalculation of tunneling deviation using WKB solutions

Determine the tunneling deviation $q - q_0$ of q from an odd integer q_0, and obtained as Eq. (18.143), by using the periodic WKB solutions below the turning points.

Solution: The turning points at z_0 and z_1 on either side of the minimum at z_+ are given by Eqs. (18.146) and (18.147). We start from Eq. (18.158), i.e.

$$y_\pm(z) = \frac{1}{2}[y_A(z) \pm \bar{y}_A(z)] = \frac{1}{2\gamma} y_{WKB}^{(l,z_0)}(z) \pm \frac{1}{2\bar{\gamma}} \bar{y}_{WKB}^{(l,z_0)}(z). \tag{18.203}$$

Different from above we now continue the solutions (in the sense of linearly matched WKB solutions) across the turning point at z_0 in the direction of the minimum of the potential at z_+. Then $((r, z_0)$ meaning to the right of z_0 and note the asymmetric factor of 2)

$$\begin{aligned}
y_\pm(z) &= \frac{1}{2\gamma} y_{WKB}^{(r,z_0)}(z) \pm \frac{1}{2\bar{\gamma}} \bar{y}_{WKB}^{(r,z_0)}(z) = \left[\frac{1}{2}qh_+^2 - \frac{1}{4}h_+^4 U(z)\right]^{-1/4} \\
&\quad \times \left\{ \frac{1}{\gamma} 2\sin \left[\int_{z_0}^z dz \left(\frac{1}{2}qh_+^2 - \frac{1}{4}h_+^4 U(z)\right)^{1/2} + \frac{\pi}{4}\right] \right. \\
&\quad \left. \pm \frac{1}{\bar{\gamma}} \cos \left[\int_{z_0}^z dz \left(\frac{1}{2}qh_+^2 - \frac{1}{4}h_+^4 U(z)\right)^{1/2} + \frac{\pi}{4}\right] \right\}. \tag{18.204}
\end{aligned}$$

**See e.g. W. Magnus and F. Oberhettinger [181], pp. 82, 93.

We also note that

$$
\frac{d}{dz} y_\pm(z) \simeq \left[\frac{1}{2}qh_+^2 - \frac{1}{4}h_+^4 U(z)\right]^{1/4} \left\{\frac{1}{\gamma}2\cos\left[\int_{z_0}^z dz\left(\frac{1}{2}qh_+^2 - \frac{1}{4}h_+^4 U(z)\right)^{1/2} + \frac{\pi}{4}\right]\right.
$$
$$
\left.\mp \frac{1}{\gamma}\sin\left[\int_{z_0}^z dz\left(\frac{1}{2}qh_+^2 - \frac{1}{4}h_+^4 U(z)\right)^{1/2} + \frac{\pi}{4}\right]\right\}.
\tag{18.205}
$$

We now apply the boundary conditions (18.132) and (18.133) at the minimum z_+ and obtain the conditions:

$$
0 = \frac{2}{\gamma}\sin\left[\int_{z_0}^{z_+} dz\left(\frac{1}{2}qh_+^2 - \frac{1}{4}h_+^4 U(z)\right)^{1/2} + \frac{\pi}{4}\right] \pm \frac{1}{\gamma}\cos\left[\int_{z_0}^{z_+} dz\left(\frac{1}{2}qh_+^2 - \frac{1}{4}h_+^4 U(z)\right)^{1/2} + \frac{\pi}{4}\right]
\tag{18.206}
$$

and

$$
0 = \frac{2}{\gamma}\cos\left[\int_{z_0}^{z_+} dz\left(\frac{1}{2}qh_+^2 - \frac{1}{4}h_+^4 U(z)\right)^{1/2} + \frac{\pi}{4}\right] \mp \frac{1}{\gamma}\sin\left[\int_{z_0}^{z_+} dz\left(\frac{1}{2}qh_+^2 - \frac{1}{4}h_+^4 U(z)\right)^{1/2} + \frac{\pi}{4}\right].
\tag{18.207}
$$

Hence

$$
\tan\left[\int_{z_0}^{z_+} dz\left(\frac{1}{2}qh_+^2 - \frac{1}{4}h_+^4 U(z)\right)^{1/2} + \frac{\pi}{4}\right] \simeq \mp\frac{\gamma}{2\gamma}
\tag{18.208}
$$

and

$$
\cot\left[\int_{z_0}^{z_+} dz\left(\frac{1}{2}qh_+^2 - \frac{1}{4}h_+^4 U(z)\right)^{1/2} + \frac{\pi}{4}\right] \simeq \pm\frac{\gamma}{2\gamma}
\tag{18.209}
$$

In the present considerations we approach the minimum of the potential at z_+ by coming from the left, i.e. from z_0. We could, of course, approach the minimum also from the right, i.e. from z_1. Then at any point $z \in (z_0, z_1)$ we expect[††]

$$
|y_{\text{WKB}}^{(r,z_0)}(z)| = |y_{\text{WKB}}^{(l,z_1)}(z)|, \quad |\overline{y}_{\text{WKB}}^{(r,z_0)}(z)| = |\overline{y}_{\text{WKB}}^{(l,z_1)}(z)|.
\tag{18.210}
$$

Choosing the point z to be z_+, this implies

$$
\left|\begin{array}{c}\sin\\\cos\end{array}\left(\int_{z_0}^{z_+} dz\left[\frac{1}{2}qh_+^2 - \frac{1}{4}h_+^4 U(z)\right]^{1/2} + \frac{\pi}{4}\right)\right|
$$
$$
= \left|\begin{array}{c}\sin\\\cos\end{array}\left(\int_{z_+}^{z_1} dz\left[\frac{1}{2}qh_+^2 - \frac{1}{4}h_+^4 U(z)\right]^{1/2} + \frac{\pi}{4}\right)\right|.
\tag{18.211}
$$

Thus e.g.

$$
\cos\left[\int_{z_+}^{z_1} dz\left(\frac{1}{2}qh_+^2 - \frac{1}{4}h_+^4 U(z)\right)^{1/2} + \frac{\pi}{4}\right]
$$
$$
= \cos\left[\int_{z_0}^{z_1}\cdots - \int_{z_0}^{z_+} + \frac{\pi}{4}\right] = \cos\left[\int_{z_0}^{z_1}\cdots - \frac{\pi}{4} - \int_{z_0}^{z_+} + \frac{\pi}{2}\right]
$$
$$
= \sin\left[\int_{z_0}^{z_1}\cdots - \frac{\pi}{4} - \int_{z_0}^{z_+}\cdots\right] = -(-1)^N \cos\left[\int_{z_0}^{z_+}\cdots + \frac{\pi}{4}\right]
$$

provided the *Bohr-Sommerfeld-Wilson quantization condition* holds, i.e.

$$
\int_{z_0}^{z_1} dz\sqrt{E - V(z)} = \int_{z_0}^{z_1} dz\left(\frac{1}{2}qh_+^2 - \frac{1}{4}h_+^4 U(z)\right)^{1/2} = (2N+1)\frac{\pi}{2}, \quad N = 1,2,3,\ldots,
\tag{18.212}
$$

[††]See e.g. A. Messiah [195], Sec. 6.2.6.

where it is understood that $qh^2/2 \simeq E_V(z_\pm)$. Similarly under the same condition

$$\sin\left[\int_{z_+}^{z_1} \cdots + \frac{\pi}{4}\right] = (-1)^N \sin\left[\int_{z_0}^{z_+} \cdots + \frac{\pi}{4}\right].$$

We rewrite the quantization condition in the present context and in view of the symmetry of the potential in the immediate vicinity of z_+ as

$$\int_{z_0}^{z_+} dz \left(\frac{1}{2}qh_+^2 - \frac{1}{4}h_+^4 U(z)\right)^{1/2} \simeq q\frac{\pi}{4}. \tag{18.213}$$

The integral on the left can be approximated by

$$\frac{h_+^2}{2}\int_{z_0}^{z_+} dz \left(\frac{2q}{h_+^2} - (z - z_+)^2\right)^{1/2} = \frac{q}{2}\sin^{-1}\frac{z_+ - z_0}{(2q/h_+^2)^{1/2}} = \frac{q}{4}\pi$$

in agreement with the right hand side (the last step following from Eq. (18.146)). We now see that Eqs. (18.209) and (18.210) assume a form as in our perturbation theory, i.e. they become

$$\tan\left\{(q+1)\frac{\pi}{4}\right\} \simeq \mp\frac{\gamma}{2\bar{\gamma}}, \quad \cot\left\{(q+1)\frac{\pi}{4}\right\} \simeq \mp\frac{\gamma}{2\bar{\gamma}}. \tag{18.214}$$

Since

$$\tan\left\{(q+1)\frac{\pi}{4}\right\} = 0 \text{ for } q = q_0 = 3, 7, 11, \ldots \text{ and } \cot\left\{(q+1)\frac{\pi}{4}\right\} = 0 \text{ for } q = q_0 = 1, 5, 9, \ldots,$$

we can expand the left hand sides about these points and thus obtain

$$q - q_0 \simeq \mp\frac{2\gamma}{\pi\bar{\gamma}} \text{ for } q_0 = 1, 3, 5, \ldots \tag{18.215}$$

in agreement with Eq. (18.143). We note here incidentally that this agreement demonstrates the significance of the factor of 2 in (18.215) which results from the factor of 2 in front of the sine in the WKB formula (18.205).

In Eq. (18.215) we can insert for $\gamma/\bar{\gamma}$ the expression given by Eqs. (18.148) and (18.149). The latter give

$$\begin{aligned}\frac{\gamma}{\bar{\gamma}} &= \exp\left(-2\int_0^{z_0} dz\left[-\frac{1}{2}qh_+^2 + \frac{1}{4}h_+^4 U(z)\right]^{1/2}\right)\\&= \exp\left(-\int_{-z_0}^{z_0} dz\left[-\frac{1}{2}qh_+^2 + \frac{1}{4}h_+^4 U(z)\right]^{1/2}\right).\end{aligned} \tag{18.216}$$

Inserting this into Eq. (18.215), the level splitting is given by

$$\begin{aligned}\triangle E(q_0, h^2) &= 2(E - E_0) = 2\frac{h^2}{\sqrt{2}}(q - q_0) = 2\frac{h^2}{\sqrt{2}}\frac{2}{\pi}\frac{\gamma}{\bar{\gamma}}\\&= \sqrt{2}h^2\frac{2}{\pi}\exp\left(-\int_{-z_0}^{z_0} dz\left[-\frac{1}{2}qh_+^2 + \frac{1}{4}h_+^4 U(z)\right]^{1/2}\right)\\&\equiv \sqrt{2}h^2\frac{2}{\pi}\exp\left(-\int_{-z_0}^{z_0} dz\sqrt{-E + V(z)}\right).\end{aligned} \tag{18.217}$$

This expression represents the WKB result for the level splitting and may therefore be called the *WKB level splitting formula*. One may note that the exponential factor here is not squared as in decay probabilities (squares of transmission coefficients). The prefactor $2\sqrt{2}h^2/\pi$ is, in fact (with our use of q_0), $2\omega/\pi$, where ω is the oscillator frequency of the wells. The formula thus agrees with that in the literature.[‡‡]

[‡‡]Cf. K. Banerjee and S. P. Bhatnagar [14], S. K. Bhattacharya [23], and L. D. Landau and E. M. Lifshitz [156].

Chapter 19

Singular Potentials

19.1 Introductory Remarks

Singular potentials have mostly been discarded in studies of quantum mechanics in view of their unboundedness from below and consequently the nonexistence of a ground state.[*] However, in an early investigation Case[†] showed that potentials of the form $r^{-n}, n \geq 2$, are not as troublesome as one might expect. In particular Case pointed out that for a repulsive singular potential the study of scattering is mathematically well-defined and useful.

Some decades ago — before the discovery of W and Z mesons which mediate weak interactions — and before the advent of quantum chromodynamics, the investigation of singular potentials in nonrelativistic quantum theory was motivated by a desire to obtain a better understanding of the (then presumed) singular nature of the nonrenormalizable weak quantum field theory interaction.[‡] The physical analogy between singular field theoretic interactions and singular potential scattering of course breaks down at short distances, since no probabilistic interpretation is available for the field theory matrix elements in virtue of creation and annihilation processes during the interaction. Nonetheless it was thought that a certain formal analogy could be seen if the field theory is supplied with Euclidean spacetime concepts at the expense of sacrificing the interpretation of the interaction in terms of particle exchange. However, there is no need to have only field

---

[*]Generally a potential more singular than the centrifugal term in the $(3 + 1)$-dimensional Schrödinger equation is described as singular. The centrifugal potential $\propto r^{-2}$ is generally considered as exceptional and is treated in detail in wellknown texts on quantum mechanics, also in P. M. Morse and H. Feshbach [198].

[†]K. M. Case [45].

[‡]For a review from this perspective with numerous references see H. H. Aly, W. Güttinger and H. J. W. Müller–Kirsten [9].

theories in mind. Singular potentials arise in various other contexts. Recently, for instance, the attractive singular potential $1/r^4$ was found to arise in the study of fluctuations about a "brane" (the $D3$ brane; roughly speaking a brane is the higher dimensional equivalent of a membrane visualized in two dimensions from which the $D3$ brane derives its name, D referring to Dirichlet boundary conditions) in 10-dimensional string theory.[§] There one could visualize this scattering off the spherically symmetric potential as a spacetime curvature effect or — with black hole event horizon zero — as that of a potential barrier surrounding the horizon (shrunk to zero at the origin). In fact, this particular singular potential plays an exceptional role in view of its relation to the Mathieu equation. It is therefore natural that we study this case here in detail. We shall see that not only can the radial Schrödinger equation be related to the *modified* or *associated Mathieu equation* (i.e. that with the hyperbolic cosine replacing the trigonometric cosine), but also that the *S*-matrix can be calculated explicitly in both the weak and strong coupling domains. This property singles this case out from many others and attributes it the role of one of very few explicitly solvable cases, which are always worth studying in view of the insight they provide into a typical case and the didactic value they possess for this reason. This chapter therefore gives the first complete solution of a Schrödinger equation with a highly singular potential. In view of the unavoidable use of Mathieu functions expanded in series of Special Functions, our presentation below is deliberately made elementary and detailed so that the reader does not shun away from it. In the concluding section we refer to additional related literature and applications, some also permitting further research.

19.2 The Potential $1/r^4$ — Case of Small h^2

19.2.1 Preliminary considerations

We consider in three space dimensions first the *repulsive potential*

$$V(r) = \frac{g^2}{r^4}. \tag{19.1}$$

The radial Schrödinger equation with this potential may then be written

$$y'' + \left[k^2 - \frac{l(l+1)}{r^2} - \frac{g^2}{r^4}\right] y = 0, \tag{19.2}$$

[§]C. G. Callan and J. Maldacena [41]; S. S. Gubser and A. Hashimoto [124]; R. Manvelyan, H. J. W. Müller–Kirsten, J.–Q. Liang and Yunbo Zhang [189]; D. K. Park, S. N. Tamaryan, H. J. W. Müller–Kirsten and Jian–zu Zhang [226].

where $E = k^2, m_0 = 1/2$ and $\hbar = c = 1$. The following substitutions are advantageous:

$$y = r^{1/2}\phi, \quad r = \gamma e^z, \quad \gamma = \frac{ig}{h}, \quad h^2 = ikg, \quad h = e^{i\pi/4}\sqrt{kg}, \qquad (19.3a)$$

and so

$$\frac{h}{g} = i\frac{k}{h} \quad \text{and} \quad e^z = \frac{h}{ig}r = \frac{kr}{h} = e^{-i\pi/4}\sqrt{\frac{k}{g}}r. \qquad (19.3b)$$

In the case of the *attractive potential* (as in the string theory context referred to above, and as we have in mind in Sec. 19.3 below) g^2 is negative and hence h^2 is real for $E > 0$. The radial Schrödinger equation then assumes the form

$$\frac{d^2\phi}{dz^2} + \left[2h^2\cosh 2z - \left(l + \frac{1}{2}\right)^2\right]\phi = 0. \qquad (19.4)$$

In the literature this equation is known as the *modified Mathieu equation* or *associated Mathieu equation*. We observe that with the replacements

$$z \to iz, \quad \left(l + \frac{1}{2}\right)^2 \to \lambda,$$

the equation converts into the periodic Mathieu equation of Chapter 17. Thus we can expect to find solutions very similar to those of the periodic case (with — for instance — cos replaced by cosh) and with the parameter ν given by the expansions we obtained previously. It may be noticed that since l is an integer, $\sqrt{\lambda}$ in the correspondence is nonintegral. This has advantages compared with higher dimensional cases,[¶] where $\sqrt{\lambda}$ can be an integer and thus leads to singularities in the expansion of ν (see e.g. Eq. (19.28) below).

In the following we study the solutions of Eq. (19.4) first for small values of h^2 and then for large values of h^2. It is clear that we draw analogies to our earlier considerations of the periodic potential. Our ultimate aim is the derivation of the S-matrix for scattering off the singular $1/r^4$ potential. We begin with the derivation of various types of weak-coupling or small-h^2 solutions which we construct again perturbatively, using the method of Dingle and Müller of Sec. 8.7 that we employed also in previous chapters. Subsequently we derive the same S-matrix from the consideration of large-h^2 expansions and calculate the *absorptivity in the case of the attractive potential*. This is rarely possible and therefore this case deserves particular attention.

[¶] R. Manvelyan, H. J. W. Müller-Kirsten, J.-Q. Liang and Yunbo Zhang [189].

19.2.2 Small h^2 solutions in terms of Bessel functions

We now develop a perturbative procedure[||] for solving the associated Mathieu equation in the domain around $h^2 = 0$. First we make the additional substitution

$$w = 2h \cosh z, \tag{19.5}$$

so that Eq. (19.4) becomes

$$\frac{d^2\phi}{dw^2} + \frac{1}{w}\frac{d\phi}{dw} + \left\{1 - \frac{(l+\frac{1}{2})^2}{w^2}\right\}\phi = \frac{2h^2}{w^2}\left\{\phi + 2\frac{d^2\phi}{dw^2}\right\}. \tag{19.6}$$

Next we define a parameter ν by the relation

$$\nu^2 = \left(l + \frac{1}{2}\right)^2 - 2\triangle h^2. \tag{19.7}$$

We shall see that this parameter is given by the same expansion as in the case of the periodic equation. Proceeding along the lines of the perturbation method of Sec. 8.7, Eq. (19.6) becomes to zeroth order $\phi^{(0)} = \phi_\nu$ of ϕ in h^2

$$D_\nu\phi_\nu = 0, \quad \text{where} \quad D_\nu := \frac{d^2}{dw^2} + \frac{1}{w}\frac{d}{dw} + \left\{1 - \frac{\nu^2}{w^2}\right\}. \tag{19.8}$$

This equation is wellknown as *Bessel's equation* or as the equation of cylindrical functions. The solutions $Z_\nu(w)$ are written $J_\nu(w), N_\nu(w), H_\nu^{(1)}(w)$, and $H_\nu^{(2)}(w)$, where these are the Bessel function of the first kind, Neumann or Bessel function of the second kind and Hankel functions of the first and second kinds respectively.

The zeroth-order approximation

$$\phi^{(0)} = Z_\nu(w) \tag{19.9}$$

leaves uncompensated on the right hand side of Eq. (19.6) terms amounting to

$$R_\nu^{(0)} = 2h^2\left[\frac{1}{w^2}Z_\nu + \frac{2}{w^2}\frac{d^2Z_\nu}{dw^2} + \frac{\triangle}{w^2}Z_\nu\right]. \tag{19.10}$$

Using the recurrence relations of cylindrical functions, i.e.

$$\frac{\nu}{w}Z_\nu = \frac{1}{2}(Z_{\nu-1} + Z_{\nu+1}), \qquad \frac{dZ_\nu}{dw} = -\frac{\nu}{w}Z_\nu + Z_{\nu-1} = \frac{1}{2}(Z_{\nu-1} - Z_{\nu+1}),$$

$$\therefore \frac{d^2Z_\nu}{dw^2} = \frac{1}{4}[Z_{\nu-2} - 2Z_\nu + Z_{\nu+2}], \tag{19.11}$$

[||]We follow here largely H. H. Aly, H. J. W. Müller–Kirsten and N. Vahedi-Faridi [10].

we can rewrite the expression (19.10) as a linear combination of various Z_ν. However, it is more convenient to use these relations in order to rewrite Eq. (19.10) in terms of functions G_ν defined by

$$G_{\nu+\alpha} = \frac{1}{w^2} Z_{\nu+\alpha}. \tag{19.12}$$

The expression (19.10) is now particularly simple:

$$R_\nu^{(0)} = h^2[(\nu, \nu - 2)G_{\nu-2} + (\nu, \nu)G_\nu + (\nu, \nu + 2)G_{\nu+2}], \tag{19.13}$$

where

$$(\nu, \nu \pm 2) = 1, \quad (\nu, \nu) = 2\triangle. \tag{19.14}$$

We observe that

$$D_\nu Z_\nu = 0, \quad D_{\nu+\alpha} Z_{\nu+\alpha} = 0, \tag{19.15}$$

but

$$D_{\nu+\alpha} = D_\nu - \left[\frac{\alpha(2\nu + \alpha)}{w^2}\right], \tag{19.16}$$

so that

$$D_\nu Z_{\nu+\alpha} = \frac{\alpha(2\nu + \alpha)}{w^2} Z_{\nu+\alpha} = \alpha(2\nu + \alpha)G_{\nu+\alpha}. \tag{19.17}$$

Thus a term $\mu G_{\nu+\alpha}$ on the right hand side of Eq. (19.6), and this means in (19.13), can be cancelled out by adding to $\phi^{(0)}$ the new contribution $\mu Z_{\nu+\alpha}/\alpha(2\nu + \alpha)$ except, of course, when α or $2\nu + \alpha = 0$. We assume in the following that $2\nu + \alpha \neq 0$ (the case of 0 has to be treated separately). The terms (19.13) therefore lead to the first order contribution

$$\phi^{(1)} = h^2[(\nu, \nu - 2)^* Z_{\nu-2} + (\nu, \nu + 2)^* Z_{\nu+2}], \tag{19.18}$$

where the starred coefficients are defined by

$$(\nu, \nu + \alpha)^* = \frac{(\nu, \nu + \alpha)}{\alpha(2\nu + \alpha)}. \tag{19.19}$$

Now $\phi^{(0)} = Z_\nu$ left uncompensated $R_\nu^{(0)}$; therefore $\phi^{(1)}$ leaves uncompensated

$$R_\nu^{(1)} = h^2[(\nu, \nu - 2)^* R_{\nu-2}^{(0)} + (\nu, \nu + 2)^* R_{\nu+2}^{(0)}]. \tag{19.20}$$

The next contribution to $\phi^{(0)} + \phi^{(1)}$ therefore becomes

$$\begin{aligned}
\phi^{(2)} = {} & h^4[(\nu, \nu - 2)^*(\nu - 2, \nu - 4)^* Z_{\nu-4} \\
& + (\nu, \nu - 2)^*(\nu - 2, \nu - 2)^* Z_{\nu-2} + (\nu, \nu + 2)^*(\nu + 2, \nu + 2)^* Z_{\nu+2} \\
& + (\nu, \nu + 2)^*(\nu + 2, \nu + 4)^* Z_{\nu+4}].
\end{aligned} \tag{19.21}$$

Proceeding in this manner we obtain the solution

$$\phi = \phi^{(0)} + \phi^{(1)} + \phi^{(2)} + \cdots$$

$$= Z_\nu + \sum_{i=1}^{\infty} h^{2i} \sum_{j=-i, j\neq 0}^{i} p_{2i}(2j) Z_{\nu+2j}, \qquad (19.22)$$

where

$$
\begin{aligned}
p_2(\pm 2) &= (\nu, \nu \pm 2)^*, \\
p_4(\pm 4) &= (\nu, \nu \pm 2)^* (\nu \pm 2, \nu \pm 4)^*, \\
p_4(\pm 2) &= (\nu, \nu \pm 2)^* (\nu \pm 2, \nu \pm 2)^*, \text{ etc..}
\end{aligned} \qquad (19.23)
$$

These coefficients may also be obtained from the recurrence relation

$$
\begin{aligned}
p_{2i}(2j) &= p_{2i-2}(2j-2)(\nu+2j-2, \nu+2j) + p_{2i-2}(2j)(\nu+2j, \nu+2j) \\
&\quad p_{2i-2}(2j+2)(\nu+2j+2, \nu+2j),
\end{aligned} \qquad (19.24)
$$

subject to the boundary conditions

$$p_{2i}(2j) = 0 \text{ for } |j| > i, \ p_0(0) = 1, \text{ and } p_0(2j \neq 0) = 0, \ p_{2i\neq 0}(0) = 0. \qquad (19.25)$$

Finally we have to consider the terms in G_ν which were left unaccounted for in $R_\nu^{(0)}, R_\nu^{(1)}, \ldots$. Adding these terms and setting the coefficient of G_ν equal to zero, we obtain

$$
\begin{aligned}
0 &= h^2(\nu, \nu) + h^4[(\nu, \nu-2)^*(\nu-2, \nu) + (\nu, \nu+2)^*(\nu+2, \nu)] \\
&\quad + h^6[(\nu, \nu-2)^*(\nu-2, \nu-2)^*(\nu-2, \nu) \\
&\quad + (\nu, \nu+2)^*(\nu+2, \nu+2)^*(\nu+2, \nu)] + \cdots.
\end{aligned} \qquad (19.26)
$$

Evaluating the first few terms, we obtain the expansion

$$
\begin{aligned}
\left(1 + \frac{1}{2}\right)^2 \equiv \lambda &= \nu^2 + \frac{h^4}{2(\nu^2-1)} + \frac{(5\nu^2+7)h^8}{32(\nu^2-1)^3(\nu^2-4)} \\
&\quad + \frac{(9\nu^4 + 58\nu^2 + 29)h^{12}}{64(\nu^2-1)(\nu^2-4)(\nu^2-9)} + O(h^{16}).
\end{aligned} \qquad (19.27)
$$

This expansion is seen to be familiar from the theory of periodic Mathieu functions where $(l+1/2)^2$ represents the eigenvalue. Reversing the expansion and setting $\alpha = (l+1/2)^2$ (not to be confused with α e.g. in Eq. (19.17)), we obtain

$$
\begin{aligned}
\nu^2 &= \alpha - \frac{h^4}{2(\alpha-1)} - \frac{(13\alpha-25)h^8}{32(\alpha-1)^3(\alpha-4)} \\
&\quad - \frac{(45\alpha^3 - 455\alpha^2 + 1291\alpha - 1169)h^{12}}{64(\alpha-1)^5(\alpha-4)^4(\alpha-9)} + O(h^{16}).
\end{aligned} \qquad (19.28)
$$

We note that this is an expansion in ascending powers of h^4, and is therefore real for both cases $g^2 > 0$ and $g^2 < 0$. Thus ν is real for small values of $|h^2|$. The solutions ϕ of the modified Mathieu equation are now completely determined, apart from a normalization factor which we have chosen (so far) such that the coefficient of Z_ν in ϕ is 1. We are still left with the question of what will happen if $2\nu + \alpha = 0$ or $\nu = \pm 1, \pm 2, \ldots$. We return to this question later but mention here that this problem has already been dealt with in Chapter 17.

19.2.3 Small h^2 solutions in terms of hyperbolic functions

For later purposes we require yet another type of solutions. Substituting (19.7) into Eq. (19.4), we can rewrite the latter as

$$\frac{d^2\phi}{dz^2} - \nu^2\phi = 2h^2(\triangle - \cosh 2z)\phi. \tag{19.29}$$

Thus to $O(0)$ in h we have

$$\phi^{(0)} = \phi_\nu = \cosh \nu z, \quad \sinh \nu z \quad \text{or} \quad e^{\pm \nu z}, \tag{19.30}$$

so that

$$D_\nu\phi_\nu = 0, \quad D_\nu = \frac{d^2}{dz^2} - \nu^2. \tag{19.31}$$

It follows that

$$D_{\nu+2n}\phi_{\nu+2n} = 0, \quad D_{\nu+2n} = D_\nu - 4n(\nu + n), \tag{19.32}$$

so that

$$D_\nu\phi_{\nu+2n} = 4n(\nu + n)\phi_{\nu+2n}. \tag{19.33}$$

Since

$$
\begin{aligned}
2\cosh 2z \cosh \nu z &= \cosh(\nu + 2)z + \cosh(\nu - 2)z, \\
2\cosh 2z \sinh \nu z &= \sinh(\nu + 2)z + \sinh(\nu - 2)z, \\
2\cosh 2z\, e^{\pm \nu z} &= e^{\pm(\nu+2)z} + e^{\pm(\nu-2)z},
\end{aligned}
$$

we may say that the first approximation $\phi^{(0)}$ leaves uncompensated terms amounting to

$$
\begin{aligned}
R_\nu^{(0)} &= 2h^2(\triangle - \cosh 2z)\phi_\nu \\
&= 2h^2\triangle\phi_\nu - h^2[\phi_{\nu+2} + \phi_{\nu-2}] \\
&= h^2[(\nu, \nu - 2)\phi_{\nu-2} + (\nu, \nu)\phi_\nu + (\nu, \nu + 2)\phi_{\nu+2}], \tag{19.34}
\end{aligned}
$$

where

$$(\nu,\nu) = 2\triangle, \quad (\nu,\nu \pm 2) = -1. \tag{19.35}$$

The form of $R_\nu^{(0)}$ is seen to be almost identical with that of the corresponding expression for solutions in terms of cylindrical functions. In fact we could have obtained the same $R_\nu^{(0)}$ by starting with the modified Mathieu equation for h^2 replaced by $-h^2$. In order to avoid confusion arising from the use of different equations, we prefer to adhere to one equation with different solutions. The use of the symbols $(\nu,\nu \pm 2)$ etc. in the present context should not be confused with the same symbols having a different meaning in the case of solutions in terms of cylindrical functions since it is generally clear which type of solutions and hence coefficients is being considered. Defining

$$(\nu,\nu + \alpha)^* = \frac{(\nu,\nu + \alpha)}{\alpha(2\nu + \alpha)}, \tag{19.36}$$

we now have the solution

$$\varphi(z,h) = \phi_\nu + \sum_{i=1}^\infty h^{2i} \sum_{j=-i, j \neq 0}^i \bar{p}_{2i}(2j)\phi_{\nu+2j}, \tag{19.37}$$

where

$$\bar{p}_{2i}(\pm 2) = (\nu,\nu \pm 2)^*, \text{ etc.}$$

For rigorous convergence and validity discussions of any of these solutions — our's here differ only in the method of derivation with the perturbation method of Sec. 8.7 — we refer to Meixner and Schäfke [193].

19.2.4   Notation and properties of solutions

We now introduce standard notation as in established literature.* The solutions of the modified Mathieu equation which we are considering here are written with a first capital letter. In particular the following notation is used for the solutions obtained above which we characterize here by their first terms:

$$
\begin{aligned}
\phi \quad &\text{with} \quad \phi_\nu = \cosh \nu z \to Ce_\nu(z,h), \\
\phi \quad &\text{with} \quad \phi_\nu = \sinh \nu z \to Se_\nu(z,h), \\
\phi \quad &\text{with} \quad \phi_\nu = \exp(\nu z) \to Me_\nu(z,h).
\end{aligned} \tag{19.38}
$$

These solutions correspond in the periodic case respectively to the solutions ce_ν, se_ν and me_ν. The solutions in terms of cylindrical functions are written

---

*J. Meixner and F. W. Schäfke [193], F. M. Arscott [11].

similarly:

$$
\begin{array}{llll}
\phi & \text{with} & \phi_\nu = J_\nu(2h\cosh z) & \to M_\nu^{(1)}(z,h), \\
\phi & \text{with} & \phi_\nu = N_\nu(2h\cosh z) & \to M_\nu^{(2)}(z,h), \\
\phi & \text{with} & \phi_\nu = H_\nu^{(1)}(2h\cosh z) & \to M_\nu^{(3)}(z,h), \\
\phi & \text{with} & \phi_\nu = H_\nu^{(2)}(2h\cosh z) & \to M_\nu^{(4)}(z,h).
\end{array}
\tag{19.39}
$$

Writing two solutions out explicitly we have for example — apart from an overall normalization constant, which, as we remarked earlier, introduces the dominant order function into higher order contributions[†] —

$$
\begin{aligned}
Me_\nu(z,h) &= \exp(\nu z) + \sum_{i=1}^\infty h^{2i} \sum_{j=-i,j\neq 0}^i p_{2i}(2j)(-1)^j \exp[(\nu+2j)z] \\
&= Ce_\nu(z,h) \pm Se_\nu(z,h)
\end{aligned}
\tag{19.40}
$$

and

$$
M_\nu^{(1)}(z,h) = J_\nu(2h\cosh z) + \sum_{i=1}^\infty h^{2i} \sum_{j=-i,j\neq 0}^i p_{2i}(2j)J_{\nu+2j}(2h\cosh z). \tag{19.41}
$$

We see immediately that

$$
Me_\nu(z+n\pi i, h) = \exp(\nu n\pi i)Me_\nu(z,h). \tag{19.42}
$$

Also from the expansion of $J_\nu(z)$ in rising powers of z, we obtain[‡]

$$
J_\nu(2h\cosh(z+in\pi)) = J_\nu(2h\cosh z \exp(in\pi)) = \exp(in\nu\pi)J_\nu(2h\cosh z),
\tag{19.43}
$$

and hence similarly

$$
M_\nu^{(1)}(z+in\pi, h) = \exp(in\nu\pi)M_\nu^{(1)}(z,h). \tag{19.44}
$$

The solutions $Me_\nu(z,h), M_\nu^{(1)}(z,h)$ are therefore proportional to each other as a comparison of Eqs. (19.42), (19.44) implies, i.e.

$$
Me_\nu(z,h) = \alpha_\nu(h)M_\nu^{(1)}(z,h), \tag{19.45}
$$

[†]Below we frequently write the normalized solution as in J. Meixner and F. W. Schäfke [193], i.e. as

$$
Me_\nu(z,h^2) = \sum_{-\infty}^\infty c_{2r}^\nu(h^2)e^{(\nu+2r)z}.
$$

As emphasized in Chapter 8, this *normalized* solution possesses contributions $\exp(\nu z)$ in higher order terms, whereas the — as yet — unnormalized solution (19.40) does not.

[‡]See e.g. W. Magnus and F. Oberhettinger [181], p. 16. For the relation of Hankel functions used below, see this reference p. 17.

where clearly

$$\alpha_\nu(h) = \frac{Me_\nu(0,h)}{M_\nu^{(1)}(0,h)}. \tag{19.46}$$

Using further properties like (19.43), i.e.

$$H_{-\nu}^{(1)}(z) = e^{i\pi\nu} H_\nu^{(1)}(z), \quad H_{-\nu}^{(2)}(z) = e^{-i\pi\nu} H_\nu^{(2)}(z),$$

we have the following for a change from ν to $-\nu$:

$$M_{-\nu}^{(3,4)} = \exp(\pm i\nu\pi) M_\nu^{(3,4)}, \quad M_\nu^{(3,4)} = M_\nu^{(1)} \pm i M_\nu^{(2)}. \tag{19.47}$$

With this equation we obtain for nonintegral values of ν:[§]

$$\pm i \sin\nu\pi \, M_\nu^{(3,4)}(z,h) = M_{-\nu}^{(1)}(z,h) - \exp(\mp i\nu\pi) M_\nu^{(1)}(z,h). \tag{19.48}$$

The series expansions of the associated Mathieu functions $M_\nu^{(j)}(z,h)$ can be shown (cf. Meixner and Schäfke [193], p.178) to be convergent for $|\cosh z| \geq 1$ but uniformly convergent only when $|\cosh z| > 1$ for otherwise complex values of z. The functions $Me_{\pm\nu}(z,h)$ on the other hand can be shown (cf. Meixner and Schäfke [193], p. 130) to converge uniformly for all finite complex values of z. These points are important in our derivation of the S-matrix below.

For later essential requirements (i.e. the explicit evaluation of the S-matrix element) we elaborate here a little on the computation of *coefficients of normalized Mathieu functions*. As mentioned earlier, we use the notation of Meixner and Schäfke [193] and so write the expansion of the function $Me_\nu(z,h^2)$:

$$Me_\nu(z,h^2) = \sum_{r=-\infty}^{\infty} c_{2r}^\nu(h^2) e^{(\nu+2r)z}. \tag{19.49}$$

Inserting this expansion into the modified Mathieu equation (19.4) we obtain

$$\sum_{r=-\infty}^{\infty} [(\nu+2r)^2 c_{2r}^\nu - \lambda c_{2r}^\nu + h^2(c_{2r-2}^\nu + c_{2r+2}^\nu)] e^{(\nu+2r)z} = 0.$$

Equating the coefficients to zero we obtain

$$[-(\nu+2r)^2 + \lambda]c_{2r}^\nu = h^2(c_{2r+2}^\nu + c_{2r-2}^\nu), \tag{19.50}$$

[§]E.g. the first relation is obtained as follows:

$$2M_{\pm\nu}^{(1)} = M_{\pm\nu}^{(3)} + M_{\pm\nu}^{(4)}, \text{ i.e. } 2M_{-\nu}^{(1)} = e^{i\nu\pi} M_\nu^{(3)} + e^{-i\nu\pi} M_\nu^{(4)},$$

$$2M_{-\nu}^{(1)} = e^{i\nu\pi} M_\nu^{(3)} + e^{-i\nu\pi}[2M_\nu^{(1)} - M_\nu^{(3)}] = 2M_\nu^{(1)} e^{-i\nu\pi} + 2i\sin\nu\pi M_\nu^{(3)}.$$

which we can rewrite as

$$h^2 \frac{c_{2r+2}^\nu}{c_{2r}^\nu} = [\lambda - (\nu + 2r)^2] - h^2 \frac{c_{2r-2}^\nu}{c_{2r}^\nu}. \qquad (19.51)$$

For ease of reading we give the steps in rewriting this. Thus now

$$\frac{c_{2r+2}^\nu}{c_{2r}^\nu} = h^{-2}[\lambda - (\nu + 2r)^2] - \frac{c_{2r-2}^\nu}{c_{2r}^\nu}$$

$$= h^{-2}[\lambda - (\nu + 2r)^2] - \frac{1}{\left(\dfrac{c_{2r}^\nu}{c_{2r-2}^\nu}\right)},$$

or

$$\frac{1}{\left(\dfrac{c_{2r}^\nu}{c_{2r-2}^\nu}\right)} = h^{-2}[\lambda - (\nu + 2r)^2] - \frac{c_{2r+2}^\nu}{c_{2r}^\nu},$$

or

$$\frac{c_{2r}^\nu}{c_{2r-2}^\nu} = \frac{1}{h^{-2}[\lambda - (\nu + 2r)^2] - \dfrac{c_{2r+2}^\nu}{c_{2r}^\nu}}$$

$$= \frac{1}{h^{-2}[\lambda - (\nu + 2r)^2] - \dfrac{1}{h^{-2}[\lambda - (\nu + 2r + 2)^2] - \dfrac{c_{2r+4}^\nu}{c_{2r+2}^\nu}}}$$

$$(19.52)$$

in agreement with Meixner and Schäfke [193], p. 117.

Alternatively taking the inverse of Eq. (19.51) we have

$$\frac{c_{2r}^\nu}{c_{2r+2}^\nu} = \frac{1}{h^{-2}[\lambda - (\nu + 2r)^2] - \dfrac{c_{2r-2}^\nu}{c_{2r}^\nu}}, \qquad (19.53)$$

or

$$\frac{c_{2r-2}^\nu}{c_{2r}^\nu} = \frac{1}{h^{-2}[\lambda - (\nu + 2r - 2)^2] - \dfrac{c_{2r-4}^\nu}{c_{2r-2}^\nu}}. \qquad (19.54)$$

This continued fraction equation can again be used to obtain the explicit expressions of coefficients of normalized modified Mathieu functions.* For

---

*Actually, see J. Meixner and F. W. Schäfke [193], Eq. (39), p. 122, $c_0^\nu = c_0^{-\nu} = 1$.

nonintegral values of ν examples obtained in this way are (see Example 19.1 for the evaluation of a typical case):

$$\frac{c_2^\nu}{c_0^\nu} = -\frac{h^2}{4(\nu+1)} - \frac{\nu^2+4\nu+7}{128(\nu+1)^2(\nu+2)(\nu-1)}h^6 + O(h^{10}),$$

$$\frac{c_4^\nu}{c_0^\nu} = \frac{h^4}{32(\nu+1)(\nu+2)} + \frac{\nu^2+5\nu+10}{768(\nu+1)^3(\nu-1)(\nu+2)(\nu+3)}h^8 + O(h^{12}),$$

$$\frac{c_6^\nu}{c_0^\nu} = -\frac{h^6}{384(\nu+1)(\nu+2)(\nu+3)} - O(h^{10}),$$

$$\cdots$$

$$\frac{c_{2r}^\nu}{c_0^\nu} = (-1)^r \frac{h^{2r}(\nu!)}{2^{2r}r!(\nu+r)!} + O(h^{2r+4}). \tag{19.55}$$

Example 19.1: Evaluation of a coefficient
Evaluate explicitly the first few terms of the coefficient c_2^ν/c_0^ν given in Eq. (19.55).

Solution: We put $r=1$ in Eq. (19.52) and obtain

$$\frac{c_2^\nu}{c_0^\nu} = \frac{1}{h^{-2}[\lambda-(\nu+2)^2] - \frac{1}{h^{-2}[\lambda-(\nu+4)^2]-\cdots}}.$$

Here we insert for λ the expansion (19.27) and truncate the continued fraction after the second step. Then

$$\frac{c_2^\nu}{c_0^\nu} = \frac{1}{h^{-2}[\nu^2+\frac{h^4}{2(\nu^2-1)}+\cdots-(\nu+2)^2] - \frac{1}{h^{-2}[\nu^2+\cdots-(\nu+4)^2]}}$$

$$\simeq \frac{h^2}{[-4(\nu+1)+\frac{h^4}{2(\nu^2-1)}]+\frac{h^4}{8(\nu+2)}} = -\frac{h^2}{4(\nu+1)-\frac{h^4(\nu^2+4\nu+7)}{8(\nu+2)(\nu^2-1)}}$$

$$= -\frac{h^2}{4(\nu+1)}\left[\frac{1}{1-\frac{h^4(\nu^2+4\nu+7)}{32(\nu+2)(\nu-1)(\nu+1)^2}}\right]$$

$$\simeq -\frac{h^2}{4(\nu+1)} - \frac{\nu^2+4\nu+7}{128(\nu+1)^2(\nu+2)(\nu-1)}h^6 + O(h^{10}),$$

which is the result expected.

19.2.5 Derivation of the S-matrix

Our next step is the derivation of the explicit form of the S-matrix for scattering off the potential $1/r^4$ since this is possibly the only singular potential permitting such a derivation in terms of known functions.[†] For this purpose

---

[†]We do this in the manner of the original derivation of R. M. Spector [257], but follow our earlier reference H. H. Aly, H. J. W. Müller–Kirsten and N. Vahedi-Faridi [10] (this paper contains several misprints which we correct here), and also R. Manvelyan, H. J. W. Müller–Kirsten, J.–Q. Liang and Yunbo Zhang [189].

we require first the regular solution y_{reg} of the radial Schrödinger equation at the origin, and then the continuation of this to infinity.

We obtain the regular solution by choosing $Z_\nu(w) = H_\nu^{(1)}(w)$ for $\Re(z) < 0$. Then

$$
\begin{aligned}
y_{\text{reg}} &= r^{1/2} M_\nu^{(3)}(z, h) \\
&= r^{1/2}\left[H_\nu^{(1)}(w) + \sum_{i=1}^{\infty} h^{2i} \sum_{j=-i, j\neq 0}^{i} p_{2i}(2j) H_{\nu+2j}^{(1)}(w) \right],
\end{aligned}
\quad (19.56)
$$

where by Eqs. (19.3a) and (19.5)

$$
w = 2h \cosh z = \left(kr + \frac{ig}{r} \right). \tag{19.57}
$$

Thus $r \to 0$ implies $|w| \to \infty$. The asymptotic behaviour of the Hankel functions $H_\nu^{(1,2)}(w)$ for $|w| \gg |\nu|, |w| \gg 1$ and $-\pi < \arg w < \pi$ is known to be given by[‡]

$$
H_\nu^{(1,2)}(w) = \left(\frac{2}{\pi w} \right)^{1/2} \exp\left[\pm i\left(w - \frac{\nu\pi}{2} - \frac{\pi}{4} \right) \right]\left[1 + O\left(\frac{1}{w} \right) \right]. \tag{19.58}
$$

The behaviour of y_{reg} near $r \approx 0$ is therefore given by

$$
\begin{aligned}
y_{\text{reg}} &\approx r\left(\frac{2}{\pi ig} \right)^{1/2} \exp\left(-\frac{g}{r} \right)\left\{ \exp\left[-i\left(\nu + \frac{1}{2} \right)\frac{\pi}{2} \right] \right. \\
&\left. + \sum_{i=1}^{\infty} h^{2i} \sum_{j=-i, j\neq 0}^{i} p_{2i}(2j) \exp\left[-i\left(\nu + 2j + \frac{1}{2} \right)\frac{\pi}{2} \right] \right\},
\end{aligned}
\quad (19.59)
$$

which tends to zero with r. In the case of the repulsive singular potential here, this wave function near the origin is the wave transmitted into this region (as distinct from the reflected and ingoing waves). The time-dependent wave function with this asymptotic behaviour is proportional to (here with $\omega = k$)

$$
e^{-i\omega t - g/r + i\pi/4}.
$$

Fixing the wave-front by setting $\varphi = -\omega t + ig/r + \pi/4 = \text{const.}$, and considering the propagation of this wave-front, we have

$$
r = \frac{ig}{\varphi + \omega t - \pi/4},
$$

[‡]W. Magnus and F. Oberhettinger [181], p. 22.

so that when $t \to \infty : r \to 0$. This means that the origin of coordinates acts as a *sink*.

In a similar manner we can define solutions $y_\pm$ by setting for $\Re(z) > 0$:

$$
\begin{aligned}
y_\pm &= r^{1/2} M_\nu^{(3,4)}(z, h) \\
&= r^{1/2} \left[H_\nu^{(1,2)}(w) + \sum_{i=1}^{\infty} h^{2i} \sum_{j=-i, j\neq 0}^{j} p_{2i}(2j) H_{\nu+2j}^{(1,2)}(w) \right]. \quad (19.60)
\end{aligned}
$$

Using the above asymptotic expressions for the Hankel functions, these solutions are seen to have the desired asymptotic behaviour for $r \to \infty$:

$$
\begin{aligned}
y_\pm &\approx \left(\frac{2}{\pi k} \right)^{1/2} e^{\pm ikr} \exp\left[\mp \frac{i\pi}{2}\left(\nu + \frac{1}{2} \right) \right] \\
&\quad \times \left[1 + \sum_{i=1}^{\infty} h^{2i} \sum_{j=-i, j\neq 0}^{i} p_{2i}(2j) \exp(\mp i\pi j) \right]. \quad (19.61)
\end{aligned}
$$

In fact, we can derive y_- from y_+ since one can show from the circuit relations of Hankel functions[§] that

$$
M_\nu^{(3)}(z + i\pi, h) = -\exp(-i\pi\nu) M_\nu^{(4)}(z, h).
$$

We now require the continuation of the regular solution y_{reg} to solutions behaving like $y_\pm$ at $r \to \infty$. From the relation $r = (ig/h)e^z$ we see that $r = 0$ corresponds to $\Re(z) \to -\infty$ and $r \to \infty$ to $\Re(z) \to \infty$. We require therefore the continuation of $M_\nu^{(3)}(z, h)$ through the entire range of $\Re(z)$.

The series $M_\nu^{(j)}(z, h)$ can be shown (cf. Meixner and Schäfke [193], p. 178) to be convergent for $|\cosh z| \geq 1$, but uniformly convergent only when $|\cosh z| > 1$ for otherwise complex values of z. Now from Eqs. (19.3a) and (19.3b)

$$
z = \ln\left(\frac{r}{r_0} \right) - i\frac{\pi}{4}, \quad r_0 = \sqrt{\frac{g}{k}}. \quad (19.62)
$$

Also with Eq. (19.57) the condition $|\cosh z| > 1$ implies

$$
|\cosh z| = \left| \frac{kr + ig/r}{2h} \right| > 1,
$$

and the square of this expression implies

$$
k^2 r^2 + \frac{g^2}{r^2} > 4hh^* = 4kg, \quad k^2 r^4 - 4kgr^2 + g^2 > 0 \quad \text{or} \quad r^4 - 4r_0^2 r^2 + r_0^4 > 0.
$$

[§] $H_\nu^{(1)}(e^{i\pi} z) = -H_{-\nu}^{(2)}(z) = -e^{-i\pi\nu} H_\nu^{(2)}(z)$. See W. Magnus and F. Oberhettinger [181], p. 17.

Hence

$$(r^2 - r_+^2)(r^2 - r_-^2) > 0 \quad \text{with} \quad r_\pm^2 = r_0^2(2 \pm \sqrt{3}).$$

This condition, i.e. $(r - r_+)(r + r_+)(r - r_-)(r + r_-) > 0$, is satisfied only for

$$r < r_- < r_+ \quad \text{or} \quad r_- < r_+ < r. \tag{19.63}$$

Thus there is a gap between the two regions of validity — i.e. the domain $r_- < r < r_+$ — as illustrated in Fig. 19.1, which has to be bridged by using another set of solutions. A suitable set is the pair of fundamental solutions $Me_{\pm\nu}$ defined by (19.38) or (19.49). These solutions converge uniformly for all finite complex values of z (cf. Meixner and Schäfke [193], p. 130).

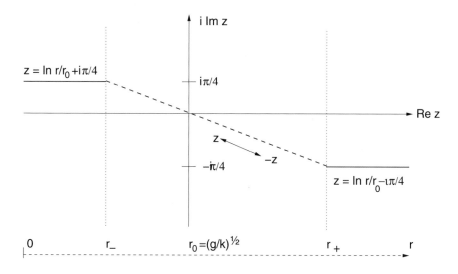

Fig. 19.1 The domains of solutions.

Originally we chose the sign of $i\pi/4$ as in Eqs. (19.3a), (19.3b) and (19.62). The one further point to observe is that the variable $w = 2h\cosh z = (kr + ig/r)$ is even under the interchange $z \to -z$. This interchange in the solution $M_\nu^{(3)}$ interchanges y_{reg} with y_+. Since the real part of z changes sign at $r = r_0$, and in order to maintain this symmetry, we have to assign different signs to the imaginary part of z in the two distant regions. Thus we choose

$$z = \ln \frac{r}{r_0} + i\frac{\pi}{4} \quad \text{for} \quad 0 < r < r_0,$$

and

$$z = \ln \frac{r}{r_0} - i\frac{\pi}{4} \quad \text{for} \quad r_0 < r < \infty.$$

Then starting from the region $r \sim 0$, we express in the domain $r_- < r < r_+$ the regular solution as a linear combination of solutions $Me_{\pm\nu}$ with coefficients α and β, and we determine these coefficients from this equality and the additional equality obtained from the derivatives. Thus we set

$$r^{1/2} M_\nu^{(3)}(r) = r^{1/2}[\alpha Me_\nu(r) + \beta Me_{-\nu}(r)],$$
$$\frac{d}{dr}[r^{1/2} M_\nu^{(3)}(r)] = \alpha \frac{d}{dr}[r^{1/2} Me_\nu(r)] + \beta \frac{d}{dr}[r^{1/2} Me_{-\nu}(r)]. \quad (19.64)$$

The right hand side of the first equation now represents the regular solution in the domain $r_- < r < r_+$.

At $r = r_0$ the real part of z vanishes and then switches from negative to positive. Thus, as just explained, at this junction we require also the imaginary part of z to change sign. Hence we match the right hand sides of Eqs. (19.64) there to another combination of solutions $Me_{\pm\nu}$ by setting:

$$r_0^{1/2}[\alpha Me_\nu(z) + \beta Me_{-\nu}(z)]_{z=i\pi/4}$$
$$= r_0^{1/2}[\alpha' Me_\nu(z) + \beta' Me_{-\nu}(z)]_{z=-i\pi/4},$$
$$\left[\alpha \frac{d}{dr}[r^{1/2} Me_\nu(z)] + \beta \frac{d}{dr}[r^{1/2} Me_{-\nu}(z)]\right]_{z=i\pi/4}$$
$$= \left[\alpha' \frac{d}{dr}[r^{1/2} Me_\nu(z)] + \beta' \frac{d}{dr}[r^{1/2} Me_{-\nu}(z)]\right]_{z=-i\pi/4}. \quad (19.65)$$

Since $Me_\nu(z, h) = Me_{-\nu}(-z, h)$ (as one can check or look up in Meixner and Schäfke [193], p. 131), these relations can be re-expressed for one and the same point $z = -i\pi/4$:

$$r_0^{1/2}[\alpha Me_{-\nu}(z) + \beta Me_\nu(z)]_{z=-i\pi/4}$$
$$= r_0^{1/2}[\alpha' Me_\nu(z) + \beta' Me_{-\nu}(z)]_{z=-i\pi/4},$$
$$\left[\alpha \frac{d}{dr}[r^{1/2} Me_{-\nu}(z)] + \beta \frac{d}{dr}[r^{1/2} Me_\nu(z)]\right]_{z=-i\pi/4}$$
$$= \left[\alpha' \frac{d}{dr}[r^{1/2} Me_\nu(z)] + \beta' \frac{d}{dr}[r^{1/2} Me_{-\nu}(z)]\right]_{z=-i\pi/4}. \quad (19.66)$$

It follows that

$$\alpha = \beta', \quad \beta = \alpha'. \quad (19.67)$$

Next we have to continue the solution beyond the point r_+ to a linear combination of solutions y_+ and y_-, i.e. to

$$r^{1/2}[A M_\nu^{(3)} + B M_\nu^{(4)}], \quad A, B \neq 0.$$

This solution can be continued into the domain below $r = r_+$ by matching to the right hand side of Eq. (19.65). Then, with the replacements of Eq. (19.67):

$$r^{1/2}[\beta Me_\nu(r) + \alpha Me_{-\nu}(r)] = r^{1/2}[AM_\nu^{(3)}(r) + BM_\nu^{(4)}(r)],$$

$$\beta \frac{d}{dr}[r^{1/2}Me_\nu(r)] + \alpha \frac{d}{dr}[r^{1/2}Me_{-\nu}(r)] = A\frac{d}{dr}[r^{1/2}M_\nu^{(3)}(r)]$$

$$+ B\frac{d}{dr}[r^{1/2}M_\nu^{(4)}(r)]. \quad (19.68)$$

From Eqs. (19.3a) and (19.3b) we infer that ($z = \ln r +$ const.)

$$\frac{d}{dr} = \frac{1}{r}\frac{d}{dz}.$$

Hence for any of the functions M_ν:

$$\frac{d}{dr}[r^{1/2}M_\nu] = \frac{1}{2r^{1/2}}M_\nu + \frac{1}{r^{1/2}}\frac{dM_\nu}{dz}. \quad (19.69)$$

Thus if we replace in the derivative relations of Eqs. (19.65) and (19.68) the derivatives with respect to r by this relation, the nonderivative parts (from the first term on the right of Eq. (19.69)) cancel out in view of the nonderivative relations, and we are left with relations expressible only in terms of z. Thus we obtain from Eqs. (19.64) the relations

$$M_\nu^{(3)}(z) = \alpha Me_\nu(z) + \beta Me_{-\nu}(z),$$

$$\frac{d}{dz}[M_\nu^{(3)}(z)] = \alpha\frac{d}{dz}[Me_\nu(z)] + \beta\frac{d}{dz}[Me_{-\nu}(z)] \quad (19.70)$$

and from Eqs. (19.68)

$$[\beta Me_\nu(z) + \alpha Me_{-\nu}(z)] = [AM_\nu^{(3)}(z) + BM_\nu^{(4)}(z)],$$

$$\beta\frac{d}{dz}[Me_\nu(z)] + \alpha\frac{d}{dz}[Me_{-\nu}(z)] = A\frac{d}{dz}[M_\nu^{(3)}(z)]$$

$$+ B\frac{d}{dz}[M_\nu^{(4)}(z)]. \quad (19.71)$$

We now determine the coefficients α, β, A and B. For the Wronskian W of two independent solutions $f(z)$ and $g(z)$ which is constant and can be evaluated at any point, we use the notation

$$W[f,g] := f(z)\frac{dg(z)}{dz} - g(z)\frac{df(z)}{dz}.$$

Thus, for instance, multiplying the first of Eqs. (19.70) by $Me'_{-\nu}$ (the prime meaning derivative), and the second of Eqs. (19.70) by $Me_{-\nu}$ and subtracting the second resulting equation from the first resulting equation, we obtain the first of the following two equations, and the second is obtained similarly:

$$\alpha = \frac{W[M_\nu^{(3)}, Me_{-\nu}]}{W[Me_\nu, Me_{-\nu}]}, \qquad \beta = \frac{W[M_\nu^{(3)}, Me_\nu]}{W[Me_{-\nu}, Me_\nu]}. \tag{19.72}$$

From Eqs. (19.71) we obtain in a similar way

$$
\begin{aligned}
A &= \frac{-W[M_\nu^{(3)}, Me_\nu]W[Me_\nu, M_\nu^{(4)}] + W[M_\nu^{(3)}, Me_{-\nu}]W[Me_{-\nu}, M_\nu^{(4)}]}{W[M_\nu^{(3)}, M_\nu^{(4)}]W[Me_\nu, Me_{-\nu}]}, \\[2mm]
B &= \frac{W[M_\nu^{(3)}, Me_\nu]W[Me_\nu, M_\nu^{(3)}] - W[M_\nu^{(3)}, Me_{-\nu}]W[Me_{-\nu}, M_\nu^{(3)}]}{W[M_\nu^{(3)}, M_\nu^{(4)}]W[Me_\nu, Me_{-\nu}]}.
\end{aligned}
$$

$$\tag{19.73}$$

We now use Eq. (19.46), i.e. $\alpha_\nu(h) = Me_\nu(0, h)/M_\nu^{(1)}(0, h)$, and Wronskians $W[M_\nu^{(i)}, M_\nu^{(j)}] \equiv [i, j]$ given in Meixner and Schäfke [193] (p. 170/171), or obtainable by substituting e.g. the leading terms of the respective cylindrical functions, i.e.¶

$$[3, 4] = -\frac{4i}{\pi}, \qquad [1, 3] = -[1, 4] = \frac{2i}{\pi}. \tag{19.74}$$

Moreover, we use the following circuit relation which can be derived like Eq. (19.48) (or cf. Meixner and Schäfke [193], p. 169)

$$M_{-\nu}^{(1)} = e^{i\nu\pi} M_\nu^{(1)} - i \sin \nu\pi \, M_\nu^{(4)}. \tag{19.75}$$

With these relations and Eq. (19.47) we obtain (α_ν defined by Eq. (19.46))

$$
\begin{aligned}
W[Me_\nu, M_\nu^{(3)}] &= \frac{2i}{\pi}\alpha_\nu, & W[Me_{-\nu}, M_\nu^{(3)}] &= \frac{2i}{\pi}e^{-i\nu\pi}\alpha_{-\nu}, \\[2mm]
W[Me_\nu, M_\nu^{(4)}] &= -\frac{2i}{\pi}\alpha_\nu, & W[Me_{-\nu}, M_\nu^{(4)}] &= -\frac{2i}{\pi}e^{i\nu\pi}\alpha_{-\nu}, \\[2mm]
W[Me_\nu, Me_{-\nu}] &= -\frac{2 \sin \nu\pi}{\pi}\alpha_\nu\alpha_{-\nu}.
\end{aligned}
$$

$$\tag{19.76}$$

With these expressions the quantities A and B are found to be

$$A = \frac{1}{2i \sin \nu\pi}\left(\frac{\alpha_\nu}{\alpha_{-\nu}} - \frac{\alpha_{-\nu}}{\alpha_\nu}\right), \quad B = \frac{1}{2i \sin \nu\pi}\left(\frac{\alpha_\nu}{\alpha_{-\nu}} - e^{-2i\nu\pi}\frac{\alpha_{-\nu}}{\alpha_\nu}\right). \tag{19.77}$$

¶Thus, using Eqs. (19.57) and (19.58) and considering large values of $|z|$, we obtain immediately $W[3, 4] \equiv W[M_\nu^{(3)}(z), M_\nu^{(4)}(z)] = 2(-i)(2/\pi w)(dw/dz) = -(4i/\pi)(2h \sinh z/2h \cosh z) \simeq -4i/\pi$.

The regular solution in this way continued to $r \sim \infty$ is then

$$
\begin{aligned}
y_{\text{reg}} &\simeq r^{1/2}[AM_\nu^{(3)}(z,h) + BM_\nu^{(4)}(z,h)] \\
&\simeq \left(\frac{2}{k\pi}\right)^{1/2}[Ae^{ikr}e^{-i(\nu+1/2)\pi/2} + Be^{-ikr}e^{i(\nu+1/2)\pi/2}].
\end{aligned} \tag{19.78}
$$

We set

$$
R = \frac{\alpha_\nu}{\alpha_{-\nu}} = \frac{M_{-\nu}^{(1)}(0,h)}{M_\nu^{(1)}(0,h)}, \tag{19.79}
$$

where the second expression follows from Eq. (19.46) together with the relation $Me_\nu(z,h) = Me_{-\nu}(-z,h)$. Then

$$
A = \frac{R^2 - 1}{2iR \sin \nu\pi}, \qquad B = \frac{R^2 - e^{-2i\nu\pi}}{2iR \sin \nu\pi}. \tag{19.80}
$$

Inserting these expressions into Eq. (19.78) we obtain

$$
y_{\text{reg}} \approx \frac{e^{-i\nu\pi/2}}{-2\sqrt{k}}[f(k,l)e^{ikr} - f(-k,l)e^{-ikr}], \tag{19.81}
$$

where

$$
\begin{aligned}
f(k,l) &= \left(\frac{2}{\pi}\right)^{1/2} \frac{R^2 - 1}{R \sin \nu\pi} e^{-i(\nu-\frac{1}{2})\frac{\pi}{2}} e^{i\nu\pi/2}, \\
f(-k,l) &= \left(\frac{2}{\pi}\right)^{1/2} \frac{R^2 - \exp(-2i\nu\pi)}{R \sin \nu\pi} e^{i(\nu-\frac{1}{2})\frac{\pi}{2}} e^{i\nu\pi/2}.
\end{aligned} \tag{19.82}
$$

The S-matrix is defined in the following way in the partial wave expansion of the scattering amplitude $f(\theta)$, as we recall from earlier considerations (e.g. in Chapter 16). The superposition of an incoming plane wave and an outgoing radial wave is written and re-expressed in terms of partial waves for $r \to \infty$ with $z = r\cos\theta$ as:

$$
e^{ikz} + f(\theta)\frac{e^{ikr}}{r} \simeq \frac{1}{2ikr}\sum_{l=0}^{\infty}[S_l e^{ikr} - (-1)^l e^{-ikr}]P_l(\cos\theta). \tag{19.83}
$$

The S-matrix element is therefore given by

$$
S_l = e^{i\pi l}\frac{f(k,l)}{f(-k,l)} = \frac{R - \frac{1}{R}}{Re^{i\nu\pi} - (Re^{i\nu\pi})^{-1}}e^{-i(l+\frac{1}{2})\pi}. \tag{19.84}
$$

We note already here that with the substitution

$$
R \equiv e^{i\pi\gamma}, \tag{19.85a}
$$

we can rewrite the S-matrix in the form

$$S_l = \frac{\sin \pi \gamma}{\sin \pi(\gamma + \nu)} e^{-i\pi(l+\frac{1}{2})}. \tag{19.85b}$$

We shall obtain the S-matrix in this form in the case of large values of h^2 later.

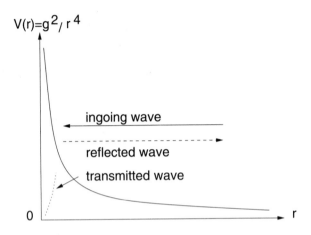

Fig. 19.2 The repulsive potential and the various waves.

We can rewrite Eq. (19.81) in another form from which we can deduce the amplitudes of reflected and transmitted waves. Thus in terms of the variable z, and recalling that y_{reg} is proportional to a function $M_\nu^{(3)}(z, h)$ (see Eqs. (19.56) to (19.58)) in the limit $r \to 0$, we can write this (multiplied by $2i \sin \nu\pi$, and with the left hand side following from Eqs. (19.56), (19.58))

$$\left(\frac{2r}{2h\pi \cosh z}\right)^{1/2} e^{-i(\nu+1/2)\pi/2} [2i \sin \nu\pi] e^{2ih \cosh z}$$

$$\overset{\Re z \to \infty}{\simeq} \left(\frac{2r}{2h\pi \cosh z}\right)^{1/2} e^{-i(\nu+1/2)\pi/2}$$

$$\times \left[\left(R - \frac{1}{R}\right) e^{2ih \cosh z} + i\left(R e^{i\nu\pi} - \frac{e^{-i\nu\pi}}{R}\right) e^{-2ih \cosh z} \right].$$

Thus, cf. Fig. 19.2, we can define respectively as amplitudes of the *incident wave*, the *reflected wave* and the *transmitted wave* the quantities:

$$A_i = R e^{i\nu\pi} - \frac{e^{-i\nu\pi}}{R} \equiv 2i \sin \pi(\gamma + \nu),$$

$$A_r = R - \frac{1}{R} \equiv 2i \sin \pi\gamma,$$

$$A_t = 2i \sin \pi\nu. \tag{19.86}$$

One can verify that for ν real (which it is here in both cases of attractive and repulsive potentials in view of Eq. (19.28)) and $R \equiv e^y$ real, unitarity is preserved, i.e. unity minus reflection probability = transmission probability, i.e.[||]

$$1 - \frac{|R - \frac{1}{R}|^2}{|Re^{i\nu\pi} - e^{-i\nu\pi}/R|^2} = \frac{|2i \sin \nu\pi|^2}{|Re^{i\nu\pi} - e^{-i\nu\pi}/R|^2}. \tag{19.87}$$

We observe that this relation remains valid if the real quantity $R \equiv e^y$ and the pure phase factor $e^{i\nu\pi}$ exchange their roles, i.e. if R becomes a pure phase factor and $e^{i\nu\pi}$ a real exponential. The latter is what happens in the S-wave case of the attractive potential, however not here, but in the 10-dimensional string theory context, in which ν is complex.[**]

19.2.6 Evaluation of the S-matrix

Our next task is to evaluate the expression (19.84) of the S-matrix. This implies basically the evaluation of the expression R of Eq. (19.79), i.e.

$$R = \frac{M_{-\nu}^{(1)}(0, h)}{M_{\nu}^{(1)}(0, h)} \equiv e^{i\pi\gamma}.$$

The function $M_{\nu}^{(1)}(0, h)$ is (cf. Eq. (19.39)) the associated Mathieu function expanded in terms of Bessel functions of the first kind, $J_{\nu+2r}$. In Meixner and Schäfke [193] (p. 178), this expansion is given as

$$M_{\nu}^{(1)}(z, h) = \frac{1}{Me_{\nu}(0, h)} \sum_{r=-\infty}^{\infty} c_{2r}^{\nu}(h^2) J_{\nu+2r}(2h \cosh z), \tag{19.88}$$

where the factor $Me_{\nu}(0, h)$ serves the use of the same coefficients as in the other expansions. By inserting in Eq. (19.88) the power expansion of the Bessel function one realizes soon that the expansion is inconvenient in view of its slow convergence. Meixner and Schäfke [193] therefore developed an expansion in terms of products of Bessel functions which is more useful in practice owing to its rapid convergence. A derivation is beyond the scope of our objectives here, therefore we cite it from Meixner and Schäfke [193], (p. 180), and then exploit it for the evaluation of the quantity R. The expansion is:

$$c_{2r}^{\pm\nu}(h^2) M_{\pm\nu}^{(1)}(z, h) = \sum_{l=-\infty}^{\infty} (-1)^l c_{2l}^{\pm\nu}(h^2) J_{l-r}(he^{-z}) J_{\pm\nu+l+r}(he^z), \tag{19.89}$$

[||] $|Re^{i\nu\pi} - (Re^{i\nu\pi})^{-1}|^2 = R^2 + \frac{1}{R^2} - 2\cos 2\nu\pi = (R - \frac{1}{R})^2 + (2\sin \nu\pi)^2$.

[**]See S. S. Gubser and A. Hashimoto [124] and R. Manvelyan, H. J. W. Müller–Kirsten, J.–Q. Liang and Yunbo Zhang [189], Eqs. (105) to (108).

so that in particular for $z = 0$

$$c_{2r}^{\pm\nu}(h^2)M_{\pm\nu}^{(1)}(0,h) = \sum_{l=-\infty}^{\infty}(-1)^l c_{2l}^{\pm\nu}(h^2)J_{l-r}(h)J_{\pm\nu+l+r}(h). \tag{19.90}$$

Here the coefficients $c_{2r}^{\pm\nu}(h^2)$ are the same as those we introduced earlier for the normalized modified Mathieu functions in Eq. (19.49). The formula (19.90) is in some sense amazing: It permits the evaluation of one and the same quantity $M_{\pm\nu}^{(1)}(0,h)$ in many different ways, i.e. by allocating different values to r, e.g. $r = 0$ and 2. As a matter of introduction we recall the definition of the coefficients with Eq. (19.49) from which we obtain for $z = 0$:

$$Me_\nu(0,h) = c_0^\nu(h^2)\sum_{r=-\infty}^{\infty}\frac{c_{2r}^\nu(h^2)}{c_0^\nu(h^2)}. \tag{19.91}$$

Inserting here the coefficients given in (19.55), one obtains

$$\begin{aligned}
Me_\nu(0,h) &= 1 + \frac{2}{\nu^2-1}\left(\frac{h}{2}\right)^2 + \frac{\nu^2+2}{(\nu^2-1)(\nu^2-4)}\left(\frac{h}{2}\right)^4 \\
&+ \frac{2(\nu^6+4\nu^4-39\nu^2-62)}{(\nu^2-1)^3(\nu^2-4)(\nu^2-9)}\left(\frac{h}{2}\right)^6 + O(h^7). \end{aligned} \tag{19.92}$$

Here in the case of the repulsive potential, h^2 is complex but ν^2 is real (cf. Eq. (19.28)). Next we evaluate $M_\nu^{(1)}(0,h)$ with the help of Eq. (19.90) and choose $r = 0$ (one can choose e.g. $r = 2$ as a check). We require the power expansion of the Bessel function of the first kind which we obtain from Tables of Functions as

$$J_\nu(z) = \left(\frac{z}{2}\right)^\nu\sum_{r=0}^{\infty}\frac{1}{r!(\nu+r)!}\left(-\frac{z^2}{4}\right)^r. \tag{19.93}$$

Thus, setting $r = 0$ in Eq. (19.90), we obtain

$$\begin{aligned}
M_{\pm\nu}^{(1)}(0,h) &= J_0(h)J_{\pm\nu}(h) - \frac{c_2^{\pm\nu}(h^2)}{c_0^{\pm\nu}(h^2)}J_1(h)J_{\pm\nu+1}(h) \\
&- \frac{c_{-2}^{\pm\nu}(h^2)}{c_0^{\pm\nu}(h^2)}J_{-1}(h)J_{\pm\nu-1}(h) + \frac{c_4^{\pm\nu}(h^2)}{c_0^{\pm\nu}(h^2)}J_2(h)J_{\pm\nu+2}(h) \\
&+ \frac{c_{-4}^{\pm\nu}(h^2)}{c_0^{\pm\nu}(h^2)}J_{-2}(h)J_{\pm\nu-2}(h) + \cdots. \end{aligned} \tag{19.94}$$

Inserting the power expansion (19.93) and the coefficient expansions (19.55) and collecting terms in ascending powers of h^2 (here for the case of nonintegral values of ν), one obtains

$$
M^{(1)}_{\pm\nu}(0,h) = \frac{1}{(\pm\nu)!}\left(\frac{h}{2}\right)^{\pm\nu}\left[1 + \frac{2}{\nu^2-1}\left(\frac{h}{2}\right)^2\right.
$$

$$
\mp\frac{2(\nu^2\mp 3\nu-7)}{(\nu\pm 1)^2(\nu\mp 1)(\nu^2-4)}\left(\frac{h}{2}\right)^4
$$

$$
\mp\frac{4(\nu^4\mp 11\nu^3-2\nu^2\pm 59\nu-23)}{(\nu\pm 1)^2(\nu\mp 1)^3(\nu^2-4)(\nu^2-9)}\left(\frac{h}{2}\right)^6
$$

$$
\left.+\cdots\right]. \tag{19.95}
$$

These expansions imply for the quantity R:

$$
R = \frac{\nu!}{(-\nu)!}\left(\frac{h}{2}\right)^{-2\nu}\frac{\left[1+\frac{2}{\nu^2-1}\left(\frac{h}{2}\right)^2+\frac{2(\nu^2+3\nu-7)}{(\nu-1)^2(\nu+1)(\nu^2-4)}\left(\frac{h}{2}\right)^4+\cdots\right]}{\left[1+\frac{2}{\nu^2-1}\left(\frac{h}{2}\right)^2-\frac{2(\nu^2-3\nu-7)}{(\nu+1)^2(\nu-1)(\nu^2-4)}\left(\frac{h}{2}\right)^4+\cdots\right]},
$$

or with the help of the inversion formula of factorials

$$
R = \frac{\nu!(\nu-1)!\sin\pi\nu}{\pi}\left(\frac{h}{2}\right)^{-2\nu}\left[1+\frac{4\nu}{(\nu^2-1)^2}\left(\frac{h}{2}\right)^4\right.
$$

$$
+\frac{2\nu(4\nu^5+15\nu^4-32\nu^3-12\nu^2+64\nu-111)}{(\nu^2-1)^4(\nu^2-4)^2}\left(\frac{h}{2}\right)^8
$$

$$
\left.+\cdots\right]. \tag{19.96}
$$

With this explicit expansion we can evaluate the S-matrix. In this procedure our attention is focussed particularly on the quantity called *absorptivity* in the case of the attractive potential. From expansion (19.96) we extract for later reference the relation

$$
\left(\frac{\sin\pi\nu}{R}\right)^2 = \frac{\pi^2}{[\nu!(\nu-1)!]^2\left[1+\frac{4\nu}{(\nu^2-1)^2}\left(\frac{h}{2}\right)^4+\cdots\right]^2}\left(\frac{h}{2}\right)^{4\nu}. \tag{19.97}
$$

This expansion will be used below in the low order approximation of the absorptivity of partial waves. We observe that this approximation involves ν and h^4 and hence is real.

19.2.7   Calculation of the absorptivity

We consider the absorptivity in a general case, and hence allow for the general case of complex parameters ν (the Floquet exponent in the case of the periodic Mathieu equation) although here in the case of small $|h^2|$ we know from Eq. (19.28) that ν is real. Thus we set

$$\nu = n + i(\alpha + i\beta) = (n - \beta) + i\alpha, \qquad (19.98)$$

where n is an integer and α and β are real and of $O(h^4)$ (cf. Eq. (19.28)). We have with Eq. (19.84):

$$S_l S_l^* = \frac{(1 - 1/R^2)(1 - 1/R^{*2})}{(e^{i\nu\pi} - e^{-i\nu\pi}/R^2)(e^{i\nu^*\pi} - e^{-i\nu^*\pi}/R^{*2})}, \qquad (19.99)$$

which can be rewritten as

$$
\begin{aligned}
S_l S_l^* = {} & e^{2\pi\alpha}\left(1 - \frac{1}{R^2}\right)\left(1 - \frac{1}{R^{*2}}\right)\left[1 - \left\{\left(\frac{1}{R^2} + \frac{1}{R^{*2}}\right)e^{2\pi\alpha}\cos 2\pi\beta\right.\right. \\
& \left.\left. + i\left\{\left(\frac{1}{R^2} - \frac{1}{R^{*2}}\right)e^{2\pi\alpha}\sin 2\pi\beta - \frac{e^{4\pi\alpha}}{R^2 R^{*2}}\right\}\right]^{-1}.
\end{aligned}
\qquad (19.100)
$$

Here we set

$$e^{2i\pi\beta} \equiv 1 + if, \qquad e^{2\pi\alpha} \equiv 1 + g, \qquad (19.101)$$

where f is complex and g is real. Then

$$\cos 2\pi\beta = 1 - \Im f, \quad \Im f \approx \frac{1}{2}(2\pi\beta)^2 \approx \frac{1}{2}(\sin 2\pi\beta)^2 \simeq 2\sin^2\pi\beta, \quad (19.102)$$

and

$$\sin 2\pi\beta = \Re f. \qquad (19.103)$$

Then

$$
\begin{aligned}
S_l S_l^* = {} & (1 + g)\left[1 - \left\{g\left(\frac{1}{R^2} + \frac{1}{R^{*2}} - \frac{2 + g}{R^2 R^{*2}}\right)\right.\right. \\
& \left.\left. - \Im f (1 + g)\left(\frac{1}{R^2} + \frac{1}{R^{*2}}\right) + (1 + g)i\Re f\left(\frac{1}{R^2} - \frac{1}{R^{*2}}\right)\right\} \right. \\
& \left. \left(1 - \frac{1}{R^2}\right)^{-1}\left(1 - \frac{1}{R^{*2}}\right)^{-1}\right]^{-1}.
\end{aligned}
\qquad (19.104)
$$

We now consider the case of $\alpha \to 0$ implying that g of Eq. (19.101) is zero. In this case $R = R^*$ and so $1/R^2 \simeq O(h^4)$. This is the case of real parameters $\nu = n - \beta$. Hence with Eq. (19.102) we obtain

$$
S_l S_l^* = \frac{1}{1 + \Im f \cdot \frac{2}{R^2}(1 - \frac{1}{R^2})^{-2}} \simeq 1 - \frac{2\Im f}{R^2(1 - \frac{1}{R^2})^{-2}}
$$

$$
\simeq 1 - 4\left(\frac{\sin \pi \beta}{R}\right)^2 = 1 - 4\left(\frac{\sin \pi \nu}{R}\right)^2. \tag{19.105}
$$

The absorptivity A is therefore given by

$$
A = 1 - S_l S_l^* \approx 4\left(\frac{\sin \pi \beta}{R}\right)^2 \tag{19.106}
$$

and evidently violates the unitarity of S. With the help of Eq. (19.97) this can be written

$$
A \approx \frac{4\pi^2}{[\nu!(\nu-1)!]^2}\left(\frac{h}{2}\right)^{4\nu}\left[1 + \frac{4\nu}{(\nu^2-1)^2}\left(\frac{h}{2}\right)^4 + O(h^8)\right]^{-2}. \tag{19.107a}
$$

In this result* we now have to insert the expansion of the parameter ν given by Eq. (19.28), i.e.

$$
\nu^2 = \left(l + \frac{1}{2}\right)^2 - \frac{h^4}{2[(l+1/2)^2 - 1]} - O(h^8).
$$

For S waves this implies an *absorptivity* given by $((-1/2)! = \sqrt{\pi})$

$$
A_{l=0} = 4h^2[1 + O(h^4)]. \tag{19.107b}
$$

Here in the *case of the attractive potential* (as remarked after Eq. (19.3b)) h^2 is real for $E = k^2 > 0$. Using the derivative of the gamma function expressed as the *psi function* $\psi(z) = \Gamma'(z)/\Gamma(z)$, and the values of $\psi(1/2)$ and $\psi(3/2)$ as given in Tables† with Euler's constant $C \simeq 0.577$, one can calculate the next term of the expansion so that

$$
A_{l=0} = 4h^2\left[1 - \frac{4h^4}{9}\{7 - 6C - 6\ln 2 - 6\ln h\} + O(h^8)\right]. \tag{19.107c}
$$

Here h^2 is actually $\sqrt{h^4}$. We can also evaluate now the amplitudes (19.86) and obtain in leading order for $l = 0$:

$$
A_i = \frac{2i}{h}[1 + O(h^2)], \quad A_r = \frac{1}{h}[1 + O(h^2)], \quad A_t = 2i[1 + O(h^2)].
$$

---

*This is the result effectively contained in R. Manvelyan, H. J. W. Müller-Kirsten, J.-Q. Liang and Yunbo Zhang [189] (there Eq. (111)) in agreement with a result in S. S. Gubser and A. Hashimoto [124]. In all comparisons with these papers — which consider predominantly the 10-dimensional string theory context — one has to keep in mind that many formulas there do not apply in the three-dimensional case we are considering here.

†W. Magnus and F. Oberhettinger [181], p.3.

19.3 The Potential $1/r^4$ — Case of Large h^2

19.3.1 Preliminary remarks

In the following we have in mind particularly the case of the attractive potential (h^2 real) and rederive the S-matrix obtained above by using the asymptotic expansions for large values of h^2, i.e. completely different solutions. This is therefore a highly interesting case which found its complete solution only recently. In addition this solvability is another aspect which singles the $1/r^4$-potential out as very exceptional and like all explicitly solvable cases it therefore deserves particular attention.[‡]

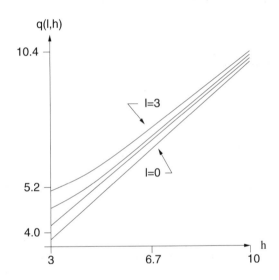

Fig. 19.3 The function $q(l, h)$ for large h and upwards $l = 0, 1, 2, 3$.

We are again concerned with the associated Mathieu equation given by Eq. (19.4) which — for some distinction from the small-h^2 treatment — we rewrite here as

$$\frac{d^2\psi}{dz^2} + [2h^2 \cosh 2z - a^2]\psi = 0, \quad a^2 = \left(l + \frac{1}{2}\right)^2. \tag{19.108}$$

For the large values of h^2 that we wish to consider here, we again make use of the replacement (17.26) and hence set

$$a^2 = -2h^2 + 2hq + \frac{\triangle(q, h)}{8}, \tag{19.109}$$

[‡]We follow here mainly D. K. Park, S. N. Tamaryan, H. J. W. Müller-Kirsten and Jian-zu Zhang [226], where the part of interest here is based on H. H. Aly, H. J. W. Müller and N. Vahedi-Faridi [8] and G.H. Wannier [278].

where q is a parameter to be determined as the solution of this equation, and $\triangle/8$ is the remainder of the large-h^2 expansion as determined perturbatively. The behaviour of q as a function of h for some values of l is shown in Fig. 19.3. The replacement (19.109) in this way enables us to obtain asymptotic expansions of solutions very analogous to those of the periodic case.

We begin, however, with an essential mathematical step, since it is clear that large h^2 considerations require a knowledge of the large-h^2 behaviour of the Floquet exponent ν.

19.3.2 The Floquet exponent for large h^2

In our treatment of the *periodic Mathieu equation* in Chapter 17, to which we have to return in this subsection, we defined the Floquet exponent ν and observed with Eq. (17.5) that this is given by the relation

$$\cos \pi\nu = \frac{y_+(\pi)}{a}, \tag{19.110}$$

where $y_+(\pi)$ is the solution evaluated at $z = \pi$, which is even about $z = 0$, and a is its normalization constant, which therefore cancels out or can be taken to be 1. We also found the boundary condition (17.21b), i.e.[§]

$$\frac{y_+(\pi)}{a} = -1 + 2\frac{y_+(\frac{\pi}{2})y'_-(\frac{\pi}{2})}{W[y_+, y_-]},$$

where $W[y_+, y_-] = ab\sqrt{\lambda}$ is the Wronskian (in leading order) of the solutions which are even and odd about $z = 0$ respectively. With the boundary conditions

$$y_+(0) = 1, \quad y_-(0) = 0, \quad y'_+(0) = 0, \quad y'_-(0) = 1, \tag{19.111}$$

we have $W[y_+, y_-] = 1$ and

$$\cos \pi\nu = -1 + 2y_+\left(\frac{\pi}{2}, h^2\right)y'_-\left(\frac{\pi}{2}, h^2\right). \tag{19.112}$$

In Chapter 17 the unnormalized even and odd large-h^2 solutions of the Mathieu equation were given by Eq. (17.63), i.e.

$$y_\pm \propto A(z)e^{2h \sin z} \pm \overline{A}(z)e^{-2h \sin z}, \tag{19.113}$$

[§] We are considering nonintegral values of ν. Therefore the second term on the right is nonzero. See also Eq. (17.21b).

and their extension to around $z = \pi/2$ by the relations

$$
y_\pm \propto \frac{B[w(z)]}{\alpha} e^{2h \sin z} \pm \frac{\overline{C}[w(z)]}{\overline{\alpha}} e^{-2h \sin z}. \tag{19.114}
$$

Our first step is therefore the determination of the normalization constants N_0, N_0' in

$$
y_+(z) = N_0 \left[A(z)e^{2h \sin z} + \overline{A}(z)e^{-2h \sin z} \right],
$$

$$
y_-(z) = N_0' \left[A(z)e^{2h \sin z} - \overline{A}(z)e^{-2h \sin z} \right]. \tag{19.115}
$$

Setting $z = 0$ we obtain in leading order

$$
y_+(0) = 2N_0 A(0) \quad \text{and} \quad y_-'(0) = 4hN_0' A(0),
$$

and hence with Eqs. (19.111) we have[¶]

$$
N_0 = \frac{1}{2^{3/2}} \left[1 + O\left(\frac{1}{h}\right) \right] \quad \text{and} \quad N_0' = \frac{1}{2^{5/2}h} \left[1 + O\left(\frac{1}{h}\right) \right] \tag{19.116}
$$

(in the first expression we have $A(z) \simeq A_q(z)$ with $A_q(z)$ given by Eq. (17.32), which is $\sqrt{2}$ at $z = 0$; in the second expression $2h$ is obtained in addition from differentiation of the exponential factor). We now obtain the expressions needed in Eq. (19.112) from evaluation of Eqs. (19.114) at $z = \pi/2$, i.e. from

$$
y_+(z) = N_0 \left[\frac{B[w(z)]}{\alpha} e^{2h \sin z} + \frac{\overline{C}[w(z)]}{\overline{\alpha}} e^{-2h \sin z} \right],
$$

$$
y_-(z) = N_0' \left[\frac{B[w(z)]}{\alpha} e^{2h \sin z} - \frac{\overline{C}[w(z)]}{\overline{\alpha}} e^{-2h \sin z} \right]. \tag{19.117}
$$

Referring back to Eq. (17.43) we see that $z = \pi/2$ implies $w(\pi/2) = 0$. Hence we require (obtained from Eq. (17.46) together with the expansion of the Hermite function from Tables of Special Functions[‖])

$$
B[w = 0] = \frac{\sqrt{\pi}}{[\frac{1}{4}(q-1)]![-\frac{1}{4}(q+1)]!} \left[1 - \frac{q}{2^5 h} + \frac{q^4 - 82q^2 - 39}{2^{14}h^2} + \cdots \right] \tag{19.118a}
$$

[¶]We emphasize again: For integral values of ν the normalization constants are different. See R. B. Dingle and H. J. W. Müller [73]. This paper contains the normalization constants of large-h^2 periodic Mathieu functions.

[‖]Or see R. B. Dingle and H. J. W. Müller [73].

and

$$\left(\frac{\partial B}{\partial z}\right)_{z=\pi/2} = \left(\frac{\partial B}{\partial w}\right)_{w=0} \underbrace{\left(\frac{\partial w}{\partial z}\right)_{z=\pi/2}}_{-2h^{1/2}} = -\frac{\sqrt{2\pi}}{[\frac{1}{4}(q-1)]![-\frac{1}{4}(q+3)]!}\left[1 - \frac{q}{2^4 h}\right.$$

$$\left. + \frac{q^4 - 106q^2 - 87}{2^{14}h^2} - \cdots\right](-2h^{1/2}). \tag{19.118b}$$

The factors $\alpha, \bar{\alpha}$ in Eq. (19.117) are known from the matching of solutions in Eqs. (17.62a) and (17.62b) (see also the explicit calculation in Example 17.3). Thus in particular

$$\alpha = \frac{(8h)^{\frac{1}{4}(q-1)}}{[\frac{1}{4}(q-1)]!}\left[1 - \frac{(3q^2 - 2q + 3)}{2^7 h} + \cdots\right].$$

Inserting these various expressions now into Eq. (19.112) we obtain

$$\cos \pi\nu + 1 = 2\frac{e^{2h}}{2^{3/2}\alpha}\frac{\sqrt{\pi}}{[\frac{1}{4}(q-1)]![-\frac{1}{4}(q+1)]!}\left[1 - \frac{q}{2^5 h} + \cdots\right]$$

$$\times \frac{e^{2h}}{2^{5/2}h\alpha}\frac{-\sqrt{2\pi}}{[\frac{1}{4}(q-1)]![-\frac{1}{4}(q+3)]!}\left[1 - \frac{q}{2^4 h} + \cdots\right](-2h^{1/2})$$

$$\simeq \frac{\pi e^{4h}}{(8h)^{q/2}[-\frac{1}{4}(q+1)]![-\frac{1}{4}(q+3)]!}$$

$$\times \left\{1 - \frac{q}{2^5 h}\right\}\left\{1 - \frac{q}{2^4 h}\right\}\left[1 + 2\frac{(3q^2 - 2q + 3)}{2^7 h}\right]$$

and hence

$$\cos \pi\nu + 1 = \frac{\pi e^{4h}}{(8h)^{q/2}[-\frac{1}{4}(q+1)]![-\frac{1}{4}(q+3)]!}\left[1 + \frac{(3q^2 - 8q + 3)}{2^6 h} + O\left(\frac{1}{h^2}\right)\right]. \tag{19.119a}$$

This formula — derived and rediscovered in Park, Tamaryan, Müller-Kirsten and Zhang [226] — is cited without proof by Meixner and Schäfke [193], p. 210. Presumably they extracted it from "between the lines" of a sophisticated paper of Langer's [159] which they cite together with others. With the help of the duplication and inversion formulas of factorials the result can be re-expressed as

$$\cos \pi\nu + 1 \simeq \frac{\cos(q\pi/2)[\frac{1}{2}(q-1)]!}{\sqrt{2\pi}(16h)^{q/2}}e^{4h}. \tag{19.119b}$$

This result gives the leading contribution of the large-h^2 expansion determining the Floquet exponent ν.

19.3.3 Construction of large-h^2 solutions

Following the procedure of Sec. 17.3.2 we insert the expression (19.109) into the equation (19.108) and set

$$\psi(q, h; z) = A(q, h; z) \exp[\pm 2hi \sinh z]. \qquad (19.120)$$

The resulting equation for the function $A(q, h; z)$ can then be written as

$$\cosh z \frac{dA}{dz} + \frac{1}{2}(\sinh z \pm iq)A = \pm \frac{1}{4ih}\left[\frac{\triangle}{8}A - \frac{d^2 A}{dz^2}\right]. \qquad (19.121)$$

We let $A_q(z)$ be the solution of this equation when the right hand side is replaced by zero (i.e. for $h \to \infty$). Then straightforward integration yields

$$A_q(z) = \frac{1}{\sqrt{\cosh z}}\left(\frac{1 + i\sinh z}{1 - i\sinh z}\right)^{\mp q/4} \overset{\Re z \to \infty}{\simeq} \frac{e^{\mp i\pi q/4}}{\sqrt{\cosh z}}. \qquad (19.122)$$

Correspondingly the various solutions $\psi(q, h; z)$ are

$$\psi(q, h; z) = A_q(z) \exp[\pm 2hi \sinh z] \overset{\Re z \to \infty}{\simeq} \frac{\exp(\pm ihe^z)}{\sqrt{\cosh z}} e^{\mp i\pi q/4},$$

$$\psi(q, h; z) = A_q(z) \exp[\pm 2hi \sinh z] \overset{\Re z \to -\infty}{\simeq} \frac{\exp(\mp ihe^z)}{\sqrt{\cosh z}} e^{\mp i\pi q/4}.$$

$$\qquad (19.123)$$

We again make the important observation by looking at Eqs. (19.120) to (19.122) that given one solution $\psi(q, h; z)$, we can obtain the linearly independent one either as $\psi(-q, -h; z)$ or as $\psi(q, h; -z)$, the expression (19.109) remaining unchanged. With the solutions as they stand, of course, $\psi(q, h; z) = \psi(-q, -h; -z)$.

In the following we require solutions $\mathrm{He}^{(i)}(z), i = 1, 2, 3, 4$, each with a specific asymptotic behaviour. We define these solutions in terms of the function

$$\mathrm{Ke}(q, h; z) := \frac{\exp[i\pi q/4]}{\sqrt{-2ih}} A_q(z) \exp[2hi \sinh z]$$

$$\equiv k(q, h)\psi(q, h; z), \quad k(q, h) = \frac{e^{iq\pi/4}}{\sqrt{-2ih}}. \qquad (19.124)$$

Since this function differs from a solution ψ by a factor $k(q, h)$, it is still a solution but not with the symmetry property $\psi(q, h; z) = \psi(-q, -h; -z)$.

Instead, in performing this cycle of replacements the function picks up a factor, i.e.

$$\mathrm{Ke}(q,h;z) = \frac{k(q,h)}{k(-q,-h)}\mathrm{Ke}(-q,-h;-z), \qquad \frac{k(q,h)}{k(-q,-h)} \simeq e^{i\frac{\pi}{2}(q\pm1)} \quad (19.125)$$

(the expression on the far right in leading order for h^2 large). One can show that the quantity Φ_0 of Wannier* (see above) is related to q by $\Phi_0 = iq\pi/2 + O(1/h)$.

In order to be able to obtain the S-matrix, we have to match a solution valid at $\Re z = -\infty$ to a combination of solutions valid at $\Re z = \infty$. This is, as we saw earlier, achieved with the help of the Floquet solutions $\mathrm{Me}_{\pm\nu}(z,h^2)$. We observed that these satisfy the relation (19.45), i.e.

$$\mathrm{Me}_\nu(z,h^2) = \alpha_\nu M_\nu^{(1)}(z,h^2), \qquad \alpha_\nu(h^2) = \frac{\mathrm{Me}_\nu(0,h^2)}{M_\nu^{(1)}(0,h^2)}. \qquad (19.126)$$

For large values of the argument $2h\cosh z$, the Bessel functions contained in the expansion of the associated Mathieu function $M_\nu^{(1)}(z,h^2)$ can be re-expressed in terms of Hankel functions, or — equivalently — by the appropriate expansion of the Bessel functions as given in Tables of Functions, i.e.

$$\begin{aligned}
J_\nu(z) &= \sqrt{\frac{2}{\pi z}}\cos\left(z - \frac{\nu\pi}{2} - \frac{\pi}{4}\right)\left[1 + O\left(\frac{1}{z^2}\right)\right] \\
&\quad - \sqrt{\frac{2}{\pi z}}\sin\left(z - \frac{\nu\pi}{2} - \frac{\pi}{4}\right)\left[O\left(\frac{1}{z^2}\right)\right]. \qquad (19.127)
\end{aligned}$$

---

*Converting to notation of Wannier [278], our h^2 is Wannier's $-k^2$ and our a^2 is his a which in terms of our parameter q becomes $a = 2k^2 + 2iqk + \triangle/8$. Wannier's parameter Φ_0 is defined by

$$\frac{1}{2}\Phi_0 = \lim_{y=\infty}[I_y - 2k\sinh y], \quad \text{where } I_y := \int_0^y d\eta\sqrt{a + 2k^2\cosh 2\eta}.$$

Inserting the expression for a, we have (using $\cosh 2\eta = 2\cosh^2\eta - 1$)

$$\begin{aligned}
I_y &= \int_0^y d\eta\left(2iqk + \frac{\triangle}{8} + 4k^2\cosh^2\eta\right)^{1/2} \simeq 2k\int_0^y d\eta\cosh\eta\left(1 + \frac{2iqk + \triangle/8}{8k^2\cosh^2\eta}\right) \\
&= 2k\sinh y + \frac{2iqk + \triangle/8}{4k}[\tan^{-1}(\sinh\eta)]_0^y.
\end{aligned}$$

Setting $\sinh\eta = \tan\theta$, we have $\eta = 0 \to \theta = 0, \pi$ and $\eta = \infty \to \theta = \pi/2$. It follows that

$$I_y = 2k\sinh y + \frac{iq\pi}{4} + \frac{\triangle\pi}{2^6 k}.$$

The factor $e^{i\pi(q+1)/2}$ on the right of Eq. (19.125) is thus seen to be identical with the proportionality factor in Eq. (44) of Wannier.

Note that the sine part is nonleading! Retaining only the dominant term of this expansion we have (with $z \to 2h \cosh z$)

$$\mathrm{Me}_{\pm\nu}(z, h^2) \simeq \sqrt{\frac{2}{\pi}} \alpha_{\pm\nu} \frac{\cos(2h \cosh z \mp \nu\pi/2 - \pi/4)}{\sqrt{2h \cosh z}}. \qquad (19.128)$$

We now define the following set of solutions of the associated Mathieu equation in terms of the function $\mathrm{Ke}(q, h, z)$:

$$
\begin{aligned}
\mathrm{He}^{(2)}(z, q, h) &:= \mathrm{Ke}(q, h, z), \\
\mathrm{He}^{(1)}(z, q, h) &:= \mathrm{He}^{(2)}(z, -q, -h), \\
\mathrm{He}^{(3)}(z, q, h) &:= \mathrm{He}^{(2)}(-z, -q, -h), \\
\mathrm{He}^{(4)}(z, q, h) &:= \mathrm{He}^{(2)}(-z, q, h).
\end{aligned}
\qquad (19.129)
$$

The solutions so defined have the following asymptotic behaviour (where $\epsilon(z) = (2h \cosh z)^{-1/2}$ and $h^2 = ikg$):

$$
\begin{aligned}
\mathrm{He}^{(1)}(z, q, h) &= \epsilon(z) \exp\left[-ihe^z - i\frac{\pi}{4}\right], \quad \Re z \gg 0, \\
&\underset{r \to \infty}{\sim} \frac{\exp[-ikr - i\pi/4]}{\sqrt{kr}}, \\
\mathrm{He}^{(2)}(z, q, h) &= \epsilon(z) \exp\left[ihe^z + i\frac{\pi}{4}\right], \quad \Re z \gg 0, \\
&\underset{r \to \infty}{\sim} \frac{\exp[ikr + i\pi/4]}{\sqrt{kr}},
\end{aligned}
\qquad (19.130a)
$$

$$
\begin{aligned}
\mathrm{He}^{(3)}(z, q, h) &= \epsilon(z) \exp\left[-ihe^{|z|} - i\frac{\pi}{4}\right], \quad \Re z \ll 0, \\
\mathrm{He}^{(4)}(z, q, h) &= \epsilon(z) \exp\left[ihe^{|z|} + i\frac{\pi}{4}\right], \quad \Re z \ll 0, \\
&\underset{r \to 0}{\sim} \frac{r^{1/2} \exp[-g/r + i\frac{\pi}{4}]}{(ig)^{1/2}}.
\end{aligned}
\qquad (19.130b)
$$

19.3.4 The connection formulas

We now return to Eq. (19.128) and consider the cosine there as composed of two exponentials whose asymptotic behaviour we identify with that of solutions of Eqs. (19.130a), (19.130b). In this way we obtain in the domain

$\Re z \gg 0$ the relations:

$$\mathrm{Me}_\nu(z, h^2) = \frac{i}{\sqrt{2\pi}} \alpha_\nu \left[e^{i\nu\pi/2} \mathrm{He}^{(1)}(z, q, h) - e^{-i\nu\pi/2} \mathrm{He}^{(2)}(z, q, h) \right],$$

$$\mathrm{Me}_{-\nu}(z, h^2) = \frac{i}{\sqrt{2\pi}} \alpha_{-\nu} \left[e^{-i\nu\pi/2} \mathrm{He}^{(1)}(z, q, h) - e^{i\nu\pi/2} \mathrm{He}^{(2)}(z, q, h) \right],$$

$$(19.131)$$

where the second relation was obtained by changing the sign of ν in the first. Changing the sign of z, we obtain in the domain $\Re z \ll 0$ the relations:

$$\mathrm{Me}_\nu(-z, h^2) = \mathrm{Me}_{-\nu}(z, h^2)$$

$$= \frac{i}{\sqrt{2\pi}} \alpha_\nu \left[e^{i\nu\pi/2} \mathrm{He}^{(3)}(z, q, h) - e^{-i\nu\pi/2} \mathrm{He}^{(4)}(z, q, h) \right],$$

$$\mathrm{Me}_\nu(z, h^2) = \frac{i}{\sqrt{2\pi}} \alpha_{-\nu} \left[e^{-i\nu\pi/2} \mathrm{He}^{(3)}(z, q, h) - e^{i\nu\pi/2} \mathrm{He}^{(4)}(z, q, h) \right].$$

$$(19.132)$$

These relations are now valid over the entire range of z. Substituting the Eqs. (19.132) into Eqs. (19.131) and eliminating $\mathrm{He}^{(3)}$ and setting again

$$\exp[i\pi\gamma] := \frac{\alpha_\nu(h^2)}{\alpha_{-\nu}(h^2)},$$

we obtain the first connection formula

$$-\sin \pi\nu \, \mathrm{He}^{(4)}(z, q, h) = \sin \pi(\gamma + \nu) \, \mathrm{He}^{(1)}(z, q, h) - \sin \pi\gamma \, \mathrm{He}^{(2)}(z, q, h).$$

$$(19.133)$$

In a similar way one obtains the connection formulas

$$-\sin \pi\nu \, \mathrm{He}^{(2)}(z, q, h) = \sin \pi(\gamma + \nu) \, \mathrm{He}^{(3)}(z, q, h) - \sin \pi\gamma \, \mathrm{He}^{(4)}(z, q, h),$$

$$\sin \pi\nu \, \mathrm{He}^{(1)}(z, q, h) = -\sin \pi\gamma \, \mathrm{He}^{(3)}(z, q, h)$$

$$+ \sin \pi(\gamma - \nu) \, \mathrm{He}^{(4)}(z, q, h). \qquad (19.134)$$

Considering now the first of relations (19.134) in the region where the function $\mathrm{He}^{(4)}(z, q, h)$ is asymptotically small (cf. Eqs. (19.130a), (19.130b)), we see that the solutions (compare with Eqs. (19.129))

$$\mathrm{He}^{(2)}(z, q, h) \equiv Ke(q, h, z) \quad \text{and} \quad \mathrm{He}^{(3)}(z, q, h) \equiv Ke(-q, -h, -z)$$

are proportional there, and thus can be matched in this region.[†] With Eq. (19.125) the proportionality factor can be seen to imply the relation

$$\exp\left[i\frac{\pi}{2}(q + 1) \right] \simeq -\frac{\sin \pi(\gamma + \nu)}{\sin \pi\nu}. \qquad (19.135)$$

[†]This means, these are exponentially increasing there, and $He^{(4)}$ exponentially decreasing.

The factor on the left, of course, is only the dominant term for large h^2.[‡]

19.3.5 Derivation of the *S*-matrix

From Eq. (19.133) we can now deduce the *S*-matrix $S_l \equiv e^{2i\delta_l}$, where δ_l is the phase shift. The latter is defined by the following large-r behaviour of the solution chosen at $r = 0$, which in our case is the solution $He^{(4)}$. Thus here the *S*-matrix is defined by (using Eq. (19.133))

$$
\begin{aligned}
- \sin \pi\nu \, He^{(4)}(z, q, h) &\stackrel{r\to 0}{\simeq} - \sin \pi\nu \frac{r^{1/2} e^{-g/r + i\pi/4}}{(ig)^{1/2}} \\
&\stackrel{r\to\infty}{\simeq} -(-1)^l \frac{\sin \pi(\gamma+\nu) e^{-i\pi/4}}{\sqrt{kr}} \left[\frac{(-1)^l \sin \pi\gamma}{\sin \pi(\gamma+\nu)} e^{i\pi/2} e^{ikr} - (-1)^l e^{-ikr} \right] \\
&\equiv \frac{e^{-i\delta_l} e^{-il\pi/2}}{2i\sqrt{r}} \left[S_l e^{ikr} - (-1)^l e^{-ikr} \right].
\end{aligned} \tag{19.136}
$$

From this we can deduce that

$$
S_l = \frac{\sin \pi\gamma}{\sin \pi(\gamma+\nu)} e^{i\pi(l+1/2)} \stackrel{(19.135)}{=} -\frac{\sin \pi\gamma}{\sin \pi\nu} e^{i\pi(l-q/2)}. \tag{19.137}
$$

We see that this expression agrees with that of Eq. (19.85b) obtained earlier. The quantity γ is now to be determined from Eq. (19.135). Squaring Eq. (19.135), i.e.

$$
\sin \pi\gamma . \cos \pi\nu + \sin \pi\nu . \exp\left[\frac{i}{2} \pi(q+1) \right] = - \cos \pi\gamma . \sin \pi\nu,
$$

and solving for $\sin \pi\gamma$ one obtains the expression

$$
\begin{aligned}
\sin \pi\gamma &= \sin \pi\nu \left\{ - i e^{iq\pi/2} \cos \pi\nu \pm \sqrt{1 + e^{iq\pi} \sin^2 \pi\nu} \right\} \\
&= -i e^{i\pi q/2} \sin \pi\nu \left\{ \cos \pi\nu \mp \sqrt{\cos^2 \pi\nu - 1 - e^{-i\pi q}} \right\}. \tag{19.138}
\end{aligned}
$$

We obtain the behaviour of the Floquet exponent for large values of h^2 from Eqs. (19.119a) and (19.119b) as

$$
1 + \cos \pi\nu \simeq \frac{e^{4h}}{(8h)^{q/2}} \left[\frac{[\frac{1}{2}(q-1)]! \cos(q\pi/2)}{\sqrt{2\pi} 2^{q/2}} + O\left(\frac{1}{h} \right) \right]. \tag{19.139}
$$

[‡]Equation (19.135) is the equation corresponding to Eq. (60) in the work of G. H. Wannier [278]. Wannier did not have the relation (19.119a) or (19.119b), and therefore remarks after his Eq. (76):"*It is not likely at this stage that an analytic relation will ever be found connecting* (what we call) ν *and* γ *to* (what we call) $(l+1/2)^2$ *and* h^2".

Since the right hand side is real for real h, i.e. for the attractive potential, the real part of ν must be an integer.[§] Since the right hand side grows exponentially with increasing h the Floquet exponent must have a large

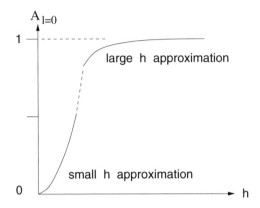

Fig. 19.4 The absorptivity of the S wave (attractive case).

imaginary part ν_I (so that 1 on the left hand side of Eq. (19.139) could actually be neglected) implying $|\cos \pi\nu| = \cosh \pi\nu_I$. This is different from the behaviour for small values of $|h^2|$, as a comparison with Eq. (19.28) shows. Using *Stirling's formula* in the form

$$z! \simeq e^{-z} z^{z+1/2} \sqrt{2\pi}, \qquad \left[\frac{1}{2}(q-1)\right]! \simeq e^{-(q-1)/2}\left(\frac{q}{2}\right)^{q/2}\sqrt{2\pi},$$

with the approximation $q \simeq h$ obtainable from Eq. (19.109), we can approximate the equation for $q \simeq h$ (i.e. irrespective of what the value of l is) by

$$\cos \pi\nu + 1 = \cos\left(\frac{h\pi}{2}\right)\left(\frac{e^7}{32}\right)^{h/2} \simeq \sqrt{e^{1.8h}}\cos\left(\frac{h\pi}{2}\right).$$

From Eq. (19.137) together with (19.138) we obtain, since $\cos \pi\nu$ is large,

$$
\begin{aligned}
S_l &= ie^{il\pi}\left[\cos \pi\nu - \sqrt{\cos^2 \pi\nu - 1 - e^{-iq\pi}}\right] \simeq ie^{il\pi}\cos \pi\nu \frac{1 + e^{-iq\pi}}{2\cos^2 \pi\nu} \\
&\simeq ie^{i(l-q/2)\pi}\left(\frac{\cos\frac{1}{2}\pi q}{\cos \pi\nu}\right).
\end{aligned}
\tag{19.140}
$$

From this we obtain the *absorptivity* $A(l, h)$ of the l-th partial wave, i.e.

$$A(l, h) := 1 - |S_l|^2 \simeq 1 - \left(\frac{\cos\frac{1}{2}\pi q}{\cos \pi\nu}\right)^2 \tag{19.141}$$

[§]$\cos \pi(\nu_R + i\nu_I) = \cos \pi\nu_R \cosh \pi\nu_I - i \sin \pi\nu_R \sinh \pi\nu_I$. This is real for ν_R an integer.

with near asymptotic behaviour (for $h^2 \to \infty$)

$$A(l,h) \overset{(19.139)}{\simeq} 1 - \frac{2\pi(16h)^q}{e^{8h}\{[\frac{1}{2}(q-1)]!\}^2} \simeq 1 - \left(\frac{32}{e^7}\right)^h. \tag{19.142}$$

Thus with Eqs. (19.107b) and (19.142) we can see the behaviour of the S-wave absorptivity as a function of h. This is sketched schematically in Fig. 19.4 (there are tiny fluctuations in the rapid approach to 1). The diminishing fluctuations of $A(l,k)$ in the approach to unity are too small to become evident here.

19.4   Concluding Remarks

In the above we have considered the solution of the Schrödinger equation for the potential r^{-4}. With the exploitation of perturbation solutions of both the periodic Mathieu equation and its associated hyperbolic form for both weak and strong coupling (more precisely h^2) together with corresponding expansions of the Floquet exponent it was possible to go as far as the explicit calculation of the S-matrix and the absorptivity. Apparently this is one of the very rare cases which permits such complete treatment. In essence, one may expect corresponding results for other singular power potentials; it seems it is not possible to guess from the above the result for the absorptivity of a singular potential with an arbitrary negative integral power. Partial or other aspects of the weak coupling (i.e. small h^2) case have been considered by some other authors.[¶] Rudimentary aspects of a singular potential with a general power n have also been considered previously,[‖] as well as other aspects, such as phase shifts.[**] Finally we should mention that the potential r^{-4} together with the associated Mathieu equation have also been studied in interesting contexts of string theory, and, in fact, in further contexts beyond those already mentioned.[††] A highly mathematical study of the potential $\sim r^{-2}$ as an emitting or absorbing centre has been given recently.[‡‡] Some of these sources can be helpful in further investigations.

[¶]J. Challifour and R. J. Eden [46], N. Dombey and R. H. Jones [78], D. Yuan–Ben [285], D. Masson [191], N. Limic [177], and G. Esposito [88].

[‖]G. Tiktopoulos and S. B. Treiman [271], L. Bertocchi, S. Fubini and G. Furlan [20].

[**]R. A. Handelsman, Y.–P. Pao and J. S . Lew [128], A. Paliov and S. Rosendorf [225].

[††]Apart from the papers already referred to, see also M. Cvetic, H. Lü, C. N. Pope and T. A. Tran [61] and M. Cvetic, H. Lü and J. F. Vazquez–Poritz [61].

[‡‡]A. E. Shabad [248].

Chapter 20

Large Order Behaviour of Perturbation Expansions

20.1 Introductory Remarks

The subject of the large order behaviour of perturbation expansions — meaning the study of the late terms of the asymptotic expansion of some function with a view to extracting information about the exact properties of the function — received wide publicity with the publication of the anharmonic oscillator studies of Bender and Wu.[*] However, the subject is much older and had been explored earlier in great detail in particular by Dingle[†] with the subsequent investigation of the behaviour of the late terms of asymptotic expansions of the eigenvalues of the Mathieu equation and others.[‡] In fact, the cosine potential is more suitable for such studies than the anharmonic oscillator since its case is simpler, e.g. the asymptotic solutions in different domains all have the same coefficients, also convergent expansions are known and a lot of literature exists on the equation. Thus in the following we shall not only obtain the large order behaviour of the coefficients of the eigenvalue expansion but also that of the coefficients of the wave functions as well as the connection between these. We shall also see explicitly, how the large order behaviour of the expansion of the eigenvalue is related to the discontinuity across the latter's cut, and how this is related to the level splitting — in fact we shall see this in both cases of the cosine potential and that of the anharmonic oscillator. This is indeed a remarkable connection.

We concentrate in this chapter on the large order behaviour of asymptotic

---

[*]C. M. Bender and T. T. Wu [18], [19].
[†]R. B. Dingle [71].
[‡]R. B. Dingle and H. J. W. Müller [74].

expansions using in particular the recurrence relation of the perturbation coefficients obtainable with the perturbation method of Sec. 8.7. Since the
method is also applicable to convergent expansions, as evident from Example 17.1 and cases in Sec. 19.2.2, one could naturally also explore the large
order behaviour of these in a similar way and expect the behaviour discussed
in Sec. 8.2 and below. Presumably this has never been done so far.

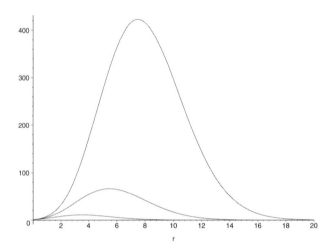

Fig. 20.1 The function $u_r = p^r/r!$ vs. r for $p = 4, 6, 8$.

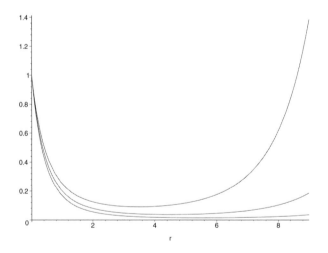

Fig. 20.2 The function $u_r = r!/p^r$ vs. r for $p = 4, 5, 6$.

It is a feature of a physicist's approach to a problem that he employs a
method of approximation which is such that the first term of the corresponding expansion yields a rough value of the desired answer. This approach leads

automatically to asymptotic series, as we discussed earlier in Chapter 8. The behaviour of the r-th term (r large) of the convergent expansion is generally of the form of a power divided by a factorial (as, for instance, in the power expansion of the exponential function), whereas in the case of the asymptotic expansion it is the opposite: a factorial in r divided by a power. The behaviour of such terms is illustrated in Figs. 20.1 and 20.2. It can be seen from these, that the maximum of the absolute value of terms in a convergent expansion and the minimum of the absolute value of terms in an asymptotic expansion are reached approximately when the value of the variable is approximately equal to the number of the term.

The formal perturbation expansion of a quantity $E(g^2)$, i.e.

$$E(g^2) = \sum_{n=0}^{\infty} E_n g^{2n} = E_0 + \sum_{n=0}^{\infty} E_{n+1} g^{2(n+1)}, \tag{20.1}$$

is an *asymptotic series* if for any N

$$\lim_{g^2 \to 0} \frac{1}{g^{2N}} \left[E(g^2) - \sum_{n=0}^{N} E_n g^{2n} \right] = 0 \tag{20.2}$$

or, equivalently for $g = 1/h$,

$$\lim_{h \to \infty} h^{2N} \left[E(h^2) - \sum_{n=0}^{N} \frac{E_n}{h^{2n}} \right] = 0. \tag{20.3}$$

Thus for any given value of g^2, the function $E(g^2)$ is given only approximately by $\sum_{n=0}^{N} E_n g^{2n}$, even if each E_n is known. The traditional method of using an asymptotic expansion is to truncate the series at the least term; this term represents the size of the error. For an authoritative discussion of these aspects we refer to Dingle [70].

A considerable amount of work has recently been devoted to the question whether it is possible to reconstruct a function exactly from its asymptotic expansion. In some cases the answer is affirmative, and the expansion is said to be *Borel summable*. If the function $E(g^2)$ can be written as the Laplace transform

$$E(g^2) = \int_0^{\infty} dz \, e^{-z/g^2} F(z) = g^2 \int_0^{\infty} dt \, e^{-t} F(g^2 t), \tag{20.4}$$

its Laplace transform $F(z)$ is called the *Borel transform* of E, and the integral $E(g^2)$ is called the *Borel sum*. The question arises: What is the information about $F(z)$ that can be obtained from the asymptotic expansion of E, i.e.

Eq. (20.1)? Asymptotic expansions originate from integrating a series (i.e. that of $F(z)$) over a domain that is larger than its region of convergence (see Chapter 8 and Dingle [70]). Thus if

$$F(z) = a_{-1}\delta(z) + \sum_{n=0}^{\infty} a_n z^n, \tag{20.5}$$

then with the integral representation of the gamma function $\Gamma(z) \equiv (z-1)!$,

$$(z-1)! = \int_0^{\infty} dt\, e^{-t} t^{z-1}, \quad \underbrace{\sum_{n=0}^{\infty} E_{n+1} g^{2(n+1)}}_{E(g^2)-E_0} = \sum_{n=0}^{\infty} a_n g^{2(n+1)} \int_0^{\infty} dt\, e^{-t} t^n,$$

we obtain

$$a_{-1} = E_0, \quad a_0 = \frac{E_1}{0!}, \dots, \quad a_n = \frac{E_{n+1}}{n!}.$$

Now if

$$E_{n+1} \simeq (-1)^n n!, \quad (\text{and } E_0 = 0),$$

we obtain

$$F(z) \simeq \sum_{n=0}^{\infty} (-z)^n = \frac{1}{1+z}. \tag{20.6}$$

The power series of $F(z)$ converges inside the unit circle about the origin at $z = 0$. The function $F = 1/(1+z)$ is an analytic function which can be continued beyond this circle to the entire positive z-axis as required by Eq. (20.4). In this ideal case the integral of Eq. (20.4) can be evaluated and $E(g^2)$ thus recovered from the asymptotic expansion by Borel summation. In general this is not so easy. In fact, many asymptotic series do not even alternate in sign (as in the case of the series in Eq. (20.6)). In any case, it is important to know the large order behaviour of the asymptotic series so that the question of its (exact or only approximate) Borel summability can be decided.

We observe that a function $E(h)$ with asymptotic series

$$E(h) = -\sum_{r=1}^{\infty} \frac{A_r}{h^r} \tag{20.7}$$

has an essential singularity at $h = 0$. We can therefore write down a dispersion relation representation choosing the cut from 0 to $+\infty$:

$$E(h) = \frac{1}{\pi} \int_0^{\infty} \frac{\Im E(h')}{h' - h} dh' = -\frac{1}{\pi} \int_0^{\infty} \sum_{r=1}^{\infty} \frac{\Im E(h')}{h} \left(\frac{h'}{h}\right)^{r-1} dh', \tag{20.8}$$

except for possible subtractions. Comparing the last two equations we obtain (by expanding in the latter the denominator in powers of h'/h)

$$A_r = \frac{1}{\pi} \int_0^\infty h'^{r-1} \Im E(h') dh'. \qquad (20.9)$$

The question is therefore: How can we determine $\Im E(h)$, or can we determine A_r independently?

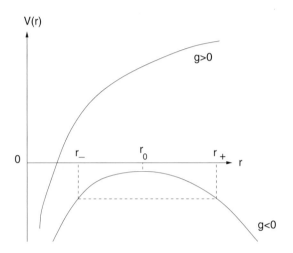

Fig. 20.3 The potential $V(r) = -\frac{1}{r} + gr^N$, $N = 1, 2, \ldots$, for $g \gtrless 0$.

When $\Im E \neq 0$, and E represents an eigenenergy, we have a quantum mechanical problem where probability can leak away, i.e. if the potential or a boundary condition is complex such that the problem is nonselfadjoint. This is the case if the real potential has a hump of some kind, e.g. if the potential is

$$V(r) = -\frac{1}{r} + gr^N, \quad N = 1, 2, \ldots, \qquad (20.10)$$

as shown in Fig. 20.3 and $g < 0$. $\Im E$ can be calculated by the WKB procedure, and with $g = 1/h$ and

$$E = \frac{p^2}{2m} + V(r),$$

it is found that[§]

$$E_n = \epsilon_0^{(n)} - \frac{i}{2}\gamma(g), \quad \text{where} \quad \gamma(g) = \gamma_0 \exp\left\{ -\frac{2}{\hbar} \int_{r_-}^{r_+} |p| dr \right\}, \qquad (20.11)$$

[§]V. S. Popov and V. M. Weinberg [228]. The potential (20.10) is considered in detail in this work.

where the frequency of particle bounces against the potential barrier is $\gamma_0 \simeq \hbar\omega_{\text{classical}}/2\pi$ and $\omega_{\text{classical}} = 2\pi/T_{\text{Kepler}} \propto 1/n^3$, as we expect on the basis of Eq. (14.76) (see also the comment on the comparison with the Kepler problem before Eq. (11.134)). Here n is the principal quantum number of the Coulomb problem (see Eq. (11.113)). Equation (20.11) is, of course, the *Breit–Wigner formula* (10.73) in the WKB or semiclassical approximation. An interesting application is provided by Example 20.1.

Example 20.1: Application to the Yukawa Potential
Evaluate $\Im E = -\gamma/2$ for the Yukawa potential

$$V(r) = -g\frac{e^{-r}}{r}$$

and insert the result into Eq. (20.9) for A_r. Evaluating the integral show that

$$E = \sum_{r=0}^{\infty} E_r g^r, \quad E_r \simeq (-1)^{r+1} r! \left(\frac{n}{2\ln r}\right)^r,$$

where n is the principal quantum number of the Coulomb problem as in Eq. (20.11).

Solution: The solution can be found in a paper of Eletsky, Popov and Weinberg.[¶] The logarithmic factor in E_r is a novel feature of this case.

20.2 Cosine Potential: Large Order Behaviour

It is instructive to investigate various methods for obtaining the large order behaviour of an asymptotic expansion. Thus in the following we apply these methods to our typical examples, the eigenvalue expansions of the cosine potential and of anharmonic oscillators, the cosine potential being the most completely investigated case, as will also be seen in the following.

The most direct way to approach the problem of determining the behaviour of the late terms of an asymptotic expansion is to consider the equation determining these coefficients. Straightforward Rayleigh–Schrödinger perturbation theory usually does not enable this in view of the unwieldy form of coefficients after very few iterations. However, the perturbation method of Sec. 8.7 with its focus on the structure of coefficients of perturbation expansions was seen to permit even the formulation of the recurrence relation of its coefficients. In the first such investigation it was sensible to consider the Schrödinger equation with the cosine potential, since this is effectively the Mathieu equation for which — for comparison purposes — extensive literature was already available, as was (and still is) not the case for the equation

---

with an anharmonic oscillator potential. Thus in the case of the Schrödinger equation with periodic cosine potential we saw (cf. Eqs. (17.58) and (17.34)) that the coefficients $P_i(q, j)$ of functions ψ_{q+4j}, $\psi_q = A_q, B_q, C_q$, arising in the i-th order of the perturbation expansion of the wave function

$$\psi = \text{const.}\left[\psi_q + \sum_{i=1}^{\infty} \frac{1}{(2^7 h)^i} \sum_{j=-i, j\neq 0}^{i} P_i(q, j)\psi_{q+4j}\right], \tag{20.12}$$

obey the recurrence relation

$$\begin{aligned} jP_i(q, j) &= (q + 4j - 1)(q + 4j - 3)P_{i-1}(q, j - 1) \\ &\quad + 2[(q + 4j)^2 + 1 + \triangle]P_{i-1}(q, j) \\ &\quad + (q + 4j + 1)(q + 4j + 3)P_{i-1}(q, j + 1). \end{aligned} \tag{20.13}$$

In the special case $j = i$ the boundary conditions stated after Eq. (17.58) eliminate the last two contributions and one obtains the exact expression

$$P_i(q, i) = \frac{4^i[2i + (q - 1)/2]!}{i![(q - 1)/2]!}.$$

We have seen earlier (cf. Eqs. (17.60), (17.61)) that the coefficients M_{2i} of the expansion of the eigenvalue $\lambda \equiv E(q, h)$, or equivalently of the quantity $\triangle$, i.e. (cf. Eqs. (17.26) and (17.55))

$$\lambda \equiv E(q, h) = -2h^2 + 2hq + \frac{\triangle}{8}, \quad -2\triangle = M_1 + \frac{1}{2^7 h}M_2 + \frac{1}{(2^7 h)^2}M_3 + \cdots,$$

are given by

$$M_{2i} = 2\sum_{j=1}^{i} j[P_i(q, j)]^2 \simeq M_{2i+1} \quad \text{for large } i. \tag{20.14}$$

Thus these coefficients can be obtained once the coefficients $P_i(q, j)$ are known. The recurrence relation (20.13) represents effectively a partial difference equation. With some simplifications for large values of i this equation can be solved approximately and the large order behaviour of the coefficients $P_i(q, j)$ and M_{2i} can be deduced. This has been done[||] and it was found that (we cite the results here since we rederive them below in detail by other methods, to avoid discussion of difference equations here)

$$P_i(q, j) \simeq \frac{(16)^i(i - 1)!(2i)^{q/2}}{[\frac{1}{4}(q - 1)]![\frac{1}{4}(q - 3)]!}\left(1 - \frac{2q^2 - 2q + 3}{8r}\right)\binom{i}{j}, \tag{20.15a}$$

[||] R. B. Dingle and H. J. W. Müller [74].

where the bracketed expression at the end is the binomial coefficient, also written ^iC_j, and from this for i even or odd it is found that

$$
M_i \simeq (16)^i i! \left(1 - \frac{2q^2 + 3}{2i}\right) \left[\frac{i^{q-1}}{\{[\frac{1}{4}(q-1)]![\frac{1}{4}(q-3)]!\}^2}\right.
$$
$$
\left. - \frac{(-1)^i i^{-q-1}}{\{[-\frac{1}{4}(q+1)]![-\frac{1}{4}(q+3)]!\}^2}\right]. \tag{20.15b}
$$

The i-th term in the expansion of the eigenvalues λ of the Schrödinger equation with cosine potential (Mathieu equation) is therefore given by

$$
i^{\mathrm{th}} \text{ term} \simeq -\frac{i!}{(8h)^{i-1}} \left(1 - \frac{2q^2 + 3}{2i}\right) \left[\frac{i^{q-1}}{\{[\frac{1}{4}(q-1)]![\frac{1}{4}(q-3)]!\}^2}\right.
$$
$$
\left. - \frac{(-1)^i i^{-q-1}}{\{[-\frac{1}{4}(q+1)]![-\frac{1}{4}(q+3)]!\}^2}\right]
$$
$$
\simeq -\frac{i! i^{q-1}}{\{[\frac{1}{4}(q-1)]![\frac{1}{4}(q-3)]!\}^2} \frac{1}{(8h)^{i-1}}, \tag{20.16a}
$$

or (with the help of the duplication formula, see also below)

$$
i^{\mathrm{th}} \text{ term} \simeq -\frac{i! i^{q-1} 2^q}{\{[\frac{1}{2}(q-1)]!\}^2 2\pi} \frac{1}{(8h)^{i-1}}. \tag{20.16b}
$$

With the help of *Stirling's formula* one can derive the following relation:

$$
\frac{i!}{(i+m)!} \overset{i \text{ large}}{\simeq} \frac{1}{i^m}\left[1 - O\left(\frac{1}{i}\right)\right], \tag{20.17}
$$

We see that assuming i is very large compared with m is here equivalent to assuming m is approximately zero. Using this relation and the *duplication formula*

$$
(z-1)!\left(z - \frac{1}{2}\right)! = \frac{(2z-1)!\sqrt{\pi}}{2^{2z-1}}, \tag{20.18}
$$

the result becomes

$$
i^{\mathrm{th}} \text{ term} \simeq -\frac{(i+2n)! 2^{2n}}{\pi(n!)^2 (8h)^{i-1}} \quad \text{where} \quad n = \frac{1}{2}(q-1). \tag{20.19}
$$

We observe that successive terms do not alternate in sign. Hence exact Borel summation is not possible.

In the following we concentrate on the cosine potential and rederive the large order behaviour obtained above by other methods. In the first method we do this by first obtaining the imaginary part of the eigenvalue with the help of nonselfadjoint boundary conditions.

20.3   Cosine Potential: Complex Eigenvalues

20.3.1   The decaying ground state

Our intention here is to consider again the cosine potential and hence the Mathieu equation with the same large-h^2 solutions obtained earlier, but with different boundary conditions. This is a very instructive exercise which shows how the eigenvalues change with a change in boundary conditions. From our previous considerations it is clear that the formal perturbation expansion of the eigenvalue $\lambda(q)$ in terms of the parameter q, when this is $q_0 = 2n+1, n = 0, 1, 2, \ldots$, remains unchanged. What changes as a result of different boundary conditions is the deviation of q from the odd integer q_0, i.e. the difference $q - q_0$. Previously, in the calculation of the level splitting as a result of tunneling, this difference was found to be real. We retain the previous boundary condition at one minimum, which means that the potential is there approximated by an infinitely high harmonic oscillator well with levels enumerated by the quantum number q_0 or (equivalently) n, but now change the boundary condition at the neighbouring minimum in such a way that the difference $q - q_0$ becomes imaginary. A means to achieve this was found by Stone and Reeve.* However, we follow here a formulation in the context of our earlier considerations of the Mathieu equation.† To enable comparison with the work of Stone and Reeve we shift the cosine potential considered previously by $\pi/2$. The potential we consider here then has the shape shown in Fig. 20.4.

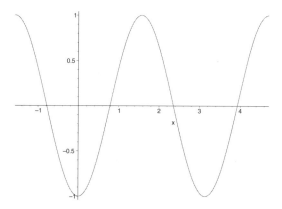

Fig. 20.4 The cosine potential shifted by $\pi/2$.

At the risk of repeating what we explained earlier, we emphasize a few points

---

*M. Stone and J. Reeve [262].
†H. J. W. Müller–Kirsten [200].

for reasons of clarity (since some considerations here may otherwise not be sufficiently clear). Effectively we consider the tunneling from the left well in Fig. 20.4 (around $z = 0$) to the one on the right (around $z = \pi$). Thus around $z = 0$ the potential behaves like that of an oscillator well with minimum there, and correspondingly we assume alternately even and odd oscillator eigenfunctions there, as we found these for this potential. The physical wave functions there are therefore the solutions which are real and exponentially decreasing. Previously this situation was the same at the neighbouring minimum, the degeneracy of eigenvalues of neighbouring oscillators being lifted by the finite height of the barrier separating them. Now we change this situation at the minimum on the right. We can distort the potential there or alternatively impose a different behaviour there on the wave function. Here we choose the second way, i.e. a boundary condition which at the face looks somewhat abstract, but achieves exactly what we want. Thus one chooses not the solution with real argument at the minimum, but that with complex argument there, and allows its imaginary part to tend to infinity. To be more precise we recall the Mathieu equation we had earlier:

$$\psi''(z) + [\lambda - 2h^2 \cos 2z]\psi(z) = 0. \qquad (20.20\text{a})$$

As stated, we shift the argument by $\pi/2$, which means replacing z by $z \pm \pi/2$, and thus obtain

$$\psi''(z) + [\lambda + 2h^2 \cos 2z]\psi(z) = 0 \qquad (20.20\text{b})$$

with potential $V(z) = -2h^2 \cos 2z$ as shown in Fig. 20.4. Around $z \sim 0$ we can expand and obtain

$$\psi''(z) + [\lambda + 2h^2 - 4h^2 z^2 \cdots]\psi(z) = 0.$$

Setting

$$w = 2h^{1/2}z, \quad \lambda + 2h^2 = 2hq + \frac{\triangle}{8}, \qquad (20.21)$$

the equation can be approximated by

$$\frac{d^2\psi(w)}{dw^2} + \left\{ \frac{1}{2}q - \frac{w^2}{4} \right\}\psi(w) \simeq 0. \qquad (20.22)$$

The solutions of this equation are the real and complex *parabolic cylinder functions*

$$D_{\frac{1}{2}(q-1)}(\pm w), \quad D_{-\frac{1}{2}(q+1)}(\pm iw).$$

In the following we find it convenient to select solutions of this type in the domains around the minima of the potential in Fig. 20.4.

As explained above, we now choose an harmonic-oscillator type of real solution around the left minimum of Fig. 20.4. The proper harmonic oscillator solution there would be that for infinitely high barriers to the left and to the right with a dependence on an integer n which enumerates the quantum states in the well from the even or symmetric ground state with $n = 0$ through alternately then odd and even states upwards with $n = 1, 2, 3, \ldots$. Here — again as before — we take the finite height of the barrier to the right of $z = 0$ into account by taking q only approximately equal to $q_0 = 2n + 1$, so that the deviation $q - q_0$ results from the finite height of the barrier taken into account with boundary conditions at the next minimum, i.e. that at $z = \pi$, such that this deviation becomes imaginary. This somewhat abstract looking boundary condition is effectively simply one imposed on a complex solution of the equation giving the oscillator eigenfunctions. One should note the difference to our earlier case of the calculation of the level splitting: There we were seeking solutions which are even or odd about the central maximum of the potential, whereas here we start off with a solution even or odd about a minimum. Since the large-h^2 solutions valid away from the minima were the solutions of types $A, \overline{A}$ (which we can loosely dub solutions of WKB-type), we naturally constructed the even and odd solutions with these (cf. Eq. 17.63), whereas here we construct these from oscillator-type solutions.

Thus we write the boundary condition at $\Re z = \pi$:

$$\psi(\pm iw) \to 0 \quad \text{for} \quad \left. \begin{array}{l} w = 2h^{1/2}z, \\ z \to \pi \pm i\infty \end{array} \right\}, \tag{20.23}$$

i.e. (where for the ground state $q = q_0 = 1$)

$$D_{-\frac{1}{2}(q+1)}(\pm i2h^{1/2}z) \to 0 \quad \text{for} \quad z \to \pm i\infty.$$

We consider here explicitly only the case of the ground state around $z = 0$. The calculation for the general case is nontrivial and so will first only be written down and will then be verified by specialization to the case of the ground state. A derivation is given subsequently, however, by a different method.

Thus, considering around $z = 0$ the ground state of the harmonic oscillator for which $q_0 = 1$, and replacing q_0 by q, the even or symmetric solution about $z = 0$ can be taken as

$$\psi_s = \frac{1}{2}[D_{\frac{1}{2}(q-q_0)}(2h^{1/2}z) + D_{\frac{1}{2}(q-q_0)}(-2h^{1/2}z)]. \tag{20.24}$$

Considering $w \to 0$ in Eq. (20.22), we see that

$$\psi_s \sim \cos\sqrt{2qh}z \quad \text{for} \quad z \sim 0.$$

Although the two parabolic cylinder functions in Eq. (20.24) are linearly independent around $z = 0$ for $q \neq q_0$, we know from the solutions of types B and C in Sec. 17.3.2 that one parabolic cylinder function is exponentially increasing in h and the other exponentially decreasing.

As before we set

$$\lambda = -2h^2 + 2hq + \frac{\triangle}{8} \equiv E_1 + E_2, \tag{20.25}$$

where in the present case

$$E_1 = -2h^2 + 2hq_0 + \frac{\triangle}{8} \quad \text{and} \quad \frac{1}{2}(q - q_0) = \frac{E_2}{4h}. \tag{20.26}$$

Thus here we can approximate ψ_s by expanding it about $E_2 = 0$:

$$\psi_s \simeq D_0(2h^{1/2}z) + E_2 \left[\frac{\partial}{\partial E_2} \frac{1}{2} \left\{ D_{E_2/4h}(2h^{1/2}z) + D_{E_2/4h}(-2h^{1/2}z) \right\} \right]_{E_2=0}. \tag{20.27}$$

The differentiation of the parabolic cylinder functions with respect to their indices is nontrivial. However, this can be handled in a few lines. The arguments of the functions appearing in Eq. (20.27) differ by an angle π. This is the feature which gives rise to an exponentially increasing contribution in the neighbourhood of $z = 0$ and hence provides the dominant contribution in going away from the origin. We see this as follows.

From Tables of Special Functions[‡] we obtain the following circuit formula of parabolic cylinder functions $D_\nu(w)$:

$$D_\nu(w) = e^{-i\nu\pi} D_\nu(-w) + \frac{\sqrt{2\pi}}{\Gamma(-\nu)} e^{-i(\nu+1)\frac{\pi}{2}} D_{-\nu-1}(iw). \tag{20.28}$$

Extracting $D_\nu(-w)$ we have

$$D_\nu(-w) = e^{i\nu\pi} D_\nu(w) - \frac{\sqrt{2\pi}}{\Gamma(-\nu)} e^{i(\nu-1)\frac{\pi}{2}} D_{-\nu-1}(iw). \tag{20.29}$$

Equation (20.27) requires the differentiation of this expression with respect to the index ν, and then evaluation at $\nu = 0$. Setting $\nu = 0$ in $\Gamma(-\nu) \equiv (-\nu-1)!$ gives infinity. Thus the only nonvanishing contribution comes from differentiation of $\Gamma(-\nu)$. The differentiated function $\Gamma(\nu)$ is in Tables of Functions expressed in terms of the so-called *psi function*, $\psi(\nu)$, or logarithmic derivative of $\Gamma(\nu)$ for which:[§]

$$\psi(\nu) = \frac{\Gamma'(\nu)}{\Gamma(\nu)}. \tag{20.30}$$

[‡]See e.g. W. Magnus and F. Oberhettinger [181], p. 92.
[§]See e.g. W. Magnus and F. Oberhettinger [181], p. 2.

From formulas given in the literature we deduce the limiting behaviour[¶]

$$\left[\frac{\psi(\nu)}{\Gamma(\nu)} \right]_{\nu \to 0} \to \left[-\frac{1}{\nu} \nu \right] = -1. \tag{20.31}$$

It follows that

$$\frac{d}{d\nu} \left[\frac{1}{\Gamma(-\nu)} \right] = -\frac{1}{\Gamma^2(-\nu)} \frac{d\Gamma(-\nu)}{d\nu} = \frac{\psi(-\nu)}{\Gamma(-\nu)} \xrightarrow{\nu \to 0} -1.$$

We now return to Eq. (20.29). Differentiation of this equation and subsequently setting $\nu = 0$ yields, with the following expression for the large-z behaviour[‖] (here $\nu = E_2/4h$)

$$D_\nu(w) \simeq w^\nu e^{-w^2/4} = e^{\nu \ln w} e^{-w^2/4} \quad \text{for} \quad |\arg w| < \frac{3}{4}\pi, \tag{20.32}$$

which we can use here since all other problems have been resolved:

$$\frac{d}{d\nu} D_\nu(-w) \Big|_{\nu=0} \overset{(20.29)}{=} (i\pi + \ln w) D_0(w) + \sqrt{2\pi} e^{-i\frac{\pi}{2}} D_{-1}(iw)$$

$$= (i\pi + \ln w) e^{-w^2/4} - \frac{i\sqrt{2\pi}}{iw} e^{+w^2/4}. \tag{20.33}$$

Inserting this result into ψ_s and retaining only the dominant contributions (i.e. dropping from the derivative part of Eq. (20.27) exponentially decreasing contributions, including the derivative of $D_{E_2/4h}(2h^{1/2}z)$), we obtain

$$\psi_s \simeq D_0(2h^{1/2}z) + \frac{1}{2}\frac{E_2}{4h}\left[O(e^{-hz^2}) - \frac{\sqrt{2\pi} e^{hz^2}}{2h^{1/2}z} \right]$$

$$= D_0(2h^{1/2}z) - \frac{E_2\sqrt{2h\pi}}{16h^2} \frac{e^{hz^2}}{z}. \tag{20.34}$$

This is the behaviour of the symmetric ground state solution away from the origin towards the limit of its domain of validity. We now go to the WKB-type solutions, i.e. those of type A, which are valid around $z \sim \pi/2$, and

---

[¶]From W. Magnus and F. Oberhettinger [181], pp. 3 - 4, we obtain the formulas

$$\frac{1}{\Gamma(\nu)} = \nu e^{C\nu} \prod_{n=1}^{\infty} \left(1 + \frac{\nu}{n} \right) e^{-\nu/n} \quad \text{and} \quad \psi(\nu) = -C - \frac{1}{\nu} + \nu \sum_{k=1}^{\infty} \frac{1}{k(\nu + k)},$$

where C = Euler's constant $= 0.577\ldots$. For $\nu \to 0$ one obtains the result (20.31). This result is not easily generalized to $q_0 > 1$. Thus calculations for $q_0 \neq 1$ are not given by M. Stone and J. Reeve [262].

[‖]See e.g. W. Magnus and F. Oberhettinger [181], p. 92.

extend these to their limit of validity to the left, where we can then match these to the solution ψ_s.

Recalling that we have shifted the potential by $\pi/2$, we obtain the required solutions from the earlier ones, i.e. Eqs. (17.29) and (17.40), by there replacing z by $z + \pi/2$. Thus in the present case and with $q \simeq q_0 = 1$ we obtain

$$\psi_A \simeq \frac{e^{2h\cos z}}{\cos\frac{z}{2}} \quad \text{and} \quad \psi_{\overline{A}} \simeq \frac{e^{-2h\cos z}}{-\sin\frac{z}{2}}. \tag{20.35}$$

We construct with constants c_1, c_2 the WKB-like linear combination

$$\psi_{\text{WKB}} = c_1\psi_A + c_2\overline{\psi}_A \simeq c_1\frac{e^{2h\cos z}}{\cos\frac{z}{2}} + c_2\frac{e^{-2h\cos z}}{-\sin\frac{z}{2}}. \tag{20.36}$$

We now want to match this solution to that of Eq. (20.34). We observe that in the matching domain (i.e. with expansion of $\cos z$ in the exponential)

$$D_0(2h^{1/2}z) = e^{-hz^2} \simeq e^{-2h}\left(\frac{e^{2h\cos z}}{\cos\frac{z}{2}}\right). \tag{20.37}$$

Comparing Eqs. (20.34) and (20.36) we can identify the dominant large-h^2 behaviour of the constant c_1 as $c_1 = e^{-2h}$. Thus the WKB-like solution becomes

$$\psi_{\text{WKB}} \simeq D_0(2h^{1/2}z) + c_2\frac{e^{-2h\cos z}}{-\sin\frac{z}{2}}$$

$$\simeq D_0(2h^{1/2}z) + c_2\frac{e^{-2h}e^{hz^2}}{-\frac{z}{2}} \tag{20.38}$$

in its limiting domain towards $z = 0$.

We now return to the WKB-like solution (20.36) and consider this at the other end of its domain of validity, i.e. that in approaching $z = \pi$. To this end we set $z' = z - \pi$. Then ψ_{WKB} becomes (with $\cos z \simeq -1 + z'^2/2$, $\cos z/2 \simeq -z'/2$)

$$\psi_{\text{WKB}} \overset{(20.36)}{\simeq} -\frac{c_1 e^{-2h}e^{hz'^2}}{\frac{z'}{2}} - c_2 e^{2h}e^{-hz'^2}. \tag{20.39}$$

Now, near $z = \pi$ the boundary condition (20.23) demands (with constant A)

$$\psi = AD_{-1}(\pm i2h^{1/2}z') \to 0 \quad \text{for} \quad z' \to \pm i\infty. \tag{20.40}$$

From Magnus and Oberhettinger [181] (p. 92) we take the formula:

$$D_\nu(w) = \frac{\Gamma(\nu+1)}{\sqrt{2\pi}}\left[e^{i\nu\pi/2}D_{-\nu-1}(iw) + e^{-i\nu\pi/2}D_{-\nu-1}(-iw)\right]. \tag{20.41}$$

Thus for $\nu = 0$ we have

$$D_0(w) = \frac{1}{\sqrt{2\pi}}[D_{-1}(iw) + D_{-1}(-iw)] \equiv \text{real},$$

and hence (adding and subtracting equal terms)

$$
\begin{aligned}
& D_{-1}(i2h^{1/2}z') \\
= \ & \frac{1}{2}[D_{-1}(i2h^{1/2}z') + D_{-1}(-i2h^{1/2}z')] \\
& + \frac{1}{2}[D_{-1}(i2h^{1/2}z') - D_{-1}(-i2h^{1/2}z')] \\
= \ & \frac{1}{2}\sqrt{2\pi}D_0(2h^{1/2}z') + \frac{1}{2}[D_{-1}(i2h^{1/2}z') - D_{-1}(-i2h^{1/2}z')] \\
= \ & \sqrt{\frac{\pi}{2}}D_0(2h^{1/2}z') + \frac{1}{2}[D_{-1}(i2h^{1/2}z') - \text{complex conjugate}] \\
\overset{(20.32)}{\simeq} \ & \sqrt{\frac{\pi}{2}}e^{-hz'^2} + \frac{2}{2}(i2h^{1/2}z')^{-1}e^{hz'^2}.
\end{aligned}
\tag{20.42}
$$

With this result Eq. (20.40) can be written

$$\psi \simeq A\sqrt{\frac{\pi}{2}}e^{-hz'^2} + A\frac{2}{2}(i2h^{1/2}z')^{-1}e^{hz'^2}. \tag{20.43}$$

Comparing now Eqs. (20.39) and (20.43) we obtain, with $c_1 = \exp(-2h)$ from above,

$$
\left.
\begin{aligned}
A &= \pm\tfrac{1}{i}4h^{1/2}e^{-4h}, \\
c_2 &= \mp\tfrac{1}{i}2(2h\pi)^{1/2}e^{-6h}.
\end{aligned}
\right\}
\tag{20.44}
$$

Hence Eq. (20.38) can be written

$$\psi_{\text{WKB}} = D_0(2h^{1/2}z) \pm \frac{4}{i}(2h\pi)^{1/2}e^{-8h}\frac{e^{hz^2}}{z}. \tag{20.45}$$

Comparing this result with ψ_s of Eq. (20.34) we obtain

$$E_2 \equiv i\Im\lambda = \pm ih^2 2^6 e^{-8h}. \tag{20.46}$$

Repeating these calculations for the general case (i.e. for both even and odd harmonic oscillator wave functions around $z = 0$ with $q_0 > 1$) is nontrivial. The general result stated by Stone and Reeve [262] is therefore presumably guessed, i.e.

$$E_2 \simeq \overset{+}{(-)} i\frac{(16h)^{q_0+1}e^{-8h}}{4\{[\frac{1}{2}(q_0 - 1)]!\}^2}. \tag{20.47}$$

For $q_0 = 1$ we recover the result (20.46). We derive the general result in the next subsection using our own method.

20.3.2   Decaying excited states

For our derivation of the generalization[**] of Eq. (20.46) we consider a half period of the cosine potential as illustrated in Fig. 20.5.

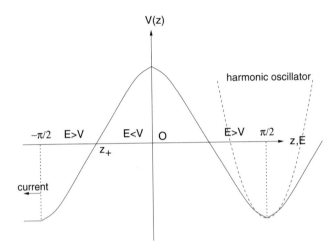

Fig. 20.5 The cosine potential from $-\pi/2$ to $\pi/2$.

We have to impose two sets of boundary conditions. At the minimum to the right of the barrier shown in Fig. 20.5, i.e. at $z = \pi/2$, we impose the usual type of boundary conditions as for the minimum of a simple harmonic oscillator there, i.e.

$$y'_+\left(\frac{\pi}{2}\right) = 0, \quad y_-\left(\frac{\pi}{2}\right) = 0, \quad y_+\left(\frac{\pi}{2}\right) \neq 0, \quad y'_-\left(\frac{\pi}{2}\right) \neq 0. \qquad (20.48)$$

These conditions provide the required quantum number $q_0 = 2n + 1$.

On the other side of the barrier at $z = -\pi/2$ we consider different boundary conditions. Considering the original equation, i.e. Eq. (20.20a), and expanding $\cos 2z$ around the point $z = -\pi/2$, this equation may be approximated there as

$$\frac{d^2\psi}{dz^2} + \left[E + 2h^2 - 4h^2\left(z + \frac{\pi}{2}\right)^2 + \cdots\right]\psi = 0.$$

Thus in the immediate neighbourhood of $z = -\pi/2$ the wave function $\psi(z)$ behaves as

$$\psi \sim e^{\pm i(E+2h^2)^{1/2}z} \sim e^{\pm i(2hq)^{1/2}z}$$

[**]We follow P. Achuthan, H. J. W. Müller–Kirsten and A. Wiedemann [4]; the result is formula (334) there.

(using Eq. (20.21)). For particles or probability to leak through the potential barrier from the trough at $z = -\pi/2$ in the direction of negative values of z where we choose the potential to be zero, we demand that

$$\psi(z)|_{z<-\pi/2} \sim e^{-i(2hq)^{1/2}z}. \tag{20.49}$$

This is a complex boundary condition which requires the coefficient of the other exponential in $\psi_{\pm}(z)$ (continued to and beyond $-\pi/2$) to vanish. For the continuation we use again the WKB formalism.

(A) Boundary conditions at the right minimum
We begin with the evaluation of the boundary conditions at $z = \pi/2$. Proceeding as in our previous cases and substituting the expression (20.21) into the Mathieu equation we obtain, as shown earlier, three pairs of solutions of types named A, B and C. Recapitulating these from Chapter 17, we have

$$\psi_A(z) = e^{2h \sin z}\left[A_q(z) + O\left(\frac{1}{h}\right)\right],$$

$$\overline{\psi}_A(q, h; z) = \psi_A(q, h; -z) = \psi_A(-q, -h; z), \tag{20.50a}$$

where (the second expression is needed for later comparison)

$$A_q(z) = \frac{\cos^{\frac{1}{2}(q-1)}(\frac{1}{2}z + \frac{1}{4}\pi)}{\sin^{\frac{1}{2}(q+1)}(\frac{1}{2}z + \frac{1}{4}\pi)} = \frac{[\tan(\frac{1}{2}z + \frac{1}{4}\pi)]^{-q/2}}{[\sin(\frac{1}{2}z + \frac{1}{4}\pi)\cos(\frac{1}{2}z + \frac{1}{4}\pi)]^{1/2}}, \tag{20.50b}$$

which are valid as asymptotically decreasing expansions in the domain away from a minimum, i.e. for

$$\left|\cos\left(\frac{1}{2}z \pm \frac{1}{4}\pi\right)\right| > O\left(\frac{1}{h^{1/2}}\right).$$

The second pair of solutions is

$$\psi_B(z) = e^{2h \sin z}\left[B_q[w(z)] + O\left(\frac{1}{h^{1/2}}\right)\right],$$

$$\overline{\psi}_B(z) = e^{-2h \sin z}\left[\overline{B}_q[w(z)] + O\left(\frac{1}{h^{1/2}}\right)\right], \tag{20.51a}$$

where

$$B_q[w(z)] = \frac{H_{\frac{1}{2}(q-1)}(w)}{2^{\frac{1}{4}(q-1)}[\frac{1}{4}(q-1)]!}, \quad \overline{B}_q[w(z)] = B_q[w(-z)],$$

$$w(z) = 4h^{1/2}\cos\left(\frac{1}{2}z + \frac{1}{4}\pi\right), \tag{20.51b}$$

the function $H_\nu(w)$ being a Hermite function. These solutions are valid asymptotically with the first in the domain around $z = \pi/2$ and the second in the domain around $z = -\pi/2$. Finally the third pair of solutions is

$$
\begin{aligned}
\psi_C(z) &= e^{2h\sin z}\left[C_q[w(-z)] + O\left(\frac{1}{h^{1/2}}\right)\right], \\
\overline{\psi}_C(z) &= e^{-2h\sin z}\left[\overline{C}_q[w(-z)] + O\left(\frac{1}{h^{1/2}}\right)\right], \\
\overline{C}_q[w(-z)] &= C_q[w(z)],
\end{aligned}
\qquad (20.52a)
$$

where

$$
C_q[w(z)] = 2^{\frac14(q+1)}\left[\frac14(q-3)\right]! \, e^{i\frac{\pi}{4}(q+1)} H_{-\frac12(q+1)}(iw). \qquad (20.52b)
$$

The first solution is valid asymptotically around $z = -\pi/2$, and the second around $z = \pi/2$.

We saw in Chapter 17 that some of these solutions can be matched in regions of common validity. Thus we have

$$
\psi_B(z) = \alpha\psi_A(z) \quad \text{and} \quad \overline{\psi}_C(z) = \overline{\alpha}\,\overline{\psi}_A(z), \qquad (20.53)
$$

where

$$
\alpha = \frac{(8h)^{\frac14(q-1)}}{\left[\frac14(q-1)\right]!}\left[1 + O\left(\frac1h\right)\right], \quad \overline{\alpha} = \frac{\left[\frac14(q-3)\right]!}{(8h)^{\frac14(q+1)}}\left[1 + O\left(\frac1h\right)\right]. \qquad (20.54)
$$

The even and odd solutions are defined as

$$
\psi_\pm(z) = \frac12[\psi_A(z) \pm \overline{\psi}_A(z)]. \qquad (20.55)
$$

Extending these solutions to the domain around the minimum at $z = \pi/2$, we obtain

$$
\psi_\pm(z) = \frac12\left[\frac1\alpha\psi_B(z) \pm \frac1{\overline{\alpha}}\overline{\psi}_C(z)\right]. \qquad (20.56)
$$

Applying the boundary conditions (20.48) and proceeding as in Chapter 17, we obtain our first transcendental equation, i.e. the relation

$$
q - q_0 \simeq \mp 2\left(\frac2\pi\right)^{1/2}\frac{(2^8 h^2)^{q/4}e^{-4h}}{\left[\frac12(q-1)\right]!}\left[1 + O\left(\frac1h\right)\right], \qquad (20.57)
$$

where $q_0 = 1, 3, 5, \ldots$. Here the upper sign refers to the even states and the lower to the odd states. We return to this condition after derivation of the second transcendental equation from the boundary condition at the other minimum.

(B) Boundary condition at the left minimum

The solutions of type A have been matched to the oscillator-type solutions at the right minimum. The solutions of type A are valid to the right as well as to the left of the maximum of the barrier. Our procedure now is to match these to exponential WKB solutions. Thereafter we extend the latter across the turning point down to periodic WKB solutions. These then provide complex exponentials like (20.49) and we can apply our boundary condition. Some nontrivial manipulations are again required to uncover the generation of important factorials.

Inserting the approximate form of the eigenvalue into the Schrödinger equation we obtain

$$\frac{d^2\psi}{dz^2} + \left[2hq + \frac{\triangle}{8} - 4h^2 \cos^2 z \right] \psi = 0. \tag{20.58}$$

Ignoring the correction $\triangle/8$, we see that the turning points at $z_{\pm}$ are given by

$$2hq - 4h^2 \cos^2 z \simeq 0,$$

i.e.

$$\cos z_{\pm} \simeq \pm \left(\frac{q}{2h} \right)^{1/2}, \quad -\frac{\pi}{2} < z_+ < \frac{\pi}{2}. \tag{20.59}$$

The WKB solutions to the right of z_+ (where $V > \Re E$ to the left of the maximum of the barrier) are given by

$$\psi_{\text{WKB}}^{(r,z_+)}(z) \simeq \frac{\exp(\int_{z_+}^z [4h^2 \cos^2 z - 2hq]^{1/2} dz)}{[4h^2 \cos^2 z - 2hq]^{1/4}},$$

$$\overline{\psi}_{\text{WKB}}^{(r,z_+)}(z) \simeq \frac{\exp(-\int_{z_+}^z [4h^2 \cos^2 z - 2hq]^{1/2} dz)}{[4h^2 \cos^2 z - 2hq]^{1/4}}. \tag{20.60}$$

We know that the solutions $\psi_A(z), \overline{\psi}_A(z)$ overlap parts of the domains of validity of these solutions. We begin therefore with the matching of $\psi_{\text{WKB}}^{(r,z_+)}(z)$ to $\psi_A(z)$ and $\overline{\psi}_{\text{WKB}}^{(r,z_+)}(z)$ to $\overline{\psi}_A(z)$. Hence it is necessary to consider the elliptic integral[††]

$$I \equiv \int_{z_+}^z [4h^2 \cos^2 z - 2hq]^{1/2} dz. \tag{20.61}$$

[††]In principle the integral could be evaluated exactly — see P. F. Byrd and M. D. Friedman [40], formula 280.01, p. 163 — but here we are interested in the WKB solution in a domain where it is proportional to a solution of type A, and we obtain the proportionality by appropriate expansion, as we shall see with the result of Eq. (20.67).

Since $\cos(z + \pi/2) = -\sin z$, we write

$$\cos^2 z = 1 - \sin^2 z = 1 - \cos^2\left(z + \frac{\pi}{2}\right) = \left(z + \frac{\pi}{2}\right)^2 + O\left[\left(z + \frac{\pi}{2}\right)^4\right].$$

Inserting this into the integral I, we obtain

$$
\begin{aligned}
I &= \int_{z_+}^{z} \left[4h^2\left(z + \frac{\pi}{2}\right)^2 - 2hq + O\left\{h^2\left(z + \frac{\pi}{2}\right)^4\right\}\right]^{1/2} dz \\
&\simeq 2h \int_{z_+}^{z} dz \left[\left(z + \frac{\pi}{2}\right)^2 - \frac{q}{2h}\right]^{1/2} = 2h \int_{z_+ + \pi/2}^{z + \pi/2} \left(z'^2 - \frac{q}{2h}\right)^{1/2} dz'.
\end{aligned}
$$

Integrating this becomes

$$
\begin{aligned}
I &= 2h\left[\frac{1}{2}z'\left(z'^2 - \frac{q}{2h}\right)^{1/2} - \frac{q}{4h}\ln\left|z' + \left(z'^2 - \frac{q}{2h}\right)^{1/2}\right|\right]_{z_+ + (\pi/2)}^{z + (\pi/2)} \\
&= h\left[\left(z + \frac{\pi}{2}\right)^2 - \frac{q}{4h} + \frac{q}{2h}\ln\left|\left(\frac{q}{2h}\right)^{1/2}\right|\right. \\
&\quad \left. -\frac{q}{2h}\ln\left|\left(z + \frac{\pi}{2}\right) + \left\{\left(z + \frac{\pi}{2}\right)^2 - \frac{q}{2h}\right\}^{1/2}\right|\right], \tag{20.62}
\end{aligned}
$$

where we set — in accordance with our expansion —

$$\left[\left(z_+ + \frac{\pi}{2}\right)^2 - \frac{q}{2h}\right]^{1/2} \simeq \left[\cos^2 z_+ - \frac{q}{2h}\right]^{1/2} \overset{(20.59)}{\simeq} 0.$$

In the same spirit we can set

$$h\left(z + \frac{\pi}{2}\right)^2 \simeq 2h\left[1 - \cos\left(z + \frac{\pi}{2}\right)\right] = 2h[1 + \sin z]. \tag{20.63}$$

We handle the logarithmic terms in Eq. (20.62) as follows, using the relation $1 + \sin z = 2\sin^2(z/2 + \pi/4)$. We have for the following expression contained in (20.62):

$$
\begin{aligned}
L : &\equiv \ln\left|\left(z + \frac{\pi}{2}\right) + \left\{\left(z + \frac{\pi}{2}\right)^2 - \frac{q}{2h}\right\}^{1/2}\right| - \ln\left|\left(\frac{q}{2h}\right)^{1/2}\right| \\
&\simeq \ln\left|\frac{2(z + \frac{\pi}{2})}{(q/2h)^{1/2}}\right| = \ln\left|\frac{2(z + \frac{\pi}{2})^2}{(q/2h)^{1/2}(z + \frac{\pi}{2})}\right| \simeq \ln\left|\frac{4(1 + \sin z)}{(q/2h)^{1/2}\cos z}\right|,
\end{aligned}
$$

and hence

$$L = \ln\left|\frac{8\sin^2(\frac{z}{2} + \frac{\pi}{4})}{(q/2h)^{1/2}2\sin(\frac{z}{2} + \frac{\pi}{4})\cos(\frac{z}{2} + \frac{\pi}{4})}\right| = \ln\left|\frac{4\tan(\frac{z}{2} + \frac{\pi}{4})}{(q/2h)^{1/2}}\right|. \tag{20.64}$$

Inserting the relations (20.63) and (20.64) into Eq. (20.62), we obtain

$$I \simeq 2h + 2h \sin z - \frac{1}{4}q - \frac{q}{2} \ln \left| 4 \left(\frac{2h}{q} \right)^{1/2} \tan \left(\frac{z}{2} + \frac{\pi}{4} \right) \right|. \qquad (20.65)$$

Hence

$$\exp \left(\int_{z_+}^{z} [4h^2 \cos^2 z - 2hq]^{1/2} dz \right) \simeq \frac{e^{2h} e^{2h \sin z} e^{-q/4} (q/2h)^{q/4}}{2^q [\tan(\frac{z}{2} + \frac{\pi}{4})]^{q/2}}. \qquad (20.66)$$

The WKB solution therefore becomes (approximating the denominator of (20.60) by $(4h^2 \cos^2 z)^{1/4}$), on using in the second step Eqs. (20.50a) and (20.50b),

$$
\begin{aligned}
\psi_{\text{WKB}}^{(r,z_+)}(z) &\simeq \frac{e^{2h} e^{2h \sin z} e^{-q/4} (q/2h)^{q/4} [\tan(\frac{z}{2} + \frac{\pi}{4})]^{-q/2}}{2^q (4h^2)^{1/4} \{2 \sin(\frac{z}{2} + \frac{\pi}{4}) \cos(\frac{z}{2} + \frac{\pi}{4})\}^{1/2}} \\
&\simeq \frac{e^{2h} e^{-q/4} (q/4)^{q/4}}{2^{q+\frac{1}{2}} (4h^2)^{1/4}} \left(\frac{2}{h} \right)^{q/4} \psi_A(z) \\
&\simeq \frac{e^{2h} [\frac{1}{4}(q-1)]!}{(2\pi)^{1/2} 2^{q+\frac{1}{2}} (4h^2)^{1/4}} \left(\frac{2}{h} \right)^{q/4} \psi_A(z). \qquad (20.67)
\end{aligned}
$$

The last expression is again obtained with the help of *Stirling's approximation*.[‡‡]

Proceeding similarly with the other WKB solution we obtain

$$
\begin{aligned}
\overline{\psi}_{\text{WKB}}^{(r,z_+)}(z) &\simeq \frac{\exp(-\int_{z_+}^{z} [4h^2 \cos^2 z - 2hq]^{1/2} dz)}{(4h^2 \cos^2 z)^{1/4}} \\
&\simeq \frac{e^{-2h} e^{-2h \sin z} e^{q/4} (q/2h)^{-q/4} [\tan(\frac{z}{2} + \frac{\pi}{4})]^{q/2}}{(4h^2)^{1/4} 2^{-q} \{2 \sin(\frac{z}{2} + \frac{\pi}{4}) \cos(\frac{z}{2} + \frac{\pi}{4})\}^{1/2}} \\
&\simeq \frac{e^{-2h} 2^{q-\frac{1}{2}}}{(4h^2)^{1/4} e^{-q/4} (q/4)^{q/4}} \left(\frac{2}{h} \right)^{-q/4} \overline{\psi}_A(z) \\
&\simeq \frac{e^{-2h} 2^{q-\frac{1}{2}} (2\pi)^{1/2}}{(4h^2)^{1/4} [\frac{1}{4}(q-3)]!} \left(\frac{2}{h} \right)^{-q/4} \overline{\psi}_A(z), \qquad (20.68)
\end{aligned}
$$

[‡‡]The use of Stirling's formula here and below — essential to obtain factorials — assumes q to be large so that corresponding corrections would have to be calculated. We ignore this here, but mention that for q an odd integer:

$$\left[\frac{1}{4}(q-1) \right]! = \sqrt{2\pi} e^{-q/4} \left(\frac{q}{4} \right)^{(q+1)/4} \left[1 + O\left(\frac{1}{q} \right) \right] \quad \text{and} \quad \left[\frac{1}{4}(q-3) \right]! \left[-\frac{1}{4}(q+1) \right]! = \pm \pi.$$

again using the Stirling formula.

We now return to the even and odd wave functions defined by Eq. (20.55) and match these to the oscillatory WKB solutions $\psi_{\text{WKB}}^{(l,z_+)}(z), \overline{\psi}_{\text{WKB}}^{(l,z_+)}(z)$ to the left of the turning point at z_+. Thus with $\psi_A(z)$ and $\overline{\psi}_A(z)$ taken from Eqs. (20.67) and (20.68) we obtain

$$
\begin{aligned}
\psi_\pm(z) &= \frac{1}{2}[\psi_A(z) \pm \overline{\psi}_A(z)] \\
&= \frac{1}{2}\left(\frac{h}{2}\right)^{q/4} \frac{e^{-2h}(2\pi)^{1/2}2^{q+\frac{1}{2}}(4h^2)^{1/4}}{[\frac{1}{4}(q-1)]!} \psi_{\text{WKB}}^{(r,z_+)}(z) \\
&\quad \pm \frac{1}{2}\left(\frac{h}{2}\right)^{-q/4} \frac{e^{2h}[\frac{1}{4}(q-3)]!2^{-q+\frac{1}{2}}(4h^2)^{1/4}}{(2\pi)^{1/2}} \overline{\psi}_{\text{WKB}}^{(r,z_+)}(z) \\
&= \frac{1}{2}\frac{(8h)^{q/4}e^{-2h}(2\pi)^{1/2}2^{1/2}(4h^2)^{1/4}}{[\frac{1}{4}(q-1)]!} \psi_{\text{WKB}}^{(l,z_+)}(z) \\
&\quad \pm \frac{1}{2}\frac{(8h)^{-q/4}e^{2h}2^{1/2}(4h^2)^{1/4}[\frac{1}{4}(q-3)]!}{(2\pi)^{1/2}} \overline{\psi}_{\text{WKB}}^{(l,z_+)}(z).
\end{aligned}
$$

Inserting here the oscillatory WKB expressions (cf. Eqs. (14.49), (14.50)), we obtain

$$
\begin{aligned}
&\simeq \frac{1}{2}\frac{(8h)^{q/4}e^{-2h}(2\pi)^{1/2}(16h^2)^{1/4}}{(2hq)^{1/4}[\frac{1}{4}(q-1)]!} \cos\left[(2hq)^{1/2}(z_+ - z) + \frac{\pi}{4}\right] \\
&\quad \pm \frac{1}{2}2\frac{(8h)^{-q/4}e^{2h}(16h^2)^{1/4}[\frac{1}{4}(q-3)]!}{(2hq)^{1/4}(2\pi)^{1/2}} \sin\left[(2hq)^{1/2}(z_+ - z) + \frac{\pi}{4}\right] \\
&\equiv \frac{(2\pi)^{1/2}(16h^2)^{1/4}}{2(2hq)^{1/4}}[T_+(\pm)\exp\{i(2hq)^{1/2}(z_+ - z)\} \\
&\qquad\qquad + T_-(\pm)\exp\{-i(2hq)^{1/2}(z_+ - z)\}],
\end{aligned} \tag{20.69}
$$

where

$$
\begin{aligned}
T_+(\pm) &= \left\{\frac{1}{2}\frac{(8h)^{q/4}e^{-2h}}{[\frac{1}{4}(q-1)]!} \pm \frac{1}{i}\frac{[\frac{1}{4}(q-3)]!e^{2h}}{2\pi(8h)^{q/4}}\right\}e^{i\pi/4} \\
T_-(\pm) &= \left\{\frac{1}{2}\frac{(8h)^{q/4}e^{-2h}}{[\frac{1}{4}(q-1)]!} \mp \frac{1}{i}\frac{[\frac{1}{4}(q-3)]!e^{2h}}{2\pi(8h)^{q/4}}\right\}e^{-i\pi/4}.
\end{aligned} \tag{20.70}
$$

Imposing the boundary condition (20.49), we obtain

$$
T_-(\pm) = 0,
$$

i.e.

$$
\frac{1}{2}\frac{2\pi(8^2h^2)^{q/4}e^{-4h}}{[\frac{1}{4}(q-1)]![\frac{1}{4}(q-3)]!} = \pm\frac{1}{i}. \tag{20.71}
$$

Using the *duplication formula*

$$\left[\frac{1}{4}(q-1)\right]! \left[\frac{1}{4}(q-3)\right]! = \left[\frac{1}{2}(q-1)\right]! \pi^{1/2} 2^{-\frac{1}{2}(q-1)},$$

this becomes

$$\left(\frac{\pi}{2}\right)^{1/2} \frac{e^{-4h}(16^2 h^2)^{q/4}}{[\frac{1}{2}(q-1)]!} = \pm\frac{1}{i}$$

or

$$\pm i \left(\frac{\pi}{2}\right)^{1/2} \frac{(16^2 h^2)^{q/4} e^{-4h} (16^2 h^2)^{q/4}}{[\frac{1}{2}(q-1)]!} = (16^2 h^2)^{q/4} \equiv (-16h)^{q/2}. \quad (20.72)$$

We observe that with the square root as on the right hand side both sides are complex, in fact, for q an integer both sides are imaginary. Equation (20.72) is our second transcendental equation.

We now have the two equations which have to be satisfied, Eqs. (20.57) and (20.72). Combining these equations by replacing $(2^8 h^2)^{q/4}$ in Eq. (20.57) by the left hand side of Eq. (20.72) we obtain

$$q - q_0 = i \frac{2(16h)^{q_0} e^{-8h}}{\{[\frac{1}{2}(q_0-1)]!\}^2} \left[1 + O\left(\frac{1}{h}\right)\right]. \quad (20.73)$$

Inserting this expression into $\lambda \equiv E(q,h)$ of Eq. (20.21) expanded about $q = q_0$, we obtain

$$
\begin{aligned}
E(q,h) &= E(q_0,h) + (q-q_0)\left(\frac{\partial E}{\partial q}\right)_{q_0} = E(q_0,h) + 2h(q-q_0)\\
&= E(q_0,h) + i \frac{(16h)^{q_0+1} e^{-8h}}{4\{[\frac{1}{2}(q_0-1)]!\}^2}\left[1 + O\left(\frac{1}{h}\right)\right]\\
&\equiv E(q_0,h) + i\Im E_{q_0}. \quad (20.74)
\end{aligned}
$$

This result agrees with that of Stone and Reeve [262], who derived it for $q_0 = 1$ and guessed the form for general values of q_0, i.e. the result (20.47).

20.3.3 Relating the level splitting to imaginary E

We can compare the results of the two problems with the cosine potential and the same solutions but with different boundary conditions. In both cases we wrote (cf. Eqs. (17.26), (20.25))

$$E \equiv \lambda = E_1 + E_2, \quad E_1 = -2h^2 + 2hq_0 + \frac{\triangle}{8}, \quad \frac{1}{2}(q-q_0) = \frac{E_2}{4h}. \quad (20.75)$$

Referring back to the calculation of the level splitting with selfadjoint boundary conditions, we recall that we obtained there for the difference $2\delta E_{q_0}$ between levels with the same oscillator quantum number q_0, cf. Eq. (17.74b),

$$2\delta E_{q_0} = 8h \left(\frac{2}{\pi} \right)^{1/2} \frac{(16h)^{q_0/2} e^{-4h}}{[\frac{1}{2}(q_0 - 1)]!} \left[1 + O\left(\frac{1}{h} \right) \right]. \tag{20.76}$$

Thus δE_{q_0} is the deviation of one of these levels from the harmonic oscillator level. Setting

$$\triangle E_{q_0} := 2i \Im E_{q_0},$$

we can obtain the imaginary part calculated above from the formula

$$4 \left(\frac{\partial E}{\partial q} \right)_{q_0} \triangle E_{q_0} = 2\pi i (\delta E_{q_0})^2. \tag{20.77}$$

One can say that this simple relation summarizes the intricate connections between the discontinuity of the eigenenergy across its cut, its tunneling properties and the large order behaviour. The same formula can be derived in the case of the double well potential.* The existence of a formula of this type was conjectured without an explicit derivation like that given here and only for the ground state, by others.† It would be interesting to see a derivation of the formula — reminiscent of an optical theorem — from first principles.

20.3.4 Recalculation of large order behaviour

Now that we have obtained $\Im E$, i.e. the discontinuity across its cut from zero to infinity, we can recalculate with the help of Eq. (20.9) the coefficients A_i of its perturbation expansion. For definiteness we recall the appropriate expansions together with the definition of coefficients A_i. We had

$$\lambda = -2h^2 + 2hq + \frac{\triangle}{8}, \quad \triangle = -\frac{1}{2} \sum_{i=0}^{\infty} \frac{M_{i+1}}{(2^7 h)^i}, \tag{20.78}$$

or

$$\lambda + 2h^2 - 2hq - \frac{M_1}{2^4} = -\frac{1}{2^4} \sum_{i=1}^{\infty} \frac{M_{i+1}}{(2^7 h)^i} = -\sum_{i=1}^{\infty} \frac{A_i}{h^i}. \tag{20.79}$$

We can calculate A_i for large i from the relation

$$A_i = \frac{1}{\pi} \int_0^{\infty} h'^{i-1} \Im E(h') dh' \tag{20.80}$$

---

*P. Achuthan, H. J. W. Müller–Kirsten and A. Wiedemann [4] and Example 20.2 below.
†E. B. Bogomol'nyi and V. A. Fateyev [26].

with $\Im E(h)$ given by Eq. (20.74). The integral is recognized to be of the type of the integral representation of the gamma function. Using again

$$n = \frac{1}{2}(q_0 - 1),$$

the integration is seen to give

$$A_i = \frac{2^{q_0+1}}{\pi 8^i 4\{[\frac{1}{2}(q_0 - 1)]!\}^2} \int_0^\infty (8h')^{q_0+i} e^{-8h'} d(8h') \simeq \frac{2^{2n}(i + 2n + 1)!}{2^{3i}(n!)^2 \pi}.$$
(20.81)

We observe that this result agrees with the result of Eq. (20.19) for the large values of i under consideration here. We also observe that these terms do not alternate in sign, so that an exact Borel summation is not possible.

20.4 Cosine Potential: A Different Calculation

A further very instructive method of obtaining the large order behaviour of a perturbation expansion was found by Dingle* in application to the case of the cosine potential or Mathieu equation. But as Dingle remarks, there is nothing special about this case, and the method can be applied in many other cases. We sketch this method here slightly modified and extend it to the final formula for comparison with our previous results.[†]

We consider again the periodic cosine potential with (unnormalized) eigenfunction expansion

$$\psi = \psi_q + \sum_{i=1}^\infty \frac{1}{(2^7 h)^i} \sum_{j=-i, j\neq 0}^i P_i(q, j)\psi_{q+4j}.$$
(20.82)

The coefficients $P_i(q, j)$ satisfy the recurrence relation (20.13) and the coefficients M_{2i} or M_{2i+1} of the eigenvalue expansion can be expressed in terms of these as in Eq. (20.14). For convenience we write Eq. (20.13) out again:

$$\begin{aligned}
jP_i(q, j) &= (q + 4j - 1)(q + 4j - 3)P_{i-1}(q, j - 1) \\
&\quad + 2[(q + 4j)^2 + 1 + \triangle]P_{i-1}(q, j) \\
&\quad + (q + 4j + 1)(q + 4j + 3)P_{i-1}(q, j + 1).
\end{aligned}$$
(20.83)

One now sets

$$p_i(\alpha) = \frac{\alpha}{(2^7 h)^i} P_i(q, j), \quad \alpha = j + \frac{1}{2}q,$$
(20.84)

---

*R. B. Dingle [74].
[†]H. J. W. Müller–Kirsten [200].

and considers large values of j for large values of i. Taking dominant terms, e.g. $(q + 4j)^2 \simeq 4j(4j + 2q)$, the recurrence relation (20.83) can be written

$$
\begin{aligned}
j\frac{2^7 h}{\alpha}p_i(\alpha) &= \frac{4j(4j + 2q - 4)}{(\alpha - 1)}p_{i-1}(\alpha - 1) + 2\frac{4j(4j + 2q)}{\alpha}p_i(\alpha) \\
&+ \frac{4j(4j + 2q + 4)}{(\alpha + 1)}p_{i-1}(\alpha + 1), \quad q \ll 4j,
\end{aligned}
$$

i.e. (with $\alpha = j + q/2$ in the coefficients)

$$
\frac{8h}{\alpha}p_i(\alpha) = p_{i-1}(\alpha - 1) + 2p_{i-1}(\alpha) + p_{i-1}(\alpha + 1). \tag{20.85}
$$

Considering now the *generating function* of the perturbation coefficients, i.e.

$$
p(\alpha, h) = \sum_i p_i(\alpha) \quad \text{with} \quad p_i(\alpha) \propto \frac{1}{h^i}, \tag{20.86}
$$

one can convince oneself that Eq. (20.85) implies

$$
p(\alpha - 1, h) + p(\alpha + 1, h) = \left[-2 + \frac{8h}{\alpha}\right]p(\alpha, h). \tag{20.87}
$$

Taking terms of $O(h^{-i+1})$ of this equation, we regain Eq. (20.85). Now, *Whittaker functions* $W_{\kappa,\mu}$, which are solutions of the differential equation known as *Whittaker's equation*, i.e.

$$
\frac{d^2 W_{\kappa,\mu}}{dz^2} + \left[-\frac{1}{4} + \frac{\kappa}{z} + \frac{\frac{1}{4} - \mu^2}{z^2}\right]W_{\kappa,\mu} = 0, \tag{20.88}
$$

satisfy among other relations the following recurrence relation which is of significance in the present context[‡]

$$
(2\kappa - z)W_{\kappa,\mu}(z) + W_{\kappa+1,\mu}(z) = \left(\mu - \kappa + \frac{1}{2}\right)\left(\mu + \kappa - \frac{1}{2}\right)W_{\kappa-1,\mu}(z). \tag{20.89}
$$

Replacing here κ by α and μ by $1/2$, we obtain

$$
(2\alpha - z)W_{\alpha,1/2}(z) + W_{\alpha+1,1/2}(z) = (1 - \alpha)\alpha W_{\alpha-1,1/2}(z). \tag{20.90}
$$

Setting

$$
V_{\alpha,1/2}(z) = \frac{W_{\alpha,1/2}(z)}{(\alpha - 1)!}, \tag{20.91}
$$

[‡]See M. Abramowitz and I. A. Stegun [1], formula 13.4.31.

we obtain

$$V_{\alpha-1,1/2}(z) + V_{\alpha+1,1/2}(z) = \left[-2 + \frac{z}{\alpha} \right] V_{\alpha,1/2}(z). \qquad (20.92)$$

Comparing Eq. (20.92) with Eq. (20.87), Dingle discovered the solutions[§]

$$p(\alpha, h) \simeq f(h) \frac{W_{\alpha,1/2}(8h)}{(\alpha - 1)!}, \quad g(h) \frac{W_{-\alpha,1/2}(-8h)}{(-\alpha - 1)!}, \qquad (20.93)$$

where $f(h), g(h)$ are functions of h which in view of Eq. (20.86) have to be chosen such that

$$p(\alpha, h) = \sum_i p_i(\alpha) \quad \text{with} \quad p_i \propto h^{-i},$$

i.e. such that $p(\alpha, h)$ is expressed in descending powers of h. We can infer the approximate behaviour of the Whittaker function for large values of z from the differential equation (20.88) which reminds us immediately of a radial Schrödinger equation with Coulomb potential. Thus the approximate large-z asymptotic behaviour is

$$W_{\kappa,\mu}(z) \sim \exp\left[i \int^z \sqrt{-\frac{1}{4} + \frac{\kappa}{z}} \, dz \right] \simeq e^{-z/2} \exp(\kappa \ln z).$$

The more precise form obtained from Tables of Special Functions[¶] is

$$\frac{W_{\kappa,\mu}(z)}{e^{-z/2} z^\kappa} \approx 1 + \sum_{n=1}^{\infty} \frac{[\mu^2 - (\kappa - \frac{1}{2})^2][\mu^2 - (\kappa - \frac{3}{2})^2] \cdots [\mu^2 - (\kappa - n + \frac{1}{2})^2]}{n! z^n},$$

which with $\kappa = \alpha$ and $\mu = 1/2$ is:

$$W_{\alpha,\frac{1}{2}}(z) \approx e^{-z/2} z^\alpha \sum_{s=0}^{\infty} \frac{(s - \alpha)!(s - \alpha - 1)!}{(-z)^s s!(-\alpha)!(-\alpha - 1)!} \quad \text{for } |z| \to \infty. \qquad (20.94)$$

Hence to insure that $p(\alpha, h)$ is expressed in descending powers of $z = 8h$, we have to remove a factor (recalling $\alpha = j + q/2$ of Eq. (20.84))

$$(8h)^{q/2} e^{-4h} \quad \text{for} \quad j < 0,$$

[§] Observe that if one solution of Whittaker's equation (20.88) is known, another solution is obtained from this by changing the signs of κ and z since this leaves the equation unchanged. In the present case this means the replacements $q \to -q, j \to -j, h \to -h$.

[¶] See W. Magnus and F. Oberhettinger [181], p. 89.

and write

$$p\left(\alpha = j + \frac{1}{2}q, h\right) \propto \frac{(8h)^{-q/2}e^{4h}}{(j + \frac{1}{2}q)!} W_{j+q/2,1/2}(8h) \quad \text{for } j < 0. \quad (20.95a)$$

Inserting (20.94) here we have a series in descending powers of h. Correspondingly we set for $j > 0$

$$p\left(\alpha = j + \frac{1}{2}q, h\right) \propto \frac{(-8h)^{q/2}e^{-4h}}{(-j - \frac{1}{2}q)!} W_{-j-q/2,1/2}(-8h) \quad \text{for } j > 0. \quad (20.95b)$$

Then for $j > 0$ we obtain from Eq. (20.94) with $s \to s - j$:

$$p\left(\alpha = j + \frac{1}{2}q, h\right) \propto \frac{(-1)^j}{(-j - \frac{1}{2}q)!} \sum_{s-j=0}^{\infty} \frac{(s + \frac{1}{2}q)!(s + \frac{1}{2}q - 1)!}{(s - j)!(j + \frac{1}{2}q)!(j + \frac{1}{2}q - 1)!} \frac{1}{(8h)^s}.$$

Using the *reflection formula*

$$(z - 1)!(-z)! = \frac{\pi}{\sin \pi z}, \quad \text{i.e.} \quad \left(-j - \frac{1}{2}q\right)!\left(j + \frac{1}{2}q - 1\right)! = (-1)^{-j-n}\pi,$$

(for $q = 2n + 1, n = 0, 1, 2, \ldots$), this becomes

$$p\left(\alpha = j + \frac{1}{2}q, h\right) \propto \frac{(-1)^{\frac{1}{2}(q-1)}}{\pi} \sum_{s-j=0}^{\infty} \frac{(s + \frac{1}{2}q)!(s + \frac{1}{2}q - 1)!}{(s - j)!(j + \frac{1}{2}q)!} \frac{1}{(8h)^s}.$$

Picking out the term in h^{-i}, we obtain the result of Dingle:

$$p_i\left(\alpha = j + \frac{q}{2}\right) \propto \frac{(i + \frac{1}{2}q)!(i + \frac{1}{2}q - 1)!}{(i - j)!(j + \frac{1}{2}q)!} \frac{1}{(8h)^i} \quad \text{for } i = j, j + 1, \ldots, j > 0,$$

$$(20.96)$$

and a similar expression for $j < 0$ with $h \to -h, q \to -q, j \to -j$. Comparing this result now with Eq. (20.84) we conclude that the original coefficients $P_i(q, j)$ are for large i given by the following expression — apart from an as yet undetermined proportionality factor c_q —

$$P_i(q, j) = c_q \frac{(2^7 h)^i}{\alpha} p_i(\alpha) = 2^{4i} \frac{c_q}{\alpha} \frac{(i + \frac{1}{2}q)!(i + \frac{1}{2}q - 1)!}{(i - j)!(j + \frac{1}{2}q)!}. \quad (20.97a)$$

We determine c_q by imposing as boundary condition the expression for $P_i(q, i)$. This expression can be obtained from the recurrence relation (20.83) and is given by[||]

$$P_i(q, i) = \frac{4^i[2i + \frac{1}{2}(q - 1)]!}{i![\frac{1}{2}(q - 1)]!}. \quad (20.97b)$$

[||]R. B. Dingle and H. J. W. Müller [74].

We observe that $P_0(q,0) = 1$ as desired (although not really enforcible for large values of i). Thus

$$c_q = \frac{(i + \frac{1}{2}q)}{2^{4i}(i + \frac{1}{2}q - 1)!}P_i(q,i) = \frac{(i + \frac{1}{2}q)}{2^{2i}[\frac{1}{2}(q-1)]!}\frac{[2i + \frac{1}{2}(q-1)]!}{i![i + \frac{1}{2}q - 1]!}.$$

Inserting this expression into Eq. (20.97a) we obtain (observe that as required this expression vanishes for $j > i$)

$$P_i(q,j) \simeq \frac{(i + \frac{1}{2}q)[2i + \frac{1}{2}(q-1)]!2^{2i}}{i![\frac{1}{2}(q-1)]!(j + \frac{1}{2}q)}\frac{[i + \frac{1}{2}q]!}{[j + \frac{1}{2}q]![i-j]!}. \tag{20.97c}$$

We assume that in a leading approximation we can cancel the factors $(i + \frac{1}{2}q)$, $(j + \frac{1}{2}q)$. Then

$$P_i(q,j) \simeq \frac{[2i + \frac{1}{2}(q-1)]!2^{2i}}{i![\frac{1}{2}(q-1)]!}\frac{[i + \frac{1}{2}q]!}{[j + \frac{1}{2}q]![i-j]!}. \tag{20.97d}$$

Then the coefficient M_{2i} of the eigenvalue expansion, i.e. of Eq. (20.14), is

$$M_{2i} = 2\sum_{j=1}^{i} j[P_i(q,j)]^2 \simeq 2\left[\frac{[2i + \frac{1}{2}(q-1)]!2^{2i}}{i![\frac{1}{2}(q-1)]!}\right]^2 \sum_{j=1}^{i} j\left(\begin{array}{c} i + \frac{1}{2}q \\ j + \frac{1}{2}q \end{array}\right)^2. \tag{20.98}$$

We evaluate the sum by approximating this to a formula given in the literature **

$$\sum_{j=0} j\left(\begin{array}{c} r \\ j \end{array}\right)\left(\begin{array}{c} R \\ j \end{array}\right) = \frac{(r + R - 1)!}{(r-1)!(R-1)!},$$

i.e.

$$\sum_{j=1}^{i} j\left(\begin{array}{c} i + \frac{1}{2}q \\ j \end{array}\right)^2 = \frac{[2i + q - 1]!}{\{[i + \frac{1}{2}q - 1]!\}^2}. \tag{20.99}$$

In the next step we use for both factorials $[2i + \frac{1}{2}(q-1)]!$ in Eq. (20.98) the *duplication formula*

$$(2z - 1)! = \frac{2^{2z-1}}{\sqrt{\pi}}(z-1)!\left(z - \frac{1}{2}\right)!,$$

and obtain

$$M_{2i} \simeq 2\left[\frac{[i + \frac{1}{4}(q-1)]![i + \frac{1}{4}(q-3)]!2^{2i+(q-1)/2}2^{2i}}{\sqrt{\pi}i![\frac{1}{2}(q-1)]!}\right]^2 \frac{[2i + q - 1]!}{\{[i + \frac{1}{2}q - 1]!\}^2}. \tag{20.100}$$

**R. B. Dingle and H. J. W. Müller [74], formula (30); note the misprint there: $(r - s)$ ought to be $(r - s)!$. A special case of the formula can be found in I. S. Gradshteyn and I. M. Ryzhik [122], formula 0.157(4), p. 5.

In the following step we use the approximate formula (20.17), i.e.

$$\frac{i!}{(i+m)!} \overset{i \text{ large}}{\simeq} \frac{1}{i^m}\left[1 - O\left(\frac{1}{i}\right)\right]. \tag{20.101}$$

This formula allows us to cancel all factorials $[i...]!$, since $\frac{1}{4}(q-1)+\frac{1}{4}(q-3) = \frac{1}{2}q-1$, i.e.

$$\frac{[i+\frac{1}{4}(q-1)]![i+\frac{1}{4}(q-3)]!}{i![i+\frac{1}{2}q-1]!} = 1.$$

Hence

$$M_{2i} \simeq 2^{8i+q}\frac{[2i+q-1]!}{\pi\{[\frac{1}{2}(q-1)]!\}^2}, \tag{20.102}$$

or, replacing $2i$ by i:

$$M_i \simeq 2^{4i+q}\frac{[i+q-1]!}{\pi\{[\frac{1}{2}(q-1)]!\}^2}. \tag{20.103}$$

Thus the result is

$$A_i \simeq \frac{M_{i+1}}{(2^7)^i} \sim \frac{1}{2^{3i}}\frac{2^q[i+q-1]!}{\pi\{[\frac{1}{2}(q-1)]!\}^2} \tag{20.104}$$

with the large i behaviour in agreement with the result (20.81) for $q = q_0 = 2n+1, n = 0, 1, 2, \ldots$.

20.5 Anharmonic Oscillators

20.5.1 The inverted double well

In Chapter 18 we obtained the imaginary part of the eigenenergy E of the Schrödinger equation for the inverted double well potential, i.e. Eq. (18.86). With this we can now derive the behaviour of the late terms in the large-h^2 perturbation expansion of the eigenvalue of the Schrödinger equation for the proper quartic anharmonic oscillator, i.e. the result of Bender and Wu [18], [19]. The imaginary part we obtained is interpreted as effectively the discontinuity of the eigenvalue across its cut from $h^2 = 0$ to infinity. We refer back to Eq. (18.86) and set (cf. Eq. (18.4))

$$\mathcal{E} \equiv \frac{2E}{h^2} = q + \frac{\triangle}{h^6} \quad \text{and} \quad y = \frac{h^6}{2c^2}, \tag{20.105}$$

the latter because we see from the eigenvalue expansion (18.34) that $\mathcal{E}$ is an expansion in ascending powers of h^6/c^2, i.e.

$$\mathcal{E} = q - \frac{3c^2}{2h^6}(q^2+1) - \frac{2c^4}{h^{12}}(4q^3 + 29q) + \cdots .$$

The Borel sum can now be formally written

$$\mathcal{E}(y) = \frac{1}{\pi} \int_0^\infty dy' \frac{\Im \mathcal{E}(y')}{1 - \frac{y'}{y}}.$$ (20.106)

With

$$\mathcal{E} = \sum_{r=0}^\infty \frac{M_r}{(-y)^r}$$ (20.107)

we obtain

$$M_r = \frac{(-1)^r}{\pi} \int_0^\infty y'^r \Im \mathcal{E}(y') dy'.$$ (20.108)

Proceeding in this way, and hence inserting here the imaginary part from Eq. (18.86), i.e. (with $E = h^2 \mathcal{E}/2$)

$$\Im \mathcal{E}(y') = \frac{2^{q_0+1}(y')^{q_0/2} e^{-y'/3}}{(2\pi)^{1/2}[\frac{1}{2}(q_0 - 1)]!},$$

and integrating with the help of the definition of the factorial or gamma function, one obtains the result

$$M_r = \frac{2^{q_0+\frac{1}{2}} 3^{r+1+\frac{q_0}{2}} (-1)^r [r + \frac{q_0}{2}]!}{\pi^{3/2}[\frac{1}{2}(q_0 - 1)]!}$$ (20.109)

or, in terms of K where $q_0 = 2K + 1$,

$$\frac{h^2}{2} M_{r-1} = \frac{h^2 2^{2K+1/2} 3^{r+K+1/2} (-1)^{r+1}}{\pi^{3/2}[K]!} \left[r + K - \frac{1}{2}\right]!.$$ (20.110)

This is the result in agreement with the large order formula (4.4) of Bender and Wu [19]. The result establishes the series unequivocally as an asymptotic series, in fact as one with Borel summability in view of the factor $(-1)^r$.

20.5.2 The double well

It is clear that one can now proceed and apply a non-selfadjoint boundary condition also in the case of the double well potential and thus obtain the imaginary part of the eigenvalue, i.e. the discontinuity across its cut. This can then be compared with the level splitting we obtained for this potential, and then the relation (20.77) can be verified. Once this has been obtained one can insert the imaginary part into the Borel transform and obtain the behaviour of the late terms in the eigenvalue expansion. We leave this to the following Example.

Example 20.2: Late term behaviour of double well eigenvalue expansion
Derive for the case of the double well potential — by application of a nonselfadjoint boundary condition (as in the case of the cosine potential) — the imaginary part of its eigenvalue and hence the large order behaviour of the asymptotic expansion of the double well eigenvalue.

Solution: For the solution we refer to the literature.[††]

20.6   General Remarks

The study of the large order behaviour of perturbation theory has become an individual direction of research.[*] A more recent status evaluation is provided by the book of LeGuillou and Zinn–Justin [161].

The relation (20.77) was conjectured without explicit derivation and only for the ground state by Bogomol'nyi and Fateyev [26]. It was later explicitly established for the level splittings and imaginary parts of arbitrary states in the cases of the cosine potential and the double well potential.[†] The relation (20.77) and its possible generality has been referred to by Brezin and Zinn–Justin [36]. Its usefulness has also been demonstrated there, in that when the quantity on one side is known, it can be used to obtain the other. An approach to arrive at the relation via instanton considerations has been examined by Zinn–Justin [292]. A relation similar to (20.77) has been obtained in the Schrödinger theory of the molecular hydrogen ion H_2^+, and is there shown to have a deep meaning (cancellation of the imaginary part of the Borel sum by explicitly imaginary terms in the perturbation expansion).[‡] The above investigation of the large order behaviour demonstrates explicitly the intimate connection between the exponentially small nonperturbative effects derivable from the perturbation expansions with boundary conditions (or, as in later chapters, from path integrals with instanton methods) and the large order behaviour of the perturbation expansions which establishes the latter clearly as asymptotic expansions.

[††]P. Achuthan, H. J. W. Müller–Kirsten and A. Wiedemann [4], Eqs. (357) to (371).

[*]Some of the literature in this field following the work of C. M. Bender and T. T. Wu [18], [19] can be traced back from articles in the Proc. of International Workshop on Perturbation Theory at Large Order (Florida, 1982) [139].

[†]P. Achuthan, H. J. W. Müller–Kirsten and A. Wiedemann [4], J.–Q. Liang and H. J. W. Müller–Kirsten [164].

[‡]R. J. Damburg, R. K. Propin, S. Graffi, V. Grecchi, E. M. Harrell, J. Cizek, J. Paldus and H. J. Silverstone [62], J. Cizek, R. J. Damburg, S. Graffi, V. Grecchi, E. M. Harrell, J. G. Harris, S. Nakai, J. Paldus, R. K. Propin and H. J. Silverstone [52].

Chapter 21

The Path Integral Formalism

21.1 Introductory Remarks

In this chapter we introduce the path integral formalism developed by Feynman.* This formalism deals with an ensemble of paths $\{x(t)\}$ rather than with wave functions and constitutes an alternative to canonical quantization in quantizing a theory. This formalism is particularly useful for evaluating scattering or transition amplitudes. The formalism is not so useful for bound-state problems (in this case the initial wave function is, in general, unknown). The path integral is, however, particularly useful in cases where the behaviour of a system is to be investigated close to its classical path, since in such cases the probability of this system choosing a path far away may be considered to be small. We illustrate the method here by rederiving the Rutherford scattering formula with this method. However, our main interest in path integrals and their uses will focus thereafter on their evaluation about solutions of classical equations. Thus we are in particular interested in rederiving the level splitting formulas obtained in earlier chapters with this method, since this illustrates the most important and most frequent use of path integrals in a wide spectrum of applications ranging from field theory to condensed matter physics. For the application of path integrals to the hydrogen atom (i.e. Coulomb problem) we refer to existing literature.† Our interest here focusses on quantum mechanics. However, we digress later a little and consider the path integral also in simple contexts of scalar field theories which can be treated like models in quantum mechanics.

---

*R. P. Feynman [93]. A standard reference is the book of R. P. Feynman and A. R. Hibbs [95]. A brief introduction is also contained in R. P. Feynman [94]. A further wellknown reference is the book of L. S. Schulman [245]. Probably the most readable introduction, which explains also difficulties, can be found in the book of B. Felsager [91], Chapters 2 and 5.

†See in particular H. Kleinert [151].

21.2   Path Integrals and Green's Functions

The one-dimensional Schrödinger equation for a particle of mass m_0 moving in a potential $V(x)$ is (with mostly $\hbar = c = 1$)

$$i\frac{\partial}{\partial t}\psi(x,t) = \left[-\frac{1}{2m_0}\frac{\partial^2}{\partial x^2} + V(x)\right]\psi(x,t). \qquad (21.1)$$

Let the wave function ψ at time $t = t_i$, the initial time, be $\psi(x,t_i)$. We are interested to know the wave function ψ at a final time $t_f > t_i$ and position coordinate $x = x_f$, i.e. we wish to know[‡]

$$\psi(x_f,t_f) = \int dx\, K(x_f,t_f;x,t_i)\psi(x,t_i), \qquad (21.2)$$

where K is a *Green's function*. Obviously

$$K(x_f,t_i;x,t_i) = \delta(x_f - x). \qquad (21.3)$$

We divide the time interval $t_f - t_i$ into $n + 1$ equal infinitesimal elements ϵ such that

$$\begin{aligned}
t_0 &= t_i, \\
t_1 &= t_i + \epsilon, \\
&\cdots \\
t_n &= t_i + n\epsilon, \\
t_f &= t_i + (n+1)\epsilon.
\end{aligned}$$

The Lagrangian L of a particle of mass m_0 moving in a potential $V(x)$ in one-time and one-space dimensions is given by

$$L = T - V = \frac{1}{2}m_0\dot{x}^2 - V(x), \qquad (21.4)$$

so that the time-sliced action S becomes

$$\begin{aligned}
S &= \int_{t_i}^{t_f} dt\, L = \lim_{n\to\infty,\epsilon\to 0}\sum_{k=1}^{n+1}\epsilon\left[\frac{1}{2}m_0\dot{x}_k^2 - V(x_k)\right] \equiv \lim_{n\to\infty,\epsilon\to 0} S_n, \\
S_n &= \sum_{k=1}^{n+1}\epsilon\left[\frac{m_0}{2\epsilon^2}(x_k - x_{k-1})^2 - V(x_k)\right]. \qquad (21.5)
\end{aligned}$$

Thus: Given $x_k(t_k)$, we can compute the action S.

[‡]In Dirac's notation we write this equation later $\langle x_f,t_f|\psi\rangle = \int dx\langle x_f,t_f|x,t_i\rangle\langle x,t_i|\psi\rangle$.

Above we considered time intervals of length ϵ. We now consider intervals in position coordinates and in particular the following integral over intermediate position coordinates x_k which corresponds to the sum of $\exp(iS/\hbar)$ over the uncountable number of possible paths from $x_i = x_0$ to $x_f = x_{n+1}$, of which one is illustrated in Fig. 21.1 (for convenience we write in the following in the argument of the exponential frequently S instead of $S/\hbar$):

$$\int_{-\infty}^{\infty} dx_1 \int_{-\infty}^{\infty} dx_2 \cdots \int_{-\infty}^{\infty} dx_n \, e^{iS/\hbar}.$$

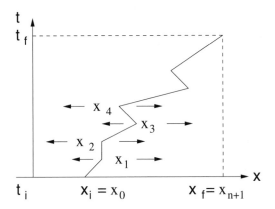

Fig. 21.1 A path in (x,t).

Thus there is a countless number of possible paths that the system may choose in propagating from its initial position at (x_i, t_i) to its final position at (x_f, t_f). The usefulness of the method therefore depends on the fact that some paths are more probable than others.

The multiple integral involves individual integrals like

$$\int_{-\infty}^{\infty} dx_k \, \exp\left[i\frac{m_0}{2\epsilon}\{(x_{k+1} - x_k)^2 + (x_k - x_{k-1})^2\}\right],$$

with x_k contained in the step up to this point and in the step away from it. The repetition of the quadratic factors in the argument of the exponential with indices differing by 1 suggests the construction of a recurrence relation, as will be shown to be possible below.

Here and in the following we need one or the other of the following integrals (cf. the Fresnel integral (13.28)):

$$\int_{-\infty}^{\infty} d\eta \, \exp\left[\frac{ia}{\epsilon}\eta^2\right] = \left(\frac{i\pi\epsilon}{a}\right)^{1/2}, \tag{21.6a}$$

$$\int_{-\infty}^{\infty} d\eta\, \eta \exp\left[\frac{ia}{\epsilon}\eta^2\right] = 0, \tag{21.6b}$$

$$\int_{-\infty}^{\infty} d\eta\, \eta^2 \exp\left[\frac{ia}{\epsilon}\eta^2\right] = \frac{i\epsilon}{2a}\left(\frac{i\pi\epsilon}{a}\right)^{1/2}. \tag{21.6c}$$

In the evaluation of these integrals we replace a by $a + i\mu, \mu > 0$. We then integrate and finally let μ go to zero.* With the first of these integrals we obtain (compare the result with the Green's function (7.37))

$$\int_{-\infty}^{\infty} dx_k \exp\left[i\frac{m_0}{2\epsilon}\{(x_{k+1} - x_k)^2 + (x_k - x_{k-1})^2\}\right]$$

$$= \int_{-\infty}^{\infty} d(x_k - x_{k+1}) \exp\left[i\frac{m_0}{2\epsilon}\{(x_k - x_{k+1})^2\right.$$

$$\left. + (x_k - x_{k+1} + x_{k+1} - x_{k-1})^2\}\right]$$

$$= \int_{-\infty}^{\infty} dx \exp\left[i\frac{m_0}{2\epsilon}\{x^2 + (x + x_{k+1} - x_{k-1})^2\}\right]$$

$$= \int_{-\infty}^{\infty} dx \exp\left[i\frac{m_0}{2\epsilon}\{2x^2 + 2x(x_{k+1} - x_{k-1}) + (x_{k+1} - x_{k-1})^2\}\right]$$

$$= \int_{-\infty}^{\infty} dx \exp\left[i\frac{m_0}{2\epsilon}\left\{2\left(x + \frac{x_{k+1} - x_{k-1}}{2}\right)^2 + \frac{1}{2}(x_{k+1} - x_{k-1})^2\right\}\right]$$

$$= \frac{1}{\sqrt{2}}N(\epsilon)\exp\left[\frac{im_0}{4\epsilon}(x_{k+1} - x_{k-1})^2\right], \quad N(\epsilon) = \left(\frac{2i\pi\epsilon}{m_0}\right)^{1/2}. \tag{21.7}$$

Thus we expect that — in view of the multiplicative $\sqrt{\epsilon}$ here in front of the phase factor — in the limit $\epsilon \to 0$

$$\int_{-\infty}^{\infty} dx_1 \int_{-\infty}^{\infty} dx_2 \ldots \int_{-\infty}^{\infty} dx_n e^{iS/\hbar} \longrightarrow 0.$$

One therefore removes the troublesome factor by *defining* (S_n being the sum in Eq. (21.5)):

$$G(x_f, t_f; x_i, t_i) := \lim_{n \to \infty} \frac{1}{[N(\epsilon)]^{n+1}} \int dx_1 \ldots \int dx_n\, e^{iS_n/\hbar}, \quad t_f > t_i. \tag{21.8}$$

Note that in spite of the n integrations, the power $n + 1$ in the denominator has to be seen in conjunction with the $n + 1$ terms in the sum of S_n in

---

*Thus, for example, with $\int_{-\infty}^{\infty} dx e^{-w^2 x^2/2} = \sqrt{2\pi/w^2}$, one has $\int_{-\infty}^{\infty} d\eta e^{i(a+i\mu)\eta^2/\epsilon} = \int_{-\infty}^{\infty} d\eta e^{-\eta^2(\mu-ia)/\epsilon} = \sqrt{\pi\epsilon/(\mu - ia)} \overset{\mu \to 0}{\longrightarrow} \sqrt{i\pi\epsilon/a}$.

Eq. (21.5). The additional factor is needed later to combine with the number n contained in the extra factor $\sqrt{n+1}$ (in Eq. (21.7) for $n = 1$) to give a time interval, as in our discretization in Eq. (21.5).[†] In the following we demonstrate that the function constructed in this way is, in fact, the Green's function K of Eq. (21.2), i.e.

$$G(x_f, t_f; x_i, t_i) = K(x_f, t_f; x_i, t_i) \quad \text{for} \quad t_f > t_i. \tag{21.9}$$

For $t_f > t_i$ we now write[‡]

$$
\begin{aligned}
G(x_f, t_f; x_i, t_i) &\equiv \int_{x_i = x(t_i)}^{x_f = x(t_f)} \mathcal{D}\{x(t)\}\, e^{iS/\hbar} \\
&= \lim_{n \to \infty} \frac{1}{[N(\epsilon)]^{n+1}} \int dx_1 \ldots \int dx_n\, e^{iS_n/\hbar}. \tag{21.10}
\end{aligned}
$$

This is the formula originally given by Feynman.[§] The factor $[N(\epsilon)]^{-(n+1)}$ associated with the multiple integration is described as its *measure*. We shall see later (with Eq. (21.26)) that — ignoring the potential — the multiple integral (21.10) can be looked at as the product of a succession of free particle Green's functions (or propagators).

The consideration of the denumerable number of possible paths connecting the particle's initial position with its final position may be handled in a number of different ways. Different constructions of these families of paths usually lead to different measures. A mathematically rigorous definition of the path integral is difficult and beyond the scope of our present aims, and will therefore not be attempted here. However, it may be noted that with a different philosophy concerning the summation over paths, we arrive at a path integral without the measure factor $N(\epsilon)$ above.[¶] These factors $N(\epsilon)$ were introduced to cancel corresponding factors arising from the Fresnel (or Gaussian) integration, i.e. summation over all possible paths. Thus one first integrates over x_1 from $-\infty$ to ∞ and obtains an exponential involving $(x_2 - x_0)^2$. Then one integrates over x_2 from $-\infty$ to ∞ and obtains an exponential involving $(x_3 - x_0)^2$, however prefixed with a new factor, and so on. These integrations are indicated by the horizontal arrows in Fig. 21.1.

[†] See the square root in front of the exponential in Eq. (21.24).

[‡] In the expression

$$\int_x^{x''} \mathcal{D}\{x\} e^{iI(x)}$$

the quantity x'' is only a symbol to indicate the endpoint of the path. It is neither a variable of integration in $\mathcal{D}\{x\} \to \prod_n dx_n$, nor the limit of integration of some variable x_i, which all vary from $-\infty$ to ∞.

[§] See e.g. B. Felsager [91], p. 183.

[¶] For discussions see e.g. B. Felsager [91], p. 183.

The factors $N(\epsilon)$ can be avoided by instead summing over piecewise linear paths. Thus consider an arbitrary continuous path connecting the endpoints and use the time divison as above but now connect the intermediate points with straight lines. Each of these linear pieces is completely characterized by its endpoints. Performing the sum now in the sense of summing and hence integrating over the vertex coordinates in Fig. 21.1 and allowing n to go to infinity, so that the piecewise linear paths fill the entire space of paths, we obtain the same as before except that there is no factor $N(\epsilon)$. One way to avoid this problem with the integration measure is to consider the ratio of two path integrals, one with the full action S, the other with that of the action of the free particle S_0. Then the measure drops out, and one is calculating the *Feynman amplitude* or *kernel* relative to that of the free theory. It is for these reasons that — unless required — the measure is frequently ignored.[||] We employ this method later in Example 23.3 in the evaluation of a specific path integral (see Eqs. (23.72) to (23.74)).

We next determine the equation satisfied by G in order to verify that this is indeed the Green's function for $t_f > t_i$. Consider a time $t = t_f + \epsilon$ for position $x_f = x_{n+2}$. Then we can construct a recurrence relation as follows. We have

$$
G(x_{n+2}, t_f + \epsilon; x_i, t_i) = \frac{1}{N(\epsilon)} \int_{-\infty}^{\infty} dx_{n+1} \exp\left[i\epsilon\left\{ \frac{m_0}{2\epsilon^2}(x_{n+2} - x_{n+1})^2 \right.\right.
$$
$$
\left.\left. -V(x_{n+2}) \right\} \right] G(x_{n+1}, t_f; x_i, t_i).
$$

Setting now $x_{n+1} - x_{n+2} \equiv \eta$, we have

$$
G(x_{n+2}, t_f + \epsilon; x_i, t_i)
$$
$$
= \frac{1}{N(\epsilon)} \int d\eta \, \exp\left[\frac{im_0}{2\epsilon}\eta^2 - i\epsilon V(x_{n+2}) \right] G(x_{n+2} + \eta, t_f; x_i, t_i).
$$
$$
(21.11)
$$

For reasons of transparency in the next few steps we replace x_{n+2} by x and suppress temporarily t_f, x_i, t_i. Then we have on the right the factor $G(x+\eta)$. Expanding this in a Taylor series around x, we have

$$
G(x + \eta) = G(x) + \eta\left(\frac{\partial G}{\partial x} \right)_{\eta=0} + \frac{1}{2}\eta^2\left(\frac{\partial^2 G}{\partial x^2} \right)_{\eta=0} + \cdots, \qquad (21.12)
$$

[||] See, however, the discussion in B. Felsager [91], pp. 187 - 188.

so that

$$
\begin{aligned}
G(x, t_f + \epsilon; x_i, t_i) &= \frac{1}{N(\epsilon)} \int_{-\infty}^{\infty} d\eta \exp\left\{ \frac{im_0}{2\epsilon} \eta^2 - i\epsilon V(x) \right\} \\
&\quad \left[G(x, t_f; x_i, t_i) + \eta \left\{ \frac{\partial}{\partial x} G(x, t_f; x_i, t_i) \right\}_{\eta=0} \right. \\
&\quad \left. + \frac{1}{2} \eta^2 \left\{ \frac{\partial^2}{\partial x^2} G(x, t_f; x_i, t_i) \right\}_{\eta=0} + \cdots \right]. \quad (21.13)
\end{aligned}
$$

We integrate with the help of the integrals (21.6a), (21.6b) and (21.6c).
Retaining only terms up to those linear in ϵ, we obtain:

$$
\begin{aligned}
G(x, t_f + \epsilon; x_i, t_i) &= \frac{1}{N(\epsilon)} e^{-i\epsilon V(x)} \left(\frac{2i\pi\epsilon}{m_0} \right)^{1/2} \\
&\quad \times \left[G(x, t_f; x_i, t_i) + \frac{i\epsilon}{2m_0} \frac{\partial^2}{\partial x^2} G(x, t_f; x_i, t_i) + O(\epsilon^2) \right] \\
&= (1 - i\epsilon V(x) + O(\epsilon^2)) \left[G(x, t_f; x_i, t_i) \right. \\
&\quad \left. + \frac{i\epsilon}{2m_0} \frac{\partial^2}{\partial x^2} G(x, t_f; x_i, t_i) + O(\epsilon^2) \right],
\end{aligned}
$$

or

$$
\begin{aligned}
G(x, t_f + \epsilon; x_i, t_i) &= G(x, t_f; x_i, t_i) - i\epsilon \left[-\frac{1}{2m_0} \frac{\partial^2}{\partial x^2} G(x, t_f; x_i, t_i) \right. \\
&\quad \left. + V(x) G(x, t_f; x_i, t_i) \right] + O(\epsilon^2). \quad (21.14)
\end{aligned}
$$

Thus

$$
\begin{aligned}
G(x, t_f + \epsilon; x_i, t_i) - G(x, t_f; x_i, t_i) &= \epsilon \frac{\partial}{\partial t_f} G + O(\epsilon^2) \\
&= -i\epsilon \left[-\frac{1}{2m_0} \frac{\partial^2 G(x, t_f; x_i, t_i)}{\partial x^2} + V(x) G(x, t_f; x_i, t_i) \right] + O(\epsilon^2),
\end{aligned}
$$

and this means

$$
i \frac{\partial}{\partial t_f} G(x, t_f; x_i, t_i) = \left[-\frac{1}{2m_0} \frac{\partial^2}{\partial x^2} + V(x) \right] G(x, t_f; x_i, t_i). \quad (21.15)
$$

We see that $G(x, t_f; x_i, t_i)$ satisfies the time-dependent Schrödinger equation.
It follows that

$$
\psi(x_f, t_f) = \int dx\, G(x_f, t_f; x, t_i) \psi(x, t_i) \quad \text{for } t_f > t_i \quad (21.16)
$$

also satisfies the time-dependent Schrödinger equation. Hence our original function K is the same as G, or

$$K(x_f, t_f; x, t_i) = G(x_f, t_f; x, t_i)\theta(t_f - t_i). \tag{21.17}$$

Differentiating K with respect to t_f we obtain

$$\frac{\partial}{\partial t_f} K(x_f, t_f; x, t_i) = \theta(t_f - t_i)\frac{\partial}{\partial t_f} G + G\delta(t_f - t_i)$$

$$= \theta(t_f - t_i)\frac{\partial}{\partial t_f} G + \delta(x_f - x)\delta(t_f - t_i), \tag{21.18}$$

since $G(x_f, t_i; x, t_i) = \delta(x_f - x)$, as may be verified by inserting this into Eq. (21.16). It follows therefore that

$$\left(i\frac{\partial}{\partial t_f} - V + \frac{1}{2m_0}\frac{\partial^2}{\partial x^2}\right) K(x_f, t_f; x, t_i) = i\delta(x_f - x)\delta(t_f - t_i). \tag{21.19}$$

Thus K is the Green's function for the problem of the Schrödinger equation. The function G which satisfies the time-dependent Schrödinger equation (21.15) is a wave function (cf. Eq. (21.1)). We have therefore established that the Feynman integral (21.10) is indeed to be identified with the time-dependent Green's function.

21.3 The Green's Function for Potential $V=0$

We now calculate the Green's function K for potential $V = 0$. The result, denoted by K_0, is the Fourier transform of the *nonrelativistic propagator*, as this is also called. In a problem where V is small it is convenient to calculate K with a perturbation method. Thus we consider here the zeroth order of the latter case and begin with the configuration space representation of this *free particle Green's function*.

21.3.1 Configuration space representation

In the case of $V = 0$ we have from the previous section for the action, now called S_0, and the Green's function K_0:

$$S_0 = \int_{t_i}^{t_f} dt\, \frac{1}{2} m\dot{x}^2, \qquad K_0 = \theta(t_f - t_i)G_0. \tag{21.20}$$

Here (cf. Eq. (21.8))

$$G_0(x_f, t_f; x_i, t_i) = \int \mathcal{D}\{x\} e^{iS_0},$$

and hence

$$
\begin{aligned}
G_0(x_f, t_f; x_i, t_i) &= \lim_{n \to \infty} \left(\frac{m_0}{2\pi i\epsilon}\right)^{(n+1)/2} \int_{-\infty}^{\infty} dx_1 \int_{-\infty}^{\infty} dx_2 \dots \\
&\quad \dots \int_{-\infty}^{\infty} dx_n \exp\left[\frac{im_0}{2\epsilon} \sum_{k=1}^{n+1} (x_k - x_{k-1})^2\right], \quad (21.21)
\end{aligned}
$$

and $t = (n+1)\epsilon = t_f - t_i$ and $\hbar = 1$. In Example 21.1 we show that

$$
\begin{aligned}
I_n(\lambda) &\equiv \int_{-\infty}^{\infty} dx_1 \int_{-\infty}^{\infty} dx_2 \dots \int_{-\infty}^{\infty} dx_n \, \exp[i\lambda\{(x_1 - x_0)^2 + \cdots \\
&\qquad\qquad\qquad\qquad \cdots + (x_{n+1} - x_n)^2\}] \\
&= \left(\frac{i\pi}{\lambda}\right)^{n/2} \frac{1}{\sqrt{n+1}} \exp\left[\frac{i\lambda}{n+1}(x_{n+1} - x_0)^2\right], \quad (21.22)
\end{aligned}
$$

where $x_i = x_0$ and $x_f = x_{n+1}$ (in agreement with Eq. (21.7) for here $n = 1$ and there $k = 1$). Then

$$
G_0(x_f, t_f; x_i, t_i) = \lim_{n \to \infty} \left(\frac{m_0}{2\pi i\epsilon}\right)^{(n+1)/2} I_n\left(\lambda = \frac{m_0}{2\epsilon}\right). \quad (21.23)
$$

We now insert our expression for I_n into Eq. (21.23) for G_0 and obtain

$$
\begin{aligned}
&G_0(x_f, t_f; x_i, t_i) \\
&= \lim_{n \to \infty} \left(\frac{m_0}{2\pi i\epsilon}\right)^{(n+1)/2} \left(\frac{2\pi i\epsilon}{m_0}\right)^{n/2} \frac{1}{\sqrt{n+1}} \exp\left[\frac{im_0}{2\epsilon(n+1)}(x_f - x_i)^2\right] \\
&= \lim_{n \to \infty} \left(\frac{m_0}{2\pi i\epsilon(n+1)}\right)^{1/2} \exp\left[\frac{im_0}{2\epsilon(n+1)}(x_f - x_i)^2\right]. \quad (21.24)
\end{aligned}
$$

Note the square root factor in front which originates as discussed after Eq. (21.8). But

$$
t = t_f - t_i = (n+1)\epsilon, \quad \text{and writing } x = x_f - x_i,
$$

we have

$$
\begin{aligned}
G_0(x_f, t_f; x_i, t_i) &\equiv G_0(x, t) = \left(\frac{m_0}{2\pi i t}\right)^{1/2} \exp\left[\frac{im_0}{2t}x^2\right], \\
K_0(x_f, t_f; x_i, t_i) &\equiv K_0(x, t) = G_0(x_f, t_f; x_i, t_i)\theta(t). \quad (21.25)
\end{aligned}
$$

The quantity $K_0(x, t)$ is the configuration space representation of the free particle ($V = 0$) nonrelativistic Green's function. We obtained the same expression previously — see Eq. (7.37) — by explicit solution of the differential equation which the Green's function satisfies.

We can now see the *group property of the Green's function* or propagator by observing that with the result (21.25) we can rewrite the Feynman path integral or kernel (21.10) in a product form, i.e. as

$$
\int_{x_i}^{x_f} \mathcal{D}\{x(t)\} e^{iS_0/\hbar} = \lim_{n\to\infty} \int_{x_i}^{x_f} \mathcal{D}\{x(t)\} \exp\left[\frac{i}{\hbar} \sum_{k=1}^{n+1} \frac{m_0}{2\epsilon}(x_k - x_{k-1})^2\right]
$$

$$
= \lim_{n\to\infty} \int \cdots \int dx_1 dx_2 \cdots dx_n \left(\frac{m_0}{2\pi i \hbar \epsilon}\right)^{1/2} e^{\frac{im_0}{2\hbar\epsilon}(x_1-x_0)^2}
$$

$$
\times \left(\frac{m_0}{2\pi i \hbar \epsilon}\right)^{1/2} e^{\frac{im_0}{2\hbar\epsilon}(x_2-x_1)^2} \cdots \left(\frac{m_0}{2\pi i \hbar \epsilon}\right)^{1/2} e^{\frac{im_0}{2\hbar\epsilon}(x_{n+1}-x_n)^2}
$$

$$
= \lim_{n\to\infty} \int \cdots \int dx_1 dx_2 \cdots dx_n K_0(x_{n+1}, t_{n+1}; x_n, t_n) \cdots
$$

$$
\cdots K_0(x_1, t_1; x_0, t_0) \tag{21.26}
$$

with n integrations and $n+1$ factors K_0. If the potential $V(x)$ of Eq. (21.5) is carried along, we have instead of S_0 and K_0 expressions S and K, the latter being a Green's function with interaction like that for the harmonic oscillator, Eq. (7.54).

Example 21.1: Evaluation of a path integral
Verify Eq. (21.22).

Solution: The result can be obtained by cumbersome integration, or else by induction. For $n = 1$ (see below)

$$
I_1 = \int_{-\infty}^{\infty} dx_1 e^{i\lambda[(x_1-x_0)^2+(x_2-x_1)^2]} = \frac{1}{\sqrt{2}}\left(\frac{i\pi}{\lambda}\right)^{1/2} e^{i\frac{\lambda}{2}(x_2-x_0)^2}.
$$

This result can be verified like Eq. (21.7) by direct integration with

$$
(x_1 - x_0)^2 + (x_2 - x_1)^2 = 2\left(x_1 - \frac{x_0 + x_2}{2}\right)^2 + \frac{1}{2}(x_0 - x_2)^2
$$

and using Eq. (21.6a). In our proof of induction we assume the result is correct for n. Then

$$
I_{n+1} = \int_{-\infty}^{\infty} I_n dx_{n+1} e^{i\lambda(x_{n+2}-x_{n+1})^2}
$$

$$
\overset{(21.22)}{=} \left[\frac{i^n \pi^n}{(n+1)\lambda^n}\right]^{1/2} \int_{-\infty}^{\infty} dx_{n+1} e^{\{\frac{i\lambda}{n+1}(x_{n+1}-x_0)^2 + i\lambda(x_{n+2}-x_{n+1})^2\}}.
$$

We set $y := x_{n+1} - x_i = x_{n+1} - x_0$ and

$$
(x_{n+2} - x_{n+1})^2 = (x_{n+2} - x_i + x_i - x_{n+1})^2 = (x_{n+2} - x_i)^2 + y^2 - 2y(x_{n+2} - x_i).
$$

Then

$$
I_{n+1} = \left[\frac{i^n \pi^n}{(n+1)\lambda^n}\right]^{1/2} \int_{-\infty}^{\infty} dx_{n+1} e^{i\lambda\{\frac{y^2}{n+1}+y^2-y^2+(x_{n+2}-x_{n+1})^2\}}
$$

$$
= \left[\frac{i^n \pi^n}{(n+1)\lambda^n}\right]^{1/2} \int_{-\infty}^{\infty} dx_{n+1} e^{i\lambda\{\frac{n+2}{n+1}y^2-2y(x_{n+2}-x_i)+(x_{n+2}-x_i)^2\}}.
$$

Next we set $z = y - \frac{n+1}{n+2}(x_{n+2} - x_i)$, so that $dz = dy = dx_{n+1}$, and

$$\frac{1}{n+2}(x_{n+2} - x_i)^2 + \frac{n+2}{n+1}z^2$$

$$= \frac{1}{n+2}(x_{n+2} - x_i)^2 + \frac{n+2}{n+1}y^2 - 2y(x_{n+2} - x_i) + \frac{n+1}{n+2}(x_{n+2} - x_i)^2$$

$$= \frac{n+2}{n+1}y^2 - 2y(x_{n+2} - x_i) + (x_{n+2} - x_i)^2.$$

Thus as required

$$I_{n+1} = \left[\frac{i^n \pi^n}{(n+1)\lambda^n}\right]^{1/2} e^{i\frac{\lambda}{n+2}(x_{n+2} - x_i)^2} \int_{-\infty}^{\infty} dz\, e^{i\lambda\frac{n+2}{n+1}z^2}$$

$$= \left[\frac{i^n \pi^n}{(n+1)\lambda^n}\right]^{1/2} e^{\frac{i\lambda}{n+2}(x_{n+2} - x_i)^2}\left[\frac{i\pi(n+1)}{\lambda(n+2)}\right]^{1/2} = \left[\frac{i^{n+1}\pi^{n+1}}{(n+2)\lambda^{n+1}}\right]^{1/2} e^{\frac{i\lambda}{n+2}(x_{n+2} - x_i)^2}$$

21.3.2 Momentum space represenation

Our next task is to derive the momentum space representation of the free particle Green's function K_0. This quantity is also described as the *nonrelativistic free particle propagator*. We pass over to momentum space with the help of the following two integrals. The first,

$$\left(\frac{\alpha}{i\pi}\right)^{1/2} \int_{-\infty}^{\infty} dp \exp\left[ipx + i\alpha p^2\right] = \exp\left[-i\frac{x^2}{4\alpha}\right], \quad (21.27)$$

can be verified by completing the square in the argument of the exponential in the integrand to $i\alpha(p + x/2\alpha)^2$ and using Eq. (21.6a). The second integral is an *integral representation of the step function*, i.e.

$$\theta(t) = \frac{1}{2\pi i} \int_{-\infty}^{\infty} d\tau \frac{e^{i\tau t}}{\tau - i\epsilon}, \quad \epsilon > 0. \quad (21.28)$$

This integral can be verified by applying Cauchy's residue theorem in the plane of complex τ and using one or the other of the contours shown in Fig. 21.2. Using Eq. (21.27) we have (with $\alpha = -t/2m_0$)

$$\exp\left[i\frac{x^2 m_0}{2t}\right] = \left(\frac{-t}{2\pi i m_0}\right)^{1/2} \int_{-\infty}^{\infty} dp \exp\left[ipx - i\frac{p^2}{2m_0}t\right]. \quad (21.29)$$

Hence with K_0 from Eq. (21.25):

$$K_0(x, t) = \left(\frac{m_0}{2\pi i t}\right)^{1/2} \exp\left[i\frac{m_0 x^2}{2t}\right]\theta(t)$$

$$= \theta(t)\left(\frac{m_0}{2\pi i t}\right)^{1/2}\left(-\frac{t}{2\pi i m_0}\right)^{1/2} \int_{-\infty}^{\infty} dp \exp\left[ipx - i\frac{p^2}{2m_0}t\right]$$

$$= \frac{1}{2\pi}\theta(t) \int_{-\infty}^{\infty} dp \exp\left[ipx - i\frac{p^2}{2m_0}t\right]. \quad (21.30a)$$

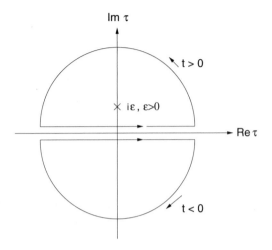

Fig. 21.2 Contours of integration.

Replacing $\theta(t)$ by its integral representation (21.28) this becomes

$$
\begin{aligned}
K_0(x,t) &= \frac{1}{i(2\pi)^2} \int_{-\infty}^{\infty} dp \int_{-\infty}^{\infty} d\tau \frac{\exp[ipx - i\frac{p^2}{2m_0}t + it\tau]}{\tau - i\epsilon} \\
&= -\frac{1}{i(2\pi)^2} \int_{-\infty}^{\infty} dp \int dE \frac{\exp[i(px - Et)]}{-E + \frac{p^2}{2m_0} - i\epsilon},
\end{aligned} \tag{21.30b}
$$

where

$$
E = \frac{p^2}{2m_0} - \tau.
$$

The expression

$$
\left[-E + \frac{p^2}{2m_0} - i\epsilon \right]^{-1} \tag{21.31}
$$

is the *nonrelativistic propagator* which describes the propagation of the free particle ($V = 0$) wave function. The generalization to three space dimensions (required below in Eq. (21.51)) is fairly self-evident.

21.4   Including V in First Order Perturbation

We now proceed to calculate the first order correction in a perturbation expansion. The first order correction contains one factor of the potential V, the second order two factors of V and so on. This is the type of expansion most people describe as perturbation theory, i.e. an expansion in rising powers of the coupling constant (contained in the potential). Thus in first order (cf. Eqs. (21.5) and (21.20)) and with $\hbar = 1$,

$$e^{iS} = \exp\left[i\left(S_0 - \int_{t_i}^{t_f} dt V(x,t)\right)\right] \simeq e^{iS_0}\left\{1 - i\int_{t_i}^{t_f} dt V(x,t)\right\}. \quad (21.32)$$

Since — we recall this for convenience —

$$\begin{aligned}
G(x_f, t_f; x_i, t_i) &= \int \mathcal{D}\{x(t)\}e^{iS} \\
&= \lim_{n\to\infty} \left(\frac{m_0}{2\pi i\epsilon}\right)^{(n+1)/2} \int_{-\infty}^{\infty} dx_1 \int_{-\infty}^{\infty} dx_2 \ldots \int_{-\infty}^{\infty} dx_n \\
&\quad \times \exp\left[\sum_{k=1}^{n+1} i\epsilon\left(\frac{m_0}{2\epsilon^2}(x_k - x_{k-1})^2 - V(x_k, t_k)\right)\right] \\
&= \int \mathcal{D}\{x(t)\} \exp\left[iS_0 - i\int_{t_i}^{t_f} dt V(x,t)\right],
\end{aligned}$$

the first order correction is $(G_1 \simeq G - G_0)$

$$\begin{aligned}
G_1(x_f, t_f; x_i, t_i) &= \int \mathcal{D}\{x(t)\}e^{iS_0} \int_{t_i}^{t_f} dt[-iV(x,t)] \\
&= \lim_{n\to\infty} -i\sum_{l=1}^{n+1}\epsilon \int \mathcal{D}\{x\}e^{iS_0}V(x, t_l = l\epsilon) = \lim_{n\to\infty} -i\sum_{l=1}^{n+1}\epsilon \left(\frac{m_0}{2\pi i\epsilon}\right)^{(n+1)/2} \\
&\quad \times \int_{-\infty}^{\infty} dx_1 \int_{-\infty}^{\infty} dx_2 \ldots \int_{-\infty}^{\infty} dx_n \exp\left[i\sum_{k=1}^{n+1}\frac{m_0}{2\epsilon}(x_k - x_{k-1})^2\right]V(x_l, l\epsilon).
\end{aligned}$$

By regrouping factors this becomes

$$\begin{aligned}
G_1 &= \lim_{n\to\infty} \sum_{l=1}^{n+1}\epsilon \int dx_n \int dx_{n-1} \ldots \int dx_{l+1} \int dx_l \int d_{l-1} \ldots dx_1 \\
&\quad \times \left(\frac{m_0}{2\pi i\epsilon}\right)^{\frac{n-(l-1)}{2}} \exp\left[i\sum_{k=l+1}^{n+1}\frac{m_0}{2\epsilon}(x_k - x_{k-1})^2\right]\{-iV(x_l, t_l)\} \\
&\quad \times \left(\frac{m_0}{2\pi i\epsilon}\right)^{l/2} \exp\left[i\sum_{k=1}^{l}\frac{m_0}{2\epsilon}(x_k - x_{k-1})^2\right].
\end{aligned}$$

With Eq. (21.21) this can be written as

$$\begin{aligned}
G_1(x_f, t_f; x_i, t_i) &= \int_{-\infty}^{\infty} dt \int_{-\infty}^{\infty} dx_l \, G_0(x_f, t_f; x_l, t_l)\{-iV(x_l, t_l)\} \\
&\quad \times G_0(x_l, t_l; x_i, t_i), \quad (21.33)
\end{aligned}$$

and hence correspondingly for the first order contribution K_1:

$$K_1(x_f, t_f; x_i, t_i) = -i \int_{-\infty}^{\infty} dt \int_{-\infty}^{\infty} dx K_0(x_f, t_f; x, t) V(x, t) K_0(x, t; x_i, t_i).$$

(21.34)

We observe that we can extend the t-integration to $-\infty$ since K_0 contains the step function $\theta(t)$. It can now be surmized that the next order contribution to $K_0 + K_1$ is

$$
\begin{aligned}
K_2 &= (-i)^2 \int dt_1 \int dt_2 \int dx_1 \int dx_2 K_0(x_f, t_f; x_1, t_1) \\
&\quad \times V(x_1, t_1) K_0(x, t_1; x_2, t_2) V(x_2, t_2) K_0(x_2, t_2; x_i, t_i) \\
&= -i \int dt_2 \int dx_2 K_1(x_f, t_f; x_2, t_2) V(x_2, t_2) \\
&\quad \times K_0(x_2, t_2; x_i, t_i).
\end{aligned}
$$

(21.35)

We also observe here that the problem of *time-ordering* arises only in perturbation theory, i.e. in writing down the expansion

$$
\begin{aligned}
\exp\left[-i \int_{t_i}^{t_f} dt V(x, t)\right] &= 1 - i \int_{t_i}^{t_f} dt V(x, t) \\
&\quad + \frac{(-i)^2}{2!} \int_{t_i}^{t_f} dt V(x, t) \int_{t_i}^{t_f} dt' V(x, t') + \cdots .
\end{aligned}
$$

(21.36)

Consider the term of second order, i.e.

$$
\begin{aligned}
&\left(\int_{t_i}^{t_1} V \, dt + \int_{t_1}^{t_f} V \, dt\right)\left(\int_{t_i}^{t_1} V \, dt' + \int_{t_1}^{t_f} V \, dt'\right) \\
&= \int_{t_i}^{t_1} V \, dt \int_{t_1}^{t_f} V \, dt' + \int_{t_i}^{t_1} V \, dt' \int_{t_1}^{t_f} V \, dt + \int_{t_i}^{t_1} V \, dt \int_{t_i}^{t_1} V \, dt' \\
&\quad + \int_{t_1}^{t_f} V \, dt \int_{t_1}^{t_f} V \, dt'.
\end{aligned}
$$

(21.37)

The last two contributions do not describe the time evolution from t_i to t_f. They must therefore be deleted. The first two terms lead to identical contributions and therefore cancel the factor $1/2!$ in the expansion. The factors of "V" originate from the series expansion of the exponential

$$\exp\left[-i \int_{t_i}^{t_f} dt V(x, t)\right].$$

Thus the term containing n times V contains a factor $1/n!$. This is cancelled by $n!$ different time orderings.

The full perturbation expansion of the Green's function K is now seen to be given by

$$K(x_f, t_f; x_i, t_i) = K_0(x_f, t_f; x_i, t_i)$$
$$-i \sum_{n=0}^{\infty} \int dt \int dx K_n(x_f, t_f; x, t) V(x, t) K_0(x, t; x_i, t_i). \quad (21.38)$$

This can also be written in the form of an integral equation which when iterated yields this expansion:

$$K(x_f, t_f; x_i, t_i) = K_0(x_f, t_f; x_i, t_i)$$
$$-i \int dt \int dx K(x_f, t_f; x, t) V(x, t) K_0(x, t; x_i, t_i). \quad (21.39)$$

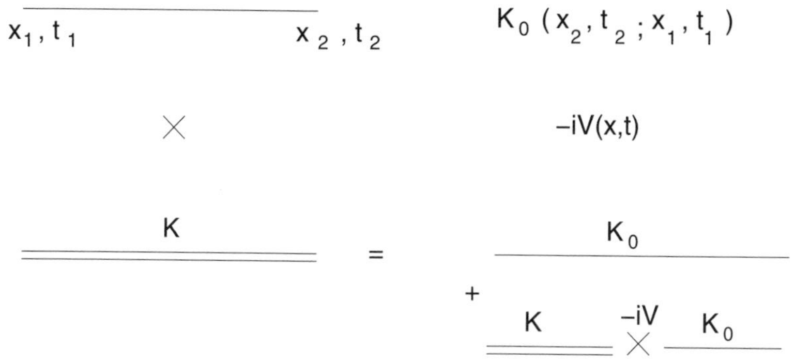

Fig. 21.3 Feynman rules and representation of the Green's function.

For various reasons — such as illustration, transparency or brevity — it is useful to introduce a diagramatic representation of individual perturbation terms. Such diagrams in the context of field theory are called *Feynman diagrams*. Here, of course, we are concerned with quantum mechanics, and hence the diagrams representing perturbation terms here are Feynman diagrams corresponding to those in field theory. The diagrams represent mathematical quantities and hence are designed on the basis of rules, called *Feynman rules*. These may be formulated in configuration space as below, or in momentum space as mostly in field theory. The Fourier transform of $V(x, t)$ is $\tilde{V}(p, E)$ which is given by (in one-time plus one-space dimensions)

$$V(x, t) = \int dp \, dE \, e^{ipx - iEt} \tilde{V}(p, E). \quad (21.40)$$

The rules for the present case are given in Fig. 21.3 along with the diagramatic representation of Eq. (21.39).

21.5 Rederivation of the Rutherford Formula

As an application of the path integral method we now consider scattering of a particle off a Coulomb potential, and this means the transition of a nonrelativistic particle from an initial state 'i' to a final state 'f' with wave functions ψ_i (or ψ_{in}) and ψ_f (or ψ_{out}). The *transition amplitude A* describing this process is defined by the Green's function sandwiched between the in- and out-state wave functions, i.e. in one-space plus one-time dimensions by

$$A = \int dx_f \int dx_i \psi_f^*(x_f, t_f) K(x_f, t_f; x_i, t_i) \psi_i(x_i, t_i). \qquad (21.41)$$

We have to pass over to three space dimensions. There the wave functions ψ_i, ψ_f have to be properly normalized, i.e. to one particle in all of space. The plane wave function normalized to one over a box of volume $V = L^3$ is, here with $E = \mathbf{k}^2/2m_0$,

$$\psi(\mathbf{x}, t) = \frac{1}{\sqrt{V}} e^{i\mathbf{k}\cdot\mathbf{x} - i\frac{Et}{\hbar}} \quad \text{with} \quad \int_V \psi^* \psi d\mathbf{x} = 1. \qquad (21.42)$$

We are interested in the total cross section defined by

$$\sigma_{\text{tot}} = \int d\mathbf{p} \frac{d\mathbf{n}}{d\mathbf{p}} \times \frac{\text{transition probability per second}}{\text{incident flux}}. \qquad (21.43)$$

Here
(1) $d\mathbf{n}/d\mathbf{p}$ = density of (plane wave) states in the box (of volume $V = L^3$) per unit three-momentum interval[**] $= V/(2\pi\hbar)^3$.
(2) The incident flux (number of particles passing through unit area in unit time) is equal to the incident velocity times the density of incident particles and this is

$$= \hbar \frac{k}{m_0} \frac{1}{V} \text{ particles per cm}^2 \text{ second,}$$

since ψ is normalized to one particle in volume V.
(3) The transition probability per second is

$$= \frac{|A|^2}{T}, \quad T = \text{duration of time for which potential is switched on.}$$

The expression $|A|^2/T$ can be related to first-order time-dependent perturbation theory in quantum mechanics. We recall for convenience from

---

[**]Note that from the wave function we obtain, with $\mathbf{p} = \hbar\mathbf{k}$, for the allowed values of $\mathbf{k}$: $k_x = 2\pi n_x/L, \mathbf{k} = 2\pi\mathbf{n}/L$, and $d\mathbf{n} = dn_x dn_y dn_z$, $d\mathbf{n} = (L/2\pi)^3 d\mathbf{k}$.

Chapter 10 the Schrödinger equation with the subdivision of its Hamiltonian into an unperturbed part and a perturbation part, and with these the following perturbation ansatz:

$$i\hbar\frac{\partial}{\partial t}\Psi = H\Psi, \quad H = H_0 + \lambda H', \quad \Psi = \sum_n a_n(t)\psi_n(\mathbf{r}, t)$$

with

$$a_n(t) = a_n^{(0)}(t) + \lambda a_n^{(1)}(t) + \cdots .$$

Then for stationary states

$$\psi_k(\mathbf{r}, t) = u(E_k, \mathbf{r})e^{-iE_k t/\hbar}, \quad H'_{ki}(t) = \int d\mathbf{r}\, u_k^*(\mathbf{r})H'u_i(\mathbf{r})$$

and

$$i\hbar a_k^{(1)} = \int_{-\infty}^t H'_{ki}(t')e^{i\frac{E_k - E_i}{\hbar}t'}\,dt'. \tag{21.44}$$

Thus the amplitude $a_k^{(1)}$ is effectively the spacetime Fourier transform of the interaction or potential $\lambda H'$ contained in the Hamiltonian. The transition probability per unit time is

$$\omega = \frac{1}{t}\sum_k |a_k^{(1)}(t)|^2.$$

It is frequently assumed that the only time dependence of $H'(t)$ is that it is switsched on at $t = 0$ and switched off at time $t > 0$. Then H'_{ki} can be taken out of the integrand in Eq. (21.44) and $\int_{-\infty}^t$ can be replaced by $\int_0^t$. Proceeding in this manner one obtains *Fermi's "golden rule"*, i.e.

$$\omega = \frac{2\pi}{\hbar}|H'_{ki}|^2\rho(k), \quad \rho(k) = \frac{dn}{dE_k} \tag{21.45}$$

(cf. Eq. (10.117) and Example 10.4). Hence ($\hbar = 1$)

$$\sigma_{\text{tot}} = \int d\mathbf{p}\frac{V}{(2\pi)^3}\frac{Vm_0}{k}\frac{|A(p, k, T)|^2}{T}. \tag{21.46}$$

We take the following expression for the time-regularized Coulomb potential, in which T is a large but finite time,

$$V(\mathbf{x}, t) = \frac{\alpha}{r}e^{-4t^2/T^2}, \quad \text{where } r = |\mathbf{x}|, \quad \alpha = \frac{e^2}{4\pi} \simeq \frac{1}{137}. \tag{21.47}$$

In calculating the Rutherford formula with the help of our previous formulas, we have to replace the one-dimensional quantities dx_f etc. by three-dimensional $d\mathbf{x}_f$ and so on.

The lowest order approximation of A, also called *Born approximation*, is now in three space dimensions

$$A \equiv A_1 = \int d\mathbf{x}_f \int d\mathbf{x}_i \psi_f^*(\mathbf{x}_f, t_f) K_1(\mathbf{x}_f, t_f; \mathbf{x}_i, t_i) \psi_i(\mathbf{x}_i, t_i), \qquad (21.48)$$

where from Eq. (21.39)

$$K_1(\mathbf{x}_f, t_f; \mathbf{x}_i, t_i) = -i \int_{-\infty}^{\infty} dt \int_{-\infty}^{\infty} d\mathbf{x}\, K_0(\mathbf{x}_f, t_f; \mathbf{x}, t) V(\mathbf{x}, t) K_0(\mathbf{x}, t; \mathbf{x}_i, t_i),$$

i.e.

$$
\begin{aligned}
A_1 &= \int d\mathbf{x}_f \int d\mathbf{x}_i \int dt \int d\mathbf{x}\, \psi_f^*(\mathbf{x}_f, t_f) K_0(\mathbf{x}_f, t_f; \mathbf{x}, t) \\
&\quad \times (-i) V(\mathbf{x}, t) K_0(\mathbf{x}, t; \mathbf{x}_i, t_i) \psi_i(\mathbf{x}_i, t_i).
\end{aligned}
\qquad (21.49)
$$

We can relate this amplitude to the amplitude $a_k^{(1)}$ of Eq. (21.44) by contracting the propagators to instantaneous point interactions. Thus with the ansatz $K_0(\mathbf{x}, t; \mathbf{x}_i, t_i) = c_0 \delta(t - t_i)\delta(\mathbf{x} - \mathbf{x}_i)$ etc., we obtain

$$A_1 = c_0^2 \delta(t_i - t_f) \int d\mathbf{x}_f (-i) V(\mathbf{x}_f, t_f) \psi^*(\mathbf{x}_f, t_f) \psi_i(\mathbf{x}_f, t_f).$$

This now has the form of the amplitude (21.44) (here in the scattering problem with initial and final energies equal).

In our previous one-dimensional considerations we had with Eq. (21.30a)

$$K_0(x_f, t_f; x_i, t_i) = \frac{\theta(t)}{2\pi} \int_{-\infty}^{\infty} dp\, e^{ipx - i\frac{p^2}{2m_0}t}. \qquad (21.50)$$

Here $x = x_f - x_i$ and $t = t_f - t_i$. The three-dimensional generalization is (taking $t_f \gg T$ and $t_i \ll -T$, T a large but finite time, so that $\theta(t)$ can be dropped)

$$K_0(\mathbf{x}_f, t_f; \mathbf{x}, t) = \frac{1}{(2\pi)^3} \int d\mathbf{q}\, e^{i\mathbf{q}\cdot(\mathbf{x}_f - \mathbf{x}) - i\frac{q^2}{2m_0}(t_f - t)}. \qquad (21.51)$$

Inserting this into Eq. (21.49) together with the following wave functions (of asymptotically free particles)

$$\psi_i(\mathbf{x}_i, t_i) = \frac{1}{\sqrt{V}} e^{i\mathbf{k}\cdot\mathbf{x}_i - i\frac{k^2}{2m_0}t_i}, \qquad \psi_f(\mathbf{x}_f, t_f) = \frac{1}{\sqrt{V}} e^{i\mathbf{p}\cdot\mathbf{x}_f - i\frac{p^2}{2m_0}t_f}, \qquad (21.52)$$

we have (for ingoing momentum $\mathbf{k}$ and outgoing momentum $\mathbf{p}$)

$$A(\mathbf{p}, \mathbf{k}, T) = \int d\mathbf{x}_f \int d\mathbf{x}_i \int dt \int d\mathbf{x} \int d\mathbf{q} \int d\mathbf{q}' \frac{1}{V} \frac{(-i)V(\mathbf{x}, t)}{(2\pi)^6}$$

$$\times \exp\left[-i\mathbf{p} \cdot \mathbf{x}_f + i\frac{\mathbf{p}^2}{2m_0} t_f + i\mathbf{q} \cdot (\mathbf{x}_f - \mathbf{x}) - i\frac{\mathbf{q}^2}{2m_0}(t_f - t) \right.$$

$$\left. + i\mathbf{q}' \cdot (\mathbf{x} - \mathbf{x}_i) - i\frac{\mathbf{q}'^2}{2m_0}(t - t_i) + i\mathbf{k} \cdot \mathbf{x}_i - i\frac{\mathbf{k}^2}{2m_0} t_i \right]. \quad (21.53)$$

The integrations with respect to $\mathbf{x}_f$ and $\mathbf{x}_i$ (including in each case a factor $(2\pi)^3$) yield delta functions $\delta(\mathbf{q} - \mathbf{p})$ and $\delta(\mathbf{k} - \mathbf{q}')$ respectively. Then integrating over $\mathbf{q}$ and $\mathbf{q}'$ we obtain

$$A(\mathbf{p}, \mathbf{k}, T) = \int dt \int d\mathbf{x} \left(\frac{-i}{V}\right) V(\mathbf{x}, t) \exp\left[-i\mathbf{p} \cdot \mathbf{x} + i\frac{\mathbf{p}^2}{2m_0} t + i\mathbf{k} \cdot \mathbf{x} - i\frac{\mathbf{k}^2}{2m_0} t \right]. \quad (21.54)$$

Now inserting the expression for the potential $V(\mathbf{x}, t)$, this becomes

$$A(\mathbf{p}, \mathbf{k}, T) = -\frac{i}{V} \int d\mathbf{x} \int dt \frac{\alpha}{r} e^{-4t^2/T^2} \exp\left[i(\mathbf{k} - \mathbf{p}) \cdot \mathbf{x} - i(\mathbf{k}^2 - \mathbf{p}^2)\frac{t}{2m_0} \right]. \quad (21.55)$$

We evaluate this integral with the help of the following integral in which we complete the square in the argument of the exponential and set $\tau = t - iaT^2/8$:

$$\int_{-\infty}^{\infty} dt\, e^{iat - \frac{4t^2}{T^2}} = e^{-\frac{a^2 T^2}{16}} \int_{-\infty}^{\infty} d\tau\, e^{-4\tau^2/T^2} = \left(\frac{\pi T^2}{4}\right)^{1/2} e^{-a^2 T^2/16}. \quad (21.56)$$

With

$$a = \frac{\mathbf{p}^2 - \mathbf{k}^2}{2m_0}, \quad (21.57)$$

we have

$$A(\mathbf{p}, \mathbf{k}, T) = -\frac{i}{V} \int d\mathbf{x} \frac{\alpha}{r} e^{i(\mathbf{k}-\mathbf{p})\cdot\mathbf{x}} \int dt\, e^{i\frac{(\mathbf{p}^2 - \mathbf{k}^2)t}{2m_0} - \frac{4t^2}{T^2}}$$

$$= -\frac{i}{V} \int d\mathbf{x} \frac{\alpha}{r} e^{i(\mathbf{k}-\mathbf{p})\cdot\mathbf{x}} \left(\frac{\pi T^2}{4}\right)^{1/2} e^{-\frac{(\mathbf{p}^2 - \mathbf{k}^2)^2}{(2m_0)^2} \frac{T^2}{16}}. \quad (21.58)$$

In doing the integration over $\mathbf{x}$ we introduce a temporary cutoff $r = 1/M$ by inserting the factor e^{-Mr} in the integrand. Then we require the following integral, which is evaluated in Example 21.2:

$$I_{\text{Yukawa}} := \int d\mathbf{x} \frac{e^{i\mathbf{q}\cdot\mathbf{x}}}{r} e^{-Mr} = \frac{4\pi}{\mathbf{q}^2 + M^2}. \quad (21.59)$$

This means: The result of Fourier transforming the *Yukawa potential* $\sim$ $\exp(-Mr)/r$ is the *propagator* (21.59), where M has the meaning of mass (in the case of the Yukawa potential the mass of the exchanged meson).

Using the result (21.59) we can re-express A as

$$A(\mathbf{p}, \mathbf{k}, T) = -\frac{i}{V}\alpha\left(\frac{\pi T^2}{4}\right)^{1/2} e^{-\frac{(\mathbf{p}^2-\mathbf{k}^2)}{(2m_0)^2}\frac{T^2}{16}}\left\{\frac{4\pi}{(\mathbf{k}-\mathbf{p})^2 + M^2}\right\}. \qquad (21.60)$$

Now, σ_{tot} contains the factor $|A(\mathbf{p}, \mathbf{k}, T)|^2/T$. This expression assumes that the integration over time t is normalized to 1. In our case this has not been done, i.e. integrating over the square of the t-dependent factor in V we have

$$\int_{-\infty}^{\infty} dt\left(e^{-4t^2/T^2}\right)^2 = \left(\frac{\pi T^2}{8}\right)^{1/2} \quad \text{instead of } T. \qquad (21.61)$$

Hence in our case $|A|^2/T$ has to be replaced by $|A|^2/(\pi T^2/8)^{1/2}$. We can now write down the probability per unit time which is

$$|A|^2\left(\frac{8}{\pi T^2}\right)^{1/2} = \frac{\alpha^2}{V^2}\frac{16\pi^2}{[(\mathbf{k}-\mathbf{p})^2+M^2]^2}\frac{\pi T^2}{4}\left(\frac{8}{\pi T^2}\right)^{1/2} e^{-\frac{(\mathbf{p}^2-\mathbf{k}^2)^2}{4m_0^2}\frac{T^2}{8}}. \qquad (21.62)$$

We want to consider the limit $T \to \infty$. For this we require the following representation of the delta function:

$$\delta(k) = \lim_{T\to\infty} T\frac{e^{-k^2T^2}}{\sqrt{\pi}}, \qquad \delta(k) = \lim_{T\to\infty} \frac{T}{\sqrt{8}}\frac{e^{-k^2T^2/8}}{\sqrt{\pi}} \qquad (21.63)$$

We establish this result for $T = 1/\epsilon$ in Example 21.3. Using Eq. (21.63) we have:

$$\text{Probability per unit time} = \frac{\alpha^2}{V^2}\frac{16\pi^2}{[(\mathbf{k}-\mathbf{p})^2+M^2]^2}2\pi\delta\left(\frac{\mathbf{p}^2-\mathbf{k}^2}{2m_0}\right) \qquad (21.64)$$

with

$$\delta\left(\frac{\mathbf{p}^2-\mathbf{k}^2}{2m_0}\right) = \frac{1}{2\pi}e^{-\frac{(\mathbf{p}^2-\mathbf{k}^2)^2T^2}{32m_0^2}}\left(\frac{\pi T^2}{2}\right)^{1/2}.$$

Inserting the result now into our expression for the total cross section, σ_{tot}, we obtain

$$\sigma_{\text{tot}} = \int d\mathbf{p}\,\frac{V}{(2\pi)^3}\frac{m_0 V}{k}\frac{32\pi^3\alpha^2\delta(\frac{\mathbf{p}^2-\mathbf{k}^2}{2m_0})}{V^2[(\mathbf{k}-\mathbf{p})^2+M^2]^2}. \qquad (21.65)$$

But

$$E = \frac{\mathbf{p}^2}{2m_0}, \quad dE = \frac{pdp}{m_0} = \frac{\sqrt{2m_0 E}\,dp}{m_0}, \qquad (21.66)$$

so that

$$
\int d\mathbf{p}\,\delta\left(\frac{\mathbf{p}^2 - \mathbf{k}^2}{2m_0}\right) = \int m_0 p\,d\left(\frac{p^2}{2m_0}\right) d\Omega\,\delta\left(\frac{\mathbf{p}^2 - \mathbf{k}^2}{2m_0}\right)
$$

$$
= \int m_0 k\,d\Omega = m_0\sqrt{2m_0 E}\int d\Omega. \qquad (21.67)
$$

The delta function here implies energy conservation, i.e. $\mathbf{p}^2 = \mathbf{k}^2$, so that

$$
(\mathbf{k} - \mathbf{p})^2 = 2\mathbf{k}^2(1 - \cos\theta) = 4\mathbf{k}^2\sin^2\frac{\theta}{2}. \qquad (21.68)
$$

Inserting (21.67) and (21.68) into Eq. (21.65) with the cutoff $M \to 0$, we obtain

$$
\sigma_{\text{tot}} = \int \frac{m_0^2 V^2\,d\Omega\,\sqrt{2m_0 E}\,32\pi^3\alpha^2}{(2\pi)^3 kV^2\,4k^4\,4\sin^4(\theta/2)},
$$

i.e. with $k = \sqrt{2m_0 E}$ we obtain the *differential cross section*

$$
\frac{d\sigma}{d\Omega} = \frac{\alpha^2 m_0^2}{4k^4\sin^4(\theta/2)}. \qquad (21.69)
$$

This is the wellknown *Rutherford formula*. This result may also be obtained from I_{Yukawa} of Eq. (21.59) by replacing there $\mathbf{q}^2$ by $(\mathbf{k}-\mathbf{p})^2$, setting $M^2 = 0$, giving the potential a coupling constant, and inserting the expression thus obtained into the final result of Example 10.4.

Example 21.2: Fourier transform of the Yukawa potential
Show that

$$
I_{\text{Yukawa}} \equiv \int d\mathbf{x}\,\frac{e^{i\mathbf{q}\cdot\mathbf{x}}}{r}e^{-Mr} = \frac{4\pi}{\mathbf{q}^2 + M^2}.
$$

Solution: We choose the z-axis along the vector $\mathbf{q}$. Then $(\mathbf{x} \equiv \mathbf{r})$

$$
I_{\text{Yukawa}} = 2\pi\int_0^\infty \frac{r^2\,dr}{r}e^{-Mr}\int_{-1}^1 d(\cos\theta)e^{iqr\cos\theta} = 2\pi\int_0^\infty \frac{r\,dr}{iqr}e^{-Mr}\left(e^{iqr} - e^{-iqr}\right)
$$

$$
= 2\pi\int_0^\infty \frac{2}{q}dr\,e^{-Mr}\sin qr = \frac{4\pi}{q}\Im\int_0^\infty dr\,e^{-(M-iq)r}
$$

$$
= \Im\frac{4\pi}{q}\left[\frac{1}{M - iq}\right] = \frac{4\pi}{\mathbf{q}^2 + M^2}.
$$

Example 21.3: A representation of the delta function
Determine the Fourier transformation of e^{-ax^2} for $a > 0$ and hence the following representation of the delta function

$$
\delta(k) = \lim_{a\to 0}\frac{1}{2\pi}\int_{-\infty}^\infty dx\,e^{-ax^2}e^{ikx} = \lim_{a\to 0}\frac{1}{2\sqrt{\pi a}}e^{-k^2/4a} = \lim_{\epsilon\to 0}\frac{e^{-k^2/\epsilon^2}}{\epsilon\sqrt{\pi}}. \qquad (21.70)
$$

The function required is

$$g(k) = \int_{-\infty}^{\infty} dx e^{-ax^2} e^{ikx}. \tag{21.71}$$

Solution: In general one uses for the evaluation of such an integral the method of contour integration. In the present case, however, a simpler method suffices. Differentiation of (21.71) with respect to k yields the same as

$$g'(k) = i \int_{-\infty}^{\infty} x dx e^{-ax^2} e^{ikx} = -\frac{i}{2a} \int_{-\infty}^{\infty} dx \frac{d}{dx}(e^{-ax^2}) e^{ikx}.$$

Partial integration of the right hand side implies

$$g'(k) = -\frac{i}{2a}\left[e^{ikx}e^{-ax^2}\Big|_{-\infty}^{\infty} - \int_{-\infty}^{\infty}\frac{d}{dx}(e^{ikx})e^{-ax^2} dx\right] = \frac{i}{2a}\int_{-\infty}^{\infty} ike^{ikx}e^{-ax^2} dx = -\frac{k}{2a}g(k).$$

Thus $g(k)$ satisfies the first order differential equation $g'(k) + (k/2a)g(k) = 0$. Direct integration yields $g(k) = Ce^{-k^2/4a}$ with the constant $C = g(0)$. Since

$$g(0) = \int_{-\infty}^{\infty} dx e^{-ax^2} = \sqrt{\frac{\pi}{a}}, \quad \text{we obtain} \quad g(k) = \sqrt{\frac{\pi}{a}}e^{-k^2/4a}. \tag{21.72}$$

From (21.70) and (21.71) we obtain a representation of the delta function:

$$\delta(k) = \lim_{a \to 0}\frac{1}{2\pi}\int_{-\infty}^{\infty} dx e^{-ax^2} e^{ikx} = \lim_{a \to 0}\frac{1}{2\pi}g(k) = \lim_{a \to 0}\frac{1}{2\sqrt{\pi a}}e^{-k^2/4a} = \lim_{\epsilon \to 0}\frac{e^{-k^2/\epsilon^2}}{\epsilon\sqrt{\pi}}.$$

21.6   Path Integrals in Dirac's Notation

It is instructive to devote a little more time on the Feynman formalism by rewriting it in terms of Dirac's bra and ket notation.

We consider again nonrelativistic quantum mechanics. We write the state vector of a particle of three-momentum $\mathbf{p}$ in the momentum space representation: $|\mathbf{p}\rangle$. In configuration space the state of the particle is written $|\mathbf{x}\rangle$. The two representations are related to one another as we saw in Chapter 4 by the following expressions or Fourier integrals:

$$|\mathbf{p}\rangle = \int d\mathbf{x}\,|\mathbf{x}\rangle\langle\mathbf{x}|\mathbf{p}\rangle, \quad |\mathbf{x}\rangle = \int \frac{d\mathbf{p}}{(2\pi)^3}|\mathbf{p}\rangle\langle\mathbf{p}|\mathbf{x}\rangle. \tag{21.73}$$

We have

$$\delta(\mathbf{p}) = \frac{1}{(2\pi)^3}\int d\mathbf{x}e^{-i\mathbf{p}\cdot\mathbf{x}}, \quad \delta(\mathbf{x}) = \frac{1}{(2\pi)^3}\int d\mathbf{p}e^{i\mathbf{p}\cdot\mathbf{x}}. \tag{21.74}$$

The normalizations

$$\langle\mathbf{q}|\mathbf{p}\rangle = (2\pi)^3\delta(\mathbf{p} - \mathbf{q}), \quad \langle\mathbf{y}|\mathbf{x}\rangle = \delta(\mathbf{y} - \mathbf{x}) \tag{21.75}$$

therefore imply

$$\langle \mathbf{x} | \mathbf{p} \rangle = e^{i\mathbf{p} \cdot \mathbf{x}}, \quad \langle \mathbf{p} | \mathbf{x} \rangle = e^{-i\mathbf{p} \cdot \mathbf{x}}. \tag{21.76}$$

These relations imply the completeness relations

$$1 = \int d\mathbf{x} |\mathbf{x}\rangle\langle\mathbf{x}|, \quad 1 = \int \frac{d\mathbf{p}}{(2\pi)^3} |\mathbf{p}\rangle\langle\mathbf{p}|. \tag{21.77}$$

The equation

$$\mathbf{P}|\mathbf{p}\rangle = \mathbf{p}|\mathbf{p}\rangle \tag{21.78}$$

is to be interpreted to say: The quantity $\mathbf{p}$ is the eigenvalue of operator $\mathbf{P}$ with eigenvector $|\mathbf{p}\rangle$. Thus

$$\langle \mathbf{q} | \mathbf{P} | \mathbf{p} \rangle = \mathbf{p}\langle \mathbf{q} | \mathbf{p} \rangle = \mathbf{p}(2\pi)^3 \delta(\mathbf{q} - \mathbf{p}). \tag{21.79}$$

Similarly we have a position operator $\mathbf{X}$ with

$$\mathbf{X}|\mathbf{x}\rangle = \mathbf{x}|\mathbf{x}\rangle. \tag{21.80}$$

Feynman's principle is expressed in terms of the propagator or transition function

$$\langle \mathbf{x}', t' | \mathbf{x}, t \rangle \equiv K(\mathbf{x}', t'; \mathbf{x}, t),$$

which describes the propagation of the particle from its initial position $\mathbf{x}_i$ at initial time t_i to its final position $\mathbf{x}_f$ at time t_f. We consider first an infinitesimal transition, i.e. $t' - t, \mathbf{x}' - \mathbf{x}$ infinitesimal. The time development is given by the Hamilton operator H, i.e.

$$|\mathbf{x}, t\rangle = e^{iHt}|\mathbf{x}, 0\rangle \equiv e^{iHt}|\mathbf{x}\rangle, \tag{21.81}$$

so that[††]

$$\frac{\partial}{\partial t}|\mathbf{x}, t\rangle = iH|\mathbf{x}, t\rangle. \tag{21.82}$$

Hence for the infinitesimal transition

$$\langle \mathbf{x}', t' | \mathbf{x}, t \rangle = \langle \mathbf{x}' | e^{iH(t-t')} | \mathbf{x} \rangle. \tag{21.83}$$

[††]Comments on signs in Eq. (21.82): The general Schrödinger equation has the state vector to the right, i.e. (with $\hbar = 1$) $\partial/\partial t|\psi\rangle = -iH|\psi\rangle$. The relation between Heisenberg picture states and Schrödinger picture states is $|\psi_t\rangle_S = e^{-iHt}|\psi\rangle_H$, so that $|\psi_{t=0}\rangle_S = |\psi\rangle_H$. The time development in the Schrödinger picture is given by $|\psi_t\rangle_S = e^{-iHt}|\psi_0\rangle_S$. We now introduce the Dirac bra and ket formalism. Then $\langle \mathbf{x}|\psi\rangle_H = \langle \mathbf{x}|\psi_0\rangle_S$. We define $|\mathbf{x}, t\rangle = e^{iHt}|\mathbf{x}\rangle$. Then the states $\{|\mathbf{x}, t\rangle\}$ constitute a "moving reference frame" in the sense that $\psi(\mathbf{x}, t) \equiv \langle \mathbf{x}, t|\psi_0\rangle_S$. If $\psi(\mathbf{x}, t)$ is to satisfy the Schrödinger equation, i.e. $\frac{\partial}{\partial t}\psi(\mathbf{x}, t) = -iH\psi(\mathbf{x}, t)$, we must have

$$\partial/\partial t\langle \mathbf{x}, t| = -iH\langle \mathbf{x}, t|, \text{ i.e. } \partial/\partial t|\mathbf{x}, t\rangle = +iH|\mathbf{x}, t\rangle,$$

as had to be shown. See also L. H. Ryder [241], pp. 159 – 160.

We can rewrite this relation as

$$
\begin{aligned}
\langle \mathbf{x}', t' | \mathbf{x}, t \rangle &= \int \frac{d\mathbf{p}'}{(2\pi)^3} \int \frac{d\mathbf{p}}{(2\pi)^3} \langle \mathbf{x}' | \mathbf{p}' \rangle \langle \mathbf{p}' | e^{iH(t-t')} | \mathbf{p} \rangle \langle \mathbf{p} | \mathbf{x} \rangle \\
&= \int \frac{d\mathbf{p}'}{(2\pi)^3} \int \frac{d\mathbf{p}}{(2\pi)^3} e^{i\mathbf{p}' \cdot \mathbf{x}'} \langle \mathbf{p}' | e^{iH(t-t')} | \mathbf{p} \rangle e^{-i\mathbf{p} \cdot \mathbf{x}}. \quad (21.84)
\end{aligned}
$$

We suppose that for a single particle of mass m_0 moving in a potential $V(\mathbf{x})$ the Hamiltonian is

$$
H = \frac{\mathbf{p}^2}{2m_0} + V(\mathbf{x}). \quad (21.85)
$$

The momentum operator $\mathbf{P}$ satisfies the relation

$$
\mathbf{P} = \int \frac{d\mathbf{p}}{(2\pi)^3} \mathbf{p} | \mathbf{p} \rangle \langle \mathbf{p} |, \quad (21.86)
$$

so that it projects out the momentum $\mathbf{q}$ of a momentum state $|\mathbf{q}\rangle$, i.e.

$$
\mathbf{P} | \mathbf{q} \rangle = \int \frac{d\mathbf{p}}{(2\pi)^3} \mathbf{p} | \mathbf{p} \rangle \langle \mathbf{p} | \mathbf{q} \rangle = \int d\mathbf{p} \, \mathbf{p} | \mathbf{p} \rangle \delta(\mathbf{p} - \mathbf{q}) = \mathbf{q} | \mathbf{q} \rangle. \quad (21.87)
$$

We also have by mapping onto configuration space

$$
\begin{aligned}
\mathbf{P} | \mathbf{p} \rangle &= \mathbf{P} \int d\mathbf{x} | \mathbf{x} \rangle \langle \mathbf{x} | \mathbf{p} \rangle = \mathbf{P} \int d\mathbf{x} | \mathbf{x} \rangle e^{i\mathbf{p} \cdot \mathbf{x}} = \mathbf{p} \int d\mathbf{x} | \mathbf{x} \rangle e^{i\mathbf{p} \cdot \mathbf{x}} \\
&= \int d\mathbf{x} | \mathbf{x} \rangle \frac{\boldsymbol{\nabla}}{i} e^{i\mathbf{p} \cdot \mathbf{x}} = \int d\mathbf{x} | \mathbf{x} \rangle \frac{1}{i} \boldsymbol{\nabla} \langle \mathbf{x} | \mathbf{p} \rangle. \quad (21.88)
\end{aligned}
$$

Thus $\mathbf{P}$ has the operator representation

$$
\mathbf{P} = \int d\mathbf{x} | \mathbf{x} \rangle \frac{1}{i} \boldsymbol{\nabla} \langle \mathbf{x} |. \quad (21.89)
$$

Using this and previous relations we obtain for the Hamiltonian (21.85) the amplitude or expectation value or position space representation

$$
\langle \mathbf{x}' | H | \mathbf{x} \rangle = -\frac{1}{2m_0} \boldsymbol{\nabla}^2 \delta(\mathbf{x} - \mathbf{x}') + V(\mathbf{x}) \delta(\mathbf{x}' - \mathbf{x}). \quad (21.90)
$$

Since

$$
\begin{aligned}
\langle \mathbf{p}' | V | \mathbf{p} \rangle &= \int d\mathbf{x}' \int d\mathbf{x} \langle \mathbf{p}' | \mathbf{x}' \rangle \langle \mathbf{x}' | V | \mathbf{x} \rangle \langle \mathbf{x} | \mathbf{p} \rangle \\
&= \int d\mathbf{x}' \int d\mathbf{x} e^{-i\mathbf{p}' \cdot \mathbf{x}'} V(\mathbf{x}) \delta(\mathbf{x}' - \mathbf{x}) e^{i\mathbf{p} \cdot \mathbf{x}} \\
&= \int d\mathbf{x} e^{-i\mathbf{p}' \cdot \mathbf{x}} V(\mathbf{x}) e^{i\mathbf{p} \cdot \mathbf{x}} \equiv \tilde{V}(\mathbf{p} - \mathbf{p}'), \quad (21.91)
\end{aligned}
$$

we have in momentum space the representation

$$\langle \mathbf{p}'|H|\mathbf{p}\rangle = \frac{\mathbf{p}^2}{2m_0}(2\pi)^3\delta(\mathbf{p}-\mathbf{p}') + \tilde{V}(\mathbf{p}-\mathbf{p}'). \tag{21.92}$$

We now return to the expression (21.84), $\langle \mathbf{x}', t'|\mathbf{x}, t\rangle$, for the amplitude of an infinitesimal transition. We have

$$
\begin{aligned}
\langle \mathbf{x}', t'|\mathbf{x}, t\rangle &= \int \frac{d\mathbf{p}'}{(2\pi)^3} \int \frac{d\mathbf{p}}{(2\pi)^3} e^{i\mathbf{p}'\cdot\mathbf{x}'} \langle \mathbf{p}'|e^{iH(t-t')}|\mathbf{p}\rangle e^{-i\mathbf{p}\cdot\mathbf{x}} \\
&= \int \frac{d\mathbf{p}'}{(2\pi)^3} \int \frac{d\mathbf{p}}{(2\pi)^3} e^{i\mathbf{p}'\cdot\mathbf{x}'} \Big\{ \langle \mathbf{p}'|\mathbf{p}\rangle \\
&\qquad + i(t-t')\langle \mathbf{p}'|H|\mathbf{p}\rangle + \cdots \Big\} e^{-i\mathbf{p}\cdot\mathbf{x}} \\
&= \cdots \Big\{ (2\pi)^3\delta(\mathbf{p}'-\mathbf{p}) + i(t-t')\Big[\frac{\mathbf{p}^2}{2m_0}(2\pi)^3\delta(\mathbf{p}-\mathbf{p}') \\
&\qquad + \tilde{V}(\mathbf{p}-\mathbf{p}') \Big] e^{-i\mathbf{p}\cdot\mathbf{x}} + O[(t-t')^2] \Big\},
\end{aligned}
$$

where we used (21.92). But

$$
\begin{aligned}
\int \frac{d\mathbf{p}'}{(2\pi)^3} e^{i\mathbf{p}'\cdot\mathbf{x}'} \tilde{V}(\mathbf{p}-\mathbf{p}') &= \int \frac{d\mathbf{p}'}{(2\pi)^3} e^{i\mathbf{p}'\cdot\mathbf{x}'} \int d\mathbf{x}'' e^{-i\mathbf{p}'\cdot\mathbf{x}''} V(\mathbf{x}'') e^{i\mathbf{p}\cdot\mathbf{x}''} \\
&= \int d\mathbf{x}'' \delta(\mathbf{x}'-\mathbf{x}'') V(\mathbf{x}'') e^{i\mathbf{p}\cdot\mathbf{x}''} \\
&= V(\mathbf{x}') e^{i\mathbf{p}\cdot\mathbf{x}'}. \tag{21.93}
\end{aligned}
$$

Hence

$$
\begin{aligned}
\langle \mathbf{x}', t'|\mathbf{x}, t\rangle &= \int \frac{d\mathbf{p}}{(2\pi)^3} e^{i\mathbf{p}\cdot\mathbf{x}'} \Big\{ 1 + i(t-t')\Big[\frac{\mathbf{p}^2}{2m_0} + V(\mathbf{x}') \Big] + \cdots \Big\} e^{-i\mathbf{p}\cdot\mathbf{x}} \\
&= \int \frac{d\mathbf{p}}{(2\pi)^3} e^{i\mathbf{p}\cdot(\mathbf{x}'-\mathbf{x})} e^{-iH(t'-t)}. \tag{21.94}
\end{aligned}
$$

In the above we have chosen the signs such that

$$\mathbf{p} = -i\boldsymbol{\nabla} \quad \text{and} \quad H = i\frac{\partial}{\partial t} \tag{21.95}$$

when applied to the bra vector $\langle \mathbf{x}, t|$, i.e. the signs are reversed when these operators are applied to the ket vector $|\mathbf{x}, t\rangle$ (cf. (21.82)).

We now consider the transition of the particle from $\mathbf{x}_i, t_i$ to $\mathbf{x}_f, t_f$ subdivided into infinitesimal transitions $\langle \mathbf{x}_{j+1}, t_{j+1}|\mathbf{x}_j, t_j\rangle$. Thus with

$$\mathbf{x}_f = \mathbf{x}_n, \quad t_f = t_n, \quad \mathbf{x}_0 = \mathbf{x}_i, \quad t_0 = t_i,$$

we obtain (it will be seen below why we do not introduce normalization factors here)

$$\langle \mathbf{x}_f, t_f | \mathbf{x}_i, t_i \rangle = \int \cdots \int d\mathbf{x}_{n-1} d\mathbf{x}_{n-2} \ldots d\mathbf{x}_1 \langle \mathbf{x}_n, t_n | \mathbf{x}_{n-1}, t_{n-1} \rangle$$
$$\times \langle \mathbf{x}_{n-1}, t_{n-1} | \mathbf{x}_{n-2}, t_{n-2} \rangle \cdots \langle \mathbf{x}_1, t_1 | \mathbf{x}_0, t_0 \rangle. \quad (21.96)$$

We define a path in configuration space by a succession of $n+1$ points $\mathbf{x}_0, \mathbf{x}_1, \ldots, \mathbf{x}_n$ which the particle passes at times $t_0, t_1, \ldots, t_n$ respectively. The integration over the $n-1$ intermediate positions, i.e. $d\mathbf{x}_1, d\mathbf{x}_2, \ldots, d\mathbf{x}_{n-1}$, then corresponds to the sum over all paths. We now insert for each of the infinitesimal intermediate transitions the amplitude (21.94) and thus obtain a phase space expression, i.e.

$$\langle \mathbf{x}_f, t_f | \mathbf{x}_i, t_i \rangle$$
$$= \int \cdots \int_{\mathbf{x}_i = \mathbf{x}(t_i)}^{\mathbf{x}_f = \mathbf{x}(t_f)} d\mathbf{x}_1 \, d\mathbf{x}_2 \, \ldots \, d\mathbf{x}_{n-1} \int \cdots \int \frac{d\mathbf{p}_1}{(2\pi)^3} \frac{d\mathbf{p}_2}{(2\pi)^3} \cdots$$
$$\times \frac{d\mathbf{p}_{n-1}}{(2\pi)^3} \frac{d\mathbf{p}_n}{(2\pi)^3} \exp\left[i\{\mathbf{p}_1 \cdot (\mathbf{x}_1 - \mathbf{x}_0) + \mathbf{p}_2 \cdot (\mathbf{x}_2 - \mathbf{x}_1)\right.$$
$$\left. + \cdots + \mathbf{p}_n \cdot (\mathbf{x}_n - \mathbf{x}_{n-1})\right]$$
$$\times \exp\left[-i\{H(\mathbf{p}_1, \mathbf{x}_1)(t_1 - t_0) + H(\mathbf{p}_2, \mathbf{x}_2)(t_2 - t_1)\right.$$
$$\left. + \cdots + H(\mathbf{p}_n, \mathbf{x}_n)(t_n - t_{n-1})\right]. \quad (21.97)$$

We can rearrange the factors here in the following form

$$\langle \mathbf{x}_f, t_f | \mathbf{x}_i, t_i \rangle = \int_{\mathbf{x}_0 = \mathbf{x}_i}^{\mathbf{x}_n = \mathbf{x}_f} \prod_{k=1}^{n-1} d\mathbf{x}_k \prod_{l=1}^{n} \left[\frac{d\mathbf{p}_l}{(2\pi)^3} \right] \exp\left[i \sum_{k=1}^{n} (t_k - t_{k-1})\right.$$
$$\left. \times \left\{ \mathbf{p}_k \cdot \left(\frac{\mathbf{x}_k - \mathbf{x}_{k-1}}{t_k - t_{k-1}} \right) - H(\mathbf{p}_k, \mathbf{x}_k) \right\} \right]. \quad (21.98)$$

We now replace $H(\mathbf{p}_k, \mathbf{x}_k)$ by

$$H(\mathbf{p}_k, \mathbf{x}_k) = \frac{\mathbf{p}_k^2}{2m_0} + V(\mathbf{x}_k).$$

Then the integration over all $\mathbf{p}_k$ can be performed by using Eq. (21.27) in which we make the replacements $x \to \epsilon \dot{x}_k, \alpha \to -\epsilon/2m_0$ and divide both sides by 2π. The resulting relation is

$$\left(\frac{m_0}{2\pi i \epsilon} \right)^{1/2} \exp\left(i\epsilon \frac{m_0 \dot{x}_k^2}{2} \right) = \int \frac{dp_k}{2\pi} \exp\left[i\epsilon \left(p_k \dot{x}_k - \frac{1}{2} \frac{p_k^2}{m_0} \right) \right]. \quad (21.99)$$

Forming products of expressions of this type we see that this relation can be generalized to the three-dimensional case, i.e. to

$$\left(\frac{m_0}{2\pi i\epsilon}\right)^{3/2}\exp\left(i\epsilon\frac{m_0\dot{\mathbf{x}}_k^2}{2}\right)=\int\frac{d\mathbf{p}_k}{(2\pi)^3}\exp\left[i\epsilon\left(\mathbf{p}_k\cdot\dot{\mathbf{x}}_k-\frac{1}{2}\frac{\mathbf{p}_k^2}{m_0}\right)\right]. \quad (21.100)$$

Hence with all $t_k - t_{k-1} = \epsilon$ we obtain

$$\int\prod_{l=1}^{n}\left[\frac{d\mathbf{p}_l}{(2\pi)^3}\right]\exp\left[i\sum_{k=1}^{n}(t_k-t_{k-1})\left\{\mathbf{p}_k\cdot\dot{\mathbf{x}}_k-\frac{\mathbf{p}_k^2}{2m_0}\right\}\right]$$

$$=\left(\frac{m_0}{2\pi i(t_k-t_{k-1})}\right)^{3n/2}\exp\left[\sum_{k=1}^{n}i(t_k-t_{k-1})\frac{m_0\dot{\mathbf{x}}_k^2}{2}\right]. \quad (21.101)$$

Then Eq. (21.98) can be rewritten as

$$\langle\mathbf{x}_f,t_f|\mathbf{x}_i,t_i\rangle=\left(\frac{m_0}{2\pi i\epsilon}\right)^{3n/2}\int\prod_{k=1}^{n-1}d\mathbf{x}_k\exp\left[i\sum_{k=1}^{n}(t_k-t_{k-1})\right.$$

$$\left.\times\left\{\frac{1}{2}m_0\dot{\mathbf{x}}_k^2-V(\mathbf{x}_k)\right\}\right]. \quad (21.102)$$

We see that the factor (with $\hbar = 1$)

$$[N(\epsilon)]^{-3}\equiv\left(\frac{m_0}{2\pi i(t_k-t_{k-1})}\right)^{3/2}=\left(\frac{m_0}{2\pi i\epsilon}\right)^{3/2}$$

is precisely the normalization factor inserted earlier for the one-dimensional case in Eq. (21.8) as $N(\epsilon)$ so that the integral does not vanish for $\epsilon \to 0$. Thus the *phase space representation of the path-integral*[*] (21.98), i.e.

$$\langle\mathbf{x}_f,t_f|\mathbf{x}_i,t_i\rangle=\int_{\mathbf{x}_i}^{\mathbf{x}_f}\mathcal{D}\{\mathbf{x}(t)\}\mathcal{D}\{\mathbf{p}(t)\}\exp\left\{i\int_{t_i}^{t_f}dt[\mathbf{p}\cdot\dot{\mathbf{x}}-H(\mathbf{x},\mathbf{p})]\right\}, \quad (21.103)$$

is another form of the path-integral for which the integration with the help of Eq. (21.100) over

$$\mathcal{D}\{\mathbf{p}\}=\lim_{n\to\infty}\prod_{l=1}^{n}\frac{d\mathbf{p}_l}{(2\pi)^3}$$

---

[*]The reader interested in a rigorous treatment of representations of solutions of Schrödinger equations in terms of Hamiltonian Feynman path integrals may like to consult O. G. Smolyanov, A. G. Tokarev and A. Truman [254].

yields automatically the correct normalization factor or metric in the config-uration space path integral[†] in agreement with Eq. (21.8), i.e.

$$\langle \mathbf{x}_f, t_f | \mathbf{x}_i, t_i \rangle = \lim_{n \to \infty} \frac{1}{N(\epsilon)^{3n}} \int_{\mathbf{x}_i}^{\mathbf{x}_f} \prod_{k=1}^{n-1} d\mathbf{x}_k \exp \left[i \int_{t_i}^{t_f} dt L(\mathbf{x}, \dot{\mathbf{x}}) \right], \quad (21.104)$$

where for the particularly simple Lagrangian L under consideration here

$$\frac{1}{2} m_0 \dot{\mathbf{x}}_k^2 - V(\mathbf{x}_k) = m_0 \dot{\mathbf{x}}_k^2 - H(\mathbf{x}_k, \mathbf{p}_k) = \mathbf{p}_k \cdot \dot{\mathbf{x}}_k - H(\mathbf{x}_k, \mathbf{p}_k) = L(\mathbf{x}_k, \dot{\mathbf{x}}_k).$$

We know from classical mechanics that this relation is a Legendre trans-form which transforms from variables $\mathbf{x}, \dot{\mathbf{x}}, t$ to $\mathbf{x}, \mathbf{p}, t$. Thus finally we have *Feynman's principle*:

$$\langle \mathbf{x}_f, t_f | \mathbf{x}_i, t_i \rangle = \int_{\mathbf{x}_i}^{\mathbf{x}_f} \mathcal{D}\{\mathbf{x}(t)\} \exp \left[i \int_{t_i}^{t_f} dt \, L(\mathbf{x}, \dot{\mathbf{x}}) \right], \quad (21.105)$$

where

$$\mathcal{D}\{\mathbf{x}\} = \lim_{n \to \infty} \frac{1}{[N(\epsilon)]^{3n}} \prod_{k=1}^{n-1} d\mathbf{x}_k.$$

In this notation the measure factor is contained in $\mathcal{D}\{\mathbf{x}\}$. Note that we have n factors $[N(\epsilon)]^3$ but only $n - 1$ integrations $d\mathbf{x}_k$; this agrees with the corresponding discrepancy in the one-dimensional case of Eq. (21.8).

We have now obtained the path integral representation of the matrix element $\langle \mathbf{x}_f, t_f | \mathbf{x}_i, t_i \rangle$ with no operator in between. Clearly for a better un-derstanding one also wants to explore the *case of sandwiched operators* in the path integral representation. Consider the following expectation value with primed and doubled-primed states denoting some intermediate states between initial and final states denoted by indices i and f:

$$\langle \mathbf{x}'', t'' | \underbrace{\mathbf{x}(t)}_{\text{operator}} | \mathbf{x}', t' \rangle. \quad (21.106)$$

In order to be able to deal with this expression we first require the momentum operator representation of the operator $\mathbf{x}(t)$. We check below that this is given by

$$\mathbf{x} = \int \frac{d\mathbf{p}}{(2\pi)^3} |\mathbf{p}\rangle i \frac{\partial}{\partial \mathbf{p}} \langle \mathbf{p}|. \quad (21.107)$$

[†]See also C. Garrod [106].

For the verification we use Eq. (21.77) and thus have

$$
\begin{aligned}
\mathbf{x}|\mathbf{x}'\rangle &= \mathbf{x}\int \frac{d\mathbf{p}}{(2\pi)^3}|\mathbf{p}\rangle\langle\mathbf{p}|\mathbf{x}'\rangle = \mathbf{x}\int \frac{d\mathbf{p}}{(2\pi)^3}|\mathbf{p}\rangle e^{-i\mathbf{p}\cdot\mathbf{x}'}, \\
\mathbf{x}|\mathbf{x}'\rangle &= \mathbf{x}'|\mathbf{x}'\rangle = \mathbf{x}'\int \frac{d\mathbf{p}}{(2\pi)^3}|\mathbf{p}\rangle e^{-i\mathbf{p}\cdot\mathbf{x}'} = \int \frac{d\mathbf{p}}{(2\pi)^3}|\mathbf{p}\rangle i\frac{\partial}{\partial\mathbf{p}}e^{-i\mathbf{p}\cdot\mathbf{x}'} \\
&= \int \frac{d\mathbf{p}}{(2\pi)^3}|\mathbf{p}\rangle i\frac{\partial}{\partial\mathbf{p}}\langle\mathbf{p}|\mathbf{x}'\rangle; \quad \therefore \mathbf{x} = \int \frac{d\mathbf{p}}{(2\pi)^3}|\mathbf{p}\rangle i\frac{\partial}{\partial\mathbf{p}}\langle\mathbf{p}|.
\end{aligned}
$$

We now consider an expectation value in which we replace the operator $\mathbf{x}$ (which is sandwiched in between exponentials containing Hamilton operators) by the expression (21.107), so that we obtain a c-number position coordinate which we can shifted across other c-numbers. Thus

$$
\begin{aligned}
\langle\mathbf{x}'',t''|\mathbf{x}|\mathbf{x}',t'\rangle &= \int \frac{d\mathbf{p}''}{(2\pi)^3}\int \frac{d\mathbf{p}'}{(2\pi)^3}e^{i\mathbf{p}''\cdot\mathbf{x}''}\langle\mathbf{p}''|e^{-iHt''}\mathbf{x}e^{iHt'}|\mathbf{p}'\rangle e^{-i\mathbf{p}'\cdot\mathbf{x}'}. \\
&= \int \frac{d\mathbf{p}''}{(2\pi)^3}\int \frac{d\mathbf{p}'}{(2\pi)^3}e^{i\mathbf{p}''\cdot\mathbf{x}''}\int d\mathbf{x}^3\int d\mathbf{x}^4\int d\mathbf{x}^5 \\
&\quad \times \int d\mathbf{x}^6\int \frac{d\mathbf{p}}{(2\pi)^3}\langle\mathbf{p}''|\mathbf{x}^3\rangle\langle\mathbf{x}^3|e^{-iHt''}|\mathbf{x}^4\rangle \times \\
&\quad \langle\mathbf{x}^4|\mathbf{p}\rangle i\frac{\partial}{\partial\mathbf{p}}\langle\mathbf{p}|\mathbf{x}^5\rangle\langle\mathbf{x}^5|e^{iHt'}|\mathbf{x}^6\rangle\langle\mathbf{x}^6|\mathbf{p}'\rangle e^{-i\mathbf{p}'\cdot\mathbf{x}'}. \quad (21.108)
\end{aligned}
$$

With the wave functions (21.76) we can rewrite the factor

$$
\langle\mathbf{x}^4|\mathbf{p}\rangle i\frac{\partial}{\partial\mathbf{p}}\langle\mathbf{p}|\mathbf{x}^5\rangle \quad \text{as} \quad \langle\mathbf{x}^4|\mathbf{p}\rangle\mathbf{x}^5\langle\mathbf{p}|\mathbf{x}^5\rangle.
$$

With contractions and integrations over delta functions the expression for the matrix element then becomes

$$
\begin{aligned}
&\langle\mathbf{x}'',t''|\mathbf{x}|\mathbf{x}',t'\rangle \\
&= \int \frac{d\mathbf{p}''}{(2\pi)^3}\int \frac{d\mathbf{p}'}{(2\pi)^3}e^{i\mathbf{p}''\cdot\mathbf{x}''}\int d\mathbf{x}^5\mathbf{x}^5\langle\mathbf{p}''|e^{-iHt''}|\mathbf{x}^5\rangle\langle\mathbf{x}^5|e^{iHt'}|\mathbf{p}'\rangle e^{-i\mathbf{p}'\cdot\mathbf{x}'} \\
&= \int d\mathbf{x}^5\mathbf{x}^5\int \frac{d\mathbf{p}''}{(2\pi)^3}\int \frac{d\mathbf{p}'}{(2\pi)^3}\langle\mathbf{x}''|\mathbf{p}''\rangle\langle\mathbf{p}''|e^{-iHt''}|\mathbf{x}^5\rangle\langle\mathbf{x}^5|e^{iHt'}|\mathbf{p}'\rangle\langle\mathbf{p}'|\mathbf{x}'\rangle,
\end{aligned}
$$

and thus

$$
\begin{aligned}
\langle\mathbf{x}'',t''|\mathbf{x}|\mathbf{x}',t'\rangle &= \int d\mathbf{x}^5\mathbf{x}^5\langle\mathbf{x}''|e^{-iHt''}|\mathbf{x}^5\rangle\langle\mathbf{x}^5|e^{iHt'}|\mathbf{x}'\rangle \\
&= \int d\mathbf{x}^5\mathbf{x}^5\langle\mathbf{x}'',t''|\mathbf{x}^5\rangle\langle\mathbf{x}^5|\mathbf{x}',t'\rangle. \quad (21.109)
\end{aligned}
$$

In the second last expression here we expand the exponentials and obtain

$$
\begin{aligned}
\langle \mathbf{x}'', t'' | \mathbf{x} | \mathbf{x}', t' \rangle \;=\; & \int d\mathbf{x}^5 \mathbf{x}^5 \big[\langle \mathbf{x}'' | \mathbf{x}^5 \rangle \langle \mathbf{x}^5 | \mathbf{x}' \rangle - it'' \langle \mathbf{x}'' | H | \mathbf{x}^5 \rangle \langle \mathbf{x}^5 | \mathbf{x}' \rangle \\
& + it' \langle \mathbf{x}'' | \mathbf{x}^5 \rangle \langle \mathbf{x}^5 | H | \mathbf{x}' \rangle + \cdots \big],
\end{aligned}
$$

and hence

$$
\langle \mathbf{x}'', t'' | \mathbf{x} | \mathbf{x}', t' \rangle = \mathbf{x}'' \delta(\mathbf{x}'' - \mathbf{x}') - it'' \mathbf{x}' \langle \mathbf{x}'' | H | \mathbf{x}' \rangle + it' \mathbf{x}'' \langle \mathbf{x}'' | H | \mathbf{x}' \rangle + \cdots .
$$
(21.110)

Recalling Eq. (21.90) we can rewrite this expression as

$$
\begin{aligned}
\langle \mathbf{x}'', t'' | \mathbf{x} | \mathbf{x}', t' \rangle &= \mathbf{x}'' \big[\delta(\mathbf{x}'' - \mathbf{x}') - i(t'' - t') \langle \mathbf{x}'' | H | \mathbf{x}' \rangle + \cdots \big] \\
&= \mathbf{x}'' \left[1 - i(t'' - t') \left[-\frac{\boldsymbol{\nabla}^2}{2m_0} + V(\mathbf{x}') \right] + \cdots \right] \delta(\mathbf{x}'' - \mathbf{x}').
\end{aligned}
$$
(21.111)

Inserting here the integral representation (21.74) of the delta function, we obtain with (21.85)

$$
\begin{aligned}
& \langle \mathbf{x}'', t'' | \mathbf{x} | \mathbf{x}', t' \rangle \\
=\; & \int \frac{d\mathbf{p}}{(2\pi)^3} \mathbf{x}'' \left[1 - i(t'' - t') \left[\frac{\mathbf{p}^2}{2m_0} + V(\mathbf{x}') \right] + \cdots \right] e^{i\mathbf{p}\cdot(\mathbf{x}'' - \mathbf{x}')} \\
=\; & \int \frac{d\mathbf{p}}{(2\pi)^3} \mathbf{x}'' e^{-i(t'' - t')H} e^{i\mathbf{p}\cdot(\mathbf{x}'' - \mathbf{x}')},
\end{aligned}
$$
(21.112)

where $\mathbf{x}''$ is a c-number. This is the expression corresponding to that of Eq. (21.94) for the case with no operator in between. We now proceed from there along parallel lines.

We now consider the transition of the expectation value of operator $\mathbf{x}$ from $\mathbf{x}_i, t_i$ to $\mathbf{x}_f, t_f$ subdivided into n transitions $\langle \mathbf{x}_{j+1}, t_{j+1} | \mathbf{x} | \mathbf{x}_j, t_j \rangle$ at equal time intervals ϵ. Thus with $\mathbf{x}_f = \mathbf{x}_n, t_f = t_n$ and $\mathbf{x}_i = \mathbf{x}_0, t_i = t_0$ as before, we obtain (here and in the following the limit $n \to \infty$ at the end being understood)

$$
\begin{aligned}
\langle \mathbf{x}_f, t_f | \mathbf{x} | \mathbf{x}_i, t_i \rangle \;=\; & \int \cdots \int d\mathbf{x}_{n-1} d\mathbf{x}_{n-2} \dots d\mathbf{x}_1 \, \langle \mathbf{x}_n, t_n | \mathbf{x} | \mathbf{x}_{n-1}, t_{n-1} \rangle \\
& \times \langle \mathbf{x}_{n-1}, t_{n-1} | \mathbf{x} | \mathbf{x}_{n-2}, t_{n-2} \rangle \cdots \langle \mathbf{x}_1, t_1 | \mathbf{x} | \mathbf{x}_0, t_0 \rangle.
\end{aligned}
$$
(21.113)

Inserting for each of the infinitesimal intermediate transitions the amplitude

(21.112), we obtain[‡]

$$\langle \mathbf{x}_f, t_f | \mathbf{x} | \mathbf{x}_i, t_i \rangle = \int \cdots \int_{\mathbf{x}_i = \mathbf{x}(t_i)}^{\mathbf{x}_f = \mathbf{x}(t_f)} d\mathbf{x}_1 \, d\mathbf{x}_2 \, \ldots \, d\mathbf{x}_{n-1} \, \mathbf{x}_1 \, \mathbf{x}_2 \, \ldots \, \mathbf{x}_{n-1} \, \mathbf{x}_n$$

$$\int \cdots \int \frac{d\mathbf{p}_1}{(2\pi)^3} \frac{d\mathbf{p}_2}{(2\pi)^3} \cdots \frac{d\mathbf{p}_{n-1}}{(2\pi)^3} \frac{d\mathbf{p}_n}{(2\pi)^3} \exp\left[-i\{ H(\mathbf{p}_1, \mathbf{x}_1)(t_1 - t_0) \right.$$

$$+ H(\mathbf{p}_2, \mathbf{x}_2)(t_2 - t_1) + \cdots + H(\mathbf{p}_n, \mathbf{x}_n)(t_n - t_{n-1})\}]$$

$$\times \exp\left[i\{ \mathbf{p}_1 \cdot (\mathbf{x}_1 - \mathbf{x}_0) + \mathbf{p}_2 \cdot (\mathbf{x}_2 - \mathbf{x}_1) + \right.$$

$$\cdots + \mathbf{p}_n \cdot (\mathbf{x}_n - \mathbf{x}_{n-1})]. \qquad (21.114)$$

The momentum integrations can be carried out as before and we obtain finally

$$\langle \mathbf{x}_f, t_f | \mathbf{x} | \mathbf{x}_i, t_i \rangle = \int_{\mathbf{x}_i}^{\mathbf{x}_f} \mathcal{D}\{\mathbf{x}(t)\} \, \mathbf{x}(t) \, \exp\left[i \int_{t_i}^{t_f} dt \, L(\mathbf{x}, \dot{\mathbf{x}}) \right], \qquad (21.115)$$

where (there is no integration over $\mathbf{x}_n$)

$$\mathcal{D}\{\mathbf{x}(t)\} = \lim_{n \to \infty} [N(\epsilon)]^{-3n} \prod_{k=1}^{n-1} d\mathbf{x}_k.$$

21.7 Canonical Quantization from Path Integrals

We saw above that in the path integral method the evolution of a system is expressed by a basic functional integral like (21.10) or (21.105). Before we can consider canonical quantization in the context of path integrals we have to investigate in more detail the role played by momenta. For simplicity we consider the case of one spatial dimension. We demonstrate first that the operator representation of P given by Eq. (21.89) is consistent with the conjugate expression

$$p = \frac{\partial L}{\partial \dot{x}}. \qquad (21.116)$$

We reconsider therefore the transition amplitude $\langle x'', t'' | x', t' \rangle$ and consider the variation

$$\delta \langle x'', t'' | x', t' \rangle \equiv \langle x'' + \delta x'', t'' | x', t' \rangle - \langle x'', t'' | x', t' \rangle = \delta x'' \nabla_{x''} \langle x'', t'' | x', t' \rangle \qquad (21.117)$$

[‡]One may observe here that if $\mathbf{x} = \mathbf{x}_f = \mathbf{x}_n$ in the integrand of Eq. (21.114), the string of factors $\mathbf{x}_1 \, \mathbf{x}_2 \, \ldots \, \mathbf{x}_{n-1} \, \mathbf{x}_n$ reduces to the single factor $\mathbf{x}_f$ and hence may be put in front of the integrals to yield $\langle \mathbf{x}_f, t_f | \mathbf{x}_f | \mathbf{x}_i, t_i \rangle = \mathbf{x}_f \langle \mathbf{x}_f, t_f | \mathbf{x}_i, t_i \rangle$.

with $\delta x'' \equiv \delta x(t'') \neq 0$ and $\delta x' \equiv \delta x(t') = 0$. For the purpose of enabling some formulations below, we separate the metric factors from $\mathcal{D}\{\mathbf{x}\}$ in Eq. (21.10) and set $\mathcal{D}\{\mathbf{x}\} \equiv \mathcal{D}^0\{\mathbf{x}\}/N(t_f - t_i)$. Then, with $N \equiv N(t'' - t')$:

$$
\begin{aligned}
\delta\langle x'', t'' | x', t' \rangle &= \delta \frac{1}{N} \int_{x'}^{x''} \mathcal{D}^0\{x\} e^{iI(x)/\hbar}, \quad I(x) = \int_{t'}^{t''} dt L(x, \dot{x}), \\
&= \frac{1}{N} \int_{x'}^{x''} \mathcal{D}^0\{x\} \left[i\delta I(x)/\hbar \right] e^{iI(x)/\hbar}.
\end{aligned} \tag{21.118}
$$

The following steps are familiar from classical mechanics:

$$
\begin{aligned}
\delta I(x) &= \delta \int_{t'}^{t''} dt L(x, \dot{x}) = \int_{t'}^{t''} dt \left[\frac{\partial L}{\partial x} \delta x + \frac{\partial L}{\partial \dot{x}} \delta \dot{x} \right] \\
&= \int_{t'}^{t''} dt \frac{\partial L}{\partial x} \delta x + \int_{t'}^{t''} dt \frac{\partial L}{\partial \dot{x}} \frac{d\delta x}{dt} \\
&= \int_{t'}^{t''} \frac{\partial L}{\partial x} \delta x dt + \left[\frac{\partial L}{\partial \dot{x}} \delta x \right]_{t'}^{t''} - \int_{t'}^{t''} \frac{d}{dt} \left(\frac{\partial L}{\partial \dot{x}} \right) \delta x dt,
\end{aligned}
$$

and hence

$$
\delta I(x) = \int_{t'}^{t''} dt \left[\frac{\partial L}{\partial x} - \frac{d}{dt} \left(\frac{\partial L}{\partial \dot{x}} \right) \right] \delta x + \left[\frac{\partial L}{\partial \dot{x}} \delta x \right]_{t'}^{t''}. \tag{21.119}
$$

The classical equation of motion follows from Hamilton's principle for which $\delta x = 0$ at $t = t', t''$. Thus in this case of classical mechanics

$$
\frac{\partial L}{\partial x} - \frac{d}{dt} \left(\frac{\partial L}{\partial \dot{x}} \right) = 0. \tag{21.120}
$$

However: In our case here $\delta x \neq 0$ for $t = t''$ (although $\delta x = 0$ at $t = t'$), and we demand the validity of the equation of motion. Hence

$$
\delta I(x) = \left(\frac{\partial L}{\partial \dot{x}} \right)_{t''} \delta x(t'') = \left(\frac{\partial L}{\partial \dot{x}} \right)_{t''} \delta x'' \tag{21.121}
$$

and — using Eq. (21.118) —

$$
\begin{aligned}
\delta\langle x'', t'' | x', t' \rangle &= \frac{1}{N} \int_{x'}^{x''} \mathcal{D}^0\{x\} \frac{i}{\hbar} \left(\frac{\partial L}{\partial \dot{x}} \right)_{t''} \delta x'' e^{iI(x)/\hbar} \\
&= \delta x'' \left\langle x'', t'' \left| \frac{i}{\hbar} \left(\frac{\partial L}{\partial \dot{x}} \right)_{t''} \right| x', t' \right\rangle \\
&\stackrel{(21.117)}{\equiv} \delta x'' \nabla_{x''} \langle x'', t'' | x', t' \rangle,
\end{aligned} \tag{21.122a}
$$

or

$$\frac{\delta}{\delta x''}\langle x'',t''|x',t'\rangle \equiv \frac{i}{\hbar}\langle x'',t''|p(t'')|x',t'\rangle. \tag{21.122b}$$

By comparison with Eq. (21.89) we see that

$$\frac{i}{\hbar}p \equiv \frac{i}{\hbar}\frac{\partial L}{\partial \dot{x}} \longrightarrow \frac{i}{\hbar}P = \nabla, \quad P = -i\hbar\nabla. \tag{21.123}$$

Having explored the appearance of *canonical momentum* in path integrals we can proceed to extract the *canonical quantization*. We consider this here in quantum mechanics in one spatial dimension but parallel to field theory.[§] The evolution of the system is expressed by the Feynman functional integral

$$\langle x_f,t_f|x_i,t_i\rangle = \frac{1}{N(\triangle)}\int_{x_i=x(t_i)}^{x_f=x(t_f)} \mathcal{D}^0\{x\}e^{iI(x)/\hbar}, \tag{21.124}$$

where with $\triangle = t_f - t_i$,

$$I(x) = \int_{\triangle} dt\, L(x,\dot{x}).$$

Here $\int_{x_i}^{x_f}\mathcal{D}^0\{x\}$ indicates that the integral is to be taken over all x with boundary conditions $x_i = x(t_i), x_f = x(t_f)$. The following properties are assumed to hold:

(1)

$$\int_{t_i}^{t_f}\mathcal{D}^0\{x\} = \sum_{x'}\int_{x_i}^{x'}\mathcal{D}^0\{x\}\int_{x'}^{x_f}\mathcal{D}^0\{x\}, \tag{21.125a}$$

(2) the completeness relation holds,

$$\sum_{x'}|x',t'\rangle\langle x',t'| = \mathbb{1}, \tag{21.125b}$$

and the measure property

(3)

$$N^{-1}(\triangle + \triangle') = N^{-1}(\triangle)N^{-1}(\triangle'). \tag{21.125c}$$

Corresponding to our relation (21.115) in three space dimensions, we now have in one spatial dimension

$$\langle x_f,t_f|x|x_i,t_i\rangle = \frac{1}{N(\triangle)}\int_{x_i}^{x_f}\mathcal{D}^0\{x\}x(t)e^{iI(x)/\hbar}. \tag{21.126}$$

[§]R. F. Streater and A. S. Wightman [264].

The time t is in general a time between t_i and t_f. But for $t = t_i$ we have (and equivalently for $t = t_f$) as we can see from Eq. (21.114)

$$\langle x_f, t_f | x(t_i) | x_i, t_i \rangle = x_i \langle x_f, t_f | x_i, t_i \rangle. \tag{21.127}$$

We can now write down an expression for a time-ordered product of operators. Thus if T denotes such ordering, we have the relation

$$\langle x_f, t_f | T(x_1 \cdots x_n) | x_i, t_i \rangle = \frac{1}{N(\triangle)} \int_{x_i}^{x_f} \mathcal{D}^0\{x\} x_1 \cdots x_n e^{iI(x)/\hbar}, \tag{21.128}$$

where t_f is later than t_i and $t_1, \ldots, t_n$ lie correspondingly in consecutive order between t_i and t_f. We consider the case $n = 2$. Then the relation follows with the help of the properties (1), (2) and (3) above. Thus

$$\langle x_f, t_f | T(x_1 x_2) | x_i, t_i \rangle = \sum_{x'} \langle x_f, t_f | x_1 | x', t' \rangle \langle x', t' | x_2 | x_i, t_i \rangle$$

$$= \frac{1}{N(\triangle)} \int_{x_i}^{x_f} \mathcal{D}^0\{x\} x_1 x_2 e^{iI(x)/\hbar}. \tag{21.129}$$

With Eq. (21.122b) we can deduce the nontrivial canonical commutation relation arising here. We have (cf. Eq. (21.127))

$$\langle x'', t'' | x'' | x', t' \rangle = x'' \langle x'', t'' | x', t' \rangle. \tag{21.130}$$

Taking the functional derivative $\delta/i\delta x''$ of this equation we obtain

$$\frac{\hbar}{i} \frac{\delta}{\delta x''} \langle x'', t'' | x'' | x', t' \rangle \overset{\text{also}}{=} \frac{\hbar}{i} \frac{\delta}{\delta x''} x'' \langle x'', t'' | x', t' \rangle$$

$$= -i\hbar \langle x'', t'' | x', t' \rangle + x'' \frac{\hbar}{i} \frac{\delta}{\delta x''} \langle x'', t'' | x', t' \rangle, \tag{21.131}$$

We can rewrite this expression as the commutator relation

$$\left(\frac{\hbar}{i} \frac{\delta}{\delta x''} x'' - x'' \frac{\hbar}{i} \frac{\delta}{\delta x''} \right) \langle x'', t'' | x', t' \rangle = -i\hbar \langle x'', t'' | x', t' \rangle, \tag{21.132}$$

or in commutator form

$$[P, x] \langle x, t | x', t' \rangle = -i\hbar \langle x, t | x', t' \rangle. \tag{21.133}$$

The generalization to higher spatial dimensions proceeds along parallel lines.

Chapter 22

Classical Field Configurations

22.1 Introductory Remarks

In field theory a perturbation expansion is generally understood to be an expansion in ascending powers of a coupling constant, such as the fine structure constant of quantum electrodynamics. Such an expansion parameter is usually known to be or assumed to be small in some sense, so that successive contributions to the first approximation do not invalidate this approximation. If the expansion is mathematically well defined, the series converges, i.e. it has some finite, nonzero radius of convergence in the domain of the coupling parameter. In physical applications such cases are rare. The vast majority of perturbation expansions discussed e.g. in the context of quantum mechanics are not convergent, but asymptotic. Such expansions are strictly speaking, i.e. in the sense of rigorous convergence tests, divergent; i.e. for sufficiently large r, the modulus of the ratio of the $(r+1)$-th term to the r-th term is larger than one, although the first few terms indicate a convergent behaviour. Expansions of this type arise, as we have seen in Chapter 8, if a function is expanded in the neighbourhood of an essential singularity* or if an expansion of the integrand of an integral representation is used (i.e. integrated over) beyond its radius of convergence. In field theory it has been known for a long time that expansions in ascending powers of a large coupling constant are fairly meaningless. It was therefore natural to search for alternative methods of expansion. Besides, it was known from quantum mechanics that Rayleigh–Schrödinger perturbation theory by itself does not yield e.g. the exponentially small level splitting which occurs in

---

*The expansion involves both positive and negative powers of the deviation; see e.g. E. T. Whittaker and G. N. Watson [283], p. 102. An essential singularity is really an ambiguity rather than an infinity. Consider e.g. $W = \exp(1/z)$, $z = \alpha + i\beta$ with $\alpha \to \infty$ or $\beta \to \infty$ or both.

the case of the symmetric double well potential or in the case of a periodic potential. There, specific boundary conditions have to be implemented in order to yield these effects which are therefore frequently termed "nonperturbative", although this terminology is not precise. In searching for other means of expansion, one considered in particular an expansion which is such that the first approximation is purely classical in a certain sense and such that this ignores quantum effects. This type of expansion is decsribed as *"semiclassical"*; it is an expansion in rising powers of a semiclassical expansion parameter which plays the role of Planck's constant h in the quantum mechanical WKB approximation. The classical, i.e. c-number, nature of the dominant approximation does not imply that this describes classical motion. On the contrary, the dominant contribution is singular (has an essential singularity) for vanishing semiclassical expansion parameter ($h \to 0$), i.e. such a series begins with a contribution proportional to (e.g.) $1/h^2$; higher order corrections are of orders

$$\frac{1}{h}, h^0, h, h^2, h^3, \ldots.$$

The study of expansions of this type has turned out to be extremely fruitful, and has led to insights which previously seemed unimaginable. In particular it enabled nonlinear problems to be studied and led to a consideration of topological properties. A further challenging task was the development of methods of quantization of theories which incorporate classical, c-number first approximations. These methods are often described as *methods of collective coordinates*, and the corresponding expansion is called the *"loop expansion"*. The procedure requires a change of variables from the original ones to *collective* and *fluctuation variables* (in analogy to centre of mass and relative coordinates), and in general this change is accompanied by certain constraints. One therefore faces the problem of quantizing a system which is subject to *constraints*. A method which achieves this is in particular Feynman's path integral procedure. The fundamental extension of the method of canonical quantization (i.e. the procedure discussed thus far) to systems with constraints was developed by Dirac [76] and is introduced in Chapter 27.

In the following we consider various typical examples, such as soliton theory. However, before we reach the stage at which quantization can be considered we have to deal with numerous other aspects such as the stability of the classical approximation, the latter's topological properties if any, the consideration of conditions which insure the existence of Green's functions of this new type of expansion procedure (which again leads to asymptotic expansions) and so on. Many of these aspects are interesting by themselves and in any case deserve detailed study.[†]

The classical first approximation in the procedure outlined above may be simply a constant (time and space independent) quantity, a static (time independent but space-dependent) solution of the classical equations of motion, or even a solution to a modified field equation (e.g. modified by going to imaginary time). It is evident that symmetries and their violation by the classical approximation play a significant role in the entire consideration.

In the following we trespass somewhat and only at very few points the sharp dividing line between quantum mechanics (which is effectively a one-dimensional field theory) and simple scalar field theories,[‡] assuming that this is acceptable to a reader with some familiarity of the basics of the more complicated field theory of electrodynamics. Thus references to field theory will be exceptional and hopefully do not irritate the reader. We shall not be concerned so much with the case of a constant first approximation. However, it is important for a better understanding of the considerations which follow, to keep this case in mind. We begin therefore with a brief recapitulation of the simple Higgs model which exemplifies this case and exhibits the phenomenon known as "*spontaneous symmetry violation*" or the Goldstone phenomenon. A main aim of this chapter is to generate some appreciation of the distinction between so-called topological and nontopological configurations (later referred to as instantons and periodic instantons and bounces respectively), and also to see these in a somewhat broader context (by comparison with higher dimensional cases).

22.2 The Constant Classical Field

We consider first the case of a constant classical field in a scalar field model called the *complex Higgs model* or *Goldstone theory*. The important aspect we want to draw attention to here is that of "*spontaneous symmetry breaking*". We write the complex scalar field $\phi(x)$ in four spacetime dimensions

$$\phi(x) = \frac{1}{\sqrt{2}}[\phi_1(x) + i\phi_2(x)].$$

The spacetime Lagrangian density of the specific theory we consider here is given by

$$\mathcal{L}[\phi, \partial_\mu \phi] = \partial_\mu \phi^* \partial^\mu \phi - V(\phi), \quad V(\phi) = m_0^2 \phi^* \phi + \frac{1}{3!}\lambda_0 (\phi^* \phi)^2, \qquad (22.1)$$

[†]The reader who wants a highly advanced overview of instantons, monopoles, vortices and kinks, may like to consult the article of D. Tong [272].

[‡]This means theories with field densities which are therefore infinite-dimensional.

where $\partial_\mu = (-\partial/\partial(ct), \boldsymbol{\nabla})$. To insure that the Hamiltonian $H = \int d\mathbf{x}\mathcal{H}$ with Hamiltonian density $\mathcal{H}(\phi(\mathbf{x}, t), \partial\mathcal{L}/\partial\dot\phi(\mathbf{x}, t))$ is bounded from below we must have $\lambda_0 > 0$. For $m_0^2 > 0$ the potential $V(\phi)$ has a single well with minimum at $|\phi| = 0$; for $m_0^2 < 0$ the potential is a double well potential with two minima at $|\phi| \neq 0$, as illustrated in Fig. 22.1. (In models of the early universe one often considers $V(|\phi|)$ with $m_0^2 > 0$ at the early stage; then after cooling with a phase transition one considers expectation values $|\phi| \neq 0$ corresponding to minima at $|\phi| \neq 0$. An even more familiar example is the *Ginzburg–Landau theory* [111] of superconductivity[§] in which the phase transition to the superconducting state implies the transition to the double well potential shape).

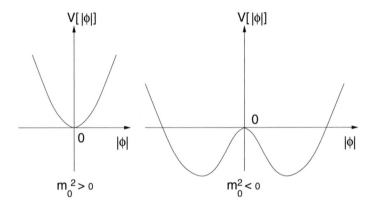

Fig. 22.1 Different potentials for different signs of m_0^2.

The Euler–Lagrange equation is

$$\partial_\mu\left(\frac{\partial\mathcal{L}}{\partial(\partial_\mu\phi^*)}\right) - \frac{\partial\mathcal{L}}{\partial\phi^*} = 0,$$

and hence here

$$[\partial_\mu\partial^\mu + m_0^2]\phi = -\frac{1}{3}\lambda_0(\phi^*\phi)\phi. \tag{22.2}$$

We ask: Are there classical (i.e. c-number) solutions $\phi(x) = \phi_c = $ const. of the equation of motion? For these we must have all derivatives of ϕ zero, i.e.

$$\left(m_0^2 + \frac{1}{3}\lambda_0\phi^*\phi\right)\phi = 0. \tag{22.3}$$

[§]See e.g. B. Felsager [91], p. 433.

For $\lambda_0 > 0, m_0^2 > 0$, the only such solution is $\phi_c = 0$. But for $\lambda_0 > 0, m_0^2 < 0$, the case we wish to consider, we have

$$\phi \to \phi_c = \frac{\lambda}{\sqrt{2}}e^{i\beta}, \quad \frac{1}{2}\lambda^2 = -\frac{3m_0^2}{\lambda_0}. \qquad (22.4)$$

Here β is a *spontaneously* chosen phase like a stick held upright along H in Fig. 22.2 when allowed to fall chooses an unpredictable phase parallel to a radius in the (ϕ_1, ϕ_2)-plane. It is important to observe that ϕ_c does not possess the rotational symmetry of the Lagrangian density $\mathcal{L}$ or of the Euler–Lagrange equation in the plane of complex fields ϕ. The $U(1)$ phase transformation $\phi \to \exp(i\alpha)\phi$ leaves both $\mathcal{L}$ and the Euler–Lagrange equation unaffected, whereas the field ϕ_c becomes

$$\phi_c \to \phi_c^{(\alpha)} = \frac{\lambda}{\sqrt{2}}e^{i(\beta+\alpha)} \neq \phi_c.$$

Every new phase defines a different solution. Thus the classical solution ϕ_c violates the $U(1)$ symmetry of $\mathcal{L}$ and of the equation of motion, like the solution $x(t)$ of the simple Newton equation $m\ddot{x}(t) = -V'(x)$ violates the invariance of this equation under time translations $t \to t + \delta t$ (i.e. the equation has this invariance, not the solution $x(t)$).

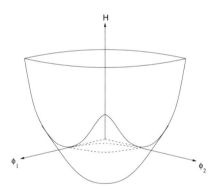

Fig. 22.2 The spontaneously chosen phase.

We can convince ourselves that ϕ_c is the field associated with a state of minimized energy. The Hamiltonian density $\mathcal{H}$ is defined by the Legendre transform

$$\mathcal{H}[\phi, \pi] = \pi\dot{\phi} - \mathcal{L}, \quad \pi = \frac{\partial\mathcal{L}}{\partial\dot{\phi}}.$$

For $\phi = \text{const.}$ with $\dot{\phi} = 0$, i.e. $\mathcal{H} = -\mathcal{L} = V$, and so

$$\mathcal{H}[\phi, \pi] = m_0^2\phi\phi^* + \frac{1}{3!}\lambda_0(\phi^*\phi)^2, \qquad (22.5)$$

the density $\mathcal{H}$ is minimized (i.e. $\partial\mathcal{H}/\partial\phi = 0$) if

$$[m_0^2 + \frac{1}{3}\phi^*\phi]\phi = 0.$$

This is the condition (22.3) obtained above.

For every value of the phase β we have a different constant c-number field configuration ϕ_c. Clearly these are the field configurations which trace out the circular bottom of the trough of the double well potential, as indicated in Fig. 22.2. Suppose we choose one such configuration, e.g. the one for $\beta = 0$, and we wish to investigate the behaviour of the field ϕ in its neighbourhood. Then we can reach some point in the neighbourhood of $\phi_c(\beta = 0)$ by travelling from ϕ_c partly along the direction of minimum configurations, i.e. along the trough of V (which we can call a *longitudinal direction*), and partly by climbing up the parabolic wall on either side, i.e. in a *transverse direction*. We examine this in more detail by setting $\phi(x)$ equal to the classical c-number configuration plus a *fluctuation field* $\eta(x)$ which is again complex like $\phi(x)$, i.e. we set

$$\phi(x) = \phi_c + \eta(x), \tag{22.6}$$

where

$$\phi_c = \frac{\lambda}{\sqrt{2}}, \quad \eta(x) = \frac{1}{\sqrt{2}}[\psi_1(x) + i\psi_2(x)].$$

Inserting this into Eq. (22.2) and separating real and imaginary parts, one obtains the equations

$$\left(\partial_\mu\partial^\mu + m_0^2 + \frac{1}{2}\lambda_0\lambda^2\right)\psi_1 = -\lambda\left(\frac{1}{3!}\lambda_0\lambda^2 + m_0^2\right) + O(\psi_1^2, \psi_1^3) \tag{22.7a}$$

and

$$\left(\partial_\mu\partial^\mu + \underbrace{m_0^2 + \frac{1}{3!}\lambda_0\lambda^2}_{0}\right)\psi_2 = O(\psi_1\psi_2, \psi_2^3). \tag{22.7b}$$

In Eq. (22.7b) the constant on the left and the coefficient of λ on the right of Eq. (22.7a) vanish on account of Eq. (22.3). We identify the coefficients of the linear terms on the left hand sides with those of the masses of the fields ψ_1, ψ_2. Then

$$m_1^2 \equiv m_0^2 + \frac{1}{2}\lambda_0\lambda^2 = m_0^2 - 3m_0^2 = -2m_0^2 > 0 \quad \text{for} \quad m_0^2 < 0,$$

$$m_2^2 \equiv m_0^2 + \frac{1}{6}\lambda_0\lambda^2 = 0.$$

We observe: The field $\psi_1(x)$ has acquired a real and positive mass whereas the field $\psi_2(x)$ is massless! We observe that the field $\psi_2(x)$ is that component of ψ which is directed along the trough of the potential, whereas $\psi_1(x)$ is

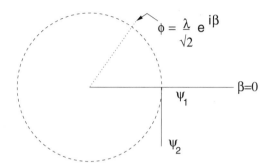

Fig. 22.3 The spontaneously chosen phase seen from above.

the component which climbs up radially outward along the profile of the potential implying a $\psi^2(x)$ term with the potential $V(\phi)$ and so in the Lagrangian density $\mathcal{L}$. We have here an example of the *Goldstone theorem* which says: If the solutions of the equation of motion do not possess a continuous symmetry of the Lagrangian, then a massless boson exists. Goldstone's theorem applies to fully relativistic field theories. In our later discussion of theories with nonconstant classical field configurations we shall encounter a similar, though not identical phenomenon. There the wave function with an associated vanishing eigenvalue is called a *"zero mode"*. This wave function is in the quantum mechanics constructed about the classical configuration ϕ_c at this point or collective coordinate of the classical path a vector in the Hilbert space pointing in the longitudinal direction, i.e. tangential to the classical path. Like the Goldstone mode in the above Higgs model it leads to a divergence in the Green's function of the theory, which, of course, has to be removed in order to allow a well defined perturbation procedure (i.e. the existence of individual perturbation contributions), irrespective of the question whether the perturbation series as such converges or not. We can see the problem here by looking at Eq. (22.7b). The appropriate Green's function G is the inhomogeneous solution of the the the equation

$$[\partial_\mu \partial^\mu + m_0^2]G(\mathbf{r}, \mathbf{r}'; t, t') = \delta(\mathbf{r} - \mathbf{r}')\delta(t - t').$$

Thus the Green's function is similar to that in electrodynamics. It is known from there that the vanishing mass of the photon together with the transversality of the electromagnetic field implies that only two of the four components of the four-vector potential A_μ are independent. The elimination of the other components results from the vanishing mass and the constraint called the *gauge-fixing condition* which can also be looked at as the condition which removes from the Green's function G the divergent contribution, i.e. equates to zero the coefficient of this would-be-divergence.

We see therefore that the classical c-number field configuration and the attempt to develop a perturbation series in its neighbourhood leads to intricate connections between symmetry properties, zero mass configurations and constraints even before the question of quantization of the fluctuation field $\eta(x)$ can be considered. In the following we will not be concerned with classical c-number solutions of the Euler–Lagrange equations which are simply constants; instead we shall consider static and time-varying solutions in important models of one spatial dimension. The principal aspects will always be very similar so that it really pays to study simple models in considerable detail.

22.3 Soliton Theories in One Spatial Dimension

We consider here the two basic soliton models known as Φ^4 and sine–Gordon theories.[*] Potentials of higher polynomial order in Φ than Φ^4 can also be considered.[†] In particular the sextic potential can be considered along parallel lines.[‡] We consider the theory (later: quantum theory) of a real, spinless field $\Phi(x,t)$ in one spatial dimension. The Lagrangian density is taken to be

$$\mathcal{L}[\Phi, \partial_\mu \Phi] = \frac{1}{2}\partial_\mu \Phi \partial^\mu \Phi - V[\Phi] = \frac{1}{2}\dot{\Phi}^2 - \frac{1}{2}\left(\frac{\partial \Phi}{\partial x}\right)^2 - V[\Phi], \qquad (22.8)$$

where $V[\Phi]$ is a self-interaction or potential of the field (to be specified later), and we choose the metric

$$\eta_{00} = +1, \quad \eta_{11} = -1 \quad (x \in \text{Minkowski manifold}).$$

To insure that the energy (see below) is positive definite we require

$$V[\Phi] \geq 0.$$

We make one more assumption concerning V which, however, we shall need only at a later stage. Later we wish to develop a perturbation series about a classical configuration of Φ, so that we are interested to have a parameter

---

[*]For a collection of many informative papers on solitons etc. see e.g. C. Rebbi and G. Soliani [237].

[†]Potentials of this type have been discussed by M. A. Lohe [179].

[‡]S. N. Behera and A. Khare [17], F. Zimmerschied and H. J. W. Müller–Kirsten [291].

which serves as the expansion parameter of the series. For this purpose we assume that $V[\Phi]$ depends on a scaling parameter g such that

$$V[\Phi] \equiv V[\Phi, g] = \frac{1}{g^2}\tilde{V}[g\Phi] \equiv \frac{1}{g^2}V[g\Phi, 1]. \qquad (22.9)$$

For instance, the so-called Φ^4-*theory* with *quartic potential* is defined by

$$V[\Phi] = \frac{m^4}{2g^2}\left[1 - \frac{g^2\Phi^2}{m^2}\right]^2 = \frac{1}{g^2}\tilde{V}[g\Phi] \equiv \frac{1}{g^2}V[g\Phi, 1], \qquad (22.10)$$

and the *sine–Gordon theory* with cosine potential by

$$V[\Phi] = \frac{m^4}{g^2}\left[1 - \cos\left(\frac{g\Phi}{m}\right)\right] = \frac{1}{g^2}\tilde{V}[g\Phi] \equiv \frac{1}{g^2}V[g\Phi, 1]. \qquad (22.11)$$

The Euler–Lagrange equation is seen to be (with $c = 1$)

$$\Box\Phi + V'[\Phi] = 0, \quad \text{i.e.} \quad \left(\frac{\partial^2}{\partial t^2} - \frac{\partial^2}{\partial x^2}\right)\Phi + V'[\Phi] = 0, \qquad (22.12)$$

where the prime denotes differentiation with respect to Φ. In a quantum theory the field Φ is an operator defined on a space of states. Since $\Phi(x, t)$ has an explicit time-dependence it is a Heisenberg-picture operator. The field $\Phi(x, t)$ is an observable (which in general — in field theory as compared with quantum mechanics — does not have to be hermitian although we assume this in the present example). Before we consider quantum aspects we study classical c-number fields which satisfy the Euler–Lagrange equation. In fact to begin with we restrict ourselves further and consider c-number fields $\Phi \to \phi$ which are static, i.e. such that

$$\frac{\partial\Phi(x, t)}{\partial t} = 0. \qquad (22.13a)$$

We write such fields $\phi(x)$. From Eq. (22.12) we see that they satisfy the Newtonian-like equation

$$\phi''(x) = V'(\phi). \qquad (22.13b)$$

We are not interested in just any solution of this equation, but in such solutions which are subject to (ignoring the dimensional aspect) reasonable physical conditions. The first such condition to be imposed is that the energy of the solution of interest must be finite, the reason for this condition being that we visualize the classical solution as representative of a *lump of energy*. The other condition which we impose (and study in detail) is that of *stability*, at least in the present case.[§]

[§]Unstable classical configurations are also important and will be considered in later chapters.

The energy $E[\Phi]$ of the system is defined to be the spatial integral of the Hamiltonian density $\mathcal{H}(\Phi, \pi)$ where π is the momentum conjugate to Φ defined by

$$\pi = \frac{\partial \mathcal{L}}{\partial \dot{\Phi}} = \dot{\Phi}, \qquad (22.14)$$

where the overdot implies differentiation with respect to time t. The energy-momentum tensor $T_{\mu\nu}$ and the conservation law it satisfies are given by the equations

$$T_{\mu\nu} = -\eta_{\mu\nu}\mathcal{L} + \frac{\partial \mathcal{L}}{\partial(\partial^\mu \Phi)}\partial_\nu\Phi, \qquad \partial_\mu T^{\mu\nu} = 0. \qquad (22.15)$$

The zero-zero component is equivalent to the Hamiltonian density $\mathcal{H}$, i.e.

$$T_{00} = -\eta_{00}\mathcal{L} + \frac{\partial \mathcal{L}}{\partial \dot{\Phi}}\dot{\Phi} = -\mathcal{L} + \pi\dot{\Phi} \equiv \mathcal{H},$$

i.e.

$$\mathcal{H} = \dot{\Phi}^2 - \mathcal{L} = \frac{1}{2}\dot{\Phi}^2 + \frac{1}{2}\left(\frac{\partial \Phi}{\partial x}\right)^2 + V[\Phi]. \qquad (22.16)$$

In the static limit we obtain therefore and write this as

$$E[\Phi] = H[\phi] \rightarrow E(\phi) = \int_{-\infty}^{\infty} dx \left[\frac{1}{2}\left(\frac{d\phi}{dx}\right)^2 + V(\phi)\right]. \qquad (22.17)$$

Since we have to integrate over all space, the classical solution will therefore have to be such that this integral is finite. We observe here already that the integrand of the integral in Eq. (22.17) is reminiscent of classical mechanics if we interpret x as time and $V(\phi)$ as the potential energy of a particle at location ϕ.

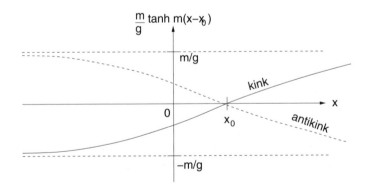

Fig. 22.4 The wellknown soliton of Φ^4-theory.

As we shall see below, it is not difficult to find classical c-number solutions ϕ_c to Eq. (22.13b) for the specific potentials (22.10) and (22.11). In fact, in the case of the first, the Φ^4-*theory*, one obtains by straightforward integration (but see also below) the *soliton configuration*

$$\phi_c(x) = \pm \frac{m}{g} \tanh m(x - x_0). \tag{22.18}$$

Here x_0 is an integration constant. The energy is correspondingly obtained — as will be shown — as

$$E(\phi_c) = \frac{4}{3} \frac{m^3}{g^2}. \tag{22.19}$$

The typical shape of the configuration given by Eq. (22.18) is depicted in Fig. 22.4.

In the *sine–Gordon theory*[¶] the classical or soliton configuration is (see below) found to be

$$\phi_c(x) = \pm 4 \frac{m}{g} \tan^{-1} \left[e^{\pm m(x - x_0)} \right] \tag{22.20}$$

with energy

$$E(\phi_c) = 8 \frac{m^3}{g^2}. \tag{22.21}$$

This sine–Gordon configuration is depicted in Fig. 22.5. Again x_0 is an integration constant.

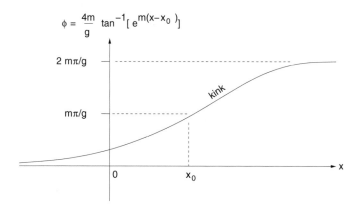

Fig. 22.5 The soliton of sine–Gordon theory.

[¶]The name sine–Gordon is derived from the analogous Klein–Gordon equation or theory, i.e. around $\phi = 0$ the sine–Gordon potential behaves like the Klein–Gordon potential proportional to ϕ^2. See J. Rubinstein [240]. The sine–Gordon theory differs from the Klein–Gordon theory in that it possesses invariance under the shift $\phi \to \phi + 2\pi$.

The solutions (22.18) and (22.20) are obtained from Eq. (22.13b) which we can write

$$\frac{d}{d\phi}\left[\frac{1}{2}(\phi')^2\right] = V'(\phi), \quad \text{or} \quad \frac{1}{2}(\phi')^2 = V(\phi) + \text{const.}$$

Here the constant is zero for $V(\phi)$ and ϕ' zero at $x = \pm\infty$. Thus

$$\pm \int_{\phi(x_0)}^{\phi(x)} \frac{d\phi}{\sqrt{2V(\phi)}} = \int_{x_0}^{x} dx = x - x_0.$$

Inserting the potential (22.10) we obtain

$$
\begin{aligned}
x - x_0 &= \pm \int_{\phi(x_0)}^{\phi(x)} d\phi \frac{1}{g}\left[\frac{m^2}{g^2}\left(1 - \frac{g^2\phi^2}{m^2}\right)\right]^{-1} \\
&= \pm\frac{1}{m}\int_{\phi(x_0)}^{\phi(x)} d\phi \frac{m}{g}\left[\frac{m^2}{g^2} - \phi^2\right]^{-1} \\
&= \pm\frac{1}{m}\left[\tanh^{-1}\left(\frac{g\phi}{m}\right)\right]_{\phi(x_0)}^{\phi(x)},
\end{aligned}
$$

i.e.

$$\phi(x) = \pm\frac{m}{g}\tanh m(x - x_0).$$

In the case of the sine–Gordon theory defined by the potential (22.11) we obtain

$$x - x_0 = \pm \int_{\phi(x_0)}^{\phi(x)} \frac{m}{g} \frac{d(g\phi/m)}{\frac{m^2}{g}\sqrt{2(1 - \cos\frac{g\phi}{m})}},$$

or

$$
\begin{aligned}
m(x - x_0) &= \pm \int_{x = g\phi/m} \frac{dx}{\sqrt{2(1 - \cos x)}} = \pm \int_{x = g\phi/m} \frac{dx}{\sqrt{4\sin^2\frac{1}{2}x}} \\
&= \pm\left\{\ln\left|\tan\frac{g\phi(x)}{4m}\right| - \ln\left|\tan\frac{g\phi(x_0)}{4m}\right|\right\},
\end{aligned}
$$

and hence

$$\phi(x) = \pm\frac{4m}{g}\tan^{-1}\left\{e^{\pm m(x - x_0)}\tan\left(\frac{g\phi(x_0)}{4m}\right)\right\}.$$

With

$$|g\phi(x_0)| = m\pi,$$

we obtain

$$\phi(x) = \pm 4\frac{m}{g}\tan^{-1}\left\{e^{\pm m(x - x_0)}\right\}.$$

From their monotonic behaviour the solutions ϕ_c in these cases derive their name as "*kinks*"; the curves rising monotonically from negative to positive values are known as *kinks*, the others as *antikinks*. The expressions (22.19), (22.21) for the energies can also be obtained from Eq. (22.17). However, instead of doing this now, we obtain these expressions later (cf. the discussion of Bogomol'nyi equations) in a simpler way. The kink solutions are also frequently described as "*domain walls*" in view of their analogy with the domain separating upward spins from downward spins as, for instance, in the one-dimensional Ising model.

22.4 Stability of Classical Configurations

The concept of stability can be complex. Naively stability means that a system does not deviate appreciably from a state of stability (or equilibrium) if it is allowed to fluctuate between neighbouring states. A state of stability is therefore associated with minimized action and/or energy. This is called more precisely "*classical stability*".[||]

The Euler–Lagrange equation is obtained by extremizing the action or Lagrangian; however, this does not require the action to be minimized. Suppose now that we demand stability in the sense of minimized energy. Then we have to demand that the functional derivative of the energy $E(\phi)$ be zero, i.e.

$$\frac{\delta E(\phi)}{\delta \phi(y)} = 0, \tag{22.22}$$

and that its second variational derivative be positive semi-definite. The relation (22.17) implies

$$
\begin{aligned}
0 &= \frac{\delta}{\delta\phi(y)} \int dx \left[\frac{1}{2}\left(\frac{d\phi}{dx}\right)^2 + V(\phi) \right] \\
&= \int dx \left[\left(\frac{d\phi}{dx}\right) \frac{\delta}{\delta\phi(y)}\left(\frac{d\phi}{dx}\right) + \frac{\delta V(\phi)}{\delta\phi(x)}\delta(x-y) \right] \\
&= \int dx \left[\left(\frac{d\phi}{dx}\right) \frac{d}{dx}\delta(x-y) + \frac{\delta V(\phi)}{\delta\phi(x)}\delta(x-y) \right],
\end{aligned}
$$

and with partial integration we obtain

$$0 = \int_{-\infty}^{\infty} dx \left[-\frac{d}{dx}\left(\frac{d\phi}{dx}\right) + \frac{\delta V(\phi)}{\delta\phi(x)} \right]\delta(x-y) + \left[\frac{d\phi}{dx}\delta(x-y) \right]_{x=-\infty}^{\infty}. \tag{22.23}$$

[||] See for instance R. Jackiw [141].

Since $y \neq \pm\infty$, the integrated contribution vanishes. Thus the vanishing of the first variational derivative yields again Eq. (22.13b) with classical solution ϕ_c. Proceeding to the second variation we obtain at $\phi = \phi_c$ the relation

$$\frac{\delta^2 E(\phi)}{\delta\phi(x)\delta\phi(y)}\bigg|_{\phi_c} = \left[-\frac{d^2}{dx^2} + V''(\phi)\right]_{\phi_c} \delta(x-y). \qquad (22.24)$$

For so-called "classical stability" the differential operator of this expression has to be positive semi-definite, i.e. its eigenvalues $w_k^2 \geq 0$. Thus we have to investigate the eigenvalue spectrum of the Schrödinger-like equation

$$\left[-\frac{d^2}{dx^2} + V''(\phi_c)\right]\psi_k(x) = w_k^2\psi_k(x). \qquad (22.25)$$

Before we begin to study this equation in detail for specific potentials we can make a very important observation on very general grounds. The original Lagrangian was, of course, written down in Lorentz-invariant form, and the Euler–Lagrange equation retains this property, in particular the invariance under spatial translations. The Newton-like equation (22.13b) retains the invariance although the solution ϕ_c does not; i.e. if $\phi_c(x)$ is a solution, the function $\phi_c(x+a), a \neq 0$ is not the same. Thus for every translational shift "a" we obtain a new solution ϕ_c in much the same way as we obtained new solutions by a change of phase in the Higgs model which we discussed before. In fact, if we apply the generator of spatial translations d/dx to the classical equation, i.e. Eq. (22.13b), we obtain

$$\phi_c'''(x) = V''(\phi_c)\phi_c', \quad \text{or} \quad \left[-\frac{d^2}{dx^2} + V''(\phi_c)\right]\phi_c'(x) = 0. \qquad (22.26)$$

Comparing this equation with Eq. (22.25), we see that the latter has one eigenfunction, namely ϕ_c'. with eigenvalue $w_k^2 = 0$. The eigenfunction $\phi_c'(x)$ is for this reason called a "*zero mode*". Since it arises from the violation of translational invariance by ϕ_c, it is also called a "*translational mode*", in fact

$$\psi_0(x) = \text{const.}\frac{d}{dx}\phi_c \qquad (22.27)$$

shows that the zero mode results from application of the generator of translations to the classical c-number field configuration. This is, in fact, a very general phenomenon which can be established as follows.

A theorem on zero modes

Let $S[U]$ be the action of some field U defined on Minkowski space. Assume that $S[U]$ is invariant under the transformations of some symmetry group G with elements

$$g = \exp(i\lambda T) \in G, \qquad (22.28)$$

where T is a generator and λ a parameter. Invariance of $S[U]$ under transformations of this group implies

$$S[U] = S[gU]. \tag{22.29}$$

Suppose now the Euler–Lagrange equation possesses the classical c-number solution U_c, and that $S[U_c]$ is finite (or equivalently the energy). Then the vanishing of the first functional variation implies that

$$\delta S = 0, \quad \text{i.e.} \quad 0 = S'[U_c] = S'[gU_c]. \tag{22.30}$$

Differentiating this with respect to λ (applying $-id/d\lambda$) and setting $\lambda = 0$, we obtain

$$0 = -i\frac{d}{d\lambda}S'[\exp(i\lambda T)U_c]|_{\lambda=0} = S''[U_c](TU_c).$$

The right hand side can be interpreted as meaning: The application of the second functional derivative of the action (at $U = U_c$) to (TU_c) is zero. Thus TU_c is a zero mode; in fact, it is an expression like (22.27). We see therefore that the occurrence of these zero modes is a very general phenomenon. We shall see later that these zero modes lead to undefined Green's functions for the semi-classical perturbation expansion unless one or more suitable constraints are imposed.

We return to our considerations of stability. As stated above, we are interested in classical c-number field configurations of finite energy, and we have seen that Eqs. (22.18) and (22.20) are such configurations in two particular models. In the one-dimensional case under discussion we can rewrite Eq. (22.13b) as

$$\frac{d}{d\phi}\left[\frac{1}{2}(\phi')^2\right] = V'(\phi). \tag{22.31}$$

The first integral of this expression is

$$\frac{1}{2}(\phi_c')^2 = V(\phi_c) + \text{const.} \tag{22.32}$$

Inserting this into Eq. (22.17) we obtain

$$E(\phi_c) = 2\int_{-\infty}^{\infty} dx[V(\phi_c) + \text{const.}].$$

Thus for $E(\phi_c)$ to be finite, this integral has to be finite. We can choose the potential V such that the constant is zero, i.e.

$$\lim_{x\to\pm\infty}[V(\phi_c) + \text{const.}] = 0 \quad \text{or} \quad \lim_{x\to\pm\infty}V(\phi_c) = 0 \text{ for const.} = 0. \tag{22.33}$$

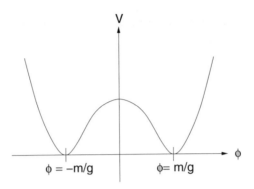

Fig. 22.6 The minima of V in Φ^4-theory at $x \to \pm\infty$.

In the Φ^4-theory defined by Eq. (22.10) we must therefore have that for $x \to \pm\infty$: $\phi_c \to \pm m/g$ as indicated in Fig. 22.6. In the sine–Gordon theory defined by Eq. (22.11) we must have correspondingly for $x \to \pm\infty$:

$$\phi_c \to \frac{m}{g} 2\pi n, \quad n = 0, \pm 1, \pm 2, \ldots .$$

Thus the classical energy or vacuum configuration is not unique; the potentials have *degenerate minima* (i.e. $V(\phi) = 0$ at different values of ϕ). For a configuration to have finite energy, the classical field ϕ_c must approach one minimum for $x \to -\infty$ and another for $x \to \infty$. Solutions which approach the same minimum for $+\infty$ and $-\infty$ are the constant solutions $\phi_c = \pm m/g$ in the Φ^4-theory. The solutions can be characterized by an integral number called the *"topological (quantum) number"*, *"topological charge"* or *"winding number"* (for an illustration see Example 22.1). This number is defined like a charge in field theory by the spatial integral of the time component of a current. In the case of the Φ^4-theory we can define a conserved current k^μ by

$$k^\mu = \epsilon^{\mu\nu}\partial_\nu\phi_c, \quad \text{since} \quad \partial_\mu k^\mu = 0, \tag{22.34}$$

in view of the antisymmetry of $\epsilon^{\mu\nu}$. The charge Q is therefore given by

$$Q = \int_{-\infty}^{\infty} dx k^0 = \int_{-\infty}^{\infty} dx \epsilon^{01}\partial_1\phi_c = [\phi_c(x)]_{-\infty}^{\infty} = \pm\frac{2m}{g}, 0. \tag{22.35}$$

The topological quantum number q is defined by

$$q = \frac{Q}{2m/g} = \pm 1, 0,$$

and it is seen that in this case q can be ± 1 or 0.

Since we call q a topological quantum number, one may ask how topology comes into the picture. We can see this as follows. First, zero is not a meaningful topological number. Thus we expect the constant solutions (which are characterized by topological number zero) to be nontopological in some sense. A property is, in general, termed topological, if it remains unchanged under continuous deformations of the (field) configuration while preserving finiteness of the energy. Thus we consider the family $\{\lambda\phi_c(x)\}$ of (field) configurations where λ is some parameter which assumes real, continuous values. For such configurations we have the energy

$$E(\lambda\phi_c) = \int_{-\infty}^{\infty} dx \left[\frac{1}{2}\lambda^2(\phi_c')^2 + V(\lambda\phi_c)\right]. \tag{22.36}$$

Consider the Φ^4-theory with $\phi_c \to \pm m/g$ for $x \to \pm\infty$. Then for $x \to \pm\infty$:

$$V(\lambda\phi_c) \xrightarrow{(22.10)} \frac{m^4}{2g^2}(1 - \lambda^2)^2 \neq 0 \quad \text{except for} \quad \lambda^2 = 1.$$

Thus even without varying E with respect to λ we can see that $E(\lambda\phi_c)$ is finite only for $\lambda = \pm 1$. Thus the condition of finiteness of energy does not permit $\lambda\phi_c$ deformations in this case other than those with

$$\lambda = e^{i\chi(\theta)} \quad \text{for} \quad \chi(\theta) = 0, \pi \tag{22.37}$$

for the one-dimensional case here, and we call this *"topological stability"*. Such a smooth deformation (in the sense that $\exp(i\chi(\theta))\phi_c$ depends smoothly on θ, and $\exp(i\chi(\theta))\phi_c$ reduces to ϕ_c when $\chi(\theta) = 0, \pi$) does not change a boundary condition

$$\lim_{x \to \pm\infty} [\lambda\phi_c(x)] = \phi_c(\pm\infty),$$

so that under such deformations the configuration remains in the same topological sector defined by the boundary condition.

In the case of the sine–Gordon theory the minima of the potential (22.11) are given by

$$\phi_c(x) \to 2\pi n\left(\frac{m}{g}\right), \quad n = 0, \pm 1 \pm 2, \ldots, \tag{22.38}$$

so that for $x \to \pm\infty$:

$$V(\lambda\phi_c) \to \frac{m^4}{g^2}[1 - \cos(2\pi n\lambda)].$$

Again we see that $E(\lambda\phi_c)$ is finite only for $\lambda = \pm 1$. The difference

$$\phi_c(\infty) - \phi_c(-\infty) = 2\pi\frac{m}{g}\triangle n$$

involves the difference $\triangle n = n_\infty - n_{-\infty} \equiv N$, where $n_{\pm\infty}$ are the integers n corresponding to the minima of the potential at $x = \pm\infty$. (For a broader view we consider briefly in Sec. 22.9 the foregoing arguments in a higher dimensional context. Thus with one space dimension higher there, the factor $\lambda = \exp[i\chi(\theta)]$ varies over a complete circle).

We observe

(a) the topological current k^μ of Eq. (22.34) is conserved independently of the equations of motion (other than Noether currents whose conservation follows from the equations of motion),**

(b) the time component k^0 depends only on ϕ but not on momenta (again in contrast to the case of Noether currents),

(c) the time component k^0 is a spatial divergence; its integral is nonzero only on account of nontrivial boundary conditions,

(d) the topological quantum number N arises in a completely classical context.

22.5   Bogomol'nyi Equations and Bounds

We have seen previously (cf. Eqs. (22.31) to (22.33)) that the classical configuration ϕ_c is the solution of the nonlinear second order differential equation (22.13b) or

$$\frac{1}{2}[\phi'(x)]^2 = V(\phi). \tag{22.39}$$

From this we construct the inequality

$$[\phi'(x) \mp \sqrt{2V(\phi)}]^2 \geq 0. \tag{22.40}$$

Since (cf. Eq. (22.17))

$$E(\phi) = \int_{-\infty}^{\infty} dx \left[\frac{1}{2}(\phi'(x))^2 + V(\phi) \right], \tag{22.41}$$

we can write

$$E(\phi) \geq \int_{-\infty}^{\infty} dx \sqrt{2V(\phi)}\phi'(x). \tag{22.42}$$

Thus the energy has a lower bound given by the right hand side of this inequality. The inequality is saturated, i.e. the energy is minimized, if

$$E_{\min}(\phi) = \int_{-\infty}^{\infty} dx \sqrt{2V(\phi)}\phi'(x). \tag{22.43}$$

**See e.g. H. J. W. Müller–Kirsten [215], pp. 425 – 428.

We observe that (cf. Eq. (22.40)) this occurs when

$$\phi'(x) \mp \sqrt{2V(\phi)} = 0. \tag{22.44}$$

This first order equation which saturates the inequality (22.40) is called the "*Bogomol'nyi equation*" since it was first introduced by Bogomol'nyi.[tt] Clearly it solves the second order equation or (22.39). In the case of the Φ^4-theory we obtain

$$
E_{\min}(\phi) = \int_{-\infty}^{\infty} dx \frac{m^2}{g}\left(1 - \frac{g^2\phi^2}{m^2}\right)\phi'(x) = \frac{m^2}{g}\int_{-\infty}^{\infty} dx \frac{d}{dx}\left(\phi - \frac{g^2\phi^3}{3m^2}\right)
$$

$$
= \frac{m^2}{g}\left[\phi - \frac{g^2\phi^3}{3m^2}\right]_{\phi(-\infty)=-m/g}^{\phi(\infty)=m/g} = 2\frac{m^3}{g^2}\left(1 - \frac{1}{3}\right) = \frac{4m^3}{3g^2},
$$

which is the expression given by Eq. (22.19).

In the sine–Gordon theory we can consider, for instance, the set of configurations ϕ_c which interpolate between $\phi(-\infty) = 0$ and $\phi(\infty) = 2\pi nm/g, n \geq 1$. Here $(V(\phi) \geq 0, \sqrt{V(\phi)} \geq 0)$

$$
E_{\min}(\phi) = \int_{-\infty}^{\infty} dx \sqrt{2V(\phi)}\phi'(x) = \int_{\phi=0}^{\phi=2\pi nm/g} d\phi\sqrt{2V(\phi)}
$$

$$
= n\int_{\phi=0}^{\phi=2\pi m/g} d\phi\sqrt{2V(\phi)}
$$

$$
= n\int_{\phi=0}^{\phi=2\pi m/g} \frac{m^2}{g}d\phi\sqrt{2\left(1 - \cos\frac{g\phi}{m}\right)},
$$

and with $x = g\phi/m$ this becomes

$$
E_{\min}(\phi) = n\frac{m^3}{g^2}\int_0^{2\pi} dx\sqrt{2(1 - \cos x)} = n\frac{m^3}{g^2}\int_0^{2\pi} dx\sqrt{4\sin^2\frac{x}{2}}
$$

$$
= n\frac{2m^3}{g^2}\left[-2\cos\frac{x}{2}\right]_0^{2\pi} = n\frac{2m^3}{g^2}[-2(-1) + 2]
$$

$$
= n\frac{8m^3}{g^2}
$$

in agreement with Eq. (22.21).

In order to obtain a better understanding of how topology comes into the picture we recall that the sine–Gordon theory has an interesting analogue in classical mechanics. We discuss this in the following Example.

[tt]E. B. Bogomol'nyi [27].

Example 22.1: Classical analogy to sine–Gordon theory

Consider a string along the x-axis. At equidistant points $x_n = na$ along this string we attach strings with pendulum bobs, all being identical, with mass m and connected with short strings to neighbouring bobs. Initially the pendulums are suspended freely in the gravitational field. Assume that the pendulums can move (i.e. rotate) only in the (y, z)-plane and only such that their elastic strings remain taut. Construct an expression for the energy of such a classical system and its Lagrangian.

Solution: Consider a string along the x-axis as described. Assuming that the pendulums can move (i.e. rotate) only in the (y, z)-plane and only such that their elastic strings remain taut the position of any one of the pendulums is determined completely by an angle $\phi(x_n, t)$ measured e.g. from the y-axis as shown in Fig. 22.7.

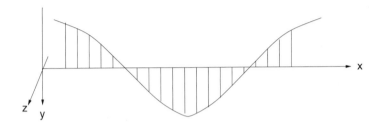

Fig. 22.7 The pendulum analogy.

We can now write down an expression for the energy of such a system of (say) n pendulums. The kinetic energy T of the bob at position x_n at time t is (recall "$mr^2\dot{\theta}^2/2$")

$$\frac{1}{2}mr^2\left(\frac{\partial\phi(x_n, t)}{\partial t}\right)^2.$$

The energy of interaction with neighbouring bobs is (recall "$kx^2/2$")

$$\frac{1}{2}k[\phi(x_{n+1}, t) - \phi(x_n, t)]^2.$$

The potential energy due to gravitation is (maximal for $\phi = \pi$) $mgr(1 - \cos\phi(x_n, t))$. Thus the total energy of the system is with the number of pendulums allowed to become infinite

$$E = \sum_{n=-\infty}^{\infty}\left[\frac{1}{2}mr^2\left(\frac{\partial\phi(x_n, t)}{\partial t}\right)^2 + \frac{1}{2}k[\phi(x_{n+1}, t) - \phi(x_n, t)]^2 + mgr(1 - \cos\phi(x_n, t))\right].$$

The continuum limit is obtained by taking[‡‡] (since $x_n = na$, $x_{n+1} - x_n = a$)

$$a \to dx, \quad \frac{m}{a} \to \mu, \quad ka \to \kappa, \quad \sum_{n=-\infty}^{\infty} \to \int_{-\infty}^{\infty},$$

so that

$$E = \int_{-\infty}^{\infty} dx\left[\frac{1}{2}\mu r^2\left(\frac{\partial\phi}{\partial t}\right)^2 + \frac{1}{2}\kappa\left(\frac{\partial\phi}{\partial x}\right)^2 + \mu gr(1 - \cos\phi)\right]. \tag{22.45}$$

We can therefore write the Lagrangian

$$L = T - V = \int_{-\infty}^{\infty} dx\left[\frac{1}{2}\mu r^2\left(\frac{\partial\phi}{\partial t}\right)^2 - \frac{1}{2}\kappa\left(\frac{\partial\phi}{\partial x}\right)^2 - \mu gr(1 - \cos\phi)\right]. \tag{22.46}$$

[‡‡]See e.g. H. Goldstein [114], Chapter XI.

Thus if the classical vacuum corresponds to the case when all pendulums are pointing downwards, each finite energy configuration beginning and ending with a classical vacuum corresponds to an integral number of rotations about the x-axis (i.e. of the continuous curve in Fig. 22.7 linking the pendulum bobs). This number, the topological quantum number, is therefore also called *"winding number"*.

22.6 The Small Fluctuation Equation

We now return to the Schrödinger-like equation (22.25) which we obtained as the condition of classical stability for eigenvalues $w_k^2 \geq 0$. It is instructive to obtain this equation by yet another method. We are interested in studying (field) configurations Φ in the neighbourhood of the classical solution ϕ_c. Hence we set

$$\Phi(x,t) = \phi_c(x) + \eta(x,t), \tag{22.47}$$

where $\eta(x,t)$ is the *fluctuation (field)*. Inserting (22.47) into (22.12) we obtain

$$\left(\frac{\partial^2}{\partial t^2} - \frac{\partial^2}{\partial x^2}\right)(\phi_c(x) + \eta(x,t)) + V'(\phi_c(x) + \eta(x,t)) = 0.$$

Since $\phi_c(x)$ obeys Eqs. (22.13a) and (22.13b) we have for small fluctuations $\eta(x,t)$:

$$\left(\frac{\partial^2}{\partial t^2} - \frac{\partial^2}{\partial x^2}\right)\eta(x,t) + V''(\phi_c(x))\eta(x,t) = 0. \tag{22.48}$$

With the ansatz

$$\eta(x,t) = \sum_k \exp(-iw_k t)\psi_k(x), \tag{22.49}$$

which represents an expansion in terms of normal modes, Eq. (22.48) becomes

$$\sum_k \left(\frac{\partial^2}{\partial x^2} + w_k^2\right)\psi_k(x)\exp(-iw_k t) = \sum_k V''(\phi_c(x))\psi_k(x)\exp(-iw_k t).$$

Equating coefficients of $\exp(-iw_k t)$, we see that $\psi_k(x)$ has to obey the equation

$$\left[-\frac{\partial^2}{\partial x^2} + V''(\phi_c(x))\right]\psi_k(x) = w_k^2\psi_k(x). \tag{22.50}$$

This equation is identical with Eq. (22.25) obtained earlier. The equation is called *stability equation* or *equation of small fluctuations*. We can understand what stability means in the present case. If some w_k^2 were negative, w_k would be imaginary and so the factor $\exp(-iw_k t)$ in $\eta(x,t)$ of Eq. (22.49) would imply an exponential growth in the future $t > 0$ or in the past $t < 0$, and so would invalidate the procedure which assumes that $\eta(x,t)$ is a small fluctuation. Thus for stability w_k^2 must not be negative. If that is necessary, is a vanishing eigenvalue $w_k^2 = 0$ acceptable for stability? Apparently for $w_k^2 = 0$ the energy is not minimized in the direction of its associated eigenfunction, i.e. the zero mode. This means that the stability is not universal but only local in a certain sense. In fact, we shall see later that the fluctuations $\eta(x, t)$ have to be orthogonal to the zero mode (which is thereby circumvented) so that the Green's function of the expansion procedure exists.

The equation of small fluctuations (22.50) has the general form of a time-independent one-dimensional Schrödinger equation. For $x \to \pm\infty$ the quantity acting as "potential", i.e. $V''(\phi_c(x))$, approaches a constant value which is nonzero in both the Φ^4-theory and the sine–Gordon theory. In fact, since we shall need the explicit form of $V''(\phi_c(x))$ we calculate this for the two cases:

(a) Φ^4-*theory*: Here

$$V(\phi) = \frac{m^4}{2g^2}\left(1 - \frac{g^2}{m^2}\phi^2\right)^2, \qquad V'(\phi) = -2m^2\left(1 - \frac{g^2}{m^2}\phi^2\right)\phi,$$

$$V''(\phi) = -2m^2 + 6g^2\phi^2.$$

Inserting ϕ_c of (22.18), we obtain

$$
\begin{aligned}
V''(\phi_c) &= -2m^2 + 6m^2\tanh^2 m(x - x_0) \\
&= -2m^2 + 6m^2(1 - \operatorname{sech}^2 m(x - x_0)) \\
&= 4m^2 - \frac{6m^2}{\cosh^2 m(x - x_0)}.
\end{aligned}
\tag{22.51}
$$

(b) *Sine–Gordon theory*: Here

$$V(\phi) = \frac{m^4}{g^2}\left[1 - \cos\left(\frac{g}{m}\phi\right)\right], \qquad V'(\phi) = \frac{m^3}{g}\sin\left(\frac{g}{m}\phi\right),$$

$$V''(\phi) = m^2\cos\left(\frac{g}{m}\phi\right).$$

Inserting ϕ_c of Eq. (22.20), we obtain

$$V''(\phi_c) = m^2\cos\left[\pm 4\tan^{-1}\left\{e^{\pm m(x - x_0)}\right\}\right].
\tag{22.52}$$

We have

$$\tan\theta := e^{\pm m(x - x_0)}, \qquad \theta = \tan^{-1}\left\{e^{\pm m(x - x_0)}\right\},$$

and need

$$\cos 4\theta = 2\cos^2(2\theta) - 1.$$

Using the formula

$$\cos 2\theta = \frac{1 - \tan^2\theta}{1 + \tan^2\theta} = \frac{1 - \exp\{\pm 2m(x - x_0)\}}{1 + \exp\{\pm 2m(x - x_0)\}} = \frac{\sinh[\mp m(x - x_0)]}{\cosh[\pm m(x - x_0)]},$$

we obtain

$$\begin{aligned}
\cos 4\theta &= 2\left(\frac{1 - \tan^2\theta}{1 + \tan^2\theta}\right)^2 - 1 \\
&= \frac{2(\cosh^2 m(x - x_0) - 1) - \cosh^2 m(x - x_0)}{\cosh^2 m(x - x_0)} \\
&= 1 - \frac{2}{\cosh^2 m(x - x_0)}.
\end{aligned}$$

Thus

$$V''(\phi_c) = m^2 - \frac{2m^2}{\cosh^2 m(x - x_0)}. \tag{22.53}$$

Hence in either case the stability equation is of the form (setting $z = m(x - x_0)$)

$$\left[-\frac{d^2}{dz^2} + l^2 - \frac{l(l + 1)}{\cosh^2 z}\right]\psi(z) = \frac{w^2}{m^2}\psi(z). \tag{22.54}$$

Regarding $-l(l + 1)/\cosh^2 z$ as the potential — it is known as the *Eckhardt* or *Pöschl–Teller potential* — and setting

$$\lambda = \frac{w^2}{m^2} - l^2$$

as the eigenvalue, we see that the equation represents a Schrödinger equation with a potential which vanishes exponentially at $\pm\infty$.[*] For such a potential the spectrum can be both discrete and continuous. The eigenfunctions of the discrete case vanish at $\pm\infty$ and are square-integrable. Those of the other case are complex and periodic at infinity. We can investigate the spectrum as follows. Consider the equation

$$\left[\frac{d^2}{dz^2} + \lambda + \frac{l(l + 1)}{\cosh^2 z}\right]\psi(z) = 0. \tag{22.55}$$

The differential operator is even in z. We can therefore construct solutions of definite parity, i.e. even and odd solutions, ψ_+, ψ_- respectively, which satisfy the boundary conditions

$$\psi_-(0) = 0, \quad \frac{d}{dz}\psi_+(0) = 0. \tag{22.56}$$

---

[*]We shall see later that the Pöschl–Teller equation is a limiting case of the Lamé equation, and as a consequence a limiting form of the small fluctuation equation of all basic potentials.

We now set (with ψ either ψ_+ or ψ_-)

$$\psi(z) = (\cosh z)^{-l}\chi(z) = (1 + \sinh^2 z)^{-l/2}\chi(z), \tag{22.57}$$

and change to the variable

$$\xi = \sinh^2 z. \tag{22.58}$$

Then the equation for $\chi(z)$ is

$$-\xi(\xi + 1)\frac{d^2}{d\xi^2}\chi + \left[(l-1)\xi - \frac{1}{2}\right]\frac{d}{d\xi}\chi - \frac{1}{4}(\lambda + l^2)\chi = 0. \tag{22.59}$$

Setting

$$\chi_+(\xi) = \sum_{n=0}^{\infty} a_n \xi^n, \qquad \chi_-(\xi) = \xi^{1/2}\sum_{n=0}^{\infty} b_n \xi^n, \tag{22.60}$$

we find the recursion formulas

$$
\begin{aligned}
a_{n+1} &= -\frac{n(n-l) + \frac{1}{4}(\lambda + l^2)}{(n + \frac{1}{2})(n+1)}a_n, \\
b_{n+1} &= -\frac{(n + \frac{1}{2})(n + \frac{1}{2} - l) + \frac{1}{4}(\lambda + l^2)}{(n+1)(n + \frac{3}{2})}b_n.
\end{aligned}
\tag{22.61}
$$

Discrete eigenvalues are obtained when the series (22.60) terminate after a finite number of terms, i.e. when in the even case

$$n(n-l) + \frac{1}{4}(\lambda + l^2) = 0,$$

i.e.

$$\lambda \to \lambda_n = -4n(n-l) - l^2 = -(2n - l)^2,$$

and in the odd case

$$\left(n + \frac{1}{2}\right)\left(n + \frac{1}{2} - l\right) + \frac{1}{4}(\lambda + l^2) = 0,$$

i.e.

$$\lambda \to \lambda_n = -(2n + 1 - l)^2.$$

Since for $\xi \to \infty$:

$$\psi_+ \sim (1 + \xi)^{-l/2}\xi^n, \qquad \psi_- \sim (1 + \xi)^{-l/2}\xi^{n+\frac{1}{2}},$$

the function ψ_+ is normalizable provided

$$\frac{1}{2}l > n, \quad \text{i.e.} \quad 2n < l,$$

and the function ψ_- is normalizable provided

$$2\left(n + \frac{1}{2}\right) < l.$$

We can summarize the results in the statement that the discrete eigenvalues are

$$\lambda = -(l - N)^2, \quad N = 0, 1, 2, \ldots, N < l, \tag{22.62}$$

where even (odd) N are associated with even (odd) eigenfunctions. We can see the continuous spectrum of Eq. (22.54) by setting

$$\lambda = k^2 \quad \text{and} \quad \psi(z) = e^{ikz} \mathcal{P}(\omega), \quad \omega = \tanh z. \tag{22.63}$$

Then

$$(1 - \omega^2)\frac{d^2}{d\omega^2}\mathcal{P} + (2ik - 2\omega)\frac{d}{d\omega}\mathcal{P} + l(l + 1)\mathcal{P} = 0. \tag{22.64}$$

This is the equation of *Jacobi polynomials*

$$\mathcal{P}_l^{(\alpha,-\alpha)}(\omega) \quad \text{with} \quad \alpha = -ik$$

of degree l in $\tanh z$.

We can now derive the spectra of the stability equation for the potentials (22.10) and (22.11).

(a) Φ^4-*theory:* Inserting (22.51) into (22.50) we obtain with $z = m(x - x_0)$ the equation

$$\left[\frac{d^2}{dz^2} + \lambda + \frac{l(l + 1)}{\cosh^2 z}\right]_{l=2, \lambda = \frac{w^2}{m^2} - l^2} \psi(z) = 0. \tag{22.65a}$$

From (22.62) we see that the equation possesses discrete eigenvalues

$$\frac{w^2}{m^2} - 2^2 = -(2 - N)^2, \quad N < 2,$$

i.e. for $N = 0, 1$:

$$w^2 = 0, \ 3m^2.$$

From (22.63) we see that the continuum starts at

$$\lambda = \frac{w^2}{m^2} - 2^2 = 0, \quad \text{i.e. } w^2 = 4m^2,$$

i.e. the continuum is $w^2 \geq 4m^2$.

(b) *Sine–Gordon theory:* Inserting (22.53) into (22.50) we obtain with $z = m(x - x_0)$ the equation

$$\left[\frac{d^2}{dz^2} + \lambda + \frac{l(l+1)}{\cosh^2 z}\right]_{l=1,\lambda=\frac{w^2}{m^2}-l^2} \psi(z) = 0. \qquad (22.65b)$$

The discrete eigenvalues are now given by

$$\frac{w^2}{m^2} - 1 = -(1 - N)^2, \quad N < 1,$$

i.e. for $N = 0$: $w^2 = 0$ and the continuum starts at $w^2 = m^2$.

Thus either spectrum contains the expected eigenvalue zero. It is now particularly interesting to look at their wave functions, the zero modes. In the Φ^4-theory we saw that (cf. Eq. (22.27))

$$\psi_0(x) \propto \frac{d}{dx}\phi_c(x) = \frac{m}{g}\frac{d}{dx}\tanh m(x - x_0) = \frac{m}{g}\frac{m}{\cosh^2 m(x - x_0)}. \qquad (22.66)$$

We see that this is a nonvanishing function for any finite value of x. Thus it has no node and so represents the eigenfunction of the lowest eigenvalue, and — as expected — this eigenfunction is even. This property can, of course, also be deduced from $d\phi_c/dx$. We know that ϕ_c is a monotonic function, so that its derivative is nowhere zero (except at $\pm\infty$). Thus $\phi_c'(x)$ has no node. The zero mode is depicted in Fig. 22.8 and we see it has the shape of a typical ground state wave function.

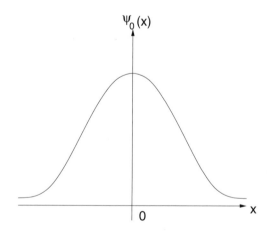

Fig. 22.8 The zero mode as typical ground state.

In the sine–Gordon theory the zero mode is proportional to

$$\psi_0(x) \propto \frac{d}{dx}\phi_c(x) \quad \propto \quad 4\frac{m}{g}\frac{d}{dx}\tan^{-1}\left[e^{\pm m(x-x_0)}\right]$$

$$= \pm 4\frac{m}{g}\frac{m}{1+e^{\pm 2m(x-x_0)}}e^{\pm m(x-x_0)}$$

$$\propto \quad \frac{1}{\cosh m(x-x_0)}.$$

Hence the conclusion is similar to that in the previous case.

Finally we consider the normalization of the zero modes. Imposing the condition

$$\int_{-\infty}^{\infty} dx [\psi_0(x)]^2 = 1, \tag{22.67}$$

we have in the Φ^4-theory:

$$\psi_0(x) = \frac{1}{M_0}\frac{d}{dx}\phi_c, \quad \phi_c = \pm\frac{m}{g}\tanh m(x-x_0), \tag{22.68}$$

and hence

$$\frac{m^3}{g^2}\frac{1}{M_0^2}\int\frac{d(mx)}{\cosh^4 m(x-x_0)} = 1,$$

i.e.

$$M_0^2 = \frac{m^3}{g^2}\left[\tanh x - \frac{1}{3}\tanh^3 x\right]_{-\infty}^{\infty} = \frac{4}{3}\frac{m^3}{g^2} = E(\phi_c). \tag{22.69}$$

The quantity M_0 here is called the *mass of the kink* (it is the mass of the soliton solution only in the classical approximation, i.e. without quantum corrections).

Similarly in the sine–Gordon theory we write

$$\psi_0(x) = \frac{1}{M_0}\frac{d}{dx}\phi_c(x),$$

so that

$$M_0^2 = \int_{-\infty}^{\infty} dx\left(\frac{d}{dx}\phi_c\right)^2 = 4\frac{m^4}{g^2}\int_{-\infty}^{\infty}\frac{dx}{\cosh^2 m(x-x_0)}$$

$$= 4\frac{m^3}{g^2}[\tanh m(x-x_0)]_{-\infty}^{\infty} = 8\frac{m^3}{g^2} = E(\phi_c). \tag{22.70}$$

22.7 Existence of Finite-Energy Solutions

So far we have been considering the case of one space and one time dimensions. We now wish to inquire whether static finite-energy configurations could exist in more than one spatial dimension. This question has important consequences, for if such configurations exist in a realistic theory with three spatial dimensions, it is important to understand their physical implications.[*] The general scaling argument used for this purpose was first introduced by Derrick[†] and is therefore referred to as *Derrick's theorem*. The arguments of Derrick were later rephrased by several authors whose line of reasoning we shall be using here.[‡] As a first example we consider the expression (22.41) for the energy of the classical configuration but now with respect to a d-dimensional position space, i.e. we consider

$$E(\phi) = \int d^d x \left[\frac{1}{2} (\boldsymbol{\nabla}\phi)^2 + V(\phi) \right], \tag{22.71}$$

which we also write as

$$E(\phi) = T(\phi) + P(\phi) \tag{22.72}$$

with

$$T(\phi) = \int d^d x \frac{1}{2} (\boldsymbol{\nabla}\phi)^2 \geq 0, \qquad P(\phi) = \int d^d x V(\phi) \geq 0.$$

We now consider the s*cale transformation* (with 'a' some number $\in \mathbb{R}$)[§]

$$\phi(x) \rightarrow \phi_a(x) = \phi(ax).$$

Under this transformation (the verification is given below)

$$E(\phi) \rightarrow E(\phi_a) = T(\phi_a) + P(\phi_a) = a^{2-d} T(\phi) + a^{-d} P(\phi), \tag{22.73}$$

since for instance

$$\begin{aligned}
T(\phi_a) &= \int d^d x \frac{1}{2} (\boldsymbol{\nabla}\phi_a)^2 = \int d^d x \frac{1}{2} (\boldsymbol{\nabla}\phi(ax))^2 \\
&= a^{2-d} \int d^d(ax) \frac{1}{2} \left[\frac{\partial}{\partial ax} \phi(ax) \right]^2 = a^{2-d} T(\phi).
\end{aligned}$$

---

[*]The considerations of this section extend in part beyond quantum mechanics and into field theory. They may help, however, to underline the fundamentally different nature of topological and nontopological finite energy configurations. This difference may not always be appreciated later in our very analogous treatment of (topological) instantons and (nontopological) bounces.

[†]G. H. Derrick [68].

[‡]See e.g. P. Goddard and D. I. Olive [113].

[§]For basic aspects of scale transformations see e.g. H. A. Kastrup [145]. One can have different types of scale transformations — see e.g. I. Affleck [5] who uses in his Eq. (2.4) $\phi(x) \rightarrow a\phi(ax)$.

We saw that the static configuration with $a = 1, d = 1$ minimizes the energy. Can the energy also be minimized for more than one spatial dimension? To see this we stationarize $E(\phi_a)$ with respect to a, i.e. we set

$$0 = \left[\frac{\partial E(\phi_a)}{\partial a}\right]_{a=1}. \tag{22.74}$$

From (22.73) we see that this implies

$$(2 - d)T(\phi) = dP(\phi). \tag{22.75}$$

Since both T and P are ≥ 0 and d is integral, we see that the stationarization of E with respect to a is possible only for $d = 1$ as in the examples we have been considering so far. In fact, since

$$\left[\frac{\partial^2 E(\phi_a)}{\partial a^2}\right]_{a=1} = 2(2 - d)T(\phi), \tag{22.76}$$

we see that $E(\phi)$ can be minimized (i.e. the second derivative is positive) only for $d = 1$. Thus this scaling argument excludes the existence of static finite-energy configurations of the given theory in more than one spatial dimension. It is also a virial argument in view of the *virial theorem* relation

$$T(\phi) = P(\phi).$$

Next we consider more complicated cases. Consider first the Lagrangian density

$$\mathcal{L}[\phi, \partial_\mu \phi] = \frac{1}{2}(\partial^\mu \phi)(\partial_\mu \phi) - V(\phi) \tag{22.77}$$

with $V(\phi) \geq 0$ and $\mathbf{x} \in \mathbb{R}^d$ and the set of minimum configurations

$$\mathcal{M}_0 := \{\phi | V(\phi) = 0\}. \tag{22.78}$$

For the energy of a static configuration, i.e. (22.71), to be finite the field $\phi \to \phi_\infty \in \mathcal{M}_0$ for $|\mathbf{x}| \to \infty$. It is convenient to discuss this limit in terms of the directions of unit vectors from the origin to the $(d - 1)$-dimensional surface in d-dimensional Euclidean space defined as the unit sphere

$$S^{d-1} = \{\mathbf{x} | \mathbf{x}^2 = 1\}, \quad \mathbf{x} = (x_1, x_2, \ldots, x_d). \tag{22.79}$$

In order to emphasize that $\phi \to \phi_\infty$ in *any* direction, we write

$$\phi_\infty\left(\frac{\mathbf{r}}{r}\right) = \lim_{r \to \infty} \phi\left(r\frac{\mathbf{r}}{r}\right) \in \mathcal{M}_0. \tag{22.80}$$

In the sine–Gordon theory $(d = 1)$ the equivalent statement is (22.38), i.e.

$$\pm 2\pi n \left(\frac{m}{g}\right) = \lim_{x \to \pm\infty} \phi_c(x) \tag{22.81}$$

with

$$\mathcal{M}_0 := \left\{ \pm 2\pi n \left(\frac{m}{g}\right) \right\},$$

and (since $d = 1$) $S = \pm 1$. Thus in this case S^{d-1} is a disconnected set consisting of two points. We see, therefore, that we have a set of *disconnected classical vacua* $(\phi_\infty, \phi_{-\infty}; \phi_\infty \ne \phi_{-\infty}$ in the case of the kink solutions) and associated with this a set of *disconnected points in space* $(\infty, -\infty$ or equivalently $S = +1, -1)$, and, in fact, one set can be mapped into the other.

Table 22.1

(A) Topologically nontrivial mapping		
manifold $S^{d-1}\|_{d=1}$	$\leftrightarrow$	manifold $\mathcal{M}_0$
$x = +\infty : S = +1$ $\quad\longleftrightarrow$	$\phi_\infty = 2\pi n \frac{m}{g}$	kink limits
$x = -\infty : S = -1$ $\quad\longleftrightarrow$	$\phi_{-\infty} = 2\pi n' \frac{m}{g}$	$(n - n' \ne 0)$
(B) Topologically trivial mapping		
$x = +\infty : S = +1$ $\quad\longleftrightarrow$	$\phi_\infty = 2\pi n \frac{m}{g}$	const.
$x = -\infty : S = -1$ $\quad\longleftrightarrow$	$\phi_{-\infty} = 2\pi n \frac{m}{g}$	$(n - n = 0)$

The *topologically trivial case* is given when the classical vacua associated with $S = +1, -1$, i.e. $x = \infty, -\infty$, are the same, i.e. $\phi_\infty = \phi_{-\infty}$, as is the case for the constant solutions. Thus in order to obtain a topologically nontrivial case we must have a different classical vacuum in association with each point at infinity. These observations are summarized in Table 22.1.

If we go to two spatial dimensions $(d = 2)$, the easiest way to visualize points at infinity is to take a circle S^1 of radius r and let $r \to \infty$. S^1, of course, is a connected region. Then each point at infinity is characterized by a different value of the polar angle θ. A correspondingly topologically nontrivial field theory would therefore have to possess a $U(1)$ symmetry as in the complex Higgs model (recall that in our considerations of Φ^4- and sine–Gordon theories we assumed Φ to be real; the Lagrangian of these theories therefore does not possess the $U(1)$ symmetry of the complex Higgs model but instead invariance under replacements $\Phi \to -\Phi$ and $\Phi \to \Phi + 2\pi$ respectively).

Suppose, for instance,

$$V(\phi(x)) = \frac{1}{4}g^2[\phi^*(x)\phi(x) - a^2]^2$$

with $(\phi(\infty) = \lim_{r\to\infty} \phi(r,\theta))$

$$\mathcal{M}_0 = \{\phi(\infty)|\phi^*(\infty)\phi(\infty) = a^2\}.$$

Thus, given any

$$\phi(\infty) = |\phi(\infty)|e^{i\theta},$$

say with $\theta \to \theta_0$, we can reach any other $\phi(\infty)$ by simple phase shifts, i.e. phase transformations $\theta_0 \to \theta_0 + \delta\theta$. If position space were one-dimensional, we would have only two points at infinity, i.e. $\infty, -\infty$ or $S = 1, -1$. Then the entire set of configurations $\phi_\infty = |\phi_\infty|\exp(i\theta)$ would correspond to, i.e. map into, the one point $S = +1$. This will be "nontrivial" in the above sense only if ϕ_∞ is independent of θ, i.e. if ϕ is real. On the other hand, if ϕ were real and position space two-dimensional, $\phi(r,\theta)$ would depend only on $r = |\mathbf{x}|$ and hence the problem would be effectively one-dimensional. We see, therefore, that if the problem is to be topologically nontrivial, the field must map position space into field space at infinity, i.e.

$$\Phi(\mathbf{r}): \quad S^1 \to \mathcal{M}_0.$$

Similarly, if we consider a theory in three spatial dimensions and so the unit sphere

$$S^2 = \{\mathbf{x}|\, x_1^2 + x_2^2 + x_3^2 = 1\},$$

the manifold of classical vacua would have to possess a similar geometry if the theory is to be topologically nontrivial, i.e. to have topologically distinct classical configurations. This can be achieved e.g. by allowing ϕ to have three components in some "internal" (isospin) space. Thus, assuming invariance of $\mathcal{L}$ under the transformations of the group $SO(3)$ in internal space, we have

$$\phi: \{\phi_i(x)\}, \quad i = 1, 2, 3 \quad (\phi \text{ real})$$

and

$$\mathcal{L} = \frac{1}{2}\sum_{i=1}^3 (\partial_\mu\phi_i)(\partial^\mu\phi_i) - V(\phi), \quad V(\phi) = \frac{1}{4}g^2\left[\sum_{i=1}^3 \phi_i^2 - a^2\right]^2 \tag{22.82}$$

Now

$$\mathcal{M}_0 := \left\{\phi(\infty)\Big|\sum_{i=1}^3 \phi_i^2 = a^2\right\}. \tag{22.83}$$

The manifold $\mathcal{M}_0$ is a sphere in a 3-dimensional Euclidean space, i.e. it is effectively a surface (and so a continuous set) S^2. Any $\phi \in \mathcal{M}_0$ can be reached from any other $\phi \in \mathcal{M}_0$ by application of an element of $SO(3)$, i.e. the appropriate rotation. In order to preserve

$$\lim_{r\to\infty} \sum_{i=1}^{3} \phi_i^2 = a^2,$$

such rotations must tend to the identity at $r = \infty$.

Summarizing we see that $\phi(\mathbf{x})$ is a mapping from S_x^{d-1} to S_ϕ^{d-1}. If, going once around S_x^{d-1} (i.e. through each point at infinity), we cover S_ϕ^{d-1} n times, the *winding number* is n. The symmetry of the potential (and Lagrangian) — i.e. invariance under replacements $\Phi \to -\Phi$ in Φ^4-theory, $\Phi \to \Phi + 2\pi$ in sine–Gordon theory, $\Phi \to \Phi \exp(i\alpha)$ in a $U(1)$ symmetric theory, $\Phi \to \Phi \exp\{i \sum_{a=1}^{3} \beta_a T_a\}$ in an $SO(3)$ invariant theory and so on — determines the transformation from one point at infinity to another (on a line, circle, sphere, ... depending on d) and correspondingly (with n complete windings) the transformation of one classical vacuum into another. In the following Example the use of Derrick's theorem is demonstrated by application to some gauged theories, a theory being described as 'gauged' if a gauge field like the electromagnetic field is involved.

Example 22.2: Derrick's theorem applied to gauged theories
Wellknown and important theories involving a gauge field (A_μ if abelian and $A_\mu^a, a = 1, 2, 3$, if $SO(3)$ nonabelian) are two models defined by the following Lagrangians which we cite here solely for the purpose of being explicit; the subsequent discussion requires only the general form of the Lagrangians, so that details of these models are irrelevant. The theories are:
(a) The Nielsen–Olesen (vortex) model (also called Abelian Higgs model)[¶] with Lagrangian density

$$\mathcal{L}[A_\mu, \phi] = -\frac{1}{4} F^{\mu\nu} F_{\mu\nu} + \frac{1}{2} |(\partial_\mu - igA_\mu)\phi|^2 - V(\phi),$$

$$V(\phi) = \frac{1}{2}\lambda g^2 \left[|\phi|^2 - \frac{\mu^2}{2\lambda g^2} \right]^2, \quad F_{\mu\nu} = \partial_\mu A_\nu - \partial_\nu A_\mu,$$

where ϕ is a complex scalar field and the spatial dimensions are 2 or 3.
(b) The Georgi–Glashow model[‖] with nonabelian gauge field in 3 spatial dimensions with Lagrangian density (V as under (a))

$$\mathcal{L}[A_\mu, \phi] = -\frac{1}{4} G_a^{\mu\nu} G_{a\,\mu\nu} + \frac{1}{2} (\mathcal{D}^\mu \phi)_a (\mathcal{D}_\mu \phi)_a - V[\phi_a],$$

where

$$G_a^{\mu\nu} = \partial^\mu A_a^\nu - \partial^\nu A_a^\mu - e\epsilon_{abc} A_b^\mu A_c^\nu \quad \text{and} \quad (\mathcal{D}^\mu \phi)_a = \partial^\mu \phi_a - e\epsilon_{abc} A_b^\mu \phi_c.$$

Write the energy $E(\phi)$ of such models (the functionals F and V containing no derivatives)

$$E(\phi, A) = T(\phi, A) + T(A) + P(\phi),$$

[¶]H. B. Nielsen and P. Olesen [220].
[‖]H. Georgi and S. L. Glashow [108].

where

$$
T(\phi, A) = \int d^d x F(\phi)(\mathcal{D}_\mu \phi)_a^\dagger (\mathcal{D}^\mu \phi)_a \geq 0, \quad T(A) = \frac{1}{4} \int d^d x G_a^{ij} G_{a\,ij} \geq 0,
$$

$$
P(\phi) = \int d^d x V(\phi) \geq 0, \tag{22.84}
$$

and investigate the existence of finite energy classical field configurations.

Solution: Each contribution in (22.84) is positive semi-definite and so must be finite by itself. We now consider scale transformations with $\lambda \in \mathbb{R}$, i.e.

$$
\phi(\mathbf{x}) \to \phi_\lambda(\mathbf{x}) = \phi(\lambda \mathbf{x}), \quad A_\mu \to A_{\mu\,\lambda}(\mathbf{x}) = \lambda A_\mu(\lambda \mathbf{x}). \tag{22.85}
$$

Here the difference in the transformations results from their difference as scalar and vector quantities which are defined by the transformations

$$
\phi'(x') = \phi(x'), \quad A'_\mu(x') = \sum_\nu \left(\frac{\partial}{\partial x_\nu} x'_\mu \right) A_\nu(x') \tag{22.86}
$$

with a scale transformation

$$
x \to x' = \lambda x.
$$

Since $\mathcal{D}_\mu \phi(x)$ transforms like a vector, we have

$$
\mathcal{D}_\mu \phi(x) \to \lambda \mathcal{D}_\mu \phi(\lambda x), \tag{22.87a}
$$

and since $G_{\mu\nu}$ transforms as a second rank tensor we must have

$$
G_{\mu\nu}(x) \to \lambda^2 G_{\mu\nu}(\lambda x). \tag{22.87b}
$$

We consider explicitly the example of $\phi \to \phi_\lambda(x) = \phi(\lambda x)$ for

$$
T^{(1)}(\phi) := \int d^d x (\partial_\mu \phi)^2.
$$

We have

$$
T^{(1)}(\phi_\lambda) = \int d^d x (\partial_\mu \phi_\lambda)^2 = \int d^d x \left[\frac{\partial \phi(\lambda x)}{\partial x_\mu} \right]^2 = \lambda^2 \int \frac{d^d(\lambda x)}{\lambda^d} \left[\frac{\partial \phi(\lambda x)}{\partial(\lambda x_\mu)} \right]^2
$$

$$
= \lambda^{2-d} \int d^d x' \left[\frac{\partial \phi(x')}{\partial x'_\mu} \right]^2 = \lambda^{2-d} T^{(1)}(\phi).
$$

Considering the contributions of (22.84) in this way one finds that the quantities involved transform as

$$
T(\phi_\lambda, A_\lambda) = \lambda^{2-d} T(\phi, A), \quad T(A_\lambda) = \lambda^{4-d} T(A), \quad P(\phi_\lambda) = \lambda^{-d} P(\phi)
$$

and so

$$
E(\phi_\lambda, A_\lambda) = \lambda^{2-d} T(\phi, A) + \lambda^{4-d} T(A) + \lambda^{-d} P(\phi). \tag{22.88}
$$

If E is to allow a static finite-energy solution, it must be stationary with respect to arbitrary field variations and so with respect to the above scale transformations also. Thus we set

$$
\left. \frac{\partial E(\phi_\lambda, A_\lambda)}{\partial \lambda} \right|_{\lambda=1} = 0, \quad \text{i.e.} \quad (2-d)T(\phi, A) + (4-d)T(A) - dP(\phi) = 0. \tag{22.89}
$$

Since $T(\phi, A), T(A), P(\phi)$ are positive semi-definite and not all zero, some contribution must be negative.

(a) Thus for $d = 1$: $P(\phi)$ must be present but $T[A]$ and so the vector field would not be required to satisfy Eq. (22.89). This case is, of course, exemplified by soliton and sine–Gordon theories.
(b) For $d = 2$: A theory with all terms would be a candidate — in fact the Nielsen–Olesen theory is an example.
(c) For $d = 3$: All three terms must be present as in the Georgi–Glashow model.
(d) For $d = 4$: This is possible only if the sign of V is reversed — this is exemplified by a so-called pure gauge theory with so-called instanton solution (which we shall encounter in a simpler context later).
(e) For $d \geq 5$: In this case it is not possible to stabilize Eq. (22.89). There are other posssibilities if we permit higher powers of derivatives as in Skyrmion models,[**] e.g. if we allow a term $T^{(4)}(\phi)$ with fourth powers of derivatives,

$$T^{(4)}(\phi_\lambda) = \lambda^{4-d}T(\phi), \tag{22.90}$$

and we would have to add to Eq. (22.89) the contribution $(4 - d)T(\phi)$. Thus such a contribution could counterbalance $T(\phi, A)$ or $P(\phi)$ for $d = 3$.

22.8   Ginzburg–Landau Vortices

We have already referred to the Nielsen–Olesen theory. In the following we consider a related theory. We first demonstrate — as a matter of interest — that the Nielsen–Olesen theory for static equilibrium (i.e. time-independent) field configurations is equivalent to the Ginzburg–Landau theory of super-conductivity.[††] The — now established — microscopic BCS theory of su-perconductivity was formulated some eight or so years after the macroscopic Ginzburg–Landau theory. We can therefore use our understanding of the BCS theory in interpreting the Ginzburg–Landau theory. From this devel-opment one knows that in the superconducting state of the metal the elec-trons combine to form pairs called "*Cooper pairs*" which then have bosonic properties. One can thus define a scalar field or wave function $\Psi(x, t)$ such that $|\Psi|^2$ describes the density of Cooper pairs. Equilibrium states of the superconductor are assumed to be described by the time-independent static wave functions $\psi(\mathbf{x})$. The Ginzburg–Landau theory assumes that the static energy density is then given by

$$\mathcal{E}(\psi) = \frac{1}{2}|\boldsymbol{\nabla}\psi(\mathbf{x})|^2 + \gamma + \frac{1}{2}\alpha|\psi(\mathbf{x})|^2 + \frac{1}{4}\beta|\psi(\mathbf{x})|^4. \tag{22.91}$$

The expression $|\psi(\mathbf{x})|$ is known as the "*order parameter*". The parameter

$$\alpha \equiv \alpha(T) = a\frac{T - T_c}{T_c}, \quad a > 0, \tag{22.92}$$

is temperature dependent and changes sign at the critical temperature T_c. We then have a situation of the potential like that described in the context

---

[**]T. H. R. Skyrme [253]. An explicit and solvable Skyrmion model in $2 + 1$ dimensions is considered in D. H. Tchrakian and H. J. W. Müller–Kirsten [268].
[††]V. L. Ginzburg and L. D. Landau [111].

of the complex Higgs model (cf. Sec. 22.2): For $T > T_c$ the overall potential has a single well (minimum), but for $T < T_c$ it has a double well shape. The parameter γ serves to put the minima of the *double well* at $V = 0$. Variation as before, i.e. setting $\delta\mathcal{E}(\psi) = 0$, leads in a *one-dimensional case* (considering this briefly) to the equation

$$\frac{d^2\psi}{dx^2} = \alpha\psi + \beta\psi^3. \tag{22.93}$$

Imposing the boundary conditions (in the domain $0 \leq x \leq \infty$)

$$\psi(0) = 0, \quad \psi(\infty) = \sqrt{\frac{-\alpha}{\beta}},$$

the solution is (in the domain $0 \leq x \leq \infty$)

$$\psi(x) = \sqrt{\frac{-\alpha}{\beta}}\tanh\left(\frac{\sqrt{-\alpha}}{2}x\right), \tag{22.94}$$

which we recognize as one half of the kink solution we obtained previously as the topological instanton solution of the Schrödinger equation with double well potential (in the present case $\psi(x) = 0$ for $x < 0$, assuming there is no superconducting material in this domain).

Returning to the three dimensional case and switching on a magnetic field (applied from outside of the superconducting material, i.e. from $x < 0$) means — as is familiar from electrodynamics — replacing

$$\boldsymbol{\nabla} \text{ by } \boldsymbol{\nabla} - i\frac{g}{\hbar}\mathbf{A}$$

(this is called *"minimal coupling"* of the electromagnetic field[‡‡] with vector potential $\mathbf{A}$ to the Cooper pair field ψ in a nonrelativistic treatment) where $g = 2e^-$ is the charge of the Cooper pair. Then the static energy density becomes (cf. Example 22.2)

$$\mathcal{E}(\psi, \mathbf{A}) = \frac{1}{2}|\mathbf{B}|^2 + \frac{1}{2}\left|\boldsymbol{\nabla} - i\frac{g}{\hbar}\mathbf{A}\right|^2 + \gamma + \frac{1}{2}\alpha|\psi|^2 + \frac{1}{4}\beta|\psi|^4. \tag{22.95}$$

This expression can be shown to be equivalent to the integrand in the Nielsen–Olesen theory. Clearly the length (see Eq. (22.94) to check the dimension)

$$x_0 = \frac{1}{\sqrt{-\alpha}}$$

is a measure of the distance from the surface of the superconductor to where

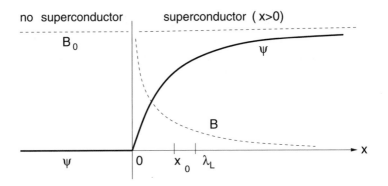

Fig. 22.9 Behaviour of ψ and $\mathbf{B}$ in the superconductor.

the order parameter $|\psi|$ attains (approximately) its asymptotic value, i.e. $\sqrt{-\alpha/\beta}$. The magnetic field on the other hand, as indicated in Fig. 22.9, exhibits a completely different behaviour — in fact, inside the superconductor it falls off exponentially away from the surface (this is the *Meissner–Ochsenfeld effect* generally described as the expulsion of lines of force from the superconductor below the critical temperature T_c). We can see this as follows. The density of the Cooper or super current is given by the typical expression of a current, i.e.

$$\mathbf{j}_s = \frac{\hbar}{2mi}[\psi^\dagger \mathbf{D}\psi - \psi \mathbf{D}^\dagger \psi^\dagger], \quad \mathbf{D} \equiv \mathbf{\nabla} - i\frac{g}{\hbar}\mathbf{A}. \tag{22.96}$$

Setting

$$\psi = \sqrt{|\rho|}e^{i\varphi}, \quad \rho = \psi^\dagger \psi,$$

we can rewrite $\mathbf{j}_s$ as

$$\mathbf{j}_s = \frac{\rho\hbar}{m}\left(\mathbf{\nabla}\varphi - \frac{g}{\hbar}\mathbf{A}\right). \tag{22.97}$$

In macroscopic electrodynamics the magnetic field strength $\mathbf{H}$ is (in SI units) defined by

$$\mathbf{H} := \frac{1}{\mu_0}\mathbf{B} - \mathbf{M}, \tag{22.98}$$

where $\mathbf{M}$ is the magnetization resulting from atomic currents, i.e. $\mathbf{\nabla} \times \mathbf{M} = \mathbf{j}_s$. In the Maxwell equation

$$\mathbf{\nabla} \times \mathbf{H} = \mathbf{j} + \frac{\partial \mathbf{D}}{\partial t},$$

---

‡‡See e.g. H. J. W. Müller–Kirsten [215], p. 450.

the quantity $\mathbf{j}$ is the density of a current applied from outside. This is zero in the present case. Also $\partial \mathbf{D}/\partial t = 0$ in a static situation like the one here. Hence

$$0 = \boldsymbol{\nabla} \times \mathbf{H} = \boldsymbol{\nabla} \times \left(\frac{1}{\mu_0} \mathbf{B} - \mathbf{M} \right),$$

i.e.

$$\boldsymbol{\nabla} \times \mathbf{B} = \mu_0 \boldsymbol{\nabla} \times \mathbf{M} = \mu_0 \mathbf{j}_s, \quad \text{and} \quad \boldsymbol{\nabla} \times (\boldsymbol{\nabla} \times \mathbf{B}) = \mu_0 \boldsymbol{\nabla} \times \mathbf{j}_s.$$

Using the relation "curl curl = grad div - div grad" and $\boldsymbol{\nabla} \cdot \mathbf{B} = 0$, this becomes

$$\triangle \mathbf{B} = -\mu_0 \boldsymbol{\nabla} \times \mathbf{j}_s. \tag{22.99}$$

Inserting (22.97) this becomes (since curl grad = 0)

$$\triangle \mathbf{B} = \mu_0 \frac{\rho g}{m} \boldsymbol{\nabla} \times \mathbf{A} = \mu_0 \frac{\rho g}{m} \mathbf{B},$$

i.e.

$$\triangle \mathbf{B} - \frac{1}{\lambda_L^2} \mathbf{B} = 0, \quad \frac{1}{\lambda_L^2} = \frac{\mu_0 \rho g}{m}.$$

Here λ_L is known as the (*London*) *penetration depth* (or length) (in practice this is of the order of 10^{-5} to 10^{-6} cm). In the one-dimensional case we have

$$B(x) = B(0) e^{-x/\lambda_L}. \tag{22.100}$$

Approximating ρ by its equilibrium value, i.e.

$$\rho = |\psi|^2 = -\frac{\alpha}{\beta},$$

we obtain

$$\lambda_L^2 = \frac{m}{\mu_0 g} \left(-\frac{\beta}{\alpha} \right), \quad \alpha < 0.$$

The ratio λ_L/x_0 is called the *Ginzburg–Landau parameter*.

There are two types of superconductors, those of the so-called type I and those of type II. Superconductors of type II are characterized by the fact that when the magnetic field is again increased and beyond a first critical value, the normal state of the metal begins to re-appear in thin vortices and not uniformly thoughout the metal as in the case of superconductors of type I. More and more vortices are formed as the magnetic field is increased further and further until only small superconducting domains remain which then disappear completely beyond a second critical value of the magnetic field. These vortices therefore carry magnetic flux which is necessarily quantized (see below). The existence of these vortices with quantized magnetic flux was predicted by Abrikosov[*] and are therefore frequently referred to as *Abrikosov vortices* (the idea was originally rejected by Landau). The existence of these vortices has been confirmed experimentally. In the Nielsen–Olesen model defined in Example 22.2 — identified as above with the Ginzburg–Landau theory — type II superconductivity occurs[†] for $\lambda > 1/4$ and type I superconductivity with the complete Meissner–Ochsenfeld effect for $\lambda < 1/4$.

22.9 Introduction to Homotopy Classes

In this section we consider topological properties of classical finite energy configurations and introduce the concept of *homotopy classes*.

In all the models we considered so far we observed that the energy E contains a contribution

$$P(V) = \int d^d x \, V(|\phi|).$$

If the energy E is to be finite — a basic requirement for the field configurations of interest to us — this integral must also be finite, i.e. we must have that

$$\lim_{|\mathbf{x}| \to \infty} V(|\phi|) = 0.$$

Thus at spatial infinity $|\phi|$ has the value of a zero of the potential. We can think of (e.g) two-dimensional space as being bounded by a circle at $r = \infty$. Finiteness of E requires $|\phi| = |\phi_\infty|$ on this circle. In the two models of Example 22.2 we saw that finiteness of $P(V)$ puts no restriction on the phase of ϕ so that

$$\phi(r, \theta) \stackrel{r \to \infty}{\Longrightarrow} e^{i\chi(\theta)} |\phi_\infty|, \tag{22.101}$$

where $\exp(i\chi(\theta))$ is an arbitrary phase factor, in fact a periodic function of the polar angle with period 2π. Thus the asymptotic field $\phi(r, \theta)|_{r \to \infty}$ represents a mapping from the circle at spatial infinity to the circle defined by $\chi(\theta)$ (or, in other words, by the element $\exp[i\chi(\theta)]$ of the internal group $U(1)$):

$$\lambda(x) = e^{i\chi(\theta)} : \; S_x^1 \to S_{\phi_\infty}^1. \tag{22.102}$$

In the one-dimensional soliton case of Eqs. (22.36), (22.37), the space consists of two diametrically opposite points on the circle, i.e. of two points on a line.

---

[*]A. A. Abrikosov [2].

[†]See the remark in H. J. de Vega and F. A. Schaposnik [69] after their equation (3.9). See also P. G. De Gennes [67].

By a *map* or *mapping* $f : X \to Y$ of a space X into a space Y we mean a single-valued continuous function from X to Y; X is called the *domain* of f and Y its *range*. The 1-sphere is the unit circle in the space of all complex numbers, i.e.

$$S^1 := \{ z \in \mathbb{C} \,|\, |z| = 1 \}.$$

In mathematical language S^1 is therefore a compact, abelian (i.e. commutative), topological group with the usual multiplication as group operation. The map $\lambda : S^1 \to S^1$ defined on $S^1 \subset \mathbb{R}^2$ by (22.102) is called the *exponential map* of S^1 onto S^1. The map satisfying $\lambda(x + y) = \lambda(x)\lambda(y)$ is said to be a *homomorphism*.

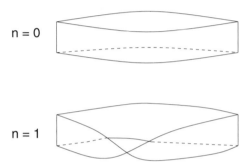

n = 0

n = 1

Fig. 22.10 States of winding number $0, 1$.

The two theories we considered above in Example 22.2 imply that such a map from a circle to a circle is associated with an integer which we called the *winding number n*. We may define this as

$$n = \frac{1}{2\pi}[\chi(\theta = 2\pi) - \chi(\theta = 0)]. \tag{22.103}$$

Later we demonstrate that *this winding number n remains unchanged under continuous deformations* (those of Eq. (22.118)). We can illustrate the simple low-dimensional example under discussion by reference to a strip with Möbius structure.[*] We can think of this as a fixed line in space which is provided at each point with an infinitesimal direction element that is free to rotate in the plane perpendicular to the line (like the set of pendulums considered in Example 22.1). Suppose $\phi, 0 \leq \phi \leq 2\pi$, defines the orientation of this line element. Then a continuous function of ϕ on $[0, 2\pi]$ remains continuous if the function assumes the same value at the two endpoints. States of the strip corresponding to winding numbers 0 and 1 are illustrated in Fig. 22.10. Naturally an integer cannot change continuously. Thus smooth deformations

---

[*]See D. Finkelstein and C. W. Misner [96].

of the fields which preserve the finiteness of the energy must preserve the winding number n. For that reason n is called a *"topological invariant"*. The continuous deformation or distortion of a set of points does not alter the topological structure of the set (e.g. if it has one twist as above or a hole, it retains this twist or hole under the continuous distortion) and is described as a *homeomorphism*. Thus two homeomorphic spaces have equivalent topological structures. Homeomorphism subdivides the possible spaces into disjoint equivalence classes (to be explained below); elements in the same class have the same topological characteristic, i.e. *connectivity*. The different kinds of connectivity of these topological spaces (e.g. no hole, one hole,...) remain the same under homeomorphic transformations and are therefore called *topological invariants*. The winding numbers n are such topological invariants. A field configuration with $n \neq 0$ therefore cannot be deformed continuously into one with winding number zero, i.e. it cannot be deformed continuously into the so-called "classical vacuum" (the latter is a constant configuration in our soliton example). We observe that since time evolution is continuous, the winding number must be a constant of the motion.

In our consideration of the Nielsen–Olesen theory we also required the gauge field A_μ to behave such that the energy is finite, i.e. we required that

$$\lim_{r\to\infty} \left| \left(\mathbf{\nabla} - \frac{ig}{\hbar}\mathbf{A} \right)\phi \right|^2 \to 0,$$

so that

$$\int d^2x \, \frac{1}{2} \left| \left(\mathbf{\nabla} - \frac{ig}{\hbar}\mathbf{A} \right)\phi \right|^2 = \text{finite}. \qquad (22.104)$$

Since $\mathbf{A}$ has direction $\mathbf{e}_\theta$, we can write

$$\left(\mathbf{\nabla} - \frac{ig}{\hbar}\mathbf{A} \right)_\theta \phi = \mathbf{e}_\theta \left(\frac{1}{r}\frac{\partial}{\partial\theta} - \frac{ig}{\hbar}A_\theta \right)\phi \xrightarrow{r\to\infty} \mathbf{e}_\theta i \left(\frac{1}{r}\frac{\partial\chi(\theta)}{\partial\theta} - \frac{g}{\hbar}A_\theta \right)\phi.$$

Thus for (22.104) to be finite we must demand that

$$\frac{g}{\hbar}A_\theta \xrightarrow{r\to\infty} \frac{1}{r}\frac{\partial\chi(\theta)}{\partial\theta}, \qquad (22.105)$$

the corrections falling off faster than $1/r$, since otherwise the energy integral would behave like

$$\int \frac{d^2x}{r^2} \propto \int \frac{dr}{r} \sim \ln r.$$

The gauge field required in the asymptotic limit (22.105) is described as *"pure gauge"* since it is determined entirely by the phase $\chi(\theta)$. In other words, it is determined by a gradient $(1/r)\partial/\partial\theta$, and we know that the gauge transformation of a static abelian field is

$$\mathbf{A} \to \mathbf{A}' = \mathbf{A} + \boldsymbol{\nabla}\chi.$$

The field which is determined entirely by $\boldsymbol{\nabla}\chi$ is therefore termed "pure gauge". Therefore, if $\mathbf{A} = \boldsymbol{\nabla}\chi$, we have

$$\mathbf{B} = \boldsymbol{\nabla} \times \mathbf{A} = \operatorname{curl}\operatorname{grad}\chi = 0.$$

It follows that the field F_{ij}, i.e. $F\mu\nu$, is also zero at $r = \infty$. Of course, for $\mathbf{B} \neq 0$, the gauge field, i.e. the vector potential $\mathbf{A}$, cannot be pure gauge everywhere, i.e. the higher order contributions in (22.105) insure that $\mathbf{B} \neq 0$. We can calculate the magnetic flux Φ through the region with

$$\Phi = \oint r d\theta \, A_\theta = \frac{\hbar}{g} \oint r d\theta \frac{1}{r} \frac{\partial\chi(\theta)}{\partial\theta} = \frac{2\pi\hbar}{g} n. \qquad (22.106)$$

Thus *the magnetic flux is quantized*, the number of flux quanta $2\pi\hbar/g$ being the winding number n. In terms of the phase factor

$$\lambda(x) = e^{i\chi(\theta)},$$

we can rewrite the relation (22.105) as

$$\frac{g}{\hbar}\mathbf{A}(x) \to -i\frac{1}{\lambda(x)}\boldsymbol{\nabla}\lambda(x) \quad (= \boldsymbol{\nabla}\chi(\theta)). \qquad (22.107)$$

The fields $\phi(r,\theta)|_\infty$ and hence the functions

$$\lambda(x) = e^{i\chi(\theta)}, \qquad (22.108)$$

map the spatial circle S^1_x into $S^1_{\phi\infty}$, the *group space* of (the phase transformations of) $U(1)$ continuously. However, going once around S^1_x we can go n times around $S^1_{\phi\infty}$. A field configuration with one loop (or twist), with $n = 1$, cannot be deformed continuously into one with two loops ($n = 2$). However, we can consider transformations which describe continuous transformations of a configuration with fixed n. Suppose two such configurations are given by

$$\lambda_n^{(0)}(\theta) = e^{i\phi_n^{(0)}(\theta)}, \quad \lambda_n^{(1)}(\theta) = e^{i\phi_n^{(1)}(\theta)}, \quad n \text{ fixed}. \qquad (22.109)$$

Then we can construct a one-parameter family of maps or transformations which transform $\phi_n^{(0)}(\theta)$ continuously into $\phi_n^{(1)}(\theta)$. An example is the mapping

$$H(t,\theta) := t\phi_n^{(1)}(\theta) + (1 - t)\phi_n^{(0)}(\theta) \quad \text{with} \quad t \in [0,1], \qquad (22.110)$$

which is continuous in the parameter t and is such that

$$H(0,\theta) = \phi^{(0)}(\theta), \qquad H(1,\theta) = \phi_n^{(1)}(\theta). \tag{22.111}$$

In principle an expression like (22.110) could be written down for configurations with different winding numbers, e.g. $n, n' \neq n$. But this type of combination is excluded sind $H(t,\theta)$ itself must be a configuration with definite winding number. A superposition of classical configurations with different winding numbers does not minimize the energy. Thus by varying t from 0 to 1 we deform $\phi_n^{(0)}(\theta)$ continuously into $\phi_n^{(1)}(\theta)$. The map $H(t,\theta)$ is called a *homotopy* (emphasis on the second syllable).[†] The homotopy relates a representative function or curve $\phi_n^{(0)}(\theta)$ to another function or curve $\phi_n^{(1)}(\theta)$ by giving a precise meaning to the idea that the one can be deformed continuously into the other. Two curves which can be related to each other in this way, like $\phi_n^{(0)}(\theta), \phi_n^{(1)}(\theta)$ in the above example, are said to be *homotopic* (emphasis on the third syllable) to each other and one writes

$$\phi_n^{(0)}(\theta) \sim \phi_n^{(1)}(\theta). \tag{22.112}$$

In fact, these two maps ϕ_n from $S_x^1 \to S_{\phi\infty}^1$ are homotopic iff there exists a homotopy connecting $\phi_n^{(0)}(\theta)$ and $\phi_n^{(1)}(\theta)$. Since the passage of time is continuous, it is clear that this is also a homotopy, i.e. time cannot make a field jump from one homotopy class (see below) to another. We can consider (22.111) as a map from the space described by θ, i.e. S_x^1, and the space of t, i.e. the unit interval $I = [0,1]$, to the internal group space of $U(1)$, i.e. the phase factors of $\phi(\theta)$, i.e.

$$H : I \otimes S_x^1 \longrightarrow S_{\phi\infty}^1 \tag{22.113}$$

with

$$H(0,\theta) = \phi_n^{(0)}(\theta) \quad \text{and} \quad H(1,\theta) = \phi_n^{(1)}(\theta).$$

The relation (22.112) is an *equivalence relation* — and as such it is defined by the three properties of (a) identity, (b) symmetry and (c) transitivity. We verify these in the following example.

Example 22.3: Homotopy an equivalence relation
Verify that the homotopy map $H(t,\theta)$ is an equivalence relation.

Solution: As stated, we have to show that the map satisfies the three properties of (a) identity, (b) symmetry and (c) transitivity.

[†]The standard texts on homotopy theory are N. Steenrod [261] and S.–T. Hu [134]. Still on the level of a mathematical text is (in spite of its title) the Argonne National Laboratory Report of G. H. Thomas [270]. A brief introduction can be found in the very readable book of W. Dittrich and M. Reuter [77].

(a) Evidently the homotopy $H(t,\theta)$ connects $\phi_n^{(0)}(\theta)$ with itself continuously.

(b) Next we check that if $\phi^{(0)} \sim \phi^{(1)}$, then $\phi^{(1)} \sim \phi^{(0)}$. We define an $\tilde{H}(t,\theta) := H(1-t,\theta)$. Then $H(t,\theta)$, being the continuous map connecting $\phi^{(0)}$ with $\phi^{(1)}$, is such that $H(0,\theta) = \phi^{(0)}(\theta)$ and $H(1,\theta) = \phi^{(1)}(\theta)$. On the other hand $\tilde{H}(0,\theta) = H(1,\theta) = \phi^{(1)}(\theta)$ and $\tilde{H}(1,\theta) = H(0,\theta) = \phi^{(0)}(\theta)$, so that $\tilde{H}$ is a homotopy connecting $\phi^{(1)}(\theta)$ with $\phi^{(0)}(\theta)$.

(c) We can demonstrate the property of transitivity as follows. Assume $\phi^{(0)} \sim \phi^{(1)}$ and $\phi^{(1)} \sim \phi^{(2)}$ for continuous maps $\phi^{(0)}, \phi^{(1)}, \phi^{(2)}$. Then there exist homotopies H_1 and H_2 connecting $\phi^{(0)}$ with $\phi^{(1)}$ and $\phi^{(1)}$ with $\phi^{(2)}$ respectively. We now define the function

$$H(t,\theta) = \begin{cases} H_1(2t,\theta) & \text{for} \quad 0 \leq t \leq \frac{1}{2}, \\ H_2(2t-1,\theta) & \text{for} \quad \frac{1}{2} \leq t \leq 1. \end{cases} \tag{22.114}$$

At

$$t = \frac{1}{2}: \quad H\left(\frac{1}{2},\theta\right) = H_1(1,\theta) = H_2(0,\theta).$$

But

$$H_1(0,\theta) = \phi^{(0)}, \quad \left.\begin{array}{l} H_1(1,\theta) = \phi^{(1)} \\ H_2(0,\theta) = \phi^{(1)} \end{array}\right\} H\left(\frac{1}{2},\theta\right), \quad H_2(1,\theta) = \phi^{(2)},$$

i.e. at $t = 1/2$ both H_1 and H_2 give $\phi^{(1)}$, so that $H(t,\theta)$ is continuous there. Moreover, since

$$H(0,\theta) = H_1(0,\theta) = \phi^{(0)} \quad \text{and} \quad H(1,\theta) = H_2(1,\theta) = \phi^{(2)},$$

H is a homotopy connecting $\phi^{(0)}$ with $\phi^{(2)}$. Hence $\phi^{(0)} \sim \phi^{(2)}$, which therefore demonstrates that the homotopy relationship is transitive.

Returning to the maps

$$\lambda(x): \quad S_x^1 \longrightarrow S_{\phi\infty}^1 \quad (\text{i.e. } U(1)),$$

we see that of the set of these maps some are equivalent and others are not. Hence the set of all such maps breaks up into *equivalence classes* or *homotopy classes*, as they are called, a class being a set of equivalent maps.[‡]

22.10 The Fundamental Homotopy Group

The set of disjoint equivalence classes in the case considered previously is denoted by the symbol

$$\pi_1[U(1)].$$

Here the subscript indicates the dimensionality of the one-dimensional sphere S^1. Two elements of $\pi_1[U(1)]$ which belong to different classes are homotopically inequivalent. In some cases the set of homotopy classes possesses group structure (satisfying the group properties); these groups are then called *homotopy groups* (cf. below). In the above case of $\pi_1[U(1)]$ we know from our earlier considerations that every equivalence class is characterized by a specific value of the winding number. Thus $\pi_1[U(1)] \equiv \pi_1[S^1]$ corresponds to (i.e. is isomorphic to) the set of integers $\mathbb{Z}$ (positive, negative or zero).

[‡]Homotopy classes are sometimes also called Chern–Pontryagin classes.

Thus there is a denumerable infinity of such homotopic classes or sectors. In order to really appreciate how topology comes into the picture here one should understand (e.g.) why $\pi_1[S^2] = 0$. How can we see this? We first sketch the idea, to make the result plausible, and then introduce some of the related mathematics.

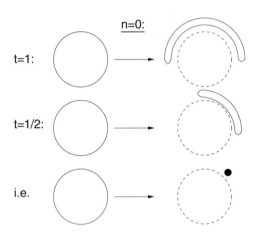

Fig. 22.11 The trivial $n = 0$ map allowing shrinkage to a point.

Sketch of idea demonstrating that $\pi_1[S^1] = \mathbb{Z}$, but $\pi_1[S^2] = 0$: S^1 is a circle, S^2 the surface of a sphere in a three-dimensional Euclidean space. In the diagrams[§] of Fig. 22.11 the circle drawn with the continuous line is the circle S_x^1, which is to be mapped into another circle $S_{\phi\infty}^1$ indicated by dashed lines. Appropriate maps, characterized by winding numbers n are for instance ($\chi_n(\theta)$ being a continuous function modulo 2π):

(a) For $n = 0$:

$$\chi_0(\theta) = 0 \qquad \text{for all } \theta.$$

This trivial map into a single point is described as a "degenerate map" and is illustrated in Fig. 22.11 for the special values of $t = 1$ and $t = 1/2$ with

$$\chi_0(\theta) = \begin{cases} t\theta & \text{for} \quad 0 \le \theta < \pi, \\ t(2\pi - \theta) & \text{for} \quad \pi \le \theta < 2\pi. \end{cases} \qquad (22.115)$$

Varying $t \in [0, 1]$ continuously to 0 one regains $\chi_0(\theta) = 0$.

(b) In the cases $n = 1$ and $n = 2$ we have

$$\chi_1(\theta) = \theta \ \text{ and } \ \chi_2(\theta) = 2\theta \quad \text{for all } \theta.$$

[§]Similar considerations are given in the wellknown book of R. Rajaraman [235], p. 52.

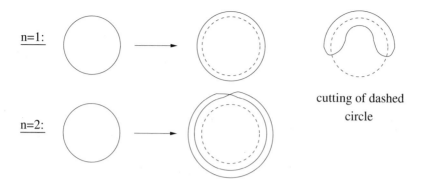

n=1:

n=2:

cutting of dashed
circle

Fig. 22.12 The $n = 1$ and $n = 2$ maps not allowing shrinkage to a point.

The maps for these cases are illustrated in Fig. 22.12. It is seen that in these cases the continuous curve can be distorted into that of case (a) only by cutting the dashed circle — thus these cases do not allow a continuous shrinkage to a point, i.e. to the case of winding number zero. In the cases $n = 1, 2$ the maps are wound around the dashed circle.

We make another observation at this point. For winding number n we have the mapping (i.e. phase factor or element of $U(1)$):

$$\lambda(x) : \lambda_n(x) = e^{in\chi(\theta)} = [\lambda_1(x)]^n, \quad 0 \le \chi(\theta) \le 2\pi. \tag{22.116}$$

In fact from this we obtain (see below)

$$n = \frac{i}{2\pi} \int_0^{2\pi} d\theta \, \lambda(x) \frac{d}{d\theta} \lambda^{-1}(x), \tag{22.117}$$

and this means

$$\text{r.h.s. of } (22.117) = \frac{i}{2\pi} \int_0^{2\pi} d\theta \, e^{in\chi(\theta)} \frac{d}{d\theta} e^{-in\chi(\theta)}$$

$$= \frac{i}{2\pi} \int_0^{2\pi} d\theta (-in) \frac{d\chi(\theta)}{d\theta} = \text{l.h.s.}$$

One can prove that the expression (22.117), and hence the winding number, is invariant under infinitesimal (continuous) deformations, i.e. the transformations[¶] (the result may be looked at as δn of expression (22.103))

$$\lambda(x) \to \lambda'(x) = \lambda(x) + i\lambda(x)\delta f(x) \equiv \lambda(x) + \delta\lambda(x), \tag{22.118}$$

where $\delta f(x)$ is an infinitesimal (continuous) real function on the circle with

$$(\delta f)_{\theta=0} = (\delta f)_{\theta=2\pi}.$$

[¶] See S. Coleman [55].

Under this transformation n changes by

$$\delta n = \frac{i}{2\pi} \int_0^{2\pi} d\theta \delta\left(\lambda \frac{d}{d\theta} \lambda^{-1}\right). \tag{22.119}$$

Now,

$$
\begin{aligned}
\delta\left(\lambda \frac{d}{d\theta} \lambda^{-1}\right) &= (\delta\lambda)\frac{d}{d\theta}\lambda^{-1} + \lambda\frac{d}{d\theta}(\delta\lambda^{-1}) = (\delta\lambda)\frac{d}{d\theta}\lambda^{-1} - \lambda\frac{d}{d\theta}\left(\frac{\delta\lambda}{\lambda^2}\right) \\
&= i\lambda\delta f(x)\frac{d}{d\theta}\lambda^{-1} - \lambda\frac{d}{d\theta}\left(\frac{i\delta f(x)}{\lambda}\right) \\
&= i\lambda\delta f(x)\frac{d}{d\theta}\lambda^{-1} - i\lambda\delta f(x)\frac{d}{d\theta}\lambda^{-1} - i\frac{d}{d\theta}(\delta f(x)) \\
&= -i\frac{d}{d\theta}(\delta f(x)).
\end{aligned}
$$

Hence

$$\delta n = \frac{1}{2\pi}\int_0^{2\pi} d\theta \frac{d}{d\theta}(\delta f(x)) = \frac{1}{2\pi}[\delta f(x)]_0^{2\pi} = 0.$$

Thus we have shown explicitly that *the winding number remains unchanged under continuous deformations.*

Considering $\pi_1(S^2)$ we recall that this symbol represents the set of disjoint equivalence classes into which the maps $\lambda(x)$ from S^1 to S^2 (circle to sphere) can be subdivided. It is intuitively clear that any circle drawn on the surface of a sphere can be shrunk to a point. Thus there is only the trivial map into a point, i.e. only one homotopy class, the identity class (the group space S^2 is simply connected, which means that a curve connecting any two points of S^2 can be continuously deformed into every other curve connecting the two points) and one writes

$$\pi_1(S^2) = 0.$$

In fact, there is a theorem which says that[||]

$$\pi_q(S^n) = 0 \quad \text{for} \quad q < n.$$

One can show that the group space of $SU(2)$ which is the sphere $S^3 \subset \mathbb{R}^4$ is simply connected and that therefore

$$\pi_1(S^3 = SU(2)) = 0,$$

whereas

$$\pi_1(SO(3)) = \mathbb{Z}_2 \quad \text{(i.e. an integer modulo 2)}$$

i.e. the cyclic group with two elements (since $SO(3)$ is doubly connected).

[||]See N. Steenrod [261], Sec. 15.8.

Chapter 23

Path Integrals and Instantons

23.1 Introductory Remarks

In Chapter 18 we considered anharmonic oscillators and calculated their eigenenergies, and in particular the exponentially small level splittings of the double well potential. In each of the cases considered we obtained the result by solving the appropriate Schrödinger equation. In the following we employ the path integral in order to derive the same result for the *ground state splitting* (only that in this chapter!). This is an important standard example of the path integral method which serves as a prototype for numerous other applications and hence will be treated in detail and, different from literature, in such a way that the factors appearing in the result are exhibited in a transparent way. Since we have not yet introduced periodic (i.e. finite Euclidean time) classical configurations (which are nontopological), our treatment in this chapter is restricted to the consideration of the asymptotically degenerate ground state of the *double well potential* which is based on the use of the *topological instanton configuration*. Together with the infinite Euclidean time limit this requires for the calculation of the tunneling contribution to the ground state energy the Faddeev–Popov method for the elimination of the zero mode. An analogous procedure will be adopted in Chapter 24 in the use of the (nontopological) bounce configuration for the calculation of the imaginary part of the ground state energy of a particle trapped in the inverted double well potential.

23.2 Instantons and Anti-Instantons

A one-dimensional field theory is a quantum mechanical problem. We consider such a theory now with action $S(\phi)$ depending on the real scalar field

$\phi(t)$, i.e.*

$$S(\phi) = \int_{t_i}^{t_f} dt \left[\frac{1}{2}\dot{\phi}^2 - V(\phi)\right],$$

$$V(\phi) = \frac{m^4}{2g^2}\left(1 - \frac{g^2\phi^2}{m^2}\right)^2 = \frac{m^4}{2g^2} - m^2\phi^2 + \frac{1}{2}g^2\phi^4, \quad g^2 > 0. \quad (23.1)$$

The potential $V(\phi)$ has — cf. Fig. 23.1 — degenerate minima (positions of classical stability or *"perturbation theory vacua"*) at positions

$$\phi(t) = \pm\frac{m}{g} \equiv \pm\phi_0. \qquad (23.2)$$

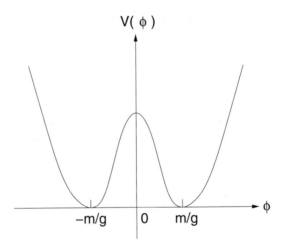

Fig. 23.1 The double well potential.

---

*For comparison with literature, in particular E. Gildener and A. Patrascioiu [110] and Chapter 18, note that here the mass of the equivalent classical particle is taken to be $m_0 = 1$. The Hamiltonian of the simple harmonic oscillator (cf. Chapter 6)

$$H = \frac{1}{2m_0}p^2 + \frac{1}{2}m_0\omega^2 q^2 \quad \text{with eigenenergies} \quad E_n = \hbar\omega\left(n + \frac{1}{2}\right), \quad n = 0, 1, 2, \dots,$$

has as the normalized ground state wave function

$$\psi_0(q) = \langle q|0\rangle = \left(\frac{m_0\omega}{\pi\hbar}\right)^{1/4} e^{-m_0\omega q^2/2\hbar}.$$

Thus for $m_0 = 1$ we have in our notation here $\omega^2 \equiv 2m^2$ (here m is the parameter in the potential $V(\phi)$ of Eq. (23.1)) and hence $E_0 = m\hbar/\sqrt{2}$, and

$$\psi^2(0) = \left(\frac{\sqrt{2}m}{\pi\hbar}\right)^{1/2}.$$

Classically the position $\phi = 0$ is a point of instability. We can develop a perturbation theory around either of the minima, as we did in Chapter 18 — but as we saw, this alone does not yield the level splitting. We know from the shape of the potential that the quantum mechanical energy spectrum is entirely discrete (no scattering), but there is tunneling between the two minima if the central hump is not infinitely high. This tunneling affects the eigenvalues, and so the question is: What sort of boundary conditions do we have to impose on the wave function other than its exponential fall-off at $\phi \to \pm\infty$? We can decide this if we observe that $S = \int dt L$ is invariant under the exchanges $\phi \to \pm\phi$. In mechanics of a particle of mass 1 with position coordinate $x(t)$, we have the Lagrangian

$$L(x, \dot{x}) = \frac{1}{2}\dot{x}^2 - V(x),$$

and in the above ϕ corresponds to x. The states of the system here must be even or odd under $x \to \pm x$, i.e. must have definite parity as in a physical situation. The corresponding wave functions $\psi_\pm(x)$ then must be peaked at the extrema of stability — as in the examples illustrated in Fig. 23.2 — and we can deduce from this the boundary conditions, e.g.

$$\psi'_+(0) = 0, \quad \psi_-(0) = 0.$$

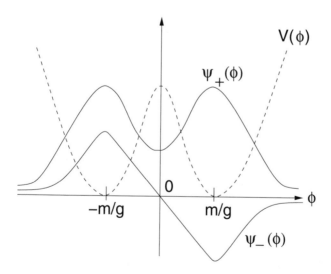

Fig. 23.2 The wave functions of the lowest states.

These boundary conditions imply nonperturbative contributions to the eigenvalues which yield the splitting of the asymptotically degenerate oscillator approximations into an even ground state and an odd first excited state. In the semiclassical path integral procedure that we want to use here, we consider the amplitude for transitions between the vacua or minima at $\phi_{f,i} = \pm\phi_0 = \pm m/g$ over a large period of time. We are therefore interested in the amplitude — called *Feynman amplitude* or *kernel* as we learned in Chapter 21 —

$$\langle \phi_0, t_f | - \phi_0, t_i \rangle = \int \mathcal{D}\{\phi\} \exp[iS(\phi)/\hbar] \tag{23.3}$$

for $t_f \to \infty, t_i \to -\infty$ (we assume the normalization factor or metric to be handled as explained in Chapter 21). For small $\hbar$ we can attempt a stationary phase method. A classical particle starting from rest ($\dot{\phi} = 0$) at $-m/g$ cannot overcome the central hump and move to $+m/g$. But quantum mechanically it can tunnel through the hump from one well to the other. The transition amplitude for this, i.e. that given by Eq. (23.3), is finite if we go to imaginary, i.e. Euclidean time $\tau = it$; classically forbidden domains then become "classically" accessible. Then

$$\langle \phi_0, \tau_f | - \phi_0, \tau_i \rangle = \int_{\phi_i=-\phi_0}^{\phi_f=\phi_0} \mathcal{D}\{\phi\} \exp[-S_E(\phi)/\hbar] \tag{23.4}$$

with the Euclidean action

$$S_E(\phi) = \int_{\tau_i}^{\tau_f} d\tau L_E = \int_{\tau_i}^{\tau_f} d\tau \left[\frac{1}{2}\left(\frac{d\phi}{d\tau}\right)^2 + V(\phi) \right] \tag{23.5}$$

and

$$iS = -S_E.$$

It follows that the equation of motion is now given by $\delta S_E(\phi) = 0$, i.e.

$$\frac{d}{d\tau}\left[\frac{\partial L_E}{\partial(d\phi/d\tau)} \right] - \frac{\partial L_E}{\partial \phi} = 0, \tag{23.6}$$

i.e.

$$\frac{d^2\phi}{d\tau^2} = V'(\phi) = -2m^2\left(1 - \frac{g^2\phi^2}{m^2}\right)\phi, \quad \frac{d^2\phi}{d\tau^2} = \frac{d}{d\phi}\left[\frac{1}{2}\left(\frac{d\phi}{d\tau}\right)^2\right], \tag{23.7}$$

so that

$$\frac{1}{2}\left(\frac{d\phi}{d\tau}\right)^2 = V(\phi) + \text{const.} \tag{23.8}$$

We observe that Eq. (23.7) resembles a Newton equation of motion with reversed sign of the potential as a result of our passage to Euclidean time. Equation (23.7) is to be solved for ϕ with the boundary conditions

$$\phi(\tau) = \frac{m}{g} \quad \text{for } \tau_f \to \infty \quad \text{and} \quad \phi(\tau) = -\frac{m}{g} \quad \text{for } \tau_i \to -\infty. \tag{23.9}$$

In fact, we see that the equation is the same as the classical equation for a static soliton (i.e. time-independent) in the $1 + 1$ dimensional Φ^4-theory which we considered previously. Thus we know the solution from there which is the *kink solution* (22.18), i.e. in the present case

$$\phi_c(\tau - \tau_0) = \frac{m}{g} \tanh m(\tau - \tau_0), \qquad (23.10)$$

where τ_0 is an integration constant. In the present context of Euclidean time the configuration ϕ_c is generally described as an "*instanton*". Its trajectory, the classical path, is depicted in Fig. 23.3.

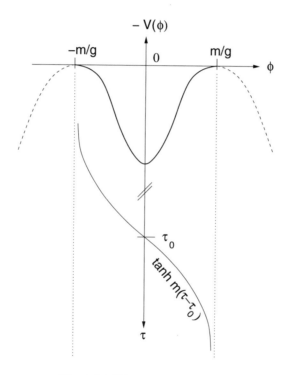

Fig. 23.3 The instanton path.

The solution $\phi_c(\tau)$ can also be looked at as describing the motion of a Newtonian particle of mass 1 in the potential $-V(\phi)$. This is a very useful description. If the constant arising in the integration of the Newton-like equation is chosen to be zero (see below), this means the total energy of the particle in Euclidean motion is zero, and hence kinetic energy gained is equal to potential energy lost. Thus we can picture the particle as starting from rest at one minimum of $V(\phi)$ or maximum of $-V(\phi)$, i.e. from $\phi_i = -m/g$, and having just enough energy to reach the other extremum at $\phi_f = m/g$ where it will again be at rest.

The Euclidean action of the instanton solution (23.10) is given by (23.5). Using Eq. (23.8) we obtain

$$S_E = \int_{\tau_i \to -\infty}^{\tau_f \to \infty} d\tau \, [2V(\phi_c) + \text{const.}]. \tag{23.11}$$

For this quantity to be finite for $\tau_i = -\infty, \tau_f = \infty$, the constant must be zero.[†] With this choice and inserting

$$V(\phi_c) = \frac{1}{2}\left(\frac{d\phi_c}{d\tau}\right)^2 = \frac{1}{2}\frac{m^4}{g^2}\frac{1}{\cosh^4 m(\tau - \tau_0)}, \tag{23.12}$$

we obtain[‡]

$$S_E = \frac{m^3}{g^2}\int_{-\infty}^{\infty}\frac{d(m\tau)}{\cosh^4 m(\tau-\tau_0)} = \frac{m^3}{g^2}\left[\tanh x - \frac{\tanh^3 x}{3}\right]_{x=-\infty}^{x=\infty} = \frac{4}{3}\frac{m^3}{g^2}. \tag{23.13}$$

We observe that this expression is identical with the *energy* of the static soliton of the $1+1$ dimensional soliton theory which we considered in Chapter 22. We also observe that the expression is singular for $g^2 \to 0$, thus indicating the nonperturbative nature of the expression.

Inserting the result (23.13) into Eq. (23.4), we obtain

$$\lim_{\tau_f \to \infty, \tau_i \to -\infty}\langle \phi_0, \tau_f | -\phi_0, \tau_i \rangle \propto \exp\left[-\frac{4}{3\hbar}\frac{m^3}{g^2}\right]. \tag{23.14}$$

But this is not yet enough. We also need to know how we can calculate the difference $\triangle E$ of the energies of the two lowest levels from this expression.

The configuration (23.10) which rises monotonically from left to right is the pseudoparticle configuration called kink in the context of soliton theory and instanton in one-dimensional field theory or quantum mechanics with topological charge 1 and Euclidean action given by (23.13). The classical equation (23.7) admits another configuration which is the negative of (23.10) and is called the *antikink* or *anti-instanton* configuration and obviously (following the arguments of Chapter 22) carries topological charge -1, but with the same Euclidean action (23.13). This anti-instanton configuration rises monotonically from right to left as shown in Fig. 23.4.

[†]For a nonzero constant we obtain the periodic instantons discussed in Chapters 25 and 26.

[‡]Note that this integral may be evaluated for finite limits $\tau = \pm T$ and presents no problem in the limit $T \to \infty$. Also note that we can re-express S_E in the form

$$S_E = \int_{-\infty}^{\infty}d\tau\left(\frac{d\phi_c}{d\tau}\right)^2 = \int_{\phi_c(-\infty)}^{\phi_c(\infty)}\frac{d\phi_c}{d\tau}d\phi_c, \quad \text{where} \quad \frac{d\phi_c}{d\tau} = \sqrt{2V(\phi_c)} = \frac{m^2}{g}\left(1 - \frac{g^2\phi_c^2}{m^2}\right).$$

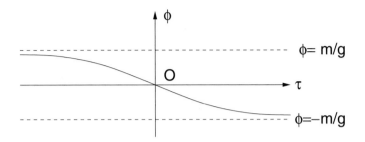

Fig. 23.4 The anti-instanton.

The terminology is motivated by analogy with the phenomenology of particles: Thus like a proton has electric charge $+1$ (in units of the fundamental charge e) and the antiproton has electric charge -1, so correspondingly one defines the instanton or kink with topological charge $+1$ and the anti-instanton or antikink with topological charge -1. Both configurations communicate between the two wells of the potential at $\phi = -m/g$ and $\phi = m/g$. Looking at the instanton (23.10) we can ask ourselves: Does it make sense to consider the sum of an instanton localized at τ_0 as indicated in Fig. 23.5, and an anti-instanton localized at (say) τ_0', e.g.

$$\phi_c^{(2)}(\tau) = \phi_c(\tau - \tau_0) + \overline{\phi}_c(\tau - \tau_0'), \quad \overline{\phi}_c(\tau) = -\phi_c(\tau), \quad (23.15a)$$

or

$$\phi_c^{(2)}(\tau) = \theta(\tau - T_0)\phi_c(\tau - \tau_0)\theta(T_1 - \tau) - \theta(\tau - T_1)\phi_c(\tau - \tau_0')\theta(T_2 - \tau). \quad (23.15b)$$

In the case of (23.15a) we have a simple superposition, in the case of (23.15b) a mending together of an instanton and an anti-instanton at $\tau = T_1$ with $-\infty = T_0 < T_1 < T_2 = \infty$ for τ_0', τ_0 far apart and $\tau_0' \gg \tau_0$. The sum (23.15b) has the shape shown in Fig. 23.5. Clearly we are here interested in configurations of type (23.15b) since we require configurations from vacuum to vacuum. Questions which arise at this point are: Is the nonmonotonic configuration (23.15b) a solution of the classical equation? Is it a topological configuration — if so, what is its associated topological charge? What is its Euclidean action? The sum of ϕ_c and $\overline{\phi}_c = -\phi_c$ defined by (23.15b) is not exactly a solution, but if τ_0, τ_0' are very far apart their sum differs from an exact solution only by an exponentially small quantity, as we can see e.g. from the sketch or from the variation of S_E. Writing

$$S_E = \int_{-\infty}^{T_1} \cdots + \int_{T_1}^{\infty} \cdots$$

and performing the variation in the usual way with $\delta\phi = 0$ at $\tau = \pm\infty$,

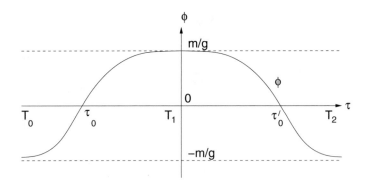

Fig. 23.5 Superposition of an instanton and a widely separated anti-instanton.

we see that in the domain $[-\infty, T_1]$ with

$$\delta \int_{-\infty}^{T_1} L_E d\tau = 0,$$

the classical equation is valid only up to a contribution

$$\left[\frac{\partial L_E}{\partial(\partial\phi/\partial\tau)}\delta\phi\right]_{-\infty}^{T_1} \propto \frac{\delta\phi(T_1)}{\cosh^2 m(T_1 - \tau_0)}.$$

At $\tau = T_1$ the variation $\delta\phi$ is finite but nonzero so that we are left with an exponentially small contribution which, of course, vanishes as $T_1 \to \pm\infty$ (of course, at ∞ the variation is $\delta\phi = 0$).

In order to answer the second question we observe that (23.15a) is a configuration which starts from $\phi = -m/g$ and ends there. Thus it belongs to the so-called vacuum sector of solutions and its topological charge (determined by the boundary conditions at plus and minus infinity) is zero. Finally, since (23.15a) is not exactly a classical solution, it also does not exactly stationarize the Euclidean action. Ignoring exponentially small contributions the Euclidean action of the configuration (23.15a) can be argued to be zero. In fact,

$$
\begin{aligned}
\frac{d\phi(\tau)}{d\tau} &= \frac{m}{g}\left[\frac{1}{\cosh^2 m(\tau - \tau_0)} - \frac{1}{\cosh^2 m(\tau - \tau_0')}\right] \\
&= \frac{m}{g}\frac{1}{\cosh^2 m(\tau - \tau_0)}\left[1 - \frac{\cosh^2 m(\tau - \tau_0)}{\cosh^2 m(\tau - \tau_0')}\right] \\
&\simeq \frac{m}{g}\frac{1}{\cosh^2 m(\tau - \tau_0)}[1 - \exp\{-2m(\tau_0' - \tau_0)\}], \quad (23.16)
\end{aligned}
$$

and looking at Eq. (23.13) and ignoring exponentially small contributions, we obtain (note the same values of the limits)

$$S_E \simeq \int_{\phi(-\infty)}^{\phi(\infty)=\phi(-\infty)} L_E \frac{d\phi}{d\phi/d\tau} = 0.$$

For the configuration (23.15b) we have with a similar argument

$$S_E = \left(\int_{-\infty}^{T_1} \cdots + \int_{T_1}^{\infty} \cdots \right) L_E d\tau = \left(\int_{-\infty}^{T_1} \cdots - \int_{-\infty}^{-T_1} \cdots \right) L_E = 0.$$

As a further example we consider a configuration consisting (approximately) of two widely separated instantons (or kinks) and one anti-instanton (or antikink) localized at $\tau_0^{(1)}, \tau_0^{(2)}$ and $\tau_0^{(3)}$ respectively,[§]

$$
\begin{aligned}
\phi_c^{(3)}(\tau) &= \theta(\tau - T_0)\phi_c(\tau - \tau_0^{(1)})\theta(T_1 - \tau) \\
&\quad -\theta(\tau - T_1)\phi_c(\tau - \tau_0^{(2)})\theta(T_2 - \tau) \\
&\quad +\theta(\tau - T_2)\phi_c(\tau - \tau_0^{(3)})\theta(T_3 - \tau)
\end{aligned}
\tag{23.17a}
$$

with $-\infty = T_0 \ll \tau_0^{(1)} \ll T_1 \ll \tau_0^{(2)} \ll T_2 \ll \tau_0^{(3)} \ll T_3 = \infty$.

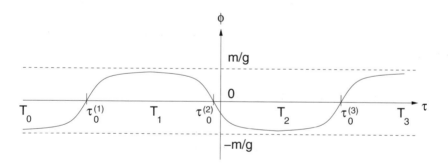

Fig. 23.6 Two widely separated instantons and one anti-instanton.

Again the configuration is non-monotonic and as in the previous case one can show that it is only approximately (with exponentially small deviations) a solution of the classical equation. In order to answer the question concerning the associated topological charge of $\phi^{(3)}(\tau)$ we recall that this is determined by the boundary conditions of the configuration; thus in the present case the configuration belongs to the instanton or kink sector with topological charge

---

[§]E.g. in the overall range of $\tau = 2T$ we can choose the locations $\tau_0^{(1)} = -2T/3, \tau_0^{(2)} = 0$ and $\tau_0^{(3)} = 2T/3$.

+1 (like the single instanton or kink). Thus for each of the four charge sectors (corresponding to the two vacua and the kink and antikink sectors) we have a multitude of approximate solutions[¶] of the classical equation as illustrated in Fig. 23.7 in which (a) and (b) depict the two constant vacuum sectors and (b) and (c) the instanton and anti-instanton sectors respectively.

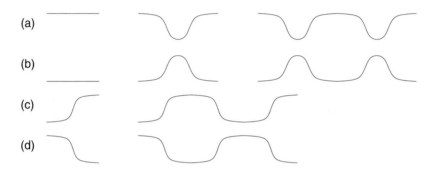

(a)

(b)

(c)

(d)

Fig. 23.7 Classical solutions in their topological sectors.

In evaluating later the Feynman amplitude about such configurations we use the completeness relation of states. Thus in the case of $\phi_c^{(3)}(\tau)$ we write

$$\langle \phi_f, \tau_f | \phi_i, \tau_i \rangle = \int_{-\infty}^{\infty} d\phi_2 \int_{-\infty}^{\infty} d\phi_1 \langle \phi_f, \tau_f | \phi_2, \tau_2 \rangle \langle \phi_2, \tau_2 | \phi_1, \tau_1 \rangle \langle \phi_1, \tau_1 | \phi_i, \tau_i \rangle.$$

(23.17b)

Evidently the evaluation of this amplitude requires two intermediate integrations.

23.3   The Level Difference

The nonrelativistic Schrödinger wave function ψ of a discrete eigenvalue problem for a potential V satisfies a homogeneous integral equation which we can write

$$\psi(x', t') = \int dx K(x', t'; x, t) \psi(x, t), \quad t' > t. \quad (23.18)$$

Here $K(x', t'; x, t)$ is a *Green's function* satisfying the differential equation

$$DK(x', t'; x, t) = i\delta(x' - x)\delta(t' - t) \quad (23.19)$$

with differential operator

$$D = i\frac{\partial}{\partial t'} - V_0(x') + \frac{1}{2m_0}\frac{\partial^2}{\partial x'^2}. \quad (23.20)$$

[¶]B. Felsager [91], p. 143, therefore calls these solutions "quasi-stationary configurations". Further discussion may be found there.

(One can convince oneself by differentiation that the integral equation is equivalent to a time-dependent Schrödinger equation). In Eq. (23.20)

$$V_0(x) = V(x) - \delta V(x), \quad V = V_0 + \delta V,$$

where $\delta V(x)$ is the deviation of the exact potential V from an approximation V_0. (Since we cannot — in general — solve a problem exactly, we have to resort to some approximation; thus we have to distinguish between the exact problem and the approximate problem which we can solve exactly; the Green's function is constructed from the eigenfunctions of the exactly solvable problem). The Green's function $K(x', t'; x, t)$ can be expanded in terms of the eigenfunctions $\psi_n^{(0)}(x)$ of the Schrödinger equation for the potential $V_0(x)$, i.e. (to be verified below)

$$K(x', t'; x, t) = \frac{i}{2\pi} \sum_n \int dE \frac{\psi_n^{(0)*}(x')\psi_n^{(0)}(x) \exp[-iE(t' - t)]}{E - E_n^{(0)}}. \quad (23.21)$$

We have (cf. Chapter 7)

$$\left[\frac{1}{2m_0} \frac{\partial^2}{\partial x'^2} - V_0(x') \right] \psi_n^{(0)}(x') = -E_n^{(0)} \psi_n^{(0)}(x'),$$

$$\sum_n \psi_n^{(0)}(x')\psi_n^{(0)}(x) = \delta(x' - x),$$

so that

$$
\begin{aligned}
DK(x', t'; x, t) &= \left(i\frac{\partial}{\partial t'} - V_0(x') + \frac{1}{2m_0}\frac{\partial^2}{\partial x'^2} \right) \\
&\quad \times \frac{i}{2\pi} \sum_n \int dE \frac{\psi_n^{(0)*}(x')\psi_n^{(0)}(x) \exp[-iE(t' - t)]}{E - E_n^{(0)}} \\
&= i \sum_n \psi_n^{(0)*}(x')\psi_n^{(0)}(x) \int \frac{dE}{2\pi} e^{-iE(t' - t)} \\
&= i\delta(x - x')\delta(t' - t), \quad (23.22)
\end{aligned}
$$

as claimed (note that the first delta function results from the completeness relation of the eigenfunctions of the exactly solvable problem). We can go one step further and integrate out the E-dependence of Eq. (23.21). Replacing $E_n^{(0)}$ in the denominator by $(E_n^{(0)} - i\epsilon)$, $\epsilon > 0$, so that the integrand of (23.21) has a simple pole at $E = E_n^{(0)} - i\epsilon$, and integrating along the contour shown in Fig. 23.8, Eq. (23.21) becomes (with the help of Cauchy's residue theorem)[||]

$$K(x', t'; x, t) = \sum_n \psi_n^{(0)*}(x')\psi_n^{(0)}(x) \exp[-iE_n^{(0)}(t' - t)]\theta(t' - t). \quad (23.23)$$

[||] This formula is discussed, for instance, in R. J. Crewther, D. Olive and S. Sciuto [60].

This is our earlier expression (7.9).

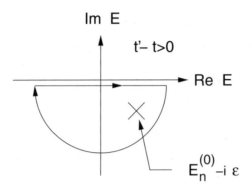

Fig. 23.8 Integration contour.

We now recall from Eqs. (21.10), (21.17), that the Green's function K can be written as a path integral, i.e. that

$$K(x_f, t_f; x_i, t_i) = \int \mathcal{D}\{x\} e^{iS/\hbar} \theta(t_f - t_i). \qquad (23.24)$$

Evidently we want to relate Eq. (23.23) to Eq. (23.24) in the case of our one-dimensional Φ^4-theory. In the derivation of Eq. (23.23) we distinguished between the exact problem with potential V and an approximate problem with potential V_0 which can be solved exactly. In an actual application such a subdivision can be a matter of convenience. Thus we could try to approximate V by taking as V_0 the harmonic oscillator part of the potential and $\delta V \propto \phi^4$; then the set $\{\psi_n^{(0)}\}$ would be the set of oscillator eigenfunctions which is wellknown. Alternatively, we could be interested in a completely different problem, e.g. one with V_0 given by Eq. (23.1) and $\delta V' \propto \phi^6$. In this case $\psi_n^{(0)}$ would be the exact eigenfunctions of the Hamiltonian for the ϕ^4-potential (23.1). It is shown for instance in scattering theory that the exact transition amplitude is characterized by poles at the exact eigenvalues (e.g. in the case of the hydrogen atom all discrete eigenvalues are simple poles of the scattering amplitude). Thus if we want to relate (23.24) to the path integral representation of the transition amplitude for the ϕ^4-potential (23.1), we must take in (23.24) the exact (real) eigenfunctions $\psi_n(x)$ and eigenvalues E_n for the potential (23.1) which, of course, we do not know.

We proceed as follows. From the above equations we obtain

$$K(\phi', t'; \phi, t) = \sum_n \psi_n(\phi') \psi_n(\phi) \exp[-iE_n(t - t')/\hbar] = \int \mathcal{D}\{\phi\} \exp[iS/\hbar]. \qquad (23.25)$$

This equivalence of the standard decomposition as a sum on the one hand, and on the other hand with the Feynman integral on the right is the reason for describing the Feynman amplitude also as *kernel*. We now consider the left hand side of this expression for Euclidean time $\tau = it$. Since $\tau' > \tau$, the expression is dominated by the lowest and next to lowest eigenvalues E_+, E_- associated with the ground and first excited state wave functions $\psi_+(\phi)$ (even) and $\psi_-(\phi)$ (odd) respectively. Thus the left hand side can then be approximated by (with $\hbar = 1$)

$$
\begin{aligned}
K(\phi', \tau'; \phi, \tau) \simeq\ & \psi_+(\phi')\psi_+(\phi) \exp[-E_+(\tau' - \tau)] \\
& + \psi_-(\phi')\psi_-(\phi) \exp[-E_-(\tau' - \tau)].
\end{aligned} \tag{23.26}
$$

We are interested in the transition with

$$
\tau' - \tau \to \infty \quad \text{and} \quad \phi' \to \phi_0, \ \phi \to -\phi_0. \tag{23.27}
$$

Moreover,

$$
\psi_+(\phi_0) = \psi_+(-\phi_0), \quad \psi_-(\phi_0) = -\psi_-(-\phi_0), \tag{23.28}
$$

so that

$$
\begin{aligned}
K(\phi', \tau'; \phi, \tau) \simeq\ & \{\psi_+(\phi_0)\}^2 \exp[-E_+(\tau' - \tau)] \\
& - \{\psi_-(\phi_0)\}^2 \exp[-E_-(\tau' - \tau)].
\end{aligned} \tag{23.29}
$$

We define $\triangle E, E_0$ by[**]

$$
\triangle E = E_- - E_+ \ \text{(level splitting)}, \quad E_0 = \frac{1}{2}(E_+ + E_-), \tag{23.30}
$$

so that

$$
E_+ = E_0 - \frac{1}{2}\triangle E, \quad E_- = E_0 + \frac{1}{2}\triangle E. \tag{23.31}
$$

Then

$$
\begin{aligned}
K(\phi_0, \tau'; -\phi_0, \tau) \simeq\ & \{\psi_+(\phi_0)\}^2 \exp\left[-\left(E_0 - \frac{1}{2}\triangle E\right)(\tau' - \tau)\right] \\
& - \{\psi_-(\phi_0)\}^2 \exp\left[-\left(E_0 + \frac{1}{2}\triangle E\right)(\tau' - \tau)\right].
\end{aligned} \tag{23.32}
$$

Here the wave functions at the minimum position ϕ_0 are comparable to (i.e. approximately given by) the corresponding ground state eigenfunctions of the harmonic oscillator and so (apart from overall normalization) (cf. Chapter 6)

$$
\psi_\pm(\phi) \propto \left\{ \exp\left[-\frac{1}{2}(\phi + \phi_0)^2\right] \pm \exp\left[-\frac{1}{2}(\phi - \phi_0)^2\right] \right\} = \pm\psi_\pm(-\phi) \tag{23.33}
$$

[**]Here $\triangle E$ denotes the splitting of the originally degenerate oscillator level. This is therefore *twice* the deviation of a split level from the originally degenerate oscillator level.

with $E_0 = m\hbar/\sqrt{2}$ (as we noted at the beginning of this chapter). Hence we can set

$$|\psi_\pm(\phi_0)| = \left|e^{-2\phi_0^2} \pm 1\right| \simeq \left|1 \pm e^{-2\phi_0^2}\right| \simeq \psi(\phi_0). \qquad (23.34)$$

Then

$$K(\phi_0, \tau'; -\phi_0, \tau)$$

$$\approx \ \{\psi(\phi_0)\}^2 \exp[-E_0(\tau' - \tau)]\left\{e^{\triangle E(\tau' - \tau)/2} - e^{-\triangle E(\tau' - \tau)/2}\right\}$$

$$= \ 2\{\psi(\phi_0)\}^2 \exp[-E_0(\tau' - \tau)/\hbar] \sinh\left\{\frac{1}{2\hbar}\triangle E(\tau' - \tau)\right\}, \quad (23.35a)$$

where we re-inserted $\hbar$ for dimensional completeness (observe that when $\triangle E(\tau' - \tau) \equiv \triangle E \triangle T \sim \hbar$ and small, the hyperbolic sine may be approximated by its argument and one returns to (23.25) for Euclidean time). The relation (23.35a) assumes fixed endpoints ϕ_0, i.e.

$$\int d\phi'' \int d\phi' \langle \psi | \phi'' \rangle \delta(\phi'' - \phi_0) \langle \phi'', \tau' | \phi', \tau \rangle \delta(\phi' + \phi_0) \langle \phi' | \psi \rangle$$

$$= \langle \psi | \phi_0 \rangle \underbrace{\langle \phi_0, \tau' | - \phi_0, \tau \rangle}_{K(\phi_0, \tau'; -\phi_0, \tau)} \langle \phi_0 | \psi \rangle. \qquad (23.35b)$$

Clearly without the delta functions the end points are not fixed. (This situation will be required in Chapter 26 in the case of certain bounce configurations). We now want to deduce the explicit expression obtained with the path integral formula and then extract $\triangle E$ by comparison. In this comparison we require $\tau' - \tau$ to be a large Euclidean time interval $(-T, T)$ with $T \to \infty$. It is clear from Eq. (23.35a) that we cannot put $T = \infty$ rightaway; thus we also require the study of a gentle approach of T to infinity, which necessarily makes the calculation more involved. We also note that since the above expression assumes quantization (discrete eigenvalues), the comparable formula must also involve quantization. Thus we must go beyond the purely classical contribution, i.e. we have to consider field fluctuations about the classical configuration.

23.4 Field Fluctuations

23.4.1 The fluctuation equation

We consider fluctuations $\eta(\tau)$ about the *instanton* (or correspondingly static kink) configuration ϕ_c:

$$\phi(\tau) = \phi_c(\tau) + \eta(\tau). \qquad (23.36)$$

Of course, the level splitting ΔE (to be extracted later) must be based on a symmetric treatment of instanton and anti-instanton. Allowing for fluctuations which are not tangential to the path one necessarily introduces more degrees of freedom — in much the same way as in allowing relative motion together with collective motion. Thus the consideration of fluctuations leads to the consideration of a larger (in fact infinite) number of degrees of freedom, i.e. to the continuum. We naturally impose on the fluctuations the boundary conditions

$$\eta(\tau = \pm T)|_{T\to\infty} = 0. \tag{23.37}$$

Clearly we have to expand the action $S_E(\phi)$ about $\phi_c(\tau)$. Inserting (23.36) into (23.5) we obtain

$$
\begin{aligned}
S_E(\phi) &= \int d\tau\, L_E = \int_{-\infty}^{\infty} d\tau \left[\frac{1}{2}\left(\frac{d\phi_c}{d\tau} + \frac{d\eta}{d\tau}\right)^2 + V(\phi_c + \eta) \right] \\
&= \int_{-\infty}^{\infty} d\tau \left[\left\{ \frac{1}{2}\left(\frac{d\phi_c}{d\tau}\right)^2 + \frac{m^4}{2g^2} - m^2\phi_c^2 + \frac{1}{2}g^2\phi_c^4 \right\} \right. \\
&\quad + \left\{ \frac{d\phi_c}{d\tau}\frac{d\eta}{d\tau} + \frac{1}{2}\left(\frac{d\eta}{d\tau}\right)^2 - 2m^2\phi_c\eta - m^2\eta^2 + 2g^2\phi_c^3\eta \right. \\
&\quad \left. \left. + 3g^2\phi_c^2\eta^2 + 2g^2\phi_c\eta^3 + \frac{1}{2}g^2\eta^4 \right\} \right].
\end{aligned}
\tag{23.38}
$$

Here the first pair of curly brackets $\{\cdots\}$ yields as before the contribution

$$S_E(\phi_c) = \frac{4}{3}\frac{m^3}{g^2}$$

(cf. Eq. (23.13)). In the second pair of curly brackets in Eq. (23.38) we use the identity

$$
\begin{aligned}
\int_{-\infty}^{\infty} d\tau \left\{ \frac{d\phi_c}{d\tau}\frac{d\eta}{d\tau} \right\} &= \frac{d\phi_c}{d\tau}\eta(\tau)\Big|_{-\infty}^{\infty} - \int_{-\infty}^{\infty} d\tau \frac{d^2\phi_c}{d\tau^2}\eta(\tau) \\
&= -\int_{-\infty}^{\infty} d\tau \frac{d^2\phi_c}{d\tau^2}\eta(\tau),
\end{aligned}
\tag{23.39}
$$

where we used Eq. (23.37). Hence

$$
\begin{aligned}
S_E(\phi) &= S_E(\phi_c) + \int_{-\infty}^{\infty} d\tau \left[\frac{1}{2}\left(\frac{d\eta}{d\tau}\right)^2 + \eta\left\{ -\frac{d^2\phi_c}{d\tau^2} - 2m^2\phi_c + 2g^2\phi_c^3 \right\} \right. \\
&\quad \left. + \eta^2\{3g^2\phi_c^2 - m^2\} + 2g^2\eta^3\phi_c + \frac{1}{2}g^2\eta^4 \right].
\end{aligned}
\tag{23.40}
$$

The term linear in η vanishes as a result of Eq. (23.7). Hence

$$S_E(\phi) = S_E(\phi_c) + \delta S_E(\eta) \tag{23.41a}$$

and

$$
\begin{aligned}
S_E(\phi) &= \int d\tau \, L_E(\phi_c + \eta) \\
&= S_E(\phi_c) + \frac{1}{2} \int d\tau \eta^2(\tau) L''_E(\phi)|_{\phi_c} + \cdots . \tag{23.41b}
\end{aligned}
$$

Here

$$\delta S_E(\eta) = \int_{-\infty}^{\infty} d\tau \left[\frac{1}{2}\left(\frac{d\eta}{d\tau}\right)^2 + \eta^2\{3g^2\phi_c^2 - m^2\} + 2g^2\phi_c\eta^3 + \frac{1}{2}g^2\eta^4 \right]. \tag{23.42}$$

Hence, since the Jacobian from $\phi(\tau)$ to $\eta(\tau)$ of Eq. (23.36) is unity (with $iS = -S_E$ and subscript 1 in the following indicating that the *single instanton case* is considered),

$$
\begin{aligned}
I_1 &:= \int \mathcal{D}\{\phi\} \exp[-S_E(\phi)/\hbar] \\
&= \int \mathcal{D}\{\phi\} \exp[-S_E(\phi_c)/\hbar] \exp[-\delta S_E(\eta)/\hbar] \\
&= \exp\left[-\frac{4}{3}\frac{m^3}{g^2\hbar} \right] \int \mathcal{D}\{\eta\} \exp[-\delta S_E(\eta)/\hbar], \tag{23.43}
\end{aligned}
$$

where we used Eq. (23.13). We consider the approximation in which we restrict ourselves in Eq. (23.42) to terms quadratic in η. This approximation is called the Gaussian or "*one-loop*" approximation.[*] Then

$$
\begin{aligned}
\delta S_E(\eta) &= \int_{-\infty}^{\infty} d\tau \left[\frac{1}{2}\left(\frac{d\eta}{d\tau}\right)^2 + \eta^2\{3g^2\phi_c^2 - m^2\} \right] \\
&\stackrel{(23.10)}{=} \int_{-\infty}^{\infty} d\tau \left[\frac{1}{2}\left(\frac{d\eta}{d\tau}\right)^2 + \eta^2 m^2\{3\tanh^2 m(\tau - \tau_0) - 1\} \right] \\
&\equiv \frac{1}{2}(\eta, M\eta), \tag{23.44}
\end{aligned}
$$

where M is an operator.[†] We set (observe that τ acts only as a parameter which parametrizes the classical path; the dynamical — real time-dependent — quantity is ξ_n)

$$\eta(\tau) = \sum_{n=0}^{\infty} \xi_n \eta_n(\tau), \tag{23.45}$$

---

[*]The word "loop" refers to internal integrations which this non-classical contribution requires.
[†]The factor $1/2$ is extracted so that the fluctuation equation assumes the form (23.54b).

where $\{\eta_n(\tau)\}$ is at every point τ of the classical path a complete set of orthonormal eigenfunctions of M with eigenvalues $\{w_n^2\}$:

$$M\eta_n(\tau) = w_n^2\eta_n(\tau), \quad \int_{-\infty}^{\infty} d\tau\eta_n(\tau)\eta_m(\tau) = \rho^2\delta_{nm}. \tag{23.46}$$

Here we have in mind $\rho^2 = 1$. However, for ease of comparison with the literature[‡] we drag the parameter ρ^2 along. Then

$$(\eta, M\eta) \overset{(23.45)}{=} \sum_{n,m} \int_{-\infty}^{\infty} d\tau(\xi_n\eta_n, M\xi_m\eta_m) = \sum_n |\xi_n|^2 w_n^2\rho^2. \tag{23.47}$$

Formally, if we insert (23.36) and (23.44) into I_1 of (23.43), we obtain

$$I_1 \equiv I_0 \exp\left[-\frac{4}{3}\frac{m^3}{g^2\hbar}\right] \tag{23.48a}$$

with

$$I_0 = \int \mathcal{D}\{\eta\} \exp\left[-\frac{1}{2}(\eta, M\eta)\right] \overset{(23.47)}{=} \int \mathcal{D}\{\eta\} \exp\left[-\frac{1}{2}\sum_n |\xi_n|^2 w_n^2\rho^2\right], \tag{23.48b}$$

where

$$\mathcal{D}\{\eta\} = \lim_{i \to \infty} \prod_{k=0}^{i} d\eta(\tau_k) \tag{23.49}$$

(in our earlier discussion (cf. Sec. 21.6): $\eta(\tau_k) \to \eta_k$, and $\eta(\tau)$ is given by Eq. (23.45)). With

$$\eta(\tau_k) = \sum_n \xi_n\eta_n(\tau_k) \quad \text{we have} \quad d\eta(\tau_k) = \sum_n d\xi_n\eta_n(\tau_k), \tag{23.50}$$

and (the determinant being the Jacobian of this transformation)

$$\prod_{k=0}^{i} d\eta(\tau_k) = \det\left(\frac{d\eta(\tau_m)}{d\xi_n}\right) \prod_{k=0}^{i} d\xi_k. \tag{23.51}$$

Hence

$$I_1 = \exp\left[-\frac{4}{3}\frac{m^3}{g^2\hbar}\right] \det\left(\frac{d\eta(\tau_m)}{d\xi_n}\right) \int \left(\prod_{k=0}^{i} d\xi_k\right) \exp\left[-\frac{1}{2}\sum_n |\xi_n|^2 w_n^2\rho^2\right]. \tag{23.52a}$$

[‡]Note that a different normalization of $\{\eta_n\}$ as in E. Gildener and A. Patrascioiu [110] (they have in our notation $\rho^2 = \hbar/\mu^2, \mu^2 = 2m^2$) introduces additional factors on the right hand side of Eq. (23.47) and hence changes some intermediate quantities.

Now,

$$\int_{-\infty}^{\infty} d\xi \, e^{-w^2\rho^2\xi^2/2} = \sqrt{\frac{2\pi}{w^2\rho^2}}, \tag{23.52b}$$

so that the integral in Eq. (23.52a) can be evaluated formally to give

$$I_1 = \exp\left[-\frac{4}{3}\frac{m^3}{g^2\hbar}\right]\det\left(\frac{d\eta(\tau_m)}{d\xi_n}\right)\prod_{k=0}^{i}\left[\frac{2\pi}{w_k^2\rho^2}\right]^{1/2}, \tag{23.52c}$$

provided no w_k is zero. We now show that this condition is not satisfied, i.e. we have one eigenvalue which is zero corresponding to the *zero mode* associated with the violation of translation invariance by the instanton configuration. This is the case if the zero mode, like all fluctuation modes, is normalized over the infinite time interval, not — however — if a finite interval of τ from $-T$ to T is considered, in which case the associated eigenvalue is nonzero (as will be shown below) and Eq. (23.52c) is well defined.

Hence first of all we show that the differential operator M is the differential operator of the *stability equation* or *small fluctuation equation*. The operator M was defined by (23.44), i.e.

$$\begin{aligned}
\frac{1}{2}(\eta, M\eta) &= \frac{1}{2}\int_{-\infty}^{\infty} d\tau \, \eta M\eta = \int_{-\infty}^{\infty} d\tau \left[\frac{1}{2}\left(\frac{d\eta}{d\tau}\right)^2 + \frac{1}{2}\eta^2 V''(\phi_c)\right] \\
&= \frac{1}{2}\int_{-\infty}^{\infty} d\tau \, \eta\left[-\frac{d^2}{d\tau^2} + V''(\phi_c)\right]\eta,
\end{aligned} \tag{23.53}$$

where in the last step we performed a partial integration and used Eq. (23.37), i.e. $\eta(\pm\infty) = 0$:

$$\int_{-\infty}^{\infty} d\tau \left(\frac{d\eta}{d\tau}\right)^2 = \frac{d\eta}{d\tau}\eta\bigg|_{-\infty}^{\infty} - \int_{-\infty}^{\infty} d\tau \, \eta\frac{d^2}{d\tau^2}\eta.$$

From Eqs. (23.44) and (23.53) we obtain

$$\begin{aligned}
M &= -\frac{d^2}{d\tau^2} + V''(\phi_c), \\
V''(\phi_c) &= 6g^2\phi_c^2 - 2m^2 = 6m^2\tanh^2 m(\tau - \tau_0) - 2m^2 \\
&= 4m^2 - 6m^2\mathrm{sech}^2 m(\tau - \tau_0).
\end{aligned} \tag{23.54a}$$

Thus Eq. (23.46) becomes

$$\left[-\frac{d^2}{d\tau^2} + V''(\phi_c)\right]\eta_n(\tau) = w_n^2\eta_n(\tau). \tag{23.54b}$$

We have considered the spectrum $\{w_n^2\}$ and the eigenfunctions $\{\eta_n\}$ of this equation previously; cf. Sec. 22.6. There we found that the spectrum consists of two discrete states (one being the zero mode) and a continuum. In Eq. (23.45) we have to sum over all of these states. We also observed earlier that the existence of the zero mode $d\phi_c/d\tau$ follows by differentiation of the classical equation, i.e.

$$\frac{d}{d\tau}\left[\frac{d^2\phi}{d\tau^2}\right] = \frac{d}{d\tau}[V'(\phi)], \quad \text{implying} \quad \left[\frac{d^2}{d\tau^2} - V''(\phi_c)\right]\left(\frac{d\phi}{d\tau}\right) = 0.$$

Thus we know that the set of eigenfunctions $\{\eta_n(\tau)\}$ of M includes one eigenfunction $\eta_0(\tau)$ with eigenvalue zero. Before we can evaluate the integral (23.52a) we therefore have to find a way to circumvent this difficulty. First, however, we consider in Example 23.1 a gentle approach to the eigenvalue zero of the zero mode by demanding the eigenfunction $\psi_0(\tau)$ to vanish not at $\tau = \infty$ but at a large Euclidean time T, as remarked earlier.[§] Some part of the evaluation of the path integral will then be performed at a large but finite Euclidean time T, and it will be seen that this large T dependence permits a clean passage to infinity.

Example 23.1: Eigenvalues for finite range normalization
Derive the particular nonvanishing eigenvalue of the fluctuation equation which replaces the zero eigenvalue of the zero mode if the latter is required to vanish not at infinity but at a large but finite Euclidean time T.

Solution: The operator of the small fluctuation equation is given by Eq. (23.54b). We change to the variable $z = m(\tau - \tau_0)$ and consider the following set of two eigenvalue problems:
(a) Our ultimate consideration is concerned with the equation

$$\left(\frac{d^2}{dz^2} + 6\,\text{sech}^2 z - n^2\right)\psi_0(z) = 0 \quad \text{with} \quad n^2 = 4 \quad \text{and} \quad \psi_0(\pm\infty) = 0.$$

However, we consider now the related equation
(b)

$$\left(\frac{d^2}{dz^2} + 6\,\text{sech}^2 z - n^2\right)\tilde{\psi}_0(z) = -\frac{\tilde{w}_0^2}{m^2}\tilde{\psi}_0(z) \quad \text{with} \quad \tilde{\psi}_0(\pm mT) = 0,$$

and our aim is to obtain $\tilde{w}_0^2$. We proceed as follows. The unnormalized solution of case (a) is (cf. Eq. (23.12))

$$\psi_0(z) \equiv \frac{d\phi_c}{d\tau} = \frac{m^2}{g}\frac{1}{\cosh^2 z}, \quad z = m(\tau - \tau_0).$$

(Note that this is the zero mode which normalized to ρ^2, i.e. $\int d\tau\eta_0^2 = \rho^2$, is $\eta_0 = \rho(d\phi_c/d\tau)/\sqrt{S_E}$, where S_E is the action (23.13)).

We multiply the equation of case (a) by $\tilde{\psi}_0(z)$ and the equation of case (b) by $\psi_0(z)$, subtract one equation from the other and integrate from $-mT$ to mT (i.e. over a total τ−range of $2T$). Then

$$\int_{-mT}^{mT} dz\left(\psi_0\frac{d^2\tilde{\psi}_0}{dz^2} - \tilde{\psi}_0\frac{d^2\psi_0}{dz^2}\right) = -\frac{\tilde{w}_0^2}{m^2}\int_{-mT}^{mT} dz\psi_0\tilde{\psi}_0.$$

[§]This corresponds to a box normalization. See also Appendix A of E. Gildener and A. Patrascioiu [110].

On the left hand side we perform a partial integration; on the right hand side we can replace (in the dominant approximation) the integral by that integrated from $-\infty$ to ∞, and since this is the integral over twice the kinetic energy (and hence $S_E(\phi_c)$ of Eq. (23.13)) we obtain (remembering the change of variable from τ to z)

$$
\left(\psi_0 \frac{d\tilde{\psi}_0}{dz} - \tilde{\psi}_0 \frac{d\psi_0}{dz} \right) \Big|_{-mT}^{mT} = \psi_0 \frac{d\tilde{\psi}_0}{dz} \Big|_{-mT}^{mT} = -\frac{\tilde{w}_0^2}{m^2} \frac{4m^4}{3g^2}. \tag{23.55a}
$$

In order to be able to evaluate the left hand side we consider WKB approximations of the solutions of the equations of cases (a) and (b) above. The WKB solution of case (a) is (c being a normalization constant and the z_0 occurring here a turning point)

$$
\psi_0(z) = \frac{c}{v^{1/4}} \exp\left(-\int_{z_0}^{z} dz v^{1/2} \right), \quad v(z) = n^2 - 6\,\mathrm{sech}^2 z \quad \text{and} \quad v(z_0) = 0.
$$

Thus for $z \to \infty$ we have $v^{1/2}(z) \to n$ and $\psi_0(z) \to 0$, as required. The corresponding WKB solution of case (b) which satisfies its required boundary condition, i.e. vanishing at $z = mT$, is similarly seen to be

$$
\tilde{\psi}_0(z) = \frac{c}{\tilde{v}^{1/4}} \left[\exp\left(-\int_{z_0}^{z} dz\tilde{v}^{1/2} \right) - \exp\left(-2\int_{z_0}^{mT} dz\tilde{v}^{1/2} \right) \exp\left(\int_{z_0}^{z} dz\tilde{v}^{1/2} \right) \right],
$$

where $\tilde{v} = n^2 - 6\,\mathrm{sech}^2 z - (\tilde{w}_0^2/m^2)$. Differentiating the last expression we obtain

$$
\frac{d}{dz}\tilde{\psi}_0(z) = -c\tilde{v}^{1/4} \left[\exp\left(-\int_{z_0}^{z} dz\tilde{v}^{1/2} \right) + \exp\left(-2\int_{z_0}^{mT} dz\tilde{v}^{1/2} \right) \exp\left(\int_{z_0}^{z} dz\tilde{v}^{1/2} \right) \right],
$$

and so

$$
\frac{d}{dz}\tilde{\psi}_0\Big|_{z=mT} \simeq -2c\tilde{v}^{1/4} \exp\left(-\int_{z_0}^{mT} dz\tilde{v}^{1/2} \right).
$$

But (see $\psi_0(z)$ above)

$$
\frac{d}{dz}\psi_0\Big|_{z=mT} \simeq -cv^{1/4} \exp\left(-\int_{z_0}^{mT} dz v^{1/2} \right).
$$

Hence (with $n^2 \simeq n^2 - (\tilde{w}_0^2/m^2)$ for $T \to \infty$)

$$
\frac{d}{dz}\tilde{\psi}_0(mT) \simeq 2\frac{d}{dz}\psi_0(mT), \quad \text{and analogously} \quad \frac{d}{dz}\tilde{\psi}_0(-mT) \simeq 2\frac{d}{dz}\psi_0(-mT).
$$

It follows that the left hand side of Eq. (23.55a) is

$$
\psi_0 \frac{d\tilde{\psi}_0}{dz} \Big|_{-mT}^{mT} \simeq 2\psi_0 \frac{d}{dz}\psi_0 \Big|_{-mT}^{mT}.
$$

But

$$
\psi_0(z) = \frac{m^2}{g} \frac{1}{\cosh^2 z} \simeq \frac{4m^2}{g} e^{\mp 2z} \quad \text{for} \quad z \to \pm\infty,
$$

so that

$$
\frac{d}{dz}\psi_0(z) \simeq \mp \frac{8m^2}{g} e^{\mp 2z} \simeq \mp 2\psi_0(z) \quad \text{for} \quad z \to \pm\infty.
$$

Hence

$$
\psi_0(\pm mT)\frac{d}{dz}\tilde{\psi}_0(\pm mT) = 2\psi_0(\pm mT)\frac{d}{dz}\psi_0(\pm mT) \simeq \mp 4\left(\frac{4m^2}{g} \right)^2 e^{-4mT}.
$$

Equation (23.55a) therefore becomes

$$-8\left(\frac{4m^2}{g}\right)^2 e^{-4mT} \simeq -\frac{\tilde{w}_0^2}{m^2}\frac{4m^4}{3g^2}.$$

We thus obtain the result

$$\tilde{w}_0^2 \simeq 96m^2 e^{-4mT}, \qquad \tilde{w}_0 \simeq \tilde{v}_0 e^{-2mT}, \qquad \tilde{v}_0 = 4\sqrt{6}m. \tag{23.55b}$$

We observe that this eigenvalue is positive; the configuration associated with the zero mode is approximated by a classically stable configuration. For $T \to \infty$ the eigenvalue vanishes. Note that here we used a Euclidean time interval of total length $2T$. This boundary *perturbation method* may be repeated for higher eigenstates which are then found to possess T-dependent eigenvalues.

23.4.2 Evaluation of the functional integral

It is clear from Eqs. (23.52a) and (23.52c) that the infinite product they contain provides a particular difficulty. It is the evaluation of this quantity that we are concerned with in this section. The highly nontrivial method was devised by Dashen, Hasslacher and Neveu* with reference to other sources.[†] We set

$$I_0 := \det\left(\frac{d\eta(\tau_m)}{d\xi_n}\right)\int\left(\prod_{k=0}^{i}d\xi_k\right)\exp\left[-\frac{1}{2}\sum_n |\xi_n|^2 w_n^2 \rho^2\right]. \tag{23.56}$$

One introduces the following mapping or so-called *Volterra transformation* for large but finite T at the lower limit

$$z(\tau) = \eta(\tau) - \int_{-T}^{\tau}\frac{\dot{N}(\tau')}{N(\tau')}\eta(\tau')d\tau', \quad z(-T) = \eta(-T) \stackrel{T\to\infty}{\longrightarrow} 0, \tag{23.57}$$

where $N(\tau)$ is the zero mode, i.e.

$$N(\tau) = \frac{d\phi_c}{d\tau} = \frac{m^2}{g}\frac{1}{\cosh^2 m(\tau - \tau_0)}. \tag{23.58}$$

One can then show — see Examples 23.2, 23.3 below — that

$$\begin{aligned}
I_0 &= \int \mathcal{D}\{z\}\left|\frac{\mathcal{D}z}{\mathcal{D}\eta}\right|^{-1}\exp\left[-\frac{1}{\hbar}\int_{-T}^{T}d\tau\left\{\frac{1}{2}\dot{z}^2(\tau)\right\}\right] \\
&= \left(\frac{1}{2\pi\hbar}\right)^{1/2}[N(T)N(-T)]^{-1/2}\left[\int_{-T}^{T}\frac{d\tau}{N^2(\tau)}\right]^{-1/2}.
\end{aligned} \tag{23.59}$$

---

*R. F. Dashen, B. Hasslacher and A. Neveu [63].

[†]Integration in functional spaces and its application in quantum physics is lucidly described in I. M. Gel'fand and A. M. Yaglom [107]. In particular this paper utilizes results of R. H. Cameron and W. T. Martin [43], [44] which are required in our context below.

Observe that this expression is independent of the normalization of the zero mode $N(\tau)$. Also observe that the transformation[‡§] (23.57) transforms effectively to a free particle Lagrangian. After evaluation in Example 23.2 of the *functional determinant* occurring here, the integration requires the introduction of a Lagrange multiplier which enforces the endpoint conditions, i.e. boundary conditions, on the fluctuation $\eta(\tau)$ and hence on $\phi(\tau)$ itself; this latter calculation is done in Example 23.3. The evaluation of I_0 with the above formula is then straightforward. We have for $T \to \infty$

$$N(T)N(-T) = \frac{m^4}{g^2} \frac{1}{\cosh^4 mT} \simeq \frac{m^4}{g^2} 2^4 e^{-4mT}$$

and

$$\int_{-T}^{T} \frac{d\tau}{N^2(\tau)} = \frac{g^2}{m^4} \frac{1}{m} \int_{-mT}^{mT} dz \, \cosh^4 z$$

$$= \frac{g^2}{m^5} \left[\frac{\sinh 4z}{32} + \frac{\sinh 2z}{4} + \frac{3z}{8} \right]_{-mT}^{mT} \simeq \frac{g^2}{m^5} \frac{1}{32} e^{4mT}.$$

We observe how the large-T dependence cancels out in I_0 and obtain the simple result

$$I_0 = \det\left(\frac{d\eta(\tau_m)}{d\xi_n} \right) \prod_{\text{all } k} \left[\frac{2\pi}{w_k^2 \rho^2} \right]^{1/2} = \left(\frac{m}{\pi\hbar} \right)^{1/2}. \qquad (23.60a)$$

[‡]Note the variable τ at the upper limit of Eq. (23.57) which implies that the transformation is a Volterra integral equation. This is an important point and should be compared with a Fredholm integral equation which has constants at both limits of integration. Consider a differential equation with second order differential operator M and Green's function $K(x, x')$, e.g.

$$(M - \lambda)u(x) = f(x), \quad MK(x, x') = \delta(x - x'), \quad a \le x \le b,$$

and with boundary conditions $B_1[u] = 0, B_2[u] = 0$. The solution $u(x)$ of the differential equation can be written

$$u(x) = h(x) + \lambda \int_a^b K(x, x')u(x')dx' \quad \text{with} \quad h(x) \equiv \int_a^b K(x, x')f(x')dx'.$$

For $f(x) \ne 0$ this integral equation is called an inhomogeneous Fredholm integral equation; for $f(x) = 0$ the equation is called a homogeneous Fredholm integral equation. If one has an integral equation where $K(x, x')$ is a function of either of the following types,

$$K(x, x') = k(x, x')\theta(x - x') \quad \text{or} \quad K(x, x') = k(x, x')\theta(x' - x),$$

the inhomogeneous Fredholm equation becomes a Volterra equation, i.e.

$$u(x) = h(x) + \lambda \int_a^x k(x, x')u(x')dx' \quad \text{or} \quad u(x) = h(x) + \lambda \int_x^b k(x, x')u(x')dx'.$$

[§]In the book of B. Felsager [91] this transformation can be found in his Eq. (5.42), p. 190.

Inserting this result into Eq. (23.52c) we obtain for the kernel I_1 the simple expression

$$I_1 = \left(\frac{m}{\pi\hbar}\right)^{1/2} \exp\left[-\frac{4}{3}\frac{m^3}{\hbar g^2}\right]. \tag{23.60b}$$

We see that this result is no-where near to allowing a comparison with our earlier result Eq. (23.35a) for the determination of $\triangle E$. The reason is that the zero mode has not really been removed, so that all fluctuations, i.e. perturbations, are still present. This has to be changed and is achieved with the Faddeev–Popov constraint insertion that we deal with in the next subsection.

Example 23.2: Evaluation of the functional determinant
Apply the shift transformation (23.57) to the fluctuation integral (23.43) and evaluate the functional determinant occurring in Eq. (23.59), i.e.

$$\left|\frac{\mathcal{D}z}{\mathcal{D}\eta}\right|.$$

Solution: We first derive the inverse of the transformation (23.57) by differentiating this which yields

$$\dot{z}(\tau) = \dot{\eta}(\tau) - \frac{\dot{N}(\tau)}{N(\tau)}\eta(\tau).$$

We can rewrite this equation as

$$\frac{\dot{z}(\tau)}{N(\tau)} = \frac{\dot{\eta}(\tau)}{N(\tau)} - \frac{\dot{N}(\tau)}{N^2(\tau)}\eta(\tau) = \frac{d}{d\tau}\left(\frac{\eta(\tau)}{N(\tau)}\right).$$

Thus

$$\frac{\eta(\tau)}{N(\tau)} = \int_{-T}^{\tau} d\tau \frac{\dot{z}(\tau)}{N(\tau)}.$$

When multiplied by $N(\tau)$, partial integration of the right hand side of this equation yields

$$\eta(\tau) = z(\tau) - N(\tau)\int_{-T}^{\tau} d\tau' \frac{d}{d\tau'}\left(\frac{1}{N(\tau')}\right)z(\tau'),$$

$$\eta(\tau) = z(\tau) + f(\tau), \quad f(\tau) = N(\tau)\int_{-T}^{\tau} d\tau' \frac{\dot{N}(\tau')}{N^2(\tau')}z(\tau'). \tag{23.61}$$

For further manipulations we observe that with $\dot{z}^2(\tau)$ (obtained by squaring the expression of the first equation above) we have by differentiating out the following expression:

$$\frac{d}{d\tau}\left\{\eta^2(\tau)\frac{\dot{N}(\tau)}{N(\tau)}\right\} = +\underbrace{\eta^2(\tau)\frac{\ddot{N}(\tau)}{N(\tau)}}_{\eta^2(\tau)V''(\phi_c)} + \underbrace{2\eta(\tau)\dot{\eta}(\tau)\frac{\dot{N}(\tau)}{N(\tau)} - \eta^2(\tau)\frac{\dot{N}^2(\tau)}{N^2(\tau)}}_{-\dot{z}^2(\tau)+\dot{\eta}^2(\tau)},$$

so that since $\ddot{N}(\tau) - V''(\phi_c)N(\tau) = 0$ (by definition $N(\tau)$ being the zero mode)

$$\dot{\eta}^2(\tau) + V''(\phi_c)\eta^2(\tau) = \dot{z}^2(\tau) + \frac{d}{d\tau}\left\{\eta^2(\tau)\frac{\dot{N}(\tau)}{N(\tau)}\right\}.$$

The derivative term will from now on be ignored since when inserted into the action integral and integrated it yields zero on account of the vanishing of the fluctuation $\eta(\tau)$ at the endpoints $\tau = \tau_0 = -T, T$. Thus

$$\delta S_E(\eta) = \frac{1}{2} \int_{-T}^{T} d\tau \left[\left(\frac{d\eta}{d\tau} \right)^2 + V''(\phi_c)\eta^2 \right] = \int_{-T}^{T} d\tau \left[\frac{1}{2} \dot{z}^2(\tau) \right].$$

The functional integral to be evaluated is I_0 of Eq. (23.48b) which excludes the classical factor. Thus

$$
\begin{aligned}
I_0 &= \int \mathcal{D}\{\eta\} \exp \left[- \int_{-T}^{T} d\tau \left\{ \frac{1}{2} \left(\frac{d\eta}{d\tau} \right)^2 + \frac{1}{2}\eta^2 V''(\phi_c) \right\} \right] \\
&= \int \mathcal{D}\{z\} \left| \frac{\mathcal{D}z}{\mathcal{D}\eta} \right|^{-1} \exp \left[- \int_{-T}^{T} d\tau \left\{ \frac{1}{2} \dot{z}^2(\tau) \right\} \right],
\end{aligned}
$$

since*

$$\mathcal{D}\{\eta\} \left| \frac{\mathcal{D}z}{\mathcal{D}\eta} \right| = \mathcal{D}\{z\}, \quad \left| \frac{\mathcal{D}z}{\mathcal{D}\eta} \right| = \left| \frac{\mathcal{D}\eta}{\mathcal{D}z} \right|^{-1}.$$

Our next step is the evaluation of the functional determinant. From Eq. (23.57) we deduce

$$\frac{\partial z(\tau)}{\partial \eta(\tau'')} = \delta(\tau - \tau'') + \int_{-T}^{\tau} K(\tau, \tau')\delta(\tau' - \tau'')d\tau' \quad \text{with} \quad K(\tau, \tau') \equiv -\frac{\dot{N}(\tau')}{N(\tau')}, \qquad (23.62)$$

where we allow for a general τ dependence in the Volterra kernel as well. Discretizing the Volterra equation (23.57), i.e.

$$z(\tau) = \eta(\tau) + \int_{-T}^{\tau} K(\tau, \tau')\eta(\tau')d\tau',$$

or more generally an integral equation of Fredholm type, by setting $\delta = \tau_{i+1} - \tau_i, i = 0, 1, \ldots, n$ with $\tau_0 = -T, \tau_n = \tau$, we can write the equation

$$z(\tau_n) = \sum_{i=1}^{n} \delta_{ni}\eta(\tau_i) + \delta \sum_{i=1}^{n} K(\tau_n, \tau_i)\eta(\tau_i). \qquad (23.63a)$$

We observe that $K(\tau_n, \tau_i) = 0$ for $i > n$. The discretization here is a very delicate point. A different and convenient — but not unique — discretization (discussed after Eq. (23.66)) is to rewrite the equation as

$$
\begin{aligned}
z(\tau_n) &= \sum_{i=1}^{n} \delta_{ni}\eta(\tau_i) + \delta \sum_{i=1}^{n} K(\tau, \tau_i)\frac{1}{2}\{\eta(\tau_i) + \eta(\tau_{i-1})\} \\
&= \sum_{i=1}^{n} \delta_{ni}\eta(\tau_i) + \frac{\delta}{2} \sum_{i=1}^{n} K(\tau, \tau_i)\eta(\tau_i) + \frac{\delta}{2} \sum_{i=0}^{n-1} K(\tau, \tau_{i+1})\eta(\tau_i). \qquad (23.63b)
\end{aligned}
$$

Equation (23.63a) can be rewritten in the matrix form of a set of simultaneous linear equations:

$$
\begin{pmatrix} z(\tau_1) \\ z(\tau_2) \\ z(\tau_3) \\ \cdot \\ z(\tau_n) \end{pmatrix} = \begin{pmatrix} 1 + \delta K(\tau_1, \tau_1) & 0 & 0 & 0 & 0 \\ \delta K(\tau_2, \tau_1) & 1 + \delta K(\tau_2, \tau_2) & 0 & 0 & 0 \\ \delta K(\tau_3, \tau_1) & \delta K(\tau_3, \tau_2) & 1 + \delta K(\tau_3, \tau_3) & 0 & 0 \\ & & & & 0 \\ \cdot & \cdot & \cdot & \cdot & 1 + \delta K(\tau_n, \tau_n) \end{pmatrix} \begin{pmatrix} \eta(\tau_1) \\ \eta(\tau_2) \\ \eta(\tau_3) \\ \cdot \\ \eta(\tau_n) \end{pmatrix}.
$$

---

*See also I. M. Gel'fand and A. M. Yaglom [107], after their Eq. (3.4).

This equation represents a set of n linear equations which has a unique solution provided the determinant D_n of the coefficient matrix on the right does not vanish. This determinant can be expanded in powers of K (more precisely in powers of a parameter λ which we attach to K). Thus

$$D_n = 1 + \delta \sum_{i=1}^{n} K(\tau_i, \tau_i) + \frac{\delta^2}{2!} \sum_{i,j=1}^{n} \begin{vmatrix} K(\tau_i, \tau_i) & 0 \\ K(\tau_j, \tau_i) & K(\tau_j, \tau_j) \end{vmatrix} + \cdots .$$

The factorials arise since every determinant of l rows and l columns appears $l!$ times when $i, j, \ldots$ are summed over all values from 1 to n, whereas the determinant arises only once in the original determinant D_n. In the limit $\delta \to 0, n \to \infty$ the expansion becomes

$$
\begin{aligned}
D &= \lim_{n \to \infty} D_n = 1 + \int_{-T}^{\tau} K(\tau_1, \tau_1) d\tau_1 + \frac{1}{2!} \int_{-T}^{\tau} d\tau_1 \int_{-T}^{\tau} d\tau_2 \begin{vmatrix} K(\tau_1, \tau_1) & 0 \\ K(\tau_2, \tau_1) & K(\tau_2, \tau_2) \end{vmatrix} + \cdots \\
&= \exp\left[\int_{-T}^{\tau} K(\tau_1, \tau_1) d\tau_1 \right].
\end{aligned} \tag{23.64}
$$

Inserting the explicit expression for the kernel given in Eq. (23.62) — and with insertion of a discretization dependent factor α still to be explained — we have in the present case

$$D = \exp\left[-\alpha \int_{-T}^{\tau} \frac{\dot{N}(\tau')}{N(\tau')} d\tau' \right] = \exp[-\alpha\{\ln N(\tau) - \ln N(-T)\}] = \left[\frac{N(\tau)}{N(-T)} \right]^{-\alpha}. \tag{23.65}$$

The quantity D is really the *Fredholm determinant* which is obtained as above by solving the Fredholm integral equation with constant integration limits. The Fredholm kernel is in general — as in our illustration above with a Green's function — a quantity with a discontinuity between the two constant integration limits, i.e. the kernels to the left and to the right would be associated with step functions $\theta(\tau - \tau')$ and $\theta(\tau' - \tau)$. Thus in our case one of the kernels is zero and we have to decide what we do when, as in Eq. (23.64), the kernel $K(\tau, \tau')$ is to be evaluated at $\tau = \tau'$. This case has been considered in the literature[†] where it is shown that if the Volterra transformation is considered as a Fredholm transformation whose kernel vanishes on one side of the discontinuity or diagonal, the value of the determinant is half (or arithmetic mean value) along the diagonal. Thus the factor α inserted above has to be taken as $1/2$.

Our final step is to relate the Volterra–Fredholm determinant to the required functional determinant. We observe that if we discretize Eq. (23.64) as above and consider the determinant of the expression, we obtain exactly the nonvanishing terms on the right of Eq. (23.63a), so that (with $\tau \to T$)

$$\left| \frac{\mathcal{D}z}{\mathcal{D}\eta} \right| \overset{(23.64)}{=} D \overset{}{=} \left[\frac{N(T)}{N(-T)} \right]^{-1/2}. \tag{23.66}$$

Finally we add the following remarks concerning the discretization (23.63b). With this specific discretization the Jacobi matrix becomes diagonal and hence we obtain with the determinant coming exclusively from the diagonal[‡]

$$\left| \frac{\mathcal{D}z}{\mathcal{D}\eta} \right| = \lim_{n \to \infty} \prod_{i=1}^{n} \left(1 + \frac{\delta}{2} K(\tau, \tau_i) \right) \tag{23.67}$$

and then

$$
\begin{aligned}
\left| \frac{\mathcal{D}z}{\mathcal{D}\eta} \right| &= \lim_{n \to \infty} \exp \ln \prod_{i=1}^{n} \left(1 + \frac{\delta}{2} K(\tau, \tau_i) \right) = \lim_{n \to \infty} \exp \left[\sum_{i=1}^{n} \ln \left(1 + \frac{\delta}{2} K(\tau, \tau_i) \right) \right] \\
&= \lim_{n \to \infty} \exp \left[\frac{\delta}{2} \sum_{i=1}^{n} K(\tau, \tau_i) \right] = \exp \left[\frac{1}{2} \int_{-T}^{\tau} K(\tau, \tau') d\tau' \right] \\
&= D \ \text{above, where } a = \frac{1}{2}.
\end{aligned} \tag{23.68}
$$

[†]See R. H. Cameron and W. T. Martin [43], [44].
[‡]This is the method explained by B. Felsager [91], p. 192.

Example 23.3: Implementation of the endpoint constraints

Using a Lagrange multiplier α insert into the functional integral (23.59) the endpoint conditions $\eta(\pm T) = 0$ and evaluate the integral.

Solution: From Eq. (23.61) we obtain the constraint on the function $z(\tau)$ which results from the endpoint constraint $\eta(\tau_f) = \eta(T) = 0$ on the fluctuations $\eta(\tau)$. Thus

$$0 = z(\tau_f) + f(\tau_f) \quad \text{or} \quad 0 = z(T) + f(T), \quad f(T) = N(T) \int_{-T}^{T} d\tau' \frac{\dot{N}(\tau')}{N^2(\tau')} z(\tau'). \tag{23.69}$$

Incorporating this condition into the functional integral I_0 with a delta function

$$1 = \int dz(T)\delta[z(T) + f(T)] = \frac{1}{2\pi} \int dz(T) \int_{-\infty}^{\infty} d\alpha e^{i\alpha[z(T)+f(T)]},$$

we have (cf. Eq. (23.59))

$$
\begin{aligned}
I_0 &= \frac{1}{2\pi} \int dz(T) \int \mathcal{D}\{z\} \int d\alpha \left|\frac{\mathcal{D}z}{\mathcal{D}\eta}\right|^{-1} \\
&\quad \times \exp\left[-\frac{1}{\hbar} \int_{-T}^{T} d\tau \frac{1}{2}\dot{z}^2(\tau) + i\frac{\alpha}{\hbar}\left(z(T) + N(T) \int_{-T}^{T} d\tau' \frac{\dot{N}(\tau')}{N^2(\tau')} z(\tau') \right) \right], \tag{23.70}
\end{aligned}
$$

and this now requires also integration over the endpoint coordinate $z(T)$. With partial integration (in the second step) we obtain (since (cf. (23.57)) $z(-T) \to 0$)

$$
\begin{aligned}
N(T) \int_{-T}^{T} d\tau' \frac{\dot{N}(\tau')}{N^2(\tau')} z(\tau') &= -N(T) \int_{-T}^{T} d\tau' \frac{d}{d\tau'}\left(\frac{1}{N(\tau')} \right) z(\tau') \\
&= N(T)\left[-\frac{z(\tau')}{N(\tau')} \right]_{\tau'=-T}^{T} + N(T) \int_{-T}^{T} d\tau' \frac{\dot{z}(\tau')}{N(\tau')} \\
&= -z(T) + \int_{-T}^{T} d\tau' \frac{N(T)}{N(\tau')} \dot{z}(\tau'). \tag{23.71}
\end{aligned}
$$

Hence (with $\hbar = 1$)

$$
\begin{aligned}
&I_0 \\
&= \frac{1}{2\pi} \int dz(T) \int \mathcal{D}\{z\} \int d\alpha \left|\frac{\mathcal{D}z}{\mathcal{D}\eta}\right|^{-1} \exp\left[-\int_{-T}^{T} d\tau \left\{ \frac{1}{2}\dot{z}^2(\tau) + i\alpha \frac{N(T)}{N(\tau)} \dot{z}(\tau) \right\} \right] \\
&= \frac{1}{2\pi} \int dz(T) \int \mathcal{D}\{z\} \int d\alpha \left|\frac{\mathcal{D}z}{\mathcal{D}\eta}\right|^{-1} \exp\left[-\int_{-T}^{T} d\tau \left\{ \frac{1}{2}\left(\dot{z}(\tau) + i\alpha \frac{N(T)}{N(\tau)} \right)^2 + \frac{\alpha^2}{2} \frac{N^2(T)}{N^2(\tau)} \right\} \right].
\end{aligned}
$$

We perform the functional integration as in Chapter 21. Briefly, the range of τ is subdivided into $n + 1$ equal elements of length ϵ with $(n + 1)\epsilon = 2T$ and $\tau_0 = -T, \tau_{n+1} = T$, so that (with $\gamma_i = \sqrt{-\alpha^2 N(T)/N(\tau_i)}$)

$$\int \mathcal{D}\{z\} \exp\left[\int_{-T}^{T} d\tau \left\{ \frac{1}{2}\left(\dot{z}(\tau) + i\alpha \frac{N(T)}{N(\tau)} \right)^2 \right\} \right] = \lim_{n\to\infty, \epsilon\to 0} I_n, \quad n \neq 0,$$

$$I_n = \int dz_1 \int dz_2 \cdots \int dz_n \exp\left[\frac{1}{\epsilon} \sum_{i=1}^{n+1} \frac{1}{2}(z_i - z_{i-1} + \epsilon\gamma_i)^2 \right]. \tag{23.72}$$

The endpoint integration $dz(T) = dz(\tau_{n+1})$ can be combined with this. In every integration $\int dz_i$ the contribution $\epsilon\gamma_i$ represents a constant translation, and hence may be ignored. Evaluating I_1, I_2 and using induction one obtains (cf. Eq. (21.22))

$$I_n = \frac{(\pi\epsilon)^{n/2}}{\sqrt{n+1}} \exp\left[\frac{1}{\epsilon(n+1)}(z_{n+1} - z_0)^2 \right] \quad n \neq 0. \tag{23.73}$$

In the required limit the argument of the exponential is proportional to $1/T$ and thus vanishes in the limit $T = (n+1)\epsilon \to \infty$. Ignoring the remaining metric factor with the reasoning explained between Eqs. (21.10) and (21.11), we are left with[§]

$$
\begin{aligned}
I_0 &= \frac{1}{2\pi}\left|\frac{\mathcal{D}z}{\mathcal{D}\eta}\right|^{-1}\int_{-\infty}^{\infty}d\alpha\,\exp\left[-\frac{\alpha^2}{2}\int_{-T}^{T}d\tau\,\frac{N^2(T)}{N^2(\tau)}\right] \\
&= \frac{1}{2\pi}\left|\frac{\mathcal{D}z}{\mathcal{D}\eta}\right|^{-1}\sqrt{2\pi}\left[N(T)\sqrt{\int_{-T}^{T}\frac{d\tau}{N^2(\tau)}}\right]^{-1} \\
&= \frac{1}{\sqrt{2\pi\hbar}}[N(T)N(-T)]^{-1/2}\left[\int_{-T}^{T}\frac{d\tau}{N^2(\tau)}\right]^{-1/2},
\end{aligned}
\tag{23.74}
$$

where we re-inserted $\hbar$. This result verifies Eq. (23.59) and agrees with that in the literature.[¶]

23.4.3   The Faddeev–Popov constraint insertion

The Faddeev–Popov constraint insertion implies the complete removal of the zero mode.[‖]   The method consists essentially in replacing the single integration $\int d\xi_0$ by a double integration involving a delta function, i.e.

$$
\int d\xi_0 = \int\int\frac{\partial F(\tau_0,\xi_0)}{\partial\tau_0}\delta[F(\tau_0,\xi_0)-\alpha\tau_0]d\tau_0 d\xi_0,\quad \alpha=\text{const.,}\tag{23.75}
$$

or more generally a three-dimensional surface integration $\int d\sigma^\nu$ by the volume integration $\int d^4x$ through the relation

$$
\int d\sigma^\nu = \int\int\frac{\partial F(x)}{\partial x_\nu}\delta[F(x)-\alpha\tau_0]d^4x,\quad \alpha=\text{const.}
$$

We started off with

$$
\phi(\tau) = \phi_c(\tau) + \eta(\tau)
$$

and set

$$
\eta(\tau) = \sum_n \xi_n\eta_n(\tau),\quad \frac{1}{\rho^2}\int_{-\infty}^{\infty}d\tau\eta_n(\tau)\eta_m(\tau)=\delta_{nm},
$$

$$
\xi_n = \frac{1}{\rho^2}\int d\tau\eta(\tau)\eta_n(\tau).\tag{23.76}
$$

Since the zero mode $\eta_0(\tau)$ is proportional to $d\phi_c(\tau)/d\tau$ with

$$
\int_{-\infty}^{\infty}d\tau\left(\frac{d\phi_c(\tau)}{d\tau}\right)^2 = S_E,
$$

[§]Recall that $\int_{-\infty}^{\infty}d\alpha e^{-\beta\alpha^2}=\sqrt{\pi/\beta}$.
[¶]E. Gildener and A. Patrascioiu [110], Eq. (3.12).
[‖]L. D. Faddeev and V. N. Popov [89].

this means that

$$\eta_0(\tau) = \frac{\rho}{\sqrt{S_E}} \frac{d\phi_c(\tau)}{d\tau}. \tag{23.77}$$

The set $\{\eta_n(\tau)\}$ constitutes a basis of the fluctuation space about the classical configuration. The number ξ_0 is the component of the fluctuation $\eta(\tau)$, e.g. at $\tau = \tau_0$, in the direction $\eta_0(\tau)$ at the point τ_0 of the parameter τ which parametrizes the trajectory of the classical configuration $\phi_c(\tau)$. A position τ_0 of this classical configuration is called the *collective coordinate*** of (the field) $\phi(\tau)$.

In analogy to (23.76) we set

$$\phi_c(\tau) = \sum_n c_n \eta_n(\tau), \quad c_n = \frac{1}{\rho^2} \int d\tau \phi_c(\tau) \eta_n(\tau). \tag{23.78}$$

Then

$$\phi(\tau) = \sum_n \tilde{\xi}_n \eta_n(\tau), \tag{23.79}$$

where

$$\tilde{\xi}_n = \xi_n + c_n. \tag{23.80}$$

We now impose (i.e. demand validity of) the *constraint* $\tilde{\xi}_0 = 0$, i.e.

$$\frac{1}{\rho^2} \int_{-\infty}^{\infty} d\tau \eta_0(\tau) \phi(\tau) = 0, \tag{23.81}$$

or

$$\frac{1}{\rho^2} \int_{-\infty}^{\infty} d\tau \eta_0(\tau) \phi_c(\tau) = -\frac{1}{\rho^2} \int_{-\infty}^{\infty} d\tau \eta_0(\tau) \eta(\tau). \tag{23.82}$$

This condition says that $\phi(\tau)$ is not to possess any component in the direction of $\eta_0(\tau)$, i.e. in the direction of the zero mode. Thus, if the set $\{\eta_n(\tau)\}$ spans the Hilbert space of fluctuation states about ϕ_c (at collective coordinate position τ), this constraint implies that we go to the subspace which is orthogonal to the zero mode.

In the integral of Eq. (23.81) the τ-dependence is integrated out and hence the integral does not provide the function $F(\tau_0, \xi_0)$ with dependence on τ_0 which the replacement (23.75) requires. We establish the function $F(\tau_0, \xi_0)$ with the following reasoning. The delta function $\delta[F(\tau_0, \xi_0) - \alpha\tau_0]$ contained in Eq. (23.75) enforces the collective coordinate τ_0 to be given by $F(\tau_0, \xi_0)/\alpha$. We consider the following integral

$$f(\tau_0) := \frac{1}{\rho} \int d\tau \eta_0(\tau) \phi_c(\tau + \tau_0) \tag{23.83}$$

**The collective coordinate in the present context is also called *"instanton time"*; in the case of the soliton of the static $1+1$ dimensional theory this is the kink coordinate.

for an infinitesimal translational shift τ_0 of the position of $\phi_c(\tau)$. Expanding $f(\tau_0)$ about $\tau_0 = 0$ we obtain (with $\phi'_c(\tau) = \eta_0(\tau)\sqrt{S_E}/\rho$)

$$
\begin{aligned}
f(\tau_0) &= \frac{1}{\rho}\int d\tau \eta_0(\tau)\phi_c(\tau) + \frac{\tau_0}{\rho}\int d\tau \eta_0(\tau)\phi'_c(\tau) \\
&= \frac{1}{\rho}\int d\tau \eta_0(\tau)\phi_c(\tau) + \sqrt{S_E}\tau_0.
\end{aligned}
\tag{23.84}
$$

Hence

$$
\frac{1}{\rho}\int d\tau \eta_0(\tau)[\phi_c(\tau+\tau_0) - \phi_c(\tau)] - \sqrt{S_E}\tau_0 = 0.
\tag{23.85}
$$

With Eq. (23.82) we can replace here

$$
-\eta_0(\tau)\phi_c(\tau) \quad \text{by} \quad \eta_0(\tau)\eta(\tau),
$$

so that with $F(\tau_0, \xi_0)$ defined as

$$
F(\tau_0, \xi_0) = \frac{1}{\rho}\int d\tau \eta_0(\tau)[\phi_c(\tau+\tau_0) + \eta(\tau)] = f(\tau_0) + \rho\xi_0,
\tag{23.86}
$$

we have

$$
F(\tau_0, \xi_0) - \sqrt{S_E}\tau_0 = 0.
\tag{23.87}
$$

We now define

$$
\triangle_{FP} := \left[\frac{\partial}{\partial \tau_0}F(\tau_0, \xi_0)\right]_{\tau_0 \sim 0}.
\tag{23.88}
$$

Then in Eq. (23.75)

$$
1 = \triangle_{FP}\int d\tau_0 \delta[(F(\tau_0, \xi_0) - \sqrt{S_E}\tau_0].
\tag{23.89}
$$

Here the factor $\triangle_{FP}$ is called the *Faddeev–Popov determinant*. We evaluate this quantity as follows in the leading approximation. We have

$$
\begin{aligned}
\triangle_{FP} &= \frac{\partial}{\partial \tau_0}F(\tau_0, \xi_0) \\
&= \frac{\partial}{\partial \tau_0}\frac{1}{\rho}\int d\tau \eta_0(\tau)[\phi_c(\tau+\tau_0) + \eta(\tau)] \\
&= \frac{1}{\rho}\left[\int d\tau \eta_0(\tau)\frac{\partial}{\partial \tau_0}\phi_c(\tau+\tau_0) + \frac{\partial}{\partial \tau_0}\int d\tau \eta(\tau)\eta_0(\tau)\right].
\end{aligned}
\tag{23.90}
$$

Now[††]

$$
\left(\frac{\partial \phi_c(\tau+\tau_0)}{\partial \tau_0}\right)_{\tau_0=0} = \left(\frac{d\phi_c(\tau)}{d\tau}\right).
\tag{23.91}
$$

[††]For more discussion of this point and related aspects see also B. Felsager [91], p. 158.

Thus a shift τ_0 in the direction of the zero mode (i.e. tangential to $\phi_c(\tau)$) leaves this and hence the square of this quantity, i.e. the "static energy" invariant. Inserting (23.91) into Eq. (23.90), we obtain (remembering that $\eta_0 = \rho d\phi_c/d\tau/\sqrt{S_E}$)

$$\triangle_{FP} \;=\; \frac{1}{\sqrt{S_E}} \int_{-T}^{T} d\tau \left(\frac{d\phi_c(\tau)}{d\tau} \right)^2 \stackrel{T\cong\infty}{=} \sqrt{S_E}. \tag{23.92}$$

We find therefore that the *value of the Faddeev–Popov determinant* is given by the square root of twice the classical kinetic energy or simply of S_E.[‡‡] The relation (23.75) now becomes

$$
\begin{aligned}
\int d\xi_0 \;&=\; \int_{\triangle T} d\tau_0 \int d\xi_0 \triangle_{FP} \delta[f(\tau_0) + \rho\xi_0 - \sqrt{S_E}\tau_0] \\
&=\; \int_{\triangle T} d\tau_0 \int d\xi_0 \triangle_{FP} \frac{1}{\rho}\delta\left[\xi_0 + \frac{f(\tau_0) - \sqrt{S_E}\tau_0}{\rho}\right] \\
&=\; \frac{1}{\rho} \int_{\triangle T} d\tau_0 \triangle_{FP},
\end{aligned} \tag{23.93}
$$

where $\triangle T = 2T \to \infty$. The factor ρ is a normalization factor which we retain here only for ease of comparison with literature and may otherwise simply be put equal to one. Thus the result is that the integration over the tangential or zero mode component ξ_0 is replaced by an integration over the collective coordinate τ_0 multiplied by the Faddeev–Popov determinant.

There is a quick way to obtain the result (23.93). Consider, using the relation (23.77) between the zero mode $d\phi_c(\tau)/d\tau$ and the fluctuation vector in its direction, $\eta_0(\tau)$,

$$d\phi_c(\tau_0) = \phi_c(\tau_0 + d\tau_0) - \phi_c(\tau_0) = d\tau_0\phi_c'(\tau_0) = d\tau_0\frac{\sqrt{S_E}}{\rho}\eta_0(\tau_0).$$

We have also

$$d\eta(\tau_0) = d\xi_0\eta_0(\tau_0) + \sum_{i\neq0} d\xi_i\eta_i(\tau_0).$$

Thus there is no fluctuation along the direction of the zero mode in the sum, i.e. in $d\phi_c(\tau_0) + d\eta(\tau_0)$, if the coefficient of $\eta_0(\tau_0)$ vanishes, i.e. if

$$d\tau_0\frac{\sqrt{S_E}}{\rho} + d\xi_0 = 0, \tag{23.94}$$

the relative sign being of no significance.

[‡‡]This is an important result which will be required later in the explicit evaluation of path integrals.

23.4.4   The single instanton contribution

We now return to the multiple integral (23.52a) and replace in this $\int d\xi_0$ by the expression (23.93), so that

$$
I_1 = \det\left(\frac{d\eta(\tau_m)}{d\xi_n}\right) \int \left(\prod_{k\neq 0} d\xi_k\right) \exp\left[-\frac{1}{2}\sum_n |\xi_n|^2 w_n^2 \rho^2\right]
$$
$$
\times \frac{1}{\rho}\triangle_{FP}\triangle T \exp\left[-\frac{4}{3}\frac{m^3}{g^2\hbar}\right]. \tag{23.95}
$$

Here $|\xi_0|^2$ in the exponential is multiplied by w_0^2 which is very small for large T and approaches zero in the limit of $T \to \infty$. We can now perform the Gaussian integrations and write the result

$$
I_1 = \lim_{T\to\infty} \frac{1}{\rho}\triangle T\triangle_{FP}\det\left(\frac{d\eta(\tau_m)}{d\xi_n}\right)\left(\prod_{k\neq 0}\sqrt{\frac{2\pi}{w_k^2\rho^2}}\right)\exp\left[-\frac{4}{3}\frac{m^3}{g^2\hbar}\right]
$$
$$
\equiv \lim_{T\to\infty}\frac{1}{\rho}\triangle T\triangle_{FP}I_0'\exp\left[-\frac{4}{3}\frac{m^3}{g^2\hbar}\right], \tag{23.96}
$$

where

$$
I_0' = \det\left(\frac{d\eta(\tau_m)}{d\xi_n}\right)\left(\prod_{k\neq 0}\sqrt{\frac{2\pi}{w_k^2\rho^2}}\right), \quad n\neq 0, \tag{23.97}
$$

is a constant still to be evaluated. This very important intermediate result is the contribution of a single instanton or kink configuration to the path integral (I_1 has been defined by (23.43)). We observe that the result (23.96) does not yet allow a comparison with (23.35a) since the dependence on $\tau_f - \tau_i = 2T$ of both expressions differs significantly. The source of lack of similarity is attributed to the fact that the contribution of a single instanton to the path integral is not enough. We have to take into account also multi-instanton configurations which almost stationarize the action integral and satisfy the same boundary conditions as the instanton.

We determine I_0' by observing that this differs from I_0 of Eq. (23.60a) in not involving the zero mode contributions. Since w_0^2 is for finite T given by $\tilde{w}_0^2$ of Eq. (23.55b), we can set

$$
I_0 = I_0'\sqrt{\frac{2\pi}{\tilde{w}_0^2\rho^2}},
$$

which with Eq. (23.60a) for I_0 and Eq. (23.55b) implies

$$
I_0' = I_0\frac{\tilde{w}_0\rho}{\sqrt{2\pi}} = \left(\frac{m}{\pi\hbar}\right)^{1/2}\frac{\tilde{\omega}_0\rho}{\sqrt{2\pi}} = \left(\frac{m}{\pi\hbar}\right)^{1/2}\frac{4\sqrt{6}m\rho}{\sqrt{2\pi}}e^{-m\triangle T}. \tag{23.98}
$$

It follows that the kernel becomes[*]

$$
\begin{aligned}
I_1 &= \lim_{T \to \infty} \frac{1}{\rho} \triangle T \triangle_{FP} I_0 \left(\frac{\tilde{\omega}_0 \rho}{\sqrt{2\pi}} \right) \exp \left[-\frac{4}{3} \frac{m^3}{g^2 \hbar} \right] \\
&= \lim_{T \to \infty} \frac{1}{\rho} \triangle T \triangle_{FP} I_0 \left(\frac{4\sqrt{6} m \rho}{\sqrt{2\pi}} e^{-m \triangle T} \right) \exp \left[-\frac{4}{3\hbar} \frac{m^3}{g^2} \right]. \quad (23.99)
\end{aligned}
$$

Inserting the explicit expressions for I_0 and $\triangle_{FP}$, the result becomes

$$
I_1 = \lim_{T \to \infty} 2mT \frac{8m^2}{\pi g \hbar^{1/2}} e^{-2mT} \exp \left[-\frac{4}{3\hbar} \frac{m^3}{g^2} \right]. \qquad (23.100)
$$

We observe that the dependence on the normalization constant ρ cancels out. Although we choose $\rho^2 = 1$, we drag the factor ρ^2 along for ease of comparison with literature.[†] Finally we rewrite the expression (23.100) in such a way that it exhibits the crucial factors of which it is composed. Thus — the subscript 1 indicating that it is the *Feynman amplitude* for the one-instanton contribution —

$$
I_1 = \lim_{T \to \infty} \triangle T \triangle_{FP} I_0 \left(\frac{\tilde{w}_0}{\sqrt{2\pi}} \right) \exp \left[-\frac{S_E(\phi_c)}{\hbar} \right], \qquad (23.101)
$$

where with $\triangle T = 2T \to \infty$ (cf. (23.55b))

$$
\tilde{w}_0 = \tilde{v}_0 e^{-m \triangle T}, \quad \tilde{v}_0 = 4\sqrt{6} m.
$$

23.4.5 Instanton–anti-instanton contributions

As already mentioned, in the present case we also have to include multi-instanton configurations in the form of one instanton plus any number of instanton–anti-instanton pairs which describe back and forth tunneling between the two wells. This means we are concerned with configurations like[‡]

$$
\begin{aligned}
\phi_c^{(2n+1)}(\tau) &= \sum_{i=1}^{2n+1} \theta(\tau - T_{i-1})(-1)^{i+1} \phi_c(\tau - \tau_0^{(i)}) \theta(T_i - \tau), \\
\phi_c^{(1)}(\tau) &= \phi_c(\tau), \\
-T = T_0 &\ll \tau_0^{(1)} \ll T_1 \ll \tau_0^{(2)} \ll \cdots \ll T_{2n+1} = T \to \infty \quad (23.102)
\end{aligned}
$$

---

[*]For comparison with the parameters of E. Gildener and A. Patrascioiu [110] make the replacements $\lambda_{GP} \leftrightarrow 2g^2$ and $\mu_{GP}^2 \leftrightarrow 2m^2$.

[†]See E. Gildener and A. Patrascioiu [110]. In view of their normalization of the normal modes, i.e. their Eq. (3.20), they have to introduce the inverse of that factor on the far right of their expression (3.32).

[‡]For the validity of the calculation the instantons or kinks have to be far apart; the approximation is then known as the *"dilute gas approximation"*.

for $n = 1, 2, \ldots$ ($n = 0$ implies the single instanton solution). Proceeding as before we now consider first in analogy with Eq. (23.36)

$$\phi(\tau) = \phi_c^{(2n+1)}(\tau) + \eta(\tau) \quad \text{with} \quad \eta(T_i) \approx 0. \tag{23.103}$$

For instance for $n = 1$ one has the configuration $\phi_c^{(3)}(\tau)$ which has the form shown in Fig. 23.6. The factor $(-1)^{i+1}$ in the sum of Eq. (23.102) distinguishes between instanton (rising from left to right) and anti-instanton (rising from right to left). Taking these configurations into account, the *total Feynman amplitude* is given by

$$\langle \phi_f, \tau_f | \phi_i, \tau_i \rangle = \sum_{n=0}^{\infty} \langle \phi_f, \tau_f | \phi_i, \tau_i \rangle^{(2n+1)}. \tag{23.104}$$

Considering the case $n = 1$ of a kink with one additional kink-anti-kink pair, this means we have to consider the amplitude (23.17b) and there perform first the integrations over the intermediate configurations whose endpoints are not fixed. Consider (with $T_0 = -T, \tau_0^{(1)} = -2T/3, T_1 = -T/3, \tau_0^{(2)} = 0, T_2 = T/3, \tau_0^{(3)} = 2T/3, T_3 = T$)

$$\begin{aligned}
I_3 &\equiv \langle \phi_f, \tau_f | \phi_i, \tau_i \rangle^{(3)} \\
&= \int_{-\infty}^{\infty} d\phi_2 \int_{-\infty}^{\infty} d\phi_1 \langle \phi_f, \tau_f | \phi_2, \tau_2 \rangle \langle \phi_2, \tau_2 | \phi_1, \tau_1 \rangle \langle \phi_1, \tau_1 | \phi_i, \tau_i \rangle \\
&\propto \int_{-\infty}^{\infty} d\phi_2 \int_{-\infty}^{\infty} d\phi_1 e^{-S_E(\phi_f, \phi_2)} e^{-S_E(\phi_2, \phi_1)} e^{-S_E(\phi_1, \phi_i)}, \tag{23.105}
\end{aligned}$$

where we write 'proportional' since only the classical instanton action depends on the endpoint configurations. Since these endpoint configurations, here ϕ_2, ϕ_1, are varied, we can expand the Euclidean actions here involved as follows (the first derivative vanishing at a solution):

$$\begin{aligned}
S_E(\phi_f, \phi_2) &= S_E(\phi_f, \phi_i) + \frac{1}{2} \frac{\partial^2 S_E}{\partial \phi_2^2} \bigg|_{\phi_i} (\phi_2 - \phi_i)^2 + \cdots, \\
S_E(\phi_2, \phi_1) &= S_E(\phi_i, \phi_1) + \frac{1}{2} \frac{\partial^2 S_E}{\partial \phi_2^2} \bigg|_{\phi_i} (\phi_2 - \phi_i)^2 + \cdots, \tag{23.106a}
\end{aligned}$$

and

$$\begin{aligned}
S_E(\phi_i, \phi_1) &= S_E(\phi_i, \phi_f) + \frac{1}{2} \frac{\partial^2 S_E}{\partial \phi_1^2} \bigg|_{\phi_f} (\phi_1 - \phi_f)^2 + \cdots, \\
S_E(\phi_1, \phi_i) &= S_E(\phi_f, \phi_i) + \frac{1}{2} \frac{\partial^2 S_E}{\partial \phi_1^2} \bigg|_{\phi_f} (\phi_1 - \phi_f)^2 + \cdots, \tag{23.106b}
\end{aligned}$$

where the first expression in Eq. (23.106b) is to be substituted into the second relation of (23.106a). It follows that — always within our approximations —

$$
\langle \phi_f, \tau_f | \phi_i, \tau_i \rangle^{(3)} \simeq \left\{ e^{-S_E(\phi_f, \phi_i)} \right\}^3 \int_{-\infty}^{\infty} d\phi_2 \exp\left[-(\phi_2 - \phi_i)^2 \frac{\partial^2 S_E}{\partial \phi_2^2}\Big|_{\phi_i} \right]
$$
$$
\times \int_{-\infty}^{\infty} d\phi_1 \exp\left[-(\phi_1 - \phi_f)^2 \frac{\partial^2 S_E}{\partial \phi_1^2}\Big|_{\phi_f} \right]
$$
$$
= \left\{ e^{-S_E(\phi_f, \phi_i)} \right\}^3 \gamma^2, \tag{23.107}
$$

with (the evaluation is given below)

$$
\gamma = \left[\frac{\pi}{(\partial^2 S_E / \partial \phi_2^2)_{\phi_i}} \right]^{1/2} = \left[\frac{\pi}{(\partial^2 S_E / \partial \phi_1^2)_{\phi_f}} \right]^{1/2} = \left(\frac{\pi}{2m} \right)^{1/2}, \tag{23.108}
$$

where the last expression is obtained as follows. With our remark after Eq. (23.12) we have

$$
S_E = \int_{\phi_c(-\infty)}^{\phi_c(\infty)} \frac{d\phi_c}{d\tau} d\phi_c = \pm \frac{m^2}{g} \int_{-\phi(-\infty)}^{\phi_c(\infty)} d\phi \left(1 - \frac{g^2 \phi^2}{m^2} \right),
$$

i.e.

$$
S_E(\phi) = \pm \frac{m^2}{g} \left[\phi - \frac{g^2 \phi^3}{3m^2} \right], \quad \text{and} \quad \frac{\partial S_E(\phi)}{\partial \phi} = \pm \frac{m^2}{g} \left[1 - \frac{g^2 \phi^2}{m^2} \right].
$$

Hence

$$
\frac{\partial^2 S_E(\phi)}{\partial \phi^2} = \pm 2g\phi, \quad \text{i.e.} \quad \left| \frac{\partial^2 S_E(\phi)}{\partial \phi^2} \right|_{\pm m/g} = 2m. \tag{23.109}
$$

The result of these integrations is that the amplitude I_{2n+1} involves $2n$ intermediate integrations, each of which implies a multiplicative factor γ, in the case of I_{2n+1} therefore γ^{2n}.

On the assumption that the kinks of the instantons are widely separated, i.e. the dilute gas approximation, each of the three intermediate kernels contained in Eq. (23.105) can in other respects be treated in analogy with the single instanton amplitude. It has to be remembered, of course, that each applies only to a fraction of the overall Euclidean time interval of $\tau = 2T$. The allowed positions of the individual kinks have to be such that their order is maintained. Thus when in each of the three cases the zero mode elimination has been performed with the help of the Faddeev–Popov method, i.e. the replacement (23.93) has been applied, we now have to integrate over three collective coordinates, and in the case depicted in Fig. 23.6 the collective coordinate $\tau_0^{(3)}$ has to be to the right of $\tau_0^{(2)}$, and $\tau_0^{(2)}$ to the right of $\tau_0^{(1)}$. This implies that in this example we have the product of integrals[§]

$$\int_{-T}^{T} d\tau_0^{(1)} \int_{-T}^{\tau_0^{(1)}} d\tau_0^{(2)} \int_{-T}^{\tau_0^{(2)}} d\tau_0^{(3)} = \frac{(2T)^3}{3!}. \tag{23.110}$$

Finally we have to recall from Eq. (23.105) that each of the three amplitudes contributes a factor $\exp[-m\triangle_i T]$ with $\sum_{i=1}^{3} \triangle_i T = \triangle T = 2T$.

In the case of $\langle \phi_f, \tau_f | \phi_i, \tau_i \rangle^{(3)}$ these considerations imply the following result parallel to that of the single instanton calculation (23.101):

$$\langle \phi_f, \tau_f | \phi_i, \tau_i \rangle^{(3)} = \lim_{T \to \infty} \frac{1}{\gamma} \gamma^3 \frac{(\triangle T)^3}{3!} \triangle_{FP}^3 I_0^3 \left(\frac{\tilde{v}_0}{\sqrt{2\pi}} \right)^3 e^{-m\triangle T}$$

$$\times \left[\exp \left(-\frac{S_E(\phi_c)}{\hbar} \right) \right]^3. \tag{23.111}$$

Finally we sum over any number of instanton–anti-instanton pair configurations and obtain for the Feynman amplitude

$$I = \lim_{T \to \infty} \sum_{n=0}^{\infty} I_{2n+1} = \lim_{T \to \infty} \frac{1}{\gamma} \sum_{n=0}^{\infty} \frac{1}{(2n+1)!} \left\{ \gamma \triangle T \triangle_{FP} I_0 \right.$$

$$\times \left. \left(\frac{\tilde{v}_0}{\sqrt{2\pi}} \right) \exp \left[-\frac{S_E(\phi_c)}{\hbar} \right] \right\}^{2n+1} e^{-m\triangle T}$$

$$= \lim_{T \to \infty} \frac{1}{\gamma} e^{-m\triangle T} \sinh \left\{ \gamma \triangle T \triangle_{FP} I_0 \left(\frac{\tilde{v}_0}{\sqrt{2\pi}} \right) \exp \left[-\frac{S_E(\phi_c)}{\hbar} \right] \right\}. \tag{23.112}$$

We now compare this result with expression (23.35a) which contains $\triangle E$ in the argument of sinh and obtain for the level splitting $\triangle^i$ resulting from consideration of the instanton plus any number of instanton–anti-instanton pairs

$$\triangle^i E = 2\hbar\gamma \triangle_{FP} I_0 \frac{\tilde{v}_0}{\sqrt{2\pi}} \exp \left(-\frac{4}{3} \frac{m^3}{g^2 \hbar} \right)$$

$$= 2\hbar \left(\frac{\pi}{2m} \right)^{1/2} \left(\frac{4}{3} \frac{m^3}{g^2} \right)^{1/2} \left(\frac{m}{\pi\hbar} \right)^{1/2} \frac{4\sqrt{6}m}{\sqrt{2\pi}} \exp \left(-\frac{4}{3} \frac{m^3}{g^2 \hbar} \right),$$

[§]With shifts $\tau_0^{(i)} \to \tau_0^{(i)} + a_i$, then sign reversals $\tau_0^{(i)} \to -\tau_0^{(i)}$, and this followed by setting $a_1 = 2T/3, a_2 = 0, a_3 = -2T/3$ this product becomes that given by E. Gildener and A. Patrascioiu [110], i.e.

$$\int_{-T/3}^{5T/3} d\tau_0^{(1)} \int_{\tau_0^{(1)}-2T/3}^{T} d\tau_0^{(2)} \int_{\tau_0^{(2)}-2T/3}^{T/3} d\tau_0^{(3)} = \frac{(2T)^3}{3!}.$$

and hence

$$\triangle^i E = \frac{8\sqrt{2}\hbar m^{5/2}}{g\sqrt{\pi\hbar}} \exp\left(-\frac{4}{3}\frac{m^3}{g^2\hbar}\right). \qquad (23.113)$$

This is (in slightly different notation) the result of Gildener and Patrascioiu [110] for consideration of the instanton plus associated pairs. Taking similarly into account the anti-instanton (since the physical level splitting does not distinguish between an instanton and its anti-instanton), by either replacing $\triangle T$ by $2\triangle T$ or doubling $\triangle^i$, we obtain $\triangle E = 2\triangle^i$ and agreement with the result (18.182) of the Schrödinger equation method provided the difference in the mass of the particle (there taken as $1/2$, here as 1) is also taken into account.¶ We see therefore that the classical Euclidean configurations ϕ_c can be considered to be responsible for the quantum mechanical tunneling which gives rise to the level splitting $\triangle E$.‖ We also see that the level splitting is a nonperturbative expression (with an essential singularity at $g^2 = 0$). Finally we realize that the computation of (23.113) is very complicated, even without explicit specification of the prefactor. We observe one more aspect of our calculation. The difference between levels, $\triangle E$, has been obtained without explicit use of canonical commutation relations (i.e. explicit quantization). In fact the equivalent to writing down canonical commutation relations is the evaluation of the Gaussian integrals which enter the prefactor of the expression in Eq. (23.113).

23.5   Concluding Remarks

In this chapter we used the path integral to obtain the level splitting of the ground state of a particle in the double well potential. We see that the method — above we included a lot of highly nontrivial details which are passed over in other literature — is considerably more complicated than the method of the Schrödinger equation. We shall see that the path integral applied to excited states is even more complicated in necessitating the evaluation of elliptic integrals, whereas the method of the Schrödinger equation in this case is of the same degree of complication as for the ground state.

¶See P. Achuthan, H. J. W. Müller–Kirsten and A. Wiedemann [4], J.–Q. Liang and H. J. W. Müller–Kirsten [163]. To help the comparison we note that — with indices indicating the authors, GP for those of E. Gildener and A. Patrascioiu [110], — the relation between the parameters is $h^4_{LMK} = 2\mu^2_{GP} = 4m^2$ and $c^2_{LMK} = \frac{1}{2}\lambda_{GP} = g^2$. With this correspondence $\triangle_1 E$ of Eq. (18.182) (with mass $= 1$ and $\hbar = 1$) becomes the following which implies twice the result (23.113) for $q_0 = 1$:

$$\triangle_1 E = 2^{q_0+2}\sqrt{\frac{2\hbar}{\pi}}\frac{m}{[\frac{1}{2}(q_0-1)]!}\left(\frac{4m^3}{g^2}\right)^{q_0/2}\exp\left(-\frac{4}{3}\frac{m^3}{g^2\hbar}\right).$$

‖For an elementary discussion see S. K. Bose and H. J. W. Müller–Kirsten [30].

Chapter 24

Path Integrals and Bounces on a Line

24.1 Introductory Remarks

In this chapter we are concerned with *nontopological classical configurations* on $\mathbb{R}^1$, and their interpretation and uses. We shall see that these classical configurations are associated with quantum mechanical instability and hence with complex eigenvalues. Since problems of instability are ubiquitous and occur in all areas of microscopic physics their understanding is of paramount importance. In order to really appreciate the difference between topological and nontopological configurations (the solitons and instantons we considered previously being of the first kind) we begin with a recapitulation of some crucial aspects of the former, however, in a somewhat different formulation.

Previously we considered in particular the quartic or symmetric double well potential for a scalar ϕ, i.e.

$$V(\phi) = \frac{m^4}{2g^2}\left(1 - \frac{g^2\phi^2}{m^2}\right)^2 \geq 0. \tag{24.1}$$

Here the two symmetrical minima of the potential are called *vacua* (classical or perturbation theory vacua as they are sometimes called because ordinary Rayleigh–Schrödinger perturbation theory would be performed around these points). Potentials of this type play a significant role for instance in models of the early inflationary (expanding) universe (as a result of its cooling). Consider for instance the temperature-dependent potential

$$V_T(\phi) = \frac{1}{2}\lambda(\phi^2 - \sigma^2)^2 + c\phi^2 T^2. \tag{24.2}$$

Here

$$\frac{dV_T(\phi)}{d\phi} = 2\lambda(\phi^2 - \sigma^2)\phi + 2cT^2\phi$$

vanishes for

$$\phi = 0, \quad \phi^2 = \sigma^2 - \frac{cT^2}{\lambda}. \qquad (24.3)$$

Since

$$\frac{d^2V_T(\phi)}{d\phi^2} = 2(cT^2 - \lambda\sigma^2) + 6\lambda\phi^2,$$

we see that

$$\left(\frac{d^2V_T(\phi)}{d\phi^2}\right)_{\phi=0} = 2(cT^2 - \lambda\sigma^2),$$

$$\left(\frac{d^2V_T(\phi)}{d\phi^2}\right)_{\phi^2=\sigma^2-cT^2/\lambda} = 4(\lambda\sigma^2 - cT^2).$$

Thus there is a critical temperature $T_c := (\lambda/c)^{1/2}\sigma$. We see that $\phi = 0$ minimizes $V_T(\phi)$ for temperatures $T > T_c$, but implies instability for $T < T_c$. Thus if we start with the system (like the early universe) at a high temperature and then allow the system to cool, a spontaneous symmetry breaking leads to the double well (i.e. at the minimum of the potential the case $\phi = 0$ changes to the case $\phi \neq 0$).

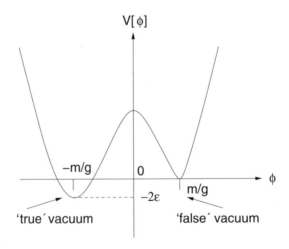

Fig. 24.1 The asymmetric double well.

We can destroy the reflection symmetry of $V(\phi)$ (i.e. its symmetry under exchanges $\phi \leftrightarrow -\phi$) by adding e.g. a linear term as in

$$V_a(\phi) = \frac{m^4}{2g^2}\left(1 - \frac{m^2\phi^2}{m^2}\right)^2 - \epsilon\left(1 - \frac{g\phi}{m}\right), \quad \epsilon > 0. \tag{24.4}$$

This is now an asymmetric double well potential as in Fig. 24.1 with

$$V(\phi = m/g) = 0, \quad V(\phi = -m/g) = -2\epsilon. \tag{24.5}$$

We see that one minimum of V now lies higher than the other. Since the true vacuum minimizes the potential, the higher minimum is called a *"false vacuum"*. False vacua play an important role in models of the inflationary universe* and elsewhere.† In the following we consider first (after some recapitulations) a simple model of a false vacuum. We calculate the classical configuration — which in this case is called a *"bounce"* — which corresponds to the soliton or instanton in our previous chapters (but is not topological) and we use this in order to calculate the lifetime of the unstable state (or equivalently the imaginary part of the energy). We shall see that:

(a) solitons (or equivalently instantons in the one-dimensional theory) describe vacuum-to-vacuum transitions, the vacuum being understood as the classical, perturbation-theory vacuum),
(b) they are minimum configurations of the energy E (instantons of the Euclidean action S_E), and
(c) determine the lowering of the energy to the true quantum mechanical ground state (i.e. the quantum mechanical vacuum).

On the other hand we shall also see that
(a) bounces describe vacuum-to-infinity transitions, which
(b) are saddle point configurations of E (or S_E), and
(c) determine $i\Im E$ of E which results from a nonvanishing probability of tunneling away from the classical vacuum only.

We recapitulate briefly a few points from our earlier detailed treatment of $(1+1)$-dimensional soliton theory. The Lagrangian density of the theory is given by

$$\mathcal{L} = \frac{1}{2}\partial_\mu\Phi\partial^\mu\Phi - U(\Phi), \quad U(\Phi) = \frac{m^4}{2g^2}\left(1 - \frac{g^2\Phi^2}{m^2}\right)^2 \geq 0. \tag{24.6}$$

The Euler–Lagrange equation is given by

$$\Box\Phi + U'(\Phi) = 0 \quad (\text{metric } +, -). \tag{24.7}$$

---

*See e.g. A. H. Guth and S.-Y. Pi [126].

†False vacua and bounces occur, for instance, also in contexts of string theory; see e.g. L. Cornalba, M. S. Costa and J. Penedones [59] and references cited there.

Static fields given by

$$\frac{\partial \Phi(x,t)}{\partial t} = 0$$

are written $\phi(x)$, so that the Euler–Lagrange equation becomes the Newton-like equation

$$\phi''(x) = U'(\phi). \tag{24.8}$$

This equation possesses the kink solution

$$\phi_c(x) = \frac{m}{g} \tanh m(x - x_0) \tag{24.9}$$

for the double-well potential given above. We observe that this kink solution is *odd* in $(x - x_0)$, which is an important point as we shall see. The energy of this classical configuration is obtained by evaluating the Hamiltonian for ϕ replaced by ϕ_c. Thus

$$E(\phi_c) = \int_{-\infty}^{\infty} dx \left[\frac{1}{2} (\phi_c')^2 + U(\phi_c) \right]. \tag{24.10}$$

Inserting ϕ_c and evaluating the integral we obtain

$$E(\phi_c) = \frac{4}{3} \frac{m^3}{g^2}. \tag{24.11}$$

We investigate the classical stability of the configuration by setting in the full t−dependent equation for small ψ_k:

$$\Phi(x,t) = \phi_c(x) + \sum_k \psi_k(x) e^{i w_k t}. \tag{24.12}$$

Since $\phi_c(x)$ solves Eq. (24.8), we see that (24.7) yields the following equation called *stability* or *small fluctuation equation*, i.e.

$$\left[-\frac{d^2}{dx^2} + U''(\phi_c(x)) \right] \psi_k(x) = w_k^2 \psi_k(x). \tag{24.13}$$

For small $w_k^2 \geq 0$, small perturbations around $\phi_c(x)$ do not grow exponentially with t (the zero eigenvalue requires special attention). We have seen previously that the same equation is also obtained from the second variational derivative of the energy functional, i.e.

$$\frac{\delta^2 E(\phi)}{\delta \phi(x) \delta \phi(y)} \bigg|_{\phi_c} = \left[-\frac{d^2}{dx^2} + U''(\phi) \right]_{\phi_c} \delta(x - y). \tag{24.14}$$

For so-called "classical stability" the differential operator has to be semi-positive definite, i.e. eigenvalues $w_k^2 \geq 0$ (actually the stability is only local in the sense that E is not minimized in the direction (in Hilbert space) of the eigenfunction associated with the zero eigenvalue, i.e. the zero mode). In the case of the double well potential (24.1) the fluctuation equation becomes

$$\left[-\frac{d^2}{dx^2} + 4m^2 - \frac{6m^2}{\cosh^2 mx} \right] \psi_k(x) = w_k^2 \psi_k(x). \qquad (24.15)$$

From our study of the solutions of this equation in Sec. 22.8 we know that this equation possesses two discrete eigenvalues (one being the eigenvalue zero, the other a positive quantity) and a continuum which are given respectively by

$$w_k^2 = 0, \, 3m^2 \quad \text{and} \quad w_k^2 \geq 4m^2. \qquad (24.16)$$

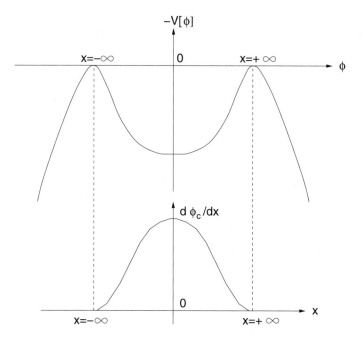

Fig. 24.2 The inverted potential; underneath the particle's velocity.

We have seen that general considerations tell us that the zero mode is given by the generator of translations applied to the classical, i.e. kink, configuration (for our ϕ^4-theory); thus the zero mode here is given by

$$\frac{d}{dx}\phi_c = \frac{d}{dx}\left[\frac{m}{g} \tanh m(x - x_0) \right] \propto \frac{1}{\cosh^2 m(x - x_0)}. \qquad (24.17)$$

We observe that this zero mode is an even function of $(x - x_0)$ in agreement with our expectations for the ground state wave function. The kink solution can now be looked at as the classical solution of a one-dimensional field theory, i.e. quantum mechanics, with Euclidean time. In this case the usual Newton equation re-appears with an opposite sign of the potential. The classical configuration is then called a pseudoparticle configuration or instanton, and — clearly — we can resort to a classical interpretation of this function. In the pseudoparticle consideration the coordinate x plays the role of time. We see that this pseudoparticle starts from rest at one peak of $-V$ at $x = -\infty$ and in losing potential energy gains just enough kinetic energy to reach the second peak at $x = +\infty$, where it is then again at rest. The associated zero mode represents the velocity of the pseudoparticle as indicated in Fig. 24.2, and also represents a nodeless even wave function — in complete agreement with our expectations for a quantum mechanical ground state wave function. Below we shall see that the configuration described as a bounce is given by an even function of x, so that its derivative representing the pseudoparticle's velocity is odd, i.e. the associated zero mode. Hence in this case there must be a lower state with a negative eigenvalue and an even (ground state) wave function. Before we consider such a theory in more detail we recall the topological aspects of a soliton or instanton. Integrating the classical equation (24.8), i.e. $\phi''(x) = U'(\phi)$, we obtain the *Bogomol'nyi inequality*

$$(\phi'(x) \mp \sqrt{2U})^2 \geq 0. \tag{24.18}$$

This inequality is saturated by the first order *Bogomol'nyi equation*

$$\phi'(x) = \pm\sqrt{2U}, \tag{24.19}$$

which has the solution $\phi_c(x)$. Then the energy satisfies

$$E(\phi_c) \geq \int_{-\infty}^{\infty} dx \sqrt{2U} \phi_c'(x - x_0) = \frac{4}{3}\frac{m^3}{g^2} \times 1. \tag{24.20}$$

The number 1 on the right is the so-called *topological charge* which is obtained as follows. We define the current

$$k^\mu := \epsilon^{\mu\nu}\partial_\nu\phi_c. \tag{24.21}$$

The charge is defined as the space integral of the time component of the current, i.e. by

$$Q = \frac{1}{2}\int_{-\infty}^{\infty} dx\left(\frac{g}{m}\epsilon_{01}\partial^1\phi_c\right) = \frac{1}{2}\frac{g}{m}[\phi_c(\infty) - \phi_c(-\infty)] = 1. \tag{24.22}$$

For reasons discussed earlier this charge is called the topological charge. It is important here to realize that the nonzero value of Q results from the different boundary conditions of the classical configuration at $+\infty$ and $-\infty$, the kink solution having a monotonic behaviour on the way from one endpoint to the other. We are here interested in these nontrivial solutions with $Q \neq 0$. The theory does, of course, possess other solutions which have $Q = 0$. These are the constant solutions $\phi = \pm m/g$ which are therefore called *nontopological* or are described as *topologically trivial*. We see already here that a configuration which has the same value at both endpoints has $Q = 0$ and so is nontopological (this will be seen to be the case for a bounce).

We can introduce the concept of *topological stability* as follows. We consider the family of configurations $\{\lambda\phi_c(x)\}$ where λ is some number. Then

$$E[\lambda\phi_c] = \int_{-\infty}^{\infty} dx \left[\frac{1}{2}\lambda^2(\phi_c')^2 + U(\lambda\phi_c)\right]. \tag{24.23}$$

In this expression we have the potential with the behaviour (see Eq. (22.37))

$$x \to \pm\infty: \quad U(\lambda\phi_c) = \frac{m^4}{2g^2}(1 - \lambda^2)^2 \neq 0 \text{ except for } \lambda^2 = 1. \tag{24.24}$$

Thus the energy is finite only for $\lambda = \pm 1$. The requirement of finite E therefore does not permit $\lambda\phi_c$ deformations. We call this topological stability.

We recall finally that the one-dimensional theory with the double well potential allows the calculation of the difference between the ground state energy and that of the first excited state. In the preceding chapter we performed this calculation with the help of the path integral method. We now turn to corresponding considerations for a potential which allows tunneling to infinity and hence the definition of a particle current. Our main objective here is in the first place the recalculation of the result of Bender and Wu[*] for the decay rate of a particle trapped in the ground state of an inverted double well potential by means of the path integral method.[†]

24.2 The Bounce in a Simple Example

The concept of bounces was introduced by Coleman.[*] In the following we discuss a simple model which allows an explicit discussion of bounces.[†]

---

[*]C. M. Bender and T. T. Wu [18], [19].

[†]J.–Q. Liang and H. J. W. Müller–Kirsten [166].

[*]S. Coleman [54]. See in particular Secs. 2.4 and 6.1 to 6.5. See also S. Coleman [56], [57] and C. G. Callan and S. Coleman [41].

[†]H. J. W. Müller–Kirsten, Jian-zu Zhang and D. H. Tchrakian [208]. In Eq. (4.11) of this reference the number 35 of 8/35 is misprinting of $3.5 = 15$.

We take as the definition of the *lifetime* of a state the inverse of the decay width Γ defined by

$$\Gamma := -2\frac{\Im E}{\hbar}. \qquad (24.25)$$

In field theory the ϕ^3 potential does not make much sense since it is unbounded from below.[‡] However, we can discuss it in the context of quantum mechanics, which we do here for purposes of illustration. We do not carry through the entire calculation to the explicit expression for the decay rate, since this will be done in detail in the next section for the inverted double well potential; the formula there obtained (i.e. the result (24.86)) is then applied to the cubic potential in Sec. 24.5 (for a WKB treatment see Example 14.6).

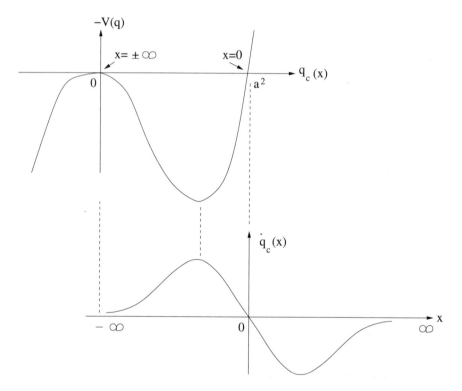

Fig. 24.3 The inverted cubic potential and velocity of the bounce.

We therefore proceed immediately to the Euclidean Euler–Lagrange equation for particle mass $m_0 = 1$

$$\frac{1}{2}\left(\frac{dq}{dx}\right)^2 = V(q). \qquad (24.26)$$

[‡]A cubic potential like the one we treat here found application in a string theory context. See e.g. K. Ohmori [222].

We take as the potential the cubic form

$$V(q) = -\frac{1}{2}b^2 q^2 (q - a^2).$$ (24.27)

For this we can solve Eq. (24.26) (e.g. with the help of elliptic functions, but the reader can convince himself by differentiation that the following expression solves (24.26)) and obtain the following solution described as a *bounce*,

$$q_c(x - x_0) = \frac{a^2}{\cosh^2[\frac{1}{2}ab(x - x_0)]}, \quad 0 \leq q \leq a^2.$$ (24.28)

We observe that this classical solution is even in $(x - x_0)$. The translational zero mode is obtained by differentiation and is seen to be

$$\dot{q} \equiv \frac{d}{dx} q_c(x - x_0) = -\frac{a^3 b \, \sinh[\frac{1}{2}ab(x - x_0)]}{\cosh^3[\frac{1}{2}ab(x - x_0)]},$$ (24.29)

which is an odd function. The bounce (24.28) satisfies the following boundary conditions for (our choice) $x_0 = 0$:

$$q_c(\pm\infty) = 0, \quad q_c(0) = a^2, \quad \text{and} \quad \dot{q}_c(\pm\infty) = 0, \quad \dot{q}_c(0) = 0.$$ (24.30)

We can understand the significance of these boundary conditions by reference to Fig. 24.3. In the figure the upper diagram shows the inverted potential, i.e. the potential which appears in Newton's equation. We see that if the pseudoparticle starts from $q_c = 0$ at time $x = -\infty$ (the infinite past), then at time $x = 0$ it reaches the position $q_c = a^2$. It then has no energy to go further and so has to return back towards $q_c = 0$, the initial position which it reaches in the infinite future. The direction reversal at $q_c = a^2$ is the reason it is called a "bounce" (it bounces back, so to speak). The consequence of this change of direction is that the derivative (i.e. the pseudoparticle's velocity) changes sign. Thus this derivative, which acts as zero mode of the stability equation (see below) is an odd function as depicted in Fig. 24.3. Since the zero mode is now the odd eigenfunction of a Schrödinger-like equation with eigenvalue zero, the corresponding ground state wave function, which has to be even, must have a negative eigenvalue. Before we turn to a consideration of the stability or small fluctuation equation, we evaluate the Euclidean action, i.e. the integral

$$S_E(q_c) = \int_{-\infty}^{\infty} dx \left[\frac{1}{2} \dot{q}_c^2 + V(q_c) \right].$$ (24.31)

Inserting (24.28) for q_c we obtain

$$S_E(q_c) \overset{(24.26)}{=} \int_{-\infty}^{\infty} dx \left(\frac{dq_c(x)}{dx}\right)^2 = 4ba^5 \int_0^{\infty} dy \frac{\sinh^2 y}{\cosh^6 y}$$

$$= \frac{8}{15} ba^5. \tag{24.32}$$

Thus the Euclidean action of the bounce is finite.

The small fluctuation equation corresponding to the *stability equation* (24.13) now becomes

$$\left[-\frac{d^2}{dx^2} + V''(q)\right]\psi_k = w_k^2 \psi_k \quad \text{with} \quad V''(q) = -3b^2 q + b^2 a^2. \tag{24.33}$$

Setting $m = ab/2$, we obtain

$$\left[-\frac{d^2}{dx^2} + 9m^2 - \frac{12m^2}{\cosh^2 m(x-x_0)}\right]\psi_k = (w_k^2 + 5m^2)\psi_k. \tag{24.34}$$

This is again a Schrödinger-like equation with a *Pöschl–Teller potential*. We studied the solutions of this equation previously in Sec. 22.6. Thus we find that the equation possesses three discrete eigenvalues and a continuum, i.e.

$$w_k^2 = -5m^2, \, 0, \, 3m^2 \quad \text{and} \quad w_k^2 = 4m^2 + k^2, \, k^2 \geq 0. \tag{24.35}$$

The one negative eigenvalue is indicative of the instability of the solution which therefore represents a *saddle point* of the action.[§]

Following our previous arguments we see that since

$$\lim_{x \to \pm\infty} q_c(x) = 0,$$

the "topological charge" of the bounce is zero. Considering the question of topological stability or instability as in the soliton case, we proceed as follows. We consider the family $\{\lambda q_c\}$ with λ some number > 0. We then find that

$$S_E(\lambda q_c) = \sqrt{\lambda}\frac{8}{15}a^5 b.$$

Thus S_E is not minimized for any $\lambda > 0$. It is topologically unstable.

In order to be able to compute the lifetime of the system trapped in the single trough of $V(q)$, we have to go to nonselfadjoint boundary conditions in solving the corresponding Schrödinger equation. Instead we follow here

---

[§]Without further calculation it is fairly clear from Eq. (24.34) that in the case of excited states the fluctuation equation is a Lamé equation like Eq. (25.44) with (therein) $n = 3$.

the procedure of the path integral method which we used in the case of the instanton for the calculation of the level splitting of the double well potential. In the path integral method the quantity $\Im E$ is also obtained by calculating the transition amplitude (cf. Eqs. (21.83) and (21.105) for Euclidean time and the normalization or measure definition discussed in Sec. 21.2)

$$\langle q_f | e^{-2TH} | q_i \rangle = \int_{q_i}^{q_f} \mathcal{D}\{q\} e^{-S_E(q)}, \qquad (24.36)$$

where $|q_i\rangle, |q_f\rangle$ are eigenstates of the position operator, H is the Hamiltonian, $2T$ is the Euclidean time interval $x_f - x_i$ or with $x \equiv \tau$ the interval $\tau_f - \tau_i$. In the neighbourhood of a classical configuration $q_c(\tau)$ of Euclidean time τ one can set

$$q(\tau) = q_c(\tau) + y(\tau), \qquad (24.37)$$

where $y(\tau)$ is the quantum fluctuation. If $q_c(\tau)$ minimizes the action $S_E(q)$, one can expand $S_E(q)$ about q_c. Setting (for $\tau_0 = $ const.)

$$y(\tau - \tau_0) = \sum_n \xi_n \psi_n(\tau - \tau_0), \qquad (24.38)$$

where $\{\psi_n\}$ is the set of eigenstates of the appropriate stability equation with spectrum $\{w_n^2\}$, one obtains in the one-loop approximation as we saw in Chapter 23 (cf. Eq. (23.52a))

$$\int_{q_i}^{q_f} \mathcal{D}\{q\} e^{-S_E(q)} = e^{-S_E(q_c)} \det\left(\frac{dy(\tau)}{d\xi_n}\right) \int \prod_k d\xi_k \exp\left[-\frac{1}{2}\sum_k w_k^2 \xi_k^2\right]. \qquad (24.39)$$

We have seen in Sec. 23.4.3 how we handle the zero mode — i.e. with the *Faddeev–Popov method*. This is therefore of no concern to us here (i.e. we apply the method here in the corresponding way). However, we recall that in evaluating the multiple integral in Eq. (24.39), the Faddeev–Popov trick exchanges ξ_0 for the *collective coordinate* of q_c (a Euclidean time τ_0); integrating this out yields the factor $2T$ of a large Euclidean time interval $(-T, T)$ and a factor $\triangle_{FP} = \sqrt{S_E}$. Thus

$$\int d\xi_0 = \triangle_{FP} \int_{-T}^{T} d\tau_0, \quad \triangle_{FP} = \sqrt{S_E}. \qquad (24.40)$$

We now observe that if we apply the same procedure in our example here with bounces as that which we applied in the case of instantons, we have the new aspect that one w_k^2, say w_{-1}^2, is negative, and hence the integral diverges, i.e. along the vector ψ_{-1} about q_c the Euclidean action is a maximum at q_c, and in the other directions it is a minimum at q_c, which means that S_E has a saddle point at q_c. We can obtain a finite value of the integral — as we must, since the decay rate is necessarily finite — if we perform the integration along a contour in the plane of complex ξ_{-1}.[¶] We assume that the dominant contribution to the imaginary part of the path integral (and hence to $\Im E$) has only a weak dependence on the detailed form of the contour as long as the integral is taken predominantly along $0 < \Im \xi_{-1} < \infty$. Thus, using Eq. (24.32), we write the imaginary part for the case of the cubic potential[‖]

$$
\begin{aligned}
i\Im & \int_{q_i}^{q_f} \mathcal{D}\{q\} e^{-S_E(q)} \\
&= 2T \triangle_{FP} e^{-S_E(q_c)} \det\left(\frac{dy(\tau)}{d\xi_n}\right) \int_0^{0+i\infty} d\xi_{-1} e^{-\frac{1}{2}\xi_{-1}^2 w_{-1}^2} \\
&\quad \int \cdots \int \left(\prod_{k\neq -1,0} d\xi_k\right) \exp\left[-\frac{1}{2}\sum_{j\neq 0} w_j^2 \xi_j^2\right] \\
&\equiv i 2T \triangle_{FP} C e^{-8ba^5/15},
\end{aligned}
\tag{24.41}
$$

where C is the product of various constants. This is the contribution of the one bounce considered to the path integral of Eq. (24.36). We now have to sum over all bounce configurations with centre in the domain of integration of the collective coordinate, i.e. the Euclidean time τ_0, in much the same way as we summed over instanton–anti-instanton contributions in the dilute gas approximation in the theory with the instanton configuration for the double well potential. This calculation is performed in Sec. 24.4 using the results of the next section.

When the hump of the potential becomes infinitely high, there is no tunneling. In this limit (24.41) vanishes. However, in that case (24.36) implies

$$
\langle q_f | e^{-2TH} | q_i \rangle = e^{-2TE_0}, \tag{24.42}
$$

where E_0 is the ground state energy associated with the potential well. In the harmonic oscillator approximation we have $E_0 = \hbar\omega/2, \omega = ab$ for mass $m_0 = 1$. Thus for a finite height of the hump the energy E_0 is replaced by

$$
E = E_0 - iC' e^{-8ba^5/15}, \quad E_0 = \frac{1}{2}ab\hbar = m\hbar, \quad C' = \text{const.}, \tag{24.43}
$$

and so

$$
\Gamma = 2C' e^{-8ba^5/15} \quad (\hbar = 1) \tag{24.44}
$$

[¶]Cf. the saddle point method explained in Example 13.3.

[‖]Recall $\int_{-\infty}^{\infty} dx e^{-w^2 x^2} = \sqrt{\pi/w^2}$.

for the cubic potential. In the following we perform the calculation in detail for the case of the inverted double well potential and return to the cubic potential in Sec. 24.4.

24.3 The Inverted Double Well: The Bounce and Complex Energy

24.3.1 The bounce solution

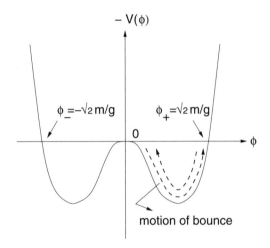

Fig. 24.4 The double well as inverse of a potential with a bounce.

The inverted double well potential provides another and important example of a bounce and hence an alternative way to derive the result (18.78) of Bender and Wu [18], [19] in the case of the ground state.[*] Suppose we consider the following potential

$$V(\Phi) = -\frac{m^4}{2g^2}\left[1 - \frac{g^2\Phi^2}{m^2}\right]^2 + \frac{m^4}{2g^2} = m^2\Phi^2 - \frac{1}{2}g^2\Phi^4, \qquad (24.45)$$

which is seen to differ in two respects from the double well potential of our Φ^4-soliton theory: The additive constant is removed and the overall sign of the potential has been reversed. Again the equation to be solved for the classical, static configuration is, with $\Phi(x,t) \to \phi(x)$ and mass $m_0 = 1$,

$$\frac{d}{d\phi}\left[\frac{1}{2}(\phi')^2\right] = V'(\phi) \quad \text{or} \quad \frac{1}{2}(\phi')^2 - V(\phi) = 0. \qquad (24.46)$$

---

[*]J.–Q. Liang and H. J. W. Müller–Kirsten [166].

Again the arbitrary additive constant is taken to be zero which means that the energy of the pseudoparticle with classical (Newtonian) interpretation is zero (for the Newtonian consideration the sign of the potential has to be reversed). Alternatively we can consider the particle as classically in motion in the potential $V(\phi)$, but in Euclidean time $\tau = it$. The solution of Eq. (24.46) can be verified to be the *even* function

$$\phi_c(x) = \overset{+}{(-)} \sqrt{2}\frac{m}{g}\mathrm{sech}[\sqrt{2}m(x - x_0)], \tag{24.47}$$

where we can write equivalently $x = \tau$ or $x = c\tau$ with $c = 1$, and correspondingly for the constant x_0. Thus

$$\phi_c'(\tau) = -\frac{2m^2}{g}\frac{\sinh[\sqrt{2}m(\tau - \tau_0)]}{\cosh^2[\sqrt{2}m(\tau - \tau_0)]}. \tag{24.48}$$

The square of this expression divided by 2 can be checked to agree with $V(\phi)$. Apart from the solution (24.47) the classical equation (24.46) has the constant solutions $\phi = 0, \pm\sqrt{2}m/g$.[†] We shall see that the even solution (24.47) is a *bounce*. The zero mode is the derivative (24.48), which — of course — is an odd function of $(\tau - \tau_0)$. Choosing $\tau_0 = 0$, we see that ϕ_c and $d\phi_c/d\tau$ satisfy the conditions:

$$\lim_{\tau\to\pm\infty}\phi_c(\tau) = 0, \quad \lim_{\tau\to 0}\phi_c(\tau) = \sqrt{2}\frac{m}{g},$$

$$\lim_{\tau\to\pm\infty}\frac{d\phi_c(\tau)}{d\tau} = 0, \quad \lim_{\tau\to 0}\frac{d\phi_c(\tau)}{d\tau} = 0. \tag{24.49}$$

These conditions are precisely the boundary conditions of a bounce, as a comparison with Eq. (24.30) demonstrates. Thus the next step is to investigate the spectrum $\{w_k^2\}$ of the *small fluctuation equation*, i.e.

$$\left[-\frac{d^2}{d\tau^2} + V''(\phi)\right]\psi_k = w_k^2\psi_k, \tag{24.50}$$

where

$$V'(\phi_c) = 2m^2\phi_c - 2g^2\phi_c^3,$$

$$V''(\phi_c) = 2m^2 - 6g^2\phi_c^2 = 2m^2 - \frac{3.4m^2}{\cosh^2[\sqrt{2}m(\tau - \tau_0)]}. \tag{24.51}$$

[†]These solutions are discussed in Sec. 24.4.

Inserting the latter into Eq. (24.50) we obtain with a change of the variable[‡] to $z = \sqrt{2}m(\tau - \tau_0)$ the Pöschl–Teller equation

$$\left[\frac{d^2}{dz^2} - 1 + \frac{w_k^2}{2m^2} + \frac{6}{\cosh^2 z}\right]\psi_k = 0. \tag{24.52}$$

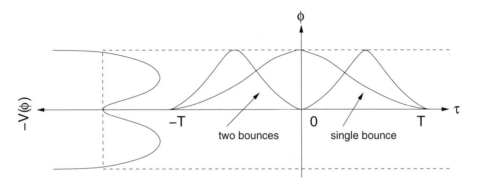

Fig. 24.5 Trajectories of one bounce and two bounces with $-V(\phi)$ on the left.

The spectrum of this equation can be read off from the result (22.62):

$$\lambda = -1 + \frac{w_k^2}{2m^2} = -(2 - N)^2, \quad N = 0, 1, \tag{24.53}$$

i.e. $w_k^2 = -6m^2, 0$ and a continuum beyond $+m^2$. Thus the equation has two discrete eigenvalues. The eigenfunction associated with the eigenvalue zero is, of course, the zero mode. We see therefore that the solution (24.48) is associated with one negative eigenvalue of the *stability equation*, i.e. it is a bounce. We can understand the significance of the bounce in terms of its Newtonian interpretation. This requires the consideration of the negative of the potential, i.e. the double well potential as shown in Fig. 24.4. The Newtonian pseudoparticle with energy zero sits at $\phi = 0$ for a long time, then (with an insignificant kick) it makes a brief excursion to ϕ_+ or ϕ_- and there reverses its direction of motion (i.e. bounces back) and finally returns to its initial position at $\phi = 0$ to remain there again for a long time. The bounce of this model is presumably the first bounce discussed in the literature;[§] the name "*bounce*" was introduced by Coleman.[¶] We can evaluate the action of

---

[‡]In frequent references to Chapter 23 please note that the variable z in Example 23.1 is defined as $m(\tau - \tau_0)$, whereas the present chapter requires — as may be seen from the solution (24.48) — the corresponding z to be $\sqrt{2}m(\tau - \tau_0)$. Thus there is a difference with $\sqrt{2}$.

[§]J. S. Langer [158]; see the discussion preceding his formula (3.30).

[¶]S. Coleman [54]; see in particular Secs. 2.4 and 6.1 to 6.5.

the bounce solution. Thus, as in the previous case,

$$
\begin{aligned}
S_E(\phi_c) &= 2\int_{-\infty}^{\infty} d\tau V(\phi_c) = 2\int_{-\infty}^{\infty} d\tau \left[m^2\phi_c^2 - \frac{1}{2}g^2\phi_c^4 \right] \\
&= 2\int_{-\infty}^{\infty} d\tau \left[2\frac{m^4}{g^2}\operatorname{sech}^2(\sqrt{2}m\tau) - 2\frac{m^4}{g^2}\operatorname{sech}^4(\sqrt{2}m\tau) \right],
\end{aligned}
$$

and hence[||]

$$
\begin{aligned}
S_E(\phi_c) &= \frac{4m^4}{\sqrt{2}mg^2}\int_{-\infty}^{\infty} d\tau \frac{\sinh^2\tau}{\cosh^4\tau} \\
&= \frac{4m^4}{\sqrt{2}mg^2}\left[\frac{\tanh^3\tau}{3} \right]_{-\infty}^{\infty} = \frac{4\sqrt{2}m^3}{3g^2}. \qquad (24.54)
\end{aligned}
$$

We now have to proceed parallel to our calculations in Chapter 23. Since the particle trapped in the central trough of the inverted double well potential can tunnel through the barriers on either side and escape to infinity, its energy is complex. Parallel to the previous chapter we consider the case of the ground state. Thus we are concerned with the evaluation of the kernel (23.25), i.e. (for $\hbar = 1$)

$$
\begin{aligned}
K(\phi_f, t_f; \phi_i, t_i) &= \sum_n \psi_n(\phi_f)\psi_n(\phi_i)\exp[-iE_n(t_f - t_i)] \\
&= \int_{\phi_i=\phi(t_i)}^{\phi_f=\phi(t_f)} \mathcal{D}\{\phi(t)\}\exp[iS(\phi)]
\end{aligned}
$$

for $i(t_f - t_i) = \tau_f - \tau_i = 2T \to \infty$ and $n = 0$, so that (re-inserting $\hbar$)

$$
\begin{aligned}
K(\phi_f, \tau_f; \phi_i, \tau_i) &\simeq \psi_0(\phi_f)\psi_0(\phi_i)\exp[-2E_0T/\hbar] \\
&= \int_{\phi_i=\phi(\tau_i)}^{\phi_f=\phi(\tau_f)} \mathcal{D}\{\phi(\tau)\}\exp[-S_E(\phi)/\hbar]. \qquad (24.55)
\end{aligned}
$$

The real part of the ground state energy E_0 is here (with the mass of the particle taken to be 1, as can be inferred from Eq. (24.46) or Chapter 23),

$$
\Re(E_0) = \frac{m\hbar}{\sqrt{2}}, \quad e^{-m\sqrt{2}T} = e^{-2T[\Re(E_0)]/\hbar}.
$$

[||]In this step we use the following formula given in Tables, such as of H. B. Dwight [81], p. 162:

$$
\int d\tau \frac{\sinh^{p-2}\tau}{\cosh^p\tau} = \frac{\tanh^{p-1}\tau}{(p-1)}.
$$

For no bounce one has simply the harmonic oscillator contribution

$$K(\phi_f, T; \phi_i, -T) \rightarrow K_0(\phi_f, T; \phi_i, -T) \simeq |\psi_0|^2 \exp[-m\sqrt{2}T]$$

$$= \left(\frac{\sqrt{2}m}{\pi\hbar}\right)^{1/2} \exp[-m\sqrt{2}T]. \qquad (24.56)$$

24.3.2 The single bounce contribution

In proceeding parallel to the evaluation of the path integral in Chapter 23 we consider first the contribution of the dominant single bounce to the Feynman amplitude. We are therefore first interested in the evaluation of ($\hbar = 1$)

$$\langle \phi_f, T | \phi_i, -T \rangle = \int_{\phi_i}^{\phi_f} \mathcal{D}\{\phi(\tau)\} \exp\left[-\int_{-T}^{T} d\tau \left\{\frac{1}{2}\left(\frac{d\phi}{d\tau}\right)^2 \right.\right.$$

$$\left.\left. + m^2\phi^2(\tau) - \frac{1}{2}g^2\phi^4(\tau)\right\}\right] \qquad (24.57)$$

in the single bounce approximation (or contribution), which we write

$$K_1(\phi_f, T; \phi_i, -T).$$

Setting

$$\phi(\tau) = \phi_c(\tau) + \eta(\tau), \qquad (24.58)$$

the Euclidean action of the single bounce can be factorized out so that

$$K_1(\phi_f, T; \phi_i, -T) = e^{-S_E(\phi_c)} \int \mathcal{D}\{\eta(\tau)\} e^{-\delta S_E} \equiv e^{-S_E(\phi_c)} I_0, \qquad (24.59)$$

where

$$\delta S_E = \frac{1}{2} \int_{-T}^{T} d\tau \, (\eta M \eta), \qquad (24.60a)$$

and

$$M = -\frac{d^2}{d\tau^2} + 2[-6m^2\text{sech}^2(\sqrt{2}m\tau) + m^2]. \qquad (24.60b)$$

Here M, the operator of the second variation of the action evaluated at the bounce, is obtained in the same way as the corresponding operator in Chapter 23 and is again restricted to terms quadratic in η, i.e. to the so-called one–loop approximation.

We let $\{\eta_n(\tau)\}$ be a complete set of eigenstates of the operator M with eigenvalues w_n^2, i.e.

$$M\eta_n = w_n^2 \eta_n \quad \text{with} \quad \int_{-T}^{T} d\tau \eta_n(\tau)\eta_m(\tau) = \delta_{nm}, \qquad (24.61)$$

again with $T \to \infty$. Expanding the fluctuation $\eta(\tau)$ in terms of these states, we can write

$$\eta(\tau) = \sum_n \xi_n \eta_n(\tau). \tag{24.62}$$

The integral I_0 in Eq. (24.59) can then be formally (i.e. for $T = \infty$) evaluated in terms of Gaussian integrals as

$$
\begin{aligned}
I_0 &= \int \cdots \int \det \left(\frac{d\eta(\tau)}{d\xi_n} \right) (\prod_k d\xi_k) \exp \left[-\frac{1}{2} \sum_n w_n^2 |\xi_k|^2 \right] \\
&= \det \left(\frac{d\eta(\tau)}{d\xi_n} \right) \prod_n \left(\frac{2\pi}{w_n^2} \right)^{1/2},
\end{aligned} \tag{24.63}
$$

where $\det(d\eta(\tau)/d\xi_n)$ is the Jacobian of the transformation (24.62). We observed earlier that the existence of the one negative eigenvalue of the fluctuation equation makes I_0 imaginary. As in Chapter 23 the zero mode is removed by using the *Faddeev–Popov technique*. This means we make in Eq. (24.63) the replacement

$$\int d\xi_0 = \triangle_{FP} \int_{-T}^T d\tau_0, \tag{24.64}$$

where τ_0 denotes the position of the *collective coordinate*, i.e. the position of the bounce, which is defined in the present case as the centre of the bounce and hence as the point where its velocity is zero, i.e. $d\phi_c/d\tau = 0$. We recall from Chapter 23 that with normalization of the fluctuation modes to unity, the *Faddeev–Popov determinant* is in the leading approximation given by the square root of the Euclidean action. Hence with Eq. (24.54)

$$\triangle_{FP} = \left(\frac{4\sqrt{2}m^3}{3g^2} \right)^{1/2}.$$

It follows therefore that the *kernel* (24.59) becomes (cf. Eq. (23.84))

$$
\begin{aligned}
K_1(\phi_f, T; \phi_i, -T) &= e^{-S_E(\phi_c)} \int d\xi_0 \det \left(\frac{d\eta(\tau)}{d\xi_n} \right) \prod_{n \neq 0} \left(\frac{2\pi}{w_n^2} \right)^{1/2} \\
&= e^{-S_E(\phi_c)} \triangle_{FP} \int_{-T}^T d\tau_0 \det \left(\frac{d\eta(\tau)}{d\xi_n} \right) \prod_{n \neq 0} \left(\frac{2\pi}{w_n^2} \right)^{1/2} \\
&= 2T \left(\frac{4\sqrt{2}m^3}{3g^2} \right)^{1/2} e^{-S_E(\phi_c)} \det \left(\frac{d\eta(\tau)}{d\xi_n} \right) \prod_{n \neq 0} \left(\frac{2\pi}{w_n^2} \right)^{1/2}.
\end{aligned} \tag{24.65}
$$

The evaluation of the functional determinant and the product factor proceeds as in Chapter 23. However, in view of various differences, we cannot simply copy from there, so that a recalculation is necessary. We deal with these steps in the next subsection.

24.3.3   Evaluation of the single bounce kernel

As in Chapter 23 we introduce the mapping

$$z(\tau) = \eta(\tau) - \int_{-T}^{\tau} d\tau' \frac{\dot{N}(\tau')}{N(\tau')} \eta(\tau'), \qquad (24.66)$$

where $N(\tau)$ is a function which satisfies the zero-mode equation

$$\ddot{N}(\tau) - 2m^2[1 - 6\,\text{sech}^2(\sqrt{2}m\tau)]N(\tau)] = 0 \qquad (24.67)$$

and is given by

$$N(\tau) = \frac{d\phi_c(\tau)}{d\tau} = -\frac{2m^2}{g} \frac{\sinh(\sqrt{2}m\tau)}{\cosh^2(\sqrt{2}m\tau)}. \qquad (24.68)$$

Taking $\eta(\tau)$ from Eq. (24.66) and substituting this into the original functional integral, i.e. Eq. (24.57), transforming to $z(\tau)$ and inserting a Lagrange multiplier along with the boundary condition that $\eta(\tau)$ has to satisfy at $\pm T$ — these steps were performed in Chapter 23 and need not be repeated here! — one obtains the (normalization independent) formula (23.59) for the quantity I_0, i.e.

$$I_0 = \left(\frac{1}{2\pi\hbar}\right)^{1/2} [N(T)N(-T)]^{-1/2} \left[\int_{-T}^{T} \frac{d\tau}{N^2(\tau)}\right]^{-1/2}. \qquad (24.69)$$

In Example 24.1 we evaluate I_0 and obtain the following finite result for $T \to \infty$:

$$I_0 = \pm i\left(\frac{m\sqrt{2}}{2\pi\hbar}\right)^{1/2} \equiv \pm i I_0'. \qquad (24.70)$$

In Example 24.2 we show that for large but finite T the eigenvalue $w_0^2 \to \tilde{w}_0^2$ of the fluctuation equation is in the present case given by

$$\tilde{w}_0^2 = 48\,m^2 e^{-2\sqrt{2}mT}. \qquad (24.71)$$

The calculation parallels the detailed calculation of Example 23.1 but a recalculation is essential here in view of the different zero mode and different numerical factors.

Example 24.1: Evaluation of I_0 for finite but large T

Show that the result (24.70) follows from a consideration of finite but large values of T.

Solution: We have, with $z = \sqrt{2m}\tau$,

$$\int_{-T}^{T} \frac{d\tau}{N^2(\tau)} = \frac{g^2}{4m^4}\int_{-T}^{T} d\tau \frac{\cosh^4(\sqrt{2m}\tau)}{\sinh^2(\sqrt{2m}\tau)} = \frac{g^2}{4m^4}\frac{1}{\sqrt{2m}}\int_{-\sqrt{2m}T}^{\sqrt{2m}T} dz \frac{\cosh^4 z}{\sinh^2 z}.$$

With a formula from Tables of Integrals** this becomes

$$\int_{-T}^{T} \frac{d\tau}{N^2(\tau)} = \frac{g^2}{4m^4}\frac{1}{\sqrt{2m}}\left[\frac{3}{2}z + \frac{1}{4}\sinh(2z) - \coth z\right]_{-\sqrt{2m}T}^{\sqrt{2m}T} \stackrel{T\to\infty}{\simeq} \frac{g^2}{m^5 2^{9/2}}e^{2\sqrt{2m}T}.$$

Also

$$N(T)N(-T) \stackrel{T\to\infty}{\simeq} -\frac{(2m)^4}{g^2}e^{-2\sqrt{2m}T}.$$

Hence

$$I_0 = \left(\frac{1}{2\pi\hbar}\right)^{1/2}[N(T)N(-T)]^{-1/2}\left[\int_{-T}^{T} \frac{d\tau}{N^2(\tau)}\right]^{-1/2} = \frac{1}{\sqrt{2\pi\hbar}}\frac{g\,m^{5/2}2^{9/4}}{\sqrt{-1}m^2 2^2 g} = \pm i\left(\frac{\sqrt{2m}}{2\pi\hbar}\right)^{1/2}.$$

The exponentials cancelled out.

Example 24.2: Eigenvalues for finite range normalization

Derive the particular nonvanishing eigenvalue of the fluctuation equation which replaces the zero eigenvalue of the zero mode if the latter is required to vanish not at infinity but for a large but finite Euclidean time T.

Solution: The operator of the small fluctuation equation is given by Eq. (24.60b). We change to the variable $z = \sqrt{2m}(\tau - \tau_0)$ and consider the following set of two eigenvalue problems:
(a) Our ultimate consideration is concerned with the equation

$$\left(\frac{d^2}{dz^2} + 6\operatorname{sech}^2 z - n^2\right)\psi_0(z) = 0 \quad \text{with} \quad n^2 = 1 \text{ and } \psi_0(\pm\infty) = 0.$$

However, we consider now the related equation (cf. Eq. (24.52))
(b)

$$\left(\frac{d^2}{dz^2} + 6\operatorname{sech}^2 z - n^2\right)\tilde\psi_0(z) = -\frac{\tilde{w}_0^2}{2m^2}\tilde\psi_0(z) \quad \text{with} \quad \tilde\psi_0(\pm\sqrt{2m}T) = 0,$$

and our aim is to obtain $\tilde{w}_0^2$. The unnormalized solution of case (a) is (cf. Eq. (24.68))

$$\psi_0(z) \equiv \frac{d\phi_c}{d\tau} = -\frac{2m^2}{g}\frac{\sinh z}{\cosh^2 z}, \quad z = \sqrt{2m}(\tau - \tau_0).$$

We multiply the equation of case (a) by $\tilde\psi_0(z)$ and the equation of case (b) by $\psi_0(z)$, subtract one equation from the other and integrate from $-\sqrt{2m}T$ to $\sqrt{2m}T$. Then

$$\int_{-\sqrt{2m}T}^{\sqrt{2m}T} dz\left(\psi_0\frac{d^2\tilde\psi_0}{dz^2} - \tilde\psi_0\frac{d^2\psi_0}{dz^2}\right) = -\frac{\tilde{w}_0^2}{2m^2}\int_{-\sqrt{2m}T}^{\sqrt{2m}T} dz\psi_0\tilde\psi_0.$$

On the left hand side we perform a partial integration; on the right hand side we can replace (in the dominant approximation) the integral by that integrated from $-\infty$ to ∞, and since this is the

---

**See I. S. Gradshteyn and I. M. Ryzhik [122], formula 40, p. 100.

integral over twice the kinetic energy (and hence $S_E(\phi_c)$ of Eq. (24.54)) we obtain (remembering the change of variable from τ to z)

$$\left(\psi_0\frac{d\tilde{\psi}_0}{dz} - \tilde{\psi}_0\frac{d\psi_0}{dz}\right)\bigg|_{-\sqrt{2mT}}^{\sqrt{2mT}} = \psi_0\frac{d\tilde{\psi}_0}{dz}\bigg|_{-\sqrt{2mT}}^{\sqrt{2mT}} = -\frac{\tilde{w}_0^2}{2m^2}\frac{4\sqrt{2}m^3}{3g^2}\sqrt{2m} = -\frac{\tilde{w}_0^2}{2m^2}\frac{8m^4}{3g^2}. \tag{24.72}$$

In order to be able to evaluate the left hand side we consider WKB approximations of the solutions of the equations of cases (a) and (b) above. The WKB solution of case (a) is (c being a constant and the parameter z_0 occurring here the position of a turning point)

$$\psi_0(z) = \frac{c}{v^{1/4}}\exp\left(-\int_{z_0}^z dz\, v^{1/2}\right), \quad v(z) = n^2 - 6\,\mathrm{sech}^2 z \quad\text{and}\quad v(z_0) = 0.$$

Thus for $z \to \infty$ we have $v^{1/2}(z) \to n$ and $\psi_0(z) \to 0$, as required. The corresponding WKB solution of case (b) which satisfies its required boundary condition, i.e. vanishing at $z = \sqrt{2mT}$, is similarly verified to be

$$\tilde{\psi}_0(z) = \frac{c}{\tilde{v}^{1/4}}\left[\exp\left(-\int_{z_0}^z dz\,\tilde{v}^{1/2}\right) - \exp\left(-2\int_{z_0}^{\sqrt{2mT}} dz\,\tilde{v}^{1/2}\right)\exp\left(\int_{z_0}^z dz\,\tilde{v}^{1/2}\right)\right],$$

where $\tilde{v} = n^2 - 6\,\mathrm{sech}^2 z - (\tilde{w}_0^2/2m^2)$. Differentiating the last expression we obtain

$$\frac{d}{dz}\tilde{\psi}_0(z) = -c\tilde{v}^{1/4}\left[\exp\left(-\int_{z_0}^z dz\,\tilde{v}^{1/2}\right) + \exp\left(-2\int_{z_0}^{\sqrt{2mT}} dz\,\tilde{v}^{1/2}\right)\exp\left(\int_{z_0}^z dz\,\tilde{v}^{1/2}\right)\right],$$

and so

$$\frac{d}{dz}\tilde{\psi}_0\bigg|_{z=\sqrt{2mT}} \simeq -2c\tilde{v}^{1/4}\exp\left(-\int_{z_0}^{\sqrt{2mT}} dz\,\tilde{v}^{1/2}\right).$$

But (see $\psi_0(z)$ above)

$$\frac{d}{dz}\psi_0\bigg|_{z=\sqrt{2mT}} \simeq -cv^{1/4}\exp\left(-\int_{z_0}^{\sqrt{2mT}} dz\, v^{1/2}\right).$$

Hence (with $n^2 \simeq n^2 - (\tilde{w}_0^2/2m^2)$ for $T \to \infty$)

$$\frac{d}{dz}\tilde{\psi}_0(\sqrt{2mT}) \simeq 2\frac{d}{dz}\psi_0(\sqrt{2mT}), \quad\text{and analogously}\quad \frac{d}{dz}\tilde{\psi}_0(-\sqrt{2mT}) \simeq 2\frac{d}{dz}\psi_0(-\sqrt{2mT}).$$

It follows that the left hand side of Eq. (24.72) is

$$\psi_0\frac{d\tilde{\psi}_0}{dz}\bigg|_{-\sqrt{2mT}}^{\sqrt{2mT}} \simeq 2\psi_0\frac{d}{dz}\psi_0\bigg|_{-\sqrt{2mT}}^{\sqrt{2mT}}.$$

But

$$\psi_0(z) = -2\frac{m^2}{g}\frac{\sinh z}{\cosh^2 z} \simeq \mp\frac{4m^2}{g}e^{\mp z} \quad\text{for } z \to \pm\infty,$$

so that

$$\frac{d}{dz}\psi_0(z) \simeq \frac{4m^2}{g}e^{\mp z} \simeq \mp\psi_0(z) \quad\text{for } z \to \pm\infty.$$

Hence

$$\psi_0(\pm\sqrt{2mT})\frac{d}{dz}\psi_0(\pm\sqrt{2mT}) \simeq \mp\psi_0^2(\pm\sqrt{2mT}) \simeq \mp\left(\frac{4m^2}{g}\right)^2 e^{-2\sqrt{2mT}}.$$

Equation (24.72) therefore becomes

$$2\left[-2\left(\frac{4m^2}{g}\right)^2 e^{-2\sqrt{2}mT}\right] \simeq -\frac{\tilde{w}_0^2}{2m^2}\frac{8m^4}{3g^2}.$$

We thus obtain the result

$$\tilde{w}_0^2 \simeq 48\,m^2 e^{-2\sqrt{2}mT} \quad \text{or} \quad \tilde{w}_0 \equiv \tilde{v}_0 e^{-\sqrt{2}mT}, \quad \tilde{v}_0 = 4\sqrt{3}m \quad \text{for} \quad T \to \infty.$$

We return to Eq. (24.65) for the evaluation of the single-bounce contribution to the Feynman kernel $K_1(\phi_f, T; \phi_i, -T)$, in which we use Eq. (24.63) and replace the vanishing w_0^2 by $\tilde{w}_0^2$, which is finite for finite values of T. Then

$$K_1(\phi_f, T; \phi_i, -T)$$
$$= e^{-S_E(\phi_c)}\left(\frac{4\sqrt{2}m^3}{3g^2}\right)^{1/2}(2T)\det\left(\frac{d\eta(\tau)}{d\xi_n}\right)\prod_{n\neq0}\left(\frac{2\pi}{w_n^2}\right)^{1/2}$$
$$= e^{-S_E(\phi_c)}\left(\frac{4\sqrt{2}m^3}{3g^2}\right)^{1/2}(2T)I_0\left(\frac{\tilde{w}_0^2}{2\pi}\right)^{1/2}. \tag{24.73}$$

We thus arrive at the formula (re-inserting $\hbar$)

$$K_1(\phi_f, T; \phi_i, -T) = (2T)\triangle_{FP}\underbrace{I_0}_{\text{imaginary}}\left(\frac{\tilde{w}_0^2}{2\pi}\right)^{1/2}\exp\left[-\frac{S_E(\phi_c)}{\hbar}\right], \tag{24.74}$$

which is formally identical with Eq. (23.101). Inserting now our results for I_0 and $\tilde{w}_0^2$, this becomes (with $I_0 \equiv \pm iI_0'$)

$$K_1(\phi_f, T; \phi_i, -T) = \pm i\,2T\triangle_{FP}\left(\frac{\tilde{v}_0^2}{2\pi}\right)^{1/2}e^{-\sqrt{2}mT}I_0'e^{-S_E(\phi_c)/\hbar}$$
$$= \pm i2T\left(\frac{4\sqrt{2}m^3}{3g^2}\right)^{1/2}\frac{4\sqrt{3}m}{\sqrt{2\pi}}\left(\frac{m\sqrt{2}}{2\pi\hbar}\right)^{1/2}e^{-\sqrt{2}mT}e^{-S_E(\phi_c)/\hbar}$$
$$= \pm i\,m2Te^{-\sqrt{2}mT}\left(\frac{16\sqrt{2}m^3}{\pi g^2}\right)^{1/2}\left(\frac{\sqrt{2}m}{\pi\hbar}\right)^{1/2}e^{-S_E(\phi_c)/\hbar}. \tag{24.75}$$

This is the contribution of a single bounce to the Feynman amplitude

$$\langle\phi_f, T|\phi_i, -T\rangle.$$

We next consider 2-bounce solutions and, more generally, n-bounce solutions.

24.3.4   Sum over an infinite number of bounces

Parallel to the summation over an infinite number of instanton-anti-instanton pairs in Chapter 23, we here have to sum over an infinite number of bounces corresponding to an arbitrary number of back and forth tunnelings in the barrier before the particle is ejected and thus escapes to infinity. For the computation of this effect it is again best to consider first the case of two bounces, perhaps more precisely the 2-bounce solution, which then enables the generalization to solutions consisting of an arbitrary number of bounces. We write the *Feynman amplitude*

$$\langle \phi_f, T | \phi_i, -T \rangle = \sum_{n=0}^{\infty} K_n(\phi_f, T; \phi_i, -T), \tag{24.76}$$

where $K_n(\phi_f, T; \phi_i, -T)$ denotes the kernel due to an n-bounce solution (K_0 that with no bounce). First we consider the kernel due to two bounces, K_2, which means we have to consider an amplitude with intermediate configurations whose endpoints are not fixed and therefore have to be integrated over. Thus, using completeness of states,

$$K_2(\phi_f, T; \phi_i, -T) = \int_{-\infty}^{\infty} d\phi_1 \langle \phi_f, T | \phi_1, T_1 \rangle \langle \phi_1, T_1 | \phi_i, -T \rangle. \tag{24.77}$$

As in Chapter 23 the two amplitudes are effectively evaluated like the single bounce amplitude above, with $T_1 = 0$ (see Fig. 24.5), but in addition K_2 requires an integration over the configurations ϕ_1. In order to deal with these intermediate integrations we write therefore (with $\hbar = 1$)

$$K_2(\phi_f, T; \phi_i, -T) \propto \int_{-\infty}^{\infty} d\phi_1 e^{-S_E(\phi_f, \phi_1)} e^{-S_E(\phi_1, \phi_i)}.$$

For the computation of the intermediate–time kernels here, the Gaussian approximation selects as optimal paths (i.e. almost stationary solutions) those for which $\phi_1 = \phi_i = \phi_f$. Thus we expand the actions as follows:

$$S_E(\phi_f, \phi_1) = S_E(\phi_f, \phi_i) + \frac{1}{2} \frac{\partial^2 S_E}{\partial \phi_1^2} \bigg|_{\phi_1 = \phi_i} (\phi_1 - \phi_i)^2 + \cdots, \tag{24.78a}$$

$$S_E(\phi_1, \phi_i) = S_E(\phi_f, \phi_i) + \frac{1}{2} \frac{\partial^2 S_E}{\partial \phi_1^2} \bigg|_{\phi_1 = \phi_f} (\phi_1 - \phi_f)^2 + \cdots. \tag{24.78b}$$

Neglecting higher order contributions we have (with $\phi_i = \phi_f$ and see the evaluation of the second derivative below)

$$
\int_{-\infty}^{\infty} d\phi_1 e^{-S_E(\phi_f, \phi_1)} e^{-S_E(\phi_1, \phi_i)}
$$

$$
= \left\{ e^{-S_E(\phi_f, \phi_i)} \right\}^2 \int_{-\infty}^{\infty} d\phi_1 \exp\left[-(\phi_1 - \phi_i)^2 \frac{\partial^2 S_E}{\partial \phi_1^2}\bigg|_{\phi_i} \right]
$$

$$
= \left\{ e^{-S_E(\phi_f, \phi_i)} \right\}^2 \gamma, \tag{24.79}
$$

with (the result is calculated below)

$$
\gamma = \left[\frac{\pi}{(\partial^2 S_E / \partial \phi_1^2)_{\phi_i}} \right]^{1/2} = \left(\frac{\pi}{\sqrt{2}m} \right)^{1/2}. \tag{24.80}
$$

Here we used

$$
S_E(\phi_1, \phi_i) = \int_{\phi_i}^{\phi_1} \left(\frac{d\phi}{d\tau} \right)^2 d\tau = \int_{\phi_i}^{\phi_1} \frac{d\phi}{d\tau} d\phi
$$

$$
= \int_{\phi_i}^{\phi_1} \sqrt{2V(\phi)} d\phi = \int_{\phi_i}^{\phi_1} \sqrt{2m^2\phi^2 - g^2\phi^4} d\phi.
$$

Hence

$$
\frac{\partial S_E}{\partial \phi_1} = \sqrt{2m^2\phi_1^2 - g^2\phi_1^4}, \qquad \frac{\partial^2 S_E}{\partial \phi_1^2} = \frac{2m^2\phi_1 - 2g^2\phi_1^3}{\sqrt{2m^2\phi_1^2 - g^2\phi_1^4}},
$$

so that (re-inserting $\hbar$)

$$
\gamma = \left[\pi\hbar \bigg/ \frac{\partial^2 S_E}{\partial \phi_1^2} \right]^{1/2}_{\phi_1 = \phi_i = \phi_f = 0} = \left(\frac{\pi\hbar}{\sqrt{2}m} \right)^{1/2}. \tag{24.81}
$$

We now apply the arguments of Chapter 23 to the present case of widely separated bounces so that a configuration consisting of n widely separated bounces extremizes approximately the action, and the collective coordinate of every bounce replaces a zero mode in the manner of the Faddeev–Popov procedure, but within an overall Euclidean time range of $2T$. The integration over the two collective coordinates $\tau_0^{(1)}, \tau_0^{(2)}$ is similar to that in the case of instantons in Chapter 23, and we have in the present case of two bounces[††]

$$
\int_{-T}^{T} d\tau_0^{(1)} \int_{-T}^{\tau_0^{(1)}} d\tau_0^{(2)} = \frac{(2T)^2}{2!}. \tag{24.82}
$$

[††]Corresponding to the remark in the instanton case, this can also be written as

$$
\int_{-T/2}^{3T/2} d\tau_0^{(1)} \int_{-T+\tau_0^{(1)}}^{T/2} d\tau_0^{(2)} = \frac{(2T)^2}{2!}.
$$

Then the approximate two-bounce configuration leads to a contribution to the Feynman kernel which we can write

$$K_2(\phi_f, T; \phi_i, -T) = \gamma(\pm i)^2 \frac{(2T)^2}{2!} e^{-\sqrt{2}mT}$$

$$\times \left[\triangle_{FP} \left(\frac{\tilde{v}_0^2}{2\pi} \right)^{1/2} I_0' e^{-S_E(\phi_c)/\hbar} \right]^2. \quad (24.83)$$

The generalization to the case of n bounces is evidently

$$K_n(\phi_f, T; \phi_i, -T) = \frac{1}{\gamma} (\pm i\gamma)^n \frac{(2T)^n}{n!} e^{-\sqrt{2}mT}$$

$$\times \left[\triangle_{FP} \left(\frac{\tilde{v}_0^2}{2\pi} \right)^{1/2} I_0' e^{-S_E(\phi_c)/\hbar} \right]^n. \quad (24.84)$$

The total Feynman amplitude therefore becomes, by including also the no-bounce contribution K_0 considered earlier (cf. Eq. (24.56)),

$$\langle \phi_f, T | \phi_i, -T \rangle = \frac{e^{-\sqrt{2}mT}}{\gamma} \sum_{n=0}^{\infty} \frac{1}{n!} \left[\pm i\gamma 2T \triangle_{FP} \left(\frac{\tilde{v}_0^2}{2\pi} \right)^{1/2} I_0' e^{-S_E(\phi_c)/\hbar} \right]^n$$

$$= \frac{e^{-\sqrt{2}mT}}{\gamma} \exp \left[\pm i\gamma 2T \triangle_{FP} \left(\frac{\tilde{v}_0^2}{2\pi} \right)^{1/2} I_0' e^{-S_E(\phi_c)/\hbar} \right]. \quad (24.85)$$

Comparing the exponential contained in this result with the exponentials in Eqs. (24.55) to (24.56), we obtain the formula

$$i\Im(E_0) = \pm i\gamma \triangle_{FP} \left(\frac{\tilde{v}_0^2}{2\pi} \right)^{1/2} I_0' e^{-S_E(\phi_c)/\hbar}. \quad (24.86)$$

Inserting the explicit expressions for the present case of the inverted double well potential, this result is

$$i\Im(E_0) = \pm i \left(\frac{\pi\hbar}{\sqrt{2}m} \right)^{1/2} \left(\frac{4\sqrt{2}m^3}{3g^2} \right)^{1/2} \frac{4\sqrt{3}m}{\sqrt{2\pi}} \left(\frac{\sqrt{2}m}{2\pi\hbar} \right)^{1/2} \exp \left[-\frac{4\sqrt{2}m^3}{3g^2\hbar} \right]$$

$$= \pm im \left(\frac{16\sqrt{2}m^3}{\pi g^2} \right)^{1/2} \exp \left[-\frac{4\sqrt{2}m^3}{3g^2\hbar} \right]. \quad (24.87)$$

This ground state result can be checked to agree with the $q_0 = 1$ (ground state) case of Eq. (18.86).[‡‡]

[‡‡]To see this one has to change from the mass $m_0 = 1/2$ case in Chapter 18 to the mass $m_0 = 1$

24.3.5   Comments

Finally we ask: Does the sine–Gordon theory permit a bounce? After the above demonstration of the existence of a bounce in Φ^4-theory (for the inverted potential) it is natural to inquire whether an appropriately modified sine–Gordon theory can similarly be shown to possess a bounce. As long as the potential remains periodic (like an infinitely long sine wave) we cannot expect a bounce since in every domain of length π the problem with periodic eigenfunctions is selfadjoint and the eigenvalues are real (though with tunneling contributions). If the eigenvalues are real there can be no bounce, since the bounce as a saddle point configuration would determine the imaginary part of the eigenvalue. Thus we cannot expect a bounce unless we truncate or distort the potential such that it allows a current to infinity.

24.4   Inverted Double Well: Constant Solutions

In the case of the constant solution $\phi = 0$ the fluctuation equation is

$$\frac{d^2\psi}{dz^2} + (w^2 - 2m^2)\psi = 0 \tag{24.88}$$

with solutions

$$\psi \sim \begin{array}{c} \sin(pz) \\ \cos(pz) \end{array} \ , \quad p^2 = w^2 - 2m^2.$$

If we demand periodicity of $\psi(z)$ with period L (later removed by letting $L \to \infty$), we have

$$\psi(z) = \psi(z+L) \quad \text{and so} \quad p^2 \to p_n^2 = \left(\frac{2\pi n}{L}\right)^2,$$

---

case here (with indices indicating mass: $2E_1 = E_{1/2}, 2h_1^4 = h_{1/2}^4, 2c_1^2 = c_{1/2}^2$). Equation (18.86) then is

$$2E = E_0(q_0, \sqrt{2}h^2) \overset{+}{(-)} i \frac{2^{q_0} \sqrt{2}h^2}{(2\pi)^{1/2}[\frac{1}{2}(q_0-1)]!} \left(\frac{\{\sqrt{2}h^2\}^3}{4c^2}\right)^{q_0/2} \exp\left[-\frac{(\sqrt{2}h^2)^3}{12c^2}\right].$$

Next the parameters of the potential have to be related: $h^2 \to 2m, c^2 \to g^2$. Then ($\hbar = 1$)

$$E = E_0 \overset{+}{(-)} i \frac{m}{\sqrt{\pi}[\frac{1}{2}(q_0-1)]!} \left(\frac{16\sqrt{2}m^3}{g^2}\right)^{q_0/2} \exp\left[-\frac{4\sqrt{2}m^3}{3g^2}\right].$$

For $q_0 = 1$ this is (with $\hbar = 1$)

$$E = E_0 \overset{+}{(-)} im \left(\frac{16\sqrt{2}m^3}{\pi g^2}\right)^{1/2} \exp\left[-\frac{4\sqrt{2}m^3}{3g^2}\right].$$

i.e.

$$w^2 = 2m^2 + p_n^2 = 2m^2 + \left(\frac{2\pi n}{L}\right)^2, \tag{24.89}$$

then

$$w^2 = 2m^2 \quad \text{for} \quad L \to \infty, \quad \text{i.e.} \quad w^2 > 0.$$

Thus, as is obvious from the potential, the configuration $\phi = 0$ is not a saddle point of the action.

Now consider the constant configurations

$$\phi_{\pm} = \pm\sqrt{2}\frac{m}{g},$$

for which (cf. Eq. (24.51))

$$V''(\phi_{\pm}) = 2m^2 - 6g^2\left(\sqrt{2}\frac{m}{g}\right)^2 = -10m^2. \tag{24.90}$$

In these cases the fluctuation equation becomes

$$\frac{d^2\psi}{dz^2} + (w_k^2 + 10m^2)\psi_k = 0. \tag{24.91}$$

Proceeding as above we obtain the eigenvalue

$$w_k^2 = -10m^2 < 0, \tag{24.92}$$

which implies that the configurations $\phi_{\pm}$ represent *saddle point configurations*, as is also clear from the potential.

24.5 The Cubic Potential and its Complex Energy

In Sec. 24.2 we considered the cubic potential for purposes of illustration. Having developed the method of calculation of complex energies in the previous section — for the case of the inverted double well potential — we can now use the general result obtained there, i.e. Eq. (24.86), to obtain the complex energy of the ground state of the cubic potential of Sec. 24.2. Clearly the quantities to be computed in the case of the cubic potential (24.27) are

$$\gamma, \quad \tilde{w}_0, \quad \text{and} \quad I_0'.$$

We leave these calculations to Examples 24.3, 24.4 and 24.5. The results are

$$\gamma = \left(\frac{\pi\hbar}{2m}\right)^{1/2}, \tag{24.93}$$

$$I_0 \equiv \pm iI_0', \quad I_0' = \left(\frac{m}{\pi\hbar}\right)^{1/2}, \tag{24.94}$$

$$\tilde{w}_0^2 \simeq 60a^2b^2 e^{-4mT}, \quad \tilde{v}_0 = 2\sqrt{15}ab, \quad T \to \infty, \tag{24.95}$$

Inserting these expressions into the formula given by Eq. (24.86), we obtain

$$i\Im E = \pm i\frac{ab\sqrt{8ba^5}}{\sqrt{\pi}}\exp\left[-\frac{8ba^5}{15}\right] = \pm i\frac{2^5 m^3\sqrt{m}}{b^2\sqrt{\pi}}\exp\left[-\frac{8(2m)^5}{15b^4}\right], \tag{24.96}$$

where $2m = ab$. This result (first version) should be compared with the WKB result (14.105). They agree, except for the usual numerical factors stemming from the non-suitability of the WKB method for a ground state, as considered here.

Example 24.3: Evaluation of γ for the cubic potential
Derive the result (24.93) from a consideration of the action.

Solution: With the help of Eq. (24.26) we can express the action in terms of the potential and consider

$$S_E(q, q_i) = \int_{q_i}^q dx\left(\frac{dq}{dx}\right)^2 = \int_{q_i}^q \frac{dq}{dx}dq = \int_{q_i}^q dq\sqrt{2V(q)} = \int_{q_i}^q dq\sqrt{b^2q^2(a^2 - q)}. \tag{24.97}$$

It follows that

$$\frac{\partial S_E}{\partial q} = \sqrt{b^2q^2(a^2 - q)}, \quad \frac{\partial^2 S_E}{\partial q^2} = \frac{2b^2q(a^2 - q) - b^2q^2}{2\sqrt{b^2q^2(a^2 - q)}}.$$

It follows that

$$\left(\frac{\partial^2 S_E}{\partial q^2}\right)_{q=0} = \frac{2b^2a^2}{2\sqrt{b^2a^2}} = ba = 2m, \tag{24.98}$$

so that

$$\gamma = \sqrt{\frac{\pi}{(\partial^2 S_E/\partial q^2)_{q=0}}} = \left(\frac{\pi\hbar}{2m}\right)^{1/2}.$$

Example 24.4: Evaluation of I_0 for large T for the cubic potential
Derive the result (24.94) from the knowledge of the zero mode given by Eq. (24.29).

Solution: The zero mode is the derivative of the classical configuration $q_c(x)$ and is given by

$$N(x) = -2a^2m\frac{\sinh mx}{\cosh^3 mx} = \dot{q}_c(x), \quad \text{with} \quad \frac{1}{2}ab \equiv m. \tag{24.99}$$

We obtain I_0 from the expression in Eq. (24.69), i.e.

$$I_0 = \left(\frac{1}{2\pi\hbar}\right)^{1/2}[N(T)N(-T)]^{-1/2}\left[\int_{-T}^T \frac{d\tau}{N^2(\tau)}\right]^{-1/2}.$$

For $T \to \infty$ we obtain readily

$$N(T)N(-T) = \left(-2a^2m\frac{\sinh mT}{\cosh^3 mT}\right)\left(2a^2m\frac{\sinh mT}{\cosh^3 mT}\right) \simeq -4a^4m^2 2^4 e^{-4mT} \quad \text{for} \quad T \to \infty. \tag{24.100}$$

Next we evaluate the integral

$$\int_{-T}^{T} \frac{dx}{N^2(x)} = \frac{1}{4a^4m^3} \int_{z=-mT}^{mT} dz \frac{\cosh^6 z}{\sinh^2 z}.$$

We use the following formulas from Tables of Integrals:[*]

$$\int dz \frac{\cosh^6 z}{\sinh^2 z} = \frac{1}{4}\frac{\cosh^5 z}{\sinh z} + \frac{5}{4}\int dz \frac{\cosh^4 z}{\sinh^2 z} \quad \text{and} \quad \int dz \frac{\cosh^4 z}{\sinh^2 z} = \frac{3}{2}z + \frac{1}{4}\sinh 2z - \coth z.$$

The dominant term is the one with the dominant exponential for $T \to \infty$. This contribution comes from the first term of the first of these integrals. Thus

$$\int_{z=-mT}^{mT} dz \frac{\cosh^6 z}{\sinh^2 z} \simeq \frac{1}{4}\left[\frac{\cosh^5 z}{\sinh z}\right]_{-mT}^{mT} \xrightarrow{T\to\infty} \frac{1}{4}\left[\frac{2e^{4mT}}{16}\right] = \frac{1}{32}e^{4mT}.$$

It follows that

$$\int_{-T}^{T} \frac{dx}{N^2(x)} \xrightarrow{T\to\infty} \frac{1}{2^7 a^4 m^3}e^{4mT}. \tag{24.101}$$

Inserting these results into the expression for I_0, we obtain

$$I_0 = \frac{[2^7 a^4 m^3 e^{-4mT}]^{1/2}}{(2\pi\hbar)^{1/2}[-4a^4m^2 2^4 e^{-4mT}]^{1/2}} = \pm i\left(\frac{m}{\pi\hbar}\right)^{1/2}.$$

Example 24.5: Finite range eigenvalues for the cubic potential

Derive the result (24.95) parallel to the calculation in Example 24.2.

Solution: The calculations parallel those of Example 24.2. Therefore we do not reproduce every intermediate step. Changing the variable in Eq. (24.34) to $z = mx$ and proceeding as in Example 24.2 we have to consider two different eigenvalue problems:
(a) The ultimate problem of the equation

$$\left(\frac{d^2}{dz^2} + 12\text{sech}^2 z - 4\right)\psi_0(z) = 0 \quad \text{with} \quad \psi_0(\pm\infty) = 0,$$

and
(b) the problem with finite range boundary conditions

$$\left(\frac{d^2}{dz^2} + 12\text{sech}^2 z - 4\right)\tilde{\psi}_0(z) = -\frac{\tilde{w}_0^2}{m^2}\tilde{\psi}_0(z) \quad \text{with} \quad \tilde{\psi}_0(\pm T) = 0.$$

We multiply the equation of case (a) by $\tilde{\psi}_0(z)$ and that of case (b) by $\psi_0(z)$ and subtract one from the other. Then with partial integration and with S_E of Eq. (24.32) one obtains

$$\psi_0\frac{d\tilde{\psi}_0}{dz}\Big|_{-mT}^{mT} = -\frac{\tilde{w}_0^2}{m^2}\int_{-mT}^{mT} dz\psi_0(z)\tilde{\psi}_0(z) = -\frac{\tilde{w}_0^2}{m^2}mS_E$$

$$\simeq -\frac{\tilde{w}_0^2}{m^2}\left(\frac{4b^2a^6}{15}\right) = -\frac{\tilde{w}_0^2}{m^2}\frac{2^8 m^6}{15b^4}. \tag{24.102}$$

Introducing WKB wave functions as in Example 24.2, but now for

$$v(z) = 4 - 12\text{sech}^2 z \quad \text{and} \quad \tilde{v}(z) = 4 - 12\text{sech}^2 z - \frac{\tilde{w}_0^2}{m^2},$$

---

[*]I. S. Gradshteyn and I. M. Ryzhik [122], formulas 2.411, p. 93 and 2.423.40, p. 100.

and proceeding as there, we obtain

$$\psi_0 \frac{d\tilde{\psi}_o}{dz}\bigg|_{-mT}^{mT} \simeq 2\psi_0 \frac{d\psi_0}{dz}\bigg|_{-mT}^{mT}. \tag{24.103}$$

But

$$\psi_0(z) = -2a^2 m \frac{\sinh z}{\cosh^3 z} \simeq \mp 8a^2 m e^{\mp 2z} \quad \text{for} \quad z \to \pm\infty,$$

and

$$\frac{d}{dz}\psi_0(z) \simeq 16a^2 m e^{\mp 2z} \simeq \mp 2\psi_0(z) \quad \text{for} \quad z \to \pm\infty.$$

It follows that

$$\left[\psi_0 \frac{d\psi_0}{dz}\right]_{z \to \pm mT} \simeq \mp 2^7 a^4 m^2 e^{-4mT}. \tag{24.104}$$

With Eq. (24.103) we then obtain from Eq. (24.102)

$$-2^8 a^4 m^2 e^{-4mT} \simeq -\frac{\tilde{w}_0^2}{m^2} \frac{2^8 m^6}{15 b^4},$$

i.e.

$$\tilde{w}_0^2 \simeq \frac{15 a^4 b^4}{m^2} e^{-4mT} = 4(15 a^2 b^2) e^{-4mT} \quad \text{for} \quad T \to \infty. \tag{24.105}$$

Chapter 25

Periodic Classical Configurations

25.1 Introductory Remarks

In the foregoing chapters we considered pseudoparticles on the space $\mathbb{R}^1$. We now want to consider these on a circle and hence as periodic configurations. Although solitons have been familiar for a long time, their consideration on a circle happened only as late as 1988 in a paper of Manton and Samols.[*] The periodic configurations arising in the *three basic cases of double well potential, cosine potential and inverted double well potential* were considered soon thereafter[†] and in particular it was shown that in each of these basic cases — in fact, also in the case of the cubic potential — the equation of small fluctuations about the periodic configuration is a standard *Lamé equation*, thus permitting a clear investigation of the spectrum with reduction to the *Pöschl–Teller equation* in the limit of an infinitely large circle.[‡] Classical periodic configurations, called *periodic instantons*[§] or *periodic bounces* or something similar, have since then been realized to be of considerable significance, particularly in condensed matter physics in the context of spin tunneling, which we consider briefly in Chapter 28. In the following we consider the three basic cases of double well potential, cosine potential and inverted double well potential in the context of theories on a circle and investigate

---

[*]N. S. Manton and T. M. Samols [187].

[†]J.–Q. Liang, H. J. W. Müller–Kirsten and D. H. Tchrakian [165].

[‡]In the case of the sextic potential the small fluctuation equation is different but assumes the form of an ellipsoidal wave equation in a certain approximation. See F. Zimmerschied and H. J. W. Müller–Kirsten [291].

[§]Periodic instantons have also been introduced by S. Y. Khlebnikov, V. A. Rubakov and P. G. Tinyakov [150].

their characteristic properties. The following chapter will then employ these configurations in the path integral derivation of level splittings and imaginary parts of eigenvalues.

25.2 The Double Well Theory on a Circle

25.2.1 Periodic configurations

We consider again the Lagrangian density[¶]

$$\mathcal{L} = \frac{1}{2}\partial_\mu\Phi\partial^\mu\Phi - U(\Phi), \quad U(\Phi) = \frac{\mu^2}{2a^2}(\Phi^2 - a^2)^2 \tag{25.1}$$

of a scalar field $\Phi(t, x)$ in $1 + 1$ dimensions, however this time the spatial coordinate x is restricted to the finite doman of the circumference of a circle. We therefore demand Φ to obey the periodicity condition

$$\Phi(t, x) = \Phi(t, x + L). \tag{25.2}$$

We are again interested in static configurations $\Phi \to \phi(x)$ for which the Euler–Lagrange equation

$$\frac{\partial^2\Phi}{\partial t^2} - \frac{\partial^2\Phi}{\partial x^2} + U'(\Phi) = 0$$

reduces to the Newton-like equation

$$\frac{\partial^2\phi(x)}{\partial x^2} = U'(\phi), \quad \text{or} \quad \frac{d}{d\phi}\left[\frac{1}{2}\left(\frac{d\phi}{dx}\right)^2\right] - U'(\phi) = 0. \tag{25.3}$$

Integrating with $C^2 = \text{const.}$ we obtain

$$\frac{1}{2}\left(\frac{d\phi}{dx}\right)^2 - U(\phi) = -\frac{1}{2}C^2. \tag{25.4}$$

We can regard $-C^2/2$ as the energy of a particle of mass $m_0 = 1$ in the potential $-U(\phi)$, the variable x in this Newtonian consideration then playing

---

[¶]For ease of comparison with literature and other chapters we note another way in which the double well potential is written, and the relation between the parameters here and there. Thus

$$U(\Phi) = \frac{\eta^2}{2}\left(\Phi^2 - \frac{m^2}{\eta^2}\right)^2 \quad \text{with} \quad \mu^2 = m^2, \quad \frac{m^2}{a^2} = \eta^2, \quad a^2 = \frac{m^2}{\eta^2} = \frac{\mu^2}{\eta^2}.$$

the role of time. Periodic motion implies that the particle moves between the two extrema of $U(\phi)$ and so its energy is subject to the bounds

$$-U(0) \le -\frac{1}{2}C^2 \le 0, \quad \text{or} \quad 0 \le \frac{1}{2}C^2 \le \frac{1}{2}\mu^2 a^2. \tag{25.5}$$

We shall see that we have to use a parameter k called the "*elliptic modulus*" of the *Jacobian elliptic functions* involved. It will be seen to be convenient to write C^2 in the form

$$C^2 = a^2 \mu^2 \left(\frac{1-k^2}{1+k^2}\right)^2, \quad 0 \le k \le 1, \tag{25.6}$$

in which case k is, in fact, this elliptic modulus. Thus, varying k over its domain is equivalent to varying the energy proportional to C^2. The double well potential and and its derivatives are

$$U(\phi) = \frac{1}{2}\frac{\mu^2}{a^2}(\phi^2 - a^2)^2, \qquad U'(\phi) = 2\frac{\mu^2}{a^2}\phi(\phi^2 - a^2),$$

$$U''(\phi) = 2\frac{\mu^2}{a^2}(3\phi^2 - a^2) \tag{25.7a}$$

with

$$U''(0) = -2\mu^2, \quad U''(\pm a) = 4\mu^2. \tag{25.7b}$$

Inserting $U'(\phi)$ into Eq. (25.4) we obtain

$$\int dx = \int \frac{d\phi}{\sqrt{2U(\phi) - C^2}} = \pm\frac{a}{\mu}\int \frac{d\phi}{\sqrt{(\phi^2 - a^2)^2 - \frac{a^2}{\mu^2}C^2}}. \tag{25.8}$$

Setting

$$s_\pm^2 = a^2 \pm C\frac{a}{\mu}, \quad s_+ = a\left(\frac{2k^2}{1+k^2}\right)^{1/2}, \quad s_- = a\left(\frac{2}{1+k^2}\right)^{1/2}, \tag{25.9}$$

(where we used Eq. (25.6)), we obtain

$$
\begin{aligned}
x - x_0 &= \pm\frac{a}{\mu}\int_{\phi_0}^{s=\phi} \frac{ds}{\sqrt{(s^2 - s_+^2)(s^2 - s_-^2)}} \\
&= \pm\frac{a}{\mu s_-}\int^{\phi/s_+} \frac{d(s/s_+)}{\sqrt{(1 - s^2/s_+^2)(1 - s^2/s_-^2)}} \\
&= \pm\frac{a}{\mu s_-}\int^{\phi/s_+} \frac{dt}{\sqrt{(1 - t^2)(1 - k^2 t^2)}}.
\end{aligned}
\tag{25.10}
$$

The integral on the right is a standard elliptic integral. It is unavoidable in the following to use Jacobian elliptic functions. However, it is not difficult to handle these with the help of tables[||] — they are then not much more difficult to handle than trigonometrical functions. The most important properties required in this text are summarized in Appendix A. The first integral we are concerned with here is in standard notation

$$u = \int_0^x \frac{dt}{\sqrt{(1-t^2)(1-k^2t^2)}} = \int_0^\varphi \frac{d\vartheta}{\sqrt{1 - k^2 \sin^2 \vartheta}}, \qquad (25.11)$$

where the second integral is obtained from the first with the substitution $t = \sin \vartheta, x = \sin \varphi$. The *Jacobian elliptic functions* written $\operatorname{sn} u$, $\operatorname{cn} u$, $\operatorname{dn} u$, are defined by

$$\begin{aligned} \operatorname{sn} u &= \sin \varphi = x, \quad \operatorname{cn} u = \cos \varphi = (1 - x^2)^{1/2}, \\ \operatorname{dn} u &= (1 - k^2 \sin^2 \varphi)^{1/2} = (1 - k^2 x^2)^{1/2}, \end{aligned} \qquad (25.12)$$

where x is the upper integration limit in Eq. (25.11). The three Jacobian elliptic functions are one-valued functions of u and are doubly periodic, one real period being

$$4K(k) = 4 \int_0^{\pi/2} \frac{d\vartheta}{\sqrt{1 - k^2 \sin^2 \vartheta}} \qquad (25.13)$$

($K(k)$ is called quarter period), i.e. $\operatorname{sn} u = \operatorname{sn}(u + 4K(k))$, etc. Important particular values of $K(k)$ are

$$K(0) = \frac{1}{2}\pi, \quad K(1) = \infty. \qquad (25.14)$$

Correspondingly if we write $\operatorname{sn} u \equiv \operatorname{sn}[u, k^2]$ (k^2 will be explicitly referred to only when important or to avoid confusion), one has

$$\begin{aligned} \operatorname{sn}[u, 0] &= \sin u, \quad \operatorname{sn}[u, 0] = \cos u, \quad \operatorname{dn}[u, 0] = 1, \\ \operatorname{sn}[u, 1] &= \tanh u, \quad \operatorname{cn}[u, 1] = \operatorname{dn}[u, 1] = \operatorname{sech} u. \end{aligned} \qquad (25.15)$$

Returning to Eq. (25.10) we see that (with Eqs. (25.11) and (25.12))

$$x - x_0 = \pm \frac{a}{\mu s_-} \operatorname{sn}^{-1}\left(\frac{\phi}{s_+}\right),$$

[||] Relations between elliptic functions can be found in most Tables of Integrals and other mathematical data, e.g. in L. S. Gradshteyn and I. M. Ryzhik [122], but also in P. F. Byrd and M. D. Friedman [40] (the most extensive compilation of elliptic integrals), L. M. Milne–Thomson [196] (which contains tables of numerical values) and E. T. Whittaker and G. N. Watson [283].

or

$$\phi = s_+ \mathrm{sn}\left[\pm \frac{\mu s_-}{a}(x - x_0)\right] = \pm s_+ \mathrm{sn}\left[\frac{\mu s_-}{a}(x - x_0)\right], \tag{25.16}$$

since $\mathrm{sn}\, u$ is an odd function of u. It is convenient to define a quantity $b(k)$ by

$$b(k) := \mu\sqrt{\frac{2}{1 + k^2}}, \quad b(0) = \mu\sqrt{2}, \quad b(1) = \mu, \tag{25.17a}$$

so that

$$\mu\frac{s_-}{a} = \mu\sqrt{\frac{2}{1 + k^2}} = b(k), \tag{25.17b}$$

and hence with (see below) $x_0 = 0$, we have the *periodic instanton configuration*

$$\phi_c(x) = \pm\frac{akb(k)}{\mu}\mathrm{sn}[b(k)x, k^2]. \tag{25.18}$$

An important aspect of this solution is its periodicity, i.e. that of

$$\mathrm{sn}\, u = \mathrm{sn}\,[u + 4K(k)] = \mathrm{sn}\,[u + 4nK(k)], \quad (n \text{ an integer}).$$

Since we require $\phi(x)$ to satisfy Eq. (25.2), i.e. $\phi(x) = \phi(x + L)$, we must have

$$\mathrm{sn}[b(k)x] = \mathrm{sn}[b(k)(x + L)],$$

i.e. the real period is

$$b(k)L = 4nK(k), \quad L \to L_n(k) = \frac{4nK(k)}{b(k)} = nL_1(k). \tag{25.19}$$

In particular for $k = 0$ and $k = 1$ we have

$$L_n(0) = \frac{4nK(0)}{b(0)} = \frac{4n}{\mu}\frac{\pi}{2\sqrt{2}} = \frac{\sqrt{2}\pi n}{\mu}, \quad L_n(1) = \frac{4nK(1)}{b(1)} = \infty. \tag{25.20}$$

The solution with period $nL_1 = L_n$ will in the following be denoted by $\phi_n(x)$, so that

$$\begin{array}{lll} \phi_1 & \text{is } \phi(x) \text{ with} & \phi_c(x) = \phi_c(x + L_1(k)), \\ \phi_2 & \text{is } \phi(x) \text{ with} & \phi_c(x) = \phi_c(x + L_2(k)), \\ \phi_n & \text{is } \phi(x) \text{ with} & \phi_c(x) = \phi_c(x + nL_1(k)). \end{array} \tag{25.21}$$

Thus we have a tower of functions of different periods. Two functions of different periods are shown schematically in Fig. 25.1. Increasing the period L indefinitely the solution $\phi_c(x)$ becomes a *one-kink configuration* in a half period as in Fig. 25.1, but a one-kink plus one antikink configuration in one complete period of the ribbon obtained by gluing the two ends together.

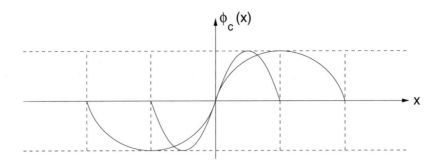

Fig. 25.1 Two solutions $\phi_c(x)$ with periods $L_1, 2L_1$.

Thus we wonder if we can regain the kink solution familiar from soliton theory on the infinitely long line. Indeed, since $L_n(1) = \infty$, we just have to increase k to 1. Then (cf. (25.15))

$$\phi_c(x) \to \pm a \tanh(\mu x). \tag{25.22}$$

The approach to $\tanh(\mu x)$ can be deduced from the approximate formula[**] for $\tanh u \sim \pm 1, u \to \pm\infty$:

$$
\begin{aligned}
\operatorname{sn}[u, k^2] \quad &\simeq \quad \tanh u + \frac{1}{4}(1-k^2)\frac{\sinh u \cosh u - u}{\cosh^2 u} \\
&\overset{u\to\infty}{\simeq} \left(1 - 2e^{-2u}\right) + \frac{1}{4}(1-k^2)\left(1 - 4ue^{-2u}\right) \\
&= \left[1 + \frac{1}{4}(1-k^2)\right] - 2e^{-2u}\left[1 + \frac{1}{2}(1-k^2)u\right]. \tag{25.23}
\end{aligned}
$$

This kind of approach is supported by the diagram of Fig. 25.2, which shows the function $\operatorname{sn} u$ for varying k^2. The $\pm$ signs in Eq. (25.22) refer, of course, to the kink and antikink configurations respectively.

On the other hand, in the limit $k^2 \to 0$

$$\operatorname{sn}[u, k^2] \to \sin u,$$

and so $\phi_c(x)$ approaches its $k = 0$ limit with a multiplicative factor k as

$$\phi_c(x) \simeq \pm\frac{akb(0)}{\mu}\sin[b(0)x], \quad \text{i.e.} \quad \phi_c(x) \simeq \pm a\sqrt{2}k\sin(\sqrt{2}\mu x). \tag{25.24}$$

As k varies from 0 to 1, the period $L_n(k) = 4K(k)n/b(k)$ increases monotonically from $L_n(0) = \sqrt{2}\pi n/\mu$ to ∞, so that $\phi_n(x)$ is a solution only for

$$L_n(0) = \frac{\sqrt{2}\pi n}{\mu} \le L_n(k) \le \infty. \tag{25.25}$$

[**]See M. Abramowitz and I. A. Stegun [1], p. 574.

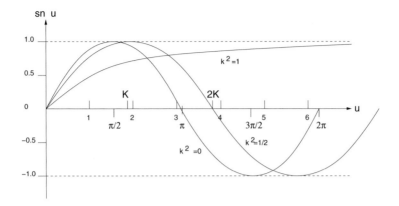

Fig. 25.2 The function sn u and its dependence on k^2.

Thus

$$\phi_1(x) \text{ is solution for } L \text{ in } [\sqrt{2}\pi/\mu, \infty],$$
$$\phi_2(x) \text{ is solution for } L \text{ in } [2\sqrt{2}\pi/\mu, \infty],$$
$$\cdots$$
$$\phi_n(x) \text{ is solution for } L \text{ in } [n\sqrt{2}\pi/\mu, \infty].$$

Thus whenever $L_n(k)$ reaches a new critical value $L_n(0)$ we have a new solution as shown schematically in Fig. 25.3.

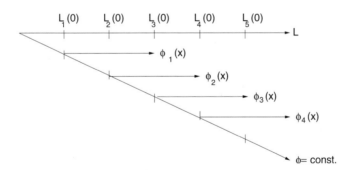

Fig. 25.3 Bifurcation of solutions at critical values of the period L.

Thus at each critical value $L_n(0)$ there is a bifurcation of static solutions, i.e. of stationary points of the static field energy given by

$$E(\phi) = \int_0^L dx \left[\frac{1}{2}\left(\frac{d\phi}{dx}\right)^2 + U(\phi) \right]. \tag{25.26}$$

At $k = 0$ the elliptic function sn u degenerates into sin u (cf. Eq. (25.15)) with quarter period $K(0) = \pi/2$, i.e. full period 2π. The solution

$$\phi_1(x) = \phi_1(x + L_1(k))$$

has the shape shown in Fig. 25.4.

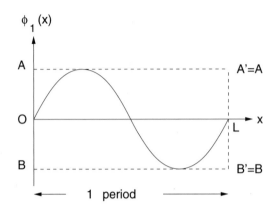

Fig. 25.4 The solution $\phi_1(x)$ on a cylinder.

Sketched on a cylinder as indicated there (with $OL = S^1$) we have to identify
the edges $AB, A'B'$. The curve is then seen to have one kink and one antikink.
Thus $\phi_1(x)$ can be looked at as the static field configuration consisting of *one
kink and one antikink*. Correspondingly n of the configurations $\phi_1(x)$ can be
looked at as describing n kinks and n antikinks, placed alternately around
the circle, but this configuration is not $\phi_n(x)$ which has period $L_n = nL_1$.

The energy $E(\phi_n)$ obtained from Eq. (25.26) can be calculated in terms
of $K(k)$, the *complete elliptic integral of the first kind* defined by Eq. (25.13),
and the following *complete elliptic integral of the second kind** (this is standard notation in the literature, thus $E(k)$ should not be confused with energy!)

$$E(k) = \int_0^1 dx \sqrt{\frac{1 - k^2 x^2}{1 - x^2}} = E\left[\frac{\pi}{2}, k\right]. \tag{25.27a}$$

It is important to keep in mind — particularly in connection with so-called
endpoint integrations in Chapter 26 — that this complete elliptic integral
is the limiting case of the following *incomplete elliptic integral of the second
kind*

$$E[\varphi, k] = \int_0^{\sin \varphi} dx \sqrt{\frac{1 - k^2 x^2}{1 - x^2}} = \int_0^{u_1} \mathrm{dn}^2 u \, du \equiv E(u_1). \tag{25.27b}$$

The evaluation of $E(\phi_n)$ is left to Example 25.1. The result[†] is

$$E(\phi_n) = n\frac{\sqrt{2}\mu a^2}{3(1 + k^2)^{3/2}}[8(1 + k^2)E(k) - (1 - k^2)(5 + 3k^2)K(k)]. \tag{25.28}$$

---

[*]P. F. Byrd and M. D. Friedman [40], see formula 110.07, p. 10.
[†]See also N. S. Manton and T. M. Samols [187].

Example 25.1: Evaluation of static field energy

Evaluate the classical static field energy $E(\phi_n)$ of the periodic instanton configuration (25.18).

Solution: The configuration we are concerned with is the function $\phi_n(x)$ of period L_n, i.e.

$$\phi_n(x) = \frac{akb(k)}{\mu}\mathrm{sn}[b(k)x] \quad \text{with} \quad \frac{d\phi_n}{dx} = \frac{akb^2(k)}{\mu}\mathrm{cn}[b(k)x]\mathrm{dn}[b(k)x].$$

This has to be inserted into Eq. (25.26), i.e.

$$E(\phi_n) = \int_0^{L_n} dx\left[\frac{1}{2}\left(\frac{d\phi_n}{dx}\right)^2 + U(\phi_n)\right], \quad \text{where} \quad L_n = \frac{4nK(k)}{b(k)} \quad \text{and} \quad b(k) = \mu\sqrt{\frac{2}{1+k^2}}$$

(cf. Eqs. (25.19) and (25.17a)). From Eqs. (25.4) and (25.6) we know that

$$U(\phi) = \frac{1}{2}\left(\frac{d\phi}{dx}\right)^2 + \frac{1}{2}C^2 \quad \text{with} \quad C^2 = a^2\mu^2\left(\frac{1-k^2}{1+k^2}\right)^2.$$

Hence

$$E(\phi_n) = \int_0^{L_n} dx\left[\left(\frac{d\phi_n}{dx}\right)^2 + \frac{1}{2}C^2\right] \quad \text{with} \quad \frac{1}{2}C^2L_n = n\frac{a^2\mu(1-k^2)^2\sqrt{2}K(k)}{(1+k^2)^{3/2}}.$$

Using the derivative of $\phi_n(x)$ given above, and the following integral[‡]

$$\int du\,\mathrm{cn}^2 u\,\mathrm{dn}^2 u = \frac{1}{3k^2}[(1+k^2)E(u) - (1-k^2)u + k^2\mathrm{sn}\,u\,\mathrm{cn}\,u\,\mathrm{dn}\,u],$$

we obtain

$$
\begin{aligned}
E(\phi_n) - \frac{1}{2}C^2L_n &= \frac{a^2k^2b^3(k)}{\mu^2}\int_0^{z=b(k)L_n=4nK(k)} dz\,\mathrm{cn}^2 z\,\mathrm{dn}^2 z \\
&= \frac{a^2k^2b^3(k)}{\mu^2}\frac{1}{3k^2}[(1+k^2)E(z) - (1-k^2)z + k^2\mathrm{sn}\,z\,\mathrm{cn}\,z\,\mathrm{dn}\,z]_0^{4nK(k)} \\
&= \frac{a^2 2^{3/2}}{3(1+k^2)^{3/2}}[(1+k^2)4nE(k) - 4nK(k)(1-k^2)].
\end{aligned}
$$

It follows that $(E(K(k)) \equiv E(k))$

$$
\begin{aligned}
E(\phi_n) &= \frac{a^2\mu\sqrt{2}n}{3(1+k^2)^{3/2}}[8(1+k^2)E(k) - 8(1-k^2)K(k) + (1-k^2)^2K(k)] \\
&= \frac{a^2\mu\sqrt{2}n}{3(1+k^2)^{3/2}}[8(1+k^2)E(k) - (1-k^2)(5+3k^2)K(k)],
\end{aligned}
$$

For $k = 1$, i.e. $L_n = \infty$, we obtain from Eq. (25.28) (since $E(1) = 1$)

$$E(\phi_n) \overset{k=1}{=} 2nE_0, \quad E_0 = \frac{4}{3}\mu a^2. \tag{25.29}$$

Here E_0 is the energy of a single kink on an infinite line (cf. S_E of Eq. (23.13) which agrees with E_0 here if the appropriate parameters of the double well

---

[‡]P. F. Byrd and M. D. Friedman [40], formula 361.03, p. 212.

potential are identified, i.e. $\mu \to m, a \to m/g$). We can obtain an approximate expression of $E(\phi_n)$ for finite but large periods L_n by using the expansions[§]

$$
\begin{aligned}
K(k) &= \ln\left(\frac{4}{\sqrt{1-k^2}}\right) + \frac{1}{4}(1-k^2)\left[\ln\left(\frac{4}{\sqrt{1-k^2}}\right) - 1\right] \\
&\quad + \frac{9}{64}(1-k^2)^2\left[\ln\left(\frac{4}{\sqrt{1-k^2}}\right) - \cdots\right] + \cdots, \\
E(k) &= 1 + \frac{1}{2}(1-k^2)\left[\ln\left(\frac{4}{\sqrt{1-k^2}}\right) - \frac{1}{2}\right] \\
&\quad + \frac{3}{16}(1-k^2)^2\left[\ln\left(\frac{4}{\sqrt{1-k^2}}\right) - \cdots\right] + \cdots. \quad (25.30)
\end{aligned}
$$

Thus

$$
\begin{aligned}
E(\phi_n) &\simeq n\frac{\mu a^2}{6}\left[16\left\{1 + \frac{1}{2}(1-k^2)\ln\left(\frac{4}{\sqrt{1-k^2}}\right)\right.\right. \\
&\qquad\left. + \frac{3}{16}(1-k^2)^2\ln\left(\frac{4}{\sqrt{1-k^2}}\right)\right\} \\
&\qquad\left. - 8(1-k^2)\left\{\ln\left(\frac{4}{\sqrt{1-k^2}}\right) + \frac{1}{4}(1-k^2)\ln\left(\frac{4}{\sqrt{1-k^2}}\right) + \cdots\right\}\right] \\
&\simeq \frac{8}{3}n\mu a^2 + \frac{n\mu a^2}{6}\left[(3-2)(1-k^2)^2\ln\left(\frac{4}{\sqrt{1-k^2}}\right) + \cdots\right] \\
&= 2n\frac{4}{3}\mu a^2 + \frac{n\mu a^2}{6}(1-k^2)^2\ln\left(\frac{4}{\sqrt{1-k^2}}\right) + \cdots. \quad (25.31)
\end{aligned}
$$

But

$$
L_n = \frac{4n\sqrt{1+k^2}}{\mu\sqrt{2}}K(k) \stackrel{k\to 1}{\simeq} \frac{4n}{\mu}\ln\left(\frac{4}{\sqrt{1-k^2}}\right),
$$

so that we see how an exponential comes about:

$$
\frac{4}{\sqrt{1-k^2}} \approx e^{\mu L_n/4n}, \quad 1 - k^2 \simeq 16\,e^{-\mu L_n/2n},
$$

and hence approximately

$$
\begin{aligned}
E(\phi_n) &\approx 2n\frac{4}{3}\mu a^2 + \frac{n\mu a^2}{6}2^8 e^{-\mu L_n/n}\left(\frac{4}{\sqrt{1-k^2}}\right) \\
&\sim 2n\frac{4}{3}\mu a^2 - \lambda_0 n\mu a^2 e^{-\mu L_n/n} + O\left(e^{-3\mu L_n/2n}\right), \quad (25.32)
\end{aligned}
$$

[§]See I. S. Gradshteyn and I. M. Ryzhik [122], pp. 905 and 906.

where λ_0 is some positive number.[¶] The first term (with factor 2n) represents the energy of n kinks and n antikinks (as independent structures), and the next term the leading contribution of — what may be interpreted as — their *interaction*.

Besides the solutions $\phi_n(x)$, the classical equation (25.3) also has important constant solutions, i.e.

$$\phi = 0, \ \pm a \quad \text{for } k = 0 \text{ and 1 respectively,} \tag{25.33}$$

the first corresponding to the local maximum of the potential and the other two to the classical vacua at the minima.

25.2.2 The fluctuation equation

As in our previous soliton considerations the small fluctuation equation is

$$\left[-\frac{d^2}{dx^2} + U''(\phi(x)) \right] \psi_k(x) = w_k^2 \psi_k(x) \tag{25.34}$$

and $\delta^2 E(\phi) \geq 0$ if $w_k^2 \geq 0$. Now, of course, we demand periodicity, i.e.

$$\psi_k(x) = \psi_k(x + L). \tag{25.35}$$

We consider three types of solutions.

(1) We consider first the *constant solutions* $\phi = \pm a$. For these $U''(\pm a) = 4\mu^2$, so that the fluctuation equation is

$$\frac{d^2 \psi}{dx^2} + (w_k^2 - 4\mu^2)\psi = 0. \tag{25.36}$$

Demanding periodicity of $\psi(x)$ we have

$$\psi \sim \begin{matrix} \sin \\ \cos \end{matrix} [p_m x] = \begin{matrix} \sin \\ \cos \end{matrix} [p_m(x + L)] \tag{25.37}$$

with

$$p_m^2 = w_m^2 - 4\mu^2, \quad p_m L = 2\pi m, \quad m = 0, 1, 2, \ldots,$$

so that

$$w_m^2 = 4\mu^2 + \frac{4\pi^2 m^2}{L^2} > 0. \tag{25.38}$$

Thus, since $w_m^2 > 0$ for all m, the constant configurations $\phi = \pm a$ have classical stability (i.e. are isolated minima of $E(\phi)$), which is not surprising since $\phi = \pm a$ are the positions of the classical vacua.

[¶]N. S. Manton and T. M. Samols [187] have $\lambda_0 = 32$.

(2) Considering next the *constant solution* $\phi = 0$, we have $U''(0) = -2\mu^2$, so that

$$\frac{d^2\psi}{dx^2} + (w_k^2 + 2\mu^2)\psi = 0.$$

Now

$$\psi \sim \begin{array}{c} \sin \\ \cos \end{array} [p_m x] = \begin{array}{c} \sin \\ \cos \end{array} [p_m(x + L)]$$

with

$$p_m^2 = w_m^2 + 2\mu^2 = \left(\frac{2\pi m}{L}\right)^2, \tag{25.39}$$

i.e.

$$w_m^2 = -2\mu^2 + \frac{4\pi^2 m^2}{L^2}. \tag{25.40}$$

At the critical values of L:

$$L^2 \to L_n^2 = \frac{2\pi^2 n^2}{\mu^2},$$

so that there

$$w_m^2 = 2\mu^2\left(\frac{m^2}{n^2} - 1\right), \quad m = 0, 1, 2, \ldots, \quad n = 1, 2, \ldots. \tag{25.41}$$

Thus for $n = 1$ (i.e. at the critical L-value L_1) we obtain for $m = 0, 1, \ldots$:

$$w_m^2 : \quad -2\mu^2, 0, 6\mu^2, 16\mu^2, \ldots$$

and for $n = 2$ (i.e. at L_2) we obtain for $m = 0, 1, 2, \ldots$:

$$w_m^2 : \quad -2\mu^2, -\frac{3}{2}\mu^2, 0, \ldots.$$

Thus negative eigenvalues occur for $n > m$ and zero eigenvalues for $n = m$. The solution $\phi = 0$ is therefore unstable, i.e. it represents a *saddle point* in the functional space about the classical configuration and is for this reason (cf. also Chapter 28) also called a "*sphaleron*" (actually the classical configuration at the top of a barrier).*

(3) Finally we consider the small fluctuation equation (25.34) for the *periodic instanton configuration* configurations (25.18). Since in this case

$$U''(\phi_c) = \frac{2\mu^2}{a^2}(3\phi_c^2 - a^2),$$

---

*The term "*sphaleron*", from Greek for "falling down", was introduced by N. S. Manton. See N. S. Manton, F. S. Klinkhamer and N. S. Manton [188].

the equation of small fluctuations is

$$\frac{d^2\psi}{dx^2} + \left[w^2 - \frac{2\mu^2}{a^2}(3\phi_c^2(x) - a^2)\right]\psi = 0,$$

i.e.

$$\psi''(x) + [w^2 + 2\mu^2 - 6k^2 b^2(k)\mathrm{sn}^2[b(k)x]]\psi(x) = 0. \tag{25.42}$$

Setting

$$z = b(k)x, \tag{25.43}$$

this becomes

$$\frac{d^2\psi(z)}{dz^2} + [\lambda - n(n+1)k^2\mathrm{sn}^2 z]\psi(z) = 0, \tag{25.44}$$

where

$$\lambda = \frac{w^2 + 2\mu^2}{b^2(k)} = \frac{w^2 + 2\mu^2}{2\mu^2}(1 + k^2), \quad n = 2. \tag{25.45}$$

Equation (25.44) is seen to be the *Lamé equation.*[†] Periodic solutions of this equation have been studied in detail in the literature.[‡] We find the equation of the zero mode ψ_0 by differentiating the classical equation (25.3), i.e.

$$\left[\frac{d^2}{dx^2} - U''(\phi)\right]\frac{d\phi}{dx} = 0.$$

Thus

$$\psi_0 \propto \frac{d\phi_n}{dx} = \frac{akb^2(k)}{\mu}\mathrm{cn}[b(k)x]\mathrm{dn}[b(k)x], \tag{25.46}$$

since

$$\frac{d}{du}\mathrm{sn}\, u = \mathrm{cn}\, u\, \mathrm{dn}\, u.$$

Since $\mathrm{sn}(bx)$ is an odd function of x, ψ_0 is an even function. But is this the ground state wave function? In order to answer this question we take a closer look at Lamé's equation (25.44). For integral values of n this equation has discrete, normalizable solutions known as *Lamé polynomials*. For $n = 2$ these are given by (in the notation of Arscott [11])[§]

$$scE_2^0(z) = \mathrm{sn}\, z\, \mathrm{cn}\, z \quad \text{with eigenvalue } \lambda_2^{(sc)0} = 4 + k^2, \tag{25.47a}$$

[†] That the small fluctuation equation of this case as well as the cases of the sine–Gordon potential and the inverted double well potential is a Lamé equation was realized in J.–Q. Liang, H. J. W. Müller–Kirsten and D. H. Tchrakian [165].

[‡] The standard reference is F. M. Arscott [11]. There is a factor of 2 missing in front of the square root of the eigenvalues $h_2^{(u)0}, h_2^{(u)1}$, p. 205. See also Example 25.2.

[§] Cf. F. M. Arscott [11], p. 205.

$$sdE_2^0(z) = \text{sn}z\,\text{dn}z \ \text{ with eigenvalue } \ \lambda_2^{(sd)0} = 1 + 4k^2, \qquad (25.47b)$$

$$cdE_2^0(z) = \text{cn}z\,\text{dn}z \ \text{ with eigenvalue } \ \lambda_2^{(cd)0} = 1 + k^2, \qquad (25.47c)$$

$$\left.\begin{array}{c} uE_2^0(z) \\ uE_2^1(z) \end{array}\right\} = \text{sn}^2 z - \frac{1 + k^2 \pm \sqrt{1 - k^2(1 - k^2)}}{3k^2}$$

with eigenvalues

$$\left.\begin{array}{c} \lambda_2^{(u)0} \\ \lambda_2^{(u)1} \end{array}\right\} = 2(1 + k^2) \mp 2\sqrt{1 - k^2(1 - k^2)}. \qquad (25.47d)$$

The third of these eigenfunctions is the *zero mode*; we can check that its eigenvalue w^2 of the small fluctuation equation is zero:

$$\lambda_2^{(cd)0} = \frac{w^2 + 2\mu^2}{2\mu^2}(1 + k^2) \overset{!}{=} (1 + k^2), \quad \text{i.e.} \quad w^2 = 0. \qquad (25.48)$$

The other eigenfunctions are those of excited states. For the first eigenfunction (25.47a) we have

$$\frac{w^2 + 2\mu^2}{2\mu^2}(1 + k^2) = 4 + k^2,$$

i.e.

$$w^2 = -2\mu^2 + 2\mu^2\left(\frac{4 + k^2}{1 + k^2}\right) = 2\mu^2\left(\frac{3}{1 + k^2}\right) > 0. \qquad (25.49)$$

For the second eigenfunction (25.47b) we have

$$\frac{w^2 + 2\mu^2}{2\mu^2}(1 + k^2) = 1 + 4k^2, \quad \text{i.e.} \quad w^2 = \frac{6k^2\mu^2}{1 + k^2} > 0. \qquad (25.50)$$

Finally for the last two cases of Eq. (25.47d) we obtain

$$\frac{w^2 + 2\mu^2}{2\mu^2}(1 + k^2) = [2(1 + k^2) \mp 2\sqrt{1 - k^2(1 - k^2)}],$$

i.e.

$$w^2 = 2\mu^2\left[1 \mp 2\frac{\sqrt{1 - k^2(1 - k^2)}}{1 + k^2}\right] > 0. \qquad (25.51)$$

Thus no eigenvalue w^2 of the equation of small fluctuations is negative and the configurations $\phi_n(x)$ therefore do not lead to saddle points of the action.

25.2.3 The limit of infinite period

In the limit of infinite period, i.e. $k \to 1$, we have

$$\text{sn}[z,1] = \tanh z, \quad \text{cn}[z,1] = \text{sech} z, \quad \text{dn}[z,1] = \text{sech} z,$$

and hence (since $b(1) = \mu$)

$$\frac{d\phi_n}{dx} = \frac{akb^2(k)}{\mu}\text{cn}[b(k)x]\text{dn}[b(k)x] \longrightarrow a\mu\text{sech}^2[\mu x].$$

This is the zero mode (22.66) we obtained previously for the double well potential on $\mathbb{R}^1$, a nicely behaving ground state wave function, the Lamé equation (25.44) in this case reducing to the Schrödinger equation with *Pöschl–Teller potential* (22.65a).

It is interesting to inquire what happens to the other eigenfunctions and eigenvalues in the limit $k \to 1$. From (25.47a) and (25.47b) we see that these merge together to yield the first excited state, i.e.

$$scE_2^0(z), \; sdE_2^0(z) \longrightarrow \tanh z \, \text{sech} z \quad \text{with eigenvalues } \lambda_2^{(sc)0}, \lambda_2^{(sd)0} = 5,$$

i.e. $w^2 = (\lambda - 2)\mu^2 = 3\mu^2$ in agreement with the discrete eigenvalue of Eq. (22.65a).

In the case of the solutions (25.47d) one finds that the first solution merges with the zero mode, the eigenvalue being zero and the solution $\tanh^2 z - 1 = -\text{sech}^2 z$. The second of these solutions has eigenvalue $w^2 = 4\mu^2$, and therefore merges with the onset of the continuum (recall the discussion after Eq. (22.65a)); in this case the solution is $\tanh^2 z - 1/3$.

Example 25.2: Verification of an eigenvalue
Verify that the eigenfunctions and eigenvalues given by Eq. (25.47d) solve the Lamé equation (25.44).

Solution: We substitute the solution $\psi(z)$ and eigenvalue λ of Eq. (25.47d) into the left hand side of the Lamé equation (25.44) and show that this yields zero. Thus

$$\frac{d\psi}{dz} = 2\text{sn} z \, \text{cn} z \, \text{dn} z, \quad \frac{d^2\psi}{dz^2} = 2[\text{cn}^2 z \, \text{dn}^2 z - \text{sn}^2 z \text{dn}^2 z - k^2\text{sn}^2 z \, \text{cn}^2 z].$$

Inserting this and λ into the left hand side of Eq. (25.44), we obtain (we have $n = 2$ here)

$$\text{l.h.s.} = 2[\text{cn}^2 z \, \text{dn}^2 z - \text{sn}^2 z \text{dn}^2 z - k^2\text{sn}^2 z \, \text{cn}^2 z] + 2[(1+k^2) \mp \sqrt{1 - k^2(1 - k^2)}] \times$$
$$\left(\text{sn}^2 z - \frac{1 + k^2 \pm \sqrt{1 - k^2(1 - k^2)}}{3k^2}\right) - 6k^2\text{sn}^2 z\left(\text{sn}^2 z - \frac{1 + k^2 \pm \sqrt{1 - k^2(1 - k^2)}}{3k^2}\right)$$

Multiplying this out and re-arranging terms with the help of the relations $1 - \mathrm{dn}^2 z = k^2 \mathrm{sn}^2 z$ and $\mathrm{sn}^2 z = 1 - \mathrm{cn}^2 z$, this becomes

$$
\begin{aligned}
\text{l.h.s.} \quad = \quad & 2\mathrm{cn}^2 z \mathrm{dn}^2 z - 2\mathrm{sn}^2 z \mathrm{dn}^2 z - 2k^2 \mathrm{sn}^2 z \mathrm{cn}^2 z + 4(1 + k^2)\mathrm{sn}^2 z - 6k^2 \mathrm{sn}^4 z \\
& - 2 \frac{(1 + k^2)^2 - \{1 - k^2(1 - k^2)\}}{3k^2} \\
= \quad & 2\mathrm{cn}^2 z \mathrm{dn}^2 z + 2\mathrm{sn}^2 z(1 - \mathrm{dn}^2 z) + 2\mathrm{sn}^2 z + 2k^2 \mathrm{sn}^2 z(1 - \mathrm{cn}^2 z) + 2k^2 \mathrm{sn}^2 z - 6k^2 \mathrm{sn}^4 z - 2 \\
= \quad & 2\mathrm{cn}^2 z \mathrm{dn}^2 z + 2k^2 \mathrm{sn}^4 z + 2\mathrm{sn}^2 z + 2k^2 \mathrm{sn}^4 z + 2k^2 \mathrm{sn}^2 z - 6k^2 \mathrm{sn}^4 z - 2.
\end{aligned}
$$

Finally using $\mathrm{dn}^2 z + k^2 \mathrm{sn}^2 z = 1$ and $1 - \mathrm{cn}^2 z = \mathrm{sn}^2 z$, we obtain

$$
\begin{aligned}
\text{l.h.s.} \quad = \quad & 2\mathrm{cn}^2 z \mathrm{dn}^2 z + 2\mathrm{sn}^2 z + 2k^2 \mathrm{sn}^2 z - 2k^2 \mathrm{sn}^4 z - 2 \\
= \quad & 2\mathrm{cn}^2 z \mathrm{dn}^2 z + 2\mathrm{sn}^2 z + 2k^2 \mathrm{sn}^2 z \mathrm{cn}^2 z - 2 = 2\mathrm{cn}^2 z + 2\mathrm{sn}^2 z - 2 = 0.
\end{aligned}
$$

25.3 The Inverted Double Well on a Circle

25.3.1 Periodic configurations

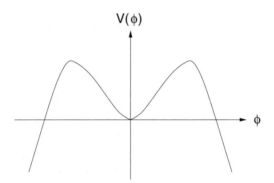

Fig. 25.5 The inverted double well potential.

Following our previous consideration of the bounce for ϕ^4-theory on a line in Sec. 24.3, we can now inquire about the properties of the corresponding configuration here when the theory is defined on a circle. We take as the potential the inverted double well with the constant subtracted out, i.e. we take

$$
U(\phi) = -\frac{\mu^2}{2a^2}(\phi^2 - a^2)^2 + \frac{\mu^2}{2a^2}a^4, \tag{25.52}
$$

and we then have to solve the equation (for a particle of mass $m_0 = 1$)

$$
\phi''(x) = U'(\phi), \quad \text{i.e.} \quad (\phi')^2 = 2U(\phi) - C^2, \tag{25.53}
$$

for the classical solution corresponding to the bounce in Chapter 24. As before we take the constant C^2 to be

$$C^2 = a^2\mu^2\left(\frac{1-k^2}{1+k^2}\right)^2. \tag{25.54}$$

For the Newtonian-like particle to have energy zero we must have $k = 1$. From Eq. (25.53) we obtain

$$\int dx = \int \frac{d\phi}{\sqrt{2U(\phi)-C^2}} = \pm\frac{a}{\mu}\int \frac{d\phi}{\sqrt{-(\phi^2-a^2)^2+a^4-a^2C^2/\mu^2}}. \tag{25.55}$$

But

$$a^4 - \frac{a^2C^2}{\mu^2} = a^4 - a^4\left(\frac{1-k^2}{1+k^2}\right)^2 = \frac{4a^4k^2}{(1+k^2)^2}. \tag{25.56}$$

Hence Eq. (25.55) becomes

$$\int^x dx = \pm\frac{a}{\mu}\int^\phi \frac{d\phi}{\sqrt{4a^4k^2/(1+k^2)^2-(\phi^2-a^2)^2}}$$

$$= \pm\int^\phi \frac{(a/\mu)d\phi}{\sqrt{\{2a^2k/(1+k^2)-\phi^2+a^2\}\{2a^2k/(1+k^2)+\phi^2-a^2\}}}. \tag{25.57}$$

Setting

$$\frac{2a^2k}{1+k^2}+a^2 = \frac{a^2(1+k)^2}{1+k^2} \equiv s_+^2, \quad \frac{2a^2k}{1+k^2}-a^2 = -\frac{a^2(1-k)^2}{1+k^2} \equiv -s_-^2,$$

i.e.

$$s_+ = \frac{a(1+k)}{\sqrt{1+k^2}}, \quad s_- = \frac{a(1-k)}{\sqrt{1+k^2}}, \tag{25.58}$$

and

$$\frac{s_-}{s_+} = \frac{1-k}{1+k}, \quad 1 - \frac{s_-^2}{s_+^2} = \frac{4k}{(1+k)^2}, \tag{25.59}$$

we obtain

$$\int^x dx = \pm\frac{a}{\mu s_+}\int^{s=\phi} \frac{d(s/s_+)}{\sqrt{(1-s^2/s_+^2)(s^2/s_+^2-s_-^2/s_+^2)}}. \tag{25.60}$$

We now put

$$t = \frac{s}{s_+} = \frac{\phi}{s_+}, \quad b \equiv \frac{s_-}{s_+}, \quad 1-b^2 \equiv \gamma^2 = \frac{4k}{(1+k)^2}, \tag{25.61}$$

and write the integral

$$\int_{x_0}^{x} dx = \pm \frac{a}{\mu s_+} \int_{\phi/s_+}^{1} \frac{dt}{\sqrt{(1-t^2)(t^2-b^2)}}. \tag{25.62}$$

From the literature on elliptic integrals of the first kind we deduce that[¶]

$$x_0 - x = \pm \frac{a}{\mu s_+} \mathrm{dn}^{-1}\left[s_+^{-1}\phi, \left|\gamma^2\right.\right] \tag{25.63}$$

for

$$b < \frac{\phi}{s_+} < 1, \quad \text{i.e.} \quad s_- < \phi < s_+,$$

so that ϕ can be zero only if we allow b to be zero, i.e. $k = 1$. We rewrite the *periodic classical solution* of Eq. (25.63) as $(s_- < \phi < s_+)$

$$\phi_c(x) = s_+ \mathrm{dn}[\beta(x-x_0), \gamma^2], \quad \beta = \frac{\mu s_+}{a}, \tag{25.64}$$

(note that in this section the elliptic modulus is γ, not k).

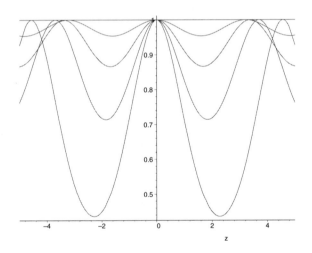

Fig. 25.6 The Jacobian function $dn(z,k)$ for $k = 0, 0.3, 0.5, 0.7, 0.9$.

This solution will be seen to play the role of the *periodic bounce*. Here γ plays the role which k played in our earlier solutions. The function $\mathrm{dn}\, u$ has real period $2K(\gamma)$. Some MAPLE-plots of the function for various values of its elliptic modulus are shown in Fig. 25.6. One can clearly see the periodic nature of the configuration. We demand again periodicity, i.e.

$$\phi_c(x) = \phi_c(x+L),$$

[¶]See e.g. L. M. Milne-Thomson [196], p. 27.

so that

$$\beta L = 2nK(\gamma), \quad L \rightarrow L_n(\gamma) = \frac{1}{\beta} 2K(\gamma), \quad n = 1, 2, \ldots. \quad (25.65)$$

We skip a discussion of these points in more detail here, and instead proceed immediately to the investigation of the small fluctuation equation.

25.3.2 The fluctuation equation

Differentiating the potential given in Eq. (25.52), we obtain

$$U'(\phi) = -\frac{2\mu^2}{a^2}\phi^3 + 2\mu^2\phi, \quad U''(\phi) = -\frac{2\mu^2}{a^2}(3\phi^2 - a^2). \quad (25.66)$$

The small fluctuation equation with eigenvalue w^2, i.e.

$$\left[-\frac{d^2}{dx^2} + U''(\phi_c) \right] \psi(x) = w^2 \psi(x), \quad (25.67)$$

then becomes

$$\frac{d^2\psi}{dx^2} + \left[w^2 + \frac{2\mu^2}{a^2}(3\phi_c^2 - a^2) \right] \psi = 0.$$

Inserting the *bounce solution* (25.64) for ϕ_c this becomes

$$\frac{d^2\psi}{dx^2} + \left[w^2 - 2\mu^2 + 6s_+^2 \frac{\mu^2}{a^2} \mathrm{dn}^2[\beta x, \gamma^2] \right] \psi = 0. \quad (25.68)$$

Using the relation

$$\mathrm{dn}^2[u, k^2] + k^2 \mathrm{sn}^2[u, k^2] = 1, \quad (25.69)$$

we can rewrite the equation as

$$\frac{d^2\psi}{dx^2} + [w^2 - 2\mu^2 + 6\beta^2 - 6\beta^2\gamma^2 \mathrm{sn}^2[\beta x, \gamma^2]]\psi = 0,$$

or with

$$z = \beta x \quad (25.70)$$

as

$$\frac{d^2\psi}{dz^2} + \left[\frac{w^2 - 2\mu^2}{\beta^2} + 6 - (2 \times 3)\gamma^2 \mathrm{sn}^2[z, \gamma^2] \right] \psi = 0. \quad (25.71)$$

We recognize this again as a *Lamé equation*:

$$\frac{d^2\psi}{dz^2} + [\lambda - n(n+1)\gamma^2 \mathrm{sn}^2[z, \gamma^2]]\psi = 0 \quad (25.72)$$

with eigenvalue λ related to the eigenvalue w^2 of the fluctuation equation by

$$\lambda = \frac{w^2 - 2\mu^2}{\beta^2} + 6, \quad n = 2. \tag{25.73}$$

The discrete eigenfunctions of this equation have been given previously by equations (25.47a) to (25.47d).

We consider first the solution which we expect to be the zero mode, i.e. the solution proportional to the derivative of the bounce, i.e.

$$\frac{d}{dz}\mathrm{dn}[z, \gamma^2] = -\gamma^2 \mathrm{sn}[z, \gamma^2]\mathrm{cn}[z, \gamma^2]. \tag{25.74}$$

We observe that this solution is an odd function of z (whereas the bounce is even). From Eq. (25.47a) we see that the eigenvalue is determined by the relation

$$\lambda^{(sc)} \equiv \frac{w^2 - 2\mu^2}{\beta^2} + 6 = 4 + \gamma^2, \tag{25.75}$$

and so

$$
\begin{aligned}
w^2 \;\; &= \;\; 2\mu^2 - 6\beta^2 + 4\beta^2 + \beta^2\gamma^2 = 2\mu^2 - 2\beta^2 + \beta^2\gamma^2 \\
&\overset{(25.61)}{=} \;\; 2\mu^2 + \beta^2\left\{\frac{4k}{(1+k)^2} - 2\right\} = 2\mu^2 - \frac{2(1+k^2)\beta^2}{(1+k)^2},
\end{aligned}
$$

and with some further manipulations

$$
\begin{aligned}
w^2 \;\overset{(25.64)}{=}\; &2\mu^2 - \frac{2(1+k^2)}{(1+k)^2}\frac{\mu^2 s_+^2}{a^2} \overset{(25.58)}{=} 2\mu^2 - \frac{2(1+k^2)}{(1+k)^2}\frac{\mu^2}{a^2}\frac{a^2(1+k)^2}{(1+k^2)} \\
&= \;\; 2\mu^2 - 2\mu^2 = 0.
\end{aligned}
$$

Thus the eigenvalue is zero as expected.

Next we consider the solution (25.47c), i.e. (observe this function is even!)

$$\psi(z) \propto \mathrm{cn}[z, \gamma^2]\mathrm{dn}[z, \gamma^2] \tag{25.76}$$

with eigenvalue equation

$$\lambda^{(cd)} = \frac{w^2 - 2\mu^2}{\beta^2} + 6 = 1 + \gamma^2,$$

from which we deduce that

$$
\begin{aligned}
w^2 \;\; &= \;\; 2\mu^2 - 5\beta^2 + \beta^2\gamma^2 = 2\mu^2 - \beta^2\left\{5 - \frac{4k}{(1+k)^2}\right\} \\
&= \;\; 2\mu^2 - \frac{\mu^2}{a^2}\frac{a^2(1+k)^2}{(1+k)^2}\left\{5 - \frac{4k}{(1+k)^2}\right\} \\
&= \;\; \frac{\mu^2}{1+k^2}[2(1+k^2) - 5(1+k)^2 + 4k],
\end{aligned}
$$

which implies

$$w^2 = -\frac{3\mu^2(1+k)^2}{(1+k^2)} < 0. \tag{25.77}$$

Thus, since this eigenvalue is negative and the associated wave function is even, one may expect this to be the ground state. However, it is best to consider first the other eigenfunctions.

Consider the solution (25.47b), i.e. (observe this function is odd!)

$$\psi(z) \propto \text{sn}[z, \gamma^2] \text{dn}[z, \gamma^2] \tag{25.78}$$

with eigenvalue equation

$$\lambda^{(sd)} = \frac{w^2 - 2\mu^2}{\beta^2} + 6 = 1 + 4\gamma^2,$$

from which we deduce that

$$
\begin{aligned}
w^2 &= 2\mu^2 - \beta^2(5 - 4\gamma^2) = 2\mu^2 - \frac{\mu^2(1+k)^2}{(1+k^2)}\left\{5 - \frac{16k}{(1+k)^2}\right\} \\
&= \frac{\mu^2}{1+k^2}\{2(1+k^2) - 5(1+k)^2 + 16k\},
\end{aligned}
$$

and hence

$$w^2 = -\frac{3\mu^2(1-k)^2}{1+k^2} \le 0. \tag{25.79}$$

We observe that this eigenvalue is also negative or zero.

There are two other eigenvalues (cf. Eqs. (25.47d)); we skip their investigation here.

25.3.3 The limit of infinite period

In order to understand the occurrence of two negative eigenvalues of the fluctuation equation here, we recall that the bounce has energy zero for $k = 1$ (or $b = 0$). In this limit $k = 1$ the eigenvalue (25.79) is zero, and in fact, the eigenfunction (25.78) coincides with the zero mode which can be seen as follows. For $k = 1$ we have $\gamma = 1$. In this limit

$$\text{sn}[z, 1] = \tanh z, \quad \text{cn}[z, 1] = \text{sech} z, \quad \text{dn}[z, 1] = \text{sech} z, \tag{25.80}$$

so that

$$\text{sn}[z, \gamma^2]\text{cn}[z, \gamma^2], \quad \text{sn}[z, \gamma^2]\text{dn}[z, \gamma^2] \xrightarrow{\gamma^2 = 1} \frac{\sinh z}{\cosh^2 z}. \tag{25.81}$$

We observe that this is the zero mode (24.48) in the case of the bounce for the inverted double well potential on the infinitely long line. We also observe that the eigenfunction associated with the negative eigenvalue in the limit $k = 1$, i.e. the eigenvalue (25.77), which is $-6\mu^2$ is (cf. Eq. (25.76))

$$\mathrm{cn}[z, \gamma^2]\mathrm{dn}[z, \gamma^2] \xrightarrow{\gamma^2 = 1} \frac{1}{\cosh^2 z}, \tag{25.82}$$

which has the typical bell-shaped form of a ground state wave function. Thus all our considerations here reduce to those for the infinitely long line when the circle or period L is allowed to expand indefinitely.

25.4 The Sine–Gordon Theory on a Circle

25.4.1 Periodic configurations

We can repeat the above procedure for the sine–Gordon potential with[||] $\Phi(x,t) \to \phi(x)$, and

$$U(\phi) = 1 + \cos\phi, \quad \phi(x) \in [-\pi, \pi], \tag{25.83}$$

as indicated in Fig. 25.7.

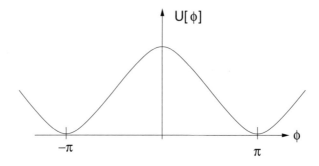

Fig. 25.7 The sine–Gordon potential.

Integrating the classical equation for static configurations $\phi(x)$, we have

$$\frac{d^2\phi}{dx^2} = U'(\phi) = -\sin\phi, \quad \frac{1}{2}\left(\frac{d\phi}{dx}\right)^2 - (1 + \cos\phi) = -2C^2. \tag{25.84}$$

Here it is convenient to set

$$C^2 \equiv C^2(k) = 1 - k^2, \quad 0 \le k \le 1. \tag{25.85}$$

[||]Here our considerations are in part analogous to those of K. Funakubo, S. Otsuki and F. Toyoda [101].

Again we demand periodicity, i.e.

$$\phi(x) = \phi(x + L).$$

The classical equation is seen to have the following *constant solutions*:

$$\left. \begin{aligned} \phi_v(x) &= \pm\pi \quad \text{with} \quad k = 1, \\ \phi_0(x) &= 0 \quad\;\; \text{with} \quad k = 0, \end{aligned} \right\} \tag{25.86}$$

the first two being vacuum solutions, the other the zero solution corresponding to the maximum between the vacua in Fig. 25.6.

The non-constant classical *periodic solution* is found to be

$$\phi_c(x) = 2 \sin^{-1}[k \operatorname{sn}[x, k^2]], \tag{25.87}$$

as we can verify by using $\operatorname{dn}^2 x = 1 - k^2 \operatorname{sn}^2 x$:

$$\left(\frac{d\phi_c(x)}{dx} \right)^2 = \left(\frac{2k \operatorname{cn}x \operatorname{dn}x}{\sqrt{1 - k^2\operatorname{sn}^2 x}} \right)^2 = 4k^2 \operatorname{cn}^2 x. \tag{25.88}$$

But $\sin(\phi_c/2) = k \operatorname{sn} x$, so that $\operatorname{cn}^2 x = 1 - \operatorname{sn}^2 x = 1 - (1/k^2)\sin^2(\phi_c/2)$, we have

$$\begin{aligned} \frac{1}{2}\left(\frac{d\phi_c}{dx} \right)^2 &= 2k^2 - 2\sin^2 \frac{\phi_c}{2} = 2k^2 - (1 - \cos\phi_c) \\ &= 2(k^2 - 1) + (1 + \cos\phi_c) = U(\phi_c) - 2C^2(k), \end{aligned}$$

in agreement with Eq. (25.84). The function $\operatorname{sn} x$ has the real period $4K(k)$ (see Appendix A), so that

$$\phi_c(x) = \phi_c(x + 4nK(k)). \tag{25.89}$$

For later convenience we note here that Eq. (25.87) implies

$$\cos\phi_c = 1 - 2\sin^2 \frac{\phi_c}{2} = 1 - 2k^2\operatorname{sn}^2 x. \tag{25.90}$$

25.4.2   The fluctuation equation

From Eq. (25.83) we obtain $U''(\phi_c) = -\cos\phi_c$, and hence the small fluctuation equation (25.67) is

$$\frac{d^2\psi}{dx^2} + [w^2 + \cos\phi_c(x)]\psi = 0, \tag{25.91}$$

or, with Eq. (25.90),

$$\frac{d^2\psi}{dx^2} + [w^2 + 1 - 2k^2\text{sn}^2x]\psi = 0. \qquad (25.92)$$

This is again a *Lamé equation*

$$\frac{d^2\psi}{dx^2} + [\lambda - n(n+1)k^2\text{sn}^2x]\psi = 0 \qquad (25.93)$$

with normalizable, periodic eigenfunctions having eigenvalues

$$\lambda = w^2 + 1 \quad \text{and} \quad n = 1. \qquad (25.94)$$

Again we can find the translational zero mode ψ_0 already by differentiating the classical solution, i.e.

$$\left(\frac{d^2}{dx^2} - U''[\phi_c]\right)\frac{d\phi_c}{dx} = 0,$$

i.e. (cf. (25.88))

$$\psi_0 \propto \frac{d\phi_c}{dx} = \pm 2k\,\text{cn}x. \qquad (25.95)$$

This is an even function of x with real period $4K(k)$. From the literature[**] we obtain the discrete normalizable solutions of Eq. (25.93). For $n = 1$ these are:

$$\begin{array}{lll} sE_1^0(x) = \text{sn}\,x & \text{with eigenvalue} & \lambda_1^{(s)0} = 1 + k^2, \\ cE_1^0(x) = \text{cn}\,x & \text{with eigenvalue} & \lambda^{(c)0} = 1, \\ dE_1^0(x) = \text{dn}\,x & \text{with eigenvalue} & \lambda_1^{(d)0} = k^2 \end{array} \right\} \qquad (25.96)$$

($\text{sn}\,x, \text{cn}\,x$ have period $4K(k)$, $\text{dn}\,x$ has period $2K(k)$). The second of these is the zero mode ψ_0 with period $4K(k)$ and eigenvalue

$$w^2 = \lambda - 1 = 0,$$

as expected. The first of the above set of three *Lamé polynomials* has $w^2 = \lambda - 1 = k^2 \geq 0$, the associated eigenfunction (with period $4K(k)$) being odd. The last of the three cases has

$$w^2 = k^2 - 1 \leq 0$$

[**]See F. M. Arscott [11], p. 205.

with eigenfunction $\operatorname{dn} x$ and period $2K(k)$. This negative eigenvalue is a new feature.[††]

In the case of the classical *solution* $\phi_0(x) = 0$, we have the stability equation

$$\frac{d^2\psi}{dx^2} + [w^2 + 1]\psi = 0,$$

i.e.

$$\psi \sim \frac{\sin}{\cos}[p_m x] = \frac{\sin}{\cos}[p_m(x + L)]$$

with

$$p_m^2 = w^2 + 1, \quad p_m L = 2\pi m, \quad m \in \mathbb{Z},$$

so that

$$w^2 \to w_m^2 = \frac{4\pi^2 m^2}{L^2} - 1.$$

At the critical values $L_n = 2\pi n, n \neq 0$:

$$w_m^2 = \frac{m^2}{n^2} - 1.$$

Thus, as in the ϕ^4-theory, we obtain for $n = 1$:

$$w_m^2 : \quad -1, 0, 3, \ldots,$$

and for $n = 2$:

$$w_m^2 : \quad -1, -\frac{3}{4}, 0, \ldots.$$

Thus again negative eigenvalues occur for $n > m$ and zero eigenvalues for $n = m$. The solution $\phi = 0$ is therefore unstable, i.e. it represents a saddle point of the function space of classical configurations and is therefore a "*sphaleron*".

25.5 Conclusions

In each of the three types of potentials considered — double well, inverted double well and cosine — the equation of small fluctuations about the classical periodic solution was found to be a Lamé equation. In each case this Lamé equation becomes a Pöschl–Teller equation in the limit of infinite period. Although not considered above, we expect the small fluctuation equation in the case of the cubic potential (cf. Sec. 24.5) also to be a Lamé equation, since its limiting form was found to be a Pöschl–Teller equation (i.e. Eq. (24.34)).

[††]The three eigenfunctions given here are also contained in the $O(3)$ nonlinear sigma model on a circle, and have been discussed in this context by K. Funakubo, S. Otsuki and F. Toyoda [101].

These observations thus reveal an interesting significance of this hitherto fairly unfamiliar equation of mathematical physics, and suggest that the potentials we considered play a basic role.

Chapter 26

Path Integrals and Periodic Classical Configurations

26.1 Introductory Remarks

Having seen in the preceding chapter where and how periodic classical configurations arise, the next natural step is to investigate these in the context of path integrals. We have already mentioned that periodic configurations are required for the study of tunneling between asymptotically degenerate excited states, which explains why we restricted ourselves in the path integral applications of Chapters 23 and 24 to ground states. Periodic configurations require a modification of the method, as we shall see. We begin again with the study of these configurations in the case of the double well potential, our aim being the derivation of the generalization of the ground state level splitting formula obtained in Chapter 23 to excited states. Our intention here is not merely to reproduce calculations from the literature, but rather to pinpoint the crucial differences between the method employed in this chapter and the method used in the foregoing chapters, by demonstrating how they lead to identical results. Thus it will be observed in this chapter that we do not introduce the Faddeev–Popov constraint as such, but nonetheless obtain the same factor and Euclidean time integral when we specialize to the ground state case considered earlier. This is a vital point. We shall also see that we arrive at fairly compact explicit formulas for the calculated Feynman amplitudes. The calculations require considerable use of Jacobian elliptic functions and elliptic integrals which are very conveniently handled with the help of Tables of Elliptic Integrals.* We consider here again the

---

*An indispensable help is the *Handbook of Elliptic Integrals for Engineers and Scientists* by P. F. Byrd and M. D. Friedman [40].

three basic examples of double well potential, cosine potential and inverted double well potential. We demonstrate also the agreement with the results of the Schrödinger method (not simply by stating this as is frequently done, but by presenting the necessary steps so that the reader can convince himself without undue efforts on his part). In each case the result has basically a simple form, namely as product of the square of the WKB normalization constant C^2 and the exponential of minus the appropriate action W. The constant C^2 has a simple meaning if we write momentum $p = m_0 d\phi/dt$ and take for simplicity $m_0 = 1, \hbar = 1$. Then, starting from the definition of C^2, we obtain[†]

$$C^2 = \left[\oint \frac{d\phi}{\sqrt{2|E - V|}} \right]^{-1} = \left[\oint \frac{d\phi}{p} \right]^{-1} = \frac{1}{\text{period}} \equiv \frac{\hbar w}{2\pi}.$$

This shows that the prefactor is essentially the frequency of attempts to penetrate the barrier.

26.2   The Double Well and Periodic Instantons

26.2.1   Periodic configurations and the double well

We consider the theory of a scalar field ϕ in one time and zero space dimensions.[‡] The Lagrangian L of this theory is given by

$$L = \frac{1}{2} \left(\frac{d\phi}{dt} \right)^2 - V(\phi), \tag{26.1}$$

where the potential $V(\phi)$ is the double well potential given by

$$V(\phi) = \frac{m^4}{2g^2} \left(1 - \frac{g^2 \phi^2}{m^2} \right)^2 = \frac{g^2}{2} \left(\phi^2 - \frac{m^2}{g^2} \right)^2. \tag{26.2}$$

The mass of the particle is seen to be taken as $m_0 = 1$. We go to Euclidean time $\tau = it$, in which case the integrated classical equation of motion, now with nonzero integration constant E_{cl}, is (as we saw also several times earlier)

$$\frac{1}{2} \left(\frac{d\phi}{d\tau} \right)^2 - V(\phi) = -E_{\text{cl}}. \tag{26.3}$$

We denote the classical solution of this equation by $\phi_c(\tau)$. Previously we considered the case $E_{\text{cl}} = 0$. Inserting here the potential (26.2) and integrating the equation (intermediate steps are akin to those in Chapter 25), one

---

[†]Compare, for instance, with Eq. (14.90) in the case of the cubic potential.
[‡]We follow here largely J.–Q. Liang and H. J. W. Müller–Kirsten [167].

obtains the solution called *periodic instanton*

$$\phi_c(\tau) = \frac{kb(k)}{g} \operatorname{sn}[b(k)\tau] \tag{26.4}$$

with an integration constant τ_0 chosen to be zero. Here k is the elliptic modulus and $b(k)$ a parameter which are related to the quantity E_{cl} and are given by

$$k^2 = \frac{1-u}{1+u}, \quad u := \frac{g}{m^2}\sqrt{2E_{\mathrm{cl}}} = \frac{1-k^2}{1+k^2}, \quad 1-k^2 = \frac{2u}{1+u},$$

$$b(k) := m\left(\frac{2}{1+k^2}\right)^{1/2} = m\sqrt{1+u}, \quad 1+k^2 = \frac{2}{1+u}. \tag{26.5}$$

In the following we will sometimes refer to the special case of $E_{\mathrm{cl}} = 0$, i.e. $u = 0, k = 1$, which implies specialization to the case we considered in Chapter 23, thus also checking the method employed here. The Jacobian elliptic function $\operatorname{sn}[u]$ appearing in $\phi_c(\tau)$ has real periods $4nK(k)$, where n is an integer and $K(k)$ is the quarter period given by the complete elliptic integral of the first kind, i.e. the integral

$$K(k) := \int_0^{\pi/2} \frac{d\vartheta}{\sqrt{1 - k^2 \sin^2 \vartheta}}.$$

The motion of the periodic instanton as a function of Euclidean time τ is

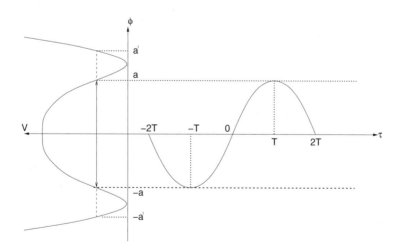

Fig. 26.1 The motion of the periodic instanton.

depicted schematically in Fig. 26.1. We also show on the left of Fig. 26.1 the double well potential so that the periodic motion of the instanton corresponds to a back and forth tunneling in the potential barrier. The *periodic instanton*[§] starts its motion (for instance) from $\phi = 0$ at $\tau = -2T$ and reaches the turning point $-a$ at time $\tau = -T$ with velocity $(d\phi_c/d\tau)_{\tau=-T} = 0$. It then returns to the origin $\phi = 0$ at $\tau = 0$ and travels on and reaches the turning point $\phi = a$ at $\tau = T$, again with zero velocity. It then returns to the origin at $\tau = 2T$, thus completing one period. The periodic instanton is therefore subject to the boundary conditions

$$\phi_c(\tau)|_{\tau=-2T} = \phi_c(\tau)|_{\tau=2T} = 0, \tag{26.6a}$$

which means $[b(k)2T] = 2K(k)$, $n = 1$, since $\mathrm{sn}[u]$ has real zeros at $u = 0, 2K(k)$.[¶] In the Newtonian picture the periodic instanton has energy E_{cl} which is entirely potential energy at the turning points, the kinetic energy being maximal at $\phi = 0$. The four turning points indicated in Fig. 26.1 and needed later follow from solving $E_{\mathrm{cl}} - V(\phi) = 0$. Thus

$$\phi \to \pm \left[\frac{m^2}{g^2} + \frac{\sqrt{2E_{\mathrm{cl}}}}{g} \right]^{1/2} \equiv \pm a' = \pm \frac{m}{g}\sqrt{1 + u},$$

$$\phi \to \pm \left[\frac{m^2}{g^2} - \frac{\sqrt{2E_{\mathrm{cl}}}}{g} \right]^{1/2} \equiv \pm a = \phi(\pm T) = \pm \frac{m}{g}\sqrt{1 - u}. \tag{26.6b}$$

26.2.2   Transition amplitude and Feynman kernel

The transition amplitude we wish to evaluate has been defined previously by Eqs. (23.35a) and (23.35b). Replacing there the ground state index "0" by the index n of an excited state, we see that the amplitude we are now interested in can be written as

$$K(\phi_f, \tau_f; \phi_i, \tau_i) = 2\psi_n^*(\phi_f)\psi_n(\phi_i)\exp[-E_n(\tau_f - \tau_i)]\sinh\left\{\frac{1}{2\hbar}\triangle E_n(\tau_f - \tau_i)\right\}. \tag{26.7a}$$

This relation (26.7a) assumes fixed endpoints ϕ_f, ϕ_i for $\tau_f = T, \tau_i = -T$. One of the main points of the following method is, as we shall see, to relax this condition by assuming variable endpoints with $\tau_f = -\tau_i$. For this reason we consider not just the kernel, but the transition amplitude that this is contained in. We encountered the general form of such a transition amplitude in Chapter 21 with Eq. (21.41) in the case of one-time plus one-space dimensions, i.e.

$$A = \int dx_f \int dx_i \psi_f^*(x_f, t_f) K(x_f, t_f; x_i, t_i)\psi_i(x_i, t_i),$$

[§]It is convenient to consider the monotonically rising part, i.e. from $\tau = -T$ to T as an "instanton" in analogy to the proper (vacuum) instanton considered earlier.

[¶]See e.g. the Table in Appendix A or L. M. Milne–Thomson [196], p. 14.

so that in the present case the amplitude to be considered becomes

$$A_{+,-} = \int d\phi_f \int d\phi_i \langle \psi_n | \phi_f, \tau_f \rangle K(\phi_f, \tau_f; \phi_i, \tau_i) \langle \phi_i, \tau_i | \psi_n \rangle.$$

We can separate off the time-dependence of the wave functions since for $\tau_f = -\tau_i$

$$\langle \psi_n | \phi_f, \tau_f \rangle = e^{-E_n \tau_f} \langle \psi_n | \phi_f \rangle, \quad \langle \phi_i, \tau_i | \psi_n \rangle = e^{-E_n \tau_i} \langle \phi_i | \psi_n \rangle.$$

A similar removal of the time dependence from the endpoint wave functions applies in Secs. 26.3 and 26.4. Hence in the present case we have the transition amplitude

$$A_{+,-} \equiv \int d\phi_f \int d\phi_i \langle \psi_n | \phi_f \rangle \underbrace{\langle \phi_f, \tau_f | \phi_i, \tau_i \rangle}_{K(\phi_f, \tau_f; \phi_i, \tau_i)} \langle \phi_i | \psi_n \rangle. \tag{26.7b}$$

The wave functions appearing here, i.e. $\langle \phi_f | \psi_n \rangle, \langle \phi_i | \psi_n \rangle$, are in the wells those of the relevant quantum mechanical state on either side of the barrier (to a first approximation of harmonic oscillator type, in the better approximation required here of WKB type). The quantity K is as before the *Feynman kernel* given by

$$K(\phi_f, \tau_f; \phi_i, \tau_i) = \int_{\phi_i}^{\phi_f} \mathcal{D}\{\phi\} \exp[-S/\hbar], \tag{26.8}$$

where for τ_f, τ_i not necessarily equal to $T, -T$ as required below,

$$S(\phi) \equiv S(\phi(\tau_f), \phi(\tau_i), \tau_f - \tau_i) = \int_{\tau_i}^{\tau_f} d\tau \left[\frac{1}{2} \left(\frac{d\phi}{d\tau} \right)^2 + V(\phi) \right]. \tag{26.9}$$

Here $\phi_f \equiv \phi(\tau_f)$ and $\phi_i \equiv \phi(\tau_i)$.

26.2.3 Fluctuations about the periodic instanton

As in Sec. 23.4.1 we begin with the consideration of the *single periodic instanton* contribution to the transition amplitude. We expand the field $\phi(\tau)$ about the periodic instanton (26.4) and set

$$\phi(\tau) = \phi_c(\tau) + \chi(\tau) \tag{26.10}$$

with the boundary conditions $\chi(\tau_f) = \chi(\tau_i) = 0$ to be imposed on the fluctuation χ. Inserting this for $\phi(\tau)$ in the kernel (26.8) we obtain

$$K = \exp[-S(\phi_c)/\hbar] \int \mathcal{D}\{\chi\} \exp[-\delta S/\hbar] \equiv \exp[-S(\phi_c)/\hbar] I_0, \tag{26.11}$$

where the remainder δS is given explicitly below. We insert the classical solution (26.4) into the action $S(\phi)$ of Eq. (26.9), the derivative of $\phi_c(\tau)$ being (cf. Appendix A)

$$\dot{\phi}_c(\tau) = \frac{kb^2(k)}{g} \operatorname{cn}[b(k)\tau] \operatorname{dn}[b(k)\tau] \equiv N(\tau), \qquad (26.12)$$

which is seen to be an even function of τ (a property of significance later). We leave the evaluation of the action to Example 26.1. The result is — with the help of Eq. (26.3) —

$$S(\phi_c) = W(\phi_c(\tau_f), \phi_c(\tau_i), E_{cl}) + E_{cl}(\tau_f - \tau_i), \qquad (26.13)$$

where

$$W(\phi_c(\tau_f), \phi_c(\tau_i), E_{cl}) = \int_{\tau_i}^{\tau_f} d\tau \dot{\phi}_c^2(\tau), \qquad (26.14a)$$

and specifically for $\tau_f \to T, \tau_i \to -T$:

$$W(\phi_c(\tau_f), \phi_c(\tau_i), E_{cl})|_{\tau_f=T,\tau_i=-T} = \frac{4m^3}{3g^2}(1+u)^{1/2}[E(k) - uK(k)]. \quad (26.14b)$$

Here $K(k)$ is the complete elliptic integral of the first kind that we encountered earlier and $E(k)$ is the complete elliptic integral of the second kind defined in Eq. (25.27a). For T not given exactly by $b(k)T = K(k)$ (cf. after Eq. (26.6a), but only approximately, the result (26.14b) involves (in general) correction terms. The equation of small fluctuations $\chi(\tau)$ about the configuration (26.4) was obtained earlier as (cf. Eq. (25.42)) the eigenvalue equation with eigenvalue w^2

$$\ddot{\chi}(\tau) + [w^2 - 2\{3g^2\phi_c^2(\tau) - m^2\}]\chi = 0. \qquad (26.15)$$

In our first steps we proceed as in Chapter 23 with the evaluation of the functional integral or kernel I_0 contained in Eq. (26.11), i.e.

$$I_0 := \int \mathcal{D}\{\chi\} \exp[-\delta S/\hbar], \quad \delta S = \int_{\tau_i}^{\tau_f} d\tau \left[\frac{1}{2}\dot{\chi}^2 + \{3g^2\phi_c^2(\tau) - m^2\}\chi^2\right]. \qquad (26.16)$$

As in Chapter 23 the integrand is converted into a quadratic form by using the following *Volterra transformation* in terms of a function $N(\tau)$ proportional to the $w^2 = 0$ (i.e. zero-eigenvalue) solution $\dot{\phi}_c(\tau)$ of Eq. (26.15), i.e.

$$y(\tau) = \chi(\tau) - \int_{\tau_i}^{\tau} d\tau' \frac{\dot{N}(\tau')}{N(\tau')}\chi(\tau'). \qquad (26.17)$$

We know from Eq. (26.12) that $\dot{\phi}_c(\tau)$ is an even function so that $N(\tau) = N(-\tau)$. The result of the transformation from $\chi(\tau)$ to $y(\tau)$ is without implementation of the endpoint constraints (cf. Example 23.3)

$$I_0^{\text{no constraints}} = \int \mathcal{D}\{y(\tau)\} \left|\frac{\mathcal{D}y}{\mathcal{D}\chi}\right|^{-1} \exp\left[-\int_{\tau_i}^{\tau_f} d\tau \frac{1}{2}\left(\frac{dy}{d\tau}\right)^2\right].$$

The boundary conditions or constraints to be imposed are $\chi(\pm T) = 0$ or, for τ_f, τ_i close to $T, -T$, $\chi(\tau_f), \chi(\tau_i) = 0$. We can express these in terms of $y(\tau)$ as (cf. Eq. (23.61) and observe that $N(T)$ would be zero)

$$y(\tau_f) + N(\tau_f) \int_{\tau_i}^{\tau_f} d\tau' \frac{\dot{N}(\tau')}{N^2(\tau')} y(\tau') \equiv y(\tau_f) + f(\tau_f) = 0.$$

We implement this constraint into the kernel I_0 with the help of a Lagrange multiplier α and the identity

$$1 = \int dy_f \delta[y(\tau_f) + f(\tau_f)] = \frac{1}{2\pi} \int dy_f \int_{-\infty}^{\infty} d\alpha e^{i\alpha[y(\tau_f)+f(\tau_f)]}.$$

Thus (with $\hbar = 1$)

$$
\begin{aligned}
I_0 &= \int_{\chi(\tau_i)}^{\chi(\tau_f)} \mathcal{D}\{\chi\} \exp[-\delta S(\chi)] \\
&= \frac{1}{2\pi} \int \mathcal{D}\{y(\tau)\} \int dy_f \int_{-\infty}^{\infty} d\alpha \left|\frac{\mathcal{D}y}{\mathcal{D}\chi}\right|^{-1} \exp\left[-\int_{\tau_i}^{\tau_f} d\tau \frac{1}{2}\left(\frac{dy}{d\tau}\right)^2\right. \\
&\quad \left. +i\alpha\left\{y(\tau_f) + N(\tau_f) \int_{\tau_i}^{\tau_f} d\tau' \frac{\dot{N}(\tau')}{N^2(\tau')} y(\tau')\right\}\right].
\end{aligned}
\tag{26.18}
$$

In Example 23.3 we evaluated this functional integral and obtained the result

$$I_0 = \frac{1}{\sqrt{2\pi}} [N(\tau_f)N(\tau_i)]^{-1/2} \left[\int_{\tau_i}^{\tau_f} \frac{d\tau}{N^2(\tau)}\right]^{-1/2}. \tag{26.19}$$

We observe that the derivation of this result applies to many similar cases. The important quantity appearing in this result is the *zero mode* $N(\tau) = \dot{\phi}_c(\tau)$. At τ_f exactly equal to T, or τ_i exactly equal to $-T$ the kernel I_0 is divergent since $N(T) = 0$, which expresses physically that the velocity of the pseudoparticle vanishes there. This is unlike the cases of the vacuum instantons and vacuum bounces we considered in Chapters 23 and 24 (where the velocity vanishes at $T = \pm\infty$, but we worked at T finite) and is the main difference between these cases and the present case. However, there is a very neat way to circumvent this apparent difficulty in order to insure that the transition amplitude is finite. We recall that with the formulation of the amplitude as in (26.7b) we do not fix the endpoints, but rather allow them to vary to effectively smear out the divergent behaviour around the turning points by expanding the integrand about these, i.e. in powers of $(\phi_f - \phi_c(T))$. We can achieve this with the help of the following formula which is verified in detail in Example 26.2:

$$\frac{\partial^2 S}{\partial \phi^2(\tau_f)} \simeq \frac{\dot{N}(\tau_f)}{N(\tau_f)} + \left[N(\tau_f)N(\tau_i) \int_{\tau_i}^{\tau_f} \frac{d\tau}{N^2(\tau)} \right]^{-1}, \quad N(\tau_{f,i}) \equiv \dot{\phi}(\tau_{f,i}).$$
$$(26.20a)$$

With the help of Eq. (26.19) we can rewrite this relation in the form

$$\frac{\partial^2 S}{\partial \phi^2(\tau_f)} - \frac{\dot{N}(\tau_f)}{N(\tau_f)} \simeq 2\pi I_0^2. \tag{26.20b}$$

Our evaluation of the Feynman amplitude for each of the three cases considered in this chapter depends on the use of this formula. We note that in the present case in which $N(\tau)$ is an even function of τ (cf. Eq. (26.12)) we have for $\tau_f = -\tau_i$: $\dot{N}(\tau_f) = \dot{N}(\tau_i)$.

Example 26.1: Action of the periodic instanton
By inserting the expressions (26.4) and (26.12) into the action given by Eq. (26.9) verify the result (26.13) with W given by Eqs. (26.14a) and (26.14b).

Solution: The evaluation is similar to that given in Example 25.1. We have for $\tau_f, -\tau_i \to T$

$$
\begin{aligned}
S(\phi_c) &= \int_{-T}^{T} d\tau \left[\frac{1}{2} \dot{\phi}_c^2(\tau) + V(\phi_c) \right] \overset{(26.3)}{=} \int_{-T}^{T} d\tau [E_{cl} + \dot{\phi}_c^2(\tau)] \\
&\overset{(26.12)}{=} \int_{-T}^{T} d\tau \left[E_{cl} + \frac{k^2 b^4(k)}{g^2} \text{cn}^2[b(k)\tau] \text{dn}^2[b(k)\tau] \right] \\
&= 2E_{cl}T + \frac{k^2 b^3(k)}{g^2} \int_{x=-b(k)T}^{x=b(k)T} dx\, \text{cn}^2[x]\text{dn}^2[x] = 2E_{cl}T + W.
\end{aligned}
$$

The incomplete elliptic integral $E(\varphi, k)$ of the second kind and its limit to the complete elliptic integral of the second kind were given in Eqs. (25.27a) and (25.27b). Here we require the additional property[||]

$$E(-\varphi, k) = -E(\varphi, k).$$

With this relation and the integral used in Example 25.1, and remembering that $\text{cn}[K(k)] = 0$, we obtain

$$
\begin{aligned}
W &= \frac{k^2 b^3(k)}{g^2}(1+k^2)\frac{1}{3k^2}\left[E(z) - \underbrace{\frac{1-k^2}{1+k^2}z}_{u} + \frac{k^2}{1+k^2}\text{sn}[z]\text{cn}[z]\text{dn}[z] \right]_{z=-K(k)}^{z=b(k)T=K(k)} \\
&= 2\frac{k^2 b^3(k)}{g^2}(1+k^2)\frac{1}{3k^2}[E(k) - uK(k)] = \frac{4m^3}{3g^2}(1+u)^{1/2}[E(k) - uK(k)].
\end{aligned}
$$

[||]See I. S. Gradshteyn and I. M. Ryzhik [122], formula 8.121(2), p. 907.

We see that when $b(k)T \neq K(k)$, but only approximately, the result is an approximation.

Example 26.2: Proof of Eq. (26.20a)
Verify the following relation, in which τ_f is close to T and τ_i close to $-T$:

$$\frac{\partial^2 S}{\partial \phi^2(\tau_f)} \simeq \frac{\dot{N}(\tau_f)}{N(\tau_f)} + \left[N^2(\tau_f) \int_{\tau_i}^{\tau_f} \frac{d\tau}{N^2(\tau)} \right]^{-1}, \quad N(\tau_{f,i}) \equiv \dot{\phi}(\tau_{f,i}).$$

At $\tau_f = T, \tau_i = -T$, i.e. at a turning point, we have, of course, $N(\pm T) = \dot{\phi}(\pm T) = 0$.

Solution: We consider the classical action S of Eq. (26.13) between Euclidean times τ_i and τ_f, i.e.

$$S(\phi(\tau_f), \phi(\tau_i); \tau_f - \tau_i) = W(\phi(\tau_f), \phi(\tau_i), E_{\text{cl}}) + E_{\text{cl}}(\tau_f - \tau_i), \quad \text{where} \quad E_{\text{cl}} = E_{\text{cl}}(\phi_f, \phi_i). \quad (26.21)$$

This relation between S and W has the character of a *Legendre transform* (like that between Hamiltonian and Lagrangian) with $\partial S / \partial E_{\text{cl}} = 0$ (verified below). In Eq. (26.14a) the functional W is expressed in terms of $\dot{\phi}_c$. Our aim now is to express the second derivative of S in terms of $\dot{\phi}_c$ (here also written $\dot{\phi}$). We first note that (recall $\dot{\phi}(\tau)|_{\pm T} = 0$ at the turning points)

$$W \stackrel{(26.14a)}{=} \int_{\tau_i}^{\tau_f} d\tau \dot{\phi}^2 = \int_{\phi(\tau_i)}^{\phi(\tau_f)} d\phi \dot{\phi}, \quad \frac{\partial W}{\partial \phi(\tau_{f,i})} = \pm \dot{\phi}(\tau_{f,i}), \quad (26.22a)$$

and (remembering that specifically in the above case of the double well potential $\tau_f = -\tau_i$ and that the solution (26.4) is an odd function and $\dot{\phi}(\tau)$ is an even function, and thus $\ddot{\phi}(\tau)$ odd)

$$\frac{\partial^2 W}{\partial \phi(\tau_i) \partial \phi(\tau_f)} = \frac{\partial}{\partial \tau_i} \left(\frac{\partial W}{\partial \phi(\tau_f)} \right) \frac{\partial \tau_i}{\partial \phi(\tau_i)} = \frac{\ddot{\phi}(\tau_i)}{\dot{\phi}(\tau_f)} = -\frac{\ddot{\phi}(\tau_f)}{\dot{\phi}(\tau_i)}, \quad \frac{\partial^2 W}{\partial \phi(\tau_f)^2} = \frac{\ddot{\phi}(\tau_f)}{\dot{\phi}(\tau_f)}. \quad (26.22b)$$

We also have

$$\dot{\phi}^2 = 2(V - E_{\text{cl}}), \quad W = \int_{\phi(\tau_i)}^{\phi(\tau_f)} d\phi \sqrt{2(V - E_{\text{cl}})},$$

and hence[**]

$$\frac{\partial W}{\partial E_{\text{cl}}} = -\int_{\phi(\tau_i)}^{\phi(\tau_f)} \frac{d\phi}{\sqrt{2(V - E_{\text{cl}})}} = -\int_{\phi(\tau_i)}^{\phi(\tau_f)} d\phi \frac{1}{\dot{\phi}} = -\int_{\tau_i}^{\tau_f} d\tau = -(\tau_f - \tau_i). \quad (26.23)$$

It follows that

$$\frac{\partial^2 W}{\partial \phi(\tau_{f,i}) \partial E_{\text{cl}}} = \mp \frac{1}{\dot{\phi}(\tau_{f,i})} \xrightarrow{\tau_{f,i} \to \pm T} \infty \quad \text{and} \quad \frac{\partial W}{\partial E_{\text{cl}}} + (\tau_f - \tau_i) = 0, \quad (26.24)$$

i.e. $\partial S / \partial E_{\text{cl}} = 0$, as is evident from Eq. (26.21). We also obtain from these relations

$$\frac{\partial^2 W}{\partial E_{\text{cl}}^2} = -\int_{\phi(\tau_i)}^{\phi(\tau_f)} \frac{d\phi}{[2(E_{\text{cl}} - V)]^{3/2}} = -\int_{\phi(\tau_i)}^{\phi(\tau_f)} \frac{d\phi}{\dot{\phi}^3} = -\int_{\tau_i}^{\tau_f} \frac{d\tau}{\dot{\phi}^2}. \quad (26.25)$$

[**]Observe that in the case of integration over one full period (needed later in Eq. (28.5))

$$\frac{\partial W}{\partial E_{\text{cl}}} = -\int_{\phi(\tau_1)}^{\phi(\tau_2)} d\phi \frac{1}{\dot{\phi}} - \int_{\phi(\tau_2)}^{\phi(\tau_1)} d\phi \frac{1}{\dot{\phi}} = 0 \quad \text{and} \quad \frac{\partial S}{\partial E_{\text{cl}}} \neq 0.$$

After these preliminary considerations we now vary the expression (26.21) for S and obtain

$$\frac{\partial S}{\partial \phi(\tau_{f,i})} = \frac{\partial W}{\partial \phi(\tau_{f,i})} + \left(\frac{\partial W}{\partial E_{\mathrm{cl}}} + (\tau_f - \tau_i) \right) \frac{\partial E_{\mathrm{cl}}}{\partial \phi(\tau_{f,i})} = \frac{\partial W}{\partial \phi(\tau_{f,i})}, \quad \text{also} \quad \frac{\partial S}{\partial (\tau_f - \tau_i)} = E_{\mathrm{cl}}. \quad (26.26)$$

From the latter of these relations and then using the former, we obtain

$$\frac{\partial E_{\mathrm{cl}}}{\partial \phi(\tau_{i,f})} = \frac{\partial}{\partial \phi(\tau_{i,f})} \frac{\partial S}{\partial (\tau_f - \tau_i)} = \frac{\partial}{\partial (\tau_f - \tau_i)} \frac{\partial S}{\partial \phi(\tau_{i,f})}$$

$$= \frac{\partial}{\partial (\tau_f - \tau_i)} \frac{\partial W}{\partial \phi(\tau_{i,f})} = \frac{\partial}{\partial (\tau_f - \tau_i)} \frac{\partial W}{\partial E_{\mathrm{cl}}} \frac{\partial W}{\partial \phi(\tau_{i,f})}. \quad (26.27)$$

Using $(\tau_f - \tau_i) = -\partial W / \partial E_{\mathrm{cl}}$ of Eq. (26.24), this is

$$\frac{\partial E_{\mathrm{cl}}}{\partial \phi(\tau_{i,f})} = -\frac{\partial}{\frac{\partial^2 W}{\partial E_{\mathrm{cl}}^2} \partial E_{\mathrm{cl}}} \left(\frac{\partial W}{\partial \phi(\tau_{i,f})} \right). \quad (26.28)$$

Differentiating Eq. (26.26) with respect to $\phi(\tau_f)$ (thereby remembering that W depends also on E_{cl}) and then using (26.28), we obtain

$$\frac{\partial^2 S}{\partial \phi(\tau_f) \partial \phi(\tau_{i,f})} = \frac{\partial^2 W}{\partial \phi(\tau_f) \partial \phi(\tau_{i,f})} + \frac{\partial^2 W}{\partial \phi(\tau_{i,f}) \partial E_{\mathrm{cl}}} \frac{\partial E_{\mathrm{cl}}}{\partial \phi(\tau_f)}$$

$$= \frac{\partial^2 W}{\partial \phi(\tau_f) \partial \phi(\tau_{i,f})} - \frac{\partial^2 W}{\partial \phi(\tau_{i,f}) \partial E_{\mathrm{cl}}} \frac{\partial}{\left(\frac{\partial^2 W}{\partial E_{\mathrm{cl}}^2} \right) \partial E_{\mathrm{cl}}} \frac{\partial W}{\partial \phi(\tau_f)}$$

$$= \frac{\partial^2 W}{\partial \phi(\tau_f) \partial \phi(\tau_{i,f})} - \frac{\partial^2 W}{\partial \phi(\tau_{i,f}) \partial E_{\mathrm{cl}}} \frac{1}{\left(\frac{\partial^2 W}{\partial E_{\mathrm{cl}}^2} \right)} \frac{\partial^2 W}{\partial E_{\mathrm{cl}} \partial \phi(\tau_f)}.$$

We are interested in this expression for $\tau_f, -\tau_i$ near T. The first term is given by Eq. (26.22b). The denominator of the second term contains the factor $\partial^2 W / \partial E_{\mathrm{cl}}^2$, which is given by Eq. (26.25). The remaining two factors in the second term are given by Eq. (26.24). Thus finally we obtain for $\tau_f, -\tau_i$ close to T:

$$\frac{\partial^2 S}{\partial \phi(\tau_f)^2} = \frac{\ddot{\phi}(\tau_f)}{\dot{\phi}(\tau_f)} + \left[\dot{\phi}^2(\tau_f) \int_{\tau_i}^{\tau_f} \frac{d\tau}{\dot{\phi}^2} \right]^{-1}. \quad (26.29)$$

It would be interesting, to find a more direct derivation of this important result.

26.2.4   The single periodic instanton contribution

We now return to the transition amplitude $A_{+,-}$ of Eq. (26.7b). With Eq. (26.11) we can write this as

$$A_{+,-} = \int d\phi_f \int d\phi_i \langle \psi_n | \phi_f \rangle K(\phi_f, \tau_f; \phi_i, \tau_i) \langle \phi_i | \psi_n \rangle$$

$$= \int d\phi_f \int d\phi_i \langle \psi_n | \phi_f \rangle \exp[- \underbrace{S(\phi_c)}_{S(\phi_f, \phi_i; \tau_f - \tau_i)} / \hbar] I_0 \langle \phi_i | \psi_n \rangle. \quad (26.30)$$

For the evaluation of this double integral we require the explicit form of the wave functions $\langle \phi_i | \psi_n \rangle, \langle \phi_f | \psi_n \rangle$. We obtain these to sufficient order with the *WKB method*. In order to construct these we recall from Eq. (14.52) (cf. also Eqs. (14.40) and (14.41)) that the WKB wave functions of exponential type (of relevance here inside the central barrier with ϕ between the two turning points indicated in Fig. 26.1, i.e. $-a < \phi < a$) are of the form (for particle mass $m_0 = 1$)

$$\sqrt{l} \exp\left[\int_\phi^{\phi(T)} \frac{d\phi}{l}\right], \quad \text{where} \quad l = \frac{\hbar}{\dot{\phi}(\tau)}, \quad \dot{\phi}(\tau) = \sqrt{2|V(\phi) - E_{\mathrm{cl}}|}. \quad (26.31)$$

We set therefore with normalization constants C_+ and C_- (integrals across the non-tunneling region, and $\hbar = 1$ for simplicity)

$$\langle \phi_f | \psi_n \rangle = \frac{C_+}{\sqrt{\dot{\phi}_f}} \exp[-\Omega(\phi_f)], \quad \langle \phi_i | \psi_n \rangle = \frac{C_-}{\sqrt{\dot{\phi}_i}} \exp[-\Omega(\phi_i)], \quad (26.32)$$

where (observe that in Fig. 26.1 the velocity $\dot{\phi}(\tau)$ — which is the gradient of the trajectory there — is always positive between $\tau = -T$ and T)

$$\Omega(\phi_f) = \int_{\phi_f}^{\phi_c(T)} \dot{\phi} d\phi, \quad \Omega(\phi_i) = \int_{\phi_c(-T)}^{\phi_i} \dot{\phi} d\phi. \quad (26.33)$$

In view of our consideration of the normalization of WKB solutions in Example 18.6, we take the formulas for the normalization constants $C_\pm$ from the literature; their agreement with the WKB normalization constants may be seen by reference to Example 18.5. Thus from the literature[*]

$$C_+ = \left[2 \int_a^{a'} \frac{d\phi}{\sqrt{2|E_{\mathrm{cl}} - V|}}\right]^{-1/2}, \quad C_- = \left[2 \int_{-a'}^{-a} \frac{d\phi}{\sqrt{2|E_{\mathrm{cl}} - V|}}\right]^{-1/2}, \quad (26.34)$$

where the integration extends across a well. We evaluate these expressions in Example 26.3 and obtain[†]

$$C_\pm = \left[\frac{m\sqrt{1+u}}{2K(\tilde{k})}\right]^{1/2} \xrightarrow{k \to 1} \left(\frac{m}{\pi}\right)^{1/2}, \quad \text{where} \quad \tilde{k}^2 = 1 - k^2 \equiv k'^2. \quad (26.35)$$

One should note that[‡]

$$K'(k) \equiv K(k'), \quad K(k=0) = K'(k=1) = \frac{\pi}{2}.$$

---

[*]See e.g. L. D. Landau and E. M. Lifshitz [156], pp. 169 – 172, or W. H. Furry [102].

[†]Observe that the same result is obtained in Example 18.6.

[‡]P. F. Byrd and M. D. Friedman [40], formulas 110.09 and 111.02, p. 10. The prime on $K(k)$ does not mean derivative! See also Appendix A here.

The integral (26.30) involves two independent integrations. We consider first the integration with respect to ϕ_f. Leaving the configuration ϕ_i untouched, we expand the action $S(\phi_f, \phi_i; \tau_f - \tau_i)$ about $\phi_f = \phi_c(T)$. Then (since the first derivative vanishes on account of $\phi_c(\tau)$ being a solution)

$$S(\phi_f, \phi_i, \tau_f - \tau_i) = \underbrace{S(\phi_c(T), \phi_i; \tau_f - \tau_i)}_{\tilde{S}} + \frac{1}{2}\left(\frac{\partial^2 S}{\partial \phi_f^2}\right)_{\phi_f=\phi_c(T)} (\phi_f - \phi_c(T))^2 + \cdots .$$

$$(26.36)$$

Correspondingly we expand the functional $\Omega(\phi_f)$. Thus

$$\Omega(\phi_f) = \int_{\phi_f}^{\phi_c(T)} \dot{\phi}\, d\phi = \underbrace{\Omega(\phi_c(T))}_{0} - (\phi_f - \phi_c(T)) \underbrace{[\dot{\phi}]_{\phi_c(T)}}_{0}$$

$$- \frac{1}{2}(\phi_f - \phi_c(T))^2 \left[\frac{\partial \dot{\phi}_f}{\partial \phi_f}\right]_{\phi_f=\phi_c(T)} + \cdots$$

$$= -\frac{1}{2}(\phi_f - \phi_c(T))^2 \left[\frac{\ddot{\phi}_f}{\dot{\phi}_f}\right]_{\phi_f=\phi_c(T)} + \cdots . \qquad (26.37)$$

Hence

$$S(\phi_f, \phi_i; \tau_f - \tau_i) + \Omega(\phi_f)$$

$$= \tilde{S} + \frac{1}{2}(\phi_f - \phi_c(T))^2 \left[\left(\frac{\partial^2 S}{\partial \phi_f^2}\right)_{\phi_f=\phi_c(T)} - \left(\frac{\ddot{\phi}_f}{\dot{\phi}_f}\right)_{\phi_f=\phi_c(T)}\right]$$

$$\overset{(26.20b)}{=} \tilde{S} + \frac{1}{2}(\phi_f - \phi_c(T))^2 [2\pi I_0^2]. \qquad (26.38)$$

Eventually we want to take the limit $\tau_f, -\tau_i \to T$. Thus we consider $\tau_f = -\tau_i$, in which case here $\dot{\phi}_i = \dot{\phi}_f$. The amplitude (26.30) then becomes in the case of the single periodic instanton (and with $\hbar = 1$ for convenience)

$$A_{+,-}^{(1)} = C_+ C_- \int \frac{d\phi_i}{\dot{\phi}_i} e^{-\tilde{S}} \exp[-\Omega(\phi_i)]$$

$$\times \int d\phi_f I_0 \exp\left[-\frac{1}{2}(\phi_f - \phi_c(T))^2 [2\pi I_0^2]\right]. \qquad (26.39)$$

Performing the integration in the second integral[§] and replacing $\dot{\phi}_i$ by $d\phi_i/d\tau$, this becomes (I_0 drops out and $\phi(\tau_i) \to \phi_c(-T)$)

$$A_{+,-}^{(1)} = \lim_{\tau_f, -\tau_i \to T} C_+ C_- \int_{\tau_i}^{\tau_f} d\tau \exp\left[-\int_{\phi_c(-T)}^{\phi(\tau_i)} \dot{\phi}\, d\phi\right] e^{-S(\phi_c(T), \phi_c(\tau_i); \tau_f - \tau_i)}$$

$$= 2T C_+^2 \exp[-S(\phi_c(T), \phi_c(-T); 2T)]. \qquad (26.40)$$

[§]Recall $\int_{-\infty}^{\infty} dx e^{-w^2 x^2/2} = \sqrt{2\pi/w^2}$.

Inserting here the expression for the action S from Eqs. (26.13) and (26.14b), as well as the expression for the WKB normalization constant C_+ from Eq. (26.35), this becomes

$$A^{(1)}_{+,-} = 2T \left[\frac{m\sqrt{1+u}}{2K(k')} \right] e^{-2E_{cl}T} \exp \underbrace{\left[-\frac{4m^3}{3g^2}(1+u)^{1/2}[E(k) - uK(k)] \right]}_{-W(u)}.$$

(26.41)

The argument of the exponential has been evaluated in Example 18.5. The result for $W(u)$ is, with $q = 2n+1, n = 0, 1, 2, \ldots$ and assuming g^2/m^3 to be small,

$$\begin{aligned}
W(u) &= \frac{4m^3}{3g^2} - q \ln \left(\frac{2^{5/2}m^{3/2}}{g\sqrt{q}} \right) - \frac{q}{2} \\
&= \frac{4m^3}{3g^2} + 2 \left(n + \frac{1}{2} \right) \ln \left(\frac{g}{4m^{3/2}} \right) - \left(n + \frac{1}{2} \right) \\
&\quad + \left(n + \frac{1}{2} \right) \ln \left(n + \frac{1}{2} \right).
\end{aligned}$$

(26.42)

With the expression (26.40) we can write down the result for the single-bounce amplitude (with $\hbar = 1$):

$$A^{(1)}_{+,-} = 2T \left[\frac{m\sqrt{1+u}}{2K(k')} \right] \left(\frac{4m^{3/2}}{g} \right)^{2n+1} \left(\frac{e}{n+\frac{1}{2}} \right)^{n+1/2} \exp \left[-\frac{4m^3}{3g^2} \right] e^{-2E_{cl}T}.$$

(26.43a)

In the case of the ground state with $n = 0$ and $E_{cl} \to 0$, which implies $u \to 0$ and hence $k^2 \to 1$, so that $K(k') = K'(k)$ is $K(0) = \pi/2$, so that this reduces to[¶]

$$A^{(1)}_{+,-}|_{n=0} = \sqrt{\frac{2e}{m}} \frac{2Tm}{\pi} \frac{4m^2}{g} \exp \left[-\frac{4m^3}{3g^2} \right] e^{-2mT}.$$

(26.43b)

Example 26.3: WKB normalization constants

Show that the normalization constants $C_\pm$ of the WKB wave functions for the double well potential (26.2) are given by

$$C_\pm = \left[\frac{m\sqrt{1+u}}{2K(\tilde{k})} \right]^{1/2}, \quad \text{where} \quad \tilde{k}^2 = 1 - k^2.$$

[¶] Cf. Eq. (23.100). The T-dependent exponential originates as in Eq. (23.55b) from the fluctuation equation which resembles a Schrödinger equation with mass $1/2$ (this mass equal to 1 would bring in a $\sqrt{2}$; cf. Sec. 23.2).

Solution: For the double well potential (26.2) we have

$$
\int_a^{a'} \frac{d\phi}{\sqrt{2|E_{\text{cl}} - V|}} = \int_a^{a'} \frac{d\phi}{\sqrt{2E_{\text{cl}} - g^2(\phi^2 - m^2/g^2)^2}}
$$

$$
= \frac{1}{g} \int_a^{a'} \frac{d\phi}{\sqrt{(a'^2 - \phi^2)(\phi^2 - a^2)}}.
$$

The integral thus obtained is of a form given in the literature[||] and is given by

$$
\int_a^{a'} \frac{d\phi}{\sqrt{(a'^2 - \phi^2)(\phi^2 - a^2)}} = \frac{1}{a'} F(\varphi, \tilde{k}),
$$

where $F(\varphi, \tilde{k})$ is the incomplete elliptic integral of the first kind and $\tilde{k}$ (we call it this and not immediately — as one finds — $k' = \sqrt{1 - k^2}$ in order to avoid confusion) is given by

$$
\tilde{k}^2 = \frac{a'^2 - a^2}{a'^2} \overset{(26.5)}{=} \frac{2u}{1+u} = 1 - k^2 \equiv k'^2.
$$

The angle φ in $F(\varphi, \tilde{k})$ is given by

$$
\sin \varphi = \left[\frac{a'^2(a'^2 - a^2)}{a'^2(a'^2 - a^2)} \right]^{1/2} = 1, \quad \varphi = \frac{\pi}{2}.
$$

Thus in this case of $\varphi = \pi/2$ the function $F(\varphi, \tilde{k})$ becomes the complete elliptic integral of the first kind,

$$
F\left(\frac{\pi}{2}, \tilde{k} \right) = K(\tilde{k}).
$$

Since $a' = m\sqrt{1+u}/g$, it follows that

$$
\int_a^{a'} \frac{d\phi}{\sqrt{2|E_{\text{cl}} - V|}} = \frac{1}{g} \frac{g K(\tilde{k})}{m\sqrt{1+u}},
$$

and hence

$$
C_+ = C_- = \left[\frac{m\sqrt{1+u}}{2K(k')} \right]^{1/2} \overset{u \to 0}{\simeq} \left(\frac{m}{\pi} \right)^{1/2}.
$$

This dominant approximation is seen to agree with the result of Example 18.6, thus verifying (26.35).

26.2.5   Sum over instanton–anti-instanton pairs

The path integral requires the sum over all possible paths. Hence in addition to the contribution of a single pseudoparticle contribution, the contributions of the infinite number of pseudoparticle–antipseudoparticle pairs must be taken into account. The transition amplitude then becomes the sum

$$
A_{+,-} = \sum_{n=0}^{\infty} A_{+,-}^{(2n+1)}, \tag{26.44}
$$

[||] P. F. Byrd and M. D. Friedman [40], formula 217.00, p. 54.

where $A^{(2n+1)}_{+,-}$ denotes the amplitude calculated for one pseudoparticle plus n (here) instanton–anti-instanton pairs. In the case of one pseudoparticle plus one pair the amplitude is

$$
A^{(3)}_{+,-} = A_{+,-}\left(T, \frac{T}{3}\right) A_{+,-}\left(\frac{T}{3}, -\frac{T}{3}\right) A_{+,-}\left(-\frac{T}{3}, -T\right). \tag{26.45}
$$

Evaluating each kernel and the end-point integrals, one then obtains (cf. Eq. (23.110))

$$
\begin{aligned}
A^{(3)}_{+,-} &= \int_{-T}^{T} d\tau_1 \int_{-T}^{\tau_1} d\tau_2 \int_{-T}^{\tau_2} d\tau \left[\frac{m\sqrt{1+u}}{2K(k')}\right]^3 e^{-3W} e^{-2TE_{\mathrm{cl}}} \\
&= \frac{(2T)^3}{3!} \left[\frac{m\sqrt{1+u}}{2K(k')}\right]^3 e^{-3W} e^{-2TE_{\mathrm{cl}}}.
\end{aligned} \tag{26.46}
$$

The expression for $A^{(2n+1)}_{+,-}$ is obtained by generalization. The sum (26.44) is then the expansion of a hyperbolic sine. Hence

$$
A_{+,-} = e^{-2ET} \sinh\left\{\frac{2Tm\sqrt{1+u}}{2K(k')} \exp\left[-\left(\frac{4m^3}{3g^2}\sqrt{1+u}[E(k) - uK(k)]\right)\right]\right\}. \tag{26.47}
$$

Comparing this expression with Eq. (26.7a), where now $\tau_f - \tau_i = 2T$, we see that the *level splitting* $\triangle^i_n$ arising from the periodic instanton and its instanton–anti-instanton pairs is given by

$$
\triangle^i E_n = \frac{m\sqrt{1+u}}{K(k')} \exp\left[-\left(\frac{4m^3}{3g^2}\sqrt{1+u}[E(k) - uK(k)]\right)\right]\hbar, \tag{26.48a}
$$

and thus by the very simple formula

$$
\triangle^i E_n = 2C_+^2 e^{-W}\hbar, \tag{26.48b}
$$

where C_+ is the WKB normalization constant. Inserting (26.42) for $W(u)$, we obtain

$$
\triangle^i E_n = \frac{2m}{\pi} \left|\frac{2^4 e\, m^3}{(n+\frac{1}{2})g^2}\right|^{n+\frac{1}{2}} e^{-\frac{4m^3}{3g^2}}. \tag{26.49}
$$

Using now the *Stirling approximation* in the form of the Furry factor f_n set equal to 1 (cf. (18.178)), i.e. $n! \simeq \sqrt{2\pi}[(n+1/2)/e]^{n+1/2}$, the expression can be written as (with $\hbar = 1$)

$$
\triangle^i E_n = \frac{2\sqrt{2}m}{\sqrt{\pi}n!} \left|\frac{2^4 m^3}{g^2}\right|^{n+\frac{1}{2}} e^{-\frac{4m^3}{3g^2}}.
$$

For $n = 0$ (ground state) this result reduces to that of Gildener and Patrascioiu [110], i.e. Eq. (23.113), obtained with vacuum instantons. Taking into account the periodic anti-instanton like the anti-instanton in Sec. 23.4, we obtain the level splitting $\Delta E_n = 2\Delta^i E_n$, i.e.

$$\Delta E_n = \frac{4\sqrt{2}m}{\sqrt{\pi n!}} \left| \frac{2^4 m^3}{g^2} \right|^{n+\frac{1}{2}} e^{-\frac{4m^3}{3g^2}}. \tag{26.50}$$

Transforming to the potential of Chapter 18 and keeping in mind that here the particle mass $m_0 = 1$, this result* can be checked to agree with that of Eq. (18.180).[†] Thus there is agreement between the path integral calculation in the one loop approximation and the Schrödinger equation calculation in leading order of boundary condition effects.

26.3   The Cosine Potential and Periodic Instantons

26.3.1   Periodic configurations and the cosine potential

We consider the Lagrangian (26.1) for a particle of mass $m_0 = 1$, i.e.

$$L = \frac{1}{2}\left(\frac{d\phi}{dt}\right)^2 - V(\phi),$$

where $V(\phi)$ is the sine–Gordon or cosine potential,[‡] here chosen in the form

$$V(\phi) = \frac{1}{g^2}[1 + \cos(g\phi)], \tag{26.51}$$

where g^2 denotes the coupling constant. The potential is shown in Fig. 26.2 and is seen to have maximum values $V(\phi_0) = 2/g^2$ at $\phi_0 = \pm n\pi/g, n = 0, 2, 4, \ldots$, and minima with value zero at $\phi_0 = \pm n\pi/g, n = 1, 3, 5, \ldots$. The classical solution $\phi_c(\tau)$ in Euclidean time $\tau = it$ which extremizes the action

---

*J.–Q. Liang and H. J. W. Müller–Kirsten [167].

[†]Comparing the double well potential $V(z)$ of Chapter 18 with $V(\phi)$ here, where mass $m_0 = 1$, i.e.

$$V(z) = -\frac{1}{4}h^4 z^2 + \frac{1}{2}c^2 z^4, \quad V(\phi) = \text{const.} - m^2\phi^2 + \frac{1}{2}g^2\phi^4,$$

one has the correspondence $c^2 \leftrightarrow g^2$, $h^2 \leftrightarrow 2m$. With these replacements in Eq. (18.180), one obtains for that perturbation result (index 1 indicating mass 1)

$$\Delta_1 E(q_0, h^2) \simeq 2\frac{\sqrt{8}m}{\sqrt{\pi n!}}\left(\frac{16m^3}{g^2}\right)^{n+1/2} e^{-4m^3/3g^2},$$

in agreement with Eq. (26.50).

[‡]We follow here the general line of J.–Q. Liang and H. J. W. Müller–Kirsten [168].

satisfies — as we have seen in Chapter 25 — the once integrated equation of motion

$$\frac{1}{2}\left(\frac{d\phi_c}{d\tau}\right)^2 - V(\phi_c) = -E_{cl} \equiv -\frac{2C^2}{g^2}, \tag{26.52}$$

where we have gone over to Euclidean time τ and take $\hbar = c = 1$ (which we use mostly throughout), and E_{cl} is the constant of integration, which we now assume to be nonzero. Inserting the potential (26.51) into (26.52) and integrating as in Section 25.4, one obtains the pseudoparticle solution called *periodic instanton* (through half a period) or *periodic bounce* (through a full period) and given by

$$\phi_c(\tau) = \frac{2}{g}\sin^{-1}[k\,\mathrm{sn}(\tau + \tau_0)], \tag{26.53}$$

where τ_0 is a constant of integration and the elliptic modulus k is given by

$$k^2 = 1 - \frac{1}{2}g^2 E_{cl} \quad \text{or} \quad C^2 = \frac{1}{2}g^2 E_{cl} = 1 - k^2 = k'^2. \tag{26.54}$$

The Jacobian elliptic function $\mathrm{sn}[\tau]$ has the real period $4K(k)$ and real zeros at 0 and $2K(k)$, $K(k)$ being the complete elliptic integral of the first kind

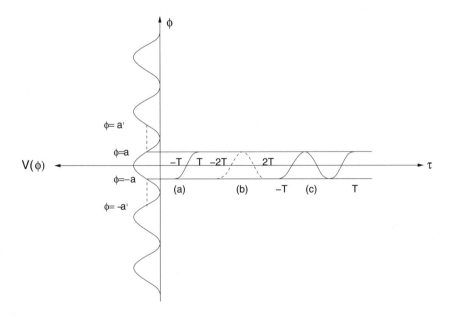

Fig. 26.2 (a) Trajectory of nonvacuum instanton (half period),
(b) of a bounce (full period) and (c) of one instanton plus a pair.

and hence a quarter period of $\phi_c(\tau)$. Thus there are critical values $[\tau]$ given by

$$[\tau] = 4n\,K(k), \quad n \text{ an integer.} \tag{26.55a}$$

Setting $[\tau] = 4T$ and taking $n = 1$, we have $4T = 4K(k)$, $T = K(k)$, $\tau_f - \tau_i \rightarrow 2K(k)$. We see that $2T$ is half the period of motion of the pseudoparticle bounce, or shortest barrier transition time of its nonvacuum instanton part, as indicated in Fig. 26.2, in which the $n = 1$ curve represents the trajectory of the single nonvacuum instanton with $\tau_0 = 0$ and value zero at $\tau = 0$. Taking $n = 3$, again with $[\tau] = 4T$, we have

$$4T = 12K(k), \quad \text{i.e. } T = 3K(k). \tag{26.55b}$$

This means the position $\tau = T$ is reached after the interval $2K(k)$ (as before) plus one period $4K(k)$ with the total range of $2T$ as before, and as indicated in Fig. 26.2. We interpret this configuration, i.e. the original configuration elongated by one full period, as that of the single periodic instanton with an instanton–anti-instanton pair (or one bounce) attached.

For zero energy $E_{\rm cl}$, i.e. $k \rightarrow 1$, the periodic solution (26.53) reduces to the vacuum instanton configuration

$$\phi_c(\tau) \xrightarrow{k=1} \frac{2}{g}\sin^{-1}[\tanh(\tau + \tau_0)]. \tag{26.56}$$

We choose $\tau_0 = 0$, so that $\phi_c(\tau)$ vanishes where $\mathrm{sn}[\tau]$ vanishes, i.e. at $\tau = 0, \pm 2K(k), \ldots$. For $\tau = \pm K(k)$ we have $\mathrm{sn}[\tau] = \pm 1$ and

$$\phi_c(\tau = \pm K(k)) = \pm\frac{2}{g}\sin^{-1}(k). \tag{26.57a}$$

These positions ϕ_c are those of the turning points $\pm a$ indicated in Fig. 26.2, since $E_{\rm cl} - V(\phi) = 0$ implies $g^2 E_{\rm cl} - 1 = \cos(g\phi) = 1 - 2\sin^2(g\phi/2)$, i.e.

$$\phi \rightarrow \pm\frac{2}{g}\sin^{-1}\sqrt{1 - \frac{g^2 E_{\rm cl}}{2}} = \pm\frac{2}{g}\sin^{-1}(k) \equiv \pm a. \tag{26.57b}$$

We interpret a configuration starting from the turning point $-a = \phi_c(-T)$ at $\tau = -T$ and ending at $a = \phi_c(T)$ at $\tau = T$ with

$$2T = (2n + 1)K(k)$$

as an instanton with n instanton–anti-instanton pairs or bounces attached, this entire configuration being a pseudoparticle configuration. The pseudoparticle oscillates between the two turning points in one of the barriers as illustrated in Fig. 26.2. For later reference we note already here the derivative of $\phi_c(\tau)$, i.e.

$$N(\tau) = \frac{d\phi_c(\tau)}{d\tau} = \frac{2k}{g}\text{cn}[\tau], \qquad (26.58)$$

an even function which is also the zero mode of the equation of small fluctuations about $\phi_c(\tau)$.

26.3.2 Transition amplitude and Feynman kernel

As in the case of the double well potential our aim is to calculate the energy splitting $\triangle E_n$ resulting from periodic instantons and instanton–anti-instanton pairs. The amplitude for a transition from a well to the left of a barrier to that on its right due to instanton tunneling at the energy E_{cl} quantized to E_n in a well can again be written[§] with $\tau_f - \tau_i = 2T$ large (cf. Eq. (23.35a))

$$A_{+,-} = \langle \psi_n | e^{-2HT} | \psi_n \rangle \sim \pm e^{-E_n(\tau_f - \tau_i)} \sinh \left\{ \frac{(\tau_f - \tau_i)\triangle E_n}{2\hbar} \right\} \qquad (26.59)$$

for negligible overlap of the wave functions which dominate in the wells, $\triangle E_n$ being here the difference $(E_- - E_+)$. In the path integral method we can write this amplitude in the form

$$A_{+,-} = \int d\phi_f \int d\phi_i \langle \psi_n | \phi_f \rangle K(\phi_f, \tau_f; \phi_i, \tau_i) \langle \phi_i | \psi_n \rangle, \qquad (26.60)$$

where the Feynman kernel K is given by

$$K(\phi_f, \tau_f; \phi_i, \tau_i) = \int_{\phi_i}^{\phi_f} \mathcal{D}\{\phi\} e^{-S/\hbar}. \qquad (26.61)$$

Here $\phi_f = \phi(\tau_f)$ and $\phi_i = \phi(\tau_i)$. We are interested in this expression, i.e. the tunneling propagator through the barrier, in the limits $\phi_i \to -a, \phi_f \to +a$, where a and $-a$ are the turning points indicated in Fig. 26.2 and given by Eqs. (26.57a) and (26.57b). The wave functions $\langle \phi_i | \psi \rangle, \langle \phi_f | \psi \rangle$ are the wave functions of the right-hand and left-hand wells and extend into the domain

---

[§]For later reference purposes compare the Hamiltonian here around a minimum at $\phi_{n_0} = \pm(2n_0 + 1)\pi/g, n_0 = 0, 1, 2, \ldots$, and for mass $m_0 = 1$, i.e.

$$H = \frac{1}{2}p^2 + V(\phi_{n_0}) + \frac{1}{2}(\phi - \phi_{n_0})^2 + \cdots,$$

with the harmonic oscillator Hamiltonian and its eigenvalues of Chapter 6. The corresponding eigenvalues of the Hamiltonian here expanded around a minimum are seen to be $E_n \simeq (n + 1/2)$ for $\hbar = 1$ (and frequency $\omega = 1$ since $V''(\phi)$ of Eq. (26.51) loses the g-dependence).

of the barrier. We consider the motion through half a period (i.e. through $2K(k)$) as the trajectory of an instanton while the motion through a full period ((i.e. $4K(k)$) is considered as the trajectory of a bounce or that of an instanton–anti-instanton pair. The Euclidean action S is, of course, again given by

$$S(\phi) = \int_{\tau_i}^{\tau_f} d\tau \left[\frac{1}{2} \left(\frac{d\phi}{d\tau} \right)^2 + V(\phi) \right]. \tag{26.62}$$

We proceed as in the previous case, but first consider the small fluctuation equation and the removal of its negative eigenmode from the Feynman kernel.

26.3.3   The fluctuation equation and its eigenmodes

As in our earlier cases we expand the field configuration $\phi(\tau)$ about its classical trajectory or *periodic instanton* $\phi_c(\tau)$ by setting

$$\phi(\tau) = \phi_c(\tau) + \chi(\tau), \tag{26.63}$$

where $\chi(\tau)$ is the deviation of $\phi(\tau)$ from $\phi_c(\tau)$ with fixed endpoints. The boundary conditions on the fluctuations $\chi(\tau)$ are that these vanish at the endpoints, i.e.

$$\chi(\tau_i) = \chi(\tau_f) = 0, \quad \tau_i \to -T = -K(k), \quad \tau_f \to T = K(k). \tag{26.64}$$

Inserting (26.63) into the kernel (26.61) and retaining only terms containing the fluctuation field $\chi(\tau)$ up to the second order (for our one loop approximation), we obtain (with $\hbar = 1$)

$$K(\phi_f, \tau_f; \phi_i, \tau_i) = \exp[-S(\phi_c)]I_0 \equiv \exp[-S(\phi_f, \phi_i; \tau_f, \tau_i)]I_0, \tag{26.65}$$

where we set

$$I_0 = \int_{\chi(\tau_i)=0}^{\chi(\tau_f)=0} \mathcal{D}\{\chi\} \exp[-\delta S(\chi)], \tag{26.66}$$

and δS is the fluctuation action, i.e. the remaining part of the action which is quadratic in the fluctuation. Evaluation of the action of the classical configuration $\phi_c(\tau)$ gives as we verify below:

$$\begin{aligned} S(\phi_c) &= \int_{-T}^{T} d\tau \left[\frac{1}{2} \left(\frac{d\phi_c}{d\tau} \right)^2 + V(\phi_c) \right] = W(\phi_c) + 2TE_{cl}, \\ W(\phi_c) &= \frac{8}{g^2}[E(k) - k'^2 K(k)]. \end{aligned} \tag{26.67}$$

Here $\phi_c(\tau_i) \to -a$ and $\phi_c(\tau_f) \to a$ with $\tau_f - \tau_i = 2T$ (the limits to the endpoints will be taken only after all integrations have been carried out so that the singularities at the turning points are avoided). The functions $K(k)$ and $E(k)$ again denote the complete elliptic integrals of the first and second kinds respectively. The result (26.67) is obtained as follows. Using Eq. (26.52) and inserting (26.58) for the derivative of $\phi_c(\tau)$, we obtain

$$
S(\phi_c) = \int_{\tau_i}^{\tau_f} d\tau \left[\left(\frac{d\phi_c}{d\tau} \right)^2 + E_{cl} \right] = \int_{-K(k)}^{K(k)} d\tau \left[\left(\frac{2k}{g} \right)^2 cn^2[\tau] + E_{cl} \right]
$$
$$
\equiv W(\phi_c) + E_{cl}(\tau_f - \tau_i).
$$

The result (26.67) then follows with the integral[¶]

$$
\int cn^2[u] \, du = \frac{1}{k^2} [E(u) - k'^2 u]
$$

and the fact that $cn[u]$ is an even function of u.

The fluctuation action can be expressed as (see for instance the step leading to Eq. (23.55b))

$$
\delta S = \frac{1}{2} \int_{-\infty}^{\infty} d\tau \left[\left(\frac{d\chi}{d\tau} \right)^2 + \chi^2 V''(\phi_c) \right]
$$
$$
= \frac{1}{2} \int_{-\infty}^{\infty} d\tau \chi \left[-\frac{d^2}{d\tau^2} + V''(\phi_c) \right] \chi \equiv \frac{1}{2} \int_{\tau_i}^{\tau_f} \chi M \chi, \quad (26.68)
$$

where M is the differential operator of the equation of small fluctuations (25.95), i.e.

$$
M\psi_m = E_m \psi_m, \quad (26.69)
$$

with (using Eq. (26.53) to convert $\cos(g\phi_c)$ into $sn[\tau]$)

$$
M = -\frac{d^2}{d\tau^2} + V''(\phi_c) = -\frac{d^2}{d\tau^2} - \cos[g\phi_c(\tau)]
$$
$$
= -\frac{d^2}{d\tau^2} - [1 - 2k^2 sn^2[\tau]]. \quad (26.70)
$$

Equation (26.69) is a *Lamé equation*, as we saw in Chapter 25, and has three discrete eigenfunctions:

$$
\begin{array}{llll}
\psi_1 = cn[\tau] & \text{with eigenvalue} & E_1 = 0, \\
\psi_2 = dn[\tau] & \text{with eigenvalue} & E_2 = k^2 - 1 \le 0, & (26.71) \\
\psi_3 = sn[\tau] & \text{with eigenvalue} & E_3 = k^2.
\end{array}
$$

[¶]See P. F. Byrd and M. D. Friedman [40], formula 312.02, p. 193.

Expanding the fluctuation field $\chi(\tau)$ in terms of the eigenmodes ψ_m of M, we set

$$\chi(\tau) = \sum_m C_m \psi_m(\tau). \tag{26.72}$$

Changing the integration variables of I_0 to $\{C_m\}$, the functional integral I_0 can be formally evaluated as a set of Gaussian integrals to yield (as we saw e.g. in Sec. 23.4.1)

$$I_0 = \left| \frac{\partial \chi(\tau)}{\partial C_m} \right| \prod_m \left(\frac{\pi}{E_m} \right)^{1/2}, \tag{26.73}$$

provided all the C_m are independent. In the present case the two boundary conditions (26.64) imply two constraint equations on the parameters C_m, i.e.

$$C_1 \psi_1(\pm K(k)) + C_2 \psi_2(\pm K(k)) + C_3 \psi_3(\pm K(k)) = 0.$$

Since (see Appendix A) $\mathrm{cn}[K(k)] = 0, \mathrm{dn}[K(k)] = k'$ and $\mathrm{sn}[K(k)] = 1$, these relations reduce to $C_2 k' + C_3 = 0, C_2 k' - C_3 = 0$, implying $C_2 = C_3 = 0$. Hence the negative eigenvalue $E_2 = k^2 - 1$ associated with $\psi_2(\tau)$ does not contribute to the functional integral, and consequently the transition amplitude and hence the level splittings are real, as expected, and as we shall see below.

26.3.4   The single periodic instanton contribution

In evaluating the transition amplitude we proceed again as in Chapter 23 and in Sec. 26.2.3. We introduce the mapping from $\chi(\tau)$ to (say) $y(\tau)$ as in Eqs. (23.59) and (26.17), i.e.

$$y(\tau) = \chi(\tau) - N(\tau) \int_{\tau_i}^{\tau} \frac{\dot{N}(\tau')}{N^2(\tau')} y(\tau') d\tau', \tag{26.74}$$

where $N(\tau)$ is the zero mode which is in the present case

$$N(\tau) = \frac{d\phi_c(\tau)}{d\tau} = \frac{2k}{g} \mathrm{cn}[\tau], \tag{26.75}$$

i.e. again an even function. The boundary conditions of $y(\tau)$ are obtained from those of $\chi(\tau)$, i.e. Eq. (26.64), as

$$y(\tau_i) = 0, \quad y(\tau_f) + f(\tau_f) = 0, \quad f(\tau_f) = N(\tau_f) \int_{\tau_i}^{\tau_f} \frac{\dot{N}(\tau')}{N^2(\tau')} y(\tau') d\tau'. \tag{26.76}$$

As in our earlier cases, the effect of the transformation is to convert the functional integral I_0 into that of a free particle propagator supplemented by the constraint of (26.76). The constraint is inserted into I_0 with the help of the identity

$$1 = \int dy_f \, \delta[y(\tau_f) + f(\tau_f)] = \frac{1}{2\pi} \int_{-\infty}^{\infty} d\alpha \int dy_f e^{i\alpha[y(\tau_f)+f(\tau_f)]}. \quad (26.77)$$

Performing in I_0 the substitutions, we obtain (for the step from χ to y see Example 23.2)

$$\begin{aligned}
I_0 &= \int_{\chi(\tau_i)=0}^{\chi(\tau_f)=0} \mathcal{D}\{\chi\} \exp[-\delta S(\chi)] \\
&= \int_{\chi(\tau_i)=0}^{\chi(\tau_f)=0} \mathcal{D}\{\chi\} \exp\left[-\frac{1}{2}\int_{\tau_i}^{\tau_f} d\tau \left\{\left(\frac{d\chi}{d\tau}\right)^2 + V''(\phi_c)\chi^2\right\}\right] \\
&= \frac{1}{2\pi} \int d\alpha \left|\frac{Dy}{D\chi}\right|^{-1} \int dy_f \int_{y_i}^{y_f} \mathcal{D}\{y\} \exp\left[-\frac{1}{2}\int_{\tau_i}^{\tau_f} d\tau \left(\frac{dy}{d\tau}\right)^2\right. \\
&\quad \left. + i\alpha\left\{y(\tau_f) + N(\tau_f)\int_{\tau_i}^{\tau_f} d\tau' \frac{\dot{N}(\tau')}{N^2(\tau')}y(\tau')\right\}\right]. \quad (26.78)
\end{aligned}$$

Following the steps in Example 23.3 this is

$$I_0 = \frac{1}{\sqrt{2\pi}} \left[N(\tau_i)N(\tau_f)\int_{\tau_i}^{\tau_f} \frac{d\tau}{N^2(\tau)}\right]^{-1/2}. \quad (26.79)$$

Our next task is the evaluation of the transition amplitude (26.60), i.e.

$$\begin{aligned}
A_{+,-} &= \int d\phi_i \int d\phi_f \langle\psi_n|\phi_f\rangle K(\phi_f, \tau_f; \phi_i, \tau_i)\langle\phi_i|\psi_n\rangle \\
&\overset{(26.65)}{=} \int d\phi_i \int d\phi_f \langle\psi_n|\phi_f\rangle \exp[-S(\phi_f, \phi_i; \tau_f - \tau_i)]I_0\langle\phi_i|\psi_n\rangle,
\end{aligned}$$
$$\qquad\qquad (26.80)$$

which we do first for the single periodic instanton, in which case we call it $A_{+,-}^{(1)}$. For this we require the wave functions $\langle\phi_i|\psi_n\rangle \equiv \psi_n(\phi_i)$ and $\langle\phi_f|\psi_n\rangle \equiv \psi_n(\phi_f)$. These are wave functions inside the barrier. We assume these to be of WKB form, i.e. we write (observe that in Fig. 26.2 the velocity $\dot{\phi}(\tau)$ — which is the gradient of the trajectory there — is positive between $\tau = -T$ and T in the case of the instanton but reverses in the case of the bounce)

$$\psi_n(\phi_i) = \frac{C}{\sqrt{\dot{\phi}_i}}\exp[-\Omega(\phi_i)], \quad \Omega(\phi_i) = \int_{-a=\phi_c(-T)}^{\phi_i} \dot{\phi}d\phi,$$

$$\psi_n(\phi_f) = \frac{C}{\sqrt{\dot{\phi}_f}}\exp[-\Omega(\phi_f)], \quad \Omega(\phi_f) = \int_{\phi_f}^{a=\phi_c(T)} \dot{\phi}d\phi. \quad (26.81)$$

Here $\pm a$ are the positions of the turning points at energy E_{cl} as indicated in Fig. 26.2. The normalization constant C is defined and evaluated in Example 26.4. The result is

$$C = \left[2 \int_a^{a'} \frac{d\phi}{\sqrt{2|E_{\text{cl}} - V(\phi)|}} \right]^{-1/2} = [4K(k)]^{-1/2}. \tag{26.82}$$

The single periodic instanton transition amplitude thus becomes

$$A_{+,-}^{(1)} = C^2 \int \frac{d\phi_i}{\sqrt{\dot{\phi}_i}} \int \frac{d\phi_f}{\sqrt{\dot{\phi}_f}} \exp[-S(\phi_f, \phi_i; \tau_f - \tau_i)]$$
$$\times I_0 \exp[-\Omega(\phi_i)] \exp[-\Omega(\phi_f)]. \tag{26.83}$$

We now expand $S(\phi_f, \phi_i; \tau_f - \tau_i)$ and $\Omega(\phi_f)$ as power series in $(\phi_f - \phi_c(T))$ with $\phi(\tau) \to a = \phi_c(T)$ up to the second order for the Gaussian appproximation. The procedure is, of course, very similar to that in Sec. 26.2, so that we do not have to repeat all details. Thus

$$S(\phi_f, \phi_i; \tau_f - \tau_i) = S(\phi(T), \phi_i; \tau_f - \tau_i) + (\phi_f - \phi_c(T)) \left(\frac{\partial S}{\partial \phi_f} \right)_{\phi_c(T)}$$
$$+ \frac{1}{2} (\phi_f - \phi_c(T))^2 \left(\frac{\partial^2 S}{\partial \phi_f^2} \right)_{\phi_c(T)} + \cdots . \tag{26.84}$$

Here the first derivative of S is zero since (cf. Example 26.2, Eq. (26.26))

$$\frac{\partial S}{\partial \phi_f} = \frac{\partial W}{\partial \phi_f} \quad \text{and} \quad W = \int_{\phi_i}^{\phi_f} d\phi \dot{\phi}, \quad \text{i.e.} \quad \frac{\partial W}{\partial \phi_f} = \dot{\phi}_f,$$

which vanishes at $\tau = T$. In the expansion of the functional $\Omega(\phi_f)$ we have $\Omega(a = \phi_c(T)) = 0$ (cf. Eq. (26.81)) and

$$\Omega'(\phi(\tau))|_{\tau=T} = -N(\phi(T)) = -\dot{\phi}_f|_{\tau=T} = 0,$$

and

$$\frac{\partial^2 \Omega(\phi_f)}{\partial \phi_f^2} = -\frac{d}{d\phi_f} \left(\frac{d\phi_f}{d\tau} \right) = -\frac{d}{d\tau} \left(\frac{d\phi_f}{d\tau} \right) \frac{d\tau}{d\phi_f} = -\frac{\dot{N}(\phi_f)}{N(\phi_f)}. \tag{26.85}$$

Hence only the second derivative remains, i.e.

$$\Omega(\phi_f) = \frac{1}{2} \frac{\partial^2 \Omega}{\partial \phi_f^2} [\phi_f - \phi_c(T)]^2 + \cdots \simeq -\frac{1}{2} \frac{\dot{N}(\phi_f)}{N(\phi_f)} [\phi_f - \phi_c(T)]^2. \tag{26.86}$$

Next we note that the zero mode $N(\phi) = \dot\phi_c(\tau)$ given by Eq. (26.75) is an even function of τ. Eventually we want to take the limit $\tau_f, -\tau_i \to T$. Thus we consider $\tau_f = -\tau_i$, in which case here $N(\tau_i) = N(\tau_f)$. We also observe that with Eqs. (26.20a) and (26.79) we have

$$\frac{\partial^2 S}{\partial \phi_f^2} + \frac{\partial^2 \Omega(\phi_f)}{\partial \phi_f^2} = \frac{\partial^2 S}{\partial \phi_f^2} - \frac{\dot N(\phi_f)}{N(\phi_f)} = \left[\dot\phi_f^2 \int_{\tau_i}^{\tau_f} \frac{d\tau}{\dot\phi^2}\right]^{-1} = 2\pi I_0^2. \qquad (26.87)$$

Returning to Eq. (26.83) we can now write the transition amplitude in the form

$$A_{+,-}^{(1)} = \lim_{\tau_f, -\tau_i \to T} C^2 \int \frac{d\phi_i}{\dot\phi_i} \exp\left[-\int_{\phi_c(-T)}^{\phi(\tau_i)} \dot\phi \, d\phi\right] I_0$$

$$\exp[-S(\phi(T), \phi_i; \tau_f - \tau_i)] \int d\phi_f \exp\left[-\frac{2\pi I_0^2}{2}(\phi_f - \phi_c(T))^2\right],$$

so that with integration of the Gaussian integral[||] and the procedure of replacing $\dot\phi_i$ by $d\phi_i/d\tau$ as in Sec. 26.2, we obtain

$$A_{+,-}^{(1)} = \frac{2T}{4K(k)} \exp[-S_c(\phi(T), \phi(-T); 2T)]$$

$$= \frac{2T}{4K(k)} e^{-W} e^{-2E_{cl}T}, \qquad (26.88)$$

where W is given by Eq. (26.67) and E_{cl} quantized is $E_n \simeq (n + 1/2)$.

Example 26.4: WKB normalization constant
Show that the normalization constants $C_\pm$ of the WKB wave functions for the periodic potential (26.51) are given by

$$C = [4K(k)]^{-1/2}.$$

Solution: We have (cf. Eq. (26.34))

$$C = \left[2\int_a^{a'} \frac{d\phi}{\sqrt{2|E_{cl} - V(\phi)|}}\right]^{-1/2}. \qquad (26.89)$$

The integral to be evaluated is (setting $\theta = g\phi$) the following across the non-tunneling region, i.e.

$$\int_a^{a'} \frac{d\phi}{\sqrt{2|E_{cl} - V(\phi)|}} = \frac{1}{\sqrt 2}\int_{\theta=ga}^{ga'} \frac{d\theta}{\sqrt{1 - g^2 E_{cl} + \cos\theta}}. \qquad (26.90)$$

The integral can be evaluated with the help of tabulated elliptic integrals. Observing that the parameters involved are such that $b \equiv 1 > |a| \equiv |1 - g^2 E_{cl}| > 0$, the relevant integral is[**]

$$\int_0^\varphi \frac{d\theta}{\sqrt{a + b\cos\theta}} = \sqrt{\frac{2}{b}} F(\xi, \tilde k), \quad 0 < \varphi \le \cos^{-1}\left(-\frac{a}{b}\right), \quad \frac{a}{b} = 1 - g^2 E_{cl},$$

[||]Recall $\int_{-\infty}^\infty dx\, e^{-w^2 x^2/2} = \sqrt{2\pi/w^2}$.
[**]See P. F. Byrd and M. D. Friedman [40], formula 290.00, p. 175.

where

$$\xi = \sin^{-1}\sqrt{\frac{b(1 - \cos\varphi)}{a + b}} \quad \text{and} \quad \tilde{k}^2 = \frac{a + b}{2b} = 1 - \frac{1}{2}g^2 E_{\text{cl}} = k^2.$$

The function $F(\xi, \tilde{k})$ is again the incomplete elliptic integral of the first kind. The turning points at $\phi = a', a$ are given by

$$\cos(g\phi) = g^2 E_{\text{cl}} - 1, \quad \text{i.e.} \quad 1 - \cos(g\phi) = 2 - g^2 E_{\text{cl}} = a + b, \quad b = 1.$$

It follows that

$$
\begin{aligned}
\int_a^{a'} \frac{d\phi}{\sqrt{2|E - V(\phi)|}} &= \int_0^{ga'} \frac{d\theta}{\sqrt{1 - g^2 E_{\text{cl}} + \cos\theta}} + \int_{ga}^0 \frac{d\theta}{\sqrt{1 - g^2 E_{\text{cl}} + \cos\theta}} \\
&= F(\xi = \sin^{-1}(1), \tilde{k}) - F(\xi = \sin^{-1}(-1), \tilde{k}) \\
&= 2F(\pi/2, \tilde{k}) = 2K(k), \quad \text{since } \tilde{k}^2 = k^2,
\end{aligned}
\tag{26.91}
$$

and we used again the relation $F(\varphi, k) = -F(-\varphi, k)$. Inserting the result into Eq. (26.89) we obtain the result (26.82).

26.3.5   Sum over instanton–anti-instanton pairs

The configuration of one instanton plus one instanton–anti-instanton pair is given by the solution with $n = 3$ in Eqs. (26.55a) and (26.55b), in which $T = 3K(k)$. Following the steps in Sec. 26.2.5 the transition amplitude for this case can be written as

$$A_{+,-}^{(3)} = \int_{-T}^T d\tau_1 \int_{-T}^{\tau_1} d\tau_2 \int_{-T}^{\tau_2} d\tau \left(\frac{1}{4K(k')}\right)^3 e^{-3W} e^{-2E_{\text{cl}}T}. \tag{26.92}$$

The total amplitude is given by the sum[*]

$$A_{+,-} = \sum_{m=0}^{\infty} A_{+,-}^{(2m+1)} = e^{-2E_{\text{cl}}T} \sinh\left\{\frac{T}{2K(k')}e^{-W}\right\}. \tag{26.93}$$

Comparing this result with Eqs. (26.59) and (23.35a), the level splitting, i.e. twice the shift of the nth asymptotic oscillator level in the wells (with $E_{\text{cl}} \simeq (n+1/2)$), as a result of tunneling through the barrier between these, is — in the case of the periodic instanton plus an infinite number of instanton–anti-instanton pairs — given by the very simple formula

$$\triangle^i E_n = \frac{1}{2K(k)}e^{-W}\hbar, \tag{26.94}$$

where (cf. (26.67))

$$W = \frac{8}{g^2}[E(k) - k'^2 K(k)]. \tag{26.95}$$

---

[*]Observe that $\int_{-T}^T d\tau_1 \int_{-T}^{\tau_1} d\tau_2 \int_{-T}^{\tau_2} d\tau = (2T)^3/3!$ and so on.

For the energy far below the barrier height we have $E_{cl} \ll 2/g^2$ implying $k^2 \to 1, k'^2 \to 0$. With the appropriate expansions of the complete elliptic integrals $E(k)$ and $K(k)$ as in Chapter 18 (cf. Eqs. (18.192), (18.193)), one obtains with (26.54), i.e.

$$k'^2 = \frac{g^2}{2} E_{cl} \to \frac{g^2}{2}\left(n + \frac{1}{2}\right),$$

the following expansion for W:

$$
\begin{aligned}
W &= \frac{8}{g^2}\left[1 - \frac{k'^2}{2}\left\{\ln\left(\frac{4}{k'}\right) + \frac{1}{2}\right\} + O(k'^4)\right] \\
&= \frac{8}{g^2} - \left(n + \frac{1}{2}\right)\ln\left[\frac{2^5}{(n + \frac{1}{2})g^2}\right] - \left(n + \frac{1}{2}\right).
\end{aligned}
\tag{26.96}
$$

Furthermore, since $K(k = 0) = \pi/2$, the result of Eq. (26.94) becomes

$$
\triangle^i E_n = \frac{2}{2\pi} e^{-8/g^2} e^{n+\frac{1}{2}}\left[\frac{2^5}{(n + \frac{1}{2})g^2}\right]^{n+\frac{1}{2}}.
\tag{26.97}
$$

With the help of the *Stirling approximation* in the form of the Furry factor f_n set equal to 1 (cf. (18.178)), i.e. $n! \simeq \sqrt{2\pi}[(n + 1/2)/e]^{n+1/2}$, the result for the contribution of the periodic instanton and its associated pairs to the level splitting becomes

$$
\triangle^i E_n = \frac{2}{\sqrt{2\pi}}\frac{1}{n!}\left(\frac{2^5}{g^2}\right)^{n+\frac{1}{2}} e^{-8/g^2}.
$$

Again taking into account the anti-instanton and its associated instanton–anti-instanton pairs (as in the case of the double well in Secs. 23.4.5 and 26.2.5), the result for the level splitting $\triangle E_n$ becomes double that, i.e.

$$
\triangle E_n = \frac{4}{\sqrt{2\pi}}\frac{1}{n!}\left(\frac{2^5}{g^2}\right)^{n+\frac{1}{2}} e^{-8/g^2}.
\tag{26.98}
$$

This result[†] agrees with the result obtained from the Schrödinger equation, as can be checked by comparison with the Eq. (17.74b).[‡]

[†]J.–Q. Liang and H. J. W. Müller–Kirsten [168].

[‡]See R. B. Dingle and H. J. W. Müller [73] and P. Achuthan, H. J. W. Müller–Kirsten and A. Wiedemann [4]. The Schrödinger equation for $m_0 = 1$ and the cosine potential (26.51) is

$$\psi'' + 2\left[E - \frac{1}{g^2}(1 + \cos gx)\right]\psi = 0.$$

26.4   The Inverted Double Well and Periodic Instantons

26.4.1   Periodic configurations and the inverted double well

We[§] use an inverted double well potential with the local minimum at $\phi = 0$, i.e. we write it[¶]

$$V(\phi) = -\frac{\mu^2}{2a^2}(\phi^2 - a^2)^2 + \frac{1}{2}\mu^2 a^2 = \mu^2\phi^2 - \frac{1}{2}\frac{\mu^2}{a^2}\phi^4. \qquad (26.99)$$

The classical solution ϕ_c which minimizes the action with Euclidean time $\tau = it$, satisfies the equation

$$\frac{1}{2}\left(\frac{d\phi}{d\tau}\right)^2 - V(\phi(\tau)) = -E_{\rm cl}. \qquad (26.100)$$

It is convenient to set (observe that $E_{\rm cl}$ is small for $k^2 \to 1$)

$$E_{\rm cl} = \frac{1}{2}a^2\mu^2 u^2, \quad u = \frac{1-k^2}{1+k^2}, \quad k^2 = \frac{1-u}{1+u}, \quad 0 \le k \le 1. \qquad (26.101)$$

This substitution defines the parameter k; below this will be shown to be related to the elliptic modulus γ and varies from 0 to 1 as $E_{\rm cl}$ varies between its extreme values, i.e.

$$-\frac{1}{2}\mu^2 a^2 \le -E_{\rm cl} \le 0.$$

The solution $\phi_c(\tau)$ of Eq. (26.100) is obtained with the help of Tables of Elliptic Integrals. Since this requires some unavoidable manipulations of

---

Setting $x = 2z/g$, the equation becomes $d^2\psi/dz^2 + (8/g^2)[E - (1/g^2)(1+\cos 2z)]\psi = 0$. Comparing this equation with the Mathieu equation of Chapter 17, $y'' + [\lambda - 2h^2\cos 2z]y = 0$, we see that we have to identify $\lambda = 8E/g^2, h = 2/g^2$. Then $\triangle\lambda(q_0)$ of Eq. (17.74b) implies

$$\triangle\lambda(q_0) = \frac{(16h)^{n+3/2}e^{-4h}}{n!\sqrt{2\pi}} \to \triangle E_n = \frac{4}{\sqrt{2\pi}n!}\left(\frac{2^5}{g^2}\right)^{n+1/2}e^{-8/g^2}.$$

[§]We follow here in part J.–Q. Liang and H. J. W. Müller–Kirsten [169].

[¶]For later reference purposes it may be noted that in the harmonic oscillator approximation the Hamiltonian to be compared with that of Chapter 6 is

$$H \simeq \frac{1}{2}p^2 + \frac{1}{2}(\sqrt{2}\mu)^2\phi^2,$$

so that the corresponding eigenvalues are here $E_n = (n + 1/2)w, w = \sqrt{2}\mu$.

parameters we leave the details to Example 26.5. The result is

$$\phi_c(\tau) \;=\; s_+(k)\mathrm{dn}[\beta(k)(\tau+\tau_0),\gamma], \quad \beta(k) = \frac{\mu}{a}s_+(k),$$

$$s_+(k) \;=\; \frac{a(1+k)}{\sqrt{1+k^2}}, \quad \gamma = \frac{4k}{(1+k)^2},$$

$$\gamma' \;=\; \sqrt{1-\gamma^2} = \frac{1-k}{1+k}, \quad 1+\gamma'^2 = \frac{2(1+k^2)}{(1+k)^2}, \qquad (26.102)$$

where $\mathrm{dn}[u]$ is (like $\mathrm{sn}[u]$ and $\mathrm{cn}[u]$) a Jacobian elliptic function and τ_0 an integration constant which we choose to be zero (see below). The Jacobian elliptic function $\mathrm{dn}[\beta(k)\tau,\gamma]$ has real periods $2K(\gamma)$, i.e.

$$\beta(k)[\tau] = 2nK(\gamma), \quad n = 1,2,3,\ldots, \qquad (26.103)$$

where $K(\gamma)$ is again the elliptic quarter period or complete elliptic integral of the first kind of elliptic modulus γ. Setting $[\tau] = 2T$ and taking $n = 1$, we have

$$\beta(k)T = K(\gamma),$$

where T is half the period of the motion of the pseudoparticle bounce as indicated in Fig. 26.3, in which the $n = 1$ curve represents the trajectory of a single

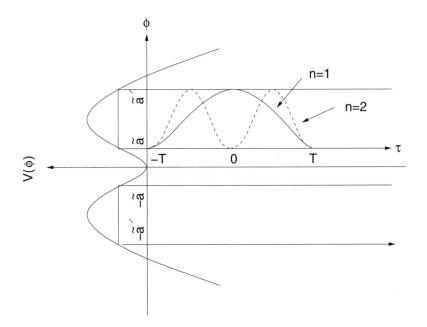

Fig. 26.3 Inverted double well and bounces for $n = 1$ and 2.

bounce with $\tau_0 \rightarrow \tau_0^{(1)} = 0$ and zeros at $\pm\beta(k)T$. For $n = 2$ in the same interval $[\tau] = 2T$ the function $\phi_c(\tau)$ has zeros at $\pm\beta(k)T/2$ and again the maximum at $\tau = 0$. With a shift $\tau_0^{(2)} = -T/2$ to the left, these zeros are shifted to $\tau = -T$ and 0, thus allowing two bounces in $\tau \in [-T, T]$, as indicated in Fig. 26.2. The pseudoparticle with position given by $\phi_c(\tau)$ oscillates in the barrier between turning points $\tilde{a}$ and $\tilde{a}'$ as indicated in Fig. 26.3. We define these turning points for the single bounce (as indicated in Fig. 26.3) as

$$\tilde{a} := \phi(-T) = \phi(T), \quad \tilde{a}' := \phi(0). \tag{26.104a}$$

Evaluating the values $\phi(\pm T)$ with (26.102) (and using that $\mathrm{dn}[K(\gamma)] = \gamma', \mathrm{dn}[0] = 1$, as in Appendix A) these turning points are given by

$$\tilde{a} = \frac{a(1 - k)}{\sqrt{1 + k^2}} = \gamma' s_+(k), \quad \tilde{a}' = \frac{a(1 + k)}{\sqrt{1 + k^2}}. \tag{26.104b}$$

As the energy E_{cl} tends to zero the solution (26.102) reduces to the usual *vacuum bounce*, i.e.

$$\phi_c(\tau) \rightarrow a\sqrt{2}\,\mathrm{sech}[\mu\sqrt{2}(\tau)]. \tag{26.105}$$

For a better distinction we dub the solution (26.102) a *"nonvacuum bounce"* or *"periodic bounce"*. On the other hand, as the energy approaches the top of the barrier, i.e. $E_{\mathrm{cl}} \rightarrow \mu^2 a^2/2$ with $k \rightarrow 0$, the solution becomes the trivial configuration $\phi_c = a$ since $\mathrm{dn}[u, \gamma] = 1$ for $\gamma = 0$. This trivial solution is called a *"sphaleron"*. The nonvacuum bounce thus interpolates between the vacuum bounce and this sphaleron. One may note that — since the configurations we are concerned with here are bounces (i.e. instanton–anti-instanton pairs or combinations as illustrated in Fig. 26.2) — the first contribution to the transition amplitude $A_{+,-}$ is that due to zero bounces, i.e. that with no pseudoparticle contribution. Comparing this case of the inverted double well potential with the previous two cases of the double well and the cosine potentials we observe already here that in the present case the classical configuration $\phi_c(\tau)$ is an even function, so that its derivative, representing the velocity, is an odd function — the opposite to the behaviour in the previous cases.

Example 26.5: The classical bounce configuration
Show that the solution of the classical Euclidean time equation of motion with the inverted double well potential is

$$\phi_c(\tau) = s_+(k)\mathrm{dn}[\beta(k)(\tau + \tau_0), \gamma], \quad \beta(k) = \frac{\mu}{a}s_+(k),$$

where γ is the elliptic modulus and $s_+(k)$ the function of k defined in Eq. (26.102).

Solution: We set

$$E_{\mathrm{cl}} = \frac{1}{2}a^2\mu^2 u^2, \quad u'^2 = 1 - u^2, \quad u = \frac{1 - k^2}{1 + k^2}, \quad 0 \le k \le 1.$$

Inserting the potential (26.99) into Eq. (26.100) we obtain

$$
\begin{aligned}
\left(\frac{d\phi}{d\tau}\right)^2 &= 2V(\phi) - 2E_{\text{cl}} = \mu^2 a^2 u'^2 - \frac{\mu^2}{a^2}(\phi^2 - a^2)^2 \\
&= \frac{\mu^2}{a^2}[a^2(u'+1) - \phi^2][\phi^2 - a^2(1-u')].
\end{aligned}
$$

Hence we obtain

$$
\frac{\mu}{a}\int d\tau = \int \frac{d\phi}{\sqrt{\{\frac{a^2(1+k)^2}{(1+k^2)} - \phi^2\}\{\phi^2 - \frac{a^2(1-k)^2}{(1+k^2)}\}}}.
$$

We have

$$
u' = \frac{2k}{1+k^2} \quad \text{and set} \quad s_+(k) = \frac{a(1+k)}{\sqrt{1+k^2}}.
$$

We evaluate the elliptic integral by using the following relation[||] with $B < x < A$:

$$
\int_x^A \frac{dt}{\sqrt{(A^2 - t^2)(t^2 - B^2)}} = \frac{1}{A} \text{dn}^{-1}[x/A, \gamma], \quad \text{where} \quad \gamma^2 = \frac{A^2 - B^2}{A^2} = \frac{4k}{(1+k)^2}.
$$

It follows that in our case we obtain

$$
\phi = s_+(k)\text{dn}\left[\frac{\mu}{a}s_+(k)(\tau - \tau_0), \gamma\right],
$$

where τ_0 is an integration constant (here we choose the bounce position such that $\tau_0 = 0$).

26.4.2 Transition amplitude and Feynman kernel

Proceeding parallel to the preceding cases we consider the transition amplitude $A_{+,-}$ from a state $|\psi\rangle$ at configuration ϕ_i to the same state $|\psi\rangle$ at configuration ϕ_f in Euclidean time $\tau_f - \tau_i = 2T \to \infty$. When there is no tunneling and the potential is that of an harmonic oscillator so that the states $|\psi\rangle$ are oscillator states $|\psi_n\rangle$, we know that the Green's function K is given by Eqs. (7.59a) and (7.59b), i.e.

$$
K(\phi_f, \tau_f; \phi_i, \tau_i) = \sum_n e^{-E_n(\tau_f - \tau_i)/\hbar} \langle\phi_f|\psi_n\rangle\langle\psi_n|\phi_i\rangle.
$$

Here $\langle\phi|\psi_n\rangle$ are oscillator eigenfunctions $\psi_n(\phi)$ with eigenvalues $E_n = (n + 1/2)\hbar w$. When there is tunneling and hence a correction to this due to additional terms in the potential, this relation holds only approximately. The transition amplitude $A_{+,-}$ of a transition from a state $|\psi_n\rangle$ to itself is then given by

$$
\begin{aligned}
A_{+,-} &= \langle\psi_n|K(\phi_f, \tau_f; \phi_i, \tau_i)|\psi_n\rangle \\
&= \int d\phi_f \int d\phi_i \langle\psi_n|\phi_f\rangle K(\phi_f, \tau_f; \phi_i, \tau_i)\langle\phi_i|\psi_n\rangle, \quad (26.106a)
\end{aligned}
$$

[||]L. M. Milne–Thomson [196], p. 27. The corresponding integral in somewhat different form in P. F. Byrd and M. D. Friedman [40] is their formula 130.04, p. 30.

which can be written

$$A_{+,-} = \langle \psi_n | e^{-E_n(\tau_f - \tau_i)/\hbar} | \psi_n \rangle = e^{-E_n(\tau_f - \tau_i)}. \tag{26.106b}$$

Thus in the following we shall find that the dominant part of the transition amplitude is an exponential in which the oscillator energy E_n is supplemented by an imaginary part. The determination of this imaginary part is our main objective here. In the Feynman representation we have again

$$K(\phi_f, \tau_f; \phi_i, \tau_i) = \int_{\phi_i}^{\phi_f} \mathcal{D}\{\phi\}\, e^{-S/\hbar}, \tag{26.107}$$

where

$$S(\phi) = \int_{\tau_i}^{\tau_f} d\tau \left[\frac{1}{2} \left(\frac{d\phi}{d\tau} \right)^2 + V(\phi) \right]. \tag{26.108}$$

26.4.3 The fluctuation equation and its eigenmodes

Considering fluctuations about the bounce $\phi_c(\tau)$ we set

$$\phi(\tau) = \phi_c(\tau) + \chi(\tau), \tag{26.109}$$

where $\chi(\tau)$ denotes the small deviation of $\phi(\tau)$ from $\phi_c(\tau)$ with the end points held fixed. Thus we demand as necessary boundary conditions

$$\chi(\tau_i) = \chi(\tau_f) = 0 \tag{26.110}$$

at inital and final Euclidean times τ_i and τ_f. Inserting the expression (26.102) for $\phi_c(\tau)$ into the Feynman kernel $K(\phi_f, \tau_f; \phi_i, \tau_i)$ (as in previous cases, see e.g. Eqs. (26.61) to (26.65)) we obtain

$$K(\phi_f, \tau_f; \phi_i, \tau_i) = I_0 \exp[-S(\phi_c)/\hbar], \tag{26.111}$$

where

$$I_0 = \int_{\chi(\tau_i)=0}^{\chi(\tau_f)=0} \mathcal{D}\{\chi\} \exp[-\delta S/\hbar]. \tag{26.112}$$

Here

$$
\begin{aligned}
S(\phi_c) &= \int_{\tau_i}^{\tau_f} d\tau \left[\frac{1}{2} \left(\frac{d\phi_c}{d\tau} \right)^2 + V(\phi_c) \right] \\
&= \int_{\tau_i}^{\tau_f} d\tau \left[\frac{1}{2} \left(\frac{d\phi_c}{d\tau} \right)^2 + \mu^2 \phi_c^2 - \frac{1}{2} \frac{\mu^2}{a^2} \phi_c^4 \right] \\
&= \int_{\tau_i}^{\tau_f} \left[\left(\frac{d\phi_c}{d\tau} \right)^2 + E_{\text{cl}} \right],
\end{aligned}
\tag{26.113}
$$

and up to terms of $O(\chi^2)$ the remainder δS is given by

$$\delta S = \frac{1}{2} \int_{\tau_i}^{\tau_f} d\tau \left[-\chi \frac{d^2\chi}{d\tau^2} + 2\chi \left(\mu^2 - \frac{3\mu^2}{a^2} \phi_c^2 \right) \chi \right] \equiv \frac{1}{2} \int_{\tau_i}^{\tau_f} d\tau \chi M \chi. \quad (26.114)$$

Here M is the operator of the equation of small fluctuations. This equation of small fluctuations about the classical configuration $\phi_c(\tau)$ is again obtained as in the previous two cases (obtained from the second variational derivative of the action evaluated at the bounce) and is given by (cf. for instance Eq. (26.69))

$$M\psi_m = E_m\psi_m \quad \text{with} \quad M = -\frac{d^2}{d\tau^2} + V''(\phi_c). \quad (26.115)$$

In the present case differentiation of the potential leads to the following operator

$$M = -\frac{d^2}{d\tau^2} + \mu^2 \left(2 - \frac{6}{a^2} \phi_c^2(\tau) \right),$$

and subsequent substitution of $\phi_c(\tau)$ from Eq. (26.102) with a change of variable leads to the Lamé equation

$$\frac{d^2\psi(z)}{dz^2} + [E_m - 6\gamma^2 \text{sn}^2[z, \gamma]]\psi(z) = 0, \quad z = \beta(k)\tau, \quad (26.116)$$

where we specify explicitly the relevant elliptic modulus γ along with the variable in order to avoid confusion with k used elsewhere. Equation (26.116) has 5 disrete eigenvalues with eigenmodes having periods $4nK(\gamma)/\beta(k)$. These have been specified previously in Eqs. (25.47a) to (25.47d) and hence will not be repeated here. As in Sec. 26.3.3 one can consider the effect of the constraints (26.110) on these eigenmodes. One can show that only one of the three discrete eigenmodes with negative eigenvalues contributes to the tunneling (apart from the zero mode). For details we refer to Ref. [169] (Appendix A). Of course, the odd zero eigenmode (derivative of $\phi_c(\tau)$) is already indicative of a lower and thus negative eigenvalue.

Differentiation of $\phi_c(\tau)$ gives the zero mode, i.e. the eigenmode of the fluctuation equation (26.116) with eigenvalue zero, i.e.

$$\frac{d\phi_c}{d\tau} = -s_+(k)\beta(k)\gamma^2 \text{sn}[\beta(k)\tau, \gamma] \, \text{cn}[\beta(k)\tau, \gamma], \quad (26.117)$$

which is seen to be an odd function of τ, and hence is indicative of a ground state with negative eigenvalue. We can insert this into Eq. (26.113) and obtain $S(\phi_c)$. Integrating this expression from $\tau_i = -T$ to $\tau_f = T$, where $T = K(\gamma)/\beta(k)$, with the help of Tables of Integrals* we obtain

$$S(\phi_c) = W(\phi(\tau_f), \phi(\tau_i), E_{cl}) + E_{cl}(\tau_f - \tau_i), \qquad (26.118)$$

where with $\phi(\tau_i) \to \tilde{a}, \phi(\tau_f) \to \tilde{a}$ (see Fig. 26.3)

$$W(\phi(\tau_f), \phi(\tau_i), E_{cl})$$

$$\to \quad \gamma^4 s_+^2(k)\beta(k) \int_{-K(\gamma)}^{K(\gamma)} du\, \mathrm{sn}^2[u,\gamma]\mathrm{cn}^2[u,\gamma]$$

$$= \quad \frac{2}{3} s_+^2(k)\beta(k)[(2 - \gamma^2)E(K(\gamma)) - 2\gamma'^2 K(\gamma)]. \qquad (26.119)$$

Here $E(u)$ is again the complete elliptic integral of the second kind.

26.4.4   The single periodic bounce contribution

We proceed as in the previous cases. We perform the transformation (26.17) from the fluctuation $\chi(\tau)$ to a function $y(\tau)$ and use formulas (26.19) (or (26.79)) along with Eqs. (26.20a), (26.20b). Apart from the specific expressions of the present case, a main difference is that in the present case the zero mode $N(\tau) = \dot{\phi}_c$ given by Eq. (26.117) is an odd function of τ. Since eventually we want to take the limit $\tau_f, -\tau_i \to T$ with $\tau_f = -\tau_i$, we now have $\dot{\phi}_i = -\dot{\phi}_f$. Thus we have

$$K(\phi_f, \tau_f; \phi_i, \tau_i) = I_0 \exp[-S(\phi_c)/\hbar] \qquad (26.120)$$

where (cf. Eq. (26.19))

$$I_0 = \frac{1}{\sqrt{2\pi}}\left[N(\phi_f)N(\phi_i)\right]^{-1/2}\left[\int_{\tau_i}^{\tau_f} \frac{d\tau}{N^2(\tau)}\right]^{-1/2} \qquad (26.121)$$

and (cf. Eqs. (26.20a), (26.20b))

$$\frac{\partial^2 S(\phi_f, \phi_i; \tau_f - \tau_i)}{\partial \phi_f^2} - \frac{\dot{N}(\tau_f)}{N(\tau_f)} \simeq 2\pi I_0^2. \qquad (26.122)$$

For the evaluation of the single bounce transition amplitude we require again the wave functions of the endpoint configurations. As before we use

---

*See P. F. Byrd and M. D. Friedman [40], formula 361.01, p. 212, i.e. the following integral with elliptic modulus k:

$$\int du\, \mathrm{sn}^2 u\, \mathrm{cn}^2 u = \frac{1}{3k^4}[(2 - k^2)E(u) - 2k'^2 u - k^2 \mathrm{sn}\, u\, \mathrm{cn}\, u\, \mathrm{dn}\, u].$$

WKB approximations, i.e. (the integer n referring to the oscillator state which the wave function is attached to inside the well)

$$\psi_n(\phi_i) \equiv \langle \phi_i | \psi_n \rangle = \frac{C}{\sqrt{N(\phi_i)}} \exp[-\Omega(\phi_i)], \quad \Omega(\phi_i) = \int_{\tilde{a}=\phi_c(-T)}^{\phi_i} \dot{\phi} d\phi,$$

$$\psi_n(\phi_f) \equiv \langle \phi_f | \psi_n \rangle = \frac{C}{\sqrt{N(\phi_f)}} \exp[-\Omega(\phi_f)], \quad \Omega(\phi_f) = \int_{\phi_f}^{\tilde{a}=\phi_c(T)} \dot{\phi} d\phi.$$

$$(26.123)$$

The WKB normalization constant is again given by

$$C = \left[2 \int_{-\tilde{a}}^{\tilde{a}} \frac{d\phi}{\sqrt{2|E - V(\phi)|}} \right]^{-1/2} \tag{26.124}$$

with the integral extending from turning point to turning point across the non-tunneling domain. Evaluating C as in Example 26.6 one obtains

$$C = \left[\frac{\mu(1+k)}{4\sqrt{1+k^2} K(\gamma')} \right]^{1/2}. \tag{26.125}$$

Next we expand the action $S(\phi_c) \equiv S(\phi_f, \phi_i; \tau_f - \tau_i)$ contained in (26.122) in powers of $(\phi_f - \phi_c(T))$ up to the second order for the Gaussian one-loop approximation. Thus we have

$$\begin{aligned} S(\phi_c) &\equiv S(\phi_f, \phi_i; \tau_f - \tau_i) \\ &= S(\phi_c(T), \phi_i; \tau_f - \tau_i) + \frac{1}{2} \frac{\partial^2 S}{\partial \phi_f^2} \bigg|_{\phi_f = \phi_c(T)} (\phi_f - \phi_c(T))^2 + \cdots . \end{aligned}$$

$$(26.126)$$

We also expand analogously $\Omega(\phi_f)$ about $\phi_c(T)$. Thus (observe that just beyond $\tau = -T$, the velocity $\dot{\phi}_c$ — which is the gradient of the trajectory in Fig. 26.3 — is positive but in approaching $\tau = T$ it is negative)

$$\begin{aligned} \Omega(\phi_f) &= \underbrace{\Omega(\phi_c(T))}_{0} - (\phi_f - \phi_c(T)) \underbrace{[\dot{\phi}]_{\phi_c(T)}}_{0} \\ &\quad - \frac{1}{2}(\phi_f - \phi_c(T))^2 \left(\frac{\ddot{\phi}}{\dot{\phi}} \right)_{\phi_c(T)} + \cdots . \end{aligned} \tag{26.127}$$

The end-point integrations can now be carried out for the single bounce contribution $A_{+,-}^{(1)}$. Thus, combining Eqs. (26.126) and (26.127) by using Eq. (26.20b), i.e. with

$$\frac{\partial^2 S}{\partial \phi^2(\tau_f)} - \frac{\dot{N}(\tau_f)}{N(\tau_f)} \simeq 2\pi I_0^2,$$

we obtain (with mostly $\hbar = 1$) for

$$A^{(1)}_{+,-} = \lim_{\tau_f, -\tau_i \to T} \int d\phi_i \int d\phi_f \langle \psi_n | \phi_f \rangle \langle \phi_i | \psi_n \rangle I_0 \exp[-S(\phi_c)/\hbar]$$

the expression

$$A^{(1)}_{+,-} \simeq \lim_{\tau_f, -\tau_i \to T} C^2 \int d\phi_i \exp\left[-\int^{\phi_i}_{\tilde{a}=\phi_c(-T)} \dot{\phi} d\phi\right] \int \frac{d\phi_f}{\sqrt{\dot{\phi}_i \dot{\phi}_f}} I_0$$

$$\times \exp[-S(\phi(T), \phi_i; \tau_f - \tau_i)] \exp\left[-\frac{2\pi I_0^2}{2}(\phi_f - \phi_c(T))^2\right]$$

$$\simeq \lim_{\tau_f, -\tau_i \to T} C^2 \int d\phi_i \exp\left[-\int^{\phi_i}_{\tilde{a}=\phi_c(-T)} \dot{\phi} d\phi\right] \int \frac{d\phi_f}{\sqrt{-\dot{\phi}_i^2}} I_0$$

$$\times \exp[-S(\phi(T), \phi_i; \tau_f - \tau_i)] \exp\left[-\frac{2\pi I_0^2}{2}(\phi_f - \phi_c(T))^2\right].$$

$$(26.128)$$

Performing the integration with respect to ϕ_f, and then replacing $\dot{\phi}_i$ by $d\phi_i/d\tau$ this expression becomes (observe that i originates from $\dot{\phi}_c$ being odd):

$$A^{(1)}_{+,-} = \lim_{\tau_f, -\tau_i \to T} C^2 \int^{\tau_f}_{\tau_i} \frac{d\phi_i}{\sqrt{-1}\dot{\phi}_i} e^{-S(\phi(T), \phi_i; \tau_f - \tau_i)}$$

$$= -i2TC^2 e^{-W} e^{-2E_{cl}T}. \qquad (26.129)$$

Inserting the expression (26.125) for C this becomes

$$A^{(1)}_{+,-} = (-i)2T e^{-W} e^{-2E_{cl}T} \frac{\mu(1+k)}{4\sqrt{1+k^2}K(\gamma')}. \qquad (26.130)$$

26.4.5 Summing over the infinite number of bounces

The path integral implies a sum over all possible paths. The single bounce contribution to the transition amplitude $A_{+,-}$ is that of the classical configuration (26.102) with period $2nK(\gamma)/\beta(k)$ (cf. Eq. (26.103)) for $n = 1$. For $n = 2$ we have two bounces in the interval $[-T, T]$, as we discussed in detail after Eq. (26.103) and may be seen from Fig. 26.3. The contribution $A^{(2)}_{+,-}$ to the transition amplitude arising from two bounces can be calculated in analogy to the contribution of an instanton–anti-instanton pair to the leading instanton contribution as discussed previously. Thus

$$
\begin{aligned}
A^{(2)}_{+,-} &= \int_{-T}^{T} d\tau \int_{-T}^{\tau} d\tau_1 e^{-2W} e^{-2E_{\rm cl}T} \left(\frac{-i\mu(1+k)}{4\sqrt{1+k^2}K(\gamma')} \right)^2 \\
&= (-i)^2 \frac{(2T)^2}{2!} \left[\frac{\mu(1+k)}{4\sqrt{1+k^2}K(\gamma')} \right]^2 e^{-2W} e^{-2E_{\rm cl}T}. \quad (26.131)
\end{aligned}
$$

The generalization to n bounces is now straightforward. Thus their contribution to the amplitude is

$$
A^{(n)}_{+,-} = (-i)^n \frac{(2T)^n}{n!} \left[\frac{\mu(1+k)}{4\sqrt{1+k^2}K(\gamma')} \right]^n e^{-nW} e^{-2E_{\rm cl}T}. \quad (26.132)
$$

The total transition amplitude $A_{+,-}$ resulting from quantum tunneling dominated by bounces is therefore obtained by summing over n and including the leading no-bounce term:

$$
A_{+,-} = \sum_{n=0}^{\infty} A^{(n)}_{+,-} = e^{-2E_{\rm cl}T} \exp\left[-i2T \frac{\mu(1+k)}{4\sqrt{1+k^2}K(\gamma')} e^{-W} \right]. \quad (26.133)
$$

This expression has to be compared with Eq. (26.106b), i.e. with

$$
A_{+,-} \simeq e^{-2(E_{\rm cl}+i\Im E)T},
$$

for the determination of the imaginary part of the energy, i.e.

$$
\Im E = \frac{\mu(1+k)}{4\sqrt{1+k^2}K(\gamma')} e^{-W}, \quad (26.134a)
$$

or

$$
\Im E \equiv C^2 e^{-W}. \quad (26.134b)
$$

Here the quantity W is given by Eq. (26.119), i.e.[†]

$$
W = \frac{2}{3} \frac{\mu}{a} \left\{ \frac{a(1+k)}{\sqrt{1+k^2}} \right\}^3 [(2-\gamma^2)E(K(\gamma)) - 2\gamma'^2 K(\gamma)], \quad (26.135)
$$

where $(1+k)/\sqrt{1+k^2} = \sqrt{2}/\sqrt{1+\gamma'^2}$. We thus have the very simple formula (26.134b) for the result.

[†]To avoid confusion we emphasize that P. F. Byrd and M. D. Friedman [40] write (cf. pp. 8 and 10):

$$
E(u_1) = \int_0^{u_1} du\, {\rm dn}^2 u \quad \text{and} \quad \int_0^{K(k)} du\, {\rm dn}^2 u = E(k),
$$

so that $E(u_1 = K(k)) = E(k)$.

Next we consider the case of low energies. This is the region in which $E_{\text{cl}} \ll a^2\mu^2/2$, i.e. small, which means that we consider low lying levels in the well, i.e. (cf. comment at the beginning of Sec. 26.4.1)

$$E_{\text{cl}} = \left(n + \frac{1}{2}\right)\mu\sqrt{2}, \quad n = 0, 1, 2, \ldots \quad . \tag{26.136}$$

Small values of E_{cl} imply (cf. (26.101), (26.102)) $k^2 \to 1$ or $\gamma^2 \to 1$ or correspondingly $\gamma'^2 \to 0$. The expansions of the elliptic integrals $E(\gamma)$ and $K(\gamma)$ in this case are those of Eqs. (18.192) and (18.193), i.e.

$$E(\gamma) = 1 + \frac{1}{2}\left[\ln\left(\frac{4}{\gamma'}\right) - \frac{1}{2}\right]\gamma'^2 + \frac{3}{16}\left[\ln\left(\frac{4}{\gamma'}\right) - \frac{13}{12}\right]\gamma'^4 + \cdots \tag{26.137}$$

and

$$K(\gamma) = \ln\left(\frac{4}{\gamma'}\right) + \frac{1}{4}\left[\ln\left(\frac{4}{\gamma'}\right) - 1\right]\gamma'^2 + \cdots \quad . \tag{26.138}$$

Inserting these expressions into Eq. (26.135) we obtain

$$W = 2^{5/2}\frac{\mu a^2}{3}\left[1 - \frac{3}{4}\gamma'^2 - \frac{3}{2}\gamma'^2\ln\left(\frac{4}{\gamma'}\right) + \cdots\right]. \tag{26.139}$$

Defining the dimensionless parameter

$$g^2 \equiv \frac{1}{\mu a^2},$$

and using the parameter u introduced in Eq. (26.101), i.e.

$$\gamma^2 = \frac{16k^2}{(1+k)^4} = \frac{(1-u^2)}{u^4}(\sqrt{1+u} - \sqrt{1-u})^4 \simeq 1 - u^2, \quad \gamma'^2 \simeq \frac{u^2}{4},$$

we can rewrite the expression for W as the expansion

$$W = \frac{4\sqrt{2}}{3g^2}\left[1 - \frac{3}{16}u^2 - \frac{3}{8}u^2\ln\left(\frac{8}{u}\right) + \cdots\right]. \tag{26.140}$$

Since $E_{\text{cl}} = a^2\mu^2 u^2/2$, we obtain from Eq. (26.136) the relation

$$u^2 = 2\sqrt{2}\left(n + \frac{1}{2}\right)g^2. \tag{26.141}$$

Inserting this into (26.140) we obtain

$$W \simeq \frac{4\sqrt{2}}{3g^2} - \left(n + \frac{1}{2}\right) - \left(n + \frac{1}{2}\right)\ln\left[\frac{2^{9/2}}{g^2\left(n + \frac{1}{2}\right)}\right]. \tag{26.142}$$

Now inserting this into (26.134a) with (for $k \to 1$) $K(\gamma') \simeq K(0) = \pi/2$, the imaginary part of the energy given by Eq. (26.134b) becomes

$$\Im E = \frac{\mu}{\pi\sqrt{2}} e^{-4\sqrt{2}/3g^2} e^{n+\frac{1}{2}} \left[\frac{16\sqrt{2}}{g^2(n+\frac{1}{2})} \right]^{n+\frac{1}{2}}. \qquad (26.143)$$

Using the *Stirling approximation* again in the form of the Furry factor f_n set equal to 1 (cf. Eq. (18.178)), i.e.

$$\left(\frac{e}{n+\frac{1}{2}} \right)^{n+\frac{1}{2}} \simeq \frac{\sqrt{2\pi}}{n!},$$

we can rewrite the expression as

$$\Im E = \frac{\mu}{\sqrt{\pi}n!} \left[\frac{16\sqrt{2}}{g^2} \right]^{n+\frac{1}{2}} e^{-4\sqrt{2}/3g^2}. \qquad (26.144)$$

This result[‡] agrees with the complex energy eigenvalue of the Schrödinger equation for the inverted double well potential obtained by Bender and Wu [18], [19] and with that of the systematic perturbation method[§] as explained in Chapter 18.[¶] As in Chapter 24 for the ground state, one can now perform a similar calculation for excited states in the case of the cubic potential and one can compare the results, and also with that of the WKB method, i.e. Eq. (14.105); this has not yet been done.

Example 26.6: Evaluation of the WKB normalization constant
Verify the result of Eq. (26.125).

Solution: We have to evaluate

$$C = \left[2 \int_{-\tilde{a}}^{\tilde{a}} \frac{d\phi}{\sqrt{2|E_{\mathrm{cl}} - V(\phi)|}} \right]^{-1/2}.$$

[‡] J.–Q. Liang and H. J. W. Müller–Kirsten [169].
[§] P. Achuthan, H. J. W. Müller–Kirsten and A. Wiedemann [4].
[¶] For the comparison with the result (18.86) of the Schrödinger equation method we first have to identify the parameters of the potentials (18.1) and (26.99). The comparison implies, with index "1" indicating mass $m_0 = 1$:

$$\mu = \frac{h_1^2}{2}, \quad a = \frac{h_1^2}{2c_1}, \quad \frac{1}{g^2} = \frac{h_1^2}{8c_1^2}.$$

Conversion from $m_0 = 1$ to $m_0 = 1/2$ (as in Chapter 18) implies

$$h_1^2 \to \frac{1}{\sqrt{2}}h^2, \quad c_1^2 \to \frac{1}{2}c^2, \quad E_{(1)} \to \frac{1}{2}E.$$

Then this result $E/2$ is obtained, i.e.

$$E_{(1)} = \frac{\mu}{\sqrt{\pi}n!} \left[\frac{16\sqrt{2}}{g^2} \right]^{n+\frac{1}{2}} e^{-4\sqrt{2}/3g^2} \longrightarrow \frac{h^2}{2\sqrt{2\pi}n!} \left(\frac{2h^6}{c^2} \right)^{n+\frac{1}{2}} e^{-h^6/6c^2} = \frac{1}{2}E \text{ of (18.86)}.$$

We can use the integrand of the integral evaluated in Example 26.5. Hence (note the modulus of the quantity under the square root)

$$
I_C \equiv \int_{-\tilde{a}}^{\tilde{a}} \frac{d\phi}{\sqrt{2|E_{\text{cl}} - V(\phi)|}} = \int_{-\tilde{a}}^{\tilde{a}} \frac{d\phi}{\sqrt{\frac{\mu^2}{a^2}\{s_+^2(k) - \phi^2\}\{\tilde{a}^2 - \phi^2\}}}
$$

$$
= \frac{2a}{\mu} \int_0^{\tilde{a}} \frac{d\phi}{\sqrt{\{s_+^2(k) - \phi^2\}\{\tilde{a}^2 - \phi^2\}}}.
$$

With the help of Tables of Integrals[||] this integral is

$$
I_c = \frac{2a}{\mu} \frac{1}{s_+(k)} \underbrace{\mathrm{sn}^{-1}[1, \gamma']}_{K(\gamma')}, \quad \text{where} \quad \gamma' = \frac{\tilde{a}}{s_+(k)}.
$$

Hence

$$
C = \left[\frac{4}{\mu} \frac{\sqrt{1 + k^2}}{(1 + k)} K(\gamma') \right]^{-1/2}.
$$

26.5   Concluding Remarks

We saw in the above that with some liberal use of the Stirling approximation (in the sense of the Furry factor set equal to 1), we obtained in each case agreement between the path integral result and that of the systematic perturbation method of matched asymptotic expansions. One can also see, that the path integral method is much more complicated. Of the three methods, WKB, perturbation and path integral, the perturbation method seems the best way to derive the results, since it is simpler, does not introduce unnatural factors like e and n^n, and does not require a separate treatment of ground states and excited states. The case of the cosine potential is the cleanest and least problematic, since in its case the Stirling approximation is never used and the parameter symmetries inherent in the original differential equation are fully exploited (it is this symmetry which is effectively restored in the path integral method by use of the Furry factor). This power of perturbation theory is not widely appreciated. Thus — as e.g. in the case of Gildener and Patrascioiu [110] — a path integral result is usually compared with that of a WKB calculation.

We mention finally that recently the semiclassical calculation of the energy levels of theories like that of the sine–Gordon model has been investigated under more complicated boundary conditions. For details we refer to a recent reference from which related investigations can be traced back.[**]

[||] Here we use L. M. Milne–Thomson [196], p. 26.

[**] See e.g. G. Mussardo, V. Riva and G. Sotkov [217].

Chapter 27

Quantization of Systems with Constraints

27.1 Introductory Remarks

In our treatment of path integrals in the preceding chapters, with the variables considered in the neighbourhood of (or, as one says, in the background of) a classical configuration with finite action or energy (depending on whether time is Euclidean or Minkowskian respectively), we required a transformation of the original (field) variables to collective and fluctuation variables, and thus to a larger number of degrees of freedom, which has the immediate consequence of the appearance of constraints. Since instantons, solitons, bounces and other such topological or nontopological classical configurations play an important role in many branches of theoretical physics, the quantization of theories with constraints is an important extension of basic quantum mechanics. Quantization around a classical configuration, i.e. around a solution of the Euler–Lagrange equation, implies also a perturbation expansion in its neighbourhood and specifically — if the path integral method is employed — the loop expansion as in the foregoing chapters. One is therefore confronted with the problem of developing such a perturbation theory for a system with constraints. This is a complicated task.[*] The fundamental steps to achieve this were discovered by Dirac [76], and have since then been developed further and extended into an individual and highly complicated direction of research.[†]

---

[*]Standard monographs, including introductions to the subject, are the books of K. Sundermeyer [265] and J. Govaerts [120].

[†]This has been pursued particularly by Russian researchers. See e.g. D. M. Gitman and I. V. Tyutin [112].

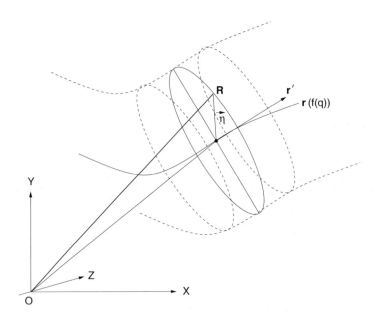

Fig. 27.1 A quantum mechanical particle near a classical path.

Frequently, there is a considerable difference between what can be done in principle and what can be done in practice. In the following we consider therefore in detail the simple but typical and very instructive example of the *rotor* to illustrate the basic problems which arise in the quantization of a system with constraints. The most straightforward example to provide a bridge between the treatment of this case and that of a theory with soliton-like configurations is that of a particle constrained to move near a classical (i.e. Newtonian) path, say $\mathbf{r}(t)$.[‡] In incorporating this path into the quantum theory of a variable $\mathbf{R}$ one sets[§]

$$\mathbf{R} = \mathbf{r}(f(q)) + \sum_a \mathbf{n}_a(f(q))\eta_a, \tag{27.1}$$

where time t has been replaced by some function $f(q)$ of a variable q which is the *collective coordinate* and parametrizes the classical path. The unit vectors $\mathbf{n}_a(f)$ together with the tangential vector $d\mathbf{r}/df$ form a moving reference frame at the point $\mathbf{r}(f)$ as depicted in Fig. 27.1. The components η_a are then the *fluctuation variables*. The substitution (27.1) corresponds exactly to the substitutions (23.36), (24.58) and (26.10) in our earlier instanton

---

[‡]This has been worked out in detail in Jian-zu Zhang and H. J. W. Müller–Kirsten [288].

[§]A very interesting Schrödinger-like theory with WKB wave functions for such systems with many degrees of freedom in a unified view of pseudoparticles and solitons has been developed by J.–L. Gervais and B. Sakita [109]. The associated equation of small fluctuations is derived in H. J. W. Müller–Kirsten and A. Wiedemann [213].

considerations. We also see a constraint coming in as the condition that the fluctuation has to be perpendicular to the classical orbit, i.e. $\mathbf{n}_a \cdot d\mathbf{r}/df = 0$, which eliminates a degree of freedom (i.e. the component of the fluctuation along the classical orbit). In our earlier considerations this is, for instance, the condition (23.81).

Since all our previous considerations of theories which involve an expansion around a classical configuration and hence to a larger number of degrees of freedom lead to constraints, our intention here is to present an introduction into this topic which in view of its extent necessarily has to be brief. There are really two basic approaches of handling constraints. In one approach, the usual one involving Dirac quantization, one proceeds to an even larger number of degrees of freedom. In the other, more recently suggested approach of Faddeev and Jackiw,[¶] the idea is to use the constraints in order to eliminate superfluous degrees of freedom. We illustrate both approaches here by application to elementary examples. The problem we want to consider can also be seen as follows. We saw that the commutators of canonical quantization satisfy the same algebra as Poisson brackets, and form a *symplectic matrix* with determinant 1, i.e. in the case of the latter

$$
\begin{pmatrix} \{q_i, q_i\} & \{q_i, p_j\} \\ \{p_j, q_i\} & \{p_j, p_j\} \end{pmatrix} = \begin{pmatrix} 0 & 1 \\ -1 & 0 \end{pmatrix}.
$$

The question is therefore: What are the obstacles in arriving at such a relation in the presence of constraints, and how can these be overcome?

27.2 Constraints: How they arise

Constraints are wellknown from classical mechanics. Easy examples which come to mind are those of particles compelled to move along a specific path, like for instance a particle of mass m_0 with coordinate $\mathbf{r}$ forced to move along a frictionless wire of parabolic shape given by

$$
f(x, y, z) = z - ar^2 = 0, \quad r = \sqrt{x^2 + y^2}, \tag{27.2a}
$$

with some initial velocity. The straight-forward application of Newton's equation of motion to such a case (i.e. without the use of constraints) is clearly inappropriate, since it is very difficult to specify the functional form of the potential which forces the particle to move along the specified path. Thus the condition or *constraint* (27.2a) is enforced by restricting the possible values of the particle's coordinates x, y, z to those of the path. If the wire

---

[¶]L. Faddeev and R. Jackiw [90]. See also R. Jackiw [142] and J. Goevaerts [121].

is placed in a gravitational field with potential $V = m_0 g z$, we would have to solve the problem of the simultaneous equations

$$m_0 \ddot{\mathbf{r}} = -\boldsymbol{\nabla} V(\mathbf{r}), \quad f(\mathbf{r}) = z - ar^2 = 0. \tag{27.2b}$$

Since the particle is to move along the wire for all time under consideration, the constraint must also hold for all times. Thus we must also have

$$\frac{d}{dt} f(\mathbf{r}) = 0,$$

and maybe higher derivatives. If the excentricity of the parabola is allowed to vary with time, i.e. $a = a(t)$, the constraint becomes explicitly time dependent, and one can see the problem becomes more complicated. In the simple case of Eq. (27.2a) we can find an easy way out by simply using the constraint to express z in terms of x and y and $\dot{z}$ in terms of $x, y, \dot{x}, \dot{y}$, and we can try to integrate Newton's equation. Thus we use the constraint to eliminate one degree of freedom, namely z. In fact, the constraint expresses the dependence of degrees of freedom.

If we go to the even simpler case of a particle constrained to move on a circle of radius a in the (x, y)-plane, we have the constraint

$$f(x, y) = r - a = 0. \tag{27.3}$$

Of course, we could write $r = \sqrt{x^2 + y^2}$ and use the constraint to eliminate one degree of freedom, say y. But it is clearly much more appropriate to exploit the symmetry of the problem and to go to polar coordinates r and θ, and then to eliminate one of these variables, i.e. r. We recall briefly how constraints are handled in classical mechanics and consider the slightly more complicated case of a mass point m_0 constrained to move freely on a sphere $\mathbb{S}^2$ of radius 1. In classical mechanics one adds to the Lagrangian L the constraint multiplied by a Lagrange multiplier λ. Thus one sets

$$L(x_i, \dot{x}_i, \lambda, \dot{\lambda}) = \frac{1}{2} m_0 \dot{x}_i^2 + \lambda(x_i^2 - 1) \tag{27.4}$$

(summation over repeated indices being understood) and treats λ as a new time-dependent degree of freedom (i.e. as a dynamical variable). The Euler–Lagrange equations for x_i and λ are seen to be

$$m_0 \ddot{x}_i - 2\lambda x_i = 0, \quad x_i^2 - 1 = 0. \tag{27.5}$$

Since the second derivative of x_i appears in the first of Eqs. (27.5), we might as well try to differentiate the second equation twice. Then

$$\mathbf{x} \cdot \dot{\mathbf{x}} = 0, \quad \dot{\mathbf{x}}^2 + \mathbf{x} \cdot \ddot{\mathbf{x}} = 0. \tag{27.6}$$

Multiplying the first of Eqs. (27.5) by x_i, we obtain

$$m_0 \mathbf{x} \cdot \ddot{\mathbf{x}} - 2\lambda \mathbf{x}^2 = 0, \quad \text{or with} \quad (27.6), (27.5) \quad m_0 \dot{\mathbf{x}}^2 + 2\lambda = 0. \quad (27.7)$$

Thus we obtain $\lambda = -m_0 \dot{\mathbf{x}}^2/2$. Having determined λ, we can insert it into the first of Eqs. (27.5) and obtain

$$m_0 \ddot{\mathbf{x}} - m_0 \dot{\mathbf{x}}^2 \mathbf{x} = 0. \quad (27.8)$$

What we have achieved is that we see from the second term of this equation that the actual potential is velocity dependent. The next part of the problem would be to solve the final equation (27.8), except that this is of little interest to us here. Instead we recall another important type of example in which constraints arise: Maxwell's electrodynamics.

The Lagrangian L of the free electromagnetic field is given by

$$L = \int d\mathbf{x} \mathcal{L}, \quad \mathcal{L}(A_\mu, \partial_\mu A_\nu) = -\frac{1}{4} F^{\mu\nu} F_{\mu\nu}, \quad F_{\mu\nu} = \partial_\mu A_\nu - \partial_\nu A_\mu. \quad (27.9)$$

The Euler–Lagrange equations are given by

$$\partial_\mu \left(\frac{\partial \mathcal{L}}{\partial(\partial_\mu A_\nu)} \right) - \frac{\partial \mathcal{L}}{\partial A_\nu} = 0, \quad \text{i.e.} \quad \partial_\mu F^{\mu\nu} = 0, \quad (27.10)$$

and may be subdivided into *equations of motion* characterized by time derivatives,

$$\partial_0 F^{0i} = -\partial_j F^{ji}, \quad (27.11)$$

and the *constraint* known as *Gauss' law*

$$\partial_k F^{k0} = 0. \quad (27.12)$$

We obtain the Gauss law only by retaining A_0 in the Lagrangian (i.e. by not starting with the gauge fixing condition $A_0 = 0$). Gauge fixing, its significance and implications play an important role in the following. The constraint (27.12) indicates, of course, that we have one or more superfluous degrees of freedom, i.e. components A_μ. Thus electrodynamics is a theory with constraints. In general one describes a gauge theory as one with a gauge field like A_μ, and one says, a theory like that of the mass point on a sphere is "gauged" by the so-called "minimal coupling prescription" of replacing the conjugate momentum

$$p_i = \frac{\partial L}{\partial \dot{x}_i}$$

by the shifted quantity $p_i - eA_i$ (e the charge, $\hbar = c = 1$), or a derivative $\partial/\partial x_i$ by the *covariant derivative*

$$D^i = \frac{\partial}{\partial x_i} + ieA^i. \tag{27.13}$$

A gauge theory is therefore one which obeys an invariance called gauge invariance. From the application of Noether's theorem to electrodynamics we know that the invariance implies the conservation of electric charge. In the case of the (academic) free electromagnetic field in space we have no charge, i.e. it is zero. This is expressed by the right hand side of the Gauss law (27.12). Thus this law or the constraint is intimately related to a gauge transformation. Thus if we have a gauge theory, we have a theory with constraints. We can ask the opposite: Is a theory with constraints also a gauge theory? We shall see that the answer is yes: Given a theory with constraints we can construct a gauge transformation — the constraints themselves appearing as the generators of this gauge transformation.

27.2.1   Singular Lagrangians

We start from a classical Lagrangian L defined for an N-dimensional configuration space with generalized coordinates q_i, i.e.

$$L = L(q_i, \dot{q}_i), \quad i, j, k \ldots = 1, \ldots, N.$$

The canonical momenta are defined by

$$p_i(t) := \frac{\partial L(q_k, \dot{q}_k)}{\partial \dot{q}_i}. \tag{27.14}$$

The transition to Hamilton's formalism is achieved with the help of a Legendre transform which implies the passage

$$q_i(t), \dot{q}_i(t) \longrightarrow q_i(t), p_i(t). \tag{27.15}$$

Equation (27.14) allows us to write

$$(M_{ij}) := \frac{\partial p_i(t)}{\partial \dot{q}_j} = \frac{\partial^2 L(q_k, \dot{q}_k)}{\partial \dot{q}_j \partial \dot{q}_i}, \tag{27.16}$$

and we can look at this $N \times N$ matrix as the Jacobian of the transformation. If this matrix is nonsingular, i.e.

$$\det(M_{ij}) \neq 0, \quad (M_{ij}) = \left(\frac{\partial^2 L(q_k, \dot{q}_k)}{\partial \dot{q}_j \partial \dot{q}_i} \right), \tag{27.17}$$

the transformation is unique, i.e. the N canonical momenta p_i are independent and the usual method of canonical quantization can be carried out. We can see the appearance of the matrix (M_{ij}) also in a set of first order equations. Looking at the Euler–Lagrange equations

$$\frac{\partial L}{\partial q_i} - \frac{d}{dt}\left(\frac{\partial L(q_k, \dot{q}_k)}{\partial \dot{q}_i}\right) = 0,$$

and performing the differentiation d/dt, we can write the equations

$$M_{ij}\ddot{q}_j = K_i(q_k, \dot{q}_k), \qquad K_i := \frac{\partial L}{\partial q_i} - \frac{\partial^2 L}{\partial \dot{q}_i \partial q_j}\dot{q}_j. \tag{27.18}$$

Setting $\dot{q}_j \equiv v_j$, we can write this equation

$$M_{ij}\dot{v}_j = K_i(q_k, v_k), \qquad v_k = \dot{q}_k. \tag{27.19}$$

Thus one obtains a set of N first order differential equations which can be solved if $\det(M_{ij}) \neq 0$. In terms of v_j the momentum becomes

$$p_i = \frac{\partial L(q_k, v_k)}{\partial v_i}. \tag{27.20}$$

Equation (27.19) together with Eq. (27.20) are sometimes called the *extended system of Euler–Lagrange equations*. Varying the momentum (27.20), we obtain

$$\delta p_i = M_{ij}\delta v_j + \frac{\partial^2 L}{\partial q_j \partial v_i}\delta q_j. \tag{27.21}$$

Thus if (M_{ij}) is singular, i.e. we cannot form its inverse, we cannot express δv_i in terms of $\delta q_j, \delta p_i$. Thus if the determinant (27.17) is zero, some momenta depend on others, and we have the case of a *singular Lagrangian*. Let R be the rank of the Jacobian (27.16) with $R < N$. Then

$$p_i = \frac{\partial L}{\partial \dot{q}_i} = f(q_j, \dot{q}_j), \tag{27.22}$$

where $i, j = 1, \ldots, N$. From the theory of matrices[||] it is known that if the $N \times N$ matrix (M_{ij}) has rank R, $R < N$, it can be cast into a form with diagonal submatrices of dimensionality R and $N - R$. Then

$$(M_{ab}) := \left(\frac{\partial^2 L}{\partial \dot{q}_a \partial \dot{q}_b}\right), \qquad a, b = 1, \ldots, R, \tag{27.23}$$

[||]See e.g. A. C. Aitken [7], pp. 66 - 67.

has $\det(M_{ab}) \neq 0$, and the $N - R$ dimensional submatrix $(M_{\alpha\beta}), \alpha, \beta = R+1, \ldots, N$, has $\det(M_{\alpha\beta}) = 0$. Now,

$$p_a = \frac{\partial L}{\partial \dot{q}_a}, \qquad \frac{\partial p_a}{\partial \dot{q}_b} = \frac{\partial^2 L(q_j, \dot{q}_j)}{\partial \dot{q}_b \partial \dot{q}_a} = M_{ba}(q_j, \dot{q}_j), \qquad (27.24)$$

and hence, different from p_i, we can have here

$$p_a = \int^{\dot{q}_b} d\dot{q}_b M_{ba}(q_j, \dot{q}_j) \equiv p_a(q_j, \dot{q}_b, \dot{q}_\alpha). \qquad (27.25)$$

Solving this equation for $\dot{q}_b$, we obtain a function of the form

$$\dot{q}_b = f_b(q_j, p_a, \dot{q}_\alpha), \qquad a, b = 1, \ldots, R, \ \alpha = R+1, \ldots, N. \qquad (27.26)$$

We can write therefore:

$$p_j = \frac{\partial L(q_j, \dot{q}_j)}{\partial \dot{q}_j} = \frac{\partial L(q_j, f_b(q_j, p_b, \dot{q}_\alpha), \dot{q}_\alpha)}{\partial \dot{q}_j} \equiv g_j(q_j, p_b, \dot{q}_\alpha). \qquad (27.27)$$

For $a = 1, \ldots, R$ and $\alpha = R+1, \ldots, N$, we have therefore

$$p_a = g_a(q_j, p_b, \dot{q}_\alpha), \quad p_\alpha = g_\alpha(q_j, p_b, \dot{q}_\alpha). \qquad (27.28)$$

But for $\alpha = R+1, \ldots, N$, we have with the following equation $\det(M_{\alpha\beta}) = 0$, i.e. that p_α is independent of $\dot{q}_\beta$:

$$\left| \frac{\partial p_\alpha}{\partial \dot{q}_\beta} \right| = \left| \frac{\partial^2 L(q, \dot{q}_a, \dot{q}_\alpha)}{\partial \dot{q}_\beta \partial \dot{q}_\alpha} \right| = 0, \quad \text{i.e.} \ \ p_\alpha = g_\alpha(q_j, p_b). \qquad (27.29)$$

Hence p_α is independent of $\dot{q}_\beta$, and we have the set of constraints called *primary constraints*:

$$\phi_\alpha := p_\alpha - g_\alpha(q_j, p_b) = 0. \qquad (27.30)$$

We observe that these primary constraints involve momenta, and therefore are not those like that of Eq. (27.2a).

Example 27.1: The Lagrangian of electrodynamics is singular
Show that the Lagrangian density of electrodynamics is singular.

Solution: The Lagrangian density of the free electromagnetic field is given by

$$\mathcal{L} = -\frac{1}{4} F_{\mu\nu} F^{\mu\nu} = -\frac{1}{4} F_{ik}^2 + \frac{1}{2} (\dot{A}_i + \partial_i A^0)^2.$$

Here

$$p_0 = \frac{\partial \mathcal{L}}{\partial \dot{A}_0} = 0, \quad p_i = \frac{\partial \mathcal{L}}{\partial \dot{A}_i} = \dot{A}_i + \partial_i A^0.$$

We see that $\partial p_0/\partial \dot{A}_\nu = 0$, i.e. one complete row or column is zero. Thus $(M_{\mu\nu})$ is singular.

Example 27.2: Singular Lagrangian of a particle moving along a circle

Show that the Lagrangian of a particle of mass m_0 which is constrained to move along a circle is singular.

Solution: In polar coordinates $(q_i = r, \theta, \lambda)$ the Lagrangian of the particle with the constraint inserted (multiplied by a Lagrange multiplier λ) is

$$L(r, \theta, \lambda; \dot{r}, \dot{\theta}, \dot{\lambda}) = \frac{1}{2}m_0(r\dot{\theta})^2 + \frac{1}{2}m_0(\dot{r})^2 + \lambda(r - a),$$

which is seen to be singular, since

$$\left| \frac{\partial^2 L}{\partial \dot{q}_i \partial \dot{q}_j} \right| = \left| \begin{matrix} m_0 r^2 & \cdot & \cdot \\ \cdot & m_0 & \cdot \\ \cdot & \cdot & \end{matrix} \right|.$$

We now convince ourselves in the first place that a blind application of the canonical quantization procedure to a system with constraints leads to ambiguities. Supposing, for example, that we have one primary constraint $\Gamma(q, p) = 0$. In keeping with the canonical quantization procedure we associate with the operator form of this equation the *null operator* $\emptyset$, i.e.

$$\Gamma(q, p) \longrightarrow \Gamma_{\text{op}} = \emptyset_{\text{op}}.$$

This then implies correspondingly that

$$\{A(q, p), \Gamma(q, p)\}_{\text{Poisson}} \longrightarrow \frac{1}{i\hbar}[A_{\text{op}}, \Gamma_{\text{op}}] = \emptyset_{\text{op}}. \tag{27.31}$$

The right hand side of this correspondence is zero since the commutator of any operator with the null operator is zero. However, the Poisson bracket on the left is in general not zero; constraints cannot be taken to be zero until after the Poisson brackets have been evaluated — if indeed they are supposed to make sense. This is already an indication that a case with constraints, i.e. with superfluous degrees of freedom, requires some new considerations. These are the ideas which Dirac [76] developed around 1950.

27.3 The Hamiltonian of Singular Systems

Following Dirac [76] we first introduce the concept of *"weak equality"*, written $\approx$: Two functions A and B of the dynamical variables $q_i(t), p_i(t), i = 1, \ldots, N$, are said to be weakly equal, i.e.

$$A(q, p) \approx B(q, p),$$

if they are equal on the hyperplane Γ_c defined by the primary constraint Γ, i.e. if

$$[A(q,p) - B(q,p)]\Big|_{\Gamma} = 0. \tag{27.32}$$

Since quantization is performed in phase space variables p and q, we have to proceed to the Hamilton formalism with Hamiltonian $H(q,p)$ which is in the nonsingular theory defined by

$$H(q,p) = \sum_{i=1}^{N} p_i \dot{q}_i - L(q_i, \dot{q}_i) \tag{27.33}$$

with

$$\frac{\partial H}{\partial q_i} = -\frac{\partial L}{\partial q_i}, \qquad \frac{\partial H}{\partial \dot{q}_i} = p_i - \frac{\partial L}{\partial \dot{q}_i}, \qquad \frac{\partial H}{\partial p_i} = \dot{q}_i. \tag{27.34}$$

In the Lagrangian formalism energy is defined by the expression

$$E := \sum_{i=1}^{N} \frac{\partial L}{\partial \dot{q}_i} \dot{q}_i - L(q_i, \dot{q}_i), \qquad L(q_i, \dot{q}_i) = \sum_{i=1}^{N} \frac{\partial L}{\partial \dot{q}_i} \dot{q}_i - E, \tag{27.35}$$

so that (replacing in Eq. (27.33) $L(q_i, \dot{q}_i)$ by the latter expression)

$$H = E + \sum_{i=1}^{N} \left[p_i - \frac{\partial L}{\partial \dot{q}_i} \right] \dot{q}_i = E + \sum_{i=1}^{N} \frac{\partial H}{\partial \dot{q}_i} \dot{q}_i. \tag{27.36}$$

Normally, in a nonsingular theory, $\partial H / \partial \dot{q}_i = 0$, and hence $H = E$. In the case of the singular theory we use Eq. (27.36) in order to define a Hamiltonian $H^{(1)}$ which depends explicitly on the velocities $\dot{q}_\alpha$. Thus we define with $i, j = 1, \ldots, N$, $a, b = 1, \ldots, R < N$, $\alpha = R+1, \ldots, N$,

$$H^{(1)}(q_j, p_j, \dot{q}_\alpha) = \left[E + \sum_{i=1}^{N} \dot{q}_i \frac{\partial H}{\partial \dot{q}_i} \right]_{\dot{q}_b = f_b(q_j, p_a, \dot{q}_\alpha)}, \tag{27.37}$$

where with Eq. (27.34)

$$\frac{\partial H}{\partial \dot{q}_i} = p_i - \frac{\partial L}{\partial \dot{q}_i} \overset{(27.27)}{=} p_i - g_i(q_j, p_a, \dot{q}_\alpha),$$

and $\dot{q}_b$ is given by Eq. (27.26). Then

$$
\overbrace{
\begin{aligned}
H^{(1)}(q_j, p_j, \dot{q}_\alpha) \;=\; & E|_{\dot{q}_b=f_b} + \sum_{a=1}^{R} f_a(q_j, p_b, \dot{q}_\alpha)[p_a - g_a(q_j, p_b, \dot{q}_\alpha)]
\end{aligned}
}^{H_{\text{can}}}
$$

$$
+ \sum_{\alpha=R+1}^{N} \dot{q}_\alpha[p_\alpha - g_\alpha(q_j, p_a)]
$$

$$
\equiv H_{\text{can}}(q_j, p_a) + \sum_{\alpha=R+1}^{N} \dot{q}_\alpha \phi_\alpha, \tag{27.38}
$$

where H_{can} is recognized to be that part of the Hamiltonian H of Eq. (27.36) which does not involve the velocities associated with the singular part of the Lagrangian. In Example 27.3 we verify that $H \to H_{\text{can}}$ does not depend on $\dot{q}_\alpha$.

Example 27.3: Canonical Hamiltonian independent of $\dot{q}_\alpha$
Check that the canonical Hamiltonian H of Eq. (27.38) is independent of the velocities $\dot{q}_\alpha$.

Solution: We have from Eq. (27.38) with E from Eq. (27.35) and f_b from Eq. (27.26):

$$
\begin{aligned}
H_{\text{can}} = & \sum_{i=1}^{N} \frac{\partial L}{\partial \dot{q}_i}\dot{q}_i - L(q_j, f_b, \dot{q}_\alpha) + \sum_{a=1}^{R}[p_a - g_a(q_j, p_b, \dot{q}_\alpha)]f_a(q_j, p_b, \dot{q}_\alpha) \\
= & \underbrace{\sum_{a=1}^{R}\left[\frac{\partial L}{\partial \dot{q}_a}\dot{q}_a - g_a(q_j, p_b, \dot{q}_\alpha)f_a(q_j, p_b, \dot{q}_\alpha)\right]}_{0} + \sum_{\alpha=R+1}^{N}\underbrace{\frac{\partial L}{\partial \dot{q}_\alpha}\dot{q}_\alpha}_{p_\alpha = g_\alpha(q_j, p_b)} \\
& -L(q_j, f_b, \dot{q}_\alpha) + \sum_{a=1}^{R} p_a f_a(q_j, p_b, \dot{q}_\alpha) \\
= & \sum_{a=1}^{R} p_a f_a(q_j, p_b, \dot{q}_\alpha) + \sum_{\alpha=R+1}^{N} \dot{q}_\alpha g_\alpha(q_j, p_a) - L(q_j, f_b, \dot{q}_\alpha), \tag{27.39}
\end{aligned}
$$

and so (cf. Eqs. (27.26), (27.27))

$$
\frac{\partial H_{\text{can}}}{\partial \dot{q}_\alpha} = \sum_{a=1}^{R}\left[p_a - \underbrace{\frac{\partial L(q_j, f_b, \dot{q}_\alpha)}{\partial f_a}}_{p_a}\right]\frac{\partial f_a(q_j, p_b, \dot{q}_\alpha)}{\partial \dot{q}_\alpha} + \left[g_\alpha(q_j, p_a) - \underbrace{\frac{\partial L(q_j, f_b, \dot{q}_\alpha)}{\partial \dot{q}_\alpha}}_{g_\alpha}\right] = 0,
$$

i.e. $H_{\text{can}} = H(q_j, p_a)$.

We observe in particular that

$$
\frac{\partial H^{(1)}}{\partial \dot{q}_\alpha} \overset{(27.38)}{=} \phi_\alpha, \tag{27.40}
$$

where $\{\phi_\alpha\}$ are the primary constraints. Obviously the velocities $\dot{q}_\alpha$ in Eq. (27.38) play the role of Lagrange multipliers. Finally we note that Hamilton's equations expressed in terms of Poisson brackets now become

$$\dot{q} = \{q, H^{(1)}\} = \frac{\partial H^{(1)}}{\partial p}, \quad \dot{p} = \{p, H^{(1)}\} = -\frac{\partial H^{(1)}}{\partial q}, \quad \phi_\alpha = 0. \quad (27.41)$$

One calls $H \to H_{\mathrm{can}}$ the *canonical Hamiltonian* and $H^{(1)}$ the *total* or *final Hamiltonian*.

27.4 Persistence of Constraints in Course of Time

For any function $F(q, p)$ we have as a consequence of Eq. (27.41)

$$\frac{d}{dt} F(q, p) = \frac{\partial F}{\partial q} \dot{q} + \frac{\partial F}{\partial p} \dot{p} = \{F(q, p), H^{(1)}(q, p)\}. \quad (27.42)$$

To insure that the constraints ϕ_α of Eq. (27.30) hold at all times, we must demand that

$$\dot{\phi}_\alpha = \{\phi_\alpha, H^{(1)}\} \overset{!}{=} 0, \quad (27.43)$$

or with Eq. (27.38)

$$\{\phi_\alpha, H_{\mathrm{can}}\} + \sum_{\beta=R+1}^{N} \lambda_\beta \{\phi_\alpha, \phi_\beta\} = 0, \quad \alpha = R+1, \ldots, N, \quad (27.44)$$

where we set $\dot{q}_\alpha = \lambda_\alpha$. Clearly we can look at this set of N minus R equations as a set of equations which determines the Lagrange multipliers λ_β. In this case, of course, we must have (see below for further distinction between the ϕ's)

$$\det|\{\phi_\alpha, \phi_\beta\}| \neq 0, \quad (27.45)$$

and we can write (with λ_β determined by Eq. (27.44))

$$H^{(1)} = H_{\mathrm{can}} - \sum_\alpha \phi_\alpha \{\phi, \phi\}_{\alpha\beta}^{-1} \{\phi_\beta, H_{\mathrm{can}}\}. \quad (27.46)$$

In case Eq. (27.45) does not hold, i.e. if the matrix of Poisson brackets of primary constraints is singular, it has rank less than its dimensionality. Then some of the λ_α remain undetermined and new constraints appear. Thus if Eq. (27.44) leads to new constraints $\phi_\rho^{(2)}, \rho = 1, \ldots, R', R' < N - R$, called *secondary constraints*, we have to repeat the procedure. These secondary constraints will have to be added to the previous Hamiltonian and we write

$$H^{(1)} \longrightarrow H^{(2)} = H_{\text{can}} + \sum_{\alpha} \lambda_{\alpha} \phi_{\alpha} + \sum_{\rho} \lambda_{\rho}^{(2)} \phi_{\rho}^{(2)}, \qquad (27.47)$$

and the entire procedure is repeated. Clearly this process has to be repeated until no further new constraints arise.

Following Dirac, one calls a function a *first class function* if its Poisson bracket with any constraint vanishes on the constraint surface. Thus in particular some additional constraint (other than primary, secondary, etc. constraints) is called *first class* if its Poisson bracket with all other constraints is ('weakly') zero.

Thus if the ϕ_{α}'s are simply the original constraints and these commute, i.e.

$$\{\phi_{\alpha}, \phi_{\beta}\} = 0,$$

expressing a symmetry, the *"gauge symmetry"*, we cannot determine the Lagrange multipliers λ_{α} from these. We must therefore demand further conditions, i.e. *gauge fixing conditions* $\chi_{\alpha} = 0$ with

$$\{\phi_{\alpha}, \chi_{\beta}\} \neq 0, \qquad (27.48)$$

which make the set of ϕ_{α} and χ_{β} *second class*, thus violating the gauge symmetry. Thus only the brackets with χ_{α}'s allow us to determine the Lagrange multipliers λ_{α} from the relation

$$\{\chi_{\alpha}, H_{\text{can}}\} + \sum_{\beta} \lambda_{\beta} \{\chi_{\alpha}, \phi_{\beta}\} = 0.$$

Before we can proceed to the quantization of theories with primary, secondary and other constraints, we have to have a deeper understanding of the significance of these constraints. In fact we show that they are intimately connected with a *local gauge invariance*. Once we have recognized this vital fact, it will be plausible that for quantization one has to demand some gauge fixing conditions, and this will then allow us to proceed to the quantization.

27.5 Constraints as Generators of a Gauge Group

We demonstrate the important fact that a constrained dynamical theory is a gauge theory by considering a simple example.* Thus we consider the case

---

*We follow here in part D. Nemeschansky, C. Preitschopf and M. Weinstein [218].

of a particle of mass m_0 which is constrained to move on a circle of radius a in the (x, y)-plane:

$$\psi := r - a = 0, \quad r = \sqrt{x^2 + y^2}. \tag{27.49}$$

This constraint is not the primary constraint we considered previously (which involves momenta). Rather it replaces a force that forces the motion along the circle. The constraint is enforced with the help of a new dynamical variable, the *Lagrange multiplier* λ, for which we can also obtain an equation of motion. Thus we start from the Lagrangian

$$L(x, y, \lambda; \dot{x}, \dot{y}, \dot{\lambda}) = \frac{1}{2}m_0(\dot{x}^2 + \dot{y}^2) - \lambda(r - a), \quad r = \sqrt{x^2 + y^2}. \tag{27.50}$$

The associated canonical Hamiltonian is

$$H(x, y, \lambda; p_x, p_y) = p_x\dot{x} + p_y\dot{y} - L = \frac{1}{2}m_0(p_x^2 + p_y^2) + \lambda(r - a). \tag{27.51}$$

Naturally we also have to demand that the constraint be preserved in the course of time. But first we exploit the symmetry of the problem and go to polar coordinates. Then the expressions become

$$L(r, \theta, \lambda; \dot{r}, \dot{\theta}, \dot{\lambda}) = \frac{1}{2}m_0(\dot{r}^2 + r^2\dot{\theta}^2) - \lambda(r - a) \tag{27.52}$$

and

$$\begin{aligned} H(r, \theta, \lambda; p_r, p_\theta, p_\lambda) &= \dot{r}p_r + \dot{\theta}p_\theta - L(r, \theta, \lambda; \dot{r}, \dot{\theta}, \dot{\lambda}) \\ &= \frac{p_r^2}{2m_0} + \frac{p_\theta^2}{2m_0r^2} + \lambda(r - a). \end{aligned} \tag{27.53}$$

Proceeding methodically we consider the Lagrangian (27.52) and observe that we have a *primary constraint* (cf. Eq. (27.30))

$$p_\lambda = \frac{\partial L}{\partial \dot{\lambda}} = 0, \quad \text{i.e.} \quad \phi_1 := p_\lambda = 0.$$

The Lagrangian is therefore singular and we incorporate the constraint into H with a Lagrange multiplier u giving the so-called *total Hamiltonian*, i.e.

$$H^{(1)} = \frac{p_r^2}{2m_0} + \frac{p_\theta^2}{2m_0r^2} + \lambda(r - a) + up_\lambda, \tag{27.54}$$

with $u = \dot{\lambda}$, as we see from Eq. (27.38). We require the primary constraint ϕ_1 to be preserved in the course of time, i.e.

$$\dot{\phi}_1 = \{\phi_1, H\} = 0. \tag{27.55}$$

It can be verified that evaluation of the Poisson bracket (with canonical variables $r, \theta, \lambda, p_r, p_\theta, p_\lambda$) yields

$$\{\phi_1, H\} = \frac{\partial \phi_1}{\partial \lambda} \frac{\partial H}{\partial p_\lambda} - \frac{\partial \phi_1}{\partial p_\lambda} \frac{\partial H}{\partial \lambda} = -(r - a) = 0.$$

This is really the equation of motion of the dynamical variable λ (i.e. if we write down the corresponding Euler–Lagrange equation this is what we obtain). Thus this is no new information. But if we proceed and treat this as a *secondary constraint*,

$$\phi_2 := r - a = 0, \tag{27.56}$$

and we require this to be preserved in the course of time, i.e.

$$\dot{\phi}_2 = \{\phi_2, H\} = 0,$$

we obtain

$$\dot{r} = \left\{ r, \frac{p_r^2}{2m_0} \right\} = \frac{p_r}{m_0} \{r, p_r\} = \frac{p_r}{m_0} = 0. \tag{27.57}$$

Thus it is required that p_r be zero here in the context of classical mechanics (one can see already here: In quantum mechanics we expect r and p_r to commute and so be simultaneously observable, but here we cannot put the Poisson bracket equal to zero, since it is 1). If we continue and set therefore

$$\phi_3 := p_r = 0,$$

we find

$$\dot{\phi}_3 = \{\phi_3, H\} = \left\{ p_r, \frac{p_\theta^2}{2m_0 r^2} + \lambda(r - a) \right\} \stackrel{!}{=} 0$$

gives

$$-\lambda + \frac{p_\theta^2}{m_0 r^3} = 0.$$

Then setting

$$\phi_4 := -\lambda + \frac{p_\theta^2}{m_0 r^3} = 0,$$

and demanding

$$\dot{\phi}_4 = \{\phi_4, H\} \stackrel{!}{=} 0,$$

we obtain again (as with Eq. (27.57)) $p_r = 0$. Thus the procedure must end here.

We can now proceed to demonstrate that the Lagrangian of the simple rotor under discussion, i.e. Eq. (27.52), has a symmetry which we can call a *gauge symmetry*. We define the following quantity Q_g, the generator of this gauge symmetry, as a linear combination of the two constraints of the problem, i.e. of $\phi_1 = p_\lambda = 0$ and $\phi_2 = r - a = 0$ ($\dot{r} = 0$ follows from ϕ_2), with appropriate coefficients which we choose such that

$$Q_g := g(t)(r - a) + \dot{g}(t)p_\lambda. \tag{27.58}$$

Thus the constraints act as generators of this gauge symmetry. The infinitesimal variation of a quantity w under this transformation is, as one learns in classical mechanics,[†] given by

$$\delta w = \{w, Q_g\}. \tag{27.59}$$

We recall as a matter of revision and as an example, that for translations in $\mathbb{R}^3$: $Q = \mathbf{a} \cdot \mathbf{p}$. Then e.g.

$$\{x, \mathbf{a} \cdot \mathbf{p}\} = \frac{\partial x}{\partial x}\frac{\partial \mathbf{a} \cdot \mathbf{p}}{\partial p_x} = a_x, \quad \text{i.e.} \quad \delta x = x' - x = a_x.$$

In the simplest form of quantum mechanics we encounter this in the operator form of the translational displacement $x \to x + a$, i.e.

$$U_a f(x)U_a^\dagger = f(x + a), \tag{27.60}$$

where

$$U_a = e^{iap} = 1 + iap + \cdots, \quad \text{with} \quad [p, x] = -i \quad (\hbar = 1).$$

Thus the quantum mechanical version of Eq. (27.59) is $\delta w = -i[w, Q_g]$. But at this stage here we are still only concerned with classical mechanics. Before we proceed, it is best to calculate some basic Poisson brackets. As a first example consider

$$\begin{aligned}\{r, Q_g\} &= \frac{\partial r}{\partial r}\frac{\partial Q_g}{\partial p_r} - \frac{\partial r}{\partial p_r}\frac{\partial Q_g}{\partial r} + \frac{\partial r}{\partial \theta}\frac{\partial Q_g}{\partial p_\theta} - \frac{\partial r}{\partial p_\theta}\frac{\partial Q_g}{\partial \theta} \\ &\quad + \frac{\partial r}{\partial \lambda}\frac{\partial Q_g}{\partial p_\lambda} - \frac{\partial r}{\partial p_\lambda}\frac{\partial Q_g}{\partial \lambda} = \frac{\partial Q_g}{\partial p_r} = 0.\end{aligned} \tag{27.61}$$

Thus $\delta r = \{r, Q_g\} = 0$, i.e. r does not change under the transformation

---

[†]See e.g. H. Goldstein [114].

generated by Q_g. Proceeding similarly with other cases, we obtain altogether

$$
\begin{aligned}
\{r, Q_g\} &= 0 &&\Longrightarrow\ \delta r = 0, \\
\{\theta, Q_g\} &= \frac{\partial Q_g}{\partial p_\theta} = 0 &&\Longrightarrow\ \delta\theta = 0, \\
\{\lambda, Q_g\} &= \frac{\partial Q_g}{\partial p_\lambda} = \dot{g}(t) &&\Longrightarrow\ \delta\lambda = \dot{g}(t), \\
\{p_r, Q_g\} &= -\frac{\partial Q_g}{\partial r} = -g(t) &&\Longrightarrow\ \delta p_r = -g(t), \\
\{p_\theta, Q_g\} &= -\frac{\partial Q_g}{\partial \theta} = 0 &&\Longrightarrow\ \delta p_\theta = 0, \\
\{p_\lambda, Q_g\} &= -\frac{\partial Q_g}{\partial \lambda} = 0 &&\Longrightarrow\ \delta p_\lambda = 0.
\end{aligned}
\tag{27.62}
$$

The gauge transformation generated by Q_g is therefore defined by these infinitesimal changes. Considering two gauge transformations with generators Q_{g_1}, Q_{g_2}, we see with Eq. (27.58), that

$$
\{Q_{g_1}, Q_{g_2}\} = 0.
\tag{27.63}
$$

This is what one expects since the symmetry with which we are concerned here is similar to that of a $U(1)$ phase transformation, which, of course, is the transformation of an abelian group for which the generators commute. We can also take the Hamiltonian of Eq. (27.53) and compute:

$$
\begin{aligned}
\delta H &= \{H, Q_g\} = \frac{1}{2m_0}\{p_r^2, Q_g\} + \{\lambda(r-a), Q_g\} \\
&= \frac{1}{2m_0}\{p_r^2, g(t)(r-a)\} + \{\lambda(r-a), \dot{g}(t)p_\lambda\},
\end{aligned}
$$

i.e.

$$
\begin{aligned}
\delta H &= \frac{g(t)}{2m_0}\{p_r^2, r\} + (r-a)\dot{g}(t)\{\lambda, p_\lambda\} \\
&= \frac{g(t)}{m_0}p_r\{p_r, r\} + (r-a)\dot{g}(t)\{\lambda, p_\lambda\} \\
&= -\frac{g(t)}{m_0}p_r + (r-a)\dot{g}(t).
\end{aligned}
\tag{27.64}
$$

Also, from Eq. (27.54) and using Eq. (27.62),

$$
\delta H^{(1)} = \delta H + \delta(\dot{\lambda}p_\lambda) = \delta H + p_\lambda\ddot{g}(t).
$$

On the other hand, if we calculate $\delta L = \{L, Q_g\}$, e.g. as δL of Eq. (27.52) with (27.62), i.e. $\delta\lambda = \dot{g}(t)$, we obtain

$$\delta L = -\delta\lambda(r - a) = -\dot{g}(t)(r - a). \tag{27.65}$$

Thus without further considerations the original Lagrangian is not invariant except on the hyperplane defined by the constraint $r - a = 0$. Of course, we want the Hamiltonian to be invariant so that the symmetry is conserved (expressed by $\{H^{(1)}, Q_g\} = 0$). We see that we have to impose the condition (27.57), i.e. $p_r = 0$. Then

$$\delta H^{(1)} \approx 0 \quad \text{with} \quad p_r = 0, \quad p_\lambda = 0. \tag{27.66}$$

In the same sense we have

$$\delta L \approx 0. \tag{27.67}$$

Can one formulate a Lagrangian which has H (as described above) for its canonically constructed Hamiltonian, and which is invariant under a class of gauge transformations which includes the transformation of Q_g above? The answer is yes, but one must use a "first order formalism" instead of the second order formalism. We therefore introduce now this first order Lagrangian formalism.

In the *"first order Lagrangian formalism"* the Lagrangian L involves only first order time derivatives, i.e. velocities to first order in return for a dependence on momenta. In the present case this Lagrangian L is obtained from Eq. (27.53) in the form

$$
\begin{aligned}
L(r, \theta, \lambda, p_r, p_\theta, p_\lambda) &= \dot{r}p_r + \dot{\theta}p_\theta - H(r, \theta, \lambda; p_r, p_\theta, p_\lambda) \\
&= \dot{r}p_r + \dot{\theta}p_\theta - \frac{p_\theta^2}{2m_0 r^2} - \frac{p_r^2}{2m_0} - \lambda\phi \\
&\equiv L_g, \quad \phi = r - a, \quad p_\lambda = 0.
\end{aligned}
\tag{27.68}
$$

This Lagrangian (see second line) has the general form

$$L(z_a, \lambda_\alpha) := \sum_a \dot{z}_a K^a - H - \sum_\alpha \lambda_\alpha \phi_\alpha, \tag{27.69}$$

and the corresponding action S is

$$S[z_a, \lambda_\alpha] = \int_{t_i}^{t_f} dt L(z_a, \lambda_\alpha). \tag{27.70}$$

The first order Lagrangian of Eq. (27.69) is obviously singular, since it is not quadratic in velocities, i.e.

$$\left| \frac{\partial^2 L}{\partial\dot{z}_a \partial\dot{z}_b} \right| = 0.$$

We shall gain a deeper insight into first order Lagrangians from other considerations later. We shall always, however, make use of the fact that L is defined uniquely only up to a total time derivative, i.e. we can always replace L by $L + (d/dt)F(z_a, \lambda_\alpha)$ with (we assume) $F(z_a, \lambda_\alpha) = 0$ at $t = t_i, t_f$. This will enable us to shift dots denoting time derivatives from some factors to others.

We can now look at L in the form (27.68) and calculate its variation under our gauge transformation. Clearly (with $\delta \dot{r} = (d/dt)\delta r$)

$$
\begin{aligned}
\delta L_g &= \dot{r}(\delta p_r) + (\delta \dot{r})p_r + (\delta \dot{\theta})p_\theta + \dot{\theta}(\delta p_\theta) - \delta H \\
&= -g(t)\dot{r} - \delta H \\
&\overset{(27.64)}{=} -g(t)\dot{r} - \dot{g}(t)(r-a) + \frac{g(t)}{m_0}p_r \\
&= -\frac{d}{dt}[g(t)(r-a)] + \frac{g(t)}{m_0}p_r.
\end{aligned}
\tag{27.71}
$$

Thus we see: For invariance we require $p_r = 0$ in order to obtain $\delta L_g = 0$ up to a total time derivative, but otherwise δL_g now vanishes strongly, i.e. not only on the hyperplane defined by the constraint surface in phase space. In this sense we have now achieved what we set out to demonstrate: We have a gauge transformation (with generator Q_g of Eq. (27.58)) which leaves the Lagrangian in its first order form invariant.

Of course, we would like to have the corresponding Hamiltonian H_g also gauge invariant, i.e.

$$\text{we want}: \quad \delta H_g = \{H_g, Q_g\} = 0.$$

This Hamiltonian H_g is the Hamiltonian associated (through a Legendre transform) with L_g for the enlarged phase space (enlarged by λ, p_λ), so that

$$
\begin{aligned}
H_g &= p_r \dot{r} + p_\theta \dot{\theta} + p_\lambda \dot{\lambda} - L_g \equiv H^{(1)} \\
&= \frac{1}{2m_0}p_r^2 + \frac{1}{2m_0 r^2}p_\theta^2 + \dot{\lambda}p_\lambda.
\end{aligned}
$$

The fact that $\delta H^{(1)} \neq 0$ as we observed earlier shows that this problem remains to be cured — this is, in fact, achieved with the BRST extension, that we cannot go into here.[‡]

[‡]The letters BRST stand for the names of the authors C. Becchi, A. Rouet and R. Stora [16] as well as V. Tyutin [274], who developed the idea of introducing anticommuting ghost degrees of freedom to obtain manifest gauge invariance.

27.6   Gauge Fixing and Dirac Quantization

The problem of the simple rotor treated in the previous section depends on the following six variables $r, \theta, \lambda, p_r, p_\theta, p_\lambda$. We have the constraints

$$\phi_1 = p_\lambda = 0, \quad \phi_2 = r - a = 0 \tag{27.72}$$

with $\{\phi_1, \phi_2\} = 0$. Thus ϕ_1, ϕ_2 commute with themselves and (all) other constraints. Clearly the two conditions (27.72) leave four variables independent. In order to achieve gauge invariance, we also imposed the additional condition

$$\phi_3 = p_r = 0. \tag{27.73}$$

Now

$$\{\phi_1, \phi_2\} = 0, \quad \{\phi_1, \phi_3\} = 0, \quad \{\phi_2, \phi_3\} = 1,$$

and the constraints (if we include ϕ_3) are no longer first but *second class* (as defined after Eq. (27.47) a constraint is first class if its Poisson bracket with all other constraints is ('locally') zero). We are then left with three remaining independent phase space variables. However, quantization requires pairs of canonically conjugate variables. Thus we still have one degree of freedom too many — this is our gauge freedom, and we can choose any function of the six variables to remove this freedom, i.e. to fix the gauge. A convenient choice is

$$\phi_4 = \lambda = 0. \tag{27.74}$$

With this (or some other appropriate choice) we obtain the following matrix of Poisson brackets:

$$(\{\phi_i, \phi_j\}) = \begin{pmatrix} 0 & 0 & 0 & -1 \\ 0 & 0 & 1 & 0 \\ 0 & -1 & 0 & 0 \\ 1 & 0 & 0 & 0 \end{pmatrix}. \tag{27.75}$$

This is a *symplectic* matrix. This matrix of "constraint brackets" plays a vital role in Dirac's method of quantization. Dirac [76] postulates that in the case of a system with constraints the transition from Poisson brackets $\{\cdots, \cdots\}$ to commutators $[\cdots, \cdots]$ is to be replaced by

$$\{A, B\}_D \longrightarrow \frac{1}{i}[A, B] \quad (\hbar = 1), \tag{27.76a}$$

where the bracket now called *Dirac bracket* is defined by

$$\{A, B\}_D := \{A, B\} - \sum_{i,j} \{A, \phi_i\} \{\phi, \phi\}_{ij}^{-1} \{\phi_j, B\}. \tag{27.76b}$$

Note that first the inverse of the matrix $(\{\phi_i, \phi_j\})$ has to be obtained, and then matrix element ij of this inverse is to be taken. We have seen previously (cf. Eq. (27.45)) that the existence of the inverse of the symplectic matrix (27.75) is in fact required for the determination of the Lagrange multipliers. We shall demonstrate this again below in a justification of Eq. (27.76b). For the time being we take (27.76b) and evaluate the various brackets. Thus we obtain:

$$
\begin{aligned}
&\{r, p_r\} = 1, \quad \{\theta, p_\theta\} = 1, \quad \{\lambda, p_\lambda\} = 1, \\
&\{r, p_\theta\} = 0, \quad \{\theta, p_r\} = 0, \quad \{r, \theta\} = 0, \\
&\{r, \phi_1\} = 0, \quad \{r, \phi_2\} = 0, \quad \{r, \phi_3\} = 1, \quad \{r, \phi_4\} = 0, \\
&\{\theta, \phi_1\} = 0, \quad \{\phi_2, p_r\} = 1, \quad \text{etc.}
\end{aligned}
\tag{27.77}
$$

We consider in particular the following cases:
(a)

$$
\{r, p_r\}_D = 1 - \sum_{i,j} \{r, \phi_i\}\{\phi, \phi\}_{ij}^{-1}\{\phi_j, p_r\}.
$$

The inverse of the symplectic matrix (27.75) can be verified to be

$$
(\{\phi_i, \phi_j\})^{-1} = \begin{pmatrix} 0 & 0 & 0 & 1 \\ 0 & 0 & -1 & 0 \\ 0 & 1 & 0 & 0 \\ -1 & 0 & 0 & 0 \end{pmatrix}.
$$

Hence

$$
\begin{aligned}
\{r, p_r\}_D &= 1 - \{r, \phi_1\}\{\phi, \phi\}_{14}^{-1}\{\phi_4, p_r\} - \{r, \phi_2\}\{\phi, \phi\}_{23}^{-1}\{\phi_3, p_r\} \\
&\quad - \{r, \phi_3\}\{\phi, \phi\}_{32}^{-1}\{\phi_2, p_r\} - \{r, \phi_4\}\{\phi, \phi\}_{41}^{-1}\{\phi_1, p_r\} \\
&= 1 - 0 - 0 - 1(+1)1 - 0 = 0.
\end{aligned}
\tag{27.78a}
$$

(b)

$$
\{\theta, p_\theta\}_D = \{\theta, p_\theta\} + 0 = 1.
\tag{27.78b}
$$

(c)

$$
\{\lambda, p_\lambda\}_D = 1 - 1 = 0.
\tag{27.78c}
$$

All other Dirac brackets vanish. Thus the theory of the *simple rotor* is quantized by setting (observe that the Poisson bracket $\{r, p_r\} = 1$, but the Dirac bracket $\{r, p_r\}_D = 0$)

$$
[r, p_r] = 0 \quad \text{and} \quad \frac{1}{i}[\theta, p_\theta] = 1,
\tag{27.79}
$$

all other commutators vanishing. This is, of course, what one expects: There is only one degree of freedom (the angle θ) with angular momentum p_θ. Thus we have quantized the theory by explicit use of its constraints and gauge fixing conditions.

27.7 The Formalism of Dirac Quantization

We have seen in the previous sections that one starts off with a *canonical Hamiltonian* $H(q,p)$ which is associated with a number of primary constraints $\phi_\alpha(q,p)$ if the original Lagrangian is singular, and one then constructs the so-called *total Hamiltonian* $H^{(1)}(q,p)$ by setting

$$H^{(1)}(q,p) := H(q,p) + \sum_{\alpha=1}^{N-R} \lambda_\alpha \phi_\alpha(q,p), \qquad (27.80)$$

where $\{\lambda_\alpha\}$ are Lagrange multipliers and ϕ_α are first class constraints, i.e. $\{\phi_\alpha, \phi_\beta\} = 0$. The presence of these constraints indicates that the problem contains some redundant or superfluous degrees of freedom like the Euler–Lagrange equations of Maxwell theory in terms of the electromagnetic vector potential A_μ: There one has four components but only two combinations, i.e. E and B are physically observable and independent; the additional two degrees of freedom were needed to enforce the gauge symmetry.[§] In Maxwell theory we know what we have to do in order to remove this redundancy: We have to "*fix the gauge*", i.e. introduce quite arbitrarily some other equation(s) which then reduce the number of degrees of freedom to those which are independent. In systems with constraints we do precisely the same: We again have to introduce arbitrarily gauge fixing conditions which effectively determine the Lagrange multipliers λ_α in Eq. (27.80). Thus we now introduce as many gauge fixing conditions χ_α as we have primary, first class constraints:

$$\chi_\alpha(q,p) = 0, \quad \alpha = 1, \ldots, N - R. \qquad (27.81)$$

The time-development of a function F of the dynamical variables q and p of the system is determined by (cf. Eq. (2.4))

$$\dot{F}(q,p) \equiv \frac{dF(q,p)}{dt} = \{F, H^{(1)}\} = \{F, H\} + \sum_{\alpha=1}^{N-R} \{F, \phi_\alpha\} \lambda_\alpha. \qquad (27.82)$$

Clearly we also have to demand the persistence of the gauge fixing conditions (27.81) in the course of time, i.e.

$$\dot{\chi}_\alpha(q,p) = 0, \qquad (27.83)$$

[§]See H. J. W. Müller–Kirsten [215], pp. 436, 437.

so that also

$$0 = \{\chi_\alpha, H\} + \sum_{\beta=1}^{N-R} \{\chi_\alpha, \phi_\beta\}\lambda_\beta. \tag{27.84}$$

This is a set of linear equations in the Lagrange multipliers λ_α. We can determine these provided

$$\det(\{\chi_\alpha, \phi_\beta\}) \equiv \det(\{\chi, \phi\}_{\alpha\beta}) \neq 0.$$

Let $C_{\alpha\gamma}$ denote the cofactor[¶] of $\{\chi_\alpha, \phi_\gamma\}$ divided by the determinant of $\{\chi_\alpha, \phi_\gamma\}$ (which defines the inverse), so that $C_{\alpha\gamma} = -C_{\gamma\alpha}$, and

$$C_{\alpha\gamma}\{\chi_\alpha, \phi_\beta\} = -C_{\gamma\alpha}\{\chi_\alpha, \phi_\beta\} = -C_{\alpha\gamma}\{\phi_\beta, \chi_\alpha\} = \delta_{\gamma\beta}. \tag{27.85}$$

Multiplying Eq. (27.84) from the left by $C_{\alpha\gamma}$, we obtain

$$0 = C_{\alpha\gamma}\{\chi_\alpha, H\} + \sum_\beta \underbrace{C_{\alpha\gamma}\{\chi_\alpha, \phi_\beta\}}_{\delta_{\gamma\beta}} \lambda_\beta = C_{\alpha\gamma}\{\chi_\alpha, H\} + \lambda_\gamma,$$

or

$$\lambda_\gamma = -C_{\alpha\gamma}\{\chi_\alpha, H\} = +C_{\gamma\alpha}\{\chi_\alpha, H\}. \tag{27.86}$$

Inserting this expression into Eq. (27.82), we obtain

$$\dot{F}(q, p) = \{F, H\} + \sum_{\alpha,\beta} \{F, \phi_\alpha\}C_{\alpha\beta}\{\chi_\beta, H\} \equiv \{F, H\}_D, \tag{27.87a}$$

where

$$\{F, H\}_D = \{F, H\} + \sum_{\alpha,\beta} \{F, \phi_\alpha\}C_{\alpha\beta}\{\chi_\beta, H\} \tag{27.87b}$$

is the *Dirac bracket* of F and H. We can rewrite this expression[∥] in a slightly different form with the following observations. From the last equality of Eq. (27.85) we obtain

$$-\{\phi_\beta, \chi_\alpha\}C_{\alpha\gamma} = \delta_{\beta\gamma}.$$

[¶]The inverse or reciprocal matrix of a nonsingular matrix A is the *adjugate* matrix of A, adj A, divided by the determinant of A, which satisfies the relation $A(\text{adj } A) = (\text{adj } A)A = |A|\mathbb{1}$. The elements of $(\text{adj } A)$ are the *cofactors* $|A_{ij}|$ of elements a_{ij} of A, but placed in transposed position. The cofactor of a_{ij} in A is the determinant obtained by suppressing the i-th row and the j-th column of A and giving it the sign $(-1)^{i+j}$. One can also use an untransposed adjugate, since the determinant of a transposed square matrix is equal to that of the matrix. See e.g. A. C. Aitken [7], pp. 39, 51 − 53.

[∥]This is the expression in the original formulation of Dirac. See P. A. M. Dirac [76] (b), Eq. (36), p. 138.

Multiplying this equation from the left by the inverse of the Poisson bracket we obtain

$$-\{\phi_\rho, \chi_\beta\}^{-1}\{\phi_\beta, \chi_\alpha\}C_{\alpha\gamma} = \{\phi_\rho, \chi_\beta\}^{-1}\delta_{\beta\gamma},$$

and hence

$$-\delta_{\rho\alpha}C_{\alpha\gamma} = \{\phi_\rho, \chi_\gamma\}^{-1} \quad \text{or} \quad -C_{\rho\gamma} = \{\phi_\rho, \chi_\gamma\}^{-1} \equiv \{\phi, \chi\}_{\rho\gamma}^{-1}.$$

Inserting this expression for the matrix C into Eq. (27.87b), we obtain the following form of the Dirac bracket:

$$\{F, H\}_D = \{F, H\} - \sum_{\alpha,\beta}\{F, \phi_\alpha\}\{\phi, \chi\}_{\alpha\beta}^{-1}\{\chi_\beta, H\} \tag{27.87c}$$

This Dirac bracket with χ_β to the right of ϕ_α is not quite the same as the one defined by Eq. (27.76b), in which the set $\{\phi_i\}$ consists of all constraints and gauge fixing conditions. However, we can relate the two versions as follows. Combining ϕ_α and χ_α into one column, we define

$$\psi_A := \begin{pmatrix} \phi_\alpha \\ \chi_\alpha \end{pmatrix}, \quad A = 1, \ldots, 2(N - R),$$

so that with $\lambda_A = 0$ for $A = N - R + 1, \ldots, 2(N - R)$, we have

$$H^{(1)}(q, p) := H(q, p) + \sum_{B=1}^{2(N-R)} \psi_B(q, p)\lambda_B.$$

Then we must have $\dot{\psi}_A(q, p) = 0$, i.e. $\{\psi_A, H^{(1)}\} = 0$, and so

$$0 = \{\psi_A, H\} + \sum_{B=1}^{N-R}\{\psi_A, \psi_B\}\lambda_B = \{\psi_A, H\} + \sum_{B=1}^{2(N-R)}\{\psi_A, \psi_B\}\lambda_B.$$

Proceeding as before we now obtain the Dirac bracket

$$\{F, H\}_D = \{F, H\} - \sum_{A,B}\{F, \psi_A\}\{\psi, \psi\}_{AB}^{-1}\{\psi_B, H\}, \tag{27.88}$$

which is (27.76b). We leave the verification of algebraic properties of the Dirac bracket to Example 27.4, and the derivation of Hamilton's equations in terms of Dirac brackets to Example 27.5.

Example 27.4: Algebraic properties of Dirac brackets
Verify that the Dirac brackets obey the properties of (a) antisymmetry, (b) linearity, and (c) the Jacobi relation.

Solution: The verification is left to whoever is interested in it.

Example 27.5: Hamilton's equations in terms of Dirac brackets

Obtain from Eq. (27.87c) Hamilton's equations in terms of Dirac brackets.

Solution: We have with $\dot{F}(q, p) = \{F, H\}$:

$$
\begin{aligned}
\{F, H\} &= \sum_i \left(\frac{\partial F}{\partial q_i} \frac{\partial H}{\partial p_i} - \frac{\partial F}{\partial p_i} \frac{\partial H}{\partial q_i} \right) = \sum_{i,j} \left(\frac{\partial F}{\partial q_i} \frac{\partial p_j}{\partial p_i} \frac{\partial H}{\partial p_j} - \frac{\partial F}{\partial p_i} \frac{\partial q_j}{\partial q_i} \frac{\partial H}{\partial q_j} \right) \\
&= \sum_{i,j} \left[\left(\frac{\partial F}{\partial q_i} \frac{\partial p_j}{\partial p_i} - \frac{\partial F}{\partial p_i} \frac{\partial p_j}{\partial q_i} \right) \frac{\partial H}{\partial p_j} + \left(\frac{\partial F}{\partial q_i} \frac{\partial q_j}{\partial p_i} - \frac{\partial F}{\partial p_i} \frac{\partial q_j}{\partial q_i} \right) \frac{\partial H}{\partial q_j} \right] \\
&= \sum_j \left(\{F, p_j\} \frac{\partial H}{\partial p_j} + \{F, q_j\} \frac{\partial H}{\partial q_j} \right). \tag{27.89}
\end{aligned}
$$

Hence for F replaced by q_i and p_i we obtain:

$$
\{q_i, H\} = \frac{\partial H}{\partial p_i} = \{q_i, p_j\} \frac{\partial H}{\partial p_j} \quad \text{and} \quad \{p_i, H\} = -\frac{\partial H}{\partial q_i} = \{p_i, q_j\} \frac{\partial H}{\partial q_j}, \tag{27.90}
$$

and for F replaced by χ_β:

$$
\{\chi_\beta, H\} = \sum_{i,j} \left(\{\chi_\beta, p_j\} \frac{\partial H}{\partial p_j} + \{\chi_\beta, q_j\} \frac{\partial H}{\partial q_j} \right). \tag{27.91}
$$

With Eq. (27.89) for $F \to q_i$ and this replacement in $\{q_i, H\}_D$, we obtain:

$$
\dot{q}_i \stackrel{(27.87a)}{=} \{q_i, H\}_D = \{q_i, p_j\}_D \frac{\partial H}{\partial p_j} + \{q_i, q_j\}_D \frac{\partial H}{\partial q_j}. \tag{27.92}
$$

Similarly

$$
\dot{p}_i \stackrel{(27.87a)}{=} \{p_i, H\}_D = \{p_i, q_j\}_D \frac{\partial H}{\partial q_j} + \{p_i, p_j\}_D \frac{\partial H}{\partial p_j}. \tag{27.93}
$$

These equations suggest that we introduce a more general notation and set $q_i = \xi_i, i = 1, \ldots, N$, and $p_i = \xi_i, i = N + 1, \ldots, 2N$. Then Eqs. (27.92) and (27.93) can be combined into

$$
\dot{\xi}_i = \{\xi_i, \xi_j\}_D \frac{\partial H}{\partial \xi_j}, \quad j = 1, \ldots, 2N. \tag{27.94}
$$

This equation becomes the Heisenberg equation of motion if the c-number observable ξ_i is replaced by the operator $\hat{\xi}_i$ and the Dirac bracket $\{\cdots, \cdots\}_D$ by the commutator $-i[\cdots, \cdots]$, i.e.

$$
\dot{\hat{\xi}}_i = \frac{1}{i} [\hat{\xi}_i, \hat{\xi}_j] \frac{\partial \hat{H}}{\partial \hat{\xi}_j}. \tag{27.95}
$$

27.7.1   Poisson and Dirac brackets in field theory

The quantum mechanical examples discussed so far differ from those in field theory mainly in as far as in the case of the latter the number of degrees of freedom is infinite. Thus as in electrodynamics, one has to distinguish between the Lagrangian L and its density in space, $\mathcal{L}$, i.e.

$$L = \int d\mathbf{x}\mathcal{L}. \tag{27.96}$$

Suppose the density $\mathcal{L}$ depends on some scalar field $b(x)$ (with spacetime x) and covariant derivatives $\partial_\mu b(x)$. Then the canonical *momentum density* $p(x)$ is defined as

$$p(x) = \frac{\partial \mathcal{L}}{\partial \dot{b}}, \quad \dot{b} = \partial_0 b(x) \quad (c = 1). \tag{27.97}$$

One can then construct the *Hamiltonian density* $\mathcal{H}(b, p)$ in the usual way. The equal time Poisson bracket of two functionals ϕ, ψ of b, p then becomes

$$\{\phi(\mathbf{x}, t), \psi(\mathbf{y}, t)\} : \;\; = \;\; \int d\mathbf{x}' \left(\frac{\partial \phi(\mathbf{x}, t)}{\partial b(\mathbf{x}', t)} \frac{\partial \psi(\mathbf{y}, t)}{\partial p(\mathbf{x}', t)} \right.$$
$$\left. - \frac{\partial \phi(\mathbf{x}, t)}{\partial p(\mathbf{x}', t)} \frac{\partial \psi(\mathbf{y}, t)}{\partial b(\mathbf{x}', t)} \right). \tag{27.98}$$

Dirac brackets are defined accordingly, i.e. for constraints ϕ_i and gauge fixing conditions χ_i we have

$$\{\phi(\mathbf{x}, t), \psi(\mathbf{y}, t)\}_D \;\; = \;\; \{\phi(\mathbf{x}, t), \psi(\mathbf{y}, t)\} - \sum_{i,j} \int d\mathbf{x}' d\mathbf{x}'' \{\phi(\mathbf{x}, t), \phi_i(\mathbf{x}', t)\}$$
$$\times \{\phi(\mathbf{x}', t), \chi(\mathbf{x}'', t)\}_{ij}^{-1} \{\chi_j(\mathbf{x}'', t), \psi(\mathbf{y}, t)\}. \tag{27.99}$$

The most immediate case to look at is electrodynamics. This is done in the next section.

27.8   Dirac Quantization of Free Electrodynamics

The Lagrangian of the free electromagnetic field is with $F_{\mu\nu} = \partial_\mu A_\nu - \partial_\nu A_\mu$,

$$L = \int d\mathbf{x}\mathcal{L}, \quad \mathcal{L} = -\frac{1}{4} F^{\mu\nu} F_{\mu\nu} \equiv \mathcal{L}(A_\mu, \partial_\mu A_\nu). \tag{27.100}$$

The Euler–Lagrange equations are given by (cf. Eq. (27.10))

$$\partial_\mu F^{\mu\nu} = 0. \tag{27.101}$$

These equations can be subdivided into *equations of motion* (observe the time derivative!) obtained for $\nu = i$:

$$\partial_0 F^{0i} = -\partial_j F^{ji} \tag{27.102}$$

and the following *constraint* (Gauss' law) obtained for $\nu = 0$:

$$\partial_k F^{k0} = 0, \quad F^{k0} = \frac{1}{c} E_k. \tag{27.103}$$

Differentiating the equations of motion, we obtain

$$\partial_i \partial_0 F^{0i} = -\partial_i \partial_j F^{ji} = 0, \quad \text{i.e.} \quad \partial_0(\partial_i F^{0i}) = 0, \tag{27.104}$$

i.e. the equations of motion lead to the vanishing of the time derivative of Gauss' law, and for the latter's integration one has to choose appropriate boundary conditions. Introducing the canonical momenta π_μ we have

$$\pi^\mu = \frac{\partial \mathcal{L}}{\partial(\partial_0 A_\mu)}, \quad \pi^0 = F^{00} = 0, \quad \pi^k = F^{k0}. \tag{27.105}$$

Thus one momentum component vanishes and the Lagrangian is therefore singular, as observed in Example 27.1. Then

$$\mathcal{L} = \frac{1}{2}\left(\frac{1}{c^2}\mathbf{E}^2 - \mathbf{B}^2\right), \quad \mathbf{B} = \mathbf{\nabla} \times \mathbf{A}. \tag{27.106}$$

Thus the two constraints obtained from the Lagrangian are the momentum constraints

$$\phi_1 := -\mathbf{\nabla} \cdot \boldsymbol{\pi}(\mathbf{x}, t) = 0, \quad \phi_2 := \pi_0(\mathbf{x}, t) = 0 \tag{27.107}$$

(ϕ_1 results from the fact that $\partial \mathcal{L}/\partial(\partial_\mu A_0) = 0$ and $\partial \mathcal{L}/\partial A_0 = 0$, and ϕ_2 expresses the fact that $\partial \mathcal{L}/\partial \dot{A}_0 = 0$). The constraint ϕ_1 is a *primary constraint* which expresses the mutual dependence of some momenta. Both ϕ_1 and ϕ_2 are in the first place *first class constraints*, i.e. their Poisson bracket is zero (no A_i involved), the Poisson bracket being given by (observe that $i = 0$ is included)

$$
\begin{aligned}
\{\phi_1(\mathbf{x}, t), \phi_2(\mathbf{y}, t)\} &= \sum_{i=0,1,2,3} \int d\mathbf{x}' \left(\frac{\partial \phi_1(\mathbf{x}, t)}{\partial A_i(\mathbf{x}', t)} \frac{\partial \phi_2(\mathbf{y}, t)}{\partial \pi_i(\mathbf{x}', t)} \right. \\
&\left. - \frac{\partial \phi_1(\mathbf{x}, t)}{\partial \pi_i(\mathbf{x}', t)} \frac{\partial \phi_2(\mathbf{y}, t)}{\partial A_i(\mathbf{x}', t)} \right) = 0. \tag{27.108}
\end{aligned}
$$

The *canonical Hamiltonian* is defined as[*] $H_c = \int d\mathbf{x} \mathcal{H}_c$ with density

$$\mathcal{H}_c = \frac{1}{\mu_0}[\pi^\rho \partial^0 A_\rho - \mathcal{L}] = -\frac{1}{\mu_0}\left[\frac{1}{c^2}\mathbf{E} \cdot \dot{\mathbf{A}} + \mathcal{L}\right] = \frac{1}{\mu_0}\left[\frac{1}{2}\left(\frac{\mathbf{E}^2}{c^2} + \mathbf{B}^2\right)\right], \quad (27.109)$$

where in the last expression we ignored a total derivative. From Eq. (27.109) one can obtain the first-order Lagrangian density

$$\mathcal{L} = -\frac{1}{c^2}\mathbf{E} \cdot \dot{\mathbf{A}} - \frac{1}{2}\left(\frac{1}{c^2}\mathbf{E}^2 + \mathbf{B}^2\right). \quad (27.110)$$

We define the general or *total Hamiltonian* H by adding to H_c the constraints multiplied by Lagrange multipliers λ_1, λ_2, i.e.

$$H = H_c + \int d\mathbf{x}\lambda_1(\mathbf{x}, t)\phi_1(\mathbf{x}, t) + \int d\mathbf{x}\lambda_2(\mathbf{x}, t)\phi_2(\mathbf{x}, t). \quad (27.111)$$

From this expression we can derive the new equations of motion. But here we are interested in the canonical quantization of the theory.

In order to exclude redundant variables we again introduce supplementary constraints or *gauge fixing conditions* χ_1, χ_2, so that some Poisson brackets of ϕ_i, χ_i do not vanish and every first class constraint becomes a second class constraint. Thus we choose

$$\chi_1 := A_3(\mathbf{x}, t) = 0, \quad \chi_2 := A_0(\mathbf{x}, t) = 0. \quad (27.112)$$

We can now calculate the following equal time Poisson brackets, of which the first is verified below:

$$\{\phi_1(\mathbf{x}, t), \chi_1(\mathbf{x}', t)\} = \partial_3\delta(\mathbf{x} - \mathbf{x}'), \qquad \{\phi_2(\mathbf{x}, t), \chi_2(\mathbf{x}', t)\} = -\delta(\mathbf{x} - \mathbf{x}'),$$
$$\{A_i(\mathbf{x}, t), \phi_1(\mathbf{x}', t)\} = -\partial_i\delta(\mathbf{x} - \mathbf{x}'), \qquad \{\chi_1(\mathbf{x}, t), \pi_j(\mathbf{x}', t)\} = \delta_{j3}\delta(\mathbf{x} - \mathbf{x}'),$$
$$\{A_i(\mathbf{x}, t), \pi_j(\mathbf{x}', t)\} = \delta_{ij}\delta(\mathbf{x} - \mathbf{x}'). \quad (27.113)$$

We verify the first, i.e.

$$\{\phi_1(\mathbf{x}, t), \chi_1(\mathbf{x}', t)\}$$
$$= \sum_{i=0,1,2,3}\int d\mathbf{y}\left(\frac{\partial\phi_1(\mathbf{x}, t)}{\partial A_i(\mathbf{y}, t)}\frac{\partial\chi_1(\mathbf{x}', t)}{\partial\pi_i(\mathbf{y}, t)} - \frac{\partial\phi_1(\mathbf{x}, t)}{\partial\pi_i(\mathbf{y}, t)}\frac{\partial\chi_1(\mathbf{x}', t)}{\partial A_i(\mathbf{y}, t)}\right)$$
$$= \sum_{i=0,1,2,3}\int d\mathbf{y}(0 + \partial_i\delta(\mathbf{x} - \mathbf{y})\delta_{3i}\delta(\mathbf{x}' - \mathbf{y})) = \partial_3\delta(\mathbf{x} - \mathbf{x}').$$

---

[*]See, for example, H. J. W. Müller–Kirsten [215], p. 430.

The Dirac bracket with $\{C_i\} \equiv \{\phi_i, \chi_i\}$ is given by

$$\{A(\mathbf{x},t), B(\mathbf{y},t)\}_D = \{A(\mathbf{x},t), B(\mathbf{y},t)\} - \sum_{i,j} \int dx' dx'' \{A(\mathbf{x},t), C_i(\mathbf{x}',t)\}$$

$$\times \{C(\mathbf{x}',t), C(\mathbf{x}'',t)\}_{ij}^{-1} \{C_j(\mathbf{x}'',t), B(\mathbf{y},t)\}. \quad (27.114)$$

We define the matrix of the Poisson brackets $(\{C(\mathbf{x}',t), C(\mathbf{x}'',t)\})$ by

$$\triangle_{ij} \equiv (\{C,C\}_{ij}) = \begin{pmatrix} \{\phi_1,\phi_1\} & \{\phi_1,\phi_2\} & \{\phi_1,\chi_1\} & \{\phi_1,\chi_2\} \\ \{\phi_2,\phi_1\} & \{\phi_2,\phi_2\} & \{\phi_2,\chi_1\} & \{\phi_2,\chi_2\} \\ \{\chi_1,\phi_1\} & \{\chi_1,\phi_2\} & \{\chi_1,\chi_1\} & \{\chi_1,\chi_2\} \\ \{\chi_2,\phi_1\} & \{\chi_2,\phi_2\} & \{\chi_2,\chi_1\} & \{\chi_2,\chi_2\} \end{pmatrix}.$$

$$(27.115a)$$

Inserting the evaluated Poisson brackets, this matrix becomes the skew expression

$$\triangle(\mathbf{x}',\mathbf{x}'') =$$

$$\begin{pmatrix} \mathbb{0}_{2\times 2} & \begin{matrix} \partial_3 \delta(\mathbf{x}'-\mathbf{x}'') & 0 \\ 0 & -\delta(\mathbf{x}'-\mathbf{x}'') \end{matrix} \\ \begin{matrix} -\partial_3 \delta(\mathbf{x}'-\mathbf{x}'') & 0 \\ 0 & \delta(\mathbf{x}'-\mathbf{x}'') \end{matrix} & \mathbb{0}_{2\times 2} \end{pmatrix}.$$

$$(27.115b)$$

The inverse $\triangle^{-1}$ of this matrix is expressible in terms of the ϵ function

$$\epsilon(x-x') := \theta(x-x') - \theta(x'-x), \quad \frac{d}{dx'}\epsilon(x-x') = -2\delta(x-x').$$

With this function we obtain by partial integration (and since $\delta(x)$ is (in the justifiable symbolical manipulation) zero at infinity and $\theta'(x) = \delta(x)$)

$$\int dx' \delta'(x'-x'')\frac{1}{2}\epsilon(x-x') = -\frac{1}{2}\int dx' \epsilon'(x-x')\delta(x'-x'')$$

$$= \int dx' \delta(x'-x'')\delta(x'-x)$$

$$= \delta(x''-x).$$

Then with $\delta_3(x-x') = \delta(x_3 - x_3')$, etc. one can verify by multiplication of $\triangle$ by $\triangle^{-1}$, that the following matrix is the inverse of $\triangle$, i.e.

$$\triangle^{-1}(\mathbf{x}',\mathbf{x}) = \delta_{1,2}(x-x')$$

$$\times \begin{pmatrix} \mathbb{0}_{2\times 2} & \begin{matrix} -\frac{1}{2}\epsilon_3(x-x') & 0 \\ 0 & \delta_3(x-x') \end{matrix} \\ \begin{matrix} \frac{1}{2}\epsilon_3(x-x') & 0 \\ 0 & -\delta_3(x-x') \end{matrix} & \mathbb{0}_{2\times 2} \end{pmatrix},$$

$$(27.116)$$

and

$$\int d\mathbf{x}' \triangle(\mathbf{x}', \mathbf{x}'') \triangle^{-1}(\mathbf{x}', \mathbf{x}) = \mathbb{1}_{4\times 4}\delta(\mathbf{x} - \mathbf{x}'').$$

Thus in the present case we obtain with Eqs. (27.113), (27.114):

$$\{A_1(\mathbf{x}, t), \pi_1(\mathbf{y}, t)\}_D$$

$$= \delta(\mathbf{x} - \mathbf{y}) - \int d\mathbf{x}' d\mathbf{x}''[-\partial_1\delta(\mathbf{x} - \mathbf{x}')][\triangle^{-1}(\mathbf{x}', \mathbf{x}'')]_{11} \times \overbrace{0}^{\{\chi_1, \pi_1\}} = 0$$

$$= \delta(\mathbf{x} - \mathbf{y}). \tag{27.117a}$$

Similarly

$$\{A_2(\mathbf{x}, t), \pi_2(\mathbf{y}, t)\}_D = \delta(\mathbf{x} - \mathbf{y}). \tag{27.117b}$$

But

$$\{A_3(\mathbf{x}, t), \pi_3(\mathbf{y}, t)\}_D - \delta(\mathbf{x} - \mathbf{y})$$

$$= -\int d\mathbf{x}' d\mathbf{x}''\{A_3(\mathbf{x}, t), \phi_1(\mathbf{x}', t)\}[\triangle^{-1}(\mathbf{x}', \mathbf{x}'')]_{11}\{\chi_1(\mathbf{x}'', t), \pi_3(\mathbf{y}, t)\}$$

$$= -\int d\mathbf{x}' d\mathbf{x}''[-\partial_3\delta(\mathbf{x} - \mathbf{x}')]\left[-\frac{1}{2}\epsilon_3(x' - x'')\delta_{1,2}(x' - x'')\right]\delta(\mathbf{x}'' - \mathbf{y})$$

$$= -\int d\mathbf{x}''\delta(\mathbf{x} - \mathbf{x}'')\delta(\mathbf{x}'' - \mathbf{y})$$

$$= -\delta(\mathbf{x} - \mathbf{y}). \tag{27.117c}$$

Hence altogether, excluding A_0,

$$\{A_1(\mathbf{x}, t), \pi_1(\mathbf{y}, t)\}_D = \delta(\mathbf{x} - \mathbf{y}),$$
$$\{A_2(\mathbf{x}, t), \pi_2(\mathbf{y}, t)\}_D = \delta(\mathbf{x} - \mathbf{y}),$$
$$\{A_3(\mathbf{x}, t), \pi_3(\mathbf{y}, t)\}_D = 0. \tag{27.117d}$$

Thus only $\{A_1, \pi_1\}_D, \{A_2, \pi_2\}_D$ are unequal zero and so the physical quantities A_1, π_1, A_2, π_2 are quantized, whereas the unphysical A_3, π_3 are not. Thus the two transverse components of A_μ are quantized whereas the other unphysical components are not and commute. The nonvanishing Dirac brackets (27.117d) therefore imply a symplectic quantization matrix like that in Sec. 27.1 with $q_i \to A_i, p_i \to \pi_i, i = 1, 2$.

The above steps of supplementing constraints by gauge fixing conditions and thus converting the system of first class constraints (with vanishing Poisson brackets) into a second class system (with some nonvanishing Poisson brackets), and then calculating the Dirac brackets (with computation of the inverse of the matrix $\triangle$) are very typical of the steps required for the quantization of a large number of models, specifically field theory models, particularly some with scalar fields and in $(1 + 1)$-spacetime, which are studied also for various other reasons.[†] We mention here without demonstration that gauge fixing is also necessary since otherwise the Green's function of the theory would not exist.[‡] Quantization of nonabelian theories along the lines of the above considerations has been discussed in the literature.[§]

27.9   Faddeev–Jackiw Canonical Quantization

An alternative to Dirac's method of quantization is the procedure suggested by Faddeev and Jackiw [90]. This is essentially a rediagonalization method which does not require the consideration of Dirac brackets; instead, the Lagrangian is rewritten in terms of transverse quantities only. One can describe the method also as one of classical Hamiltonian reduction of dynamics. In principle the idea is to use the constraints to eliminate superfluous degrees of freedom which also leads to an invertible symplectic matrix.[*] It is very instructive to consider this method in some detail and to compare it with the method of Dirac, and to demonstrate the equivalence.

27.9.1   The method of Faddeev and Jackiw

We employ the first-order Lagrangian introduced earlier. In this case we write the Lagrangian L[†]

$$L = p_\alpha \dot{q}_\alpha - H(p_\alpha, q_\alpha), \tag{27.118}$$

where p_α, q_α have $\alpha = 1, 2, \ldots, 2N$. Since L is uniquely defined only up to a total time derivative (which gives only a vanishing contribution to the action $S = \int L dt$), we can rewrite L as

$$L = \frac{1}{2}(p_\alpha \dot{q}_\alpha - q_\alpha \dot{p}_\alpha) - H(p_\alpha, q_\alpha), \tag{27.119}$$

[†]For such typical examples and calculations see e.g. Usha Kulshreshtha, D. S. Kulshreshtha and H. J. W. Müller–Kirsten [155], from where related literature can also be traced back to.

[‡]L. Maharana, H. J. W. Müller–Kirsten and A. Wiedemann [183].

[§]P. P. Srivastava [259].

[*]Various aspects to further clarify and illustrate the application of the method have been discussed by G. V. Dunne, R. Jackiw and C. A. Trugenberger [80], Jian–zu Zhang and H. J. W. Müller–Kirsten [289], [290].

[†]Note that the system described by this Lagrangian does not possess (primary) constraints since $p = \partial L/\partial \dot{q} = p$, i.e. $p - p = 0$ does not yield a constraint. See Example 27.6 for a Lagrangian which does possess primary constraints.

where we dropped the contribution $d(pq/2)/dt$. Next we introduce the following notation for the $4N$ commuting phase space variables (commuting here meaning bosonic):

$$(\xi_\alpha) \equiv (\xi_{1\alpha}, \xi_{2\alpha}) = \overbrace{\underbrace{p_1, p_2, \ldots, p_N}_{\xi_{1i}}, \ldots, \underbrace{p_{2N}}_{\xi_{1a}}}^{\xi_{1\alpha}}, \overbrace{\underbrace{q_1, q_2, \ldots, q_N}_{\xi_{2i}}, \ldots, \underbrace{q_{2N}}_{\xi_{2a}}}^{\xi_{2\alpha}} \equiv (p_\alpha, q_\alpha)$$

with index α attached to ξ assuming the values $1, 2, \ldots, 4N$, and attached to ξ_1 and ξ_2 the values $1, 2, \ldots, 2N$. Thus if $(\xi_{1\alpha}, \xi_{2\alpha}) \equiv (p, q)$, and if f^0 is the symplectic matrix

$$f^0 = \begin{pmatrix} 0 & 1 \\ -1 & 0 \end{pmatrix} \quad \text{with} \quad f^{0-1} = \begin{pmatrix} 0 & -1 \\ 1 & 0 \end{pmatrix},$$

we have

$$\frac{1}{2}(\xi_{1\alpha}, \xi_{2\alpha}) \begin{pmatrix} 0 & \mathbb{1}_{2N \times 2N} \\ -\mathbb{1}_{2N \times 2N} & 0 \end{pmatrix} \begin{pmatrix} \dot{\xi}_{1\alpha} \\ \dot{\xi}_{2\alpha} \end{pmatrix} \equiv \frac{1}{2} \xi f^0 \dot{\xi} = \frac{1}{2}(p\dot{q} - q\dot{p}).$$

$$(27.120)$$

Hence the canonical one-form of the Lagrangian L can be rewritten as

$$L = \frac{1}{2} \xi f^0 \dot{\xi} - H(\xi). \tag{27.121}$$

The equations of motion follow from $\delta L = 0$. This variation of the expression in Eq. (27.121) implies

$$\begin{aligned}
0 &= \delta L = \frac{1}{2} \delta\xi f^0 \dot{\xi} + \frac{1}{2} \xi f^0 \delta\dot{\xi} - \delta\xi H'(\xi) \\
&= \frac{1}{2} \delta\xi f^0 \dot{\xi} + \frac{1}{2} \frac{d}{dt}(\xi f^0 \delta\xi) - \frac{1}{2} \dot{\xi} f^0 \delta\xi - \delta\xi H'(\xi),
\end{aligned}$$

and hence $(f^{0T} = -f^0)$

$$0 = \frac{1}{2} \delta\xi f^0 \dot{\xi} - \frac{1}{2} \delta\xi f^{0T} \dot{\xi} - \delta\xi H'(\xi) = \delta\xi[f^0 \dot{\xi} - H'(\xi)]. \tag{27.122}$$

Now if the $\delta\xi_\alpha$ are independent (this is crucial!), we obtain

$$f^0 \dot{\xi} = H'(\xi), \tag{27.123}$$

i.e.

$$\dot{\xi}_\alpha = (f^0)^{-1}_{\alpha\beta} H'_\beta(\xi) = (f^0)^{-1}_{\alpha\beta} \frac{\partial H}{\partial \xi_\beta}. \tag{27.124}$$

This equation expresses in a compact form Hamilton's equations, i.e.

$$\dot{p} = -\frac{\partial H}{\partial q}, \quad \dot{q} = \frac{\partial H}{\partial p}$$

(for anticommuting (i.e. Grassmann) quantities ξ_α, both equations have the same sign).

We now consider the problem in a more general form.* We assume a Lagrangian of the form

$$L = A_\alpha(\xi)\dot{\xi}_\alpha - H(\xi), \quad \alpha = 1, \ldots, 4N, \tag{27.125}$$

where $A_\alpha = A_\alpha(\xi)$. Varying L as before we obtain

$$
\begin{aligned}
\delta L &= \delta\xi_\beta \frac{\partial A_\alpha}{\partial \xi_\beta}\dot{\xi}_\alpha + A_\alpha \delta\dot{\xi}_\alpha - \delta\xi H'(\xi) \\
&= \delta\xi_\beta \frac{\partial A_\alpha}{\partial \xi_\beta}\dot{\xi}_\alpha + \frac{d}{dt}(A\delta\xi) - \dot{A}\delta\xi_\alpha - \delta\xi H'(\xi) \\
&= \delta\xi_\beta \frac{\partial A_\alpha}{\partial \xi_\beta}\dot{\xi}_\alpha - \dot{\xi}_\beta \frac{\partial A_\alpha}{\partial \xi_\beta}\delta\xi_\alpha - \delta\xi H'(\xi) \\
&= \delta\xi_\beta \left[\left(\frac{\partial}{\partial \xi_\beta}A_\alpha - \frac{\partial}{\partial \xi_\alpha}A_\beta \right)\dot{\xi}_\alpha - \frac{\partial H}{\partial \xi_\beta} \right].
\end{aligned}
\tag{27.126}
$$

Again, if the $\delta\xi_\beta$ are independent, we can equate to zero their coefficients here, so that

$$f_{\beta\alpha}\dot{\xi}_\alpha = \frac{\partial H}{\partial \xi_\beta}, \tag{27.127}$$

where (observe the analogy with the field tensor of electrodynamics!)

$$f_{\beta\alpha} = \frac{\partial}{\partial \xi_\beta}A_\alpha - \frac{\partial}{\partial \xi_\alpha}A_\beta = -f_{\alpha\beta}. \tag{27.128}$$

If f^{-1} exists, we can rewrite Eq. (27.127) as

$$\dot{\xi}_\alpha = (f^{-1})_{\alpha\beta}\frac{\partial H}{\partial \xi_\beta}, \tag{27.129}$$

which is again the compact form of Hamilton's equations.

Our next step is to express this equation in terms of Poisson brackets. The (bosonic) Poisson brackets are defined by[†]

$$\{A, B\}_P := -\sum_\alpha \left(\frac{\partial A}{\partial \xi_{1\alpha}}\frac{\partial B}{\partial \xi_{2\alpha}} - \frac{\partial A}{\partial \xi_{2\alpha}}\frac{\partial B}{\partial \xi_{1\alpha}} \right). \tag{27.130}$$

---

*D. S. Kulshreshtha and H. J. W. Müller–Kirsten [154].

[†]Observe that since $\xi_{1\alpha}$ represents momentum components, the minus sign in front of the sum ensures that the definition agrees with that in Chapter 2.

Thus one obtains for these ξ_α:

$$\{\xi_1, \xi_2\}_P = -1, \quad \{\xi_2, \xi_1\}_P = 1, \quad \{\xi_2, \xi_2\}_P = 0, \quad \{\xi_1, \xi_1\}_P = 0.$$

Hence for ξ_α:

$$
\begin{aligned}
\{H, \xi_\alpha\}_P &= -\sum_\kappa \left(\frac{\partial H}{\partial \xi_{1\kappa}} \frac{\partial \xi_\alpha}{\partial \xi_{2\kappa}} - \frac{\partial H}{\partial \xi_{2\kappa}} \frac{\partial \xi_\alpha}{\partial \xi_{1\kappa}} \right) \\
&= -\sum_{\beta, \kappa} \frac{\partial H}{\partial \xi_\beta} \left(\frac{\partial \xi_\beta}{\partial \xi_{1\kappa}} \frac{\partial \xi_\alpha}{\partial \xi_{2\kappa}} - \frac{\partial \xi_\beta}{\partial \xi_{2\kappa}} \frac{\partial \xi_\alpha}{\partial \xi_{1\kappa}} \right) \\
&= \sum_\beta \frac{\partial H}{\partial \xi_\beta} \{\xi_\beta, \xi_\alpha\}_P = -\dot{\xi}_\alpha,
\end{aligned}
$$

the last equality holding provided (cf. Eq. (27.129))

$$\{\xi_\beta, \xi_\alpha\}_P = -(f^{-1})_{\alpha\beta}, \tag{27.131}$$

if, of course, the inverse of f exists, which is guaranteed if the ξ_α's, $\alpha = 1, 2, \ldots, 4N$, are independent, and so are not subjected to constraints.

In the method of Faddeev and Jackiw Eq. (27.129) is written

$$\dot{\xi}_\alpha = -\frac{\partial H}{\partial \xi_\beta} \{\xi_\beta, \xi_\alpha\}_{FJ}, \tag{27.132}$$

where the *Faddeev–Jackiw bracket* $\{\cdots, \cdots\}_{FJ}$ is defined by

$$\{\xi_\beta, \xi_\alpha\}_{FJ} := -(f^{-1})_{\alpha\beta}. \tag{27.133}$$

We illustrate the use of this method by application to a specific case in Example 27.6. One should note that since

$$\det(f_{\alpha\beta}) = \det(f_{\alpha\beta}^T) = \det(f_{\beta\alpha}),$$

and in view of the antisymmetry of $f_{\alpha\beta}$, we have

$$\det(f_{\alpha\beta}) = \det(-f_{\beta\alpha}) = (-1)^{n_d} \det(f_{\beta\alpha}),$$

where n_d is the dimensionality of the matrix $(f_{\alpha\beta})$. Thus either

$$\det(f_{\alpha\beta}) = 0 \quad \text{or} \quad n_d \quad \text{is even.}$$

If n_d is odd and so $\det(f_{\alpha\beta}) = 0$, we have to impose another condition (the gauge fixing condition) in order to be able to invert $(f_{\alpha\beta})$. In Example 27.6

one does not have to impose such a condition since the dimensionality of $(f_{\alpha\beta})$ has been constructed to be even.[‡]

Example 27.6: Faddeev–Jackiw method versus Dirac's method

Demonstrate the equivalence of the Faddeev–Jackiw and Dirac methods of canonical quantization in the case of the theory defined by the following Lagrangian:

$$L = \frac{1}{2}qC\dot{q} - H(q,p), \quad \text{with} \quad C = -C^T. \tag{27.134}$$

Solution: Since

$$p_\alpha = \frac{\partial L}{\partial \dot{q}_\alpha} = \frac{1}{2}(qC)_\alpha, \quad (\therefore \ q = 2pC^{-1}),$$

the problem possesses primary constraints

$$\Gamma_\alpha := p_\alpha - \frac{1}{2}(qC)_\alpha = 0, \quad \alpha = 1, \dots, 2N. \tag{27.135}$$

Since $\alpha = 1, \dots, 2N$, only $2N$ of the $4N$ canonical variables q_α, p_α are independent. Thus we can use the constraints (27.135) to eliminate all p_α by expressing these in terms of q_α, or we could eliminate all q_α and retain only variables p_α, or we can eliminate half of each. Each possibility has its own quantization conditions. We consider these now one after the other.
(a) In the first case the momenta $p_\alpha \equiv \xi_{1\alpha}$ are re-expressed in terms of $q_\alpha \equiv \xi_{2\alpha}$, and hence the quantity $A(\xi)$ of Eq. (27.125) is given by (recall from Eq. (27.120) the correspondence $q_\alpha \leftrightarrow \xi_{2\alpha}$)

$$A_\alpha(\xi)\dot{\xi}_\alpha = A_\alpha(\xi)(\dot{\xi}_{1\alpha}, \dot{\xi}_{2\alpha}) \equiv \frac{1}{2}qC\dot{q} \equiv \frac{1}{2}\xi_{2\alpha}C_{\alpha\beta}\dot{\xi}_{2\beta}, \tag{27.136}$$

and so the matrix $(f_{\beta\alpha})$ of Eq. (27.128) becomes the $2N \times 2N$ matrix (ξ_α here $\xi_{2\alpha}$)

$$f_{\beta\alpha} = \frac{\partial}{\partial \xi_\beta}\left(\frac{1}{2}\xi_{2\gamma}C_{\gamma\alpha}\right) - \frac{\partial}{\partial \xi_\alpha}\left(\frac{1}{2}\xi_{2\rho}C_{\rho\beta}\right) = \frac{1}{2}(C_{\beta\alpha} - C_{\alpha\beta}) = C_{\beta\alpha}, \quad \alpha, \beta = 1, \dots, 2N. \tag{27.137}$$

Since (see Eq. (27.133)) the inverse of f determines the Faddeev–Jackiw brackets, we obtain according to Eq. (27.133)

$$\{q_\alpha, q_\beta\}_{FJ} = -C_{\beta\alpha}^{-1} = -C_{\alpha\beta}^{-1T} = C_{\alpha\beta}^{-1}. \tag{27.138}$$

This is one set of the brackets, and hence one set of possible quantization conditions. We can obtain the other alternatives as follows.
(b) Using the constraint (27.135) to replace q_α's by p_α's, we can also write (since $\dot{q} = 2\dot{p}C^{-1}$)

$$\begin{aligned} L &= 2p_\alpha(\dot{p}C^{-1})_\alpha - H(q,p) = 2\dot{p}C^{-1}p - H(q,p) \\ &= -2pC^{-1}\dot{p} - H(q,p) + \text{total time derivative.} \end{aligned} \tag{27.139}$$

In this case we have by comparison with Eqs. (27.125) and (27.128) (and $p_\alpha \leftrightarrow \xi_{1\alpha}$ here ξ_α)

$$A\dot{\xi} = -2pC^{-1}\dot{p} \quad \text{and} \quad f_{\alpha\beta} = \frac{\partial}{\partial \xi_\alpha}(-2\xi_{1\gamma}C_{\gamma\beta}^{-1}) - \frac{\partial}{\partial \xi_\beta}(-2\xi_{1\rho}C_{\rho\alpha}^{-1}) = -4C_{\alpha\beta}^{-1}.$$

Thus now

$$\{p_\alpha, p_\beta\}_{FJ} = \frac{1}{4}C_{\beta\alpha} = \frac{1}{4}C_{\alpha\beta}^T. \tag{27.140}$$

[‡]D. S. Kulshreshtha and H. J. W. Müller–Kirsten [154].

(c)   With the help of the constraint (27.135) we can also write L as

$$L = p_\alpha \dot{q}_\alpha - H(q,p), \quad \alpha = 1, \dots, 2N,$$

and use the $2N$ constraints to eliminate N of the p's and N of the q's, so that the remaining N p's and N q's are independent. For convenience we choose the matrix C with vanishing $N \times N$ submatrices in the diagonal, i.e. we choose it skew symmetric as

$$C = \begin{pmatrix} \mathbb{0}_{ij} & G_{ib} \\ -G_{aj} & \mathbb{0}_{ab} \end{pmatrix}, \quad C_{ia} = -C_{ia}^T. \tag{27.141}$$

We express $p_\alpha \dot{q}_\alpha$ in terms of $p_i, q_j, i, j = 1, \dots, N$. Using Eq. (27.141) and allowing a, b to assume values $N+1, \dots, 2N$, we have

$$p_i = \frac{1}{2} q_\alpha C_{\alpha i} = \frac{1}{2} q_j \overbrace{C_{ji}}^{\mathbb{0}_{ij}} + \frac{1}{2} q_a C_{ai} = \frac{1}{2} q_a C_{ai}, \quad \dot{q}_a = 2\dot{p}_i C_{ia}^{-1},$$

and similarly

$$p_a = \frac{1}{2} q_j C_{ja}. \tag{27.142}$$

Then

$$p_a \dot{q}_a = \frac{1}{2}(q_j C_{ja})(2\dot{p}_i C_{ia}^{-1}) = q_j C_{ja} C_{ai}^{-1T} \dot{p}_i = -q_j C_{ja} C_{ai}^{-1} \dot{p}_i = -q_i \dot{p}_i. \tag{27.143}$$

Thus L can now be written

$$L = p_i \dot{q}_i + p_a \dot{q}_a - H(q,p) = p_i \dot{q}_i - q_i \dot{p}_i - H(q,p), \quad i = 1, \dots, N. \tag{27.144}$$

Hence with $p_i \equiv \xi_{1i}, q_i \equiv \xi_{2i}, i = 1, 2, \dots, N$ ($\xi_\alpha = (\xi_{1i}, \xi_{2i}), \alpha, \beta = 1, \dots, 2N$) as independent variables (out of the total of $2N$ of each of q_α and p_α) we have by comparison with Eq. (27.125):

$$A\dot{\xi} \equiv (p_i \dot{q}_i - q_i \dot{p}_i) = (\xi_{1i}, \xi_{2i}) \begin{pmatrix} \mathbb{0} & \mathbb{1}_{ij} \\ -\mathbb{1}_{ij} & \mathbb{0} \end{pmatrix} \begin{pmatrix} \dot{\xi}_{1j} \\ \dot{\xi}_{2j} \end{pmatrix} \equiv \xi_\alpha f^0_{\alpha\beta} \dot{\xi}_\beta, \tag{27.145a}$$

i.e.

$$f_{\alpha\beta} = \frac{\partial}{\partial \xi_\alpha}(\xi_\gamma f^0_{\gamma\beta}) - \frac{\partial}{\partial \xi_\beta}(\xi_\rho f^0_{\rho\alpha}) = 2f^0_{\alpha\beta}, \tag{27.145b}$$

with

$$f^0 = \begin{pmatrix} \mathbb{0} & \mathbb{1}_{N \times N} \\ -\mathbb{1}_{N \times N} & \mathbb{0} \end{pmatrix}, \quad f^{-1} = \frac{1}{2} \begin{pmatrix} \mathbb{0} & -\mathbb{1}_{N \times N} \\ \mathbb{1}_{N \times N} & \mathbb{0} \end{pmatrix}, \tag{27.146}$$

and so with (27.133)

$$\{\xi_\alpha, \xi_\beta\} = -(f^{-1})_{\beta\alpha} = \frac{1}{2} \begin{pmatrix} \mathbb{0} & -\mathbb{1}_{N \times N} \\ \mathbb{1}_{N \times N} & \mathbb{0} \end{pmatrix}_{\alpha\beta},$$

$$\{p_i, q_j\}_{FJ} = -\frac{1}{2}\delta_{ij}, \quad \{q_i, p_j\}_{FJ} = \frac{1}{2}\delta_{ij}. \tag{27.147}$$

The brackets (27.138), (27.140), (27.146) are the *Faddeev–Jackiw brackets* which correspond to Dirac brackets $\{\cdots, \cdots\}_D$, but are not introduced as such here.

It is now interesting to compare the results above, i.e. for the Faddeev–Jackiw brackets, with the brackets of Dirac's method. In the latter the constraints and gauge fixing conditions $\Gamma_\alpha(q,p)$ have to be second class, so that

$$\{\Gamma_\alpha(q,p), \Gamma_\beta(q,p)\}_P \equiv \triangle_{\alpha\beta} \neq 0, \tag{27.148}$$

so that the inverse $\triangle^{-1}$ exists. Then, as we have seen, the Dirac brackets are defined by

$$\{A, B\}_D = \{A, B\}_P - \sum_{\alpha,\beta} \{A, \Gamma_\alpha\}_P (\triangle^{-1})_{\alpha\beta} \{\Gamma_\beta, B\}_P. \tag{27.149}$$

In the above with constraints (27.135) we have the Poisson brackets

$$\triangle_{\alpha\beta} = \{\Gamma_\alpha, \Gamma_\beta\}_P = \sum_\rho \left(\frac{\partial \Gamma_\alpha}{\partial q_\rho} \frac{\partial \Gamma_\beta}{\partial p_\rho} - \frac{\partial \Gamma_\beta}{\partial q_\rho} \frac{\partial \Gamma_\alpha}{\partial p_\rho} \right)$$

$$= \left(-\frac{1}{2} C_{\rho\alpha} \right) \delta_{\beta\rho} - \left(-\frac{1}{2} C_{\rho\beta} \right) \delta_{\alpha\rho} = C_{\alpha\beta},$$

$$\{q_\alpha, \Gamma_\beta\}_P = \delta_{\alpha\beta}, \quad \{p_\alpha, \Gamma_\beta\}_P = -\left(-\frac{1}{2} C_{\rho\beta} \right) \delta_{\alpha\rho} = \frac{1}{2} C_{\alpha\beta}, \tag{27.150}$$

and hence

$$\{q_\alpha, q_\beta\}_D = 0 - \sum_{\alpha',\beta'} \{q_\alpha, \Gamma_{\alpha'}\}_P (C^{-1})_{\alpha'\beta'} \{\Gamma_{\beta'}, q_\beta\}_P = -(\delta_{\alpha\alpha'})(C^{-1})_{\alpha'\beta'}(-\delta_{\beta'\beta}) = C^{-1}_{\alpha\beta}$$

$$\tag{27.151}$$

in agreement with Eq. (27.138). Similarly we have

$$\begin{aligned}
\{p_\alpha, p_\beta\}_D &= 0 - \sum_{\alpha',\beta'} \{p_\alpha, \Gamma_{\alpha'}\}_P (C^{-1})_{\alpha'\beta'} \{\Gamma_{\beta'}, p_\beta\}_P \\
&= -\left(\frac{1}{2} C_{\alpha\alpha'} \right)(C^{-1})_{\alpha'\beta'} \left(\frac{1}{2} C_{\beta'\beta} \right) = \frac{1}{4} C_{\beta\alpha} \tag{27.152}
\end{aligned}$$

in agreement with Eq. (27.140). Finally we have

$$\begin{aligned}
\{p_\alpha, q_\beta\}_D &= -\delta_{\alpha\beta} - \sum_{\alpha',\beta'} \{p_\alpha, \Gamma_{\alpha'}\}_P (C^{-1})_{\alpha'\beta'} \{\Gamma_{\beta'}, q_\beta\}_P \\
&= -\delta_{\alpha\beta} - \left(\frac{1}{2} C_{\alpha\alpha'} \right)(C^{-1})_{\alpha'\beta'}(-\delta_{\beta\beta'}) \\
&= -\delta_{\alpha\beta} + \frac{1}{2} \delta_{\alpha\beta} = -\frac{1}{2} \delta_{\alpha\beta} \tag{27.153}
\end{aligned}$$

in agreement with Eq. (27.146). The transition to quantum mechanics is now achieved with the replacement of Dirac brackets by (bosonic) operator commutation relations according to the recipe

$$\{A, B\}_D \longrightarrow \frac{1}{i} [A, B] \quad (\hbar = 1). \tag{27.154}$$

Chapter 28

The Quantum-Classical Crossover as Phase Transition

28.1 Introductory Remarks

Up to this point the major part of this text has been concerned with the exploitation of the two calculational methods in quantum mechanics characterized by the Schrödinger equation in the one case and the path integral in the other. Our comparison of the methods in deriving the same results with either method required in the path integral method the approximation known as the *dilute gas approximation*, i.e. the assumed independence of instantons and anti-instantons in the summation of the effects of an infinite number of these. We also observed that (with some minor deficiency) the WKB approximation also reproduces these results. This means, one achieves agreement by considering relatively low-lying quantum states, i.e. those close to the ground state, and on the assumption of high barriers. Thus the decay rates of metastable states are determined by tunneling processes whose dynamics is described by periodic instantons or bounces. Solutions of the associated classical equation can be obtained at all levels of a barrier. In considering what happens higher up in a barrier where thermal activation becomes more and more important and close to the top of a barrier beyond some critical or crossover temperature T_c (where thermal activation becomes the decisive mechanism), the above methods break down, i.e. the Schrödinger equation and the dilute gas approximation, but one can still find use of the classical configurations, i.e. the periodic configurations up to the *sphaleron*, by relating their decreasing period to increasing temperature. This is the

753

aspect we explore in this chapter on an introductory level, which is therefore on the border of the contents of the foregoing and indicates the limitations of the latter. In continuation of our previous considerations it is thereby convenient and instructive to consider again the quartic and the cosine potentials, however with an extension originating from models of large spins, which we can only briefly refer to here since there was no space in earlier chapters to enter into the subject of spin or intrinsic angular momentum of particles. However, in the following at most a general appreciation of such spins is all that is required along with the view that a conglomeration of a large number of such spins when somehow aligned may be looked at as an almost macroscopic system with a large spin. In recent years considerable progress was made in understanding macroscopic spin systems with degenerate vacua. This progress was achieved largely independently of developments in field theory where very similar methods were motivated by the realization that classical field configurations play an important role in our understanding of fundamental particle phenomena. Thus instanton methods wellknown in theoretical particle physics for more than two decades appeared in spin tunneling investigations about ten years later.* Configurations corresponding to the *periodic instantons* found in field theory contexts[†] were called *thermons* in independent spin tunneling contexts.[‡] Since then progress in the study of spin systems uncovered a host of model theories with degenerate vacua permitting explicit evaluation of periodic instantons and their investigation. In all nontrivial cases these classical configurations involve again elliptic functions.

Very recently the study of macroscopic spin systems aroused again new interest with the discovery[§] that they provide examples which exhibit first order phase transitions (in the plot of action S versus temperature T_e) of which simple examples were not known. In view of the possibility of experimental verification of such a transition in decay rates of certain spin systems, and their interpretation as a crossover from classical to quantum behaviour, such systems are also of fundamental interest. The characteristic way in which phase transitions appear in quantum-mechanical tunneling processes was first worked out by Chudnovsky [50], and was elaborated on by Gorokhov and Blatter [117], [118]. There are very few models which al-

---

*E. M. Chudnovsky [50], M. Enz and R. Schilling [84].

[†]J.–Q. Liang, H. J. W. Müller–Kirsten and D. H. Tchrakian [165]. This means field theory models effectively reduced to quantum mechanics, like those discussed in earlier chapters.

[‡]E. M. Chudnovsky [50]. The thermon defined by Chudnovsky is a configuration travelling through one complete period, whereas the periodic instanton, in keeping with its name (i.e. the analogy with the vacuum instanton) is defined as a configuration over half the complete period, so that the action of the thermon is twice the action of the corresponding periodic instanton.

[§]D. A. Garanin and E. M. Chudnovsky [104], E. M. Chudnovsky and D. A. Garanin [51].

low an explicit and analytic investigation,[¶] so that these few models serve as very instructive prototypes for the study of the crossover from quantum to classical behaviour. The most straightforward of these models can be related to the double well and cosine potentials which figured prominently in this text, so that it is natural to include their discussion here. This is therefore our intention here. We can relate the period of the periodic instantons to the reciprocal of temperature, the harmonic oscillations around the sphaleron configuration at the peak of a barrier then providing a critical temperature. We then investigate the behaviour of the action as a function of temperature. The double well and cosine potentials we discussed earlier will be shown to possess smooth *second order crossover transitions*. Spin models which can be related to these but possess in addition a dependence of the mass on the classical (field) configuration, will be shown to exhibit *first order crossover transitions* provided this dependence is not too weak. We show that these crossover transitions are indeed very similar to phase transitions in the thermodynamics of (e.g.) a van der Waals gas.

28.2 Relating Period to Temperature

In this section we relate the period $P(E_{cl})$ of the periodic instantons that we studied previously to temperature T_e in their approach to the peak of the potential barrier. All the periodic instantons are saddle point solutions of the classical equation of motion in imaginary time, as we discussed in detail in earlier chapters, and hence are associated with a complex energy whose imaginary part determines the rate of escape of the particle from a potential well through a barrier of finite height. In the *Breit–Wigner formula* (10.73) the complex energy was written

$$E = E^{(0)} - i\frac{\gamma}{2}, \tag{28.1}$$

where $E^{(0)} = E_n$ is the energy of the particle in the well (i.e. with no possibility for escape in the dominant approximation). We also saw that in the WKB or semiclassical approximation the quantity $\gamma \to \gamma_n$ is given by

$$\hbar\gamma_n = \frac{\hbar\omega(E_n)}{2\pi} \exp\left[-\frac{1}{\hbar}\oint |p|dq\right] = \frac{\hbar\omega(E_n)}{2\pi} \exp\left[-\frac{S}{\hbar}\right], \tag{28.2}$$

[¶]J.–Q. Liang, H. J. W. Müller–Kirsten, D. K. Park and F. Zimmerschied [170]; S. Y. Lee, H. J. W. Müller–Kirsten, D. K. Park and F. Zimmerschied [160].

where p is the momentum and S the classical action of the classical configuration, and the frequency ω is given by (see e.g. Sec. 26.1)

$$\left[\oint \frac{dq}{p} \right]^{-1} = \frac{\hbar \omega}{2\pi}$$

(compare, for instance, with Eqs. (14.89) and (14.90)).

We encountered the *canonical distribution* of statistical mechanics earlier briefly with Eqs. (5.43) and (7.57), i.e. the averaging factor

$$z_n = \frac{1}{Z} e^{-\beta E_n}, \quad \beta = \frac{1}{k_B T_e}, \tag{28.3}$$

where $Z = \sum_n \exp(-\beta E_n)$ is the partition function associated with the particle in the well, and $\{E_n\}$ are its eigenenergies therein. Hence the *mean decay rate* Γ *at temperature* T_e is given by

$$\Gamma = \sum_n z_n \gamma_n = \frac{1}{Z} \sum_n \frac{\omega(E_n)}{2\pi} \exp\left[-\frac{S}{\hbar} - \beta E_n \right]. \tag{28.4}$$

One now approximates this expression by its value at the maximum of the exponential with respect to energy E. Thus we solve ($E \leftrightarrow E_n$)

$$\frac{d}{dE}\left[-\frac{S}{\hbar} - \beta E \right] = -\frac{1}{\hbar}\frac{\partial S}{\partial E} - \beta = 0.$$

In a periodic case with period $P(E_{\text{cl}})$ and (cf. e.g. Eq. (26.21))

$$S(\phi_c) = W + E_{\text{cl}} \underbrace{(\tau_f - \tau_i)}_{-P(E_{\text{cl}})} \quad \text{and} \quad \frac{\partial W}{\partial E_{\text{cl}}} = 0,$$

(see the comment to Eq. (26.23)) the function W is independent of E_{cl}. Hence

$$\frac{\partial S}{\partial E_{\text{cl}}} = -P(E_{\text{cl}}), \tag{28.5}$$

and

$$\frac{P(E_{\text{cl}})}{\hbar} - \beta = 0, \quad \text{i.e.} \quad P(E_{\text{cl}}) = \hbar\beta = \frac{\hbar}{k_B T_e}. \tag{28.6}$$

28.3   Crossover in Previous Cases

For the demonstration of the method of investigation we consider first familiar cases examined in detail in earlier chapters. Thus we consider specifically the two cases of double well and cosine potentials and show that (with no new changes) these provide examples of *second order phase transitions*, this classification being taken over from thermodynamics, as we explain later.

28.3.1 The double well and phase transitions

Since we studied the double well in detail in Chapters 18 and 26 it is convenient to begin with this case, as this enables us also to exploit earlier results.
Thus the double well potential defined by Eq. (26.2) is (with mass $m_0 = 1$)

$$V(\phi) = \frac{g^2}{2} \left(\phi^2 - \frac{m^2}{g^2} \right)^2. \tag{28.7}$$

The solution of the Euclidean time classical equation was found to be the
result (26.4), i.e.

$$\phi_c(\tau) = \frac{kb(k)}{g} \mathrm{sn}[b(k)\tau], \tag{28.8}$$

$$b(k) = \frac{m\sqrt{2}}{\sqrt{1+k^2}}, \quad u = \frac{g}{m^2}\sqrt{2E_{\mathrm{cl}}} = \frac{1-k^2}{1+k^2}, \tag{28.9}$$

and k is the elliptic modulus of the Jacobian elliptic function. Here E_{cl} is
the energy of the pseudo-classical configuration, i.e. the periodic instanton,
and is given by Eq. (26.3), i.e.

$$\frac{1}{2} \left(\frac{d\phi}{d\tau} \right)^2 - V(\phi) = -E_{\mathrm{cl}}. \tag{28.10}$$

The finite maximum and the minima of the potential at $\phi = 0$ and $\phi_\pm = \pm m/g$ respectively determine the range of E_{cl} which is of relevance here.
Thus with the corresponding values of the elliptic modulus we have

$$\underbrace{0}_{k^2=1} \le E_{\mathrm{cl}} \le \underbrace{\frac{m^4}{2g^2} \equiv E_0}_{k^2=0}. \tag{28.11}$$

Here E_0 is the energy of the pseudoparticle at the top of the barrier; this
configuration is the so-called *sphaleron*. The periodic instanton is the configuration which performs periodic back and forth tunneling in the potential
barrier, i.e. between one turning point and the other. At the turning points
its velocity is zero. Assuming the periodic instanton starts from $\phi = 0$ at
Euclidean time $\tau = -2T$ and reaches the turning point $\phi = -a$ at time
$\tau = -T$, and then travels back to $\phi = 0$ which it reaches at time $\tau = 0$, and
so on, the configuration $\phi_c(\tau)$ has to satisfy the boundary conditions

$$\phi_c(\tau)\Big|_{\tau=-2T} = \phi_c(\tau)\Big|_{\tau=2T} = 0.$$

Since sn[u] has real zeros at $u = 0, 2K(k)$ (cf. Eq. (26.6a)), we must have

$$b(k)2T = 2K(k),$$

or with the full period $4T \equiv P(E_{cl})$:

$$b(k)P(E_{cl}) = 4K(k), \quad P(E_{cl}) = \frac{4}{b(k)}K(k) = \frac{4\sqrt{1+k^2}}{m\sqrt{2}}K(k). \quad (28.12)$$

The quarter period $K(k)$ is a monotonically increasing function as shown in Fig. 28.1.[||]

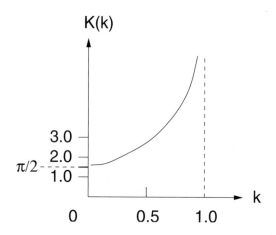

Fig. 28.1 The behaviour of the quarter period $K(k)$.

It follows that in the energy range (28.11) from 0 to E_0 the period $P(E_{cl})$ is monotonically decreasing as shown in Fig. 28.2. Thus the equation $P'(E_{cl}) = 0$ does not have a solution anywhere in this range (i.e. no extremum). With the identification

$$P(E_{cl}) = \frac{\hbar}{k_B T_e}, \quad (28.13)$$

where T_e means temperature, we see that the temperature increases with the energy.

We next consider the action of the thermon, S_{T_e} (or of the pair of a periodic instanton and an anti-instanton, as illustrated in Fig. 26.1). We obtain this action from Eqs. (26.13) and (26.14b) (with $4T = P(E_{cl}) = \hbar/k_B T_e = 1/T_e$ with $\hbar = k_B = 1$) as

$$S_{T_e} = \frac{E_{cl}}{T_e} + \frac{8m^3}{3g^2}(1+u)^{1/2}[E(k) - uK(k)], \quad (28.14)$$

[||]See also P. F. Byrd and M. D. Friedman [40], Fig. 6, p. 17.

where $E(k)$ is the complete elliptic integral of the second kind. The action at the top of the barrier, i.e. of the sphaleron configuration, is that for $k^2 = 0$ and is called *"thermodynamic action"* S_0, i.e. (since $E(0) = K(0) = \pi/2$)

$$S_0 = \frac{m^4}{2g^2} \frac{1}{T_e}. \tag{28.15}$$

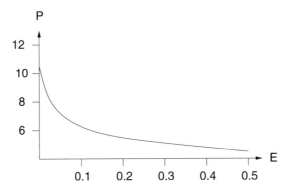

Fig. 28.2 The double well period in the relevant range of E_{cl} for $m = 1$ and $g = 1$.

Figure 28.3 displays the behaviour of S_{T_e} and S_0 versus temperature T_e for $m = 1, g = 1$. One can clearly see the typically smooth behaviour of a *second order transition* from the *thermal to the quantum regime*, i.e. from $T_e = 0.5$ to zero, i.e. as the temperature is lowered. Conversely we can argue that as the temperature is increased the number of periodic instantons and anti-instantons or thermons increases and crowd together, and the dilute gas approximation breaks down. The critical temperature at which the periodic instantons and anti-instantons condense and disorder the system can be shown to be that of harmonic oscillations around the sphaleron.[**]

28.3.2 The cosine potential and phase transitions

In the case of the sine–Gordon potential (26.51)

$$V(\phi) = \frac{1}{g^2}[1 + \cos(g\phi)], \tag{28.16}$$

the periodic instanton is the configuration (26.53), i.e.

$$\phi_c(\tau) = \frac{2}{g} \sin^{-1}[k\,\mathrm{sn}(\tau)] \tag{28.17}$$

[**]E. M. Chudnovsky [50].

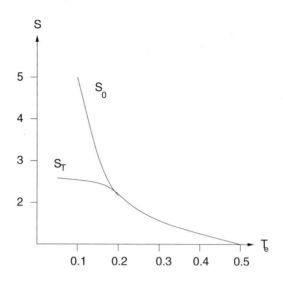

Fig. 28.3 Thermodynamic and thermon actions S_0, S_{T_e} versus temperature T_e.

with an integration constant τ_0 chosen to be zero. In this case the period is (cf. Eq. (26.55a))

$$P(E_{\rm cl}) = 4K(k), \quad k^2 = 1 - \frac{1}{2}g^2 E_{\rm cl}. \tag{28.18}$$

Again the derivative of the period with respect to the energy can be shown not to possess a solution in the domain $0 < E_{\rm cl} < 2/g^2$. Proceeding as above and using the results of Sec. 26.3, the thermon and thermodynamic actions are obtained respectively as

$$S_{T_e} = \frac{E_{\rm cl}}{T_e} + \frac{16}{g^2}[E(k) - (1 - k^2)K(k)], \quad S_0 = \frac{2}{g^2 T_e}. \tag{28.19}$$

Again one finds a second order transition from the thermal to the quantum regime similar to the double well case so that we do not present the corresponding figures.

28.4   Crossover in a Simple Spin Model

We now consider in a semiclassical approximation a large spin like that of an almost macroscopic (e.g. ferromagnetic) particle in the context of a model with degenerate vacua. Loosely speaking, we will study a problem in which the double well appears around the circumference of a circle, so that the problem is indeed very analogous to the type of problems we considered previously. Quantum tunneling between macroscopically distinct states is of great interest both theoretically and experimentally. Since the topic has developed into an individual area of research, our intention here is — and can only be — to supplement our earlier detailed studies with a glimpse into this field by considering the most immediate analogous example.

As a simple introduction consider a *classical* three-dimensional vector $\mathbf{S}$ in Euclidean space with Hamiltonian[*]

$$H^{\text{cl}} := -AS_x^2 + BS_z^2 - hS_y, \tag{28.20}$$

where A, B, h (h an applied magnetic field) are real coefficients. In spherical polar coordinates H^{cl} becomes, with $S = |\mathbf{S}|$,

$$H^{\text{cl}}(\theta, \phi) = -AS^2 \cos^2 \phi \sin^2 \theta + BS^2 \cos^2 \theta - hS \sin \phi \sin \theta. \tag{28.21}$$

One can see that this expression is minimized for $\theta_0 = \pi/2$, where it becomes

$$H^{\text{cl}}_{\theta_0=\pi/2} = -AS^2 \cos^2 \phi - hS \sin \phi = -AS^2 + AS^2 \sin^2 \phi - hS \sin \phi. \tag{28.22}$$

Setting the derivative of this expression with respect to $\sin \phi$ equal to zero (and excluding the unphysical case of $\cos \phi = 0$) one obtains the two degenerate positions (for $0 \leq h < 2AS$) given by

$$\sin \phi_0 = \frac{h}{2AS} \equiv -\sin(\phi_0 \pm \pi), \quad \text{i.e.} \quad \phi_0 = \begin{cases} \mp\pi - \sin^{-1}(\frac{h}{2AS}) \\ \sin^{-1}(\frac{h}{2AS}) \end{cases} \tag{28.23}$$

(for $h = 0$ the classical vacua lie diametrically opposite) with energy

$$E_0^{\text{cl}} = -\left[1 + \left(\frac{h}{2AS}\right)^2\right] AS^2. \tag{28.24}$$

Thus this model is analogous to that of the double well problem in having two degenerate minima at the positions ϕ_0 given by Eq. (28.23). However, the analytical calculation of the level splitting was not immediately successful as is recalled in the literature.[†] Thus, proceeding to the physical formulation, in which $\mathbf{S}$ represents a *spin operator* and h an applied magnetic field, one can expect the problem to be even more complicated. In the following we attempt to present a simple introduction which can serve as a first step into the literature on a large number of similar models with various directions of an applied magnetic field (probably all possible cases have by now been considered somewhere in the literature).

---

[*]M. Enz and R. Schilling [84], first paper.
[†]See M. Enz and R. Schilling [84], first paper, who cite the earlier investigations.

The first idea is to formulate the problem in such a way that in a leading approximation the operator form of the Hamiltonian, i.e.

$$\hat{H} = -A\hat{S}_x^2 + B\hat{S}_z^2 - h\hat{S}_y, \qquad (28.25)$$

reproduces the classical form (28.20). Here the operators $\hat{S}_x, \hat{S}_y, \hat{S}_z, \hat{S}_+, \hat{S}_-$ have to be looked at as operators satisfying an algebra very much akin to that of the angular momentum operators l_x, l_y, l_z, l_+, l_- of Eqs. (11.54) to (11.56). Once this has been accepted, the following arguments are not too difficult. Since our aim is a semiclassical expansion with the classical expression as the leading term, a standard representation of the operators is inappropriate. Thus a new representation of the operators is needed. This representation with components which satisfy the algebra of Eqs. (11.54) to (11.56) is a representation discovered by Villain.[‡] This *Villain representation* is given by

$$\hat{S}_+ = \exp(i\hat{\phi})[S(S+1) - \hat{S}_z(\hat{S}_z + 1)]^{1/2}, \quad \hat{S}_- = (\hat{S}_+)^\dagger \qquad (28.26)$$

with the canonical commutation relation

$$[\hat{\phi}, \hat{S}_z] = i. \qquad (28.27)$$

These Villain operators, with e.g. $\hat{S}_\pm = \hat{S}_x \pm i\hat{S}_y$ etc., can be verified to satisfy the algebra of Eqs. (11.54) to (11.56). The commutation relation (28.27) is, effectively, the relation $[\theta, p_\theta] = i$ we encountered earlier in Eq. (27.79), and expresses simply the quantization of this angular momentum, which is classically $S_z = S \cos\theta$ (as in Eq. (28.21)). Thus all operators $\hat{S}_x, \hat{S}_y, \hat{S}_z, \hat{S}_+, \hat{S}_-$ can now be re-expressed in terms of the operators $\hat{\phi}$ and $\hat{S}_z$.

The next step is now[§] to replace the operators in the operator Hamiltonian (28.25) by the Villain operators, then to replace the canonical operators $\hat{\phi}, \hat{S}_z$ by their classical equivalents $\phi, S \cos\theta$, and to expand the expressions in powers of $1/S$ and S_z/S. In this way one obtains for $\hat{S}_x^2$ the following *semiclassical* expression:

$$\hat{S}_x^2 \to \cos^2\phi[S(S+1) - S_z(S_z + 1)],$$

and with $S_z = S \cos\theta$,

$$
\begin{aligned}
\hat{S}_x^2 &= \cos^2\phi[S(S+1) - S\cos\theta(S\cos\theta + 1)] \\
&= \cos^2\phi[S^2 \sin^2\theta + S(1 - \cos\theta)] \\
&= \cos^2\phi S^2\left[1 + \frac{1}{S} - \left(\frac{S_z}{S}\right)^2 - \left(\frac{S_z}{S^2}\right)\right] \\
&\simeq \cos^2\phi S^2\left[\sin^2\theta + \frac{1}{S}\right]. \qquad (28.28)
\end{aligned}
$$

[‡]J. Villain [277].

[§]In case of doubt see also the steps as explained by M. Enz and R. Schilling [84].

We see that the leading term reproduces the corresponding term in the classical Hamiltonian (28.21). We also observe that the contribution $\cos\theta/S$ is neglected. This term violates in any case the invariance of the original operator Hamiltonian under rotations about the y-axis (replacements $\theta \to \theta + \pi, \phi \to \phi + \pi$). In a similar way a symmetry violating $\cos\theta$ term has to be ignored in the semiclassical expansion of the expression obtained from $\hat{S}_y$, in order to maintain that symmetry at every order of the large-S expansion. We can now see that in this way one obtains the semiclassical Hamiltonian (up to the order of terms maintained)

$$H^{\mathrm{scl}}(\theta,\phi) = H^{\mathrm{cl}}(\theta,\phi) - AS\cos^2\phi - \frac{h}{2}\sin\phi\sin\theta \quad \text{with} \quad S\cos\theta = \frac{p}{\hbar}, \quad (28.29)$$

where the latter expression insures that the Hamiltonian is expressed in terms of the appropriate canonical variables, i.e. ϕ and the conjugate momentum p.

In the remaining part of this chapter we restrict ourselves predominantly to the case without a magnetic field (i.e. $h = 0$) with

$$
\begin{aligned}
H^{\mathrm{scl}} &= -A\cos^2\phi S^2\left(1 - \frac{p^2}{\hbar^2 S^2}\right) - AS\cos^2\phi + \frac{Bp^2}{\hbar^2} \\
&= -AS(S+1)\cos^2\phi + \frac{p^2}{\hbar^2}(A\cos^2\phi + B).
\end{aligned}
$$

In order to obtain the Lagrangian $L(\phi,\dot{\phi})$, we perform a Legendre transform, for which

$$\dot{\phi} = \frac{\partial H^{\mathrm{scl}}(\theta,\phi)}{\partial p} = \frac{2}{\hbar^2}(A\cos^2\phi + B)p, \quad p = \frac{\hbar^2}{2}\frac{\dot{\phi}}{(A\cos^2\phi + B)}. \quad (28.30)$$

Then

$$L(\phi,\dot{\phi}) = -p\dot{\phi} + H^{\mathrm{scl}}, \quad H^{\mathrm{scl}} = -AS(S+1)\cos^2\phi + \frac{\hbar^2\dot{\phi}^2}{4(A\cos^2\phi + B)}. \quad (28.31)$$

Transforming to Euclidean time $\tau = it$, we then obtain

$$L\left(\phi,\frac{d\phi}{d\tau}\right) = \frac{1}{2}m(\phi)\left(\frac{d\phi}{d\tau}\right)^2 + V(\phi), \quad (28.32)$$

where

$$m(\phi) = \frac{\hbar^2}{2(A\cos^2\phi + B)} \quad (28.33)$$

and

$$V(\phi) = -AS(S+1)\cos^2\phi. \quad (28.34)$$

The corresponding Hamiltonian and classical energy E_{cl} now follow as

$$H(\phi(\tau), p(\tau)) = \frac{1}{2}m(\phi)\left(\frac{d\phi}{d\tau}\right)^2 - V(\phi) = -E_{\mathrm{cl}}. \tag{28.35}$$

In the following and for ease of comparison with the literature it is convenient to redefine the constants A and B in the following way:

$$A + B \equiv K_1 \quad \text{and} \quad A \equiv K_2 \quad \text{with} \quad \lambda = \frac{K_2}{K_1}. \tag{28.36}$$

Then[*]

$$m(\phi) = \frac{\hbar^2}{2K_1(1 - \lambda\sin^2\phi)}, \quad V(\phi) = \underbrace{-K_2 S(S+1)}_{\text{below ignored}} + K_2 S(S+1)\sin^2\phi. \tag{28.37}$$

In Eq. (28.35) E_{cl} is a constant of integration (of the second order differential equation) and may be viewed as the classical energy of the pseudoparticle configuration. The maximum value of this energy, $E_0 = K_2 S(S+1)$, is the energy of the pseudoparticle configuration there, i.e. that of the *sphaleron*, $\phi_{\mathrm{sph}} = \pi/2$. Integration of Eq. (28.35) yields the following periodic instanton configuration as is shown in Example 28.1:

$$\phi_c(\tau) = \sin^{-1}\left[\frac{1 - k^2\mathrm{sn}^2(\omega\tau)}{1 - \lambda k^2\mathrm{sn}^2(\omega\tau)}\right]^{1/2}, \tag{28.38}$$

where the elliptic modulus k and the other parameters are given by the following relations:

$$k^2 = \frac{\nu^2 - 1}{\nu^2 - \lambda} = \frac{K_2 S(S+1) - E_{\mathrm{cl}}}{K_2 S(S+1) - E_{\mathrm{cl}}\lambda}, \quad \nu^2 = \frac{K_2 S(S+1)}{E_{\mathrm{cl}}},$$

$$\omega^2 = \omega_0^2\left(1 - \frac{\lambda}{\nu^2}\right) = 4K_2[K_1 S(S+1) - E_{\mathrm{cl}}], \quad \omega_0^2 = 4K_1 K_2 S(S+1). \tag{28.39}$$

---

[*]It may be observed here that the theory defined by Eq. (28.25) without applied magnetic field ($h = 0$) leads to a periodic potential and hence to a quantum mechanical level splitting. The latter has been investigated in J.–Q. Liang, H. J. W. Müller–Kirsten and Jian-Ge Zhou [171]. With inclusion of the magnetic field h Eqs. (28.33) and (28.34) are [84]

$$\begin{aligned} m(\phi) &= \hbar^2[2(A\cos^2\phi + B) + h\sin\phi/(S+1/2)]^{-1}, \\ V(\phi) &= -[AS(S+1)\cos^2\phi + h(S+1/2)\sin\phi]. \end{aligned}$$

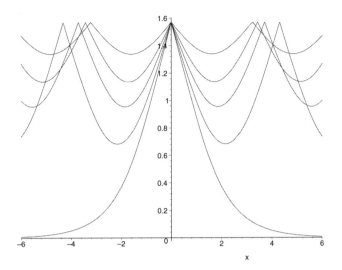

Fig. 28.4 The periodic instanton (28.38) for $k^2 = 0.1, 0.3, 0.5, 0.75, 1.0$
from top to bottom, $\lambda = 0.5$ and $x = \omega\tau = -6$ to 6.

The behaviour of the solution (28.38) as a function of $x = \omega\tau$ is shown by
the MAPLE-plot in Fig. 28.4 for several values of the elliptic modulus k. We
infer from the periodic solution (28.38) that its period $P(E)$ is

$$P(E) = \frac{4K(k)}{\omega} = \frac{2}{\sqrt{K_1}} \frac{1}{\sqrt{K_2 S(S+1) - E_{\text{cl}}\lambda}} K(k). \qquad (28.40)$$

We consider again the derivative of the period with respect to the energy
and inquire about a nontrivial solution with an energy in the domain $0 < E_{\text{cl}} < E_0 \equiv K_2 S(S+1)$. Such a solution would be different from that with
the monotonically decreasing behaviour of the period $P(E_{\text{cl}})$ observed in the
examples of Sec. 28.3. Using the formula[†]

$$\frac{dK(k)}{dk} = \frac{1}{k}\left[\frac{E(k)}{1 - k^2} - K(k)\right], \qquad (28.41)$$

one obtains for the vanishing of the derivative of $P(E_{\text{cl}})$ the equation

$$K(k) - \frac{K_2 S(S+1)}{E_{\text{cl}}} E(k) = 0. \qquad (28.42)$$

The solution of this equation can be investigated both numerically and with
approximation analytically.[‡] In the case of the second method, we expand

---

[†]P. F. Byrd and M. D. Friedman [40], formula 710.00, p. 282.

[‡]J.–Q. Liang, H. J. W. Müller–Kirsten, D. K. Park and F. Zimmerschied [170].

$K(k)$ and $E(k)$ in rising powers of k^2 around $\pi/2$.[§] Taking into account only terms up to $O(k^2)$ and, using Eq. (28.39), this implies that

$$\left[1 + \frac{k^2}{4}\right] = \nu^2\left[1 - \frac{k^2}{4}\right] \quad \text{or} \quad \nu^2 = \frac{5\nu^2 - (4\lambda + 1)}{3\nu^2 - (4\lambda - 1)}$$

with solutions

$$\nu_+^2 = \frac{4\lambda + 1}{3}, \quad \nu_-^2 = 1,$$

so that one obtains two possible solutions, i.e.

$$E_{\text{cl}} = K_2 S(S+1) \quad \text{or} \quad E_{\text{cl}} = \frac{3K_2 S(S+1)}{1 + 4\lambda}. \tag{28.43}$$

Since $E_{\text{cl}} < E_0 = K_2 S(S+1)$, the nontrivial solution is obtained for

$$\frac{3K_2 S(S+1)}{1 + 4\lambda} < K_2 S(S+1), \quad \text{i.e.} \quad \lambda > \frac{1}{2}. \tag{28.44}$$

Thus one expects a crossover of the first order type for $\lambda > 1/2$. This is, in fact, what one finds as is shown below and as illustrated in Fig. 28.5.

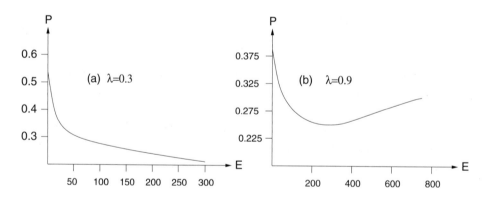

Fig. 28.5 The period of the periodic instanton for $S = \sqrt{1000}$ and $K_1 = 1$. In (a) for $\lambda = 0.3$, in (b) for $\lambda = 0.9$.

As in the case of our earlier examples, we next consider the two types of action. The action of the periodic instanton configuration (28.38) can be obtained in much the same way as in earlier chapters, i.e. from evaluation of the following relation, in which τ_i and τ_f are the initial and final Euclidean

---

[§]P. F. Byrd and M. D. Friedman [40], formulae 900.00 and 900.07, pp. 298, 299. The first few terms are: $K(k) = \frac{\pi}{2}[1 + \frac{1}{4}k^2 + \frac{9}{64}k^4 + \cdots]$ and $E(k) = \frac{\pi}{2}[1 - \frac{1}{4}k^2 - \frac{3}{64}k^4 + \cdots]$.

times τ of the pseudoparticle's path from one turning point at ϕ_i to the other at ϕ_f, i.e.

$$S_c = \int_{\tau_i}^{\tau_f} [m(\phi_c)\dot{\phi}_c^2 + E_{cl}]d\tau \equiv W_c + E_{cl}(\tau_f - \tau_i). \qquad (28.45)$$

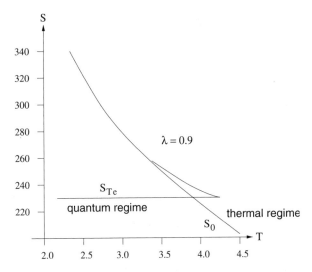

Fig. 28.6 Actions S_0, S_{T_e} as functions of temperature T_e.

Here

$$\phi_i = \sin^{-1}\left[\frac{E_{cl}}{K_2 S(S+1)}\right]^{1/2}, \quad \phi_f = \pi - \sin^{-1}\left[\frac{E_{cl}}{K_2 S(S+1)}\right]^{1/2}. \qquad (28.46)$$

We leave the calculation to Example 28.2 below and cite the result for W_c which is also cited in the literature:[*]

$$W_c = \frac{\omega}{\lambda K_1}[K(k) - (1 - \lambda k^2)\Pi(K(k), \lambda k^2)], \qquad (28.47)$$

where ω is given by Eq. (28.39) and $\Pi(K(k), \lambda k^2)$ is the *complete elliptic integral of the third kind*. Here it must be remembered that this is the expression for the periodic instanton which — as explained earlier — is to be considered as half a thermon or half a pair of an instanton and an anti-instanton. Thus, in using the concept of thermons, we have to introduce in

---

[*]J.–Q. Liang, Y.–B. Zhang, H. J. W. Müller–Kirsten, J.–G. Zhou, F. Zimmerschied and F.–C. Pu [175], equation (32).

the action an appropriate factor of 2. Proceeding as in the earlier examples
we have the thermodynamic action ($\hbar = k_B = 1$)

$$S_0 = \frac{K_2 S(S+1)}{T_e},\tag{28.48}$$

and the thermon action (as twice the action of the periodic instanton)

$$S_T = \frac{E_{cl}}{T_e} + 2W_c.\tag{28.49}$$

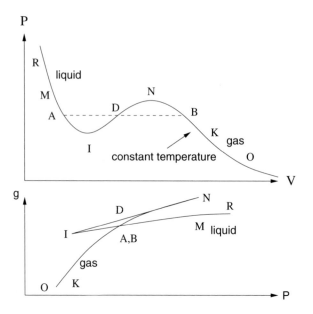

Fig. 28.7 For comparison: Van der Waals gas: (a) Equation of state,
(b) enthalpy/mole, g, vs. pressure, P.

The plots of these actions are shown in Fig. 28.6, again plotted for $\lambda = 0.9$.
Only the two lowest branches, marked S_0 and S_{T_e}, are physical. The convention of describing such a sharp crossover as a *first order transition* is taken
over from the analogous behaviour of phase transitions in thermodynamics.
For the comparison we show in Fig. 28.7(a) the equation of state of a van
der Waals gas, and in Fig. 28.7(b) a plot of the free enthalpy per mole versus
pressure as one finds this in textbooks on thermodynamics.[†] We observe a
very analogous behaviour to the first order transition in our case here. A
similar behaviour for appropriate values of the parameters involved has in

---

[†]See e.g. F. Reif [238], Sec. 8.6 on phase transitions and equation of state.

the mean time been found in many other models, including in particular those with applied magnetic fields as we mentioned at the beginning of this chapter. A natural question that arises is that for criteria for the occurrence of crossover transitions of one or the other type. This question has also been investigated by various authors but will not be considered here.[‡]

Example 28.1: Derivation of the periodic instanton
Derive the periodic instanton solution (28.38) by integration of its equation (28.35).

Solution: Inserting the explicit expressions for $m(\phi)$ and $V(\phi)$, separating the integrations with respect to $d\tau$ and $d\phi$ and setting $z = \sin^2 \phi$, the resulting integral becomes

$$\tau - \tau_0 = -\frac{1}{\alpha} \int_y^1 \frac{dz}{\sqrt{z(1-z)(z_1 - z)(z - z_2)}}$$

with

$$\alpha^2 = 16\lambda K_1 K_2 S(S+1), \quad z_1 = \frac{1}{\lambda}\frac{K_1}{K_2}, \quad z_2 = \frac{E_{cl}}{K_2 S(S+1)}.$$

With the parameters within the ranges of the inequality $z_1 > 1 > y \geq z_2 > 0$ the integral can be evaluated with the help of an integral given in Tables.[§] Then

$$(\tau - \tau_0) = -\frac{g}{\alpha}\mathrm{sn}^{-1}(\sin \varphi), \quad \sin^2 \varphi = \frac{(z_1 - z_2)(1 - y)}{(1 - z_2)(z_1 - y)}, \quad k^2 = \frac{z_1(1 - z_2)}{(z_1 - z_2)}, \quad g = \frac{2}{\sqrt{z_1 - z_2}}.$$

Extracting y from the second relation and then using the first, we obtain (with $\tau_0 = 0$)

$$y = \sin^2 \phi = \frac{(z_1 - z_2) - z_1(1 - z_2)\sin^2 \varphi}{(z_1 - z_2) - (1 - z_2)\sin^2 \varphi} = \frac{1 - k^2 \sin^2 \varphi}{1 - \lambda k^2 \sin^2 \varphi} = \frac{1 - k^2\mathrm{sn}^2(\omega\tau)}{1 - \lambda k^2\mathrm{sn}^2(\omega\tau)}.$$

Example 28.2: Derivation of the action of the periodic instanton
Derive the expression for W_c given by Eq. (28.47).

Solution: We use the notation of Example 28.1. We have to evaluate the expression

$$W_c = \int_{\tau_i}^{\tau_f} d\tau m(\phi_c)\dot{\phi}_c^2 = \int_{\phi_i}^{\phi_f} m(\phi_c)\dot{\phi}_c d\phi_c.$$

From Eqs. (28.35) and (28.37) we obtain

$$\dot{\phi}_c^2 = \frac{4}{\hbar^2} K_1(1 - \lambda \sin^2 \phi_c)\underbrace{[K_2 S(S+1)\sin^2 \phi_c - E_{cl}]}_{V(\phi_c)}.$$

We see that the turning points are given by

$$\sin^2 \phi_c = \frac{E_{cl}}{K_2 S(S+1)} \equiv z_2 \ll 1, \quad \phi_c = \sin^{-1}(\pm\sqrt{z_2}),$$

i.e. $\quad \phi_i = \sin^{-1}\sqrt{z_2}, \quad \phi_f = \pi - \sin^{-1}\sqrt{z_2}.$

[‡]Criteria for first order transitions have been investigated in (a) D. A. Gorokhov and G. Blatter [117], [118], (b) H. J. W. Müller–Kirsten, D. K. Park and J. M. S. Rana [214], and (c) J.–Q. Liang, Y.–B. Zhang, H. J. W. Müller–Kirsten, S.–P. Kou, X.–B. Wang and F.–C. Pu [176].
[§]P. F. Byrd and M. D. Friedman [40], formula 255.00, p. 116.

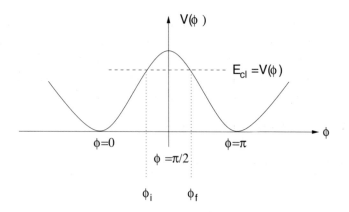

Fig. 28.8 The relative positions of turning points and sphaleron.

It follows now that W_c can be written as twice the integral from one turning point to the sphaleron position, as indicated in Fig. 28.8. Thus

$$W_c = 2 \int_{\sin^{-1}\sqrt{z_2}}^{\pi/2} m(\phi_c)\dot\phi_c d\phi_c = \frac{2\hbar}{\sqrt{K_1}} \int_{\sin^{-1}\sqrt{z_2}}^{\pi/2} \frac{d\phi_c \sqrt{K_2 S(S+1)\sin^2\phi_c - E_{cl}}}{\sqrt{1-\lambda\sin^2\phi_c}}.$$

With $z = \sin^2\phi_c, z_1 = 1/\lambda$, as in Example 28.1 this becomes

$$W_c = \frac{\hbar\sqrt{K_2 S(S+1)}}{\sqrt{K_1}\sqrt{\lambda}} \int_{y=z_2}^{1} \frac{dz\sqrt{z-z_2}}{\sqrt{z}\sqrt{1-z}\sqrt{z_1-z}} \quad \text{with} \quad z_1 > 1 > y \geq z_2 > 0.$$

We evaluate this integral with the help of Tables of Integrals:[¶]

$$W_c = \hbar\sqrt{S(S+1)}(1-z_2)g w_c, \quad \text{where} \quad g = \frac{2}{\sqrt{z_1-z_2}}, \quad k^2 = \frac{z_1(1-z_2)}{z_1-z_2},$$

and

$$w_c = \frac{1}{\alpha^2}[K(k) - (1-\alpha^2)\Pi(K(k),\alpha^2)], \quad \alpha^2 = \frac{1-z_2}{z_1-z_2} = \lambda k^2,$$

and $\Pi(K(k),\alpha^2)$ is the *complete elliptic integral of the third kind*.[||] Hence

$$W_c = \frac{2\hbar\sqrt{S(S+1)}(1-z_2)}{\sqrt{z_1-z_2}\lambda k^2}[K(k) - (1-\lambda k^2)\Pi(K(k),\alpha^2)].$$

With

$$\nu^2 - 1 = \frac{1-z_2}{z_2}, \quad \sqrt{\frac{z_1}{z_2}} = \sqrt{\frac{K_1}{K_2}}\nu, \quad z_1 - z_2 = \frac{z_1(1-z_2)}{k^2},$$

we obtain (using also the relations (28.39), e.g. for ω)

$$\frac{2\hbar\sqrt{S(S+1)}(1-z_2)}{\sqrt{z_1-z_2}\lambda k^2} = \frac{2\hbar\sqrt{S(S+1)}}{\sqrt{z_1}\lambda k} = \frac{2\hbar\sqrt{S(S+1)}\sqrt{\nu^2-1}\sqrt{K_2}}{\lambda k\sqrt{K_1}\nu}$$

$$= \frac{2\hbar\sqrt{S(S+1)}\sqrt{\nu^2-1}\sqrt{K_2 K_1}}{K_2 k\sqrt{K_1}\nu} = \frac{\omega\hbar}{\lambda K_1}.$$

Hence we obtain the result of Eq. (28.47):

$$W_c = \frac{\omega\hbar}{\lambda K_1}[K(k) - (1-\lambda k^2)\Pi(K(k),\lambda k^2)].$$

[¶] P. F. Byrd and M. D. Friedman [40] formula 255.19, p. 118 and formula 338.01, p. 202.
[||] P. F. Byrd and M. D. Friedman [40], p. 223.

28.5 Concluding Remarks

In the above we considered a specific model defined by the operators in Eq. (28.25). It is clear that a multitude of similar models can be constructed, also with inclusion of applied magnetic fields in arbitrary directions. Most of these models have — presumably — been investigated by now, in one way or another. As the example of Eq. (28.25) demonstrates, these models permit not only the investigation of quantum-classical transitions, but also numerous other aspects. In particular one can — for instance — investigate also the level splitting and its dependence on the field dependence of the effective mass, and one can do this with the path integral method* or, in some cases, with the Schrödinger method.† Also, in the mean time some other methods have been developed for the evaluation of path integrals and can therefore be explored.‡ A vital and convincing aspect is, of course, the application to realistic cases, such as ferromagnetic particles with large spin, and comparison of calculated quantities with experimental observation. A huge number of papers has appeared over the last 10 to 15 years in the field of spin tunneling and related phenomena which it is impossible to enter here in more detail.§ Thus we limit ourselves here to this brief discussion and to mentioning a few references which may provide an entrance into more specialized directions.

---

*J.–Q. Liang, H. J. W. Müller–Kirsten and Jian–Ge Zhou [171]. Another spin model investigation along the lines considered here, including level splitting, can be found in J.–Q. Liang, Y.–B. Zhang, H. J. W. Müller–Kirsten, Jian–Ge Zhou, F. Zimmerschied and F.–C. Pu [175].

†For the model of M. Enz and R. Schilling [84] see J.–Q. Liang, H. J. W. Müller–Kirsten, A. V. Shurgaia and F. Zimmerschied [172]. For related aspects see J.–Q. Liang, L. Maharana and H. J. W. Müller–Kirsten [173] and J.–Q. Liang, H. J. W. Müller–Kirsten and J. M. S. Rana [174].

‡V. V. Ulyanov and O. B. Zaslavskii [275].

§A very useful general reference to work in this field is the collection of articles in the *Proceedings of the Meeting on Quantum Tunneling of Magnetization* QTM'94 [125].

Chapter 29

Summarizing Remarks

The considerations in the preceding text demonstrate that with perturbation theory one can go significantly beyond the examples which are usually treated in elementary texts on quantum mechanics, like the harmonic oscillator, the Coulomb potential and several other simple potentials, and — unlike any other method — one can handle with it such diverse cases as periodic potentials, anharmonic oscillators, screened Coulomb potentials, and even a typical singular potential. In most cases the expansions of interest in physics are asymptotic, in a few cases supplemented by convergent expansions (as in the case of the cosine potential). A crucial point in the formulation of the perturbation method is the full exploitation of the symmetries provided by the parameters of the respective Schrödinger equation. If these are taken into account, the perturbation method can be extended and employed even for the calculation of the behaviour of its large order terms. This is not widely appreciated, and therefore one encounters repeatedly attempts to find a better way with some kind of convergent expansion. The critical test of any other method, if designed as alternative or superior procedure, would be the calculation of the level splitting or analogous exponentially small imaginary quantity for each of the cases considered here, particularly for the cases of the cosine and elliptic potentials. Since at least formally as shown, a level splitting can be related to the discontinuity of a cut which enters the Borel transform for the determination of the large order behaviour (and is found to be that typical of asymptotic expansions), it is hard to envisage how that can be achieved with convergent expansions. In the above, the WKB method played only a supplementary role; however, we pointed out that with consideration of higher order WKB corrections, the eigenvalue expansion for screened Coulomb potentials had been shown in the literature to be identical with that of perturbation theory. The study of Schrödinger equations

773

with perturbation methods is thus basically an extension of the study of fundamental differential equations of mathematical physics.

The Feynman path integral evolves from the free particle Green's function of the time-dependent Schrödinger equation, and thus does not, *a priori*, take into account the parameter symmetries of the Schrödinger wave equation. As a result the calculated level splittings, for instance, involve unnatural factors (like first order WKB), and only by application of the Furry factor (the factor originally established by Furry to improve a WKB result for near ground state levels), full agreement is in each case achieved with the perturbation theory result. A noteworthy feature of the path integral expansion about the pseudoparticle configuration for the basic potentials we considered here, is that in each such case the equation of small fluctuations about this configuration is a Lamé equation. In spite of the complexity of the path integral derivation of level splittings and decay widths, the pseudoparticle configuration is and remains a very useful quantity, even beyond the domain of validity of the Schrödinger equation, as the examples of transitions from the quantum regime to the thermal regime illustrate. These latter connections were recognized and investigated only recently. This applies similarly to extensive investigations into the quantization of systems with constraints, which now allow us — as demonstrated — to see the treatment of waves (i.e. light) completely parallel to that of particles, with Dirac brackets replacing Poisson brackets. Thus the attempts to gain a deeper understanding of some specific aspects of more complicated contexts, such as quantum field theory, can also open the view to a simpler understanding of aspects of basic quantum mechanics.

Appendix A

Properties of Jacobian Elliptic Functions

In this appendix we collect some useful formulas for *Jacobian elliptic functions*, some of which are needed at various points of the text.[*] It can be seen that the manipulation of these functions is not much more difficult than that of trigonometric functions.

Notation and definitions: The three basic Jacobian elliptic functions are

$$\text{sn}[u, k], \quad \text{cn}[u, k], \quad \text{dn}[u, k]. \tag{A.1}$$

Here

$$u = \int_0^x \frac{dt}{[(1 - t^2)(1 - k^2 t^2)]^{1/2}} = \int_0^\phi \frac{d\theta}{(1 - k^2 \sin^2 \theta)^{1/2}} \equiv \text{sn}^{-1} x \tag{A.2}$$

with

$$x = \text{sn}[u] \equiv \sin \phi, \quad \text{cn}[u] = \cos \phi, \quad \text{dn}[u] = (1 - k^2 \sin^2 \phi)^{1/2}. \tag{A.3}$$

Here u is called the *argument* of the functions and k the *elliptic modulus* with *complementary modulus* k' given by $k'^2 = 1 - k^2$. Hence the more complete notation in Eq. (A.1). The three basic Jacobian elliptic functions are one-valued functions of the argument u and are doubly periodic. The following *complete elliptic integral of the first kind* defines the so-called real *quarter period* $K(k)$ (in the case of trigonometric functions equivalent to $\pi/2$):

$$K(k) = \int_0^{\pi/2} \frac{d\theta}{(1 - k^2 \sin^2 \theta)^{1/2}}, \quad K'(k) = \int_0^{\pi/2} \frac{d\theta}{[1 - (1 - k^2) \sin^2 \theta]^{1/2}}. \tag{A.4}$$

---

[*]A handy reference is the small book of L. M. Milne–Thomson [196]. Much more complete references are the Tables of F. Byrd and M. D. Friedman [40] and L. S. Gradshteyn and I. M. Ryzhik [122].

One defines as imaginary quarter period $iK'(k)$.

Important special values:

$$\text{sn}[u, 0] = \sin u, \quad \text{cn}[u, 0] = \cos u, \quad \text{dn}[u, 0] = 1;$$
$$\text{sn}[u, 1] = \tanh u, \quad \text{cn}[u, 1] = \text{sech}\, u, \quad \text{dn}[u, 1] = \text{sech}\, u,$$
$$K(k = 0) = K'(k = 1) = \frac{\pi}{2}, \quad K(k = 1) = K'(k = 0) = \infty. \quad \text{(A.5)}$$

Reciprocal and ratio functions: Just as in trigonometry in addition to the functions $\sin u$ and $\cos u$ it is usual to consider their reciprocals $\csc u$ and $\sec u$, and their ratios $\tan u$ and $\cot u$, so with the Jacobian elliptic functions one defines reciprocals and ratios written:

$$\text{ns}\, u = \frac{1}{\text{sn}\, u}, \quad \text{nc}\, u = \frac{1}{\text{cn}\, u}, \quad \text{nd}\, u = \frac{1}{\text{dn}\, u};$$
$$\text{sc}\, u = \frac{\text{sn}\, u}{\text{cn}\, u}, \quad \text{cs}\, u = \frac{\text{cn}\, u}{\text{sn}\, u}, \quad \text{ds}\, u = \frac{\text{dn}\, u}{\text{sn}\, u};$$
$$\text{sd}\, u = \frac{\text{sn}\, u}{\text{dn}\, u}, \quad \text{dc}\, u = \frac{\text{dn}\, u}{\text{cn}\, u}, \quad \text{cd}\, u = \frac{\text{cn}\, u}{\text{dn}\, u}. \quad \text{(A.6)}$$

Thus there are altogether twelve Jacobian elliptic functions. The function $\text{sn}[u, k]$ is an odd function of u, whereas the functions $\text{cn}[u, k], \text{dn}[u, k]$ are even functions of u.

Relations between squares of Jacobian functions: The following relations are frequently needed and very useful, and are similar to wellknown relations of trigonometrical functions:

$$\text{sn}^2 u + \text{cn}^2 u = 1, \quad \text{dn}^2 u + k^2 \text{sn}^2 u = 1, \quad \text{dn}^2 u - k^2 \text{cn}^2 u = k'^2. \quad \text{(A.7)}$$

There are more relations involving ratio and reciprocal functions, also double and half argument formulas as for trigonometric functions, but these are not needed in the text, so that we do not cite them here. Important are, however, derivatives.

Derivatives of Jacobian elliptic functions: Derivatives of various functions are needed in the text. These are contained in the following set:

$$\frac{d}{du}\text{sn}[u, k] = \text{cn}[u, k]\text{dn}[u, k], \quad \frac{d}{du}\text{cn}[u, k] = -\text{sn}[u, k]\text{dn}[u, k],$$
$$\frac{d}{du}\text{dn}[u, k] = -k^2\text{sn}[u, k]\text{cn}[u, k]. \quad \text{(A.8)}$$

In particular one has, for instance,

$$\frac{d}{du}\ln(\text{dn}[u, k] + k\text{cn}[u, k]) = -k\text{sn}[u, k], \quad \text{(A.9)}$$

and

$$\int sn[u, k]du = \frac{1}{k} \ln(dn[u, k] - kcn[u, k]). \qquad (A.10)$$

Further properties of the basic functions: Some further properties of the basic Jacobian elliptic functions needed in the text are collected in the following Tables:

Table A.1 Periods and Zeros

	$sn\,u$	$cn\,u$	$dn\,u$
Periods	$4K, 2iK'$	$4K, 2K + 2iK'$	$2K, 4iK'$
Zeros	$0, 2K$	$K, 3K$	$K + iK', K + 3iK'$

Table A.2 Changes of the argument

Argument	sn	cn	dn
u	$sn\,u$	$cn\,u$	$dn\,u$
$-u$	$-sn\,u$	$cn\,u$	$dn\,u$
$u + K$	$cd\,u$	$-k'sd\,u$	$k'nd\,u$
$u - K$	$-cd\,u$	$k'sd\,u$	$k'nd\,u$
$K - u$	$cd\,u$	$k'sd\,u$	$k'nd\,u$
$u + 2K$	$-sn\,u$	$-cn\,u$	$dn\,u$
$u - 2K$	$-sn\,u$	$-cn\,u$	$dn\,u$

Table A.3 Special values of the argument

u	$sn\,u$	$cn\,u$	$dn\,u$
0	0	1	1
$\frac{1}{2}K$	$(1 + k')^{-1/2}$	$k'^{1/2}(1 + k')^{-1/2}$	$k'^{1/2}$
K	1	0	k'
$2K$	0	-1	1

Bibliography

[1] M. Abramowitz and I. A. Stegun, *Handbook of Mathematical Functions* (Dover, 1970).

[2] A. A. Abrikosov, *Soviet Physics, JETP* **5** (1957) 1174.

[3] N. I. Achieser and L. M. Glasman, *Theorie der linearen Operatoren im Hilbert–Raum* (Akademie–Verlag, 1968).

[4] P. Achuthan, H. J. W. Müller–Kirsten and A. Wiedemann, *Fortschr. Phys.* **38** (1990) 77.

[5] I. Affleck, *Nucl. Phys.* **B 191** (1981) 429.

[6] A. Ahmadzahdeh, P. G. Burke and C. Tate, *Phys. Rev.* **131** (1963) 1315.

[7] A. C. Aitken, *Determinants and Matrices* (Oliver and Boyd, 1954).

[8] H. H. Aly, H. J. W. Müller and N. Vahedi–Faridi, *Lett. Nuovo Cimento* **2** (1969) 485.

[9] H. H. Aly, W. Güttinger and H. J. W. Müller–Kirsten, *Singular Interactions in Nonrelativistic Quantum Theory* in H. H. Aly (Ed.), *Lectures in Theoretical High Energy Physics* (Gordon and Breach, 1970), p.247.

[10] H. H. Aly, H. J. W. Müller–Kirsten and N. Vahedi–Faridi, *J. Math. Phys.* **16** (1975) 961.

[11] F. M. Arscott, *Periodic Differential Equations* (Pergamon, 1964).

[12] M. A. Ashbaugh and E. M. Harrell, *Perturbation Theory for Shape Resonances and Large Barrier Potentials*, University of Tennessee Report, 1981.

[13] C. Bachas, G. Lazarides, Q. Shafi and G. Tiktopoulos, Report No. CERN-TH. 6183/91, BA:91-19 (unpublished).

[14] K. Banerjee and S. P. Bhatnagar, *Phys. Rev.* **D 18** (1978) 4767.

[15] G. Barton, *Ann. Phys.* **166** (1986) 322.

[16] C. Becchi, A. Rouet and R. Stora, *Phys. Lett.* **B 52** (1974) 344.

[17] S. N. Behera and A. Khare, *Pramana (J. of Physics)*, **15** (1980) 245.

[18] C. M. Bender and T. T. Wu, *Phys. Rev. Lett.* **21** (1968) 406; **27** (1971) 461; *Phys. Rev.* **184** (1969) 1231.

[19] C. M. Bender and T. T. Wu, *Phys. Rev.* **D7** (1973) 1620.

[20] L. Bertocchi, S. Fubini and G. Furlan, *Nuovo Cimento* **35** (1965) 633.

[21] H. A. Bethe and T. Kinoshita, *Phys. Rev.* **128** (1962) 1418.

[22] S. K. Bhattacharya and A. R. P. Rau, *Phys. Rev.* **A 26** (1982) 2315.

[23] S. K. Bhattacharya, *Phys. Rev.* **A 31** (1985) 1991.

[24] S. N. Biswas, K. Datta, R. P. Saxena, P. K. Srivastava, *Phys. Rev.* **D** **4**(1971) 3617.

[25] R. Blankenbecler, M. L. Goldberger, N. N. Khuri and S. B. Treiman, *Ann. Phys.* **10** (1960) 62.

[26] E. B. Bogomol'nyi and V. A. Fateyev, *Phys. Lett.* **B71** (1977) 93.

[27] E. B. Bogomol'nyi, *Sov. J. Nucl. Phys.* **24** (1976) 449.

[28] S. K. Bose, G. E. Hite and H. J. W. Müller-Kirsten, *J. Math. Phys.* **20** (1979) 1878.

[29] S. K. Bose and H. J. W. Müller–Kirsten, *J. Math. Phys.* **20** (1979) 2471.

[30] S. K. Bose and H. J. W. Müller–Kirsten, *Phys. Lett.* **A 162** (1992) 79.

[31] A. Bottino, A. M. Longoni and T. Regge, *Nuovo Cimento* **25** (1962) 242.

[32] A. Brachner and R. Fichtner, *Quantenmechanik für Lehrer und Studenten* (A. Schroedel Verlag, 1976).

[33] B. Booss and D. D. Bleecker, *Topology and Analysis* (Springer, 1985), p. 26.

[34] J. I. Boukema, *Physica* **30** (1965) 1320.

[35] J. I. Boukema, *Physica* **30** (1965) 1909.

[36] E. Brezin and J. Zinn–Justin, *J. de Physique – Lettres* **40** (1979) 511.

[37] L. Brillouin, *J. Phys. Radium* **7** (1926) 353.

[38] D. M. Brink and G. R. Satchler, *Angular Momentum* (Oxford University Pres, 1971).

[39] L. de Broglie, *Sur le parallelisme entre la dynamique du point materiel et l'optique géométrique, J. de Phys.* **7** (1926) 1.

[40] P. F. Byrd and M. D. Friedman, *Handbook of Elliptic Integrals for Engineers and Scientists* (Springer, 1971).

[41] C. G. Callan and S. Coleman, *Phys. Rev.* **D 16** (1977) 1762.

[42] C. G. Callan and J. M. Maldacena, *Nucl. Phys.* **B 513** (1998) 198, hep-th/9805140.

[43] R. H. Cameron and W. T. Martin, *Bull. Amer. Math. Soc.* **51** (1945) 73.

[44] R. H. Cameron and W. T. Martin, *Trans. Amer. Math. Soc.* **58** (1945) 184.

[45] K. M. Case, *Phys. Rev.* **80** (1950) 797.

[46] J. Challifour and R. J. Eden, *J. Math. Phys.* **4** (1063) 359.

[47] H. Cheng and T. T. Wu, *Phys. Rev.* **D5** (1972) 3189.

[48] G. F. Chew, *S-Matrix Theory of Strong Interactions* (W. A. Benjamin, 1962).

[49] Y. Cho, N. Kan and K. Shiraishi, *Compactification in deconstructed gauge theory with topologically non-trivial link fields*, hep–th/0306012.

[50] E. M. Chudnovsky, *Phys. Rev.* **A 46** (1992) 8011.

[51] E. M. Chudnovsky and D. A. Garanin, *Phys. Rev. Lett.* **79** (1997) 4469.

[52] J. Cizek, R. J. Damburg, S. Graffi, V. Grecchi, E. M. Harrell, J. G. Harris, S. Nakai, J. Paldus, R. K. Propin and H. J. Silverstone, *Phys. Rev.* **A 33** (1986) 12.

[53] C. Cohen–Tannoudji, B. Diu and F. Laloa, *Quantum Mechanics* I, II (Wiley-Interscience, 1977).

[54] S. Coleman in *The Whys of Subnuclear Physics*, ed. by A. Zichichi (Plenum Press, 1979), pp. 805 – 916.

[55] S. Coleman, *The Uses of Instantons*, 1977 International School of Subnuclear Physics, Ettore Majorana.

[56] S. Coleman, *Phys. Rev.* **D 15** (1977) 2929; *Phys. Rev.* **D 16**(E) (1977).

[57] S. Coleman, *Nucl. Phys.* **B 298** (1988) 178.

[58] M.–C. Combourieu and H. Rauch, *The Wave-Particle Dualism in 1992: A Summary* in *Foundations in Physics* **22** (1992) 1403.

[59] L. Cornalba, M. S. Costa and J. Penedones, *Coleman meets Schwinger*, hep-th/0501151.

[60] R. J. Crewther, D. Olive and S. Sciuto, *Riv. Nuovo Cimento* **2** (No. 8) (1979) 1.

[61] M. Cvetic, H. Lü, C. N. Pope and T. A. Tran, *Phys. Rev.* **D 59** (1999) 126002, hep-th/9901002; M. Cvetic, H. Lü and J. F. Vazquez–Poritz, *Phys. Lett.* **B 462** (1999) 62, hep-th/9904135 and hep-th/0002128.

[62] R. J. Damburg, R. K. Propin, S. Graffi, V. Grecchi, E. M. Harrell, J. Cizek, J. Paldus and H. J. Silverstone, *Phys. Rev. Lett.* **52** (1984) 1112.

[63] R. F. Dashen, B. Hasslacher and A. Neveu, *Phys. Rev.* **D 10** (1974) 4114.

[64] C. Davisson and L. H. Germer, *Nature* 16th April (1927), *Phys. Rev.* **30** (1927) 706.

[65] P. C. W. Davies and J. R. Brown, *Der Geist im Atom* (Birkhäuser, 1988).

[66] J. D. de Deus, *Phys. Rev.* **D 26** (1982) 2782.

[67] P. G. De Gennes, *Superconductivity of Metals and Alloys* (Benjamin, 1966).

[68] G. H. Derrick, *J. Math. Phys.* **5** (1964) 1252.

[69] H. J. de Vega and F. A. Schaposnik, *Phys. Rev.* **D 14** (1976) 1100.

[70] R. B. Dingle, *Asymptotic Expansions: Their Derivation and Interpretation* (Academic Press, 1973).

[71] R. B. Dingle, *Proc. Roy. Soc.* **A 244** (1958) 456, 476, 484; **A 249** (1959) 270, 285, 293. *Buletinul Institutului Politehnic din Iasi* **8** (1962) 53.

[72] R. B. Dingle, *The Role of Difference Equations in Perturbation Theory*, University of St. Andrews Report (1964).

[73] R. B. Dingle and H. J. W. Müller, *J. reine angew. Math.* **211** (1962) 11.

[74] R. B. Dingle and H. J. W. Müller, *J. reine angew. Math.* **216** (1964) 123.

[75] P. A. M. Dirac, *The Principles of Quantum Mechanics* (Oxford University Press, 1944), 3rd ed.

[76] P. A. M. Dirac, (a) *Lectures on Quantum Mechanics* (Yeshiva University, New York, 1964); (b) *Canad. J. Math.* **2** (1950) 129; (c) *Phys. Rev.* **114** 924 (1959).

[77] W. Dittrich and M. Reuter, *Classical and Quantum Dynamics — from Classical Paths to Path Integrals* (Springer, 1992).

[78] N. Dombey and R. H. Jones, *J. Math. Phys.* **9** (1968) 986.

[79] G. V. Dunne and K. Rao, *J. High Energy Phys.* **01** (2000) 019.

[80] G. V. Dunne, R. Jackiw and C. A. Trugenberger, *Phys. Rev.* **D 41** (1990) 661.

[81] H. B. Dwight, *Tables of Integrals and Other Mathematical Data* (Macmillan, 1957).

[82] E. Eichten, K. Gottfried, T. Kinoshita, K. D. Lane and T.–M. Yan, *Phys. Rev. Lett.* **34** (1975) 369 and **36** (1976) 500.

[83] V. L. Eletsky, V. S. Popov and V. M. Weinberg, *Phys. lett.* **A 84** (1981) 235.

[84] M. Enz and R. Schilling, *J. Phys.* **C 19** (1986) 1765; **C19** (1986) L711.

[85] A. Erdélyi, *Phil. Mag.* **205** (1941) 123.

[86] A. Erdélyi, W. Magnus, F. Oberhettinger and F. G. Tricomi, (Editors, Bateman Manuscript Project), *Higher Transcendental Functions*, Vol. II (MacGraw–Hill, 1955).

[87] A. Erdélyi, W. Magnus, F. Oberhettinger and F. G. Tricomi, (Editors, Bateman Manuscript Project), *Higher Transcendental Functions*, Vol. III (MacGraw–Hill, 1955).

[88] G. Esposito, *Scattering from Singular Potentials in Quantum Mechanics*, J. Phys. **A 31** (1998) 9493, hep-th/9807018.

[89] L. D. Faddeev and V. N. Popov, *Phys. Lett.* **B 25** (1967) 29; L. D. Faddeev, *Teor. Mat. Fiz.* **1** (1969) 3. See also L. D. Faddeev in *Methods in Field Theory* (Les Houches Lectures, 1976).

[90] L. Faddeev and R. Jackiw, *Phys. Rev. Lett.* **60** (1988) 1692.

[91] B. Felsager, *Geometry, Particles and Fields* (Odense University Press, 1981).

[92] E. Fermi, *Nuclear Physics*, Notes compiled by J. Orear, A. H. Rosenfeld and R. A. Schluter, revised ed. (University of Chicago Press, 1949/50).

[93] R. P. Feynman, *Rev. Mod. Phys.* **20** (1948) 367, *Phys. Rev.* **84** (1951) 108.

[94] R. P. Feynman, *Statistical Mechanics* (W. A. Benjamin, 1972).

[95] R. P. Feynman and A. R. Hibbs, *Quantum Mechanics and Path Integrals* (McGraw–Hill, 1965).

[96] D. Finkelstein and C. W. Misner, *Ann. Phys.* **6** (1959) 230.

[97] S. C. Frautschi, *Regge Poles and S-Matrix Theory* (W. A. Benjamin, 1963).

[98] R. Friedberg and T. D. Lee, *New ways to solve the Schrödinger equation*, Ann. Phys. **308** (2003) 263, quant-ph/0407207; T. D. Lee, *A new approach to solve the low-lying states of the Schrödinger equation*, quant-ph/0501054.

[99] N. Fröman and P.-O. Fröman, JWKB Approximation (North-Holland, 1965).

[100] N. Fröman, P.-O. Fröman, U. Myhrman and R. Paulsson, *Ann. Phys.* **74** (1972) 314; R. Paulsson and N. Fröman, *Ann. Phys.* **163** (1985) 227; R. Paulsson, *Ann. Phys.* **163** (1985) 245; N. Fröman and U. Myhrman, *Arkiv för Fysik* **40** (1970) 497; M. Lakshmanan, F. Karlsson and P.-O. Fröman, *Phys. Rev.* **D 24** (1981) 2586.

[101] K. Funakubo, S. Otsuki and F. Toyoda, *Progr. Theor. Phys.* **83** (1990) 118.

[102] W. H. Furry, *Phys. Rev.* **71** (1947) 360.

[103] A. Ganguly, *Associated Lamé and various other new classes of elliptic potentials from sl(2, ℝ) and related orthogonal potentials*, University of Calcutta Report (2003).

[104] D. A. Garanin and E. M. Chudnovsky, *Phys. Rev.* **B 56** (1997) 11 102.

[105] A. Garg, *Amer. J. Phys.* **68** (2000) 430.

[106] C. Garrod, *Rev. Mod. Phys.* **38** (1966) 483.

[107] I. M. Gel'fand and A. M. Yaglom, *J. Math. Phys.* **1** (1960) 48.

[108] H. Georgi and S. L. Glashow, *Phys. Rev.* **D 6** (1972) 2977.

[109] J.–L. Gervais and B. Sakita, *Phys. Rev.* **D 16** (1977) 3507.

[110] E. Gildener and A. Patrascioiu, *Phys. Rev.* **D 16** (1977) 423.

[111] V. L. Ginzburg and L. D. Landau, *Zh. Eksp. Teor. Fiz.* **20** (1950) 1064.

[112] D. M. Gitman and I. V. Tyutin, *Quantization of Fields with Constraints* (Springer, 1990).

[113] P. Goddard and D. I. Olive, *Rep. Progr. Phys.* **41** (1978) 1357.

[114] H. Goldstein, *Classical Mechanics* (Addison–Wesley, 1959) 6th ed.

[115] S. Goldstein, *Trans. Camb. Phil. Soc.* **23** (1927) 303.

[116] S. Goldstein, *Proc. Roy. Soc. Edin.* **49** (1929) 210.

[117] D. A. Gorokhov and G. Blatter, *Phys. Rev.* **B 56** (1997) 3130.

[118] D. A. Gorokhov and G. Blatter, *Phys. Rev.* **B 57** (1998) 3586.

[119] P. Gosdzinsky and R. Tarrach, *Learning Quantum Field Theory from Elementary Quantum Mechanics*, University of Barcelona Report, UB-ECM-PF (December, 1989).

[120] J. Govaerts, *Hamiltonian Quantization and Constrained Dynamics* (Leuven University Press, 1991).

[121] J. Govaerts, *Int. J. Mod. Phys.* **A 5** (1990) 3625.

[122] I. S. Gradshteyn and I. M. Ryzhik, *Table of Integrals, Sums and Products* (Academic Press, 1965).

[123] Z.-Y. Gu and S.-W. Qian, *J. Phys. A: Math. Gen.* **21** (1988) 2573.

[124] S. S. Gubser and A. Hashimoto, *Commun. Math. Phys.* **203** (1999) 325, hep-th/9805140.

[125] *Proceedings of the Meeting on Quantum Tunneling of Magnetization* QTM'94, edited by L. Gunther and B. Barbara, NATO ASI, Ser. E, Vol.301 (Kluwer, 1995).

[126] A. H. Guth and S.-Y. Pi, *Phys. Rev.* **D 23** (1985) 1899.

[127] W. Güttinger, *Fortschr. Physik* **14** (1966) 483.

[128] R. A. Handelsman, Y.-P. Pao and J. S. Lew, *Nuovo Cimento* **LV A**(1968) 453.

[129] E. M. Harrell and B. Simon, *Duke Math. J.* **47** (1980) 845.

[130] E. M. Harrell, *On the Effect of the Boundary Conditions on the Eigenvalues of Ordinary Differential Equations*, Johns Hopkins University Report, 1981.

[131] E. M. Harrell, *Com. Math. Phys.* **119** (1979) 351. This paper deals specifically with "Double Wells".

[132] E. M. Harrell, *On Estimating Tunneling Phenomena*, Johns Hopkins University Report, 1981, contained in [139].

[133] W. Heisenberg, *Z. Physik* **120** (1943) 513, 673.

[134] Sze–Tsen Hu, *Homotopy Theory* (Academic Press, 1959).

[135] G. J. Iafrate and L. B. Mendelsohn, *Phys. Rev.* **182** (1969) 244 and **A2** (1970) 561.

[136] E. L. Ince, *Proc. Roy. Soc. Edin.* **46** (1926) 316.

[137] E. L. Ince, *Proc. Roy. Soc. Edin.* **47** (1927) 294.

[138] E. L. Ince, *Proc. Roy. Soc. Edin.* **A 60** (1939) 47, 83.

[139] Proc. of International Workshop on Perturbation Theory at Large Order, Florida, *Int. J. Quantum Chem.* (1) **21** (1982) 1.

[140] J. D. Jackson, *Lectures on the New Particles*, Lawrence Berkeley Report LBL-5500 (1976), also published in Proceedings, 1976 Summer Institute on Particle Physics, edited by M. C. Zipf, SLAC, Stanford, California.

[141] R. Jackiw, *Rev. Mod. Phys.* **49** (1977) 681.

[142] R. Jackiw, *"Constrained Quantization without Tears"*, Proceedings, Workshop on Constraint Theory and Quantization Methods, Montepulciano (World Scientific, 1993).

[143] E. Jahnke and F. Emde, *Tables of Functions* (Dover, 1945).

[144] H. Jeffreys, *Proc. Lond. Math. Soc.* **23** (1924) 428.

[145] H. A. Kastrup, *Ann. Physik* **9** (1962) 388.

[146] R. S. Kaushal, *J. Phys. A: Math. Gen.* **12** (1979) L253.

[147] R. S. Kaushal and H. J. W. Müller–Kirsten, *J. Math. Phys.* **20** (1979) 2233.

[148] A. Khare and U. Sukhatme, *PT-Invariant Periodic Potentials with a Finite Number of Band Gaps*, math-ph/0505027.

[149] Y. S. Kim and M. E. Noz, *Harmonic Oscillators as Bridges between Theories: Einstein, Dirac and Feynman*, quant–ph/0411017.

[150] S. Y. Khlebnikov, V. A. Rubakov and P. G. Tinyakov, *Nucl. Phys.* **B 367** (1991) 334.

[151] H. Kleinert, *Path Integrals in Quantum Mechanics, Statistics, Polymer Physics and Financial Markets*, 3rd edition (World Scientific, 2004).

[152] H. Koppe, *Quantenmechanik*, Mimeographed Lecture Notes, University of Munich (1964).

[153] H. A. Kramers, *Z. Phys.* **39** (1926) 828.

[154] D. S. Kulshreshtha and H. J. W. Müller–Kirsten, *Phys. Rev.* **D 43** (1991) 3376.

[155] Usha Kulshreshtha, D. S. Kulshreshtha and H. J. W. Müller–Kirsten, *Helv. Phys. Acta* **66** (1993) 743, 752.

[156] L. D. Landau and E. M. Lifshitz, *Quantum Mechanics*, Vol. 3 of *Course of Theoretical Physics*, third revised edition (Pergamon, 1977), p. 183.

[157] L. D. Landau and E. M. Lifshitz, *Relativistic Quantum Theory* (Pergamon, 1971), p. 288.

[158] J. S. Langer, *Ann. Phys.* **41** (1967) 198.

[159] R. E. Langer *Trans. Amer. Math. Soc.* **36** (1934) 637 – 695.

[160] S. Y. Lee, H. J. W. Müller–Kirsten, D. K. Park and F. Zimmerschied, *Phys. Rev.* **B 58** (1998 - I) 5554.

[161] J. C. LeGuillou and J. Zinn–Justin, *Large-Order Behaviour of Perturbation Theory* (North-Holland, 1990).

[162] H. Lehmann, *Nuovo Cimento* **10** (1958) 579.

[163] J.-Q. Liang and H. J. W. Müller–Kirsten, *Anharmonic Oscillator Equations: Treatment Parallel to Mathieu Equation*, quant-ph/0407235.

[164] J.-Q. Liang and H. J. W. Müller–Kirsten, *Nonvacuum Pseudoparticles, Quantum Tunneling and Metastability*, Proc. of Conference on Theories of Fundamental Interactions, Maynooth, Ireland (May 1995), ed. D. H. Tchrakian (World Scientific, 1995), 54.

[165] J.-Q. Liang, H. J. W. Müller–Kirsten and D. H. Tchrakian, *Phys. Lett.* **B 282** (1992) 105.

[166] J.-Q. Liang and H. J. W. Müller–Kirsten, *Phys. Rev.* **D 45** (1992) 23963; the correction of a number of factor of 2 errors is given in *Phys. Rev.* **D 49** (1993) (E) 105.

[167] J.-Q. Liang and H. J. W. Müller–Kirsten, *Phys. Rev.* **D 46** (1992) 4685. This paper treats the double well at excited states.

[168] J.-Q. Liang and H. J. W. Müller–Kirsten, *Phys. Rev.* **D 51** (1995) 718. This paper treats the sine–Gordon potential at excited states.

[169] J.-Q. Liang and H. J. W. Müller–Kirsten, *Phys. Rev.* **D 50** (1994) 6519. This paper treats the inverted double well potential at excited states.

[170] J.-Q. Liang, H. J. W. Müller–Kirsten, D. K. Park and F. Zimmerschied, *Phys. Rev. Lett.* **81** (1998) 216.

[171] J.-Q. Liang, H. J. W. Müller–Kirsten and Jian–Ge Zhou, *Z. Phys.* **B 102** (1997) 525. Note that the factor '3' in equations (30) and (32) there are misprints and have to be removed.

[172] J.-Q. Liang, H. J. W. Müller–Kirsten A. V. Shurgaia and F. Zimmer-schied, *Phys. Lett.* **A 237** (1998) 169.

[173] J.-Q. Liang, L. Maharana and H. J. W. Müller–Kirsten, *Physica* **B271** (1999) 28.

[174] J.-Q. Liang, H. J. W. Müller–Kirsten and J. M. S. Rana, *Phys. Lett.* **A 231** (1997) 255.

[175] J.-Q. Liang, Y.-B. Zhang, H. J. W. Müller–Kirsten, J.-G. Zhou, F. Zimmerschied and F.-C. Pu, *Phys. Rev.* **B 57** (1998) 529.

[176] J.-Q. Liang, Y.-B. Zhang, H. J. W. Müller–Kirsten, S.-P. Kou, X.-B. Wang and F.-C. Pu, *Phys. Rev.* **B 60** (1999) 12886.

[177] N. Limič, *Nuovo Cimento* **26** (1962) 581.

[178] J. J. Loeffel, A. Martin, B. Simon and A. Wightman, *Phys. Lett.* **B 30** (1969) 656.

[179] M. A. Lohe, *Phys. Rev.* **D 20** (1979) 3120.

[180] C. Lovelace and D. Masson, *Nuovo Cimento* **26** (1962) 472.

[181] W. Magnus and F. Oberhettinger, *Formulas and Theorems for the Functions of Mathematical Physics*, (Chelsea Pub. Co., 1954). The extended version of this book is: W. Magnus, F. Oberhettinger and R. P. Soni, *Formulas and Theorems for the Special Functions of Mathematical Physics*, (Springer, 1960).

[182] B. P. Mahapatra, N. Santi and N. B. Pradhan, *A New General Approximation Scheme in Quantum Theory: Application to Anharmonic and Double Well Oscillators*, quant-ph/0406036.

[183] L. Maharana, H. J. W. Müller–Kirsten and A. Wiedemann, *Nuovo Cimento* **97 A** (1987) 535.

[184] S. Mandelstam, *Phys. Rev.* **112** (1958) 1344.

[185] S. Mandelstam, *Ann. Phys.* (N. Y.) **19** (1962) 254.

[186] H. M. M. Mansour and H. J. W. Müller–Kirsten, *J. Math. Phys.* **23** (1982) 1835.

[187] N. S. Manton and T. M. Samols, *Phys. Lett.* **B 207** (1988) 179.

[188] N. S. Manton, *Phys. Rev.* **D 28** (1983) 2019; F. S. Klinkhamer and N. S. Manton, *Phys. Rev.* **D 30** (1984) 2212.

[189] R. Manvelyan, H. J. W. Müller-Kirsten, J.-Q. Liang and Yunbo Zhang, *Nucl. Phys.* **B 579** (2000) 177, hep-th/0001179.

[190] H. Margenau and G. M. Murphy, *The Mathematics of Physics and Chemistry*, (D. van Nostrand, 1943), German: *Mathematik für Physik und Chemie* (Deutsch, 1965).

[191] D. Masson, *Nuovo Cimento* **35** (1965) 125, Appendix.

[192] J. Mauss, *On Matching Principles*, in *Asymptotic Analysis*, Lecture Notes in Mathematics, edited by A. Dold and B. Eckmann (Springer, 1979), p. 1.

[193] J. Meixner and F. W. Schäfke, *Mathieusche Funktionen und Sphäroid–Funktionen* (Springer, 1954). All page number references to these authors in the text refer to this book. For completeness we mention the supplementary volume they published later: J. Meixner, F. W. Schäfke and G. Wolf, *Mathieu Functions and Spheroidal Functions and their Mathematical Foundations: Further Studies* (Springer, 1980).

[194] E. Merzbacher, *Quantum Mechanics* (Wiley, 1975), 2nd ed.

[195] A. Messiah, *Quantum Mechanics*, Vols.I, II (North–Holland, 1960).

[196] L. M. Milne–Thomson, *Jacobian Elliptic Functions* (Dover, 1950).

[197] P. Mittelstaedt, *Klassische Mechanik*, BI-Hochschultasschenbuch Nr. 500/500* (Bibliogr. Institut, 1970).

[198] P. M. Morse and H. Feshbach, *The Methods of Theoretical Physics* (McGraw–Hill, 1953), Vol. 2, Chapter 12.

[199] N. F. Mott, *Elementary Quantum Mechanics* (Wykeham, 1972).

[200] H. J. W. Müller–Kirsten, *Fortschr. Physik* **34** (1986) 775.

[201] H. J. W. Müller, *Ann. Physik* **15** (1965) 395.

[202] H. J. W. Müller, *Physica* **31** (1965) 688.

[203] H. J. W. Müller and K. Schilcher, *J. Math. Phys.* **9** (1968) 155.

[204] H. J. W. Müller, *Ann. Physik* **16** (1965) 255.

[205] H. J. W. Müller, *Math. Nachr.* **31** (1966) 89, **32** (1966) 49, **32** (1966) 157 (split up at the request of the then editor). The first two papers are concerned with the derivation of 3 pairs of solutions, the third paper considers their matching and the calculation of the level splitting. The different sign of the power of $(1 + k)/(1 - k)$ in Eqs. (17.87), (17.89) corrects a minus sign error in the exponential factor in the latter (compare the solutions (2) in the third paper with those inserted in Eq. (17) there).

[206] H. J. W. Müller–Kirsten, Jian-zu Zhang and Yunbo Zhang, *JHEP* **11** (2001) 011.

[207] H. J. W. Müller, *J. reine angew. Math.* **211** (1962) 33, **212** (1963) 26.

[208] H. J. W. Müller–Kirsten, Jian–zu Zhang and D.H. Tchrakian, *Int. J. Mod. Phys.* **A 5** (1990) 1319.

[209] H. J. W. Müller–Kirsten, *An Introduction to J-Plane Models of Subnuclear Reactions* in *Studies in Particles and Fields*, ed. H. H. Aly (College of Science Press, Bagdad, 1978), 128.

[210] H. J. W. Müller–Kirsten, *Perturbation Approach for Regular Interactions* in *Lectures in Theoretical High Energy Physics*, ed. H. H. Aly (Wiley Interscience, 1968), 371.

[211] H. J. W. Müller, J. Math. Phys. **11** (1970) 355.

[212] H. J. W. Müller–Kirsten and N. Vahedi–Faridi, *J. Math. Phys.* **14** (1973) 1291.

[213] H. J. W. Müller–Kirsten and A. Wiedemann, *Phys. Lett.* **B 132** (1983) 169.

[214] H. J. W. Müller–Kirsten, D. K. Park and J. M. S. Rana, *Phys. Rev.* **B 60** (1999) 6662, cond-mat/9902184.

[215] H. J. W. Müller–Kirsten, *Electrodynamics, An Introduction including Quantum Effects* (World Scientific, 2004).

[216] H. J. W. Müller, *J. reine angew. Math.* **211** (1962) 33; **212** (1963) 26 and *Z. angew. Math. Mech.* **44** (1964) 371; **45** (1965) 29.

[217] G. Mussardo, V. Riva and G. Sotkov, *Semiclassical Energy Levels of Sine–Gordon Model on a Strip with Dirichlet Boundary Conditions*, hep-th/0406246.

[218] D. Nemeschansky, C. Preitschopf and M. Weinstein, *Ann. Phys.* **183** (1988) 226.

[219] R. G. Newton, *Scattering Theory of Waves and Particles* (Springer, 1982). Check reference in text!

[220] H. B. Nielsen and P. Olesen, *Nucl. Phys.* **B 61** (1973) 45.

[221] M. Nieto, V. P. Gutschik, C.M. Bender, F. Cooper and D. Strottman, *Phys. Lett.* **B 163** (1985) 336.

[222] K. Ohmori, *A Review on Tachyon Condensation in Open String Field Theories*, hep-th/0102085.

[223] R. Omnés and M. Froissart, *Mandelstam Theory and Regge Poles* (W. A. Benjamin, 1963).

[224] B. A. Orfanopoulos, *Physics Essays* **3** (1990) 368.

[225] A. Paliov and S. Rosendorf, *J. Math. Phys.* **8** (1967) 1829.

[226] D. K. Park, S. N. Tamaryan, H. J. W. Müller–Kirsten and Jian-zu Zhang, *Nucl. Phys.* **B 594** (2001) 243, hep-th/0005165.

[227] W. Pauli in *Probleme der Modernen Physik, A. Sommerfeld zum 60. Geburtstag* (Hirzel, 1928).

[228] V. S. Popov and V. M. Weinberg, *ITEP* (Moscow) (1982) Report number 101. The potential (20.10) is considered in detail in this work.

[229] E. Predazzi and T. Regge, *Nuovo Cimento* **24** (1962) 518.

[230] C. Quigg and J. L. Rosner, *Fermilab Reports* 77/82, 77/90, 77/106.

[231] C. Quigg and J. L. Rosner, *Phys. Reports* **56** (1979) 168.

[232] C. Quigg and J. L. Rosner, *Phys. Lett.* **B 71** (1977) 153.

[233] C. Quigg and J. L. Rosner, *Phys. Lett.* **B 72** (1978) 462.

[234] A. Rae, *Quantum Physics: Illusion or Reality?* (Cambridge University Press, 1986).

[235] R. Rajaraman, *Solitons and Instantons* (North–Holland, 1982).

[236] M. Razavy, *Quantum Theory of Tunneling* (World Scientific, 2003).

[237] C. Rebbi and G. Soliani, *Solitons and Particles* (World Scientific, 1984).

[238] F. Reif, *Fundamentals of Statistical and Thermal Physics* (McGraw–Hill, 1965).

[239] R. P. van Royen and V. F. Weisskopf, *Nuovo Cimento* **A 50** (1967) 617.

[240] J. Rubinstein, *J. Math. Phys.* **11** (1970) 258.

[241] L. H. Ryder, *Quantum Field Theory* (Cambridge Univ. Press, 1987).

[242] M. Salem and T. Vachaspati, *Phys. Rev.* **D 66** (2002) 025003, hep-th/0203037.

[243] L. Schiff, *Quantum Mechanics* (McGraw-Hill, 1955), second ed.

[244] E. Schrödinger, *Ann. d. Physik* **79** (1926) 489, 734.

[245] L. S. Schulman, *Techniques and Applications of Path Integrals* (Interscience, 1981).

[246] F. Schwabl, *Quantenmechanik* (Springer, 1988).

[247] N. Schwind, *Transactions Amer. Math. Soc.* **37** (1935) 339.

[248] A. E. Shabad, *Black Hole Approach to the Singular Problem of Quantum Mechanics*, hep-th/0403177.

[249] L. K. Sharma and H. J. W. Müller–Kirsten, *J. Math. Phys.* **23** (1982) 367.

[250] E. V. Shuryak, *Nucl. Phys.* **B 302** (1988) 621.

[251] S. Simons, *Introduction to the Principles of Quantum Mechanics* (Logos Press, 1968).

[252] V. Singh, *Phys. Rev.* **127** (1962) 632.

[253] T. H. R. Skyrme, *Proc. Roy. Soc.***A 247** (1958) 260; **A 252** (1959) 236; **A 260** (1961) 127; **A 262** (1961) 237.

[254] O. G. Smolyanov, A. G. Tokarev and A. Truman, *J. Math. Phys.* **43** (2002) 5161.

[255] I. N. Sneddon, *Special Functions of Mathematical Physics and Chemistry*, 3rd ed. (Longmans, 1980).

[256] A. Sommerfeld, *Partial Differential Equations in Physics* (Academic Press, 1949).

[257] R. M. Spector, *J. Math. Phys.* **5** (1964) 1185.

[258] E. J. Squires, *Complex Angular Momenta and Particle Physics* (W. A. Benjamin, 1963).

[259] P. P. Srivastava, *Nuovo Cimento* **64 A** (1981) 259.

[260] S. G. Starling and A. J. Woodall, *Physics* (Longmans, 1952).

[261] N. Steenrod, *The Topology of Fibre Bundles* (Princeton University Press, 1951).

[262] M. Stone and J. Reeve, *Phys. Rev.* **D18** 4746.

[263] M. Stone, *Phys. Rev.* **D18** 4752.

[264] R. F. Streater and A. S. Wightman, *PCT, Spin and Statistics, and All That* (W. A. Benjamin, 1964).

[265] K. Sundermeyer, *Constrained Dynamics* (Springer, 1982).

[266] G. Süssmann, *Einführung in die* Quantenmechanik (Bibliogr. Inst., Hochschulbücherverlag, 1963).

[267] J. R. Taylor, *A New Rigorous Approach to Coulomb Scattering*, Imperial College Report ICTP/73/7 (1973), *Nuovo Cimento* **B 23** (1974) 313.

[268] D. H. Tchrakian and H. J. W. Müller–Kirsten, *Phys. Rev.* **D 44** (1991) 1204.

[269] H. B. Thacker, C. Quigg and J. L. Rosner, *Phys. Lett.* **B 74** (1978) 350.

[270] G. H. Thomas, *Introductory Lectures on Fibre Bundles and Topology for Physicists*, Argonne National Laboratory Report (1978) ANL–HEP–PR–78–23.

[271] G. Tiktopoulos and S. B. Treiman, *Phys. Rev.* **134** (1964) 844.

[272] D. Tong, *TASI Lectures on Solitons*, hep-th/0509216.

[273] A. V. Turbiner, *Anharmonic Ocillator and Double Well Potentials: Approximating Eigenfunctions*, math-ph/0506033; to be publ. in *Lett. Math. Phys.* (Special Issue in Memory of F. A. Berezin, 2005).

[274] V. Tyutin, *Lebedev Report*, No. FIAN-39 (1975) unpublished.

[275] V. V. Ulyanov and O. B. Zaslavskii, *Phys. Rep.* **216** (1992) 179.

[276] A. D. de Veigy and S. Ouvry, *Phys. Lett.* **B 307** (1993) 91.

[277] J. Villain, *J. de Phys.* **35** (1974) 27.

[278] G. H. Wannier, *Quart. Appl. Math.* **11** (1953) 33.

[279] A. E. A. Warburton, *Nuovo Cimento* **41** (1966) 360.

[280] G. N. Watson, *Proc. Roy. Soc.* (London) **95** (1918) 83.

[281] M. Weinstein, *Adaptive Perturbation Theory I: Quantum Mechanics*, hep-th/0510159; *Adaptive Perturbation Theory: Quantum Mechanics and Field Theory*, hep-th/0510160.

[282] G. Wentzel, *Z. Phys.* **38** (1926) 518.

[283] E. T. Whittaker and G. N. Watson, *A Course of Modern Analysis*, fourth ed. (Cambridge University Press, 1958).

[284] R. M. Wilcox, *J. Math. Phys.* **8** (1967) 962.

[285] D. Yuan–Ben, *Sci. Sinica* (Peking) **13** (1964) 1319.

[286] E. Zauderer, *J. Math. Phys.* **13** (1972) 42.

[287] Jian–zu Zhang, H. J. W. Müller–Kirsten and J. M. S. Rana, *J. Phys. A: Math. Gen.* **31** (1998) 7291.

[288] Jian–zu Zhang and H. J. W. Müller–Kirsten, *Phys. Rev.* **D 50** (1994) 6531.

[289] Jian–zu Zhang and H. J. W. Müller–Kirsten *Phys. Lett.* **A 200** (1995) 243.

[290] Jian–zu Zhang and H. J. W. Müller–Kirsten *Phys. Lett.* **A 202** (1995) 241.

[291] F. Zimmerschied and H. J. W. Müller–Kirsten, *Phys. Rev.* **D 49** (1994) 5387.

[292] J. Zinn–Justin, *J. Math. Phys.* **25** (1984) 549.

[293] J. Zinn–Justin and U. D. Jentschura, *Ann. Phys.* **313** (2004) 197, 269.

[294] B. Zwiebach, *A First Course in String Theory* (Cambridge University Press, 2004).

Index